Passive Microwave Components and Antennas

Edited by

Vitaliy Zhurbenko

Passive Microwave Components and Antennas
Edited by Vitaliy Zhurbenko

Republished by InTech
Janeza Trdine 9, 51000 Rijeka, Croatia

Additional hard copies can be obtained from orders@intechopen.com

Passive Microwave Components and Antennas,
Edited by Vitaliy Zhurbenko
p. cm.
ISBN 978-953-307-083-4

For EU product safety concerns :
IN TECH d . o . o . , Prolaz Marije Krucifikse Kozulić 3 , 51000 Rijeka , Croatia ,
info@intechopen . com or visit our website at intechopen . com .

Meet the editor

Vitaliy Zhurbenko obtained the B.Sc. and M.Sc. degrees from the Kharkiv National University of Radio Electronics, Kharkiv, Ukraine, in 2000 and 2001, respectively, and the Ph.D. degree from the Technical University of Denmark, Copenhagen, Denmark, in 2008, all in electrical engineering. From November 2000 to June 2005 he was a Metrology Engineer with the Laboratory of Metrology, Kharkiv, Ukraine. In 2004 he became a Junior Member of the Teaching Staff with the Kharkiv National University of Radio Electronics. In 2005 he joined the Technical University of Denmark, where he is currently an Assistant Professor. His current teaching and research interests include wireless communications, microwave and millimeter-wave sensing for biomedical and security applications, microwave and millimeter-wave devices and integrated circuits for instrumentation applications, antenna and passive circuit design and characterization, terahertz technologies.

Preface

State-of-the-art microwave systems always require higher performance and lower cost microwave components. Constantly growing demands and performance requirements of industrial and scientific applications often make employing traditionally designed components impractical. For that reason, the design and development process remains a great challenge today. This problem motivated intensive research efforts in microwave design and technology, which is responsible for a great number of recently appeared alternative approaches to analysis and design of microwave components and antennas. This book highlights these new trends focusing on passive components such as novel resonators, filters, diplexers, power dividers, directional couplers, impedance transformers, waveguides, transmission lines and transitions as well as antennas, metamaterial-based structures, and various electromagnetic analysis and design techniques.

Modelling and computations in electromagnetics is a quite fast-growing research area. The recent interest in this field is caused by the increased demand for designing complex microwave components, modeling electromagnetic materials, and rapid increase in computational power for calculation of complex electromagnetic problems. The first part of this book is devoted to the advances in the analysis techniques such as method of moments, finite-difference time-domain method, boundary perturbation theory, Fourier analysis, mode-matching method, and analysis based on circuit theory. These techniques are considered with regard to several challenging technological applications such as those related to electrically large devices, scattering in layered structures, photonic crystals, and artificial materials.

The second part of the book deals with waveguides, transmission lines and transitions. This includes microstrip lines (MSL), slot waveguides, substrate integrated waveguides (SIW), vertical transmission lines in multilayer media as well as MSL to SIW and MSL to slot line transitions.

Impedance matching is an important aspect in the design of microwave circuitry since impedance mismatches may severely deteriorate performance of the overall system. Different techniques for wideband matching are presented in the third part of this book. The design of compact microwave resonators and filters is also covered in this part. Compact, high-performance microwave filters are essential for high-efficiency miniaturized microwave systems. The filter circuit size is large in traditionally designed planar bandpass filters due to a high number of large area resonators. The rejection level in the upper stopband of the filters is usually degraded by the spurious response at twice the passband frequency. Several types of resonators have been designed to overcome these problems, such as miniaturized hairpin resonators, stepped-impedance hairpin resonators, and slow-wave open-loop resonators.

Miniaturized resonators lead to a reduced filter size, but not always improve the spurious response. Another method relies on various resonator combinations within one filter structure to reduce the circuit size, such as the loop resonator or hairpin resonator incorporated with one or several open stubs.

Recently, microwave filters based on electromagnetic bandgap structures and artificial materials have attracted a great deal of interest because of improved characteristics in comparison to traditional filter design. Such artificial materials can be realized using periodic inclusion of variously shaped metals into a host medium. The most prominent candidate for such structures has been the split-ring resonator. In addition to the split-ring resonator there are several alternative realizations based on lumped elements, quasi-lumped LC resonators and other planar microwave resonators which are in details discussed in the fourth part of this book.

Antennas are key components in most microwave devices and systems. They are used everywhere where a transformation between a guided wave and a free-space wave (or vice versa) is required. The final part of the book is dedicated mainly to the design and applications of planar antennas and arrays including metamaterial-based antennas, monopoles, slot antennas, reflector antennas and arrays.

The book concludes with a chapter considering accuracy aspects of antenna gain measurements.

Editor

Vitaliy Zhurbenko

Contents

1

Boundary Perturbation Theory for Scattering in Layered Rough Structures

Pasquale Imperatore, Antonio Iodice and Daniele Riccio
University of Naples "Federico II",
Department of Biomedical, Electronic and Telecommunication Engineering,
Napoli, Italy

1. Introduction

The electromagnetic wave interaction with layered structures constitutes a crucial topic of current interest in theoretical and experimental research. Generally speaking, several modelling and design problems, encountered, for instance, in SAR (*Synthetic Aperture Radar*) application, GPR (*Ground Penetrating Radar*) sensing, radar altimeter for planetary exploration, microstrip antennas and MMICs (*Monolithic Microwave Integrated Circuits*), radio-propagation in urban environment for wireless communications, through-the-wall detection technologies, optics, biomedical diagnostic of layered biological tissues, geophysical and seismic exploration, lead to the analysis of the electromagnetic wave interaction with multilayered structure, whose boundaries can exhibit some amount of roughness.

This chapter is aimed primarily at providing a comprehensive analytical treatment of electromagnetic wave propagation and scattering in three-dimensional multilayered structures with rough interfaces. The emphasis is placed on the general formulation of the scattering problem in the analytic framework of the *Boundary Perturbation Theory (BPT)* developed by Imperatore et al. A systematic perturbative expansion of the fields in the layered structure, based on the gently rough interfaces assumption, enables the transferring of the geometry randomness into a non-uniform boundary conditions formulation. Subsequently, the fields' expansion can be analytically evaluated by using a recursive matrix formalism approach encompassing a proper scattered field representation in each layer and a matrix reformulation of non-uniform boundary conditions. A key-point in the development resides in the appropriate exploitation of the *generalized reflection/transmission* notion, which has strong implications in order to make the mathematical treatment manageable and to effectively capture the physics of the problem. Two relevant compact closed-form solutions, derived in the first-order limit of the perturbative development, are presented. They refer to two complementary *bi-static* configurations for the scattering, respectively, from and through layered structures with arbitrary number of rough interfaces. The employed formalism is fully-*polarimetric* and suitable for applications. In addition, it is demonstrated how the symmetrical character of the *BPT* formalism reflects the inherent conformity with the *reciprocity* theorem of the electromagnetic theory.

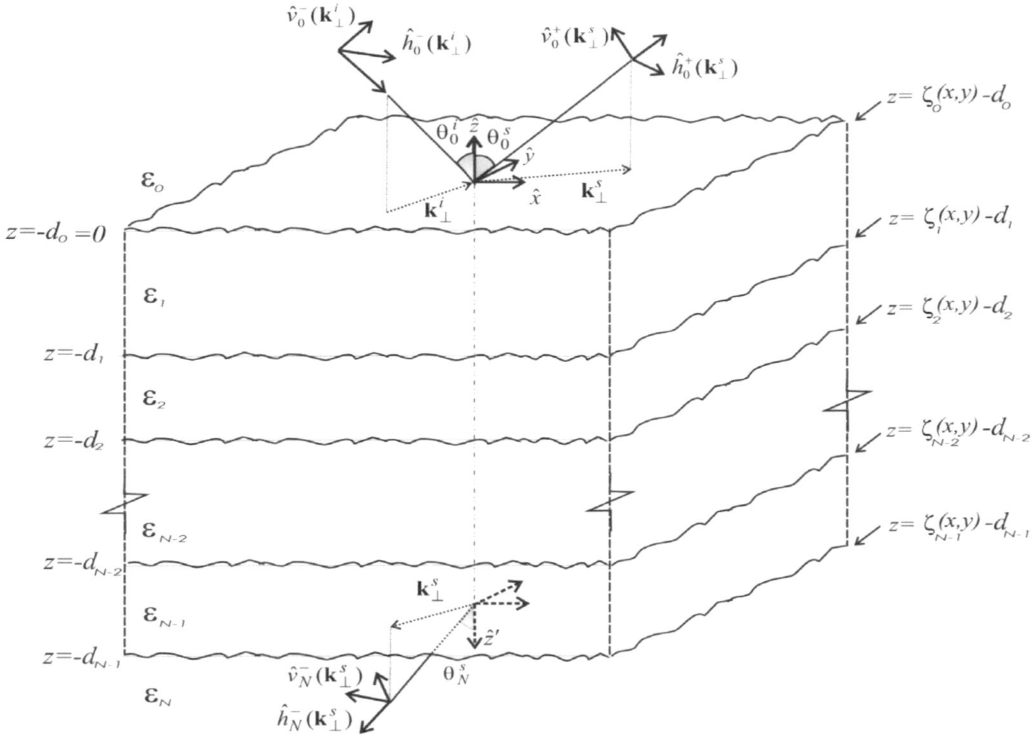

Fig. 1. Geometry for an N-rough boundaries layered medium

2. Statement of the problem

When stratified media with rough interfaces are concerned, the possible approaches to cope with the EM scattering problem fall within three main categories. First, the numerical approaches do not permit to attain a comprehensive understanding of the general functional dependence of the scattering response on the structure parameters, as well as do not allow capturing the physics of the involved scattering mechanisms. In addition, the numerical approach turns out to be feasible for non-fully 3D geometry or configurations in which a very limited number of rough interfaces is accounted for. Layered structures with rough interfaces have been also treated resorting to *radiative transfer theory* (RT). However, coherent effects are not accounted for in RT theory and could not be contemplated without employing full wave analysis, which preserves phase information. Another approach relies on the full-wave methods. Although, to deal with the electromagnetic propagation and scattering in complex random layered media, several analytical formulation involving some idealized cases and suitable approximations have been conducted in last decades, the relevant solutions usually turn out to be too complicated to be generally useful in applications, even if simplified geometries are accounted for. The proliferation of the proposed methods for the simulation of wave propagation and scattering in stratified media and the continuous interest in this topic are indicative of the need of appropriate modelling and interpretation of the complex physical phenomena that take place in layered structures. Indeed, the availability of accurate, sound physical and manageable models turns out still to be a strong necessity, in perspective to apply them, for instance, in retrieving of add-valued information from the data acquired by microwave sensors.

Generally speaking, an exact analytical solution of *Maxwell* equations can be found only for a few idealized problems. Subsequently, appropriate approximation methods are needed. Regarding the perturbative approaches, noticeable progress has been attained in the analytic investigation on the extension of the classical SPM (small perturbation method) solution for the scattering from rough surface to specific layered configurations. Most of previous existing works analyze different layered configurations in the first-order limit, using procedures, formalisms and final solutions that can appear of difficult comparison (Yarovoy et al., 2000), (Azadegan and Sarabandi, 2003), (Fuks, 2001). All these formulations, which refer to the case of a single rough interface, have been recently unified in (Franceschetti et al, 2008). On the other hand, solution for the case of two rough boundaries has also been proposed in (Tabatabaeenejad and Moghaddam, 2006).

Methodologically, we underline that all the previously mentioned existing perturbative approaches, followed by different authors in analyzing scattering from simplified geometry, imply an inherent analytical complexity, which precludes the treatment to structures with more than one (Fuks, 2001) (Azadegan et al., 2003) (Yarovoy et al., 2000) or two (Tabatabaeenejad er al., 2006) rough interfaces.

The general problem we intend to deal with here refers to the analytical evaluation of the electromagnetic scattering from and through layered structure with an arbitrary number of rough interfaces (see Fig.1). As schematically shown in fig.1, an arbitrary polarized monochromatic plane wave

$$\mathbf{E}_0^i(\mathbf{r}) = [E_0^{ih}\hat{h}_0^-(\mathbf{k}_\perp^i) + E_0^{iv}\hat{v}_0^-(\mathbf{k}_\perp^i)]\, e^{j(\mathbf{k}_\perp^i \cdot \mathbf{r}_\perp - k_{z0}^i z)} \tag{1}$$

is considered to be incident on the layered medium at an angle θ_0^i relative to the $\hat{z}$ direction from the upper half-space, where in the field expression a time factor $exp(-j\omega t)$ is understood, and where, using a spherical frame representation, the incident vector wave direction is individuated by θ_0^i, φ_0^i:

$$k_0\hat{k}_0^i = \mathbf{k}_0^i = \mathbf{k}_\perp^i - \hat{z}k_{z0}^i = k_0(\hat{x}\sin\theta_0^i\cos\varphi_0^i + \hat{y}\sin\theta_0^i\sin\varphi_0^i - \hat{z}\cos\theta_0^i), \tag{2}$$

with

$$\hat{h}_0^-(\mathbf{k}_\perp^i) = \frac{\hat{k}_0^i \times \hat{z}}{\left|\hat{k}_0^i \times \hat{z}\right|} = \sin\varphi_0^i\hat{x} - \cos\varphi_0^i\hat{y}, \tag{3}$$

$$\hat{v}_0^-(\mathbf{k}_\perp^i) = \hat{h}_0^-(\mathbf{k}_\perp^i)\times\hat{k}_0^i = (\hat{x}\cos\varphi_0^i + \hat{y}\sin\varphi_0^i)\cos\theta_0^i + \hat{z}\sin\theta_0^i, \tag{4}$$

where $\mathbf{k}_\perp^i = k_x^i\hat{x} + k_y^i\hat{y}$ is the two-dimensional projection of incident wave-number vector on the plane $z=0$. The parameters pertaining to layer m with boundaries $-d_{m-1}$ and $-d_m$ are distinguished by a subscript m. Each layer is assumed to be homogeneous and characterized by arbitrary and deterministic parameters: the *dielectric relative permittivity* ε_m, the *magnetic relative permeability* μ_m and the *thickness* $\Delta_m=d_m-d_{m-1}$. With reference to Fig.1, it has been assumed that in particular, $d_0=0$. In the following, the symbol $\perp$ denotes the projection of the corresponding vector on the plan $z=0$. Here $\mathbf{r} = (\mathbf{r}_\perp, z)$, so we distinguish the transverse

spatial coordinates $\mathbf{r}_\perp = (x, y)$ and the longitudinal coordinate z. In addition, each mth rough interface is assumed to be characterized by a zero-mean two-dimensional *stochastic process* $\zeta_m = \zeta_m(\mathbf{r}_\perp)$ with normal vector $\hat{n}_m$. No constraints are imposed on the degree to which the rough interfaces are correlated.
A general methodology has been developed by *Imperatore et al.* to analytically treat EM bistatic scattering from this class of layered structures that can be described by small changes with respect to an idealized (unperturbed) structure, whose associated problem is exactly solvable. A thorough analysis of the results of this theoretical investigation (BPT), which is based on perturbation of the boundary condition, will be presented in the following, methodologically emphasizing the development of the several inherent aspects.

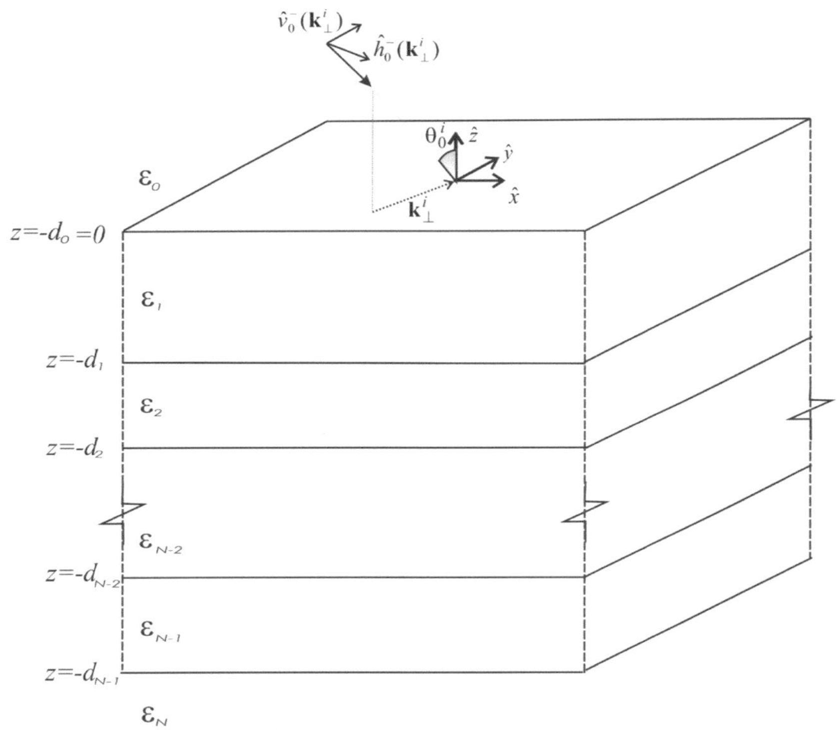

Fig. 2. Geometry for a flat boundaries layered medium

3. Basic definition and notations

This section is devoted preliminary to introduce the formalism used in the following of this chapter. The Flat Boundaries layered medium (*unperturbed structure*) is defined as a stack of parallel slabs (Fig.2), sandwiched in between two half-spaces, whose structure is shift invariant in the direction of x and y (infinite lateral extent in x-y directions). With the notations $T^p_{m|m+1}$ and $R^p_{m|m+1}$, respectively, we indicate the *ordinary transmission* and *reflection coefficients* at the interface between the regions m-1 and m+1,

$$R^h_{m|m+1} = \frac{\mu_{m+1}k_{zm} - \mu_m k_{z(m+1)}}{\mu_{m+1}k_{zm} + \mu_m k_{z(m+1)}}, \tag{5}$$

$$R^v_{m|m+1} = \frac{\varepsilon_{m+1}k_{zm} - \varepsilon_m k_{z(m+1)}}{\varepsilon_{m+1}k_{zm} + \varepsilon_m k_{z(m+1)}}, \tag{6}$$

$$T^h_{m|m+1} = \frac{2\mu_{m+1}k_{zm}}{\mu_{m+1}k_{zm} + \mu_m k_{z(m+1)}}, \tag{7}$$

$$T^v_{m|m+1} = \frac{2\varepsilon_{m+1}k_{zm}}{\varepsilon_{m+1}k_{zm} + \varepsilon_m k_{z(m+1)}}, \tag{8}$$

with the superscript $p\in\{v, h\}$ indicating the polarization state for the incident wave and may stand for *horizontal* (h) or *vertical* (v) polarization (Tsang et al., 1985) (Imperatore et al. 2009a), and where

$$k_{zm} = \sqrt{k_m^2 - |\mathbf{k}_\perp|^2} = k_m \cos\theta_m, \tag{9}$$

where $k_m = k_0\sqrt{\mu_m\varepsilon_m}$ is the *wave number* for the electromagnetic medium in the mth layer, with $k_0 = \omega / c = 2\pi / \lambda$, and where $\mathbf{k}_\perp = k_x\hat{x} + k_y\hat{y}$ is the two-dimensional projection of vector wave-number on the plane $z=0$. In addition, we stress that:

$$T^p_{i|j} = 1 + R^p_{i|j}, \qquad R^p_{i|j} = -R^p_{j|i}, \qquad i=j\pm 1. \tag{10}$$

3.1 Generalized reflection formalism

The *generalized reflection coefficients* $\Re^p_{m-1|m}$ at the interface between the regions (m-1) and m, for the p-polarization, are defined as the ratio of the amplitudes of *upward-* and *downward*-propagating waves immediately above the interface, respectively. They can be expressed by recursive relations as in (Chew W. C., 1997) (Imperatore et al. 2009a):

$$\Re^p_{m-1|m} = \frac{R^p_{m-1|m} + \Re^p_{m|m+1} e^{j2k_{zm}\Delta_m}}{1 + R^p_{m-1|m}\Re^p_{m|m+1} e^{j2k_{zm}\Delta_m}}. \tag{11}$$

Likewise, at the interface between the regions (m+1) and m, $\Re^p_{m+1|m}$ is given by:

$$\Re^p_{m+1|m} = \frac{R^p_{m+1|m} + \Re^p_{m|m-1} e^{j2k_{zm}\Delta m}}{1 + R^p_{m+1|m}\Re^p_{m|m-1} e^{j2k_{zm}\Delta_m}}, \tag{12}$$

Furthermore, it should be noted that the factors

$$\vec{M}_m^p(k_\perp) = 1 - R^p_{m|m-1}\,\Re^p_{m|m+1}\,e^{j2k_{zm}\Delta_m}\,, \tag{13}$$

$$\overleftarrow{M}_m^p(k_\perp) = 1 - \Re^p_{m|m-1}\,R^p_{m|m+1}\,e^{j2k_{zm}\Delta_m}\,, \tag{14}$$

$$\overleftrightarrow{M}_m^p(k_\perp) = 1 - \Re^p_{m|m-1}\,\Re^p_{m|m+1}\,e^{j2k_{zm}\Delta_m}\,, \tag{15}$$

take into account the multiple reflections in the mth layer.

3.2 Generalized transmission formalism

The *generalized transmission coefficients* in *downward* direction $\Im^p_{0|m}$ can be defined as:

$$\Im^p_{0|m}(k_\perp) = \exp\left[j\sum_{n=1}^{m-1} k_{zn}\Delta_n\right]\prod_{n=0}^{m-1} T^p_{n|n+1}\left[\prod_{n=1}^{m}\vec{M}_n^p\right]^{-1}, \tag{16}$$

where $p\in\{v, h\}$. The *generalized transmission coefficients* in *upward* direction $\Im^p_{m|0}$ are then given by

$$\Im^p_{m|0} = \begin{cases} \Im^p_{0|m}\dfrac{\mu_0 k_{zm}}{\mu_m k_{z0}} & for\ p = h \\ \Im^p_{0|m}\dfrac{\varepsilon_0 k_{zm}}{\varepsilon_m k_{z0}} & for\ p = v \end{cases} . \tag{17}$$

As a counterpart of (17), we have

$$\Im^p_{m+1|N} = \begin{cases} \Im^p_{N|m+1}\dfrac{\mu_N k_{z(m+1)}}{\mu_{m+1} k_{zN}} & for\ p = h \\ \Im^p_{N|m+1}\dfrac{\varepsilon_N k_{z(m+1)}}{\varepsilon_{m+1} k_{zN}} & for\ p = v \end{cases} . \tag{18}$$

Equations (17) and (18) formally express the *reciprocity* of the generalized transmission coefficients for an arbitrary flat-boundaries layered structure (Imperatore et al. 2009b).
Here we introduce notion of layered *slab*, which refers to a layered structure sandwiched between two half-spaces. Accordingly, the *generalized transmission coefficients* in *downward* direction for a layered slab between two half-spaces (*0, N*), $\Im^{p(slab)}_{0|N}$, can be defined as:

$$\Im^{p(slab)}_{0|N}(k_\perp) = \exp\left[j\sum_{n=1}^{N-1} k_{zn}\Delta_n\right]\prod_{n=0}^{N-1} T^p_{n|n+1}\left[\prod_{n=1}^{N-1}\vec{M}_n^p\right]^{-1}. \tag{19}$$

It should be noted that the parenthesized superscript *slab* indicates that both the media *0* and *N* are half-space. Similarly, the *generalized transmission coefficients* in *downward* direction for the layered slab between two half-spaces (*m+1, N*), $\Im^{p(slab)}_{m+1|N}$, are defined as:

$$\Im^{p(slab)}_{m+1|N}(k_\perp) = \exp\left[j \sum_{n=m+2}^{N-1} k_{zn}\Delta_n \right] \prod_{n=m+1}^{N-1} T^p_{n|n+1} \left[\prod_{n=m+2}^{N-1} \vec{M}^p_n \right]^{-1} . \tag{20}$$

Note also that

$$\Im^p_{m+1|N} = [\vec{M}^p_{m+1}]^{-1} \Im^{p(slab)}_{m+1|N} . \tag{21}$$

Moreover, we consider the *generalized transmission coefficients* in *upward* direction for the layered *slab* between two half-spaces (*m*, 0), $\Im^{p(slab)}_{m|0}$, which are defined as

$$\Im^{p(slab)}_{m|0}(k_\perp) = \exp\left[j \sum_{n=1}^{m-1} k_{zn}\Delta_n \right] \prod_{n=0}^{m-1} T^p_{n+1|n} \left[\prod_{n=1}^{m-1} \vec{M}^p_n \right]^{-1} . \tag{22}$$

Note also that

$$\Im^p_{m|0}(k_\perp) = [\vec{M}^p_m(k_\perp)]^{-1} \Im^{p(slab)}_{m|0}(k_\perp) . \tag{23}$$

The *generalized transmission coefficients* in *downward* direction for the layered *slab* between two half-spaces (*0*, *m*), $\Im^{p(slab)}_{0|m}$, can be defined as

$$\Im^{p(slab)}_{0|m}(k_\perp) = \exp\left[j \sum_{n=1}^{m-1} k_{zn}\Delta_n \right] \prod_{n=0}^{m-1} T^p_{n|n+1} \left[\prod_{n=1}^{m-1} \vec{M}^p_n \right]^{-1} . \tag{24}$$

It should be noted that the $\Im^p_{0|m}$ are distinct from the coefficients $\Im^{p(slab)}_{0|m}$, because in the evaluation of $\Im^p_{0|m}$ the effect of all the layers under the layer *m* is taken into account, whereas $\Im^{p(slab)}_{0|m}$ are evaluated referring to a different configuration in which the intermediate layers *1…m* are bounded by the half-spaces *0* and *m*.

We stress that generalized reflection and transmission coefficients do not depend on the direction of $\mathbf{k}_\perp$. In the following, we shown how the employing the generalized reflection/transmission coefficient notions not only is crucial in obtaining a compact closed-form perturbation solution, but it also permit us to completely elucidate the obtained analytical expressions from a physical point of view, highlighting the role played by the *equivalent reflecting interfaces* and by *the equivalent slabs*, so providing the inherent connection between the global scattering response.

4. Stochastic characterization for the 3-D geometry description

In this section, the focus is on stochastic description for the geometry of the investigated structure, and the notion of wide-sense stationary process is detailed. First of all, when the description of a rough interface by means of deterministic function $\zeta_m(\mathbf{r}_\perp)$ is concerned, the corresponding *ordinary* 2-D *Fourier Transform* pair can be defined as

$$\tilde{\zeta}_m(\mathbf{k}_\perp) = (2\pi)^{-2} \iint d\mathbf{r}_\perp \, e^{-j\mathbf{k}_\perp \cdot \mathbf{r}_\perp} \zeta_m(\mathbf{r}_\perp), \tag{25}$$

$$\zeta_m(\mathbf{r}_\perp) = \iint d\mathbf{k}_\perp \, e^{j\mathbf{k}_\perp \cdot \mathbf{r}_\perp} \tilde{\zeta}_m(\mathbf{k}_\perp). \tag{26}$$

Let us assume now that $\zeta_m(\mathbf{r}_\perp)$, which describes the generic (mth) rough interface, is a 2-D stochastic process satisfying the conditions

$$< \zeta_m(\mathbf{r}_\perp) > = 0, \tag{27}$$

$$< \zeta_m(\mathbf{r}_\perp + \boldsymbol{\rho}) \zeta_m(\mathbf{r}_\perp) > = B_{\zeta_m}(\boldsymbol{\rho}), \tag{28}$$

where the *angular bracket* denotes statistical ensemble averaging, and where $B_{\zeta_m}(\boldsymbol{\rho})$ is the interface *autocorrelation* function, which quantifies the similarity of the spatial fluctuations with a displacement ρ. Equations (27)-(28) constitute the basic assumptions defining a *wide sense stationary* (*WSS*) stochastic process: the statistical properties of the process under consideration are invariant to a spatial shift. Similarly, concerning two mutually correlated random rough interfaces ζ_m and ζ_n, we also assume that they are *jointly WSS*, i.e.

$$< \zeta_m(\mathbf{r}_\perp + \boldsymbol{\rho}) \zeta_n(\mathbf{r}_\perp) > = B_{\zeta_m \zeta_n}(\boldsymbol{\rho}), \tag{29}$$

where $B_{\zeta_m \zeta_n}(\boldsymbol{\rho})$ is the corresponding *cross-correlation function* of the two random processes. It can be readily derived that

$$B_{\zeta_m \zeta_n}(\boldsymbol{\rho}) = B_{\zeta_n \zeta_m}(-\boldsymbol{\rho}). \tag{30}$$

The integral in (25) is a *Riemann* integral representation for $\zeta_m(\mathbf{r}_\perp)$, and it exists if $\zeta_m(\mathbf{r}_\perp)$ is piecewise continuous and *absolutely integrable*. On the other hand, when the spectral analysis of a stationary random process is concerned, the integral (25) does not in general exist in the framework of theory of the ordinary functions. Indeed, a WSS process describing an interface $\zeta_m(\mathbf{r}_\perp)$ of infinite lateral extension, for its proper nature, is not *absolutely integrable*, so the conditions for the existence of the Fourier Transform are not satisfied. In order to obtain a spectral representation for a *WSS* random process, this difficulty can be circumvented by resorting to the more general *Fourier-Stieltjes* integral (Ishimaru, 1978); otherwise one can define space-truncated functions. When a finite patch of the rough interface with area A is concerned, the space-truncated version of (25) can be introduced as

$$\tilde{\zeta}_m(\mathbf{k}_\perp; A) = (2\pi)^{-2} \iint_A d\mathbf{r}_\perp \, e^{-j\mathbf{k}_\perp \cdot \mathbf{r}_\perp} \zeta_m(\mathbf{r}_\perp), \tag{31}$$

subsequently, $\tilde{\zeta}_m(\mathbf{k}_\perp) = \lim_{A \to \infty} \tilde{\zeta}_m(\mathbf{k}_\perp; A)$ is not an ordinary function. Nevertheless, we will use again the (25)-(26), regarding them as symbolic formulas, which hold a rigorous mathematical meaning beyond the ordinary function theory (generalized Fourier Transform). We underline that by virtue of the condition (27) directly follows also that $< \tilde{\zeta}_m(\mathbf{k}_\perp) > = 0$. Let us consider

$$<\zeta_m(\mathbf{r}'_\perp)\zeta_n^*(\mathbf{r}''_\perp)> = \iint d\mathbf{k}'_\perp \iint d\mathbf{k}''_\perp \, e^{j(\mathbf{k}'_\perp\cdot\mathbf{r}'_\perp - \mathbf{k}''_\perp\cdot\mathbf{r}''_\perp)} <\tilde{\zeta}_m(\mathbf{k}'_\perp)\tilde{\zeta}_n^*(\mathbf{k}''_\perp)>, \quad (32)$$

where the asterisk denotes the complex conjugated, and where the operations of average and integration have been interchanged. When *jointly WSS* processes ζ_m and ζ_n are concerned, accordingly to (29), the LHS of (32) must be a function of $\mathbf{r}'_\perp - \mathbf{r}''_\perp$ only; therefore, it is required that

$$<\tilde{\zeta}_m(\mathbf{k}'_\perp)\tilde{\zeta}_n^*(\mathbf{k}''_\perp)> = W_{mn}(\mathbf{k}'_\perp)\delta(\mathbf{k}'_\perp - \mathbf{k}''_\perp), \quad (33)$$

where $\delta(\cdot)$ is the *Dirac* delta function, and where $W_{mn}(\boldsymbol{\kappa})$ is called the (spatial) *cross power spectral density* of two interfaces ζ_m and ζ_n, for the spatial frequencies of the roughness. Equation (33) states that the different spectral components of the two considered interfaces must be uncorrelated. This is to say that the (generalized) Fourier transform of jointly WSS processes are *jointly non stationary* white noise with average power $W_{mn}(\mathbf{k}'_\perp)$. Indeed, by using (33) into (32), we obtain

$$<\zeta_m(\mathbf{r}'_\perp)\zeta_n(\mathbf{r}''_\perp)> = \iint d\mathbf{k}''_\perp \, e^{j\cdot\mathbf{k}''_\perp\cdot(\mathbf{r}'_\perp - \mathbf{r}''_\perp)} W_{mn}(\mathbf{k}''_\perp), \quad (34)$$

where the RHS of (34) involves an (ordinary) 2D Fourier Transform. Note also that as a direct consequence of the fact that $\zeta_n(\mathbf{r}_\perp)$ is real we have the relation $\tilde{\zeta}_n(\mathbf{k}_\perp) = \tilde{\zeta}_n^*(-\mathbf{k}_\perp)$. Therefore, setting $\boldsymbol{\rho} = \mathbf{r}'_\perp - \mathbf{r}''_\perp$ in (34), we have

$$B_{\zeta_m\zeta_n}(\boldsymbol{\rho}) = \iint d\boldsymbol{\kappa} \, e^{j\boldsymbol{\kappa}\cdot\boldsymbol{\rho}} \, W_{mn}(\boldsymbol{\kappa}) \, . \quad (35)$$

The *cross-correlation function* $B_{\zeta_m\zeta_n}(\boldsymbol{\rho})$ of two interfaces ζ_m and ζ_n is then given by the (inverse) 2D Fourier Transform of their *(spatial) cross power spectral density*, and Equation (35) together with its Fourier inverse

$$W_{mn}(\boldsymbol{\kappa}) = (2\pi)^{-2} \iint d\boldsymbol{\rho} \, e^{-j\boldsymbol{\kappa}\cdot\boldsymbol{\rho}} B_{\zeta_m\zeta_n}(\boldsymbol{\rho}) \, , \quad (36)$$

may be regarded as the (generalized) *Wiener-Khinchin* theorem. In particular, when *n=m,* (33) reduces to

$$<\tilde{\zeta}_m(\mathbf{k}'_\perp)\tilde{\zeta}_m^*(\mathbf{k}''_\perp)> = W_m(\mathbf{k}'_\perp)\delta(\mathbf{k}'_\perp - \mathbf{k}''_\perp), \quad (37)$$

where $W_m(\boldsymbol{\kappa})$ is called the *(spatial) power spectral density* of *n*th corrugated interface ζ_m and can be expressed as the *(ordinary)* 2D Fourier transform of *n*-corrugated interface autocorrelation function, i.e., satisfying the transform pair:

$$W_m(\boldsymbol{\kappa}) = (2\pi)^{-2} \iint d\boldsymbol{\rho} \, e^{-j\boldsymbol{\kappa}\cdot\boldsymbol{\rho}} B_{\zeta_m}(\boldsymbol{\rho}) \, , \quad (38)$$

$$B_{\zeta_m}(\boldsymbol{\rho}) = \iint d\boldsymbol{\kappa} \, e^{j\boldsymbol{\kappa}\cdot\boldsymbol{\rho}} \, W_m(\boldsymbol{\kappa}) \, , \quad (39)$$

which is the statement of the classical *Wiener-Khinchin* theorem. We emphasize the physical meaning of $W_m(\boldsymbol{\kappa})d\boldsymbol{\kappa} = W_m(\kappa_x,\kappa_y)d\kappa_x d\kappa_y$: it represents the power of the spectral components of the mth rough interface having spatial wave number between κ_x and $\kappa_x + d\kappa_x$ and κ_y and $\kappa_y + d\kappa_y$, respectively, in x and y direction. Furthermore, from (30) and (36) it follows that

$$W_{mn}(\boldsymbol{\kappa}) = W_{nm}^{*}(\boldsymbol{\kappa}). \tag{40}$$

This is to say that, unlike the power spectral density, the cross power spectral density is, in general, neither real nor necessarily positive. Furthermore, it should be noted that the *Dirac*'s delta function can be defined by the integral representation

$$\delta(\boldsymbol{\kappa}) = (2\pi)^{-2} \iint d\boldsymbol{\rho}\, e^{-j\boldsymbol{\kappa}\cdot\boldsymbol{\rho}} = \lim_{A\to\infty} \delta(\boldsymbol{\kappa}; A). \tag{41}$$

By using in (37) and (33) the relation $\delta(0;A) = A/(2\pi)^2$, we have, respectively, that the (spatial) power *spectral density* of nth corrugated interface can be also expressed as

$$W_m(\boldsymbol{\kappa}) = (2\pi)^2 \lim_{A\to\infty} \frac{1}{A} < \left|\tilde{\zeta}_m(\boldsymbol{\kappa}; A)\right|^2 >, \tag{42}$$

and the (spatial) *cross power spectral density* of two interfaces ζ_m and ζ_n is given by

$$W_{mn}(\boldsymbol{\kappa}) = (2\pi)^2 \lim_{A\to\infty} \frac{1}{A} < \tilde{\zeta}_m(\boldsymbol{\kappa}; A)\tilde{\zeta}_n^{*}(\boldsymbol{\kappa}; A) >. \tag{43}$$

It should be noted that the domain of a rough interface is physically limited by the illumination beamwidth. Note also that the different definitions of the Fourier transform are available and used in the literature: the sign of the complex exponential function are sometimes exchanged and a multiplicative constant $(2\pi)^{-2}$ may appear in front of either integral or its square root in front of each expression (25)-(26). Finally, we recall that the theory of random process predicts only the averages over many realizations.

5. Boundary Perturbation Theory (BPT)

In this section, we first introduce the general perturbative expansion on which the BPT formulation is based. A systematic matrix reformulation, which enables the formal evaluation of pertinent scattered field solutions, is then presented.

5.1 Perturbative formulation

With reference to the geometry of Fig.1, in order to obtain a solution valid in each region of the structure, we have to enforce the continuity of the tangential fields:

$$[\hat{n}_m \times \Delta\mathbf{E}_m]_{z=\zeta_m(\mathbf{r}_\perp)-d_m} = 0, \tag{44}$$

$$[\hat{n}_m \times \Delta\mathbf{H}_m]_{z=\zeta_m(\mathbf{r}_\perp)-d_m} = 0, \tag{45}$$

where $\Delta\mathbf{E}_m = \mathbf{E}_{m+1} - \mathbf{E}_m$, $\Delta\mathbf{H}_m = \mathbf{H}_{m+1} - \mathbf{H}_m$, and the surface normal vector is given by:

$$\hat{n}_m = \frac{\hat{z} - \boldsymbol{\gamma}_m}{\sqrt{1-\gamma_m^2}}, \tag{46}$$

with the slope vector $\boldsymbol{\gamma}_m$:

$$\boldsymbol{\gamma}_m = \nabla_\perp \zeta_m = \left[\frac{\partial}{\partial x}\hat{x} + \frac{\partial}{\partial y}\hat{y}\right]\zeta_m, \tag{47}$$

and where $\nabla_\perp$ is the *nabla* operator in the *x-y* plane. In order to study the fields $\mathbf{E}_m$ and $\mathbf{H}_m$ within the generic *m*th layer of the structure, we assume then that, for each *m*th rough interface, the deviations and slopes of the interface, with respect to the reference mean plane $z=-d_m$, are small enough in the sense of (Ulaby et al, 1982) (Tsang et al., 1985), so that the fields can be expanded about the reference mean plane. Assume that the fields can be expanded about the reference mean plane $z=-d_m$ as:

$$\Delta\mathbf{E}_m(z) = \Delta\mathbf{E}_m\Big|_{z=-d_m} + \frac{\partial\Delta\mathbf{E}_m}{\partial z}\Big|_{z=-d_m}(z+d_m) + \frac{1}{2}\frac{\partial^2\Delta\mathbf{E}_m}{\partial z^2}\Big|_{z=-d_m}(z+d_m)^2 + \dots, \tag{48}$$

$$\Delta\mathbf{H}_m(z) = \Delta\mathbf{H}_m\Big|_{z=-d_m} + \frac{\partial\Delta\mathbf{H}_m}{\partial z}\Big|_{z=-d_m}(z+d_m) + \frac{1}{2}\frac{\partial^2\Delta\mathbf{H}_m}{\partial z^2}\Big|_{z=-d_m}(z+d_m)^2 + \dots, \tag{49}$$

where the dependence on $\mathbf{r}_\perp$ is understood. Then (48), (49) are the fields expansions in perturbative orders of the fields and their derivatives at the interfaces of the structure; they can be injected into the boundary conditions (44-45). Retaining only up to the first-order terms with respect to ζ_m and γ_m, we obtain:

$$\hat{z}\times\Delta\mathbf{E}_m\Big|_{z=-d_m} = \nabla_\perp\zeta_m\times\Delta\mathbf{E}_m\Big|_{z=-d_m} - \zeta_m\,\hat{z}\times\frac{\partial\Delta\mathbf{E}_m}{\partial z}\Big|_{z=-d_m}, \tag{50}$$

$$\hat{z}\times\Delta\mathbf{H}_m\Big|_{z=-d_m} = \nabla_\perp\zeta_m\times\Delta\mathbf{H}_m\Big|_{z=-d_m} - \zeta_m\,\hat{z}\times\frac{\partial\Delta\mathbf{H}_m}{\partial z}\Big|_{z=-d_m}. \tag{51}$$

The field solutions can then be represented formally as

$$\mathbf{E}_m(\mathbf{r}_\perp, z) \approx \mathbf{E}_m^{(0)} + \mathbf{E}_m^{(1)} + \mathbf{E}_m^{(2)} + \dots, \tag{52}$$

$$\mathbf{H}_m(\mathbf{r}_\perp, z) \approx \mathbf{H}_m^{(0)} + \mathbf{H}_m^{(1)} + \mathbf{H}_m^{(2)} + \dots. \tag{53}$$

where the parenthesized superscript refers to the perturbation field of order *n*: $\mathbf{E}_m^{(0)}, \mathbf{H}_m^{(0)}$ is the unperturbed solution and $\mathbf{E}_m^{(1)}, \mathbf{H}_m^{(1)}$ is correction to the first-order of ζ_m and γ_m. It should be noted that the unperturbed solution represents the field existing in flat boundaries stratification, and satisfying:

$$\hat{z} \times \Delta \mathbf{E}_m^{(0)}\Big|_{z=-d_m} = 0\,, \qquad \hat{z} \times \Delta \mathbf{H}_m^{(0)}\Big|_{z=-d_m} = 0\,. \tag{54}$$

The fields expansion (52)-(53) can be then injected into the boundary conditions (50)-(51), so that, retaining only up to the first-order terms, the following *nonuniform boundary conditions* can be obtained (Imperatore et al. 2008a) (Imperatore et al. 2008b) (Imperatore et al. 2009a)

$$\hat{z} \times \Delta \mathbf{E}_m^{(1)}\Big|_{z=-d_m} = \nabla_\perp \zeta_m \times \Delta \mathbf{E}_m^{(0)}\Big|_{z=-d_m} - \zeta_m\, \hat{z} \times \frac{\partial \Delta \mathbf{E}_m^{(0)}}{\partial z}\Bigg|_{z=-d_m}, \tag{55}$$

$$\hat{z} \times \Delta \mathbf{H}_m^{(1)}\Big|_{z=-d_m} = \nabla_\perp \zeta_m \times \Delta \mathbf{H}_m^{(0)}\Big|_{z=-d_m} - \zeta_m\, \hat{z} \times \frac{\partial \Delta \mathbf{H}_m^{(0)}}{\partial z}\Bigg|_{z=-d_m}. \tag{56}$$

Therefore, the boundary conditions from each mth rough interface can be transferred to the associated equivalent flat interface. In addition, the right-hand sides of Eqs. (55) and (56) can be interpreted as effective magnetic ($\mathbf{J}_{Hm}^{p(1)}$) and electric ($\mathbf{J}_{Em}^{p(1)}$) surface current densities, respectively, with p denoting the incident polarization; so that we can identify the first-order fluctuation fields as being excited by these effective surface current densities imposed on the unperturbed interfaces. Accordingly, the geometry randomness of each corrugated interfaces is then translated in random current sheets imposed on each reference mean plane (z=-d_m), which radiate in an unperturbed (flat boundaries) layered medium. As a result, within the first-order approximation, the field can be than represented as the sum of an unperturbed part $\mathbf{E}_n^{(0)}, \mathbf{H}_n^{(0)}$ and a random part, so that $\mathbf{E}_n(\mathbf{r}_\perp, z) \approx \mathbf{E}_n^{(0)} + \mathbf{E}_n^{(1)}$, $\mathbf{H}_n(\mathbf{r}_\perp, z) \approx \mathbf{H}_n^{(0)} + \mathbf{H}_n^{(1)}$. The first is the primary field, which exists in absence of surface boundaries roughness (flat-boundaries stratification), detailed in (Imperatore et al. 2009a); whereas $\mathbf{E}_n^{(1)}, \mathbf{H}_n^{(1)}$ can be interpreted as the superposition of single-scatter fields from each rough interface. In order to perform the evaluation of perturbative development, the scattered field in each region of the layered structure is then represented as the sum of *up-* and *down-going* waves, and the first-order scattered field in each region of the layered structure can be then characterized by adopting the following field *spectral representation* in terms of the unknown coefficients $S_m^{\pm q(1)}(\mathbf{k}_\perp)$:

$$\mathbf{E}_m^{(1)} = \mathbf{E}_m^{-(1)} + \mathbf{E}_m^{+(1)}\,, \tag{57}$$

$$\mathbf{H}_m^{(1)} = \mathbf{H}_m^{-(1)} + \mathbf{H}_m^{+(1)}\,, \tag{58}$$

with

$$\mathbf{E}_m^{\pm(1)} = \sum_{q=h,v} \iint d\mathbf{k}_\perp e^{j\mathbf{k}_\perp \cdot \mathbf{r}_\perp}\, \hat{q}_m^{\pm}(\mathbf{k}_\perp) S_m^{\pm q(1)}(\mathbf{k}_\perp) e^{\pm j k_{zm} z}\,, \tag{59}$$

$$\mathbf{H}_m^{\pm(1)} = \sum_{q=h,v} \frac{1}{Z_m} \iint d\mathbf{k}_\perp e^{j\mathbf{k}_\perp \cdot \mathbf{r}_\perp}\, \hat{k}_m^{\pm} \times \hat{q}_m^{\pm}(\mathbf{k}_\perp) S_m^{\pm q(1)}(\mathbf{k}_\perp) e^{\pm j k_{zm} z}\,. \tag{60}$$

where Z_m is the intrinsic impedance of the medium m, and where

$$\hat{h}_m^{\pm}(\mathbf{k}_\perp) = \hat{k}_\perp \times \hat{z} = \hat{h} \tag{61}$$

$$\hat{v}_m^{\pm}(\mathbf{k}_\perp) = \mp \frac{k_{zm}}{k_m}\hat{k}_\perp + \frac{k_\perp}{k_m}\hat{z} \tag{62}$$

is a basis for the horizontal/vertical polarization vectors.

5.2 Matrix reformulation

In this section, we reformulate the *non-uniform boundaries condition* (55, 56), reducing the scattering problem to the formal solution of a linear system of equations; the unknowns are the scalar amplitudes, $S_m^{\pm q(1)}(\mathbf{k}_\perp)$, of the scattered fields. Eqs. (55, 56) can be rewritten by using their spectral representation:

$$\hat{z} \times \Delta\mathbf{E}_m^{(1)}\Big|_{z=-d_m} = \iint d\mathbf{k}_\perp\, e^{j\mathbf{k}_\perp \cdot \mathbf{r}_\perp} \widetilde{\mathbf{J}}_{Hm}^{p(1)}(\mathbf{k}_\perp, \mathbf{k}_\perp^i), \tag{63}$$

$$\hat{z} \times \Delta\mathbf{H}_m^{(1)}\Big|_{z=-d_m} = \iint d\mathbf{k}_\perp\, e^{j\mathbf{k}_\perp \cdot \mathbf{r}_\perp} \widetilde{\mathbf{J}}_{Em}^{p(1)}(\mathbf{k}_\perp, \mathbf{k}_\perp^i), \tag{64}$$

where the spectral densities $\widetilde{\mathbf{J}}_{Em}^{p(1)}, \widetilde{\mathbf{J}}_{Hm}^{p(1)}$ are the two-dimensional *Fourier transform* (2D-FT), with respect to $\mathbf{k}_\perp$, of the right-hand sides of (55) and (56), respectively, so that:

$$\widetilde{\mathbf{J}}_{Hm}^{p(1)}(\mathbf{k}_\perp, \mathbf{k}_\perp^i) = \widetilde{\zeta}_m(\mathbf{k}_\perp - \mathbf{k}_\perp^i)\left\{ j(\mathbf{k}_\perp - \mathbf{k}_\perp^i) \times \Delta\widetilde{\mathbf{E}}_m^{(0)}\Big|_{z=-d_m} - \hat{z} \times \frac{\partial \Delta\widetilde{\mathbf{E}}_m^{(0)}}{\partial z}\bigg|_{z=-d_m} \right\}, \tag{65}$$

$$\widetilde{\mathbf{J}}_{Em}^{p(1)}(\mathbf{k}_\perp, \mathbf{k}_\perp^i) = \widetilde{\zeta}_m(\mathbf{k}_\perp - \mathbf{k}_\perp^i)\left\{ j(\mathbf{k}_\perp - \mathbf{k}_\perp^i) \times \Delta\widetilde{\mathbf{H}}_m^{(0)}\Big|_{z=-d_m} - \hat{z} \times \frac{\partial \Delta\widetilde{\mathbf{H}}_m^{(0)}}{\partial z}\bigg|_{z=-d_m} \right\}, \tag{66}$$

where $\widetilde{\zeta}_m(\mathbf{k}_\perp)$ is the *spectral representation* (2D-FT) of the corrugation $\zeta_m(\mathbf{r}_\perp)$, and where $\Delta\widetilde{\mathbf{E}}_m^{(0)} = e^{-j\mathbf{k}_\perp^i \cdot \mathbf{r}_\perp}\Delta\mathbf{E}_m^{(0)}, \Delta\widetilde{\mathbf{H}}_m^{(0)} = e^{-j\mathbf{k}_\perp^i \cdot \mathbf{r}_\perp}\Delta\mathbf{H}_m^{(0)}$, $p \in \{v, h\}$ is associated with the incident field polarization, and where we have taken into account that the 2D-FT of $\nabla_\perp \zeta_m(\mathbf{r}_\perp)$ is $j\mathbf{k}_\perp \widetilde{\zeta}_m(\mathbf{k}_\perp)$, and that the 2D-FT of $\zeta_m(\mathbf{r}_\perp)e^{j\mathbf{k}_\perp^i \cdot \mathbf{r}_\perp}$ is $\widetilde{\zeta}_m(\mathbf{k}_\perp - \mathbf{k}_\perp^i)$.

Therefore, a solution valid in each region of the layered structure can be obtained from (57)-(62) taking into account the non uniform boundary conditions (63)-(64). In order to solve the scattering problem in terms of the unknown expansion coefficients $S_m^{\pm q(1)}(\mathbf{k}_\perp)$, we arrange their amplitudes in a single vector according to the notation:

$$\mathbf{S}_m^{q(1)}(\mathbf{k}_\perp, d_m) = \begin{bmatrix} S_m^{+q(1)}(\mathbf{k}_\perp)e^{-jk_{zm}d_m} \\ S_m^{-q(1)}(\mathbf{k}_\perp)e^{+jk_{zm}d_m} \end{bmatrix}. \tag{67}$$

Subsequently, the *nonuniform boundary conditions* (63)-(64) can be reformulated by employing a suitable matrix notation, so that for the (*q=h*) *horizontal* polarized scattered wave we have (Imperatore et al. 2008a) (Imperatore et al. 2009a):

$$\mathbf{S}_m^{h(1)}(\mathbf{k}_\perp, d_m) + \boldsymbol{\Theta}_m^p(\mathbf{k}_\perp, \mathbf{k}_\perp^i) = \boldsymbol{N}_{m|m+1}^h(k_\perp)\mathbf{S}_{m+1}^{h(1)}(\mathbf{k}_\perp, d_m), \tag{68}$$

where

$$\boldsymbol{\Theta}_m^p(\mathbf{k}_\perp, \mathbf{k}_\perp^i) = \begin{bmatrix} -\dfrac{k_0 Z_0 \mu_m}{2k_{zm}}(\hat{k}_\perp \times \hat{z}) \cdot \widetilde{\mathbf{J}}_{Em}^{p(1)} + \dfrac{1}{2}\hat{k}_\perp \cdot \widetilde{\mathbf{J}}_{Hm}^{p(1)} \\ +\dfrac{k_0 Z_0 \mu_m}{2k_{zm}}(\hat{k}_\perp \times \hat{z}) \cdot \widetilde{\mathbf{J}}_{Em}^{p(1)} + \dfrac{1}{2}\hat{k}_\perp \cdot \widetilde{\mathbf{J}}_{Hm}^{p(1)} \end{bmatrix} \tag{69}$$

is the term associated with the *effective* source distribution, where the expressions of the effective currents $\widetilde{\mathbf{J}}_{Em}^{p(1)}$ and $\widetilde{\mathbf{J}}_{Hm}^{p(1)}$, imposed on the (flat) unperturbed boundary $z = -d_m$, for an incident polarization $p \in \{v, h\}$ are detailed in (Imperatore et al. 2009a); and where Z_0 is the intrinsic impedance of the vacuum. Furthermore, the fundamental *transfer matrix operator* is given by:

$$\boldsymbol{N}_{m-1|m}^q(k_\perp) = \frac{1}{T_{m-1|m}^q}\begin{bmatrix} 1 & R_{m-1|m}^q \\ R_{m-1|m}^q & 1 \end{bmatrix}, \tag{70}$$

with the superscripts $q \in \{v, h\}$ denoting the polarization. Moreover, it should be noted that on a (*k*th) flat interface Eq. (68) reduces to the *uniform boundary conditions,* thus getting:

$$\mathbf{S}_k^{h(1)}(\mathbf{k}_\perp, d_k) = \boldsymbol{N}_{k|k+1}^h(k_\perp)\mathbf{S}_{k+1}^{h(1)}(\mathbf{k}_\perp, d_k). \tag{71}$$

We emphasize that equations (68) state in a simpler form the problem originally set by Eqs. (55)-(56): as matter of fact, solving Eq. (68) $\forall m$ implies dealing with the determination of unknown scalar amplitudes $S_m^{\pm q(1)}(\mathbf{k}_\perp)$ instead of working with the corresponding vector unknowns $\mathbf{E}_m^{(1)}, \mathbf{H}_m^{(1)}$. Therefore, the scattering problem in each *m*th layer is reduced to the algebraic calculation of the unknown expansion scattering coefficients vector (67). As a result, when a structure with rough interfaces is considered, the enforcement of the original non uniform boundary conditions through the stratification (*m=0, ..., N-1*) can be addressed by writing down a linear system of equations with the aid of the matrix formalism (68)-(69) with *m=0, ..., N-1.* As a result, the formulation of *non-uniform boundary conditions* in matrix notation (68)-(69) enables a systematic method for solving the scattering problem: For the *N*-layer stratification of Fig.1, we have to find *2N* unknown expansion coefficients, using *N* vectorial equations (68), i.e., *2N* scalar equations. It should be noted that, for the considered configuration, the relevant scattering coefficients $S_N^{+q(1)}(\mathbf{k}_\perp), S_0^{-q(1)}(\mathbf{k}_\perp)$ are obviously supposed to be zero. The scattering problem, therefore, results to be reduced to a formal solution of a linear equation system. It can be demonstrated that, making use of a *recursive approach* involving the key-concept of generalized transmission/reflection (see also Sect. 3), the system of equations (68, 69) is susceptible of a straightforward closed form solution, so that the first-order perturbation fields anywhere in the upper half-space that arise from each *m*th rough interface can be formally found (Imperatore et al. 2009a) (Imperatore et al. 2009b).

In conclusion, the derivation of scattering field contribution, due to each rough interface, for instance in the upper or the lower half-space, can be then accomplished by avoiding the use of the cumbersome *Green functions* formalism.
We finally emphasize that here we are interested in the scattering from and through the stratification; therefore, the determination of the pertinent unknown expansion coefficients $S_0^{+q(1)}(\mathbf{k}_\perp)$ and $S_N^{-q(1)}(\mathbf{k}_\perp^s)$ of the scattered wave, respectively, into the upper and the lower half-space, is required only. Full expressions for these coefficients are reported in (Imperatore et al. 2009a) (Imperatore et al. 2009b).

6. *BPT* closed-form solutions

The aim of this section is to present the relevant BPT solutions for the scattering from and through the 3-D layered rough structure pictured schematically in Fig.1. We underline that the corresponding first-order solutions refer to two complementary bistatic configuration: in the first case, both the transmitter and the receiver are into the same half-space, whereas, in the second case, each one is located in a different half-space.

6.1 Scattering from layered structure with an arbitrary number of rough interfaces

First we consider the case of one rough interface embedded in the layered structure. The field scattered upward in the upper half-space in the first-order limit can be written in the form (see (57)-(60)):

$$\mathbf{E}_0^{(1)}(\mathbf{r}) = \sum_{q=h,v} \iint d\mathbf{k}_\perp e^{j\mathbf{k}_\perp \cdot \mathbf{r}_\perp} \hat{q}_0^+(\mathbf{k}_\perp) S_0^{+q(1)}(\mathbf{k}_\perp) e^{jk_{z0}z} \,. \tag{72}$$

By employing the *method of stationary phase,* we evaluate the integral (72) in the *far field* zone, obtaining:

$$\mathbf{E}_0^{(1)}(\mathbf{r}) \cdot \hat{q}_0^+(\mathbf{k}_\perp^s) \cong -j2\pi\, k_0 \cos\theta_0^s \frac{e^{j k_0 r}}{r} S_0^{+q(1)}(\mathbf{k}_\perp^s)\,, \tag{73}$$

with $q \in \{v, h\}$ is the polarization of the scattered field. Taking into account the expressions for the unknowns expansion coefficients $S_0^{+q(1)}(\mathbf{k}_\perp^s)$ (Imperatore et al. 2009a), we get

$$\mathbf{E}_0^{(1)}(\mathbf{r}) \cdot \hat{q}_0^+(\mathbf{k}_\perp^s) = \pi\, k_0^2 \frac{e^{jk_0 r}}{r} \tilde{\alpha}_{qp}^{m,m+1}(\mathbf{k}^s, \mathbf{k}^i) \tilde{\zeta}_m(\mathbf{k}_\perp^s - \mathbf{k}_\perp^i)\,, \tag{74}$$

wherein

$$\begin{aligned}\tilde{\alpha}_{hh}^{m,m+1} &= (\varepsilon_{m+1} - \varepsilon_m) \frac{k_{z0}^s}{k_{zm}^s} (\hat{k}_\perp^s \cdot \hat{k}_\perp^i) \\ & e^{jk_{zm}^s \Delta_m} \mathfrak{I}_{m|0}^h (k_\perp^s)[1 + \mathfrak{R}_{m|m+1}^h (k_\perp^s)]\, e^{jk_{zm}^i \Delta_m} \mathfrak{I}_{0|m}^h (k_\perp^i)[1 + \mathfrak{R}_{m|m+1}^h (k_\perp^i)]\,,\end{aligned} \tag{75}$$

$$\tilde{\alpha}_{vh}^{m,m+1} = (\varepsilon_{m+1} - \varepsilon_m)\frac{k_{z0}^s}{k_0}\hat{z}\cdot(\hat{k}_\perp^i \times \hat{k}_\perp^s)$$
$$e^{jk_{zm}^s\Delta_m}\mathfrak{I}_{m|0}^v(k_\perp^s)[1-\mathfrak{R}_{m|m+1}^v(k_\perp^s)]\,e^{jk_{zm}^i\Delta_m}\mathfrak{I}_{0|m}^h(k_\perp^i)[1+\mathfrak{R}_{m|m+1}^h(k_\perp^i)] \quad , (76)$$

$$\tilde{\alpha}_{hv}^{m,m+1} = (\varepsilon_{m+1} - \varepsilon_m)\frac{k_{z0}^s k_{zm}^i}{k_0 k_{zm}^s \varepsilon_m}\hat{z}\cdot(\hat{k}_\perp^i \times \hat{k}_\perp^s) \quad (77)$$
$$e^{jk_{zm}^s\Delta_m}\mathfrak{I}_{m|0}^h(k_\perp^s)[1+\mathfrak{R}_{m|m+1}^h(k_\perp^s)]\,e^{jk_{zm}^i\Delta_m}\mathfrak{I}_{0|m}^v(k_\perp^i)[1-\mathfrak{R}_{m|m+1}^v(k_\perp^i)] \, ,$$

$$\tilde{\alpha}_{vv}^{m,m+1} = (\varepsilon_{m+1} - \varepsilon_m)\frac{k_{z0}^s}{k_0^2 k_{zm}^s \varepsilon_m}e^{jk_{zm}^s\Delta_m}\mathfrak{I}_{m|0}^v(k_\perp^s)e^{jk_{zm}^i\Delta_m}\mathfrak{I}_{0|m}^v(k_\perp^i)$$
$$\{[1+\mathfrak{R}_{m|m+1}^v(k_\perp^s)][1+\mathfrak{R}_{m|m+1}^v(k_\perp^i)]\frac{\varepsilon_m}{\varepsilon_{m+1}}k_\perp^i k_\perp^s \quad (78)$$
$$-[1-\mathfrak{R}_{m|m+1}^v(k_\perp^s)][1-\mathfrak{R}_{m|m+1}^v(k_\perp^i)]k_{zm}^s k_{zm}^i(\hat{k}_\perp^s\cdot\hat{k}_\perp^i)\} \quad ,$$

where $k_{zm}^i = k_{zm}(k_\perp^i)$, $k_{zm}^s = k_{zm}(k_\perp^s)$, $\mathfrak{I}_{0|m}^p$ and $\mathfrak{I}_{m|0}^p$ are, respectively, the generalized transmission coefficients in *downward* direction and the generalized transmission coefficients in *upward* direction (see (16)-(17)), and $\mathfrak{R}_{m|m+1}^p$ are the generalized reflection coefficients (see eq. (11)). The coefficients $\tilde{\alpha}_{qp}^{m,m+1}$ are relative to the p-polarized incident wave impinging on the structure from upper half space 0 and to the q-polarized scattering contribution from structure into the upper half space, originated from the rough interface between the layers m, $m+1$. Finally, we emphasize that the total scattering from the N-rough interfaces layered structure can be straightforwardly obtained, in the first-order approximation, by superposition of the different contributions pertaining each rough interface:

$$\mathbf{E}_0^{(1)}(\mathbf{r})\cdot\hat{q}_0^+(\mathbf{k}_\perp^s) = \pi k_0^2 \frac{e^{jk_0 r}}{r}\sum_{m=0}^{N-1}\tilde{\alpha}_{qp}^{m,m+1}(\mathbf{k}^s,\mathbf{k}^i)\tilde{\zeta}_m(\mathbf{k}_\perp^s - \mathbf{k}_\perp^i)\cdot \quad (79)$$

6.2 Scattering through layered structure with an arbitrary number of rough interfaces

Similarly, when one rough interface embedded in the layered structure is concerned, the field scattered into the last half-space, through the 3-D layered structure, in the first-order limit can be then written in the form:

$$\mathbf{E}_N^{(1)}(\mathbf{r}) = \sum_{q=h,v}\iint d\mathbf{k}_\perp e^{j\mathbf{k}_\perp\cdot\mathbf{r}_\perp}\hat{q}_N^-(\mathbf{k}_\perp)S_N^{-q(1)}(\mathbf{k}_\perp)e^{-jk_{zN}z} \, . \quad (80)$$

In order to evaluate the integral (80) in *far field* zone, we firstly consider a suitable change of variable $\mathbf{r}' = (\mathbf{r}_\perp, z')$, with $z' = -z - d_{N-1}$:

$$\mathbf{E}_N^{(1)}(\mathbf{r}') = \sum_{q=h,v} \iint d\mathbf{k}_\perp e^{j\mathbf{k}_\perp \cdot \mathbf{r}_\perp} \hat{q}_N^-(\mathbf{k}_\perp) S_N^{-q(1)}(\mathbf{k}_\perp) e^{jk_{zN} d_{N-1}} e^{jk_{zN} z'}, \tag{81}$$

then we use the method of *stationary phase* and obtain:

$$\mathbf{E}_N^{(1)}(\mathbf{r}') \cdot \hat{q}_N^-(\mathbf{k}_\perp^s) \cong -j2\pi\, k_N \cos\theta_N^s \frac{e^{jk_N r'}}{r'} S_N^{-q(1)}(\mathbf{k}_\perp^s) e^{jk_{zN}^s d_{N-1}}, \tag{82}$$

with $q \in \{v, h\}$. Taking into account the expressions for the unknowns expansion coefficients $S_N^{-q(1)}(\mathbf{k}_\perp^s)$ (Imperatore et al. 2009b), we get

$$\mathbf{E}_N^{(1)}(\mathbf{r}') \cdot \hat{q}_N^-(\mathbf{k}_\perp^s) = \pi\, k_0^2 \frac{e^{jk_N r'}}{r'} \, {}_N^0\tilde{\beta}_{qp}^{m,m+1}(\mathbf{k}^s, \mathbf{k}^i)\, \tilde{\zeta}_m(\mathbf{k}_\perp^s - \mathbf{k}_\perp^i), \tag{83}$$

wherein

$$\begin{aligned} {}_N^0\tilde{\beta}_{hh}^{m,m+1} &= \left(\varepsilon_{m+1} - \varepsilon_m\right) \frac{k_{zN}^s}{k_{z(m+1)}^s} (\hat{k}_\perp^s \cdot \hat{k}_\perp^i) \\ &\quad [1 + \Re_{m+1|m}^h(k_\perp^s)] e^{jk_{z(m+1)}^s \Delta_{m+1}} \Im_{m+1|N}^h(k_\perp^s) [1 + \Re_{m|m+1}^h(k_\perp^i)] e^{jk_{zm}^i \Delta_m} \Im_{0|m}^h(k_\perp^i), \end{aligned} \tag{84}$$

$$\begin{aligned} {}_N^0\tilde{\beta}_{vh}^{m,m+1} &= (\varepsilon_{m+1} - \varepsilon_m) \frac{k_{zN}^s}{k_0 \sqrt{\varepsilon_N}} \hat{z} \cdot (\hat{k}_\perp^s \times \hat{k}_\perp^i) \\ &\quad [1 - \Re_{m+1|m}^v(k_\perp^s)] e^{jk_{z(m+1)}^s \Delta_{m+1}} \Im_{m+1|N}^v(k_\perp^s) [1 + \Re_{m|m+1}^h(k_\perp^i)] e^{jk_{zm}^i \Delta_m} \Im_{0|m}^h(k_\perp^i), \end{aligned} \tag{85}$$

$$\begin{aligned} {}_N^0\tilde{\beta}_{hv}^{m,m+1} &= (\varepsilon_{m+1} - \varepsilon_m) \frac{k_{zN}^s k_{zm}^i \sqrt{\varepsilon_0}}{k_{z(m+1)}^s k_0 \varepsilon_m} \hat{z} \cdot (\hat{k}_\perp^i \times \hat{k}_\perp^s) \\ &\quad [1 + \Re_{m+1|m}^h(k_\perp^s)] e^{jk_{z(m+1)}^s \Delta_{m+1}} \Im_{m+1|N}^h(k_\perp^s) [1 - \Re_{m|m+1}^v(k_\perp^i)] e^{jk_{zm}^i \Delta_m} \Im_{0|m}^v(k_\perp^i), \end{aligned} \tag{86}$$

$$
{}^{0}_{N}\widetilde{\beta}^{m,m+1}_{vv} = (\varepsilon_{m+1}-\varepsilon_m)\frac{k^s_{zN}\sqrt{\varepsilon_0}}{k^s_{z(m+1)}k_0^2\varepsilon_m\sqrt{\varepsilon_N}}e^{jk^s_{z(m+1)}\Delta_{m+1}}\Im^v_{m+1|N}(k^s_\perp)\,\Im^v_{0|m}(k^i_\perp)e^{jk^i_{zm}\Delta_m}
$$
$$
\left\{ [1+\Re^v_{m+1|m}(k^s_\perp)][1+\Re^v_{m|m+1}(k^i_\perp)]k^i_\perp k^s_\perp + [1-\Re^v_{m+1|m}(k^s_\perp)][1-\Re^v_{m|m+1}(k^i_\perp)](\hat{k}^s_\perp\cdot\hat{k}^i_\perp)k^i_{zm}k^s_{z(m+1)} \right\},
$$

(87)

where $k^i_{zm} = k_{zm}(k^i_\perp)$, $k^s_{zm} = k_{zm}(k^s_\perp)$, $\Im^p_{0|m}$ and $\Im^p_{m+1|N}$ are, respectively, the generalized transmission coefficients in *downward* direction and the generalized transmission coefficients in *downward* direction given, respectively, by (16) and (21), and $\Re^p_{m|m+1}$ are the generalized reflection coefficients (see (11)). The coefficients ${}^{0}_{N}\widetilde{\beta}^{m,m+1}_{qp}$ are relative to the *p*-polarized incident wave impinging on the structure from half-space *0* and to *q*-polarized scattering contribution, originated from the rough interface between the layers *m* and *m+1*, through the structure into last half-space *N*. Finally, we emphasize that the total scattering through the *N*-rough interfaces layered structure can be straightforwardly obtained, in the first order approximation, by superposition of the different contributions pertaining each rough interface:

$$
\mathbf{E}^{(1)}_N(\mathbf{r}')\cdot\hat{q}^-_N(\mathbf{k}^s_\perp) = \pi k_0^2\frac{e^{jk_N r'}}{r'}\sum_{m=0}^{N-1}{}^{0}_{N}\widetilde{\beta}^{m,m+1}_{qp}(\mathbf{k}^s,\mathbf{k}^i)\,\widetilde{\zeta}_m(\mathbf{k}^s_\perp-\mathbf{k}^i_\perp)\cdot \tag{88}
$$

As a result, the relevant final solutions (79) and (88) turn out formally identical, provided that the coefficients $\widetilde{\alpha}^{m,m+1}_{qp}$ are replaced with the complementary ones ${}^{0}_{N}\widetilde{\beta}^{m,m+1}_{qp}$.

7. Bi-static scattering cross sections

In this section, we calculate the bi-static scattering cross sections of the layered structure arising from the *BPT* solutions, which have been derived in the first-order approximation in the previous sections. The estimate of the mean power density can be obtained by averaging over an ensemble of statistically identical interfaces.

7.1 Scattering Cross Section of an arbitrary layered structure with an embedded rough interface

In this section, we focus on the scattering property of a single rough interface embedded in the layered structure. The bi-static scattering *cross section* of a generic (*n*th) rough interface embedded in the layered structure can be then defined as

$$
\widetilde{\sigma}^0_{qp,n} = \lim_{r\to\infty}\lim_{A\to\infty}\frac{4\pi r^2}{A}<\left|\mathbf{E}^{(1)}_0(\mathbf{r})\cdot\hat{q}^+_0(\mathbf{k}^s_\perp)\right|^2>, \tag{89}
$$

where < > denotes ensemble averaging, where $q \in \{v, h\}$ and $p \in \{v, h\}$ denote, respectively, the polarization of scattered field and the polarization of incident field, and where A is the illuminated surface area. Therefore, by substituting (74) into (89) and considering that the (spatial) *power spectral density* $W_n(\mathbf{\kappa})$ of nth corrugated interface is defined as in (42), the *scattering cross section* relative to the contribution of the nth corrugated interface, according to the formalism used in [Franceschetti et al. 2008], can be expressed as

$$\tilde{\sigma}^0_{qp,n} = \pi k_0^4 \left| \tilde{\alpha}_{qp}^{n,n+1}(\mathbf{k}^s, \mathbf{k}^i) \right|^2 W_n(\mathbf{k}_\perp^s - \mathbf{k}_\perp^i) \ , \tag{90}$$

with $p, q \in \{v, h\}$ denoting, respectively, the incident and the scattered polarization states, which may stand for *horizontal* polarization (h) or *vertical* polarization (v); Furthermore, we stress when the backscattering case ($\hat{k}_\perp^s \times \hat{k}_\perp^i = 0$) is concerned, our cross-polarized scattering coefficients (75)-(78) evaluated in the plane of incidence vanish, in full accordance with the classical first-order SPM method for a rough surface between two different media (Ulaby et al, 1982) (Tsang et al., 1985).

7.2 Scattering Cross Section into last half-space of an Arbitrary Layered Structure with an Embedded Rough Interface

As counterparts of the configuration considered in the last subsection, we now refer to the complementary one in which the scattering through the structure is concerned. The bi-static scattering cross section into last half-space of the structure with one embedded (nth) rough interface can be defined as

$$\tilde{\sigma}^0_{qp,n} = \lim_{r \to \infty} \lim_{A \to \infty} \frac{4\pi r'^2}{A} < \left| \mathbf{E}_N^{(1)}(\mathbf{r}') \cdot \hat{q}_{\bar{N}}(\mathbf{k}_\perp^s) \right|^2 > \mathrm{Re}\left\{ \sqrt{\frac{\varepsilon_N}{\varepsilon_0}} \right\}, \tag{91}$$

where < > denotes ensemble averaging, where the index $q \in \{v, h\}$ index $p \in \{v, h\}$ and denote, respectively, the polarization of scattered field and the polarization of incident field, A is the surface area, and where we have considered the *Poynthing* power density of the transmitted wave in Nth region normalized to the power density of the incident wave. Therefore, by substituting (82) into (91) and considering that the (spatial) *power spectral density* $W_n(\mathbf{\kappa})$ of nth corrugated interface is defined as in (42), as final result, we obtain:

$$\tilde{\sigma}^0_{qp,n} = \pi k_0^4 \left| {}_N^0\tilde{\beta}_{qp}^{n,n+1}(\mathbf{k}^s, \mathbf{k}^i) \right|^2 W_n(\mathbf{k}_\perp^s - \mathbf{k}_\perp^i) \mathrm{Re}\left\{ \sqrt{\frac{\varepsilon_N}{\varepsilon_0}} \right\}. \tag{92}$$

7.3 Scattering Cross Section of a Layered Structure with N-rough interfaces

We now show that the solutions, given by the expressions (90) and (92) respectively, are susceptible of a straightforward generalization to the case of arbitrary stratification with N-rough boundaries. Taking into account the contribution of each nth corrugated interface (see (79)), the global *bi-static scattering cross section* of the N-rough interface layered media can be expressed as:

$$\tilde{\sigma}_{qp}^{0} = \pi k_0^4 \sum_{n=0}^{N-1} \left| \tilde{\alpha}_{qp}^{n,n+1}(\mathbf{k}^s,\mathbf{k}^i) \right|^2 W_n(\mathbf{k}_\perp^s - \mathbf{k}_\perp^i) + \pi k_0^4 \sum_{i \neq j} \mathrm{Re}\left\{ \tilde{\alpha}_{qp}^{i,i+1}(\mathbf{k}^s,\mathbf{k}^i) \left[\tilde{\alpha}_{qp}^{j,j+1}(\mathbf{k}^s,\mathbf{k}^i) \right]^* \right\} W_{ij}(\mathbf{k}_\perp^s - \mathbf{k}_\perp^i), \tag{93}$$

with $p, q \in \{v, h\}$, where the asterisk denotes the complex conjugated, where $\tilde{\alpha}_{qp}^{i,i+1}$ are given by (75)-(78), and where the *cross power spectral density* W_{ij}, between the interfaces i and j, for the spatial frequencies of the roughness is given by (43).

Likewise, the solution given by the expression (92), is susceptible of a straightforward generalization to the case of arbitrary stratification with N-rough boundaries. Taking into account the contribution of each nth corrugated interface (see (88)), the global *bi-static scattering cross section* into last half-space of the N-rough interface layered media can be expressed as:

$$\tilde{\sigma}_{qp}^{0} = \pi k_0^4 \,\mathrm{Re}\left\{ \sqrt{\frac{\varepsilon_N}{\varepsilon_0}} \right\} \sum_{n=0}^{N-1} \left| {}_N^0\tilde{\beta}_{qp}^{n,n+1}(\mathbf{k}^s,\mathbf{k}^i) \right|^2 W_n(\mathbf{k}_\perp^s - \mathbf{k}_\perp^i) + \pi k_0^4 \,\mathrm{Re}\left\{ \sqrt{\frac{\varepsilon_N}{\varepsilon_0}} \right\} \sum_{i \neq j} \mathrm{Re}\left\{ {}_N^0\tilde{\beta}_{qp}^{i,i+1}(\mathbf{k}^s,\mathbf{k}^i) \left[{}_N^0\tilde{\beta}_{qp}^{j,j+1}(\mathbf{k}^s,\mathbf{k}^i) \right]^* \right\} W_{ij}(\mathbf{k}_\perp^s - \mathbf{k}_\perp^i), \tag{94}$$

where $p, q \in \{v, h\}$, where the asterisk denotes the complex conjugated, ${}_N^0\tilde{\beta}_{qp}^{i,i+1}$ are given by (84)-(87), and where the *cross power spectral density* W_{ij}, between the interfaces i and j, for the spatial frequencies of the roughness is given by (43).

Some final considerations are now in order. As a matter of fact, the presented closed-form solutions permit the *polarimetric* evaluation of the scattering for a *bi-static* configuration, *from* or *through* the layered rough structure, once the *three-dimensional* layered structure's parameters (shape of the roughness spectra, layers thickness and complex permittivities), the incident field parameters (frequency, polarization and direction) and the observation direction are been specified. As a result, an elegant closed form solution is established, which takes into account parametrically the dependence of scattering properties on structure (geometric and electromagnetic) parameters. Therefore, BPT formulation leads to solutions which exhibit a direct functional dependence (no integral evaluation is required) and, subsequently, permit to show that the scattered field can be parametrically evaluated considering a set of parameters: some of them refer to an unperturbed structure configuration, i.e. intrinsically the physical parameters of the smooth boundary structure, and others which are determined exclusively by (random) deviations of the corrugated boundaries from their reference position. Note also that the coefficients $\tilde{\alpha}_{qp}^{m,m+1}$ and ${}_N^0\tilde{\beta}_{qp}^{i,i+1}$ depend parametrically on the unperturbed structure parameters only. Procedurally, once the *generalized reflection/transmission coefficients* are recursively evaluated, the coefficients

$\tilde{\alpha}_{qp}^{i,i+1}$ and/or ${}_{N}^{0}\tilde{\beta}_{qp}^{i,i+1}$ can be than directly computed, so that the scattering cross sections (93) and/or (94) for the pertinent structure with rough interfaces can be finally predicted. Furthermore, the scattering from or through the rough layered media is sensitive to the correlation between rough profiles of different interfaces. In fact, a real layered structure will have interfaces cross-correlation somewhere between two limiting situations: perfectly correlated and uncorrelated roughness. Consequently, the degree of correlation affects the phase relation between the fields scattered by each rough interface. Obviously, when the interfaces are supposed to be uncorrelated, the second terms respectively in (93) and (94) vanish and accordingly, in the first-order approximation, the total scattering from or through the structure arises from the incoherent superposition of radiation scattered from each interface. We emphasize that the effects of the interaction between the rough interfaces can limited be treated, in the first-order approximation, only when the rough interfaces exhibit some correlation. In addition, it has been demonstrated that the proposed global solution turns out to be completely interpretable with basic physical concepts, clearly discerning the physics of the involved scattering mechanisms (Imperatore et al 2008c) (Imperatore et al. 2009c). Finally, it should be noted that the method to be applied needs only the classical gently-roughness assumption, without any further approximation.

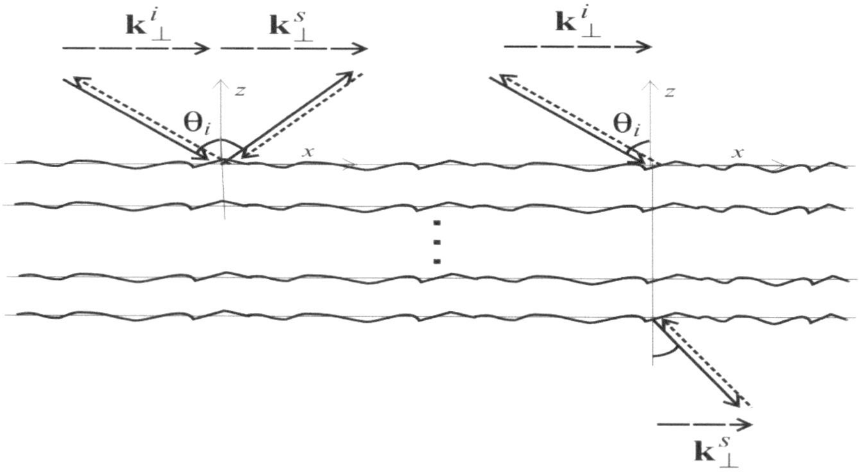

Fig. 3. Reciprocity for scattering from and through a layered structure with rough interfaces.

8. Reciprocal character of the BPT solutions

In this section, the emphasis is placed on the reciprocal character of the final BPT scattering solutions, which evidently constitutes a crucial point in the formal framework of the BPT. Generally speaking, the reciprocity principle is a statement that expresses some form of symmetry in the laws governing a physical system. Analytically speaking, both the BPT final solutions (79) and (88), respectively, from and through the layered structure with N-rough interfaces can be expressed in a common formal frame exhibiting a symmetric nature:

$$\tilde{\alpha}_{qp}^{m,m+1}(\mathbf{k}^s,\mathbf{k}^i)=\tilde{\alpha}_{pq}^{m,m+1}(-\mathbf{k}^i,-\mathbf{k}^s), \tag{95}$$

$${}_N^0\tilde{\beta}_{qp}^{m,m+1}(\mathbf{k}^s,\mathbf{k}^i)=-{}_0^N\tilde{\beta}_{pq}^{m+1,m}(-\mathbf{k}^i,-\mathbf{k}^s). \tag{96}$$

These formal relations are not only a mere matter of aesthetic; in fact their symmetry inherently reflects the conformity with the reciprocity principle of the electromagnetic theory. We emphasize that the relations (95), (96) imply that the wave amplitude for the scattering process $\mathbf{k}^i \to \mathbf{k}^s$ equals that of reciprocal scattering process $-\mathbf{k}^s \to -\mathbf{k}^i$ .Therefore, (95) and (96) are also *reciprocity* relationships for the scattering, respectively, from and through a layered structure with an (mth) embedded rough interface. This is to say that for the presented scattering solutions the role of the source and the receiver can be exchanged (see Fig.3), in conformity with the reciprocity principle of the electromagnetic theory. It should be noted that when the N-rough interfaces structure is concerned the properties (95)-(96) are satisfied as well, since the solutions in first-order limit are obtainable by superposition of the contribution of each (mth) rough interface. In order to provide general demonstration of these fundamental relationships, we found a more compact expression for (75)-(78) and (84)-(87), respectively. First, we introduce the following suitable notation:

$$\xi_{0\to m}^{\pm p}(k_\perp)=\Im_{0|m}^{p}(k_\perp)e^{jk_{zm}\Delta_m}[1\pm\Re_{m|m+1}^{p}(k_\perp)], \tag{97}$$

$$\xi_{N\to m+1}^{\pm p}(k_\perp)=\Im_{N|m+1}^{p}(k_\perp)e^{jk_{z(m+1)}\Delta_{m+1}}[1\pm\Re_{m+1|m}^{p}(k_\perp)]. \tag{98}$$

Next, when the solution for the scattering from the layered structure with an embedded rough interface is concerned, substituting relations (17) into (75)-(78), we obtain the alternative and more compact expressions for the relevant solution:

$$\tilde{\alpha}_{vv}^{m,m+1}(\mathbf{k}^s,\mathbf{k}^i)=(\varepsilon_{m+1}-\varepsilon_m)\left[\frac{\varepsilon_m}{\varepsilon_{m+1}}\frac{k_\perp^s}{k_0\varepsilon_m}\xi_{0\to m}^{+v}(k_\perp^s)\frac{k_\perp^i}{k_0\varepsilon_m}\xi_{0\to m}^{+v}(k_\perp^i) - \hat{k}_\perp^s\cdot\hat{k}_\perp^i\frac{k_{zm}^s}{k_0\varepsilon_m}\xi_{0\to m}^{-v}(k_\perp^s)\frac{k_{zm}^i}{k_0\varepsilon_m}\xi_{0\to m}^{-v}(k_\perp^i)\right], \tag{99}$$

$$\tilde{\alpha}_{vh}^{m,m+1}(\mathbf{k}^s,\mathbf{k}^i)=(\varepsilon_{m+1}-\varepsilon_m)\hat{z}\cdot(\hat{k}_\perp^i\times\hat{k}_\perp^s)\frac{k_{zm}^s}{k_0\varepsilon_m}\xi_{0\to m}^{-v}(k_\perp^s)\xi_{0\to m}^{+h}(k_\perp^i), \tag{100}$$

$$\tilde{\alpha}_{hv}^{m,m+1}(\mathbf{k}^s,\mathbf{k}^i)=(\varepsilon_{m+1}-\varepsilon_m)\hat{z}\cdot(\hat{k}_\perp^i\times\hat{k}_\perp^s)\xi_{0\to m}^{+h}(k_\perp^s)\frac{k_{zm}^i}{k_0\varepsilon_m}\xi_{0\to m}^{-v}(k_\perp^i), \tag{101}$$

$$\tilde{\alpha}_{hh}^{m,m+1}(\mathbf{k}^s,\mathbf{k}^i)=(\varepsilon_{m+1}-\varepsilon_m)\hat{k}_\perp^s\cdot\hat{k}_\perp^i\,\xi_{0\to m}^{+h}(k_\perp^s)\xi_{0\to m}^{+h}(k_\perp^i). \tag{102}$$

Then, by direct inspection of (99)-(102) we ultimately find Eq. (95).

On the other hand, when the solution for the scattering through the layered structure with an embedded rough interface is concerned, we proceed similarly as done previously. Substituting relations (18) into (84)-(87), we obtain the alternative and more compact expressions for the relevant solution:

$$
{}_N^0\tilde{\beta}_{vv}^{m,m+1}(\mathbf{k}_\perp^s,\mathbf{k}_\perp^i)=(\varepsilon_{m+1}-\varepsilon_m)\frac{\sqrt{\varepsilon_0\varepsilon_N}}{k_0^2\varepsilon_{m+1}\varepsilon_m}
\left\{k_\perp^s k_\perp^i\,\xi_{N\to m+1}^{+v}(k_\perp^s)\xi_{0\to m}^{+v}(k_\perp^i)+(\hat{k}_\perp^s\cdot\hat{k}_\perp^i)k_{z(m+1)}^s k_{zm}^i\,\xi_{N\to m+1}^{-v}(k_\perp^s)\xi_{0\to m}^{-v}(k_\perp^i)\right\}.
\tag{103}
$$

$$
{}_N^0\tilde{\beta}_{vh}^{m,m+1}(\mathbf{k}^s,\mathbf{k}^i)=(\varepsilon_{m+1}-\varepsilon_m)\,\hat{z}\cdot(\hat{k}_\perp^s\times\hat{k}_\perp^i)\frac{\sqrt{\varepsilon_N}\,k_{z(m+1)}^s}{k_0\varepsilon_{m+1}}\xi_{N\to m+1}^{-v}(k_\perp^s)\,\xi_{0\to m}^{+h}(k_\perp^i)\,,
\tag{104}
$$

$$
{}_N^0\tilde{\beta}_{hv}^{m,m+1}(\mathbf{k}^s,\mathbf{k}^i)=(\varepsilon_{m+1}-\varepsilon_m)\,\hat{z}\cdot(\hat{k}_\perp^i\times\hat{k}_\perp^s)\frac{\sqrt{\varepsilon_0}\,k_{zm}^i}{k_0\varepsilon_m}\xi_{N\to m+1}^{+h}(k_\perp^s)\,\xi_{0\to m}^{-v}(k_\perp^i),
\tag{105}
$$

$$
{}_N^0\tilde{\beta}_{hh}^{m,m+1}(\mathbf{k}^s,\mathbf{k}^i)=(\varepsilon_{m+1}-\varepsilon_m)(\hat{k}_\perp^s\cdot\hat{k}_\perp^i)\,\xi_{N\to m+1}^{+h}(k_\perp^s)\,\xi_{0\to m}^{+h}(k_\perp^i),
\tag{106}
$$

Then, by direct inspection of (103)-(106) we ultimately find Eq. (96). This is to say that BPT formalism satisfies reciprocity.

9. Conclusion

The problem of electromagnetic scattering in 3D layered rough structures can be analytical treated by relying on effective results of the *Boundary Perturbation Theory* (*BPT*), whose formulation has been introduced by P. Imperatore and his coauthors in many different papers. A structured presentation of the pertinent theoretical body of results has been provided in this chapter. The first-order scattering models obtained in the framework of the *BPT* allow us to polarimetrically deal with the (bi-static) scattering, from and through three-dimensional layered structures with an arbitrary number of gently rough interfaces.

Analytically speaking, two relevant closed-form solutions, obtained for two different configurations, respectively, for the scattering from and through the structure, are presented in a common formal frame. As a matter of fact, beyond a certain economy and mathematical elegance in the final analytical solutions, their inherent symmetry is intimately related to the reciprocity.

Some remarkable considerations on the meaning of the *BPT* solutions are in order. It can be demonstrated that, beyond the technicalities of the *BPT* formulation, the pertinent analytical results are also susceptible of a powerful physical interpretation; so that the fundamental interactions contemplated by the *BPT* can be revealed, gaining a coherent explanation and a neat picture of the physical meaning of the *BPT* theoretical construct (Imperatore et al 2008c)

(Imperatore et al. 2009c). The consequent phenomenological implications on the practical applications are then considerable.
Therefore, the formally symmetric, physically revealing, and fully polarimetric *BPT solutions* are amenable of direct and parametric numerical evaluation and, therefore, can be effectively applied to several practical situations of interest. We underline that it can be also demonstrated that all the previous existing perturbative scattering models, introduced by other authors to deal with simplified layered geometry with one (Yarovoy et al., 2000), (Azadegan and Sarabandi, 2003), (Fuks, 2001) or two rough interfaces (Tabatabaeenejad and Moghaddam, 2006), can be all rigorously regarded as a special cases of the general BPT solutions (see also Franceschetti et al, 2008). This analytical consistency also provides a unifying perspective on the perturbative approaches. Finally, the body of the *BPT* theoretical results can be also regarded as a generalization to the case of layered media with rough interfaces of the classical *SPM* for rough surface.

10. References

Franceschetti G., Imperatore P., Iodice A., Riccio D., and Ruello G. (2003), Scattering from a Layered Medium with One Rough Interface: Comparison and Physical Interpretation of Different Methods, *Proceedings of IEEE International Geoscience and Remote Sensing Symposium (IGARSS 2003), Toulouse, France,* pp. 2912–2914, Jul. 2003.

Franceschetti G., Imperatore P., Iodice A., Riccio D., and Ruello G. (2008). Scattering form Layered Strucutures with One rough Interface: A Unified Formulation of Perturbative Solutions. *IEEE Trans. on Geoscience and Remote Sensing,* vol.46, no.6, June 2008, pp.1634-1643.

Imperatore P., Iodice A., Riccio D. (2008a), Perturbative Solution for the Scattering from Multilayered Structure with Rough Boundaries, *Proceedings of Microwave Radiometry and Remote Sensing of the Environment* (MICRORAD 2008), pp.1-4, 11-14 March 2008, Florence, Italy.

Imperatore P., Iodice A., Riccio D. (2008b), Small Perturbation Method for Scattering From Rough Multilayers, *Proceedings of IEEE International Geoscience and Remote Sensing Symposium,* 2008 (IGARSS 2008), vol.5, pp.V-271-V-274, Boston 7-11 July 2008.

Imperatore P., Iodice A., Riccio D. (2008c), Interpretation of Perturbative Solution for the Scattering From Layered Structure With Rough Interfaces, *Proceedings of IEEE International Geoscience and Remote Sensing Symposium,* 2008 (IGARSS 2008), vol.4, pp. IV -1141-IV-1144, Boston 7-11 July 2008.

Imperatore P., Iodice A., Riccio D. (2009a). Electromagnetic Wave Scattering From Layered Structures with an Arbitrary Number of Rough Interfaces. *IEEE Trans. on Geoscience and Remote Sensing,* vol.47, no.4, pp.1056-1072, April 2009.

Imperatore P., Iodice A., Riccio D. (2009b). Transmission Through Layered Media With Rough Boundaries: First-Order Perturbative Solution. *IEEE Transaction on Antennas and Propagation,* vol.57, no.5, pp.1481-1494, May 2009.

Imperatore P., Iodice A., Riccio D. (2009c). Physical Meaning of Perturbative Solutions for Scattering from and Through Multilayered Structure with Rough Interfaces. *submitted to IEEE Trans. on Antennas and Propagation.*

Azadegan R. and Sarabandi K. (2003), Analytical formulation of the scattering by a slightly rough dielectric boundary covered with a homogeneous dielectric layer, *Proceedings of IEEE AP-S Int. Symp.*, Columbus, OH, Jun. 2003, pp. 420–423.

Fuks I. M.(1998), Radar contrast polarization dependence on subsurface sensing, *Proceedings of IEEE IGARSS'98*, vol. 3, Seattle, WA, USA, July 6–10, 1998, pp. 1455–1459.

Fuks I.M. (2001). Wave diffraction by a rough boundary of an arbitrary plane-layered medium. *IEEE Transaction on Antennas and Propagation*, vol. 49, no.4, Apr. 2001. pp.630–639.

Yarovoy A.G., de Jongh R.V., Ligthard L.P. (2000). Scattering properties of a statistically rough interface inside a multilayered medium. *Radio Science*, vol.35, n.2, 2000, pp. 445–462.

Tabatabaeenejad A., Moghaddam M. (2006). Bistatic scattering from three-dimensional layered rough surfaces. *IEEE Trans. on Geoscience and Remote Sensing*, vol. 44, no. 8, pp. 2102–2114, Aug. 2006.

Chew W. C. (1997). *Waves and Fields in Inhomogeneous Media*. IEEE Press, 1997.

Ishimaru A. (1978). *Wave Propagation and Scattering in Random Media*, vol.2. Academic Press, New York, 1978.

Tsang L., Kong J. A, and Shin R. T. (1985). *Theory of Microwave Remote Sensing*. New York: Wiley, 1985.

Ulaby F. T., Moore R. K., and Fung A. K. (1982). *Microwave Remote Sensing*, vol. I, II, II. Reading, MA: Addison-Wesley, 1982.

2

Numerical Analysis of Planar Periodic Multilayer Structures by Method of Moments

Stanislav Gona
Tomas Bata University in Zlin
Czech Republic

1. Introduction

Numerical analysis of planar periodic multilayer structures is often carried out with the aid the method of moments (MoM). The advantage of the technique against the other, more general methods (like FEM and FDTD) is in faster computation of reflection and transmission properties of a periodic structure. Besides of computation speed, usage of MoM typically results in lower memory requirements when compared to FEM and FDTD. The method of moments can be either formulated in the spatial or spectral domain. For analysis of periodic structures the spectral formulation is more advantageous - a discrete space spectrum (Scott, 1989). Thus, the original integral equations reduce into algebraic ones (that is double summations are being used instead of surface integrals). The disadvantage of the spectral formulation is that double summations arising in MoM formulation are slowly convergent and a high number of Floquet modes is needed for analysis of periodic structures having fine patch details inside the periodic cell or densely stacked structures. Simple spectral domain MoM codes for analysis periodic structures consider uniform mesh of cells and utilize FFT to accelerate the double summations (Cwick & Mittra, 1987; Wan & Encinar, 1995). These simple MoM codes also typically use small domain basis functions (like rooftops (Cwick & Mittra, 1987), or triangular (RWG) basis functions (Kipp & Chan, 1994)). If more general geometry is to be analyzed, then a non-uniform mesh of rectangular or quad shaped cells (Kolundzija, 1998) must be used. From the computational point of view it also highly desirable to consider use of higher order large domain basis functions for representation of surface currents. Then, the conductive currents in patches may be accurately described with a small number of unknown expansion coefficients.
Solution of reflection/transmission properties of a multilayer periodic structure can be performed either directly (Wu, 1995) or by the use of a cascade approach (Mittra et. al, 1988) or (Wan & Encinar, 1995). The cascade approach is suitable for periodic structures which with a large or a medium electrical thickness of individual dielectric layers. When the electrical thickness of a particular dielectric layer becomes too small then a large number of Floquet modes must be used during the cascade process (Wan & Encinar, 1995). In such a case, which is a typical for periodic structures with microscopically thin layers (e.g. carbon fibre composite materials) the direct approach is being used.

The chapter is organized as follows. In section 2, the formulation of MoM in the spectral domain is described. The formulation is based on (Wu, 1995) and in the further text it will be referenced as a direct approach. Section 3 describes approximation of surface currents with small domain rooftops defined over a non-uniform mesh of rectangular cells and use of higher order domain basis functions defined over quadrilaterals. Computation of reflection and transmission coefficients is described in section 4. Numerical examples are given in section 5. These examples fall into two different groups. First group belongs to the area of frequency selective surfaces, where results from our in-house MoM code called FSSMQ are compared with those existing in literature. Second group of examples spans into the area of composite materials, where FSSMQ code is used for prediction of shielding effectiveness of a composite with partially conductive carbon fibres. Finally, concluding remarks are given in section 6.

2. Spectral domain method of moments

In this section, a detail overview of the spectral domain method of moments is given. The formulation and notation being used is based on (Wu, 1995) and (Mittra et al., 1988). The formulation is based in the immitance approach developed originally by (Itoh, 1980) and is slightly different from the notation being used by other authors (Scott, 1989). As a new, when compared to existing papers related to the spectral domain MoM, incorporation of non-uniform mesh rectangular cells and mesh of quadrilateral cells is presented in section 3.

2.1 Multilayer problem – formulation (EFIE)

Let's consider a planar periodic multilayer structure according to Fig. 1. As an example structure with 2 dielectric layers is given. Metal layers are denoted as M_0, M_1 and M_2. The top of the dielectric structure is a placed at position $z = 0$.

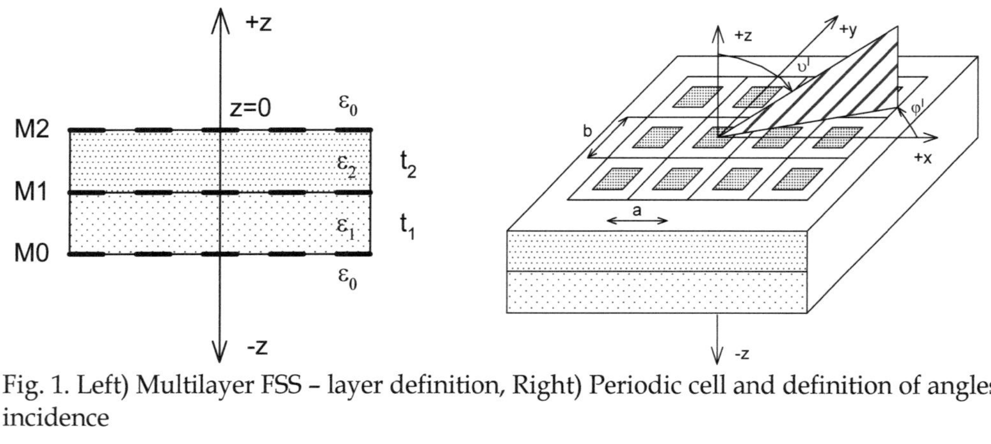

Fig. 1. Left) Multilayer FSS - layer definition, Right) Periodic cell and definition of angles of incidence

The reflection/transmission from the multilayer structure is solved within a single periodic cell dimensioned a and b. The structure is illuminated by the TE or TM polarized plane wave with the angles of incidence ϑ^I and φ^I. The formulation of MoM in the spectral domain starts with expressing an electric intensity of the incident and scattered wave in space and spectral domain (equations 1 and 2). Then, the electric field integral equation (EFIE), (Scott, 1989) and (Wu, 1995) is formulated in the spectral domain (3).

$$\mathbf{E}_{ii}^{inc}(x,y)e^{j(\alpha_0 x+\beta_0 y)}+\mathbf{E}_{ii}^{scat}(x,y)=Z_s\mathbf{J}_{ii}(x,y) \tag{1}$$

where ii stands for the index of metal layer (ii=0,1,2, ...), $\mathbf{E}_{ii}^{inc}(x,y)$ is electric intensity of the incident wave at the ii^{th} metal layer (but with all metal layers removed). The components of intensity $\mathbf{E}_{ii}^{inc}(x,y)$ can be calculated according to a procedure given in section 4. The symbol $\mathbf{E}_{ii}^{scat}(x,y)$ stands for the electric intensity at ii^{th} layer arising from conductive currents from ii^{th} and all remaining metal layers placed above and beneath the ii^{th} metal layer. As metal layers may be lossy, with a finite conductivity σ_{ii} [S/m], then the total tangential electric field is not completely vanishing on the conductors. Such a situation may be approximated with the so called surface impedance Z_s. The surface impedance is determined from the sheet resistance R_s (Cwick & Mittra, 1987) as $Z_s = (1+j)R_s$, where R_s is calculated as $R_s=\sqrt{\omega\mu_{ii}/2\sigma_{ii}}$. The concept of sheet resistance is valid when skin depth is less than the metal thickness. Last two symbols α_0 and β_0 to be explained, stand for x and y components of the wave vector of the incident wave. Time dependence of $e^{j\omega t}$ is assumed in this chapter.

In the subsequent paragraphs, double index ii or jj will used for the index of a metal layer, while the single index i will stand for an integer index of a dielectric layer.

$$\mathbf{E}_{ii}^{inc}(m,n)+\mathbf{E}_{ii}^{scat}(m,n)=Z_s\mathbf{J}_{ii}(m,n)\ \ldots \text{ on metal parts of } ii^{th} \text{ metal layer} \tag{2}$$

$$\mathbf{E}_{ii}^{inc}(m,n)+\sum_{jj=0}^{N_m}\begin{bmatrix} G_{xx}^{ii\leftarrow jj} & G_{xy}^{ii\leftarrow jj} \\ G_{yx}^{ii\leftarrow jj} & G_{yy}^{ii\leftarrow jj}\end{bmatrix}\begin{bmatrix}\mathbf{J}_{jj,x}(m,n) \\ \mathbf{J}_{jj,y}(m,n)\end{bmatrix}=Z_s\mathbf{J}_{ii}(m,n)\ ;\text{ii}=0,1,\ldots,\text{N}_\text{m} \tag{3}$$

with $G_{xx}^{ii\leftarrow jj}$, $G_{xy}^{ii\leftarrow jj}$, $G_{yx}^{ii\leftarrow jj}$ and $G_{yy}^{ii\leftarrow jj}$ standing for components of the self and mutual impedance matrix $[G^{ij}]$ which relates the scattered electric field at ii^{th} metal layer through the conductive currents at jj^{th} layer. N_m is index of a maximal metal layer in the analyzed multilayer structure. Symbols m and n have a meaning of integer indexes (Floquet harmonics). These indexes range from $-M$ to $+M$, resp. $-N$ to $+N$, where M, N represent the maximal Floquet harmonics.

$$\mathbf{E}_{ii}=[G^{ij}]\mathbf{J}_{jj} \tag{4}$$

Components of $[G^{ij}]$ are evaluated by the immitance approach (Itoh, 1980) and (Wu, 1995). The evaluation of $[G^{ij}]$ is different for the case with $ii=jj$ (self-impedance matrix) and $ii \neq jj$ (mutal impedance matrix).

2.2 Self impedance matrixes

In the case of self-impedance matrixes, the total admittance looking up and down from the *ii*^th^ layer is evaluated in series of steps described below (Wu, 1995). As an example, situation for the two layer dielectric structure is given in Fig . 2.

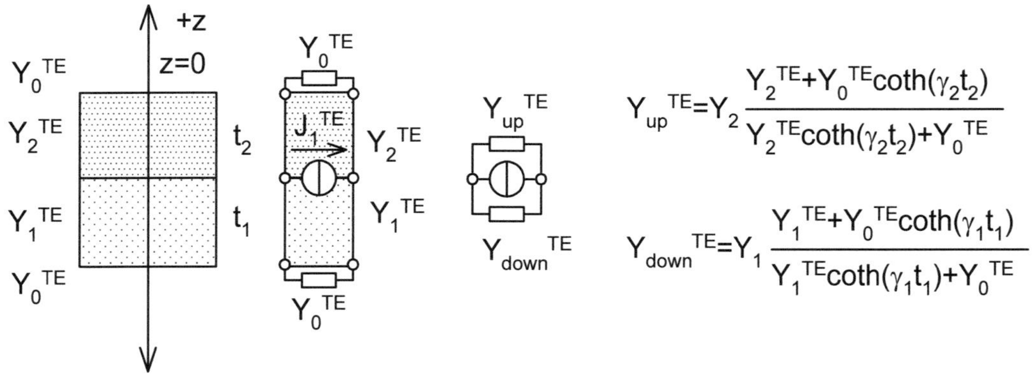

Fig. 2. Multilayer FSS - example of evaluation of up and down looking admittance for the metal layer M1.

Evaluation of up and looking admittance is performed with the aid of the equation (5) known the transmission line theory

$$Y_i^{TE,TM} = Y_i^{TE,TM} \frac{Y_i^{TE,TM} + Y_L \coth(\gamma_i t_i)}{Y_i^{TE,TM} \coth(\gamma_i t_i) + Y_L} \tag{5}$$

where Y_i^{TE} and Y_i^{TM} are TE and TM admittances of the *ii*^th^ dielectric layer. According to (Wu, 1995) these impedances can be expressed as

$$Y_i^{TE} = \frac{-\gamma_i}{j\omega\mu_i} \tag{6a}$$

$$Y_i^{TM} = \frac{-j\omega\varepsilon_i}{\gamma_i} \tag{6b}$$

Symbol ω stands for the angular frequency, the complex permittivity and permeability of the *ii*^th^ dielectric layer are denoted as ε_i and μ_i. Last symbol γ_i being used in (6a,b) represents the complex propagation constant in the z-direction (for the *i*^th^ dielectric layer).

$$\gamma_i = \sqrt{\alpha_m^2 + \beta_n^2 - k_i^2} \tag{7}$$

The propagation constant γ_i is determined upon the transversal propagation constant $k_{xy,i} = \sqrt{\alpha_m^2 + \beta_n^2}$ and the wave number k_i valid for the *i*^th^ dielectric layer.

Components of the transversal propagation constant $k_{xy,i}$ are called as Floquet harmonics and may be expressed as

$$\alpha_m(m,n) = \alpha_0 + \frac{2\pi}{a} m \tag{8a}$$

$$\beta_n(m,n) = \beta_0 + \frac{2\pi}{b} n \tag{8b}$$

Wave numbers α_0 and β_0 are the fundamental (zeroth order) spectral harmonics and they are directly linked with the incident angles ϑ^I and φ^I by equations (9a) and (9b)

$$\alpha_0 = k_0 \sin(\vartheta^I)\cos(\varphi^I) \tag{9a}$$

$$\beta_0 = k_0 \sin(\vartheta^I)\sin(\varphi^I) \tag{9b}$$

where $k_0 = 2\pi f \sqrt{\mu_0 \varepsilon_0}$ and $k_i = 2\pi f \sqrt{\mu_i \varepsilon_i}$

The total TE and TM impedances connected to the ii^{th} metal layer are then

$$Z_{g,ii}^{TE} = \frac{1}{Y_{down,ii}^{TE} + Y_{up,ii}^{TE}} \tag{10a}$$

$$Z_{g,ii}^{TM} = \frac{1}{Y_{down,ii}^{TM} + Y_{up,ii}^{TM}} \tag{10b}$$

Finally, components of $[G^{ii}]$ can be written (Wu, 1995) as

$$[G^{ii}] = \begin{bmatrix} G_{xx,ii} & G_{xy,ii} \\ G_{yx,ii} & G_{yy,ii} \end{bmatrix} \tag{11}$$

$$G_{xx,ii} = Z_{g,ii}^{TE} \sin^2\Theta + Z_{g,ii}^{TM} \cos^2\Theta \tag{12a}$$

$$G_{xy,ii} = (Z_{g,ii}^{TM} - Z_{g,ii}^{TE}) \sin\Theta \cos\Theta \tag{12b}$$

$$G_{yx,ii} = G_{xy,ii} \tag{12c}$$

$$G_{yy,ii} = Z_{g,ii}^{TE} \cos^2\Theta + Z_{g,ii}^{TM} \sin^2\Theta \tag{12d}$$

where

$$\sin\Theta = \frac{\beta}{\sqrt{\alpha^2 + \beta^2}} \tag{13a}$$

$$\cos\Theta = \frac{\alpha}{\sqrt{\alpha^2 + \beta^2}} \tag{13b}$$

2.3 Mutual impedance matrixes

Mutual impedance matrixes $[G^{ij}]$ are evaluated with the aid of self impedance matrix for the ii^{th} layer and the transfer impedance Z_t (Wu, 1995).

$$[G^{ij}] = \begin{bmatrix} G_{xx,ij} & G_{xy,ij} \\ G_{yx,ij} & G_{yy,ij} \end{bmatrix} \tag{14}$$

$$G_{xx,ij} = Z_{g,ij}^{TE} \sin^2\Theta + Z_{g,ij}^{TM} \cos^2\Theta \tag{15a}$$

$$G_{xy,ij} = (Z_{g,ij}^{TM} - Z_{g,ij}^{TE}) \sin\Theta\cos\Theta \tag{15b}$$

$$G_{yx,ij} = G_{xy,ij} \tag{15c}$$

$$G_{yy,ij} = Z_{g,ij}^{TE} \cos^2\Theta + Z_{g,ij}^{TM} \sin^2\Theta \tag{15d}$$

$$Z_{g,ij}^{TE} = Z_{g,ii}^{TE} Z_{t,ii\leftarrow jj}^{TE} \tag{16a}$$

$$Z_{g,ij}^{TM} = Z_{g,ii}^{TM} Z_{t,ii\leftarrow jj}^{TM} \tag{16b}$$

The transfer impedances $Z_{t,ii\leftarrow jj}^{TE}$ and $Z_{t,ii\leftarrow jj}^{TM}$ are calculated with application of the cascade matrix. As known from the transmission line theory, the matrix relates input and output voltages and currents on the section (or several sections) of a transmission line. When the source and destination metal layers are distanced by one dielectric layer (that is $|ii - jj| = 1$), equation (17) applies to the case

$$Z_{t,ii\leftarrow jj}^{TE,TM} = \frac{Y_i^{TE,TM}}{Y_i^{TE,TM} \cosh(\gamma_i t_i) + Y_{L,ii}^{TE,TM} \sinh(\gamma_i t_i)} \tag{17}$$

with $Y_i^{TE,TM}$ being admittances of the dielectric layer between the ii^{th} and jj^{th} metal layers. Admittance $Y_{L,ii}^{TE,TM}$ is the total load admittance connected to the ii^{th} metal layer, when looking from jj^{th} to ii^{th} layer.

Finally, the global impedance matrix [Z] valid for the structure from Fig. 2 (2 dielectric and 3 metal layers) is then

$$[Z] = \begin{bmatrix} G_{00} & G_{01} & G_{02} \\ G_{10} & G_{11} & G_{12} \\ G_{20} & G_{21} & G_{22} \end{bmatrix} \tag{18}$$

which links total currents densities in metal layers 0,1 and 2 and the scattered electric intensities at these layers. Matrix [Z] actually represents the matrix written in the EFIE according to equation (3).

$$\begin{bmatrix} E_{00}^{scat} \\ E_{11}^{scat} \\ E_{22}^{scat} \end{bmatrix} = [Z] \begin{bmatrix} J_{00} \\ J_{11} \\ J_{22} \end{bmatrix} \tag{19}$$

3. Approximation of surface currents

This section discusses use of two different current expansion basis functions. First, non-uniform rooftops and their Fourier transform is given. Second, large domain basis functions defined over quad elements are presented. As a last subsection of this section, solution of the EFIE via the Galerking method (method of moments) is briefly mentioned.

3.1 Roof-top basis functions

Roof-top basis functions are most common basis functions being used for approximation of currents within simple spectral domain codes. Typically, uniform rooftops are being incorporated in order to use FFT for acceleration of double summations (Cwick & Mittra, 1987) or (Wan & Encinar, 1995). If non-uniform rooftops are incrorporated, then FFT cannot be used. As an advantage a wider class geometries can analyzed without the restriction of snapping into the uniform grid. This is especially valuable when tuning characteristics of the periodic structure or making optimization of the structures with gradient methods (e.g. Newton one).

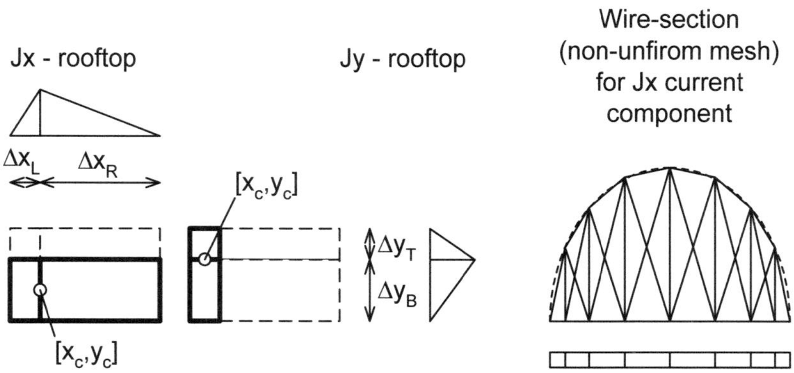

Fig. 3. Non-uniform roof-tops – definition and example of approximation of surface current density

Non-uniform rooftop basis functions are defined in Fig. 3. For example the rooftops being used for approaximation of x-directed currents have lengths of their left and right part denoted as Δx_L and Δx_R . The width of the J_x curent cell is denoted as Δy. The y-directed rooftops are defined analogically with indexes L and R being replaced by B (bottom) and T (top). Fouried transform of the non-uniform rooftops is given in equations (20a) and (20b)

$$F\{B_x(\alpha,\beta)\} = [\Delta_y \sin c(\beta \Delta_y / 2) e^{-j\beta y_c}][\frac{(\Delta x_L + \Delta x_R) - (\Delta x_L e^{-j\alpha \Delta x_R} + \Delta x_R e^{j\alpha \Delta x_L})}{\Delta x_L \Delta x_R \alpha^2} e^{-j\alpha x_c}] \quad (20a)$$

$$F\{B_y(\alpha,\beta)\} = [\frac{(\Delta y_B + \Delta y_T) - (\Delta y_B e^{-j\beta \Delta y_T} + \Delta y_T e^{j\beta \Delta y_B})}{\Delta y_B \Delta y_T \beta^2} e^{-j\beta y_c}][\Delta_x \sin c(\alpha \Delta_x / 2) e^{-j\alpha x_c}] \quad (20b)$$

where α and β stand for the space frequency. Symbols x_c and y_c represent x and y co-ordinates of the centre of the J_x or J_y rooftop.

Global approximation of surface currents on conductive patches of the periodic structure, is typically written as

$$J_x(x,y) = \sum_{n_x=1}^{N_x} I_{x,nx} B_{x,nx}(x,y) \quad (21a)$$

$$J_y(x,y) = \sum_{n_y=1}^{N_y} I_{y,ny} B_{y,ny}(x,y) \quad (21b)$$

where $B_{x,nx}$ and $B_{y,ny}$ are J_x and J_y rooftop basis functions and $I_{x,nx}$, $I_{y,ny}$ are unknown current expansion coefficients.

Rooftops have generally advantage in easier program implementation and provide good accuracy when analyzing "Manhattan shaped" geometries. Their disadvantage is in the inability to model general or curves patches. Rooftops also provide accurate modelling of amplitude characteristics of FSS. While, the convergence of phase is very slow. These elements are inefficient when analyzing phase shifters in planar reflector antennas (Gona, 2004). In this case, suitable large domain basis functions (e.g. composed from sinusoidal and Tchebyschev functions) must be used.

3.2 Higher-order basis functions defined over quad elements

As a large domain basis functions, historically sinusoidal and cosinusoidal functions were used to represent J_x and J_y current densities over the rectangular patch (Scott, 1989). Alternatively combination of sinusoidal and Tchebyshev functions (Mittra et. al, 1988) may be used. In this case, singular edge currents are well modelled for with low order of expansion functions (typically 2 to 3). During 1990's (Kolundzija, 1998) and (Notaros et al., 2001) extended the use of entire domain basis functions into quadrilaterals (equation 22). These functions satisfy conditions for a local continuity of density and are successfully being used within the modern spatial domain MoM calculations. Their use the spectral domain codes, is less common.

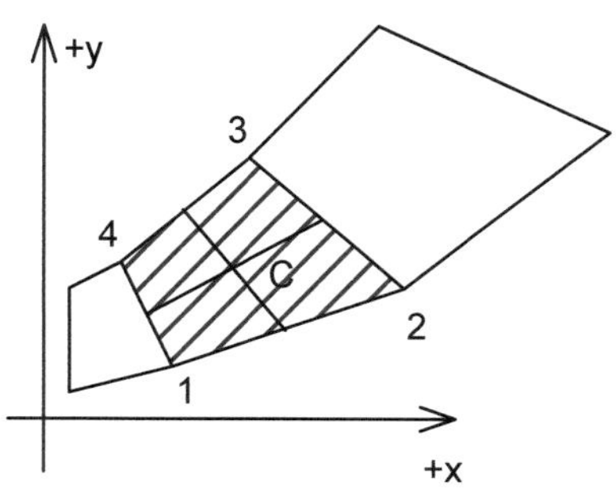

$$[x_c, y_c] = \frac{1}{4}(\mathbf{r}_1 + \mathbf{r}_2 + \mathbf{r}_3 + \mathbf{r}_4)$$

$$\mathbf{r}(u,v) = [\mathbf{r}_1(1-u)(1-v) + \mathbf{r}_2(u+1)(1-v) + \mathbf{r}_3(u+1)(v+1) + \mathbf{r}_4(1-u)(1+v)]/4$$

$$\mathbf{r}(u,v) = \mathbf{r}_c + \mathbf{r}_u u + \mathbf{r}_v v + \mathbf{r}_{uv} uv$$

$$x(u,v) = r_{cx} + r_{ux} u + r_{vx} v + r_{uvx} uv$$

$$y(u,v) = r_{cy} + r_{uy} u + r_{vy} v + r_{uvy} uv$$

Fig. 4. Quadrilateral element - definition

$$\mathbf{J}_u(u,v) = \mathbf{e}_u \Big(\sum_{i=0}^{Nu}\sum_{j=0}^{Nv} a_{uij} \begin{Bmatrix} 1-u, & i=0 \\ u+1, & i=1 \\ u^i - 1, & i=2,4,\ldots,N_u \\ u^i - u, & i=3,5,\ldots,N_u - 1 \end{Bmatrix}\Big) v^j \tag{22}$$

where u,v = <-1;1>

Large domain basis functions defined according to equation (22) are orthogonal and their Fourier Transform is as following

$$F_x\{J_u(\alpha,\beta)\} = \int_{v=-1}^{v=+1}\int_{u=-1}^{v=+1} J_u(u,v)\cdot e_{ux}(u,v)\cdot e^{-j(\alpha\cdot x(u,v)+\beta\cdot y(u,v))}\left|J_1(u,v)\right| du dv \tag{23a}$$

$$F_y\{J_u(\alpha,\beta)\} = \int_{v=-1}^{v=+1}\int_{u=-1}^{v=+1} J_u(u,v)\cdot e_{uy}(u,v)\cdot e^{-j(\alpha\cdot x(u,v)+\beta\cdot y(u,v))}\left|J_1(u,v)\right| du dv \tag{23b}$$

where $J_1(u,v)$ is Jacobian. The Jacobian can be expressed by equations (23b)

$$J_1(u,v) = c_0 + c_1 u + c_2 v \tag{23c}$$

Constants c_0, c_1, c_2 are found as

$$c_0 = \frac{1}{8}((r_{4x} - r_{2x}) * (r_{1y} - r_{3y}) + (r_{3x} - r_{1x}) * (r_{4y} - r_{2y})) \tag{23d}$$

$$c_1 = \frac{1}{8}((r_{4x} - r_{3x}) * (r_{3y} - r_{1y}) + (r_{1x} - r_{2x}) * (r_{4y} - r_{3y})) \tag{23e}$$

$$c_2 = \frac{1}{8}((r_{4x} - r_{1x}) * (r_{2y} - r_{3y}) + (r_{3x} - r_{2x}) * (r_{4y} - r_{1y})) \tag{23f}$$

where r_{1x}, r_{1y} ; r_{2x}, r_{2y}; r_{3x}, r_{3y}; r_{4x}, r_{4y} are x and y co-ordinates of quadrilateral vertices.
Further details about analytical or semianalytical expressions for the Fourier transform of the J_u current component are beyond the extent of the chapter.

3.3 Solution of current expansion coefficients

Once the suitable current approximation (e.g. rooftops) is selected and the metal patches in all layers are divided into cells, the unknown current expansion coefficients can be determined by the Galerkin method (Scott, 1989) or (Wu, 1995). As a result of the Galerkin testing procedure, a system of linear equations is obtained and unknown current expansion coefficients are solved by the matrix inversion. After solution of the system of equations, current density is evaluated in the spectral domain and scattered electric intensity $\mathbf{E}^{scat}$ in all layers is calculated using equation (4).

Generation of elements a matrix of equations for large number of unknowns is time consuming due to the presence of double summations in the electric field equation. Several methods were introduced during last 20 years. One of them is acceleration of double summations by use of a Poission summation formula or more newly by using hybrid spatial-spectral domain approach (Kipp & Chan, 1994).

3.4 Reflection and transmission coefficients

Prior definition of reflection and transmission coefficients, let's assume that the periodic structure is independently illuminated by a plane wave with parallel (TM) and orthogonal (TE) polarizations (transverse to z). The x and y components of the incident electric intensity are then

$$\mathbf{E}^{inc,TM} = [\cos(\varphi^I) \quad \sin(\varphi^I)] \tag{24a}$$

$$\mathbf{E}^{inc,TE} = [-\sin(\varphi^I) \quad \cos(\varphi^I)] \tag{24b}$$

To compute the reflection and transmission coefficients for a multilayer periodic structure, the electric intensity in the most top (reflected wave $\boldsymbol{E}^{refl}$) and the electric intensity at the most bottom (transmitted wave $\boldsymbol{E}^{trans}$) air to dielectric interfaces must be calculated. These intensities are given by equations (25)

$$\mathbf{E}^{refl,TM}_{mm\_max} = \mathbf{E}^{scat,TM}_{mm\_max} + \mathbf{E}^{inc,TM}(1+\Gamma^{TM}) \tag{25a}$$

$$\mathbf{E}^{refl,TE}_{mm\_max} = \mathbf{E}^{scat,TE}_{mm\_max} + \mathbf{E}^{inc,TE}(1+\Gamma^{TE}) \tag{25b}$$

$$\mathbf{E}^{trans,TM}_{mm=0} = \mathbf{E}^{scat,TM}_{mm=0} + \mathbf{E}^{inc,TM}\mathrm{T}^{TM} \tag{25c}$$

$$\mathbf{E}^{trans,TE}_{mm=0} = \mathbf{E}^{scat,TE}_{mm=0} + \mathbf{E}^{inc,TE}\mathrm{T}^{TE} \tag{25d}$$

where coefficients Γ^{TM}, Γ^{TE} and T^{TM}, T^{TE} are computed according to the procedure described in section 4.

Reflection and transmission coefficients are defined differently by different authors. Most common definition is upon the z-directed potentials (Cwick & Mittra, 1987) or the tangential components of the electric intensity (Wan & Encinar, 1995). In this section, the definition from thesis (Gona, 2004) will be used. The definition is being used in physics related textbooks and assumes that the electric intensity of the incident and reflected wave are expressed by E_ϑ and E_φ components (Fig. 5).

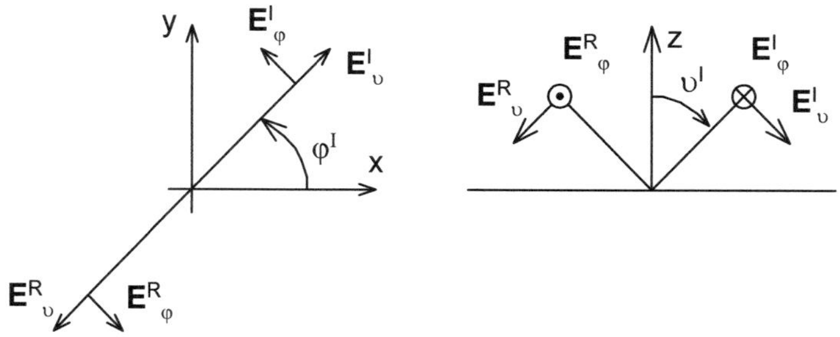

Fig. 5. Definition of reflection coefficients

Reflection coefficients defined by spherical components of the electric intensity are given by the matrix (26).

$$[\rho]=\begin{bmatrix}\rho_{\vartheta\vartheta} & \rho_{\vartheta\varphi}\\ \rho_{\varphi\vartheta} & \rho_{\varphi\varphi}\end{bmatrix} \tag{26}$$

The advantage of the definition is that the cross-polarized reflection coefficients $\rho_{\vartheta\varphi}$ and $\rho_{\varphi\vartheta}$ are equal.

The definition of reflection coefficients by the tangential components of the electric intensity is more common (Wan & Encinar, 1995) and is given by matrix (27), with cross-polarized terms being different

$$[R]=\begin{bmatrix}R_{TMTM} & R_{TMTE}\\ R_{TETM} & R_{TETE}\end{bmatrix} \tag{27}$$

A relation between definitions (26) and (27) as given in the thesis (Gona, 2004) and is as follows

$$R_{TMTM}=-\rho_{\vartheta\vartheta} \tag{28a}$$

$$R_{TETM}=-\rho_{\varphi\vartheta}\frac{1}{\cos(\vartheta)} \tag{28b}$$

$$R_{TMTE}=-\rho_{\vartheta\varphi}\cos(\vartheta) \tag{28c}$$

$$R_{TETE}=-\rho_{\varphi\varphi} \tag{28d}$$

Once the tangential components of the electric intensity of incident and transmitted wave are evaluated with the aid of equations (25) and (4) then the fundamental mode ((0,0) Floquet mode) reflection transmission coefficients can be obtained by equations (29).

$$\rho_{\vartheta\vartheta}=\frac{1}{\left|E_{TM}^{inc}\right|}\left(-\frac{\cos(\varphi^I)}{\cos(\vartheta^I)}E_{x,mm\_\max}^{refl,TM}-\frac{\sin(\varphi^I)}{\cos(\vartheta^I)}E_{y,mm\_\max}^{refl,TM}\right) \tag{29a}$$

$$\rho_{\varphi\vartheta} = \frac{1}{\left|E_{TM}^{inc}\right|}(\sin(\varphi^I)E_{x,mm\_\max}^{refl,TM} - \cos(\varphi^I)E_{y,mm\_\max}^{refl,TM}) \tag{29b}$$

where $\left|E_{TM}^{inc}\right| = 1/\cos(\vartheta^I)$ and $E_{x.mm\_\max}^{refl,TM}$, $E_{y,mm\_\max}^{refl,TM}$ are components of the total electric intensity of the reflected wave at the most top layer of the periodic structure (Parallel polarization of the incident plane wave is assumed). According to the example given in Fig. 1, the most top layer is M2. Similarly the coefficients for the orthogonal polarization (TE) are defined

$$\rho_{\vartheta\varphi} = \frac{1}{\left|E_{TE}^{inc}\right|}(-\frac{\cos(\varphi^I)}{\cos(\vartheta^I)}E_{x,mm\_\max}^{refl,TE} - \frac{\sin(\varphi^I)}{\cos(\vartheta^I)}E_{y,mm\_\max}^{refl,TE}) \tag{29c}$$

$$\rho_{\varphi\varphi} = \frac{1}{\left|E_{TE}^{inc}\right|}(\sin(\varphi^I)E_{x,mm\_\max}^{refl,TE} - \cos(\varphi^I)E_{y,mm\_\max}^{refl,TE}) \tag{29d}$$

where $\left|E_{TE}^{inc}\right| = 1$.

Calculation of transmission coefficients is also done with equations (29) but intensities for reflected wave are replaced by intensities from transmitted fields and spherical angles ϑ^I and φ^I are replaced with $-\vartheta^I$ and $\varphi^I + \pi$.

4. Reflection and transmission from a multilayer dielectric structure

The derivation of reflection and transmission coefficients for a multilayer dielectric structure can be found in (Wu, 1995), for example. The derivation makes use of the unitary z-components of the vector potentials $\mathbf{A}_z$ and $\mathbf{F}_z$.

$$\mathbf{A}^{TM} = \mathbf{u}_z A_z = \mathbf{u}_z \psi^{TM} = \mathbf{u}_z\, e^{j(\alpha_0 x + \beta_0 y)} e^{\gamma_0 z} \tag{30a}$$

$$\mathbf{F}^{TE} = \mathbf{u}_z F_z = \mathbf{u}_z \psi^{TE} = \mathbf{u}_z\, e^{j(\alpha_0 x + \beta_0 y)} e^{\gamma_0 z} \tag{30b}$$

As known from the electromagnetic theory, the TE and TM polarized plane waves can be derived from the unit $\mathbf{A}_z$ and $\mathbf{F}_z$ potentials by application of a complete set Maxwell equations (that is Maxwell equations with both electric and magnetic currents).

$$\mathbf{E}^{TM} = \frac{1}{j\omega\varepsilon}(\nabla\nabla + k^2 I)\mathbf{A}^{TM} \tag{30a}$$

$$\mathbf{H}^{TM} = \nabla \times \mathbf{A}^{TM} \tag{30b}$$

$$\mathbf{E}^{TE} = -\nabla \times \mathbf{F}^{TE} \tag{30c}$$

$$\mathbf{H}^{TE} = \frac{1}{j\omega\mu}(\nabla\nabla + k^2 I)\mathbf{F}^{TE} \tag{30d}$$

By expressing tangential components E_x and H_y with applications of equations (31) and enforcing continuity of components E_x and H_y at all dielectric and air-to dielectric interfaces, the TM/TE reflection coefficients may be found by solution of matrix equations (32).

$$E_x^{TM} = \frac{1}{j\omega\varepsilon}\frac{\partial^2\psi^{TM}}{\partial x\partial z} \tag{31a}$$

$$H_y^{TM} = -\frac{\partial\psi^{TM}}{\partial x} \tag{31b}$$

$$E_x^{TE} = -\frac{\partial\psi^{TE}}{\partial y} \tag{31c}$$

$$H_y^{TE} = \frac{1}{j\omega\mu}\frac{\partial^2\psi^{TE}}{\partial y\partial z} \tag{31d}$$

$$[M^{TM}][x^{TM}] = [b^{TM}] \tag{32a}$$

$$[M^{TE}][x^{TE}] = [b^{TE}] \tag{32b}$$

with $[x^{TM}] = [\Gamma^{TM} \quad C_{11} \quad C_{12} \quad C_{21} \quad C_{22} \quad C_{31} \quad C_{32} \quad T^{TM}]^T$

$$[b^{TM}] = [-\frac{\gamma_0}{\varepsilon_0} \quad 0 \quad 0 \quad 0 \quad -1 \quad 0 \quad 0 \quad 0]^T$$

and $[x^{TE}] = [\Gamma^{TE} \quad C_{11} \quad C_{12} \quad C_{21} \quad C_{22} \quad C_{31} \quad C_{32} \quad T^{TE}]^T$

$$[b^{TE}] = [-1 \quad 0 \quad 0 \quad 0 \quad \gamma_0\frac{\mu_1}{\mu_0} \quad 0 \quad 0 \quad 0]^T,$$

where TM/TE reflection and transmission coefficients are denoted as Γ^{TM}, Γ^{TE} and T^{TM}, T^{TE} respectively. As an example, the matrixes [M^{TE}] and [M^{TM}] for the three-layer dielectric structure (Fig. 6) are given. From the tables 1 and 2 the structure of the [M^{TE}] and [M^{TM}] matrixes for a dielectric structure with general number of layers can be easily derived.

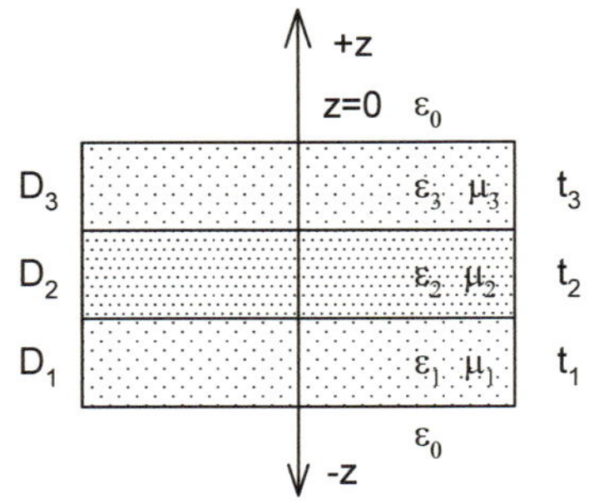

Fig. 6. The 3-layer dielectric structure

$-\gamma_0/\varepsilon_0$	$-\frac{\gamma_1}{\hat{\varepsilon}_1}\sinh(\gamma_1 t_1$	$-\frac{\gamma_1}{\hat{\varepsilon}_1}\cosh(\gamma_1 t$	0	0	0	0	0
0	0	$\gamma_1/\hat{\varepsilon}_1$	$-\frac{\gamma_2}{\hat{\varepsilon}_2}\sinh(\gamma_2 t_2$	$-\frac{\gamma_2}{\hat{\varepsilon}_2}\cosh(\gamma_2 t_2$	0	0	0
0	0	0	0	$\gamma_2/\hat{\varepsilon}_2$	$-\frac{\gamma_3}{\hat{\varepsilon}_3}\sinh(\gamma_3 t_3$	$-\frac{\gamma_3}{\hat{\varepsilon}_3}\cosh(\gamma_3 t_3$	0
0	0	0	0	0	0	$\gamma_3/\hat{\varepsilon}_3$	$-\gamma_0/\varepsilon_0$
+1	$-\cosh\gamma_1 t_1)$	$-\sinh(\gamma_1 t_1)$	0	0	0	0	0
0	+1	0	$-\cosh(\gamma_2 t_2)$	$-\sinh(\gamma_2 t_2)$	0	0	0
0	0	0	+1	0	$-\cosh(\gamma_3 t_3)$	$-\sinh(\gamma_3 t_3)$	
0	0	0	0	0	+1	0	-1

Table 1. Matrix M^{TM}(3 layer – dielectric structure)

+1	$-\cosh(\gamma_1 t_1)$	$-\sinh(\gamma_1 t_1)$	0	0	0	0	0
0	+1	0	$-\cosh(\gamma_2 t_2)$	$-\sinh(\gamma_2 t_2)$	0	0	0
0	0	0	+1	0	$-\cosh(\gamma_3 t_3)$	$-\sinh(\gamma_3 t_3)$	0
0	0	0	0	0	+1	0	-1
$\gamma_0\hat{\mu}_1/\mu_0$	$\gamma_1\sinh(\gamma_1 t_1)$	$\gamma_1\cosh(\gamma_1 t_1)$	0	0	0	0	0
0	0	$\gamma_1\hat{\mu}_2/\hat{\mu}_1$	$-\gamma_2\sinh(\gamma_2 t_2)$	$-\gamma_2\cosh(\gamma_2 t_2$	0	0	0
0	0	0	0	$\gamma_2\hat{\mu}_3/\hat{\mu}_2$	$-\gamma_3\sinh(\gamma_3 t_3)$	$-\gamma_3\cosh(\gamma_3 t_3$	0
0	0	0	0	0	0	$\gamma_3\hat{\mu}_0/\hat{\mu}_3$	$-\gamma_0$

Table 2. Matrix M^{TE} (3 layer – dielectric structure)

5. Numerical examples

This section presents numerical results for reflection and transmission coefficients for periodic structures with different elements (patches). Where available the numerical results are compared with results obtained by other authors. All numerical examples given, were calculated by our in-house periodic MoM code called FSSMQ. The code was written upon the theory presented in the previous sections. At the present, the code allows analysis of periodic structures having up to 4 dielectric and 5 metal layers. At each metal layer, arbitrarily shaped patches may be present. These patches are meshed with uniform or non-uniform rectangular elements depending on the graphical editor being used. As current approximation basis functions, non-uniform rooftops are used. The FSSMQ code is based on a direct approach, that is the global impedance matrix, which relates interactions among all metal layers, is being used for description of interactions among the layers. The program is written in Matlab and can be used for analysis of any planar periodic structures. Its target applications are frequency selective surfaces, composite materials and potentially metamaterials. During solution of a system of linear equations for unknown current expansion coefficients, Matlab's matrix inversion is being used. With this type of solver and speed optimized program architecture, multilayer periodic structures with several hundreds of unknowns may be analyzed within few seconds per frequency (Celeron M, 1.6 GHz). Program allows also interactive postprocessing of currents (surface and vector plots) and total electric intensities at selected layers.

5.1 Jerusalem cross

Jerusalem Cross is one of the oldest elements. It was developed to produce stable reflection coefficients within a large range of angles of incidences (Mittra et al., 1988). The element was analyzed thoroughly by (Cwick & Mittra, 1987) and (Mittra at al., 1988). In this subsection more newer results from (Weile & Michielssen, 2001) are being compared with results from the FSSMQ code.

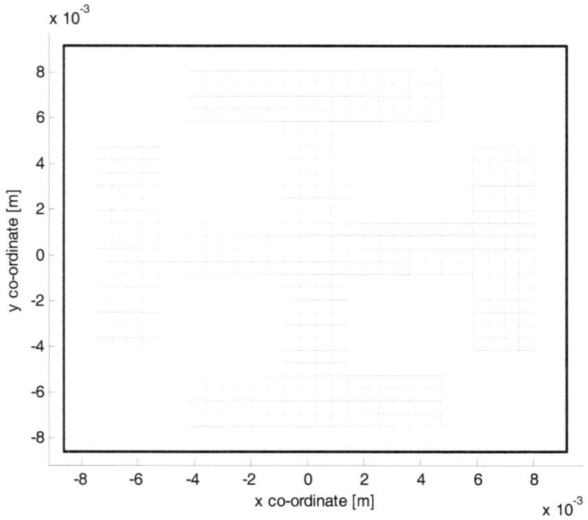

Fig. 7. Jerussalem cross element. Dimensions of the periodic cells $a = b = 17.8$ mm. Discretization 32x32 cells. Relative permittivity 1.0. Maximal number of Floquet harmonics $M = N = 18$

Geometry of the Jerusalem cross which was analyzed is shown in Fig. 7. Dimensions of the periodic cell are $a = b = 17.8$ mm. Periodic cell is divided into 32x32 cells. It is considered that the element resides in free-space and the structure is illuminated by the 40 degree TM polarized wave. Frequency dependence of the TM power reflection coefficient is shown in Fig. 8. It can be seen, that the FSSMQ produces identical results with those given in (Weile & Michielssen, 2001) where resonance arising for oblique incidence is well captured.

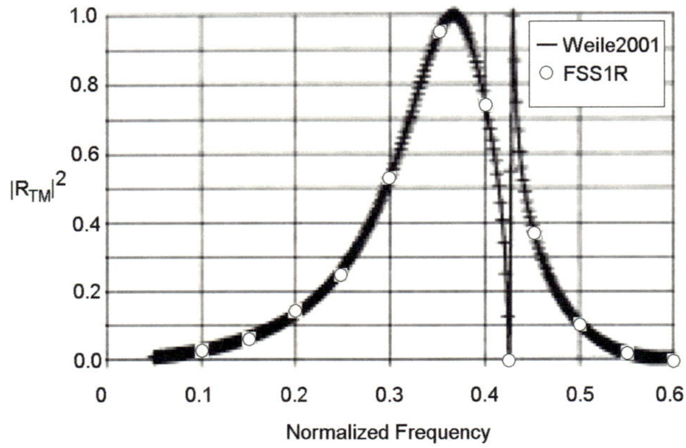

Fig. 8. Power reflection coefficient versus frequency (parallel polarization) , normalization frequency was f_n = c/a = 16.842 GHz, υ^I = 40 deg, φ^I = 0 deg

5.2 Cross element

Cross element is often used in practise in design or reflector antennas. However it does not show as superior performance as loop elements (e.g. Fourlegged or square ring element). The element was analyzed by many authors. In this subsection results from (Wan & Encinar, 1995) serve as a reference. In figures 9 and 10, reflection coefficient for a single and a double layer cross element FSS are compared. For the purpose of validation of the FSSMQ code for a single and multilayer case, the same number of Floquet modes as in (Wan, 1995) was selected. In both cases, excellent agreement is observed.

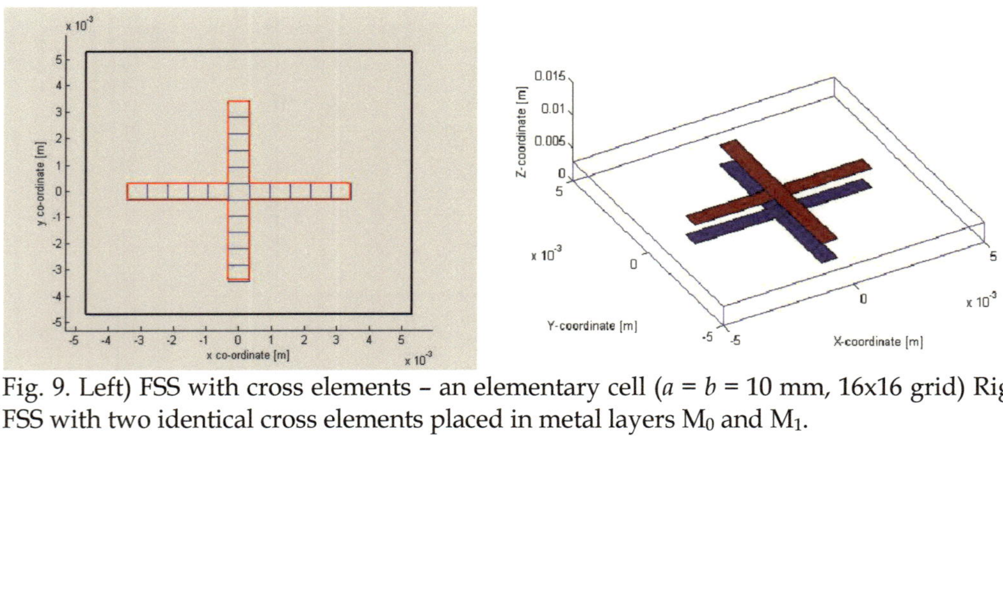

Fig. 9. Left) FSS with cross elements - an elementary cell ($a = b = 10$ mm, 16x16 grid) Right) FSS with two identical cross elements placed in metal layers M_0 and M_1.

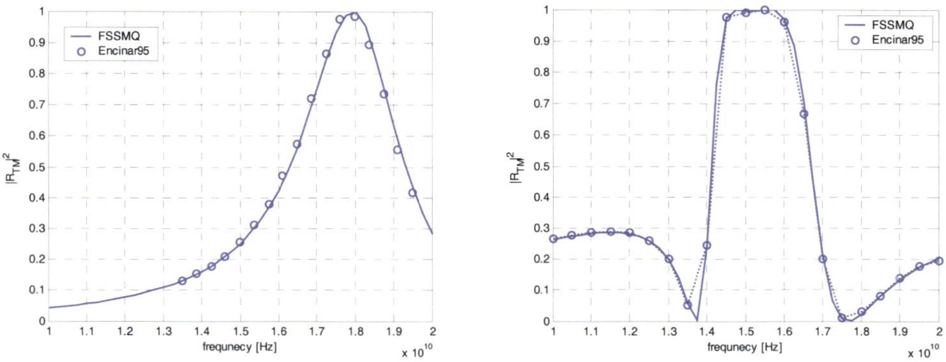

Fig. 10. Power reflection coefficient versus frequency (FSS with cross elements) - comparison of results produced by FSSMQ code and results from (Wan & Encinar, 1995) - cell period $a = b = 10$mm, length of cross dipole $a' = b' = 6.875$ mm, width of cross dipole $w_x = w_y = 0.625$ mm, 16x16 grid). Left) single dielectric and single metal layer, substrate thickness $d = 0.5$ mm, Relative permittivity was 2.0. Right) double cross element, $d = 2.362$ mm, $\varepsilon_r = 2.58$

5.3 Four-legged element

This element is reported as one of the best elements in terms of stability of reflection coefficient with respect to the angle of incidence (Munk, 2000). If properly designed in combination with suitable selection of the relative permittivity of a dielectric substrate, reflection coefficient stays stable well beyond the 45 degrees for all frequencies within the operating band.

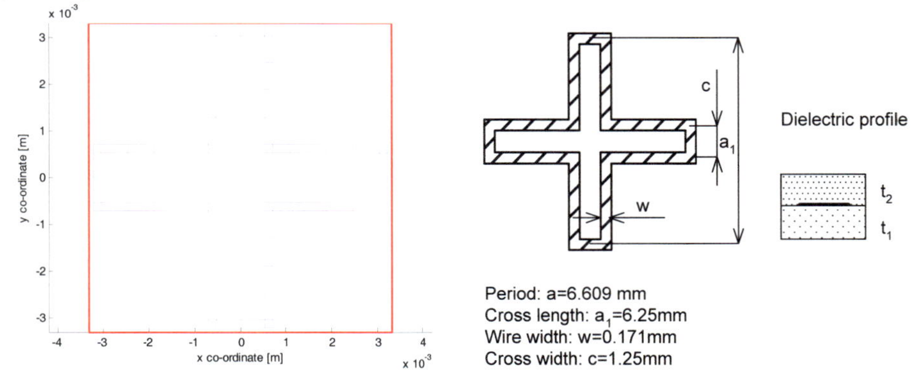

Fig. 11. Left) FSS with four-legged element (Munk, 2000) – non-uniformly meshed with rectangular cells, Right) element dimensions and the dielectric profile ($t_1 = t_2 = 0.508$ mm, $\varepsilon_{r1} = \varepsilon_{r2} = 2.2$)

The geometry of the four-legged element being analyzed is shown in Fig. 11. Its dimensions and properties of the dielectric profile were overtaken from (Munk, 2000). The element was meshed with non-uniform rooftops (Fig. 11, left), contrary to the (Munk, 2000) where large domain basis functions were used. The total number of unknown current expansion coefficients was $N_x+N_y = 26+26 = 52$. The calculation of reflection properties was performed from 1 to 20 GHz, considering an oblique incidence (45 degrees). As seen from Fig. 12, a very good agreement was obtained between the two approaches. The slight disagreements are attributed to readout errors from the graphs given in (Munk, 2000).

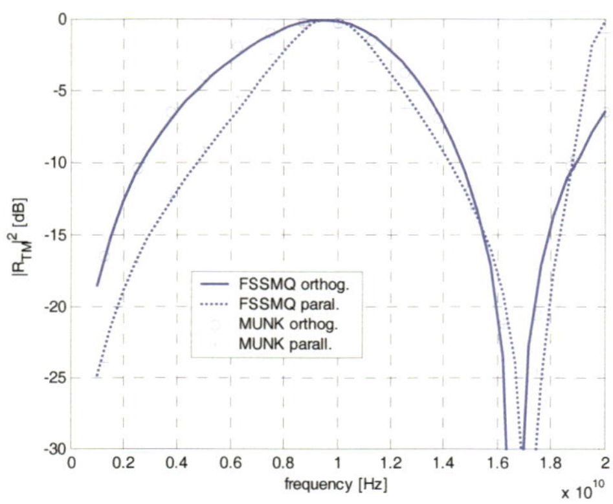

Fig. 12a. FSS with four-legged element (TM reflection coefficient versus frequency)

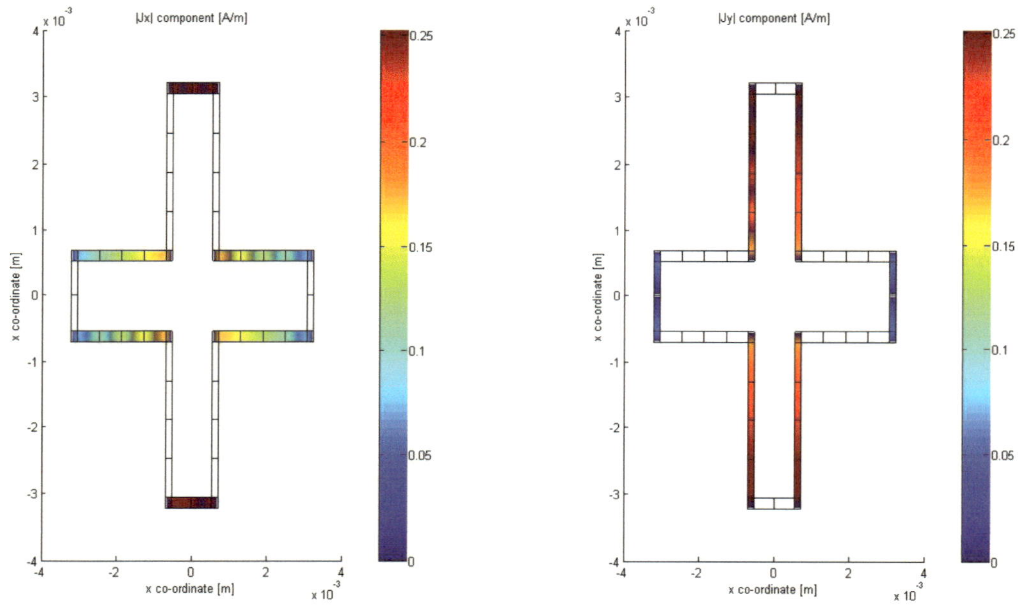

Fig. 12b. FSS with four-legged element (current density at resonance (10 GHz), TM incidence)

5.4 Composite materials – reflectivity study

In this subsection a study of reflection properties of a composite material is presented. The composite material consists from 10 micrometers wide flat fibres (which approximate real rounded ones). Fibres are placed in the y direction with a period of a = 20 micrometers (Fig. 13). Different composite materials have fibres with conductivity ranging from 100 to 10000 S/m. For more details about physical properties of composite materials see for example paper (Jayasree at el, 2008). In this example conductivity γ = 10000 S/m was selected. Three different simulations were performed. First, the composite with a single layer of wires was analyzed assuming the infinite wire conductivity (Fig. 14). Second, the effect of the finite conductivity was studied. It has been found that shielding effectiveness at 1 GHz was reduced from -85 to -18 dB. Third, three layers of lossy wires distanced 100um apart were analyzed. In this case, the shielding effectiveness improved by 10 dB at 1 GHz. Simulations were performed at frequency range 1-18 GHz. Since at frequencies 1-18 GHz the skin depth was larger than the thickness of the flat wire, the sheet resistance Rs could be approximated as Rs = 1/(γ*t).

At each metal layer, flat wire was meshed by 50 cells. Number of J_x and J_y current expansion coefficients per one layer was 40 + 50 = 90. For a three layer structure, the global impedance matrix sized 270x270 was assembled. Solution time per one frequency was about 5 seconds (Celeron M, 1.6 GHz).

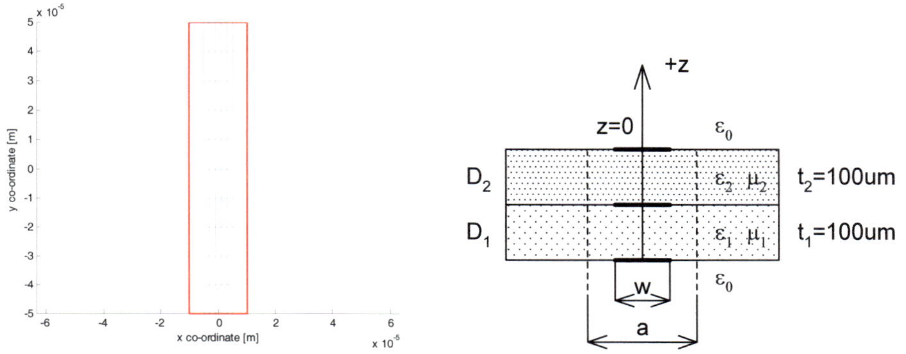

Fig. 13. Composite material with carbon fibres (a = 20um, w = 10um, t = 10um) Left) Meshed fibre, Right) Profile of the composite ($t_1 = t_2$ = 100 um, $\varepsilon_{r1} = \varepsilon_{r2} = 1$)

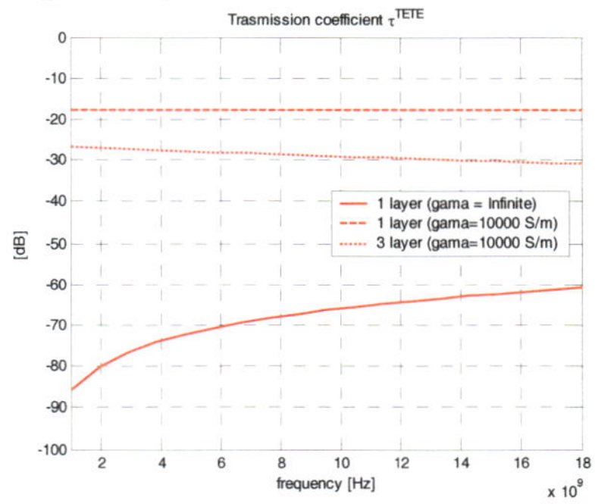

Fig. 14. Transmission coefficient of the composite structure (Number of Floquet harmonics $M = N = 18$), angle of incidence υ^I=0 deg, φ^I = 90 deg (electric intensity parallel with wire).

6. Conclusions

In this chapter, numerical analysis of planar periodic multilayer structures by the spectral domain method was addressed. Compared to other authors, use of non-uniform rooftops and a direct approach of analysis periodic structure with the global impedance matrix, was presented. Briefly, large domain basis functions defined over quadrilateral elements were outlined. Based, on the theory described in previous sections, capabilities of written FSSMQ simulation program were demonstrated on several examples. A very good agreement with results presented by other authors was obtained with the use of the code.

7. References

Itoh, T. (1980). Spectral Domain Immitance Approach for Disperion Chracteristics of Generalized Printed Transmission Lines. *IEEE transactions on Microwave Theory and Techniques,* Vol. 28, No. 7, July 1980, pp. 733-736.

Liu, C., C.; Hessel, A.; Hanfling, J., D., Usoff, J. M. (1985). Plane wave reflection from microstrip patch arrays - Theory and Experiment. *IEEE transactions on Antennas and Propagation,* Vol. 33, No. 4, April 1985, pp. 426-435.

Cwick, T., A.; Mittra, R. (1987). Scattering from a Periodic Array of Free-Standing Arbitrarily Shaped Perfectly Conducting or Resistive Patches. *IEEE transactions on Antennas and Propagation,* Vol. 35, No. 11, Nov 1987, pp. 1226-1234.

Mittra, R.; Chi, H., C.; Cwick, T. (1988). Techniques for Analyzing Frequnecy Selective Surfaces - A Review. *IEEE Proceedings,* Vol. 76, No. 2, Dec 1988, pp. 1593-1615.

Scott, G. (1989). *Spectral domain method in electromagnetics,* Artech House, New York.

Wu, T., K. (1995). *Frequency selective Surfaces and Grid Arrays,* John Wiley & Sons, ISBN 0-471-31189-8, New York.

Kipp, R., A.; Chan, C., H. (1994). A numerically Efficient Technique for the Method of Moments Solution for Planar Periodic Structures in Layered Media. *IEEE Transactions on Microwave Theory and Technique,* Vol. 42, No. 4, Apr 1994, pp. 635-642.

Wan, C.; Encinar, J., A. (1995). Efficient Computation of Generalized Scattering Matrix For Analyzing Multilayered Periodic Structures. *IEEE Transactions on Antennas and Propagation,* Vol. 43, No. 11, Dec 1995, pp. 1233-1242.

Kolundzija, B. (1998). On the Locally Continuous Formulation of Surface Doublets. *IEEE Transactions on Antennas and Propagation,* Vol. 46, No. 12, Dec 1998, pp. 1879-1883.

Gona, S.; Raida, Z. (1999). Hybrid FE/SD MoM analysis of frequency selective surfaces with arbitrarily shaped elements. *In proceedings of the international conference on electronics in advanced applications (ICEAA 99), Sept* 13-17, 1999, Torino, Italy.

Munk, B., A. (2000). *Frequency selective Surfaces – Theory and Design,* John Wiley & Sons, ISBN 0-471-37047-9, New York.

Notaros, M., B.; Popovic, B., D.; Weem, J., P.; Brown, R., A.; Popovic, Z. Efficient Large-Domain MoM Solutions to Electrically Large Practical EM Problems. *IEEE Transactions on Microwave Theory and Techniques,* Vol. 49, No. 1, Jan 2001, pp. 151-159.

Weile, D., S.; Michielssen, E. (2001); Galivan, K. Reduced-Order Modeling of Multiscreen Frequency-Selective Surfaces Using Krylov-Based Rational Interpolationn. *IEEE Transactions on Antennas and Propagation,* Vol. 49, No. 5, May 2001, pp. 801-813.

Gona, S. (2004). *Analysis and design of planar reflector antennas,* Ph.D Thesis. Brno University of Technology, Brno.

Gona, S.; Kresalek, V. (2008). Development of a Versatile Planar Periodic Structure Simulator in MATLAB. *In proceedings of the international conference on microwave techniques (COMITE 2008),* Apr23-24 2008, pp. 335-338. ISBN 978-1-4244-2137-4. Prague.

Jayasree, P., V., Y.; Baba, V., S., S., N.; Rao, B., P. (2008). Shielding Effectiveness of Laminated Shields. *Radioengineering magazine,* Vol. 17, No. 4, Dec 2008.

3

The High-Order Symplectic Finite-Difference Time-Domain Scheme

Wei E.I. Sha[a], Xian-liang Wu[b], Zhi-xiang Huang[b], and Ming-sheng Chen[c]
a. Department of Electrical and Electronic Engineering, The University of Hong Kong, Pokfulam Road, Hong Kong, China
Email: wsha@eee.hku.hk
b. Key Laboratory of Intelligent Computing & Signal Processing, Anhui University, Feixi Road 3, Hefei 230039, China
c. Department of Physics and Electronic Engineering, Hefei Teachers College, Jinzhai Road 327, Hefei 230061, China

Abstract
The book chapter will aim at introducing the background knowledge, basic theories, supporting techniques, numerical results, and future research for the high-order symplectic finite-difference time-domain scheme. The theories of symplectic geometry and Hamiltonian are reviewed in Section 2 followed by the symplectiness of Maxwell's equations presented in Section 3. Next, the numerical stability and dispersion analyses are given in Section 4. Then, in Section 5, we will make a tour of the supporting techniques but do not discuss them in detail. These techniques involve source excitation, perfectly matched layer, near-to-far-field transformation, inhomogeneous boundary treatments, and parameter extractions. The numerical results on propagation, scattering, and guided-wave problems are shown in Section 6. The high-order symplectic finite-difference time-domain scheme demonstrates the powerful advantages and potentials for the time-domain solution of Maxwell's equations, especially for electrically-large objects and for long-term simulation. Finally, the conclusion and future research are summarized in Section 7.

Keywords: Symplectic Finite-Difference Time-Domain Scheme; High-Order Techniques; Symplectic Geometry and Hamiltonian; Numerical Stability and Dispersion; Maxwell's Equations.

1. Introduction

The traditional finite-difference time-domain (FDTD) method [1-4] , which is explicit second-order-accurate in both space and time, has been widely applied to electromagnetic computation and simulation. The main advantages of the FDTD-based techniques for solving electromagnetic problems are computational simplicity and low operation count. Furthermore, it is well suited to analyze transient problems and is good at modeling inhomogeneous geometries. Most important of all, the method can readily be implemented on the massive computers.

However, the FDTD method has two primary drawbacks, one is the inability to accurately model the curved complex surfaces and material discontinuities by using the staircasing approach with structured grids, and another is the significant accumulated errors from numerical instability, dispersion and anisotropy. Hence fine grids are required to obtain satisfying numerical results, which leads to vast memory requirements and high computational costs, especially for electrically-large domains and for long-term simulation.

For the first pitfall, a variety of alternative methods in conjugation with unstructured grids were proposed to reduce the inaccuracy owing to the staircase approximation, including the finite-volume time-domain (FVTD) [5], finite-element time-domain (FETD) [6], and discontinuous Galerkin time-domain (DGTD) methods [7]. Although the methods are easy to treat boundaries, they are less efficient than the traditional FDTD method. Meanwhile, for the traditional FDTD method, a variety of conformal [8-11] and subgridding strategies [12] were proposed also.

To overcome the second problem, other high-order spatial discretization strategies were developed. The multi-resolution time-domain (MRTD) [13] and pseudo-spectral time-domain (PSTD) [14] methods reduce the spatial sampling rate drastically, but they are difficult to handle the material interface for modeling the three-dimensional complex objects [15, 16]. Another approach is the staggered fourth-order FDTD method [17-21], which retains the simplicity of the original Yee algorithm and can save computational resources with coarse grids compared to the traditional FDTD method. However, the approach must set lower Courant-Friedrichs-Levy (CFL) number to comply with the stability criterion. Furthermore, the high-order compact difference [22, 23] is easier to treat the inhomogeneous boundaries, but it requires the sparse matrix inversion for each time step.

Except for the solvers in space direction, novel solvers in time direction were proposed as well. The Runge-Kutta (R-K) method used in [3, 22] can achieve the high-order accuracy. However, it will consume additional memory and has amplitude error. The alternative direction implicit time-stepping strategy [24-26] is unconditionally stable, but it suffers from the intolerable numerical dispersion once the CFL number is too large. Moreover, the strategy will consume more CPU times caused by the sparse matrix inversion. For the time direction, does a high-order-accurate and energy-conserving solver with low computational costs exist? Surprisingly, Yes!

Most physical and chemical phenomenons can be modeled by Hamiltonian differential equations whose time evolution is symplectic transform and flow conserves the symplectic structure [27-29]. The symplectic schemes include a variety of different temporal discretization strategies designed to preserve the global symplectic structure of the phase space for a Hamiltonian system. They have demonstrated their advantages in numerical computation for the Hamiltonian system, especially for long-term simulation. Since Maxwell's equations can be written as an infinite-dimensional Hamiltonian system, a stable and accurate solution can be obtained by using the symplectic schemes, which preserve the energy of the Hamiltonian system constant. The symplectic schemes can be explicit or implicit and can be generalized to high-order with controllable computational complexity. Recently, researchers from computational electromagnetics society have focused on the symplectic schemes for solving Maxwell's equations. Symplectic finite-difference time-domain (**SFDTD**) scheme [30-41], symplectic discrete singular convolution method [42], symplectic pseudo-spectral time-domain approach [43], symplectic wave equation strategy [44], and multi-symplectic method [45, 46] were proposed and studied. This chapter we will focus on the explicit high-order symplectic integration schemes with the high-order staggered spatial differences for solving the Maxwell's equations.

2. Mathematical foundations

The partial mathematical proofs are cited from [28, 29, 47].

Definition 1.1. For $p_{2n}^0, q_{2n}^0 \in R_{2n}$, real-symplectic inner product can be defined as

$$\varpi(p^0, q^0) = p^0 J (q^0)^T \tag{1}$$

where T denotes transpose and $J = \begin{bmatrix} \{0\}_{n\times n} & I_{nxn} \\ -I_{nxn} & \{0\}_{n\times n} \end{bmatrix}_{2n\times 2n}$ which satisfies skew-symmetric and orthogonal properties, i.e. $J = -J^T$, $J^{-1} = J^T = -J$.

The real-symplectic inner product has the following properties:

(1) Bilinear property:

$\varpi(p^0 + q^0, r^0 + s^0) = \varpi(p^0, r^0) + \varpi(p^0, s^0) + \varpi(q^0, r^0) + \varpi(q^0, s^0)$,

$\varpi(\lambda p^0, \eta q^0) = \lambda\eta\varpi(p^0, q^0)$, $r^0, s^0 \in R_{2n}$, and $\lambda, \eta \in R$;

(2) Skew-symmetric property: $\varpi(p^0, q^0) = -\varpi(q^0, p^0)$;

(3) Non-degenerate property: $\forall q^0 \neq 0$, $\exists p^0$, $\varpi(p^0, q^0) = 0 \Rightarrow p^0 = 0$.

Definition 1.2. If V is a vector space defined on R_{2n} and the mapping $\varpi : V \times V \rightarrow R$ is real-symplectic, (V, ϖ) is called real-symplectic space and ϖ is called real-symplectic structure.

Definition 1.3. A linear transform $T : V \rightarrow V$ is called real-symplectic transform, if it meets $\varpi(Tp^0, Tq^0) = \varpi(p^0, q^0)$, $\forall p^0, q^0 \in R_{2n}$.

Definition 1.4. The matrix T is called real-symplectic matrix if $T^T J T = J$ and $\varpi(Tp^0, Tq^0) = \varpi(p^0, q^0)$. The group including all the real-symplectic matrices is called real-symplectic group. We sign it as $Sp(2n, R)$.

Definition 1.5. B is an infinitesimally real-symplectic matrix if $B^T J + JB = 0$. The infinitesimally real-symplectic matrices can be composed of Lie algebra via anti-commutable Lie Poisson bracket $[A, B] = AB - BA$.

Theory 1. B is an infinitesimally real-symplectic matrix $\Rightarrow$ $\exp(B) \in Sp(2n, R)$.

Above mentioned definitions and theory can be extended to complex space.

Definition 2.1. For $p_{2n}^0, q_{2n}^0 \in C_{2n}$, complex-symplectic inner product can be defined as

$$\varpi(p^0, q^0) = p^0 J (q^0)^H \tag{2}$$

where H denotes complex conjugate transpose or adjoint.
The complex-symplectic inner product has the following properties:
(1) Conjugate bilinear property:
$\varpi(p^0+q^0,r^0+s^0)=\varpi(p^0,r^0)+\varpi(p^0,s^0)+\varpi(q^0,r^0)+\varpi(q^0,s^0)$,

$\varpi(\lambda p^0,\eta q^0)=\lambda\overline{\eta}\,\varpi(p^0,q^0)$, $r^0,s^0\in C_{2n}$, $\lambda,\eta\in C$, and $\overline{\eta}$ is the conjugate complex of η ;

(2) Skew-Hermitian property: $\varpi(p^0,q^0)=-\overline{\varpi(q^0,p^0)}$;
(3) Non-degenerate property: $\forall q^0\neq 0$, $\exists p^0$, $\varpi(p^0,q^0)=0 \Rightarrow p^0=0$.

Definition 2.2. If V is a vector space defined on C_{2n} and the mapping $\varpi: V\times V\to C$ is complex-symplectic, (V,ϖ) is called complex-symplectic space and ϖ is called complex-symplectic structure.

Definition 2.3. A linear transform $T:V\to V$ is called complex-symplectic transform, if it meets $\varpi(Tp^0,Tq^0)=\varpi(p^0,q^0)$, $\forall p^0,q^0\in C_{2n}$.

Definition 2.4. The matrix T is called complex-symplectic matrix if $T^H JT=J$ and $\varpi(Tp^0,Tq^0)=\varpi(p^0,q^0)$. The group including all the complex-symplectic matrices is called complex-symplectic group. We sign it as $Sp(2n,C)$.

Definition 2.5. B is an infinitesimally complex-symplectic matrix if $B^H J+JB=0$. The infinitesimally complex-symplectic matrices can be composed of Lie algebra via anti-commutable Lie Poisson bracket $[A,B]=AB-BA$.

Theory 2. B is an infinitesimally complex-symplectic matrix $\Rightarrow \exp(B)\in Sp(2n,C)$.

Definition 3. If $p^0=(p_1,p_2,\cdots p_n)$, $q^0=(q_1,q_2,\cdots q_n)$, $(p^0,q^0)\in\Omega\subseteq R_{2n}$, and $t_0\in I$, the Hamiltonian canonical equations can be written as

$$\frac{dp_i}{dt_0}=-\frac{\partial H}{\partial q_i},\quad \frac{dq_i}{dt_0}=+\frac{\partial H}{\partial p_i},\quad i=1,2,\cdots n \tag{3}$$

where $H(p^0,q^0,t_0)$ is the Hamiltonian function, Ω is the phase space, and $\Omega\times I$ is the extended phase space.

Theory 3. If the solution of (3) at any time t_* is (p^*,q^*) and (p^*,q^*) still satisfies the equation (3), the Jacobi matrix Θ is a symplectic matrix

$$\Theta^T J\Theta=J \tag{4}$$

where $\Theta=\dfrac{\partial(p^*,q^*)}{\partial(p^0,q^0)}=\begin{pmatrix}\partial p^*/\partial p^0 & \partial p^*/\partial q^0\\ \partial q^*/\partial p^0 & \partial q^*/\partial q^0\end{pmatrix}$.

Theory 4. If the time evolution operator of (3) from t_0 to t_* is $\Psi(t_*, t_0)$ and $(p^*, q^*) = \Psi(t_*, t_0)(p^0, q^0)$, the operator conserves the symplectic structure

$$\Psi(t_*, t_0)^* \varpi^* = \varpi^0 \tag{5}$$

where $\varpi^* = dp^* \wedge dq^*$, $\varpi^0 = dp^0 \wedge dq^0$, and $\Psi(t_*, t_0)^*$ is the conjugate operator of $\Psi(t_*, t_0)$. The time evolution operator is also called Hamiltonian flow or symplectic flow.

Theory 5. The matrix $L = \begin{bmatrix} 0 & A \\ -A & 0 \end{bmatrix} \Rightarrow \exp(L) = \begin{bmatrix} \cos(A) & \sin(A) \\ -\sin(A) & \cos(A) \end{bmatrix}$.

Theory 6. If the matrix $L = \begin{bmatrix} 0 & A \\ -A & 0 \end{bmatrix}$ and $A = A^T$, we have: (1) L is skew-symmetric, i.e. $L = -L^T$; (2) $\exp(L)$ are both orthogonal and real-symplectic matrices. We call $\exp(L)$ symplectic-orthogonal matrix.

Theory 7. If the matrix $L = \begin{bmatrix} 0 & A \\ -A & 0 \end{bmatrix}$ and $A = A^H$, we have: (1) L is skew-Hermitian, i.e. $L = -L^H$; (2) $\exp(L)$ are both unitary and complex-symplectic matrices. We call $\exp(L)$ symplectic-unitary matrix.

3. Symplectic framework of Maxwell's equations

A Helicity generating function [48] for Maxwell's equations in free space is introduced as

$$G(\mathbf{H}, \mathbf{E}) = \frac{1}{2}\left(\frac{1}{\varepsilon_0}\mathbf{H} \cdot \nabla \times \mathbf{H} + \frac{1}{\mu_0}\mathbf{E} \cdot \nabla \times \mathbf{E}\right) \tag{6}$$

where $\mathbf{E} = (E_x, E_y, E_z)^T$ is the electric field vector, $\mathbf{H} = (H_x, H_y, H_z)^T$ is the magnetic field vector, and ε_0 and μ_0 are the permittivity and permeability of free space. The differential form of the Hamiltonian is

$$\frac{\partial \mathbf{H}}{\partial t} = -\frac{\delta G}{\delta \mathbf{E}}, \quad \frac{\partial \mathbf{E}}{\partial t} = \frac{\delta G}{\delta \mathbf{H}} \tag{7}$$

According to the variation principle, we can derive Maxwell's equations of free space from (7)

$$\frac{\partial}{\partial t}\begin{pmatrix} \mathbf{H} \\ \hat{\mathbf{E}} \end{pmatrix} = L \begin{pmatrix} \mathbf{H} \\ \hat{\mathbf{E}} \end{pmatrix} \tag{8}$$

$$L = \begin{pmatrix} \{0\}_{3\times3} & -\dfrac{1}{\sqrt{\mu_0\varepsilon_0}}R_{3\times3} \\ \dfrac{1}{\sqrt{\mu_0\varepsilon_0}}R_{3\times3} & \{0\}_{3\times3} \end{pmatrix}, \ \hat{\mathbf{E}} = \sqrt{\frac{\varepsilon_0}{\mu_0}}\mathbf{E} \tag{9}$$

$$R = \begin{pmatrix} 0 & -\dfrac{\partial}{\partial z} & \dfrac{\partial}{\partial y} \\ \dfrac{\partial}{\partial z} & 0 & -\dfrac{\partial}{\partial x} \\ -\dfrac{\partial}{\partial y} & \dfrac{\partial}{\partial x} & 0 \end{pmatrix} = \nabla\times \tag{10}$$

where $\{0\}_{3\times3}$ is the 3×3 null matrix and R is the three-dimensional curl operator. However, the Helicity generating function has little physical meaning.
It is known however that the total stored energy of electromagnetic field is constant in an energy conserving system. Hence, the total stored energy is taken to be the Hamiltonian

$$G = \iiint_V \left(\frac{1}{2}\mathbf{E}\cdot\mathbf{D} + \frac{1}{2}\mathbf{B}\cdot\mathbf{H}\right)dV \tag{11}$$

It is well known that

$$\mathbf{H} = \frac{1}{\mu_0}\nabla\times\mathbf{A} \tag{12}$$

$$\mathbf{E} = -\nabla\Phi - \frac{\partial\mathbf{A}}{\partial t} \tag{13}$$

where $\mathbf{A}$ and Φ are the vector and scalar potentials and can be uniquely defined by using a Lorentz gauge or a Coulomb gauge. If we define the conjugate momentum and coordinate as

$$\mathbf{\Pi} = \varepsilon_0(\nabla\Phi + \frac{\partial\mathbf{A}}{\partial t}) \tag{14}$$

$$\mathbf{Q} = \mathbf{A} + \nabla\varphi \tag{15}$$

The Hamiltonian can be rewritten as

$$G = \iiint_V \left[\frac{1}{2\varepsilon_0}\mathbf{\Pi}\cdot\mathbf{\Pi} + \frac{1}{2\mu_0}\nabla\times\mathbf{Q}\cdot\nabla\times\mathbf{Q}\right]dV \tag{16}$$

The equations of motion is to be

$$\frac{\delta G}{\delta\mathbf{\Pi}} = \frac{\partial\mathbf{Q}}{\partial t}, \ \frac{\delta G}{\delta\mathbf{Q}} = -\frac{\partial\mathbf{\Pi}}{\partial t} \tag{17}$$

If we define $\Phi = \partial\varphi / \partial t$ and $\nabla A = -\mu_0\varepsilon_0 \frac{\partial \Phi}{\partial t}$ (Lorentz gauge), the above is equivalent to Maxwell's equations.

The time evolution of (8) from $t = 0$ to $t = \Delta_t$ can be written as

$$\begin{pmatrix} \mathbf{H} \\ \hat{\mathbf{E}} \end{pmatrix}(\Delta_t) = \exp(\Delta_t L)\begin{pmatrix} \mathbf{H} \\ \hat{\mathbf{E}} \end{pmatrix}(0) \tag{18}$$

where $\exp(\Delta_t L)$ is the time evolution matrix (TEMA) or symplectic flow of Maxwell's equations.

For infinite-dimensional real space, we define the inner product

$$< F(t,\mathbf{r}), G(t,\mathbf{r}) >= \int_{-\infty}^{\infty} F(t,\mathbf{r}) \cdot G(t,\mathbf{r}) d\mathbf{r} \tag{19}$$

where $\mathbf{r} = x\mathbf{e}_\mathbf{x} + y\mathbf{e}_y + z\mathbf{e}_z$ is the position vector and t is the time variable. According to the identity both in the generalized distribution space and in the Hilbert space

$$< \partial F / \partial\delta, G >= - < F, \partial G / \partial\delta >,\ \delta = x, y, z \tag{20}$$

we can know $\partial / \partial\delta$ is a skew-symmetric operator. Hence R is a symmetric operator, i.e. $R = R^T$. Based on **Theory 6**, the TEMA of Maxwell's equations is a symplectic-orthogonal matrix in real space.

For infinite-dimensional complex space, we define the inner product

$$< F(t,\mathbf{r}), G(t,\mathbf{r}) >= \int_{-\infty}^{\infty} F(t,\mathbf{r}) \cdot \overline{G(t,\mathbf{r})} d\mathbf{r} \tag{21}$$

The forward and inverse Fourier transforms for electromagnetic field components are respectively

$$\widetilde{F}(t,\mathbf{k_0}) = \frac{1}{\sqrt{2\pi}} \int_{-\infty}^{\infty} F(t,\mathbf{r}) \exp(j_0 \mathbf{k_0} \cdot \mathbf{r}) d\mathbf{r} \tag{22}$$

$$F(t,\mathbf{r}) = \frac{1}{\sqrt{2\pi}} \int_{-\infty}^{\infty} \widetilde{F}(t,\mathbf{k_0}) \exp(-j_0 \mathbf{k_0} \cdot \mathbf{r}) d\mathbf{k_0} \tag{23}$$

where j_0 is the imaginary unit and $\mathbf{k_0} = k_x\mathbf{e}_\mathbf{x} + k_y\mathbf{e}_y + k_z\mathbf{e}_z$ is the wave vector. For simplicity, we use the shorthand notations $\widetilde{F} = \psi F$ and $F = \psi^{-1}\widetilde{F}$.

In the beginning, with the help of Parseval theorem

$$< \psi F, \widetilde{G} >=< F, \psi^{-1}\widetilde{G} > \tag{24}$$

we know that the Fourier operator ψ is a unitary operator, i.e. $\psi^{-1} = \psi^H$.

Next, using the differential property of Fourier transform $\frac{\partial F}{\partial \delta} \leftrightarrow -j_0 k_\delta \widetilde{F}, \delta = x, y, z$, we can obtain the spectral-domain form of Maxwell's equations

$$\frac{\partial}{\partial t}\begin{pmatrix}\widetilde{\mathbf{H}}\\ \widetilde{\hat{\mathbf{E}}}\end{pmatrix}=\begin{pmatrix}\{0\}_{3\times3} & -\frac{1}{\sqrt{\mu_0\varepsilon_0}}\widetilde{R}_{3\times3}\\ \frac{1}{\sqrt{\mu_0\varepsilon_0}}\widetilde{R}_{3\times3} & \{0\}_{3\times3}\end{pmatrix}\begin{pmatrix}\widetilde{\mathbf{H}}\\ \widetilde{\hat{\mathbf{E}}}\end{pmatrix} \tag{25}$$

$$\widetilde{R}_{3\times3}=\begin{pmatrix}0 & j_0k_z & -j_0k_y\\ -j_0k_z & 0 & j_0k_x\\ j_0k_y & -j_0k_x & 0\end{pmatrix} \tag{26}$$

where $\widetilde{R}$ is a Hermitian matrix, i.e. $\widetilde{R}^H = \widetilde{R}$.
Finally, considering the unitary property of the Fourier operator, we can convert the spectral-domain form (25) into the spatial-domain form (27)

$$\frac{\partial}{\partial t}\begin{pmatrix}\mathbf{H}\\ \hat{\mathbf{E}}\end{pmatrix}==\begin{pmatrix}\{0\}_{3\times3} & -\frac{1}{\sqrt{\mu_0\varepsilon_0}}\Psi^H_{3\times3}\widetilde{R}_{3\times3}\Psi_{3\times3}\\ \frac{1}{\sqrt{\mu_0\varepsilon_0}}\Psi^H_{3\times3}\widetilde{R}_{3\times3}\Psi_{3\times3} & \{0\}_{3\times3}\end{pmatrix}\begin{pmatrix}\mathbf{H}\\ \hat{\mathbf{E}}\end{pmatrix} \tag{27}$$

where $\Psi_{3\times3}=diag(\psi\psi\psi)$. It is easy to show that $R=\psi^H_{3\times3}\tilde{R}_{3\times3}\psi_{3\times3}$ is a Hermitian matrix, i.e. $R = R^H$. Based on **Theory 7**, the TEMA of Maxwell's equations is a symplectic-unitary matrix in complex space.
It is well known that the total energy of electromagnetic field in free space can be represented as

$$\Sigma^V=\frac{1}{2}\mu_0\left(<\mathbf{H},\mathbf{H}>+<\hat{\mathbf{E}},\hat{\mathbf{E}}>\right)=\iiint\left(\frac{1}{2}\mu_0\,|\,\mathbf{H}\,|^2+\frac{1}{2}\varepsilon_0\,|\,\mathbf{E}\,|^2\right)dV \tag{28}$$

No matter in complex space or in real space, the TEMA $\exp(\Delta_t L)$ accurately conserves the total energy of electromagnetic field. In other words, the TEMA $\exp(\Delta_t L)$ only rotates the electromagnetic field components (**Theory 5**). In addition, if an algorithm can accurately conserve the total energy of electromagnetic field, it is to be unconditionally stable.
Both in complex space and in real space, we can split L into U and V

$$L=U+V \tag{29}$$

$$U=\begin{pmatrix}\{0\}_{3\times3} & -\frac{1}{\sqrt{\mu_0\varepsilon_0}}R_{3\times3}\\ \{0\}_{3\times3} & \{0\}_{3\times3}\end{pmatrix},\ V=\begin{pmatrix}\{0\}_{3\times3} & \{0\}_{3\times3}\\ \frac{1}{\sqrt{\mu_0\varepsilon_0}}R_{3\times3} & \{0\}_{3\times3}\end{pmatrix} \tag{30}$$

The split TEMA can be approximated by the explicit m-stage pth-order symplectic integration scheme [32, 49]

$$\exp\left(\Delta_t (U+V)\right) = \prod_{l=1}^{m} \exp(d_l \Delta_t V)\exp(c_l \Delta_t U) + O\left(\Delta_t^{\ p+1}\right) \qquad (31)$$

where c_l and d_l are the symplectic integrators and satisfy the time-reversible [49] or symmetric relations [50], i.e.

$$c_l = c_{m-l+1}(1 \le l \le m),\ d_l = d_{m-l}(1 \le l \le m-1),\ d_m = 0 \qquad (32)$$

$$d_l = c_{m-l+1}(1 \le l \le m) \qquad (33)$$

Table 1 lists the three-order symmetric symplectic integrators and the fourth-order time-reversible symplectic integrators [40]. The time-stepping diagram for the five-stage fourth-order symplectic scheme [32] is shown in Fig. 1.

Order (p)	Stage (m)	c_1	c_2	c_3	d_1	d_2
3	3	0.26833010	-0.18799162	0.91966152		
4	5	0.16537923	1.35491814	-2.04059474	0.51541261	-0.01541261

Table 1. The three-order symmetric symplectic integrators and the fourth-order time-reversible symplectic integrators.

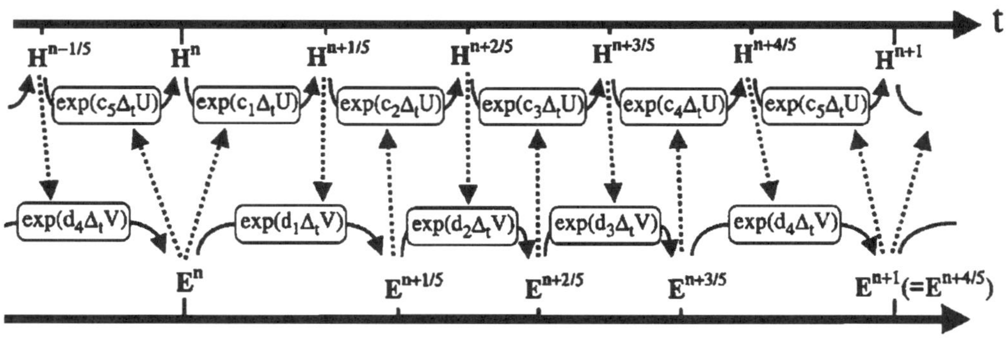

Fig. 1. Time-stepping diagram for the five-stage fourth-order symplectic scheme.

For real space, $R = R^T$ and therefore U and V are the infinitesimally real-symplectic matrices. Likewise, for complex space, $R = R^H$ and therefore U and V are the infinitesimally complex-symplectic matrices. In particular, we have: (1) U and V can be composed of Lie algebra semicolon at Line 11. (2) $\exp(d_l \Delta_t V)$ and $\exp(c_l \Delta_t U)$ are the symplectic matrices.

Although the orthogonal properties can not be retained by the two matrices $\exp(d_l \Delta_t V)$ and $\exp(c_l \Delta_t U)$, the determinants of them are equal to 1 [51]. Thus the explicit symplectic integration scheme is conditionally stable and does not have amplitude error.

4. Numerical stability and dispersion analyses

We first present the numerical stability and dispersion analyses for the one-dimensional problem, then extend them to the three-dimensional problem.

Given the field components $\mathbf{F}^{\mathbf{n}} = (H_y^n, E_x^n)$ at the n-th time step, the field components $\mathbf{F}^{\mathbf{n+1}} = (H_y^{n+1}, E_x^{n+1})$ at the (n+1)-th time step can be represented as

$$\mathbf{F}^{\mathbf{n+1}} = S\mathbf{F}^{\mathbf{n}} \tag{34}$$

where S is the amplification matrix.

The well-known plane wave expansions are

$$F(x, y, z, t) = f_0 \exp\left(-j_0(i\Delta_x k_x + j\Delta_y k_y + k\Delta_z k_z)\right) \tag{35}$$

$$k_x = k_0 \sin\theta \cos\varphi,\ k_y = k_0 \sin\theta \sin\varphi,\ k_z = k_0 \cos\theta \tag{36}$$

where k_0 is the numerical wave number, and θ and φ are spherical angles.

Using the q-th staggered differences to approximate the spatial first-order derivatives, we get

$$\begin{aligned}
\frac{\partial F}{\partial z} &\approx \sum_{r=1}^{q/2} W_r \frac{F(i, j, k + r - 1/2) - F(i, j, k - r + 1/2)}{\Delta_z} \\
&= \sum_{r=1}^{q/2} W_r \frac{\exp[-j_0(r - 1/2)k_z\Delta_z] - \exp[j_0(r - 1/2)k_z\Delta_z]}{\Delta_z} \cdot F \\
&= \eta_z F
\end{aligned} \tag{37}$$

where $\eta_z = \sum_{r=1}^{q/2} W_r \frac{\exp[-j_0(r - 1/2)k_z\Delta_z] - \exp[j_0(r - 1/2)k_z\Delta_z]}{\Delta_z}$, and W_r are the spatial difference coefficients [40] as shown in Table 2.

The continuous Maxwell's equations

$$\frac{\partial}{\partial t}\begin{pmatrix} H_y \\ E_x \end{pmatrix} = \begin{pmatrix} 0 & -\frac{1}{\mu}\frac{\partial}{\partial z} \\ -\frac{1}{\varepsilon}\frac{\partial}{\partial z} & 0 \end{pmatrix} \begin{pmatrix} H_y \\ E_x \end{pmatrix} \tag{38}$$

can be semi-discretized as

$$\frac{\partial}{\partial t}\begin{pmatrix} H_y \\ E_x \end{pmatrix} = \begin{pmatrix} 0 & -\frac{1}{\mu}\eta_z \\ -\frac{1}{\varepsilon}\eta_z & 0 \end{pmatrix}\begin{pmatrix} H_y \\ E_x \end{pmatrix} \tag{39}$$

If we set $U = \begin{pmatrix} 0 & -\frac{1}{\mu}\eta_z \\ 0 & 0 \end{pmatrix}$, $V = \begin{pmatrix} 0 & 0 \\ -\frac{1}{\varepsilon}\eta_z & 0 \end{pmatrix}$, and use the symplectic integration scheme for approximating the TEMA of Maxwell's equations, the amplification matrix S can be written as

$$S = \prod_{l=1}^{m}\begin{pmatrix} 1 & 0 \\ -\frac{1}{\varepsilon}\eta_z d_l \Delta_t & 1 \end{pmatrix}\begin{pmatrix} 1 & -\frac{1}{\mu}\eta_z c_l \Delta_t \\ 0 & 1 \end{pmatrix} \tag{40}$$

Each stage of (40) is the symplectic transform, and therefore $\det[\exp(c_l \Delta_t U)] = 1$ and $\det[\exp(d_l \Delta_t V)] = 1$ [51], which can be easily testified by (40). As a result, $\det S = 1$. The amplification matrix is

$$S = \begin{bmatrix} S_{11} & S_{12} \\ S_{21} & S_{22} \end{bmatrix} \tag{41}$$

and its eigenvalues λ satisfy the following equation

$$\lambda^2 - (S_{11} + S_{22})\lambda + (S_{11}S_{22} - S_{12}S_{21}) = 0 \tag{42}$$

Notice that $tr(S) = S_{11} + S_{22}$ and $(S_{11}S_{22} - S_{12}S_{21}) = \det S = 1$, (42) can be rewritten as

$$\lambda^2 - tr(S)\lambda + 1 = 0 \tag{43}$$

and its solutions $\lambda_{1,2} = \dfrac{tr(S) \pm j_0\sqrt{4 - tr(S)^2}}{2}$. A stable algorithm requires $|\lambda_{1,2}| = 1$, which says that $|tr(S)| \le 2$.

The right side of (40) is multiplied term by term, then we get

$$tr(S) = 2 + \sum_{l=1}^{m} g_l \left(v_0{}^2 \Delta_t^2 \eta_z^2\right)^l \tag{44}$$

$$g_l = \sum_{1\le i_1 \le j_1 < i_2 \le j_2 < \cdots < i_l \le j_l \le m} c_{i_1} d_{j_1} c_{i_2} d_{j_2} \cdots c_{i_l} d_{j_l} + \sum_{1\le i_1 < j_1 \le i_2 < j_2 \le \cdots \le i_l < j_l \le m} d_{i_1} c_{j_1} d_{i_2} c_{j_2} \cdots d_{i_l} c_{j_l} \tag{45}$$

where $v_0 = \dfrac{1}{\sqrt{\mu\varepsilon}}$ is the velocity of light.

For the three-dimensional problem, the continuous-time discrete-space Maxwell's equations can be written as

$$\frac{\partial}{\partial t}\begin{pmatrix}\mathbf{H}\\ \mathbf{E}\end{pmatrix}=\begin{pmatrix}0 & \frac{-1}{\mu}(\eta_x\mathbf{e}_x+\eta_y\mathbf{e}_y+\eta_z\mathbf{e}_z)\times \\ \frac{1}{\varepsilon}(\eta_x\mathbf{e}_x+\eta_y\mathbf{e}_y+\eta_z\mathbf{e}_z)\times & 0\end{pmatrix}\begin{pmatrix}\mathbf{H}\\ \mathbf{E}\end{pmatrix} \tag{46}$$

Notice $(\eta_x^2+\eta_y^2+\eta_z^2)<0$, (46) can be rewritten as the tensor form

$$\frac{\partial}{\partial t}\begin{pmatrix}\mathbf{H}\\ \mathbf{E}\end{pmatrix}=\begin{pmatrix}0 & \frac{-\sqrt{-(\eta_x^2+\eta_y^2+\eta_z^2)}}{\mu}\overline{\mathbf{K}}\cdot \\ \frac{\sqrt{-(\eta_x^2+\eta_y^2+\eta_z^2)}}{\varepsilon}\overline{\mathbf{K}}\cdot & 0\end{pmatrix}\begin{pmatrix}\mathbf{H}\\ \mathbf{E}\end{pmatrix} \tag{47}$$

where $\overline{\mathbf{K}}$ is the tensor matrix defined by the spherical angles [52]. Although (47) is a 6×6 matrix, it has only two independent eigenvalues related to TE and TM waves. Hence, (47) and (39) are isomorphic. Using the similar technique, we can get

$$tr(S)=2+\sum_{l=1}^{m}g_l\left\{v_0{}^2\Delta_t^2\left(\eta_x^2+\eta_y^2+\eta_z^2\right)\right\}^l \tag{48}$$

Generally speaking, the stability limit $CFL_{\max}$ for a time-domain solver can be written as [16, 40]

$$CFL_{\max}=\frac{\lambda_T}{\lambda_S} \tag{49}$$

where λ_S is the spatial stability factor which can be defined as

$$\lambda_S=\sqrt{d}W_s \tag{50}$$

where $d=1,2,3$ are the spatial dimensions, and W_s is the summation of the spatial difference coefficients as shown in Table 2. λ_T is the time stability factor, which can be obtained by the spatial stability factor and the constraint $|tr(S)|\le 2$. The form (49) decoupling the spatial stability factor from the time stability factor is more flexible and convenient for analyzing the stability limits of SFDTD(p,q) schemes, where p is the order for the time-stepping scheme and q is the order for the spatial differences. The stability limits [40] for the time-domain solvers are listed in Table 3.

Order (q)	W_1	W_2	W_3	W_4	W_S
2	1				2
4	9/8	-1/24			7/3
6	75/64	-25/384	3/640		149/60
8	1225/1024	-245/3072	49/5120	-5/7168	2161/840

Table. 2. The spatial difference coefficients.

Algorithms	CFL number
FDTD(2,2)	0.577
FDTD(2,4)	0.495
J-Fang(4,4)	0.577
R-K(4,4)	0.700
SFDTD(4,4)	0.858

Table. 3. The stability limits for different algorithms.

The disperision relation can be written as

$$\omega\Delta t = \arccos[tr(S)/2] \tag{51}$$

and the phase velocity error can be defined as

$$Err = 20\log_{10}\left|\frac{v_p - v_0}{v_0}\right| \tag{52}$$

where $v_p = \dfrac{\omega}{k_0}$ is the numerical phase velocity. The phase velocity error as a function of points per wavelength (PPW) is shown in Fig. 2. The SFDTD(4,4) scheme is superior to the traditional FDTD(2,2) method, FDTD(2,4) approach [18], and R-K(4,4) [3] strategy. Although the J-Fang(4,4) method [17] is the best solver, but it suffers from the intractable boundary treatments.

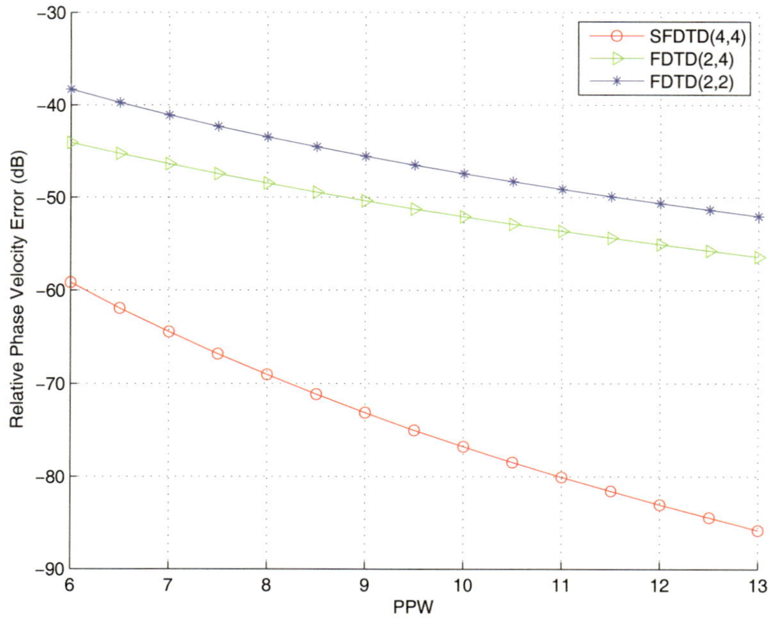

(a) CFL=0.4

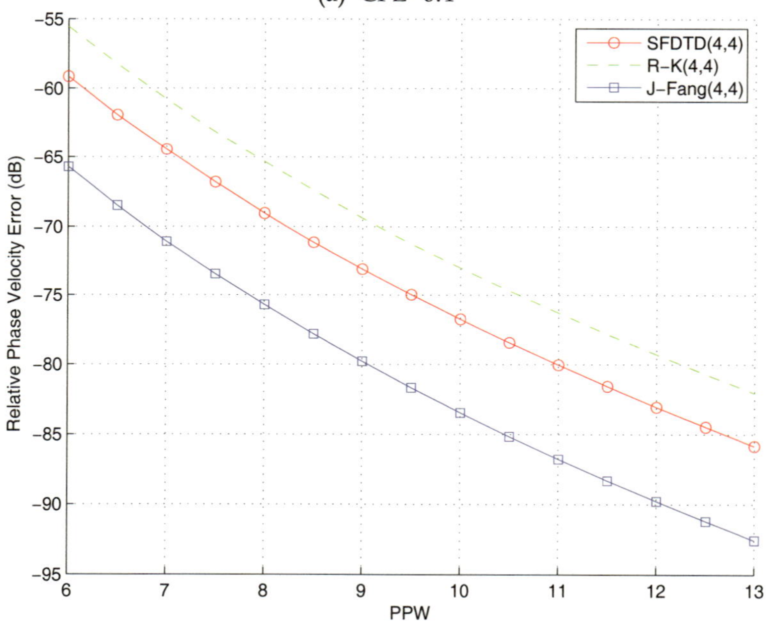

(b) CFL=0.5

Fig. 2. Numerical dispersion comparisons. Dispersion curves for a plane wave traveling at $\theta = 60^\circ$ and $\varphi = 30^\circ$.

5. Supporting techniques

The basic formulations of the high-order SFDTD scheme are presented in [32, 38]. The perfectly matched layer (PML) absorbing boundary conditions are given in [31, 41-43]. The total field and scattered field techniques are developed in [34, 53]. The near-to-far-field transformation is put forward in [38]. The high-order subcell and the high-order conformal strategies are proposed in [38, 39, 54, 55, 56]. The parameter extraction and source excitation techniques are discussed in [41].

A function of space and time evaluated at a discrete point in the Cartesian lattice and at a discrete stage in the time step can be notated as

$$F(x,y,z,t)=F^{n+l/m}(i\Delta_x, j\Delta_y, k\Delta_z, (n+\tau_l)\Delta_t) \tag{53}$$

where Δ_x, Δ_y, and Δ_z are, respectively, the lattice space increments in the x, y, and z coordinate directions, Δ_t is the time increment, i, j, k, n, l, and m are integers, $n+l/m$ denotes the $l-th$ stage after n time steps, m is the total stage number, and τ_l is the fixed time with respect to the $l-th$ stage.

Take the SFDTD(p,4) scheme for example, the update equation for the scaled electric field component is given by

$$\begin{aligned}
\hat{E}_x^{n+l/m}\left(i+\frac{1}{2},j,k\right) &= \hat{E}_x^{n+(l-1)/m}\left(i+\frac{1}{2},j,k\right)+\frac{1}{\varepsilon_r\left(i+\frac{1}{2},j,k\right)} \\
&\times\left\{\alpha_{y1}\times\left[H_z^{n+l/m}\left(i+\frac{1}{2},j+\frac{1}{2},k\right)-H_z^{n+l/m}\left(i+\frac{1}{2},j-\frac{1}{2},k\right)\right]\right. \\
&-\alpha_{z1}\times\left[H_y^{n+l/m}\left(i+\frac{1}{2},j,k+\frac{1}{2}\right)-H_y^{n+l/m}\left(i+\frac{1}{2},j,k-\frac{1}{2}\right)\right] \\
&+\alpha_{y2}\times\left[H_z^{n+l/m}\left(i+\frac{1}{2},j+\frac{3}{2},k\right)-H_z^{n+l/m}\left(i+\frac{1}{2},j-\frac{3}{2},k\right)\right] \\
&\left.-\alpha_{z2}\times\left[H_y^{n+l/m}\left(i+\frac{1}{2},j,k+\frac{3}{2}\right)-H_y^{n+l/m}\left(i+\frac{1}{2},j,k-\frac{3}{2}\right)\right]\right\}
\end{aligned} \tag{54}$$

$$\alpha_{y1}=\frac{9}{8}d_l\times CFL_y \qquad \alpha_{z1}=\frac{9}{8}d_l\times CFL_z \tag{55}$$

$$\alpha_{y2}=\frac{-1}{24}d_l\times CFL_y \qquad \alpha_{z2}=\frac{-1}{24}d_l\times CFL_z \tag{56}$$

$$CFL_y=\frac{1}{\sqrt{\mu_0\varepsilon_0}}\frac{\Delta_t}{\Delta_y} \qquad CFL_z=\frac{1}{\sqrt{\mu_0\varepsilon_0}}\frac{\Delta_t}{\Delta_z} \tag{57}$$

where ε_r is the relative permittivity. For the cubic grid, $\Delta_x = \Delta_y = \Delta_z = \Delta_\delta$ and $CFL_x = CFL_y = CFL_z = CFL_\delta$.

The source conditions for $\hat{E}_x$ field at the plane $k = k_2$ are given as follows

$$\hat{E}_x^{n+l/m}\left(i+\frac{1}{2},j,k_2-1\right) = \hat{E}_x^{n+l/m}\left(i+\frac{1}{2},j,k_2-1\right) + \alpha_{z2} \times H_{y,inc}^{n+l/m}\left(k_2+\frac{1}{2}\right) \quad (58)$$

$$\hat{E}_x^{n+l/m}\left(i+\frac{1}{2},j,k_2\right) = \hat{E}_x^{n+l/m}\left(i+\frac{1}{2},j,k_2\right) + \alpha_{z1} \times H_{y,inc}^{n+l/m}\left(k_2-\frac{1}{2}\right) + \alpha_{z2} \times H_{y,inc}^{n+l/m}\left(k_2-\frac{3}{2}\right) \quad (59)$$

$$\hat{E}_x^{n+l/m}\left(i+\frac{1}{2},j,k_2+1\right) = \hat{E}_x^{n+l/m}\left(i+\frac{1}{2},j,k_2+1\right) + \alpha_{z2} \times H_{y,inc}^{n+l/m}\left(k_2-\frac{1}{2}\right) \quad (60)$$

where $H_{y,inc}$ is the one-dimensional incident wave source.

The discretized y subcomponent of $\hat{E}_x$ field in the PML region can be deduced as

$$\hat{E}_{xy}^{n+l/m}\left(i+\frac{1}{2},j,k\right) = \exp(-\xi) \times \hat{E}_{xy}^{n+(l-1)/m}\left(i+\frac{1}{2},j,k\right) + \frac{1-\exp(-\xi)}{\xi} \times \left\{\alpha_{y1} \times \left[H_z^{n+l/m}\left(i+\frac{1}{2},j+\frac{1}{2},k\right) - H_z^{n+l/m}\left(i+\frac{1}{2},j-\frac{1}{2},k\right)\right] + \alpha_{y2} \times \left[H_z^{n+l/m}\left(i+\frac{1}{2},j+\frac{3}{2},k\right) - H_z^{n+l/m}\left(i+\frac{1}{2},j-\frac{3}{2},k\right)\right]\right\} \quad (61)$$

$$\xi = \frac{d_l \Delta_t \sigma_y\left(i+\frac{1}{2},j,k\right)}{\varepsilon_0} \quad (62)$$

where σ_y is the local electric conductivity at $\left(i+\frac{1}{2},j,k\right)$ in the PML region. Polynomial conductivities are employed varying from zeros at the vacuum-layer interface to $\sigma_{y,\max}$ at the outer side of the PML layer, i.e.

$$\sigma_y(\Lambda) = \sigma_{y,\max}\left(\frac{\Lambda}{\Gamma}\right)^\kappa \quad (63)$$

where Γ is the layer thickness, Λ is the distance from the interface, and κ is the polynomial order. When $\kappa = 3$, $\sigma_{y,\max}$ can be set as

$$\sigma_{y,\max} = \frac{0.08}{\sqrt{\mu_r^b \varepsilon_r^b}\,\Delta_y} \tag{64}$$

where ε_r^b and μ_r^b are the permittivity and permeability of the background media. For the free space, $\varepsilon_r^b = \mu_r^b = 1$ .Considering the electric and magnetic fields are interleaved in the space lattice at intervals of half space increments, we must use efficient interpolation method to obtain the values of the scattered field components at the same locations. At one virtual plane $k = k_1$ on the rectangular locus, the one-dimensional fourth-order cubic interpolation formula for the electric field can be defined as

$$\begin{aligned}\overline{\hat{E}}_x^{n+l/m}\left(i+\frac{1}{2}, j+\frac{1}{2}, k_1\right) &= \frac{-1}{16}\times\left[E_x^{n+l/m}\left(i+\frac{1}{2}, j-1, k_1\right)\right.\\ &\left.+\hat{E}_x^{n+l/m}\left(i+\frac{1}{2}, j+2, k_1\right)\right]+\frac{9}{16}\times\left[\hat{E}_x^{n+l/m}\left(i+\frac{1}{2}, j, k_1\right)\right.\\ &\left.+\hat{E}_x^{n+l/m}\left(i+\frac{1}{2}, j+1, k_1\right)\right]\end{aligned} \tag{65}$$

where $\overline{\hat{E}}_x^{n+l/m}$ is the averaged value of the scaled electric field component $\hat{E}_x^{n+l/m}$. The two-dimensional interpolation formula for the magnetic field can be expressed in the form

$$\begin{aligned}\overline{H}_x^{n+l/m}\left(i+\frac{1}{2}, j+\frac{1}{2}, k_1\right) &= \frac{1}{256}\times\left[H_x^{n+l/m}\left(i-1, j+\frac{1}{2}, k_1-\frac{3}{2}\right)\right.\\ &+H_x^{n+l/m}\left(i+2, j+\frac{1}{2}, k_1-\frac{3}{2}\right)+H_x^{n+l/m}\left(i-1, j+\frac{1}{2}, k_1+\frac{3}{2}\right)\\ &\left.+H_x^{n+l/m}\left(i+2, j+\frac{1}{2}, k_1+\frac{3}{2}\right)\right]+\frac{-9}{256}\times\left[H_x^{n+l/m}\left(i, j+\frac{1}{2}, k_1-\frac{3}{2}\right)\right.\\ &+H_x^{n+l/m}\left(i+1, j+\frac{1}{2}, k_1-\frac{3}{2}\right)+H_x^{n+l/m}\left(i-1, j+\frac{1}{2}, k_1-\frac{1}{2}\right)\\ &+H_x^{n+l/m}\left(i+2, j+\frac{1}{2}, k_1-\frac{1}{2}\right)+H_x^{n+l/m}\left(i-1, j+\frac{1}{2}, k_1+\frac{1}{2}\right)\\ &+H_x^{n+l/m}\left(i+2, j+\frac{1}{2}, k_1+\frac{1}{2}\right)+H_x^{n+l/m}\left(i, j+\frac{1}{2}, k_1+\frac{3}{2}\right)\\ &\left.+H_x^{n+l/m}\left(i+1, j+\frac{1}{2}, k_1+\frac{3}{2}\right)\right]+\frac{81}{256}\times\left[H_x^{n+l/m}\left(i, j+\frac{1}{2}, k_1-\frac{1}{2}\right)\right.\\ &+H_x^{n+l/m}\left(i+1, j+\frac{1}{2}, k_1-\frac{1}{2}\right)+H_x^{n+l/m}\left(i, j+\frac{1}{2}, k_1+\frac{1}{2}\right)\\ &\left.+H_x^{n+l/m}\left(i+1, j+\frac{1}{2}, k_1+\frac{1}{2}\right)\right]\end{aligned} \tag{66}$$

where $\overline{H}_x^{n+l/m}$ is the averaged value of the x component of magnetic field $\mathbf{H}^{n+l/m}$.

6. Numerical results

a. One-dimensional propagation problem

A Gaussian pulse can be defined by $\exp\left[-4\pi\left(\frac{t-t_0}{\tau}\right)^2\right]$ with $t_0 = 10^{-8}s$ and $\tau = 1.33\times10^{-8}s$. The space increment is set as $\Delta_z = 0.1m$, and the CFL number is chosen to be 0.5. The time-domain waveforms are recorded in Fig. 3 after the pulse travels 10000 cells. Compared with the traditional FDTD(2,2) method and the staggered FDTD(2,4) method, the SFDTD(4,4) scheme agrees with the analytical solution very well.

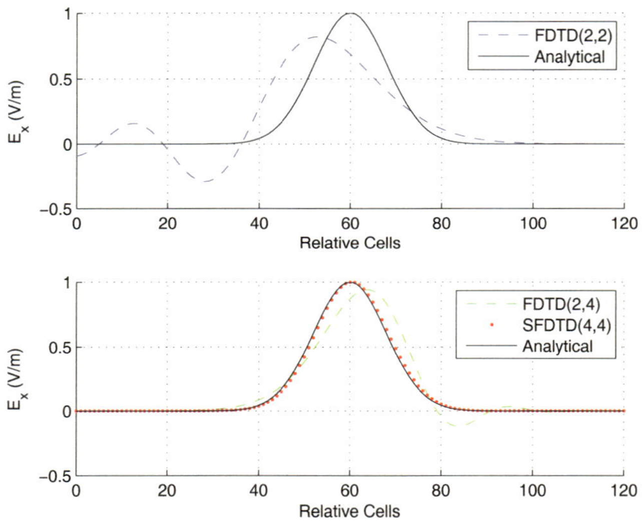

Fig. 3. The time-domain waveforms of the Gaussian pulse by the traditional FDTD(2,2) method, the staggered FDTD(2,4) method, and the SFDTD(4,4) scheme.

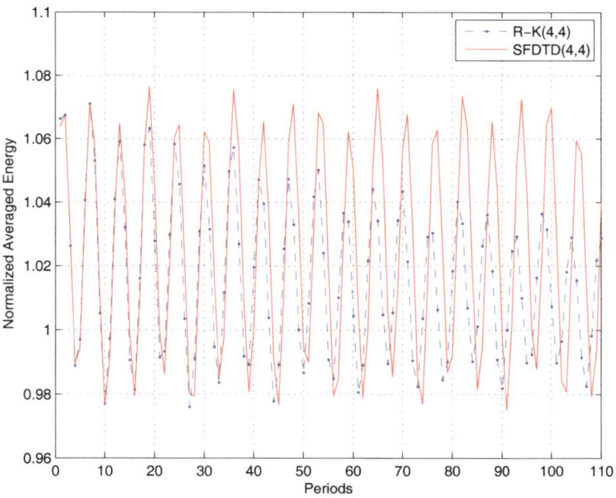

Fig. 4. The normalized averaged energy of two-dimensional waveguide resonator calculated by the R-K(4,4) approach and the SFDTD(4,4) scheme.

b. Two-dimensional waveguide resonator problem

A two-dimensional waveguide resonator with size $2.286cm \times 1.016cm$ is driven in $\mathbf{TE}_{21}$ mode. Calculated by the above mentioned SFDTD(4,4) scheme and the R-K (4,4) approach, the normalized averaged energy per three periods is drawn in Fig. 4. The uniform space increment $\Delta_{\delta} = 1.27mm$, the CFL number is chosen to be 0.797, and the time step $n = 5100$. To obtain high-order accuracy, we use the analytical solution to treat the perfect electric conductor (PEC) boundary. Compared with the SFDTD(4,4) scheme, the R-K (4,4) approach has obvious amplitude error. Furthermore, within given numerical precision, the required memory of the R-K approach is four times more than that of the symplectic scheme.

c. Three-dimensional waveguide resonator problem

The resonant frequency is analyzed for a rectangular waveguide cavity. The size of the waveguide resonator is $a \times b \times c = 19.050mm \times 9.525mm \times 14.288mm$. Other parameters are taken as $\Delta_{\delta} = 2.381mm$, CFL=0.4, and $n_{\max} = 10000$. The frequency of the cosine-modulated Gaussian pulse ranges form 12GHz to 21GHz. Within the frequency range, all possible resonant modes include $\mathbf{TE}_{101}$, $\mathbf{TE}_{110}$ ($\mathbf{TM}_{110}$), $\mathbf{TE}_{011}$, and $\mathbf{TE}_{111}$ ($\mathbf{TM}_{111}$). In particular, the PEC boundary is treated with the image theory [15] for the SFDTD(3,4) scheme. Fig. 5 shows the curves of the normalized total energy and their peaks correspond to the resonant frequencies. One can see that compared with the high-order FDTD(2,4) approach and the traditional FDTD(2,2) method, the SFDTD(3,4) scheme can find the resonant frequencies better.

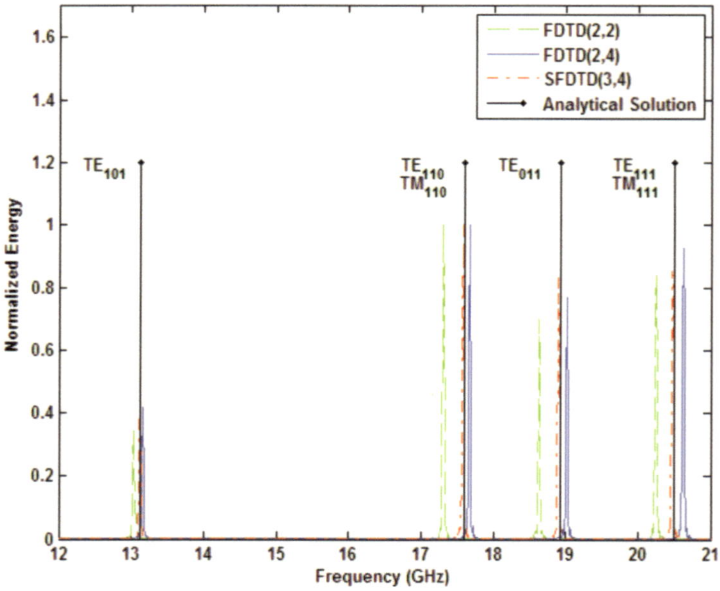

Fig. 5. The resonant frequencies of the rectangular waveguide cavity.

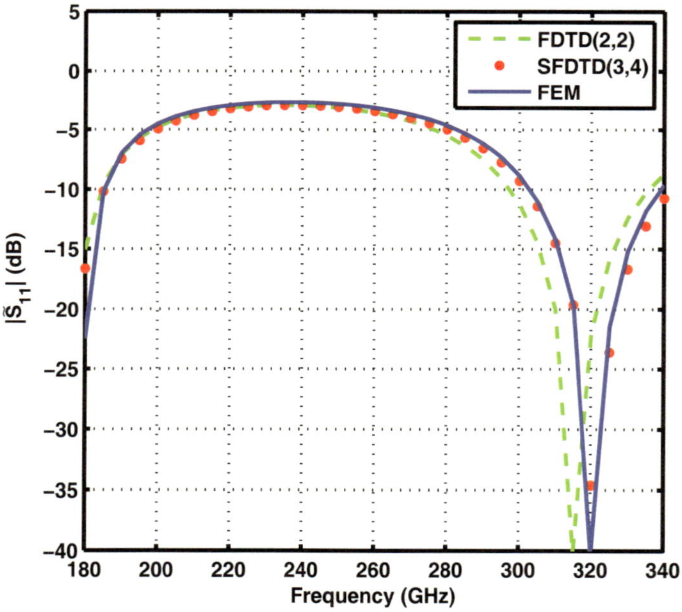

Fig. 6. The scattering parameter of the dielectric-loaded waveguide.

d. Three-dimensional waveguide discontinuity problem

Partially filled with a dielectric of permittivity 3.7, the WR-3 waveguide is driven in $\mathbf{TE}_{10}$ dominant-mode. The size of the waveguide is $0.8636mm \times 0.4318mm$, and the length of the loaded-dielectric is $0.504mm$. The settings are taken as $\Delta_{\delta} = 0.072mm$ and CFL=0.5. The ten layered PML is used to truncate the two waveguide ports, and the sinusoidal-modulated Gaussian pulse is employed as the excitation source. In particular, the PEC boundary is treated with the image technique [15], and the air-dielectric interface is modeled by the scheme proposed in [38]. As shown in Fig. 6, the wide-band scattering parameter is extracted after 5000 time steps. Compared with the traditional FDTD(2,2) method, the SFDTD(3,4) scheme can obtain satisfying numerical solution under the coarse grid condition.

e. Three-dimensional scattering problem of electrically-large sphere

The next example considered is the scattering from a electrically-large conducting sphere of diameter 14 wavelengths. In particular, we use only 7 PPW to model the curved surfaces. From Fig. 7 and Fig. 8, compared with the low-order conformal (LC)-FDTD(2,2) method [8] and the High-order staircased (HS)-SFDTD(3,4) approach, the high-order conformal (HC)-SFDTD(3,4) scheme [55, 56] agrees with the analytical solution very well. The relative two-norm errors of the bistatic RCS by different methods in the E-plane and H-plane are listed in Table 4. The numerical error of the HC-SFDTD(3,4) scheme is controlled by 1%. It can be clearly seen that the locations of the error peaks for the HS-SFDTD(3,4) and the LC-FDTD(2,2) methods are different. The error by the HS-SFDTD(3,4) method is due to the staircase approximation, while the error by the LC-FDTD(2,2) method is due to the numerical dispersion. Within the same relative two-norm errors bound (1%), we change the settings of the space step and the CFL number, and the CPU time and memory consumed by different algorithms are recorded in Table 5. From the table, the HC-SFDTD(3,4) scheme saves considerable memory and CPU time.

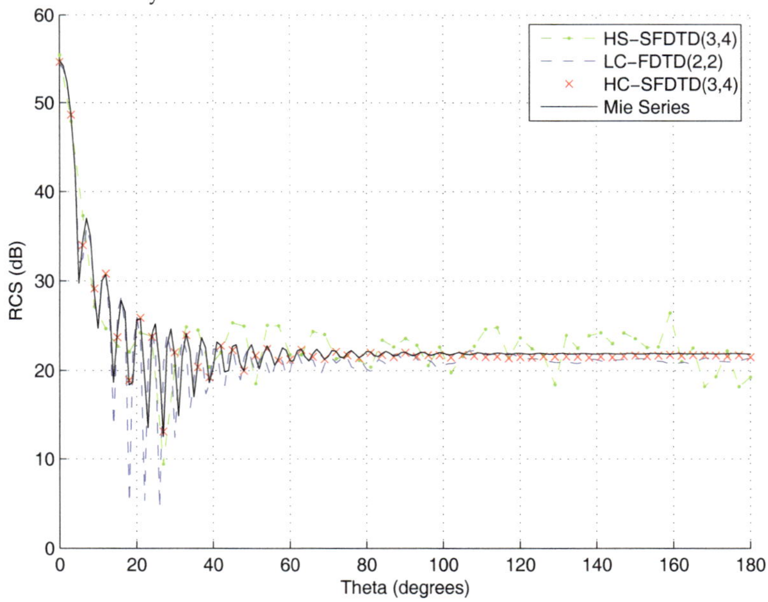

Fig. 7. The E-plane bistatic RCS of the conducting sphere.

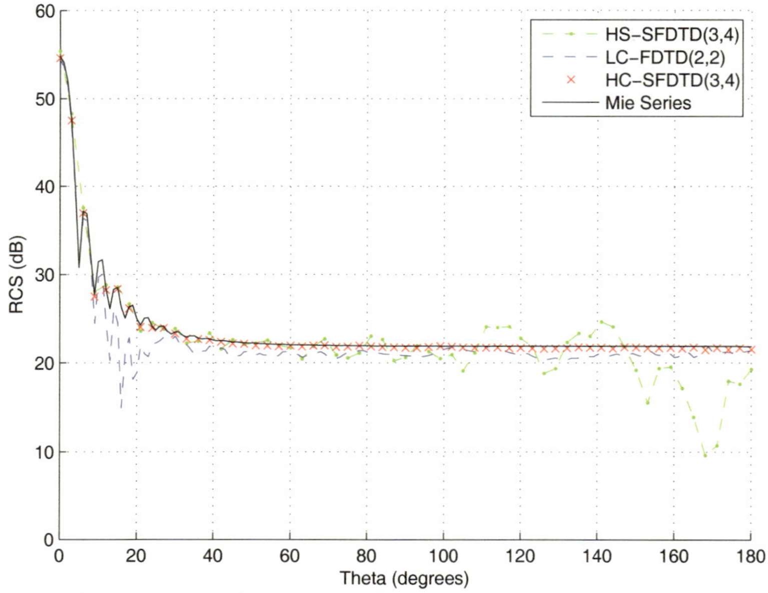

Fig. 8. The H-plane bistatic RCS of the conducting sphere.

Error	HS-SFDTD(3,4)	LC-FDTD(2,2)	HC-SFDTD(3,4)
E-plane	9.89%	11.08%	1.35%
H-plane	12.33%	7.31%	0.85%

Table 4. The relative two-norm errors of bistatic RCS. Seven points per wavelength are adopted.

Algorithms	PPW	CFL	Time (s)	Memory (MB)
HC-SFDTD(3,4)	7	0.50	5891	258
HS-SFDTD(3,4)	16	1.00	56279	1318
LC-FDTD(2,2)	13	0.20	23359	820

Table 5. The consumed CPU time and memory under the same relative two-norm errors condition.

7. Conclusion and future work

The SFDTD scheme, which is explicit high-order accurate in both space and time, is energy-conserving, highly stable, and efficient. On one hand, the scheme can achieve high-order accuracy by using the high-order spatial differences with the simple Yee lattice. On the other hand, by using the symplectic integrators, the scheme demonstrates satisfying numerical performances under long-term simulation. Finally, with the supporting techniques, the scheme is suitable for the electromagnetic modeling of complex structures and media. The future work will focus on the following aspects: (1) The other symplectic integrators, such as composite symplectic integrators [57], can be introduced and optimized for computational electromagnetics; (2) The symplectic integration scheme can be combined with other spatial discretization methods, such as multi-resolution expansion method; (3) The high-order implicit symplectic scheme can be developed for some engineering applications; (4) The symplectic integration scheme is a general solver for a variety of Hamiltonian systems and can be applied to the multi-physics simulation.

8. Acknowledgement

The work was supported by the National Natural Science Foundation of China. (Key Program, No. 60931002.)

9. References

[1] K. S. Yee, "Numerical solution of initial boundary value problems involving Maxwell's equations in isotropic media," IEEE Transactions on Antennas and Propagation, vol. 14, pp. 302-307, 1966.

[2] A. Taflove, Computational Electrodynamics: the Finite-Difference Time-Domain Method. Norwood, MA: Artech House, 1995.

[3] A. Taflove, etc., Advances in Computational Electrodynamics: The Finite-Difference Time-Domain Method. Norwood, MA: Artech House, 1998.

[4] D. M. Sullivan, Electromagnetic Simulation Using the FDTD Method. New York: IEEE Press, 2000.

[5] V. Shankar, A. H. Mohammadian, and W. F. Hall, "A time-domain, finite-volume treatment for the Maxwell equations," Electromagnetics, vol. 10, pp. 127-145, Jan 1990.

[6] J. F. Lee, R. Lee, and A. Cangellaris, "Time-domain finite-element methods," IEEE Transactions on Antennas and Propagation, vol. 45, pp. 430-442, Mar 1997.

[7] T. Lu, W. Cai, and P. W. Zhang, "Discontinuous Galerkin time-domain method for GPR simulation in dispersive media," IEEE Transactions on Geoscience and Remote Sensing, vol. 43, pp. 72-80, Jan 2005.

[8] S. Dey and R. Mittra, "A locally conformal finite-difference time-domain algorithm for modeling three-dimensional perfectly conducting objects," IEEE Microwave and Guided Wave Letters, vol. 7, pp. 273-275, Sep 1997.

[9] W. H. Yu and R. Mittra, "A conformal FDTD algorithm for modeling perfectly conducting objects with curve-shaped surfaces and edges," Microwave and Optical Technology Letters, vol. 27, pp. 136-138, Oct 2000.

[10] I. A. Zagorodnov, R. Schuhmann, and T. Weiland, "A uniformly stable conformal FDTD-method in Cartesian grids," International Journal of Numerical Modelling-Electronic Networks Devices and Fields, vol. 16, pp. 127-141, Feb 2003.

[11] T. Xiao and Q. H. Liu, "Enlarged cells for the conformal FDTD method to avoid the time step reduction," IEEE Microwave and Wireless Components Letters, vol. 14, pp. 551-553, Dec 2004.

[12] M. Okoniewski, E. Okoniewska, and M. A. Stuchly, "Three-dimensional subgridding algorithm for FDTD," IEEE Transactions on Antennas and Propagation, vol. 45, pp. 422-429, Mar 1997.

[13] M. Krumpholz and L. P. B. Katehi, "MRTD: new time-domain schemes based on multiresolution analysis," IEEE Transactions on Microwave Theory and Techniques, vol. 44, pp. 555-571, 1996.

[14] Q. H. Liu, "PSTD algorithm: A time-domain method requiring only two cells per wavelength," Microwave and Optical Technology Letters, vol. 15, pp. 158-165, 1997.
[15] Q. S. Cao, Y. C. Chen, and R. Mittra, "Multiple image technique (MIT) and anistropic perfectly matched layer (APML) in implementation of MRTD scheme for boundary truncations of microwave structures," IEEE Transactions on Microwave Theory and Techniques, vol. 50, pp. 1578-1589, Jun 2002.
[16] S. Zhao and G. W. Wei, "High-order FDTD methods via derivative matching for Maxwell's equations with material interfaces," Journal of Computational Physics, vol. 200, pp. 60-103, Oct 2004.
[17] J. Fang, "Time domain finite difference computation for Maxwell's equations," Ph.D. dissertation, Univ. California, Berkeley, CA 1989.
[18] A. Yefet and P. G. Petropoulos, "A staggered fourth-order accurate explicit finite difference scheme for the time-domain Maxwell's equations," Journal of Computational Physics, vol. 168, pp. 286-315, Apr 2001.
[19] S. V. Georgakopoulos, C. R. Birtcher, C. A. Balanis, and R. A. Renaut, "Higher-order finite-difference schemes for electromagnetic radiation, scattering, and penetration, Part I: Theory," IEEE Antennas and Propagation Magazine, vol. 44, pp. 134-142, Feb 2002.
[20] S. V. Georgakopoulos, C. R. Birtcher, C. A. Balanis, and R. A. Renaut, "Higher-order finite-difference schemes for electromagnetic radiation, scattering, and penetration, Part 2: Applications," IEEE Antennas and Propagation Magazine, vol. 44, pp. 92-101, Apr 2002.
[21] S. V. Georgakopoulos, C. R. Birtcher, C. A. Balanis, and R. A. Renaut, "HIRF penetration and PED coupling analysis for scaled fuselage models using a hybrid subgrid FDTD(2,2)/FDTD(2,4) method," IEEE Transactions on Electromagnetic Compatibility, vol. 45, pp. 293-305, 2003.
[22] J. L. Young, D. Gaitonde, and J. S. Shang, "Toward the construction of a fourth-order difference scheme for transient EM wave simulation: staggered grid approach," IEEE Transactions on Antennas and Propagation, vol. 45, pp. 1573-1580, Nov 1997.
[23] J. S. Shang, "High-order compact-difference schemes for time-dependent Maxwell equations," Journal of Computational Physics, vol. 153, pp. 312-333, Aug 1999.
[24] T. Namiki, "New FDTD algorithm based on alternating-direction implicit method," IEEE Transactions on Microwave Theory and Techniques, vol. 47, pp. 2003-2007, Oct 1999.
[25] F. H. Zhen, Z. Z. Chen, and J. Z. Zhang, "Toward the development of a three-dimensional unconditionally stable finite-difference time-domain method," IEEE Transactions on Microwave Theory and Techniques, vol. 48, pp. 1550-8, Sep 2000.
[26] H. De Raedt, J. S. Kole, K. F. L. Michielsen, and M. T. Figge, "Unified framework for numerical methods to solve the time-dependent Maxwell equations," Computer Physics Communications, vol. 156, pp. 43-61, Dec 2003.
[27] K. Feng, "Difference-schemes for Hamiltonian-formalism and symplectic-geometry," Journal of Computational Mathematics, vol. 4, pp. 279-289, Jul 1986.
[28] J. M. Sanz-Serna and M. P. Calvo, Numerical Hamiltonian Problems. London, U.K.: Chapman & Hall, 1994.
[29] K. Feng and M. Z. Qin, Symplectic Geometric Algorithm for Hamiltonian Systems. Hangzhou: Zhejiang Science & Technology Press, 2003.

[30] T. Hirono, W. W. Lui, and K. Yokoyama, "Time-domain simulation of electromagnetic field using a symplectic integrator," IEEE Microwave and Guided Wave Letters, vol. 7, pp. 279-281, Sep 1997.
[31] T. Hirono, W. W. Lui, and S. Seki, "Successful applications of PML-ABC to the symplectic FDTD scheme with 4th-order accuracy in time and space," IEEE MTT-S International Microwave Symposium Digest, vol. 3, pp. 1293-1296, 1999.
[32] T. Hirono, W. Lui, S. Seki, and Y. Yoshikuni, "A three-dimensional fourth-order finite-difference time-domain scheme using a symplectic integrator propagator," IEEE Transactions on Microwave Theory and Techniques, vol. 49, pp. 1640-1648, Sep 2001.
[33] M. Kusaf, A. Y. Oztoprak, and D. S. Daoud, "Optimized exponential operator coefficients for symplectic FDTD method," IEEE Microwave and Wireless Components Letters, vol. 15, pp. 86-88, Feb 2005.
[34] P. W. Zhai, G. W. Kattawar, P. Yang, and C. H. Li, "Application of the symplectic finite-difference time-domain method to light scattering by small particles," Applied Optics, vol. 44, pp. 1650-1656, Mar 2005.
[35] Z. X. Huang, W. Sha, X. L. Wu, and M. S. Chen, "A novel high-order time-domain scheme for three-dimensional Maxwell's equations," Microwave and Optical Technology Letters, vol. 48, pp. 1123-1125, Jun 2006.
[36] M. Kusaf and A. Y. Oztoprak, "Higher stability limits for the symplectic FDTD method by making use of Chebyshev polynomials," IEEE Microwave and Wireless Components Letters, vol. 16, pp. 579-581, Nov 2006.
[37] Z. X. Huang, X. L. Wu, W. Sha, and M. S. Chen, "Optimal symplectic integrators for numerical solution of time-domain Maxwell's equations," Microwave and Optical Technology Letters, vol. 49, pp. 545-547, Mar 2007.
[38] W. Sha, Z. X. Huang, X. L. Wu, and M. S. Chen, "Application of the symplectic finite-difference time-domain scheme to electromagnetic simulation," Journal of Computational Physics, vol. 225, pp. 33-50, Jul 2007.
[39] W. Sha, X. L. Wu, M. S. Chen, and Z. X. Huang, "Application of the high-order symplectic FDTD scheme to the curved three-dimensional perfectly conducting objects," Microwave and Optical Technology Letters, vol. 49, pp. 931-934, Apr 2007.
[40] W. Sha, Z. X. Huang, M. S. Chen, and X. L. Wu, "Survey on symplectic finite-difference time-domain schemes for Maxwell's equations," IEEE Transactions on Antennas and Propagation, vol. 56, pp. 493-500, Feb 2008.
[41] W. Sha, X. L. Wu, Z. X. Huang, and M. S. Chen, "Waveguide simulation using the high-order symplectic finite-difference time-domain scheme," Progress In Electromagnetics Research B, vol. 13, pp. 237-256, 2009.
[42] Z. Shao, Z. Shen, Q. He, and G. Wei, "A generalized higher order finite-difference time-domain method and its application in guided-wave problems," IEEE Transactions on Microwave Theory and Techniques, vol. 51, pp. 856-861, Mar 2003.
[43] Y. Shi and C. H. Liang, "Multidomain pseudospectral time domain algorithm using a symplectic integrator," IEEE Transactions on Antennas and Propagation, vol. 55, pp. 433-439, Feb 2007.
[44] S. Wang, Z. H. Shao, and G. J. Wen, "A modified high order FDTD method based on wave equation," IEEE Microwave and Wireless Components Letters, vol. 17, pp. 316-318, May 2007.

[45] S. Reich, "Multi-symplectic Runge-Kutta collocation methods for Hamiltonian wave equations," Journal of Computational Physics, vol. 157, pp. 473-499, Jan 2000.

[46] T. J. Bridges and S. Reich, "Multi-symplectic integrators: numerical schemes for Hamiltonian PDEs that conserve symplecticity," Physics Letters A, vol. 284, pp. 184-193, Jun 2001.

[47] W. Sha, X. L. Wu, Z. X. Huang, and M. S. Chen, "Maxwell's equations, symplectic matrix, and grid" Progress In Electromagnetics Research B, vol. 8, pp. 115-127, 2008.

[48] N. Anderson and A. M. Arthurs, "Helicity and variational principles for Maxwell's equations," International Journal of Electronics, vol. 54, pp. 861-864, Jun 1983.

[49] H. Yoshida, "Construction of higher order symplectic integrators," Physica D: Nonlinear Phenomena, vol. 46, pp. 262-268, Nov 1990.

[50] S. K. Gray and D. E. Manolopoulos, "Symplectic integrators tailored to the time-dependent Schrödinger equation," Journal of Chemical Physics, vol. 104, pp. 7099-7112, May 1996.

[51] F. M. Dopico and C. R. Johnson, "Complementary bases in symplectic matrices and a proof that their determinant is one," Linear Algebra and Its Applications, vol. 419, pp. 772-778, Dec 2006.

[52] W. C. Chew, Waves and Fields in Inhomogenous Media. New York: Van Nostrand Reinhold, 1990.

[53] W. Sha, Z. X. Huang, X. L. Wu, and M. S. Chen, "Total field and scattered field technique for fourth-order symplectic finite difference time domain method," Chinese Physics Letters, vol. 23, pp. 103-105, Jan 2006.

[54] W. Sha, X. L. Wu, and M. S. Chen, "A diagonal split-cell model for the high-order symplectic FDTD scheme," PIERS Online, vol. 2, pp. 715-719, 2006.

[55] W. Sha, X. L. Wu, Z. X. Huang, and M. S. Chen, "A new conformal FDTD(2,4) scheme for modeling three-dimensional curved perfectly conducting objects," IEEE Microwave and Wireless Components Letters, vol. 18, pp. 149-151, Mar 2008.

[56] W. Sha and X. L. Wu, "High-Order Conformal Symplectic FDTD Scheme," in APMC2008 Hong Kong, 2008.

[57] S. A. Chin, "Symplectic integrators from composite operator factorizations," Physics Letters A, vol. 226, pp. 344-348, March 1997.

4

Nonlocal Electromagnetic Media: A Paradigm for Material Engineering

Said M. Mikki
Royal Millitary College, Kingston
Canada

Ahmed A. Kishk
University of Mississippi, University
USA

Abstract

In this paper, we provide a general Fourier transform formalism suitable for studying the electromagnetic response of material media. This approach can handle media that exhibit natural optical activity, magnetoelectric effects, spatial dispersion, etc. Moreover, it is a powerful method in addressing the impact of electromagentic systems on the spatial structure of the field, particulary at the nanoscale (e.g., near-field nano-optics, subwavelength imaging, etc.) The formalism is employed to analyze the localization of electromagnetic energy around radiating sources and also to provide a new paradigm for thinking about metamaterials.

1. Introduction

Traditionally, the research area known under the label "artificial materials," or what has become popular nowadays as "metamatrials," is based on the idea of mimicking the way natural media respond to an applied electromagnetic field. The mechanism responsible of the electromagnetic character of the medium, for example the optical properties, can be applied to repeat the whole process artificially in the sense that the atomic constituents of matter are individually manipulated and controlled in order to achieve a desired electromagnetic profile.
The conventional approach to describe material responses rely on assuming that the external field induce multipole electric and magnetic moments in the medium, giving rise to polarization and magnetization density vectors. This approach, as will be demonstrated in this paper, has its merits although theoretically problematic. It provides an extremely simple mathematical model that is adequate for a very wide range of applications. However, on the other hand, with the exploding progress in nanotechnology and experimental research, it is becoming increasingly pressing to employ a more general mathematical formalism that allows us to explore new dimensions in the material response that go beyond the traditional multipole description.
It is the vision of the present authors that a large proportion of the future research in the field of artificial and metamaterials should be invested in studying the *spatial* degrees of freedom of the medium response, a space hitherto unexplored in depth with few notable exceptions (1),

(2). The purely spatial effects, for example spatial dispersion, has been often neglected because natural materials happen to have very small interatomic spacing to operating wavelength ratio, which implies that when a macroscopic field measurement is employed, all microscopic spatial information are washed away. However, there is nothing in principle prohibiting designing artificial media with arbitrary spatial response profile.[1] The possibility of controlling wave propagation through a given device by manipulating both the temporal and spatial dispersion was already proposed in conjunction with realizing the so-called negative refraction metamaterials (4).

In this paper, we review a Fourier-space formalism suitable for modeling the spatial effects of a given natural or artificial medium. The formalism is compatible with the traditional multipole approach but is conceptually easier to understand. The Fourier transform method we introduce here is inspired by techniques developed in the physics community to attack plasma problems (1), (2).

There is a plethora of advantages in employing this particular point of view in this setting. The chief advantage is that the Fourier-space formalism is more general in its applicability to fluctuating fields with higher frequencies. Also, it naturally provides a complete characterization of the field in both space and time. Finally, being a spectral method, it allows for deeper understanding of localization phenomena and coupling mechanisms.

Some of the disadvantages is that it requires an additional mathematical background that is not usually part of the training of professional electromagnetic engineers. It also does not apply to static field problems. In general, the Fourier formalism does not conform with the conventional literature standards of notation and usage. For this reason, the current paper will provide, in a pedagogically illuminating way, a review of the space formalism starting from the ground, Maxwell's equations, and building up to advanced applications in the concluding section.

2. Maxwell's Equation

We start with the fundamental equations governing the free Maxwellian fields **B** and **E**. These are

$$\nabla \times \mathbf{E} = -\frac{\partial \mathbf{B}}{\partial t}, \tag{1}$$

$$\nabla \times \mathbf{B} = \mu_0 \mathbf{J} + \left(1/c^2\right) \frac{\partial \mathbf{E}}{\partial t}, \tag{2}$$

$$\nabla \cdot \mathbf{E} = \frac{\rho}{\varepsilon_0}, \tag{3}$$

$$\nabla \cdot \mathbf{B} = 0, \tag{4}$$

where c is the speed of light and $\varepsilon_0 = 8.854 \times 10^-12\,\mathrm{F/m}$ and $\mu_0 = 4\pi \times 10^-7\,\mathrm{H/m}$ are the permittivity and permeability of free space, respectively.

We notice that these set of Maxwell's equations are complete since they capture everything related to electromagnetic interactions. However, in order to solve Maxwell's equations *in the presence of matter*, one has to supply suitable decompositions of the source terms appearing in (2) and (3) in the following manner

$$\rho = \rho_{\text{ext}} + \rho_{\text{ind}} \tag{5}$$

[1] The implementation of a particular solution of Maxwell's equations coupled with a suitable mechanical model is a technological problem, not a theoretical one. In this sense, the present paper should be viewed as a theoretical contribution.

and

$$\mathbf{J} = \mathbf{J}_{\text{ext}} + \mathbf{J}_{\text{ind}}, \tag{6}$$

where ρ_{ext} and $\mathbf{J}_{\text{ext}}$ are the imposed sources supplied externally. Matter will interact with the fields radiated by these sources and respond by generating induced sources ρ_{ind} and $\mathbf{J}_{\text{ind}}$. These induced sources cannot be deduced from Maxwell's equations themselves. They must be found upon constructing an appropriate mechanical model for matter in the radiation field.

2.1 The Continuity Equation and Energy Conservation

By imposing the conservation of electric charge density $\rho(t,\mathbf{r})$, the equation of continuity for electromagnetism takes the following form

$$\frac{\partial \rho}{\partial t} + \nabla \cdot \mathbf{J} = 0. \tag{7}$$

Energy conservation is already built into the structure of Maxwell's equations in continuous media. Indeed, it is possible to directly derive the following relation

$$\frac{\partial}{\partial t}\left(\frac{1}{2}\varepsilon_0 |\mathbf{E}|^2 + \frac{1}{2}|\mathbf{B}|^2 \Big/ \mu_0\right) + \nabla \cdot \left(\frac{1}{\mu_0}\mathbf{E} \times \mathbf{B}\right) = -\mathbf{J} \cdot \mathbf{E}. \tag{8}$$

Let us supplement this equation with Lorentz force law

$$\mathbf{F} = q\mathbf{E} + \mathbf{v} \times \mathbf{B}. \tag{9}$$

One can carefully build the interpretation of the terms appearing in the RHS of (8) starting from the basic law of force (9). As it turns out, the time rate of the volume density of the work done by the electric current $\mathbf{J}$ on the electric field $\mathbf{E}$ is given by $-\mathbf{J} \cdot \mathbf{E}$. This provides us with an interpretation of the RHS of (8). Now, in order to interpret (8) as a continuity equation, we observe that, in vacuum, the quantities $\varepsilon_0 |\mathbf{E}|^2 \big/ 2$ and $|\mathbf{B}|^2 \big/ 2\mu_0$ can straightforwardly be interpreted as volume densities of electric and magnetic energies, respectively, stored in free space. It follows then that the last term, that of $\mathbf{E} \times \mathbf{B} / \mu_0$, can be easily interpreted to stand for the volume density of the power flow, or the electromagnetic flux.

3. Fourier Transform Approach to the Greens Functions

3.1 Maxwell's Equations in the Spectral Domain

As we are going to formulate the entire problem in terms of Fourier transform, the usual spatio-temporal form of Maxwell's equations must be transformed into the spectral domain. In this section, we handle the problem of a source radiating in infinite isotropic and homogenous medium. Maxwell's equations (1)-(4) can be written in the Fourier transform domain as

$$\mathbf{k} \times \mathbf{E}(\omega,\mathbf{k}) = \omega \mathbf{B}(\omega,\mathbf{k}), \tag{10}$$

$$i\mathbf{k} \times \mathbf{B}(\omega,\mathbf{k}) = -i\omega \mathbf{E}(\omega,\mathbf{k}) \big/ c^2 + \mu_0 \mathbf{J}(\omega,\mathbf{k}), \tag{11}$$

$$\mathbf{k} \cdot \mathbf{E}(\omega,\mathbf{k}) = -i\rho(\omega,\mathbf{k}) / \varepsilon_0, \tag{12}$$

$$\mathbf{k} \cdot \mathbf{B}(\omega,\mathbf{k}) = 0. \tag{13}$$

The equation of continuity (7) can be also Fourier transformed into the form

$$\omega \rho(\omega,\mathbf{k}) = \mathbf{k} \cdot \mathbf{J}(\omega,\mathbf{k}). \tag{14}$$

The reader must notice that these equations cannot be used to describe static fields, which may be tackled on their own by applying the Coloumb gauge. Therefore, throughout this paper, we restrict ourselves to the case $\omega \neq 0$.
The program of performing calculations in electromagnetism using the Fourier transform method can be elucidated in the following manner

1. Express the magnetic field in terms of the electric field using Maxwell's equation (10)

$$\mathbf{B}(\omega,\mathbf{k}) = \mathbf{k} \times \mathbf{E}(\omega,\mathbf{k})/\omega. \tag{15}$$

2. Express the charge density in terms of the current density using the equation of continuity (14)

$$\rho(\omega,\mathbf{k}) = \frac{1}{\omega}\mathbf{k} \cdot \mathbf{J}(\omega,\mathbf{k}). \tag{16}$$

3. End up with a single equation in one unknown, $\mathbf{E}(\omega,\mathbf{k})$, and forcing term $\mathbf{J}(\omega,k)$; i.e., solve

$$\frac{\omega^2}{c^2}\mathbf{E}(\omega,\mathbf{k}) + \mathbf{k} \times [\mathbf{k} \times \mathbf{E}(\omega,\mathbf{k})] = -i\omega\mu_0\mathbf{J}(\omega,\mathbf{k}). \tag{17}$$

Therefore, the program of solving Maxwell's equations reduces to solving a single algebraic equation in terms of the electric field $\mathbf{E}(\omega,\mathbf{k})$. All the other field and source components can obtained from the solution of the electric field together with the given form of the source.

3.2 The Greens Function Tensor in the Spectral Domain

To obtain the Greens function in the Fourier domain, we first put equation (17) in a suitable form. We use the following identity

$$\mathbf{A} \times \mathbf{B} = \epsilon_{ijk} A_j B_k, \tag{18}$$

where ϵ_{ijk} is the permutation tensor.[2] Therefore, we have

$$\mathbf{k} \times \mathbf{E} = \epsilon_{ijk} k_j E_k. \tag{19}$$

Iterating, we obtain

$$\mathbf{k} \times \mathbf{k} \times \mathbf{E} = \epsilon_{ijk} k_j \epsilon_{kj'k'} k_{j'} E_{k'} = \epsilon_{ijk} \epsilon_{kj'k'} k_j k_{j'} E_{k'}. \tag{20}$$

We use the following basic identity

$$\begin{aligned}\epsilon_{abc}\epsilon_{ijk} = \delta_{ai}\delta_{bj}\delta_{ck} + \delta_{ak}\delta_{bi}\delta_{cj} + \delta_{aj}\delta_{bk}\delta_{ci} \\ -\delta_{bi}\delta_{aj}\delta_{ck} - \delta_{ak}\delta_{ai}\delta_{cj} - \delta_{bj}\delta_{ak}\delta_{ci}.\end{aligned} \tag{21}$$

Therefore, we have

$$\epsilon_{iab}\epsilon_{ijk} = \delta_{aj}\delta_{bk} - \delta_{ak}\delta_{bj}. \tag{22}$$

Using this identity in (17), we arrive to

$$\left[\left(\frac{\omega^2}{c^2} - k^2\right)\delta_{nm} + k_n k_m\right] E_m(\omega,\mathbf{k}) = -i\omega\mu_0 J_n(\omega,\mathbf{k}). \tag{23}$$

[2] Throughout this paper, the Einstein (repeated) summation index is used. That is, whenever an index is repeated in a given expression, summation is implied with respect to these indices.

The Greens function tensor is defined to satisfy the following equation

$$\left[\left(\frac{\omega^2}{c^2}-k^2\right)\delta_{nm}+k_nk_m\right]G_{ml}(\omega,\mathbf{k})=-i\omega\mu_0\delta_{nl}(\omega,\mathbf{k}). \tag{24}$$

Therefore, by inverting the matrix operator appearing in the equation above, the Greens function tensor is readily obtained in the following compact closed form

$$G_{nm}(\omega,\mathbf{k})=\frac{-i\omega\mu_0}{\omega^2/c^2-k^2}\left(\delta_{nm}-\frac{c^2}{\omega^2}k_nk_m\right) \tag{25}$$

Finally, we notice that it is possible to separate the field into two components, one transverse to the direction of the wave vector **k** (transverse mode), and another perpendicular to this direction, which we call longitudinal mode. The longitudinal mode is not involved in the radiation and is related to the near field. It contributes directly to the structure of the field surrounding the source.

4. Review of the Traditional Description of Electromagnetic Materials in terms of Multipole Moments

The conventional old description of electromagnetic materials involves the introduction of two quantities to calculate the induced charge and current distributions. We review here the traditional view and show how it can be derived by a Fourier transform approach to the multipole expansion of the source.

The conventional idea is to assume that a given medium responds to both electric and magnetic fields by generating an induced *polarization* density **P** and *magnetization* **M**. However, this description is strictly valid when both the electric and magnetic responses can be unambiguously separated from each other. This is possible only when the fields are static; otherwise, it should be viewed as an approximation. Indeed, if rapid field fluctuations at the microscopic scale are taken into consideration, then the separation becomes ill-defined and problematic.

Let us see how **P** and **M** arise from the Fourier transform perspective. Consider an arbitrary charge and current distribution

$$\rho(t,\mathbf{k})=\int d^3re^{-i\mathbf{k}\cdot\mathbf{r}}\rho(t,\mathbf{r}), \tag{26}$$

$$\mathbf{J}(t,\mathbf{k})=\int d^3re^{-i\mathbf{k}\cdot\mathbf{r}}\mathbf{J}(t,\mathbf{r}). \tag{27}$$

Expand the exponential in Taylor series

$$e^{-i\mathbf{k}\cdot\mathbf{r}}=1-i\mathbf{k}\cdot\mathbf{r}+\frac{1}{2}(i\mathbf{k}\cdot\mathbf{r})^2+\cdots. \tag{28}$$

Inserting (28) into (26), we obtain

$$\begin{aligned}
\rho(t,\mathbf{k}) &= \int d^3r\rho(t,\mathbf{r})\left[1 - i\mathbf{k}\cdot\mathbf{r} + \tfrac{1}{2}(i\mathbf{k}\cdot\mathbf{r})^2 + \cdots\right] \\
&= \int d^3r\rho(t,\mathbf{r}) - \int d^3ri\mathbf{k}\cdot\mathbf{r}\rho(t,\mathbf{r}) + \tfrac{1}{2}\int d^3r(i\mathbf{k}\cdot\mathbf{r})^2\rho(t,\mathbf{r}) + \cdots \\
&= -\int d^3rik_nr_n\rho(t,\mathbf{r}) - \tfrac{1}{2}\int d^3rk_nr_nk_mr_m\rho(t,\mathbf{r}) + \cdots \\
&= -ik_n\int d^3rr_n\rho(t,\mathbf{r}) - \tfrac{1}{2}k_nk_m\int d^3rr_nr_m\rho(t,\mathbf{r}) + \cdots \\
&= -i\mathbf{k}\cdot\mathbf{p}(t) - \tfrac{1}{2}k_nk_mq_{nm}(t) + \cdots,
\end{aligned} \tag{29}$$

where

$$p_n(t) = \int d^3rr_n\rho(t) \tag{30}$$

and

$$q_{nm}(t) = \int d^3rr_nr_m\rho(t) \tag{31}$$

are the dipole and quadrable moments, respectively. We also used the assumption that the charge distribution is neutral $\int d^3r\rho(t,\mathbf{r}) = 0$. Similarly, by inserting (28) into (27), one obtains

$$\begin{aligned}
J_n(t,\mathbf{k}) &= \int d^3rJ_n(t,\mathbf{r})\left[1 - i\mathbf{k}\cdot\mathbf{r} + \tfrac{1}{2}(i\mathbf{k}\cdot\mathbf{r})^2 + \cdots\right] \\
&= \int d^3rJ_n(t,\mathbf{r}) - \int d^3r(i\mathbf{k}\cdot\mathbf{r})J_n(t,\mathbf{r}) + \tfrac{1}{2}\int d^3r(i\mathbf{k}\cdot\mathbf{r})^2J_n(t,\mathbf{r}) + \cdots \\
&= \underbrace{\int d^3rJ_n(t,\mathbf{r})}_{\mu_n(t)} - ik_m\underbrace{\int d^3rr_mJ_n(t,\mathbf{r})}_{\mu_{mn}(t)} - \tfrac{1}{2}k_lk_m\int d^3rr_lr_mJ_n(t,\mathbf{r}) + \cdots \\
&= \tfrac{\partial}{\partial t}p_n(t) - ik_m\tfrac{1}{2}\tfrac{\partial}{\partial t}q_{mn}(t) - i\epsilon_{mns}k_mm_s(t) + \cdots \\
&= \tfrac{\partial}{\partial t}p_n(t) - ik_m\tfrac{1}{2}\tfrac{\partial}{\partial t}q_{mn}(t) + i\epsilon_{nms}k_mm_s(t) + \cdots,
\end{aligned} \tag{32}$$

where equations (110) and (120) (see Appendix) were utilized in obtaining the fourth equality, and the relation $\epsilon_{nms} = -\epsilon_{mns}$ is employed in the writing last equality. By ignoring all quadrable and higher terms in (29) and (32), we find

$$\rho(t,\mathbf{k}) = -i\mathbf{k}\cdot\mathbf{p}(t), \tag{33}$$

$$\mathbf{J}(t,\mathbf{k}) = \frac{\partial}{\partial t}\mathbf{p}(t) + i\mathbf{k}\times\mathbf{m}(t). \tag{34}$$

Define the polarization and magnetization densities **P** and **M**, respectively, by the following relations

$$\mathbf{p}(t) = \int d^3r\mathbf{P}(t,\mathbf{r}) \tag{35}$$

and

$$\mathbf{m}(t) = \int d^3r\mathbf{M}(t,\mathbf{r}). \tag{36}$$

Inserting (33) and (34) into (26) and (27), it follows

$$\rho(t,\mathbf{k}) = \int d^3r\left[-i\mathbf{k}\cdot\mathbf{P}(t,\mathbf{r})\right], \tag{37}$$

$$\mathbf{J}(t,\mathbf{k}) = \int d^3r \left[\frac{\partial}{\partial t}\mathbf{P}(t,\mathbf{r}) + i\mathbf{k} \times \mathbf{M}(t,\mathbf{r}) \right]. \tag{38}$$

Therefore, by inverting the Fourier transforms (37) and (38), we obtain

$$\rho_{\text{ind}}(t,\mathbf{r}) = -\nabla \cdot \mathbf{P}(t,\mathbf{r}), \tag{39}$$

$$\mathbf{J}_{\text{ind}}(t,\mathbf{r}) = \frac{\partial}{\partial t}\mathbf{P}(t,\mathbf{r}) + \nabla \times \mathbf{M}(t,\mathbf{r}). \tag{40}$$

As can be seen now, this derivation ignores higher-order multipole without providing a clear-cut criterion for when and why this approximation is valid. Since we are attempting to construct a general theory for *both* near and far fields in the context of material response, it is important to employ a formulation that does not involve approximations that may not hold in certain media. Some other difficulties relate to the question of the convergence of the multipole expansion that is seldom addressed in literature. Finally, there is the incompleteness issue in the expansion (28), which includes only terms with zero trace.

5. Material Response Through the Fourier Transform Approach

We will now carefully introduce the equivalent representation of the electromagnetic material response in terms of the Fourier transform of the fields, not the actual field in space and time. There are several advantages in this approach that are worthy detailed considerations in themselves. First, notice that this approach does not apply to static fields, which are better addressed by the classical **P-M** approach. On other hand, certain complex electromagnetic effects, like spatial dispersion (nonlocality) magnetoelectric responses and optical activity, can be regarded as special case of spatial dispersion.

It appears to the authors that operating directly on material systems with a formalism tailored especially to handle spatial dispersion is very advantageous. Besides its ability to deal with complex media exhibiting phenomena like magnetoelectric effects and optical activity, it can also provide a natural window to probe near-field interactions. Although we are still trying to mathematically identify the meaning of the near field, remember that one of the most immediate features that come to mind when thinking about fields in the near zone (close to the radiator or the scatterer) is that they tend to be localized, or, equivalently, contain short wavelength components that contribute significantly to the field structure. In this case, one is looking naturally for a mathematical device that characterize electromagnetic wave phenomena in terms of the Fourier spatial modes, i.e., the **k**-component. Therefore, the formalism should look for information about the response of the system to particular wavevecotrs **k**. This is essentially the goal of integrating spatial dispersion in the theoretical description of material media.

Let us try to address in more details some of the difficulties in the traditional approach to electromagnetic material response. By Fourier transforming equation (40) in time, we obtain

$$\mathbf{J}_{\text{ind}}(\omega,\mathbf{k}) = -i\omega\mathbf{P}(\omega,\mathbf{k}) + i\mathbf{k} \times \mathbf{M}(\omega,\mathbf{k}). \tag{41}$$

The problem here is that there exists no general *a priori* method to tell how the individual contributions of the quantities **P** and **M** divide in forming the total induced current. In this sense, one can view these two vectors as mere calculational tools, auxiliary devices used to compute the actually observed induced current $\mathbf{J}_{\text{ind}}$. In particular, there seems to be no harm in just setting the magnetization density **M** to zero and considering only a polarization density **P** contributing to the induced charge and current densities.

As we have just observed in Section 3.1, the program of calculating the fields through Maxwell's equations can be reduced to the solution of a single equation, namely (17), which contains a single unknown, the electric field vector $\mathbf{E}(\omega,\mathbf{k})$ itself. If the relation between the induced current density $\mathbf{J}_{\text{ind}}(\omega,\mathbf{k})$ and the electric field is known, then this relation, together with the master equation (17), can be used to completely solve the problem of light-matter interaction. It seems natural then to introduce a *single* material response tensor

$$(J_{\text{ind}})_m(\omega,\mathbf{k}) = \sigma_{mn}(\omega,\mathbf{k})\,E_n(\omega,\mathbf{k}), \tag{42}$$

where the matrix $\sigma_{mn}(\omega,\mathbf{k})$ is called the *conductivity tensor*. After solving for the electric field, all the remaining quantities, the magnetic field $\mathbf{B}(t,\mathbf{r})$ and the charge density $\rho(t,\mathbf{r})$, can be calculated from the knowledge of the total current and the electric field.
One can replace the conductivity tensor by different equivalent representations that may turn out to be handy in some applications. In particular, we discuss here the polarization tensor $\alpha_{nm}(\omega,\mathbf{k})$ and the equivalent dielectric constant $\varepsilon^{\text{eq}}_{nm}(\omega,\mathbf{k})$, defined by the following equation

$$\begin{aligned}\varepsilon^{\text{eq}}_{nm}(\omega,\mathbf{k}) &= \delta_{nm} + \tfrac{i}{\omega\varepsilon_0}\sigma_{nm}(\omega,\mathbf{k})\\ &= \delta_{nm} + \tfrac{1}{\omega^2\varepsilon_0}\alpha_{nm}(\omega,\mathbf{k})\\ &= \delta_{nm} + \chi_{nm}(\omega,\mathbf{k}).\end{aligned} \tag{43}$$

The reader should notice that the equivalent dielectric function $\varepsilon_0\varepsilon^{\text{eq}}_{nm}(\omega,\mathbf{k})$ is *not* the same as the conventional dielectric function defined in terms of the polarization and magnetization densities appearing in equation (40). In terms of the new dielectric function $\varepsilon^{\text{eq}}_{nm}(\omega,\mathbf{k})$, we write

$$D_n(\omega,\mathbf{k}) = \sum_m \varepsilon_0\varepsilon^{\text{eq}}_{nm}(\omega,\mathbf{k})\,E_m(\omega,\mathbf{k}). \tag{44}$$

It follows that in the Fourier transform approach to the material response, we effectively kill the magnetization vector $\mathbf{M}$ and collect all relevant physical processes into a single vector, the effective polarization density $\mathbf{P}$.

6. Comparison between the Traditional Multipole and the Fourier Transform Approaches to the Material Response

Within the miltipole approach to the material response, two new fields are traditionally introduced, the electric induction $\mathbf{D}$ (the electric displacement vector), and the magnetic field strength $\mathbf{H}$. These are defined by the relations

$$\mathbf{D} \equiv \varepsilon_0\mathbf{E} + \mathbf{P}, \tag{45}$$

$$\mathbf{H} \equiv \frac{1}{\mu_0}\mathbf{B} - \mathbf{M}. \tag{46}$$

The electric susceptibility χ^e and the magnetic susceptibility χ^m are defined by the following equations

$$\mathbf{P} = \varepsilon_0\chi^e\mathbf{E}, \tag{47}$$

$$\mathbf{M} = \frac{1}{\mu_0}\chi^m\mathbf{B}. \tag{48}$$

The effective dielectric constant, or electric permittivity ε, and the magnetic permeability μ, can now be defined in terms of the quantities above as

$$\mathbf{D}(\omega,\mathbf{k}) = \varepsilon\mathbf{E}(\omega,\mathbf{k}), \tag{49}$$

$$\mathbf{H}(\omega,\mathbf{k}) = \frac{1}{\mu}\mathbf{B}(\omega,\mathbf{k}). \tag{50}$$

We now proceed to derive the equivalence between this traditional approach and the Fourier formalism of Section 5. First, the current distribution is decomposed into two parts, one due to external (applied) sources, $\mathbf{J}_{\text{ext}}$, and the other, $\mathbf{J}_{\text{ind}}$ due to the interaction between the medium and the electromagnetic fields. We write

$$\mathbf{J}(t,\mathbf{r}) = \mathbf{J}_{\text{ext}}(t,\mathbf{r}) + \mathbf{J}_{\text{ind}}(t,\mathbf{r}). \tag{51}$$

The induced current is written using the conductivity tensor introduced in (42) and the result is substituted to the master equation (17). After simple re-arranging of terms, we find

$$\frac{\omega^2}{c^2}\mathbf{E}(\omega,\mathbf{k}) + \mathbf{k}\times\mathbf{k}\times\mathbf{E}(\omega,\mathbf{k}) + i\omega\mu_0\bar{\bar{\sigma}}(\omega,\mathbf{k})\cdot\mathbf{E}(\omega,\mathbf{k}) = -i\omega\mu_0\mathbf{J}_{\text{ext}}(\omega,\mathbf{k}). \tag{52}$$

Now let us calculate by means of the ϵ-μ method. In this case, the induced current is written in terms of both the polarization and magnetization current densities **P** and **M** as shown in (40). Using (47) and (48) in (41), we find

$$\mathbf{J}_{\text{ind}}(\omega,\mathbf{k}) = -i\omega\varepsilon_0\chi^e\mathbf{E}(\omega,\mathbf{k}) + i\mathbf{k}\times\frac{\chi^m}{\mu_0}\mathbf{B}(\omega,\mathbf{k}). \tag{53}$$

But from Maxwell's equations in the Fourier domain, specifically (10), we know that

$$i\mathbf{k}\times\mathbf{B}(\omega,\mathbf{k}) = \frac{i}{\omega}\mathbf{k}\times\mathbf{k}\times\mathbf{E}(\omega,\mathbf{k}). \tag{54}$$

The induced current in (53) becomes then

$$\mathbf{J}_{\text{ind}}(\omega,\mathbf{k}) = -i\omega\varepsilon_0\chi^e\mathbf{E}(\omega,\mathbf{k}) + i\frac{\chi^m}{\omega\mu_0}\mathbf{k}\times\mathbf{k}\times\mathbf{E}(\omega,\mathbf{k}). \tag{55}$$

Combining (51) and (55) and substituting the result into (17), we arrive after some rearranging to

$$\begin{aligned}&\tfrac{\omega^2}{c^2}\mathbf{E}(\omega,\mathbf{k}) + \mathbf{k}\times\mathbf{k}\times\mathbf{E}(\omega,\mathbf{k})\\ &\quad+\tfrac{\omega^2}{c^2}\chi^e\mathbf{E}(\omega,\mathbf{k}) - \chi^m\mathbf{k}\times\mathbf{k}\times\mathbf{E}(\omega,\mathbf{k}) = -i\omega\mu_0\mathbf{J}_{\text{ext}}(\omega,\mathbf{k})\end{aligned} \tag{56}$$

By comparing (52) and (56), we conclude that we must have

$$i\omega\mu_0\bar{\bar{\sigma}}(\omega,\mathbf{k})\cdot\mathbf{E}(\omega,\mathbf{k}) = \frac{\omega^2}{c^2}\chi^e\mathbf{E}(\omega,\mathbf{k}) - \chi^m\mathbf{k}\times\mathbf{k}\times\mathbf{E}(\omega,\mathbf{k}). \tag{57}$$

In tensor form, equation (57) becomes

$$i\omega\mu_0\sigma_{nl}(\omega,\mathbf{k})E_l(\omega,\mathbf{k}) = \frac{\omega^2}{c^2}\chi^e E_n(\omega,\mathbf{k}) - \chi^m\left[k_nk_l - k^2\delta_{nl}\right]E_l(\omega,\mathbf{k}). \tag{58}$$

Since the equality holds for arbitrary E_l, we obtain

$$\sigma_{nl}(\omega,\mathbf{k}) = \frac{1}{i\omega\mu_0}\left\{\frac{\omega^2}{c^2}\chi^e\delta_{nl} - \chi^m\left[k_nk_l - k^2\delta_{nl}\right]\right\}. \tag{59}$$

From (43), we reach to

$$\begin{aligned}\varepsilon_{nl}^{\text{eq}}(\omega,\mathbf{k}) &= \delta_{nl} + \tfrac{i}{\omega\varepsilon_0}\tfrac{1}{i\omega\mu_0}\left\{\tfrac{\omega^2}{c^2}\chi^e\delta_{nl} - \chi^m\left[k_nk_l - k^2\delta_{nl}\right]\right\} \\ &= \delta_{nl} + \chi^e\delta_{nl} - \tfrac{c^2}{\omega^2}\chi^m\left(k_nk_l - k^2\delta_{nl}\right) \\ &= (1+\chi^e)\,\delta_{nl} - \tfrac{c^2}{\omega^2}\chi^m\left(k_nk_l - k^2\delta_{nl}\right)\end{aligned} \tag{60}$$

Finally, we use the definitions (45), (46), (47), (48) to write

$$\varepsilon_{nm}^{\text{eq}}(\omega,\mathbf{k}) = \left(\frac{\varepsilon}{\varepsilon_0}\right)\delta_{nm} - \frac{c^2}{\omega^2}\left(1 - \frac{\mu_0}{\mu}\right)\left(k_nk_m - k^2\delta_{nm}\right). \tag{61}$$

This is the main equation we are looking for. It shows that a medium which is magnetic in the ε-μ approach translates into spatial dispersion in the Fourier approach. It follows also that the two dielectric constants are the same only if there is no spatial dispersion.

7. General Properties of the Material Response Tensors

The requirement that the electromagnetic fields should by themselves satisfy Maxwell's equations cannot fully specify how the very same fields will behave in a material environment. Such behavior is dictated by a more complex structure consisting of the mechanical response coupled with the electromagnetic fields. In this section, we survey and present rigorously the most important *non*-electromagnetic restrictions imposed on the material tensor. Such restrictions can be conveniently gathered under the heading 'General properties of the Material Tensor' since they involve quite broad characteristics that are wider than the particular dynamical laws encapsulated by the Maxwell's equations.

Our main equations will be the relation between the electric flux density vector and the electric field in both the spatio-temporal and spectral domain. These are, respectively,

$$D_n(\omega,\mathbf{k}) = \sum_m \varepsilon_0\varepsilon_{nm}^{\text{eq}}(\omega,\mathbf{k})\,E_m(\omega,\mathbf{k}), \tag{62}$$

$$D_n(t,\mathbf{r}) = \varepsilon_0\int dt'\int d^3r'\sum_m \varepsilon_{nm}^{\text{eq}}(t-t',\mathbf{r}-\mathbf{r}')\,E_m(t',\mathbf{r}'). \tag{63}$$

These equations describe electromagnetic processes in homogenous, isotropic or anisotropic media. It is important to keep in mind that within the Fourier-space formalism the equivalent dielectric tensor is *inherently* a tensor; even when the medium under consideration is isotropic, the dielectric function is still generally a tensor. Also, the reader may notice from (63) that the field induced at particular time t and location $\mathbf{r}$ depends generally on the applied field at *different* times and locations. We say that the medium exhibit "memory" in both the temporal and spatial sense. The spatial sense of the this memory, which is going to be the main concern for us here, is called *nonlocality*.[3]

[3] Whenever there is no risk of confusion, we drop the superscript 'eq' from $\varepsilon_{nm}^{\text{eq}}(\omega,\mathbf{k})$ and refer to the equivalent dielectric function as merely the *dielectric tensor*.

7.1 The Reality of the Fields

Since the fields appearing in equation (63) are all real, the properties of the Fourier transform dictate that the negative and positive frequencies appearing in the spectrum of the fields are both essentially equivalent to each other. Formally, we express this requirement in the following relation that any material tensor describing the responses of the medium to real quantities must satisfy

$$\varepsilon^{*}_{nm}(\omega,\mathbf{k}) = \varepsilon_{nm}(-\omega,-\mathbf{k}). \tag{64}$$

7.2 Dissipative and Non-Dissipative Processes

The material tensorial response is the Fourier transform of a real quantity and hence generally complex. The real part and the imaginary part of this tensor are usually interpreted as those responsible for dispersion and losses (dissipation), respectively. In this section, we provide the mathematical evidence in support of this interpretation.

We start be decomposing an arbitrary response tensor into hermitian and antihermitian parts

$$\varepsilon_{nm}(\omega,\mathbf{k}) = \varepsilon^{\mathrm{H}}_{nm}(\omega,\mathbf{k}) + \varepsilon^{\mathrm{A}}_{nm}(\omega,\mathbf{k}), \tag{65}$$

where

$$\varepsilon^{\mathrm{H}}_{nm}(\omega,\mathbf{k}) = \frac{1}{2}\left[\varepsilon_{nm}(\omega,\mathbf{k}) + \varepsilon^{*}_{mn}(\omega,\mathbf{k})\right], \tag{66}$$

$$\varepsilon^{\mathrm{A}}_{nm}(\omega,\mathbf{k}) = \frac{1}{2}\left[\varepsilon_{nm}(\omega,\mathbf{k}) - \varepsilon^{*}_{mn}(\omega,\mathbf{k})\right]. \tag{67}$$

It is obvious that the two parts satisfy

$$\varepsilon^{\mathrm{H}*}_{nm}(\omega,\mathbf{k}) = \varepsilon^{\mathrm{H}*}_{mn}(\omega,\mathbf{k}), \tag{68}$$

$$\varepsilon^{\mathrm{A}*}_{nm}(\omega,\mathbf{k}) = -\varepsilon^{\mathrm{A}*}_{mn}(\omega,\mathbf{k}). \tag{69}$$

We now recall our interpretation in Section 2.1 of the term $-\mathbf{J}\cdot\mathbf{E}$ as the density of the rate of energy transfer by the current $\mathbf{J}$ into the electric field $\mathbf{E}$. The current can be decomposed into external and induced parts as $\mathbf{J} = \mathbf{J}_{\mathrm{ex}} + \mathbf{J}_{\mathrm{ind}}$. Thus, the total work done by the *medium* on the electric field is given by integrating $-\mathbf{J}_{\mathrm{ind}}\cdot\mathbf{E}$ in both time and space as

$$-\int dt \int d^3r \mathbf{J}_{\mathrm{ind}}(t,\mathbf{r})\cdot\mathbf{E}(t,\mathbf{r}) = \int \frac{d\omega d^3k}{(2\pi)^4}\mathbf{J}_{\mathrm{ind}}(\omega,\mathbf{k})\cdot\mathbf{E}^{*}(\omega,\mathbf{k}), \tag{70}$$

where the power theorem of Fourier transforms was used in writing the equality. We now have

$$\begin{aligned}&\int \tfrac{d\omega d^3k}{(2\pi)^4}\mathbf{J}_{\mathrm{ind}}(\omega,\mathbf{k})\cdot\mathbf{E}^{*}(\omega,\mathbf{k})\\ &\quad= \int \tfrac{d\omega d^3k}{(2\pi)^4}\tfrac{1}{2}\left[\mathbf{J}^{*}_{\mathrm{ind}}(\omega,\mathbf{k})\cdot\mathbf{E}(\omega,\mathbf{k}) + \mathbf{J}_{\mathrm{ind}}(\omega,\mathbf{k})\cdot\mathbf{E}^{*}(\omega,\mathbf{k})\right]\end{aligned} \tag{71}$$

In deriving this, the integral was first divided into its negative and positive frequency parts, and then a transformation of variables was applied to the negative frequencies integral. Finally, the symmetry condition (reality condition) given in (64) was applied. Employing equation (42) in (71), we can write

$$\begin{aligned}&-\int \tfrac{d\omega d^3k}{(2\pi)^4}\mathbf{J}_{\mathrm{ind}}(\omega,\mathbf{k})\cdot\mathbf{E}^{*}(\omega,\mathbf{k})\\ &= \int \tfrac{d\omega d^3k}{(2\pi)^4}\tfrac{1}{2}\left[\sigma_{nm}(\omega,\mathbf{k})E_m(\omega,\mathbf{k})E^{*}_n(\omega,\mathbf{k}) + \sigma^{*}_{mn}(\omega,\mathbf{k})E^{*}_n(\omega,\mathbf{k})E_m(\omega,\mathbf{k})\right]\\ &= \int \tfrac{d\omega d^3k}{(2\pi)^4}\sigma^{H}_{nm}(\omega,\mathbf{k})E_m(\omega,\mathbf{k})E^{*}_n(\omega,\mathbf{k}).\end{aligned} \tag{72}$$

Therefore, it is the hermitian part of the conductivity tensor which contributes to the dissipation of energy by the medium. Equivalently, by considering the relation between the conductivity and the equivalent dielectric tensor (43), we find that it is the antihermitian part of the dielectric tensor that contributes to energy dissipation by the medium.

7.3 Onsager Relations

Since any material responses tensor is ultimately based on a mechanical model, of which the dynamical equations must satisfy certain space-time symmetry transformations, there exists certain general restrictions on the mathematical form of a physically realizable material tensor. In order to give the reader some idea about such requirement, we list the classical dynamical equation for the particle motion, namely the Lorentz force (9). By writing the force as $\mathbf{F} = d\mathbf{p}/dt$, where $\mathbf{p}$ is the linear momentum, it is an easy matter to verify that the equation of motion is invariant under the transformations

$$t \to -t, \ \mathbf{p} \to -\mathbf{p}, \ \mathbf{B} \to -\mathbf{B} \tag{73}$$

The same conclusion can be obtained if the Lorentz force law is replaced by the Schrodinger equation.
Notice that a time-reversal corresponds to the substitution $\omega \to -\omega$ in the Fourier domain. The reversal of the sign of the momentum corresponds to reversing the sign of the wavevector. Finally, the reversal of the sign of the magnetic field is shown explicitly in the following standard form of the Onsager relations[4]

$$\varepsilon_{nm}^{\mathrm{eq}}\left(\omega,-\mathbf{k}\right)\Big|_{-\mathbf{B}} = \varepsilon_{mn}^{\mathrm{eq}}\left(\omega,\mathbf{k}\right)\Big|_{\mathbf{B}}. \tag{74}$$

The Onsager relations places severe restrictions on the physically allowable form of the material response. We discuss below particular examples of isotropic spatially dispersed media.
Let us focus on materials that don't respond to the magnetic field. In this case, the Onsager relations reduces to the situation in which the tensorial responses is required to be invariant under the transformation

$$\mathbf{k} \to -\mathbf{k}, \ n \leftrightarrow m. \tag{75}$$

First, notice that in the Fourier transform approach, even when the medium is isotropic, the response is still described by a tensorial quantity, c.f. equation (61). For isotropic media that is spatially dispersive, we can analyze the situation by pure matrix theoretic arguments. Indeed, the only available vector in this case is k_m, while the only available tensors are δ_{nm} and ϵ_{nml}. It can be shown that the Onsager relations leads to the result that we can construct only three independent second-rank tensors. A popular choice in the condensed-matter physics literature is the following

$$\varepsilon_{nm}^{\mathrm{eq}}\left(\omega,\mathbf{k}\right) = \varepsilon^{L}\left(\omega,k\right)\kappa_n\kappa_m + \varepsilon^{T}\left(\omega,k\right)\left(\delta_{nm} - \kappa_n\kappa_m\right) + i\varepsilon^{R}\left(\omega,k\right)\epsilon_{nml}k_l, \tag{76}$$

where

$$\kappa_m = k_m/k, \ \ k = |\mathbf{k}|. \tag{77}$$

Here, the quantities $\varepsilon^L(\omega,k)$, $\varepsilon^T(\omega,k)$, $\varepsilon^R(\omega,k)$ are the longitudinal, transverse, and rotational permittivities, respectively. The rotatory parts can be ignored in media that don't exhibit optical activity. Notice that for media in which both the longitudinal and transverse parts happen to be equal to each other, the equivalent dielectric tensor reduces to the scalar case.

[4] The symmetry relations (64) are used to simplify the final form.

7.4 The Kramers–Kronig Relations

The fact that the dielectric tensor is a response function imposes a restriction on the relationship between the real and imaginary part. This restriction is due to causality and can be rigourously derived by standard techniques in the theory of complex functions.[5] Kramers-Kronig relations say that the real and imaginary parts of the Fourier transform of a function that is causal (i.e., a function that its inverse Fourier transform is identically zero for a time interval in the form $-\infty < t < t_0$) satisfy

$$\varepsilon_{nm}^{\mathrm{eq,H}}(\omega,\mathbf{k}) - \delta_{nm} = \frac{i}{\pi}\wp\int_{-\infty}^{\infty} d\omega' \frac{\varepsilon_{nm}^{\mathrm{eq,A}}(\omega',\mathbf{k}) - \delta_{nm}}{\omega - \omega'}, \tag{78}$$

$$\varepsilon_{nm}^{\mathrm{eq,A}}(\omega,\mathbf{k}) = \frac{i}{\pi}\wp\int_{-\infty}^{\infty} d\omega' \frac{\varepsilon_{nm}^{\mathrm{eq,H}}(\omega',\mathbf{k})}{\omega - \omega'}, \tag{79}$$

where $\wp$ symbolizes the Cauchy principal value.[6] Equations (78) and (79) show that if dissipation is known, then dispersion can be uniquely determined (and vice versa) by applying the Hilbert transform operator to the available data.

One can see that when spatial dispersion is present, then in the case of non-dissipative medium, i.e., a medium with negligible losses which, as can be seen from Section 7.2, corresponds to $\varepsilon_{nm}^{\mathrm{eq,A}}(\omega,\mathbf{k}) = 0$, the dispersion behavior dictated by $\varepsilon_{nm}^{\mathrm{eq,H}}(\omega,\mathbf{k})$ is restricted to only the class of functions of ω which has zero Hilbert transform. It can be shown that such functions take the basic form $1/(\omega - \omega_m)$ with constant ω_m. This explains partially why such basic form pops out very frequently in practice. However, they also demonstrate the power of Kramers-Kronig relations in being able to severely restrict the allowable functional form of the dispersion in lossless media.

The general lesson we learn from taking casuality into consideration when thinking about designing artificial media is that once the losses is neglected for the entire frequency range $-\infty < \omega < \infty$ (or the medium is designed to have small losses globally), the *global* form of dispersion is no more a free degree of freedom but, instead, takes a particular form. However, in practice we seldom achieve or require particular specifications of the losses and/or dispersion to hold for the entire frequency range. Noticing that the Hilbert transform relations in (78) and (79) are global operators, i.e., they involve integration over the entire frequency range in order to know the value at a single frequency (nonlocal or memory-dependent in frequency), we need just to restrict ourselves to a finite frequency and wavenumber range upon which the desired losses and dispersion characteristics are required to apply. By this restrictions, the Kramers–Kronig relations cannot impose a serious restriction on the design and analysis of artificial media.

8. Advanced Properties of the Material Tensor

In this Section, we look at the material tensor through the point of view of complex analysis. The motivation for such study is that certain characteristics of signals excited in media, like

[5] The causality restriction translates formally to the following setting. Imagine that the medium is excited by an applied electric field **E**. The material responses, for example through (47), will appear in the form of a forced (induced) quantity, here the polarization density **P**. If the applied field was zero for time $t < 0$, then causality implies that there must be no induced polarization in this time interval.

[6] These relations represent a Hilbert transform relation between the hermitian and antihermitian parts, which play the role of real and imaginary parts, respectively, in the case of matrices (linear operators).

short-term disturbances and damped waveforms are best understood analytically if viewed using the mathematical device of Laplace transform instead of the familiar Fourier transform, the latter being best suited ideally to analyze the steady-state behavior of a given system. As will be shown below, there are general restrictions on the mathematical form of the response functions when viewed in the complex plane. Knowledge of such global restrictions is vital in the theory and practice of meta-materials.

8.1 Stability Restrictions

From the physical point of view, a passive medium cannot generate energy and hence all propagating signals must be damped or decaying when the distance goes to infinity.[7] Mathematically, this translates to the requirement that all poles are located in the LHP. We call the following the statement of the stability condition of material media

All poles of the material tensor must be located in the LHP. (80)

To see why this should be the case, just (Laplace) invert a spectral component in the form $1/(\omega - \omega_0 + i\gamma/2)$ and notice the sign of the resulting exponential factor. For signals to exponentially decay, instead of growing, the algebraic sign of the factor γ must be positive.

8.2 Causality Restrictions

Although we have already looked at casuality in the study of Kramers -Kronig relations, we want to understand here this topic at a deeper level. Consider the Fourier transform of a causal function $f(t)$ given by

$$f(\omega) = \int_0^\infty dt\, f(t)\, e^{i\omega t}. \tag{81}$$

Let us study the asymptotic behavior of this function when $t \to \infty$. We first notice that when $\mathrm{Im}\{\omega\} > 0$, the integral in (81) has a finite value since the integrand approaches zero as t grows to infinity. Moreover, on repeatedly differentiating this integral, we conclude also that all derivatives of $f(\omega)$ are finite. Therefore, the function $f(\omega)$ is analytic in the upper half complex plane. We have then

A causal function is analytic in the UHP. (82)

An immediate corollary is that

A causal function has no poles or branch points in the UHP. (83)

This principle forms the mathematical background behind the derivation of Kramers-Kronig relations.

8.3 Landau Condition

The Laplace transform of a signal is defined as

$$F(s) \equiv \int_0^\infty dt\, f(t)\, e^{-st}. \tag{84}$$

[7] Notice that for a range that is bounded, both growing and decaying signals are possible. For example, consider a multilayered medium. In one intermediate layer both growing and decaying waves are permitted.

Therefor, the s-plane and the complex ω-plane are related by $s = i\omega$, which means that 'upper' and 'lower' in one plane translates into 'right' and 'left', respectively, in the other plane.
The inverse Laplace transform is given by the equation

$$f(t) = \frac{1}{2\pi i} \int_{\Gamma - i\infty}^{\Gamma + i\infty} ds\, F(s)\, e^{st}, \tag{85}$$

where Γ specifies how the integration contour should be chosen. Landau condition states that

The contour in (85) is to the left of all singularites in the s-plane. (86)

Therefore, the integration contour must be above all singularities in the complex ω-plane. It can be shown then that the resulting function does not depend on the particular path provided it satisfies the Landau condition.

9. Wave Propagation

9.1 Dispersion Relations

By wave modes or wave propagation we mean electromagnetic disturbances that can propagate in a source-free medium. In our case, the medium response is described by the nonlocal model of the Fourier approach.
Equation (52) is the *inhomogeneous* wave equation in our medium. From the definition (42), the induced current in terms of the vector potential (temporal gauge) is expressed as follows

$$J_{\text{ind},m}(\omega, \mathbf{k}) = \alpha_{mn}(\omega, \mathbf{k})\, A_n(\omega, \mathbf{k}). \tag{87}$$

In tensor form, we can write then (52) as

$$\Xi_{nm}(\omega, \mathbf{k})\, A_m(\omega, \mathbf{k}) = -\frac{\mu_0 c^2}{\omega^2} J_{\text{ext},n}(\omega, \mathbf{k}), \tag{88}$$

where

$$\Xi_{nm}(\omega, \mathbf{k}) = \frac{c^2}{\omega^2}\left(k_n k_m - k^2 \delta_{nm}\right) + \chi_{nm}(\omega, \mathbf{k}). \tag{89}$$

If the source term in (88) is set to zero, we obtain the *homogenous* wave equation describing the propagation of waves in a source-free environment, i.e., the eigenmodes. However, as we found in Section 7.2, the antihermitian part of the tensor $\Xi_{nm}(\omega, \mathbf{k})$ is responsible of dissipation or energy generation in the medium. Such term must be omitted from the final homogeneous equation describing pure wave propagation. The desired equation of motion is therefore given by

$$\Xi^{\text{H}}_{nm}(\omega, \mathbf{k})\, A_m(\omega, \mathbf{k}) = 0, \tag{90}$$

where $\Xi^{\text{H}}_{nm}(\omega, \mathbf{k})$ describes the hermitian part of the tensor $\Xi_{nm}(\omega, \mathbf{k})$. The reader should notice that there is a thermodynamic hypothesis implicit in the derivation of this fundamental equation. That is, dissipation is treated as equivalent to source, and so the antihermitian part is removed even when it describes only a passive medium. Such hypothesis, equivalence of source and sink, is an additional postulate that cannot be derived from Maxwell's equations and should be supplied by an external theory, in this case thermodynamics of continuous media.

Notice that (90) is a matrix equation. From linear algebra, the necessary and sufficient condition for the existence of a nontrivial solution is that the determinant of the hermitian matrix $\Xi^{\mathrm{H}}_{nm}(\omega,\mathbf{k})$ is identically zero. That is, the dispersion relation is given by

$$\det\left[\Xi^{\mathrm{H}}_{nm}(\omega,\mathbf{k})\right] = 0. \tag{91}$$

In general, this dispersion relation has potentially many solutions, each is called a *mode* or *branch*. We write the solution of the lth mode of the dispersion equation (91) as

$$\omega = \omega_l(\mathbf{k}). \tag{92}$$

For each mode, there corresponds a vector A_n satisfying equation (90). Such vector is called the *polarization* of the wave mode. within the scheme of Fourier-space electromagnetics, there exists a detailed theory of how to obtain and classify polarization in various types of media, which is based on direct application of results from tensor calculus. However, we omit such details for the limitations of space.

9.2 The Greens Function

The solution of the inhomogeneous wave equation (88) can be formally written as

$$A_n(\omega,\mathbf{k}) = -\frac{\mu_0 c^2}{\omega^2} G_{nm}(\omega,\mathbf{k})\, J_{\mathrm{ext},n}(\omega,\mathbf{k}), \tag{93}$$

where $G_{nm}(\omega,\mathbf{k})$ is the Greens function dyad in the spectral domain. From matrix theory, an expression of this dyad can be immediately written as

$$G_{nm}(\omega,\mathbf{k}) = \frac{\mathrm{cof}_{nm}\left[\Xi_{n'm'}(\omega,\mathbf{k})\right]}{\det\left[\Xi_{nm}(\omega,\mathbf{k})\right]}, \tag{94}$$

where cof_{nm} is the cofactor matrix. In deriving this result, only the hermitian part of the operator $\Xi_{nm}(\omega,\mathbf{k})$ is used, and therefore the Greens dyad as it stands here is hermitian. However, when inverting the Fourier transform to calculate the fields in the spatio-temporal domain, a singularity in the spectral domain is encountered around $\omega = \omega_l(\mathbf{k})$. The traditional solution of this problem is to carefully enforce suitable causality conditions. Technically, the determinant is expanded in the following form

$$\det\left[\Xi_{nm}(\omega,\mathbf{k})\right] \approx \frac{\partial}{\partial\omega}\det\left[\Xi_{nm}(\omega,\mathbf{k})\right]\Big|_{\omega=\omega_l(\mathbf{k})} \left\{\omega - \omega_l(\mathbf{k}) + i0\right\}, \tag{95}$$

where the expansion illustrated here is taken around the lth mode pole. By formally inverting the Fourier transform using the Dirac delta function, we obtain the following expression for the antihermitian part of the Greens function

$$D^{\mathrm{A}}_{nm}(\omega,\mathbf{k}) = \sum_l -i\pi\omega_l(\mathbf{k})\,\mathbf{e}^*_{l,m}(\mathbf{k})\,\mathbf{e}_{l,n}(\mathbf{k})\,F_l\left[\Xi_{nm}(\omega,\mathbf{k})\right]\delta\left(\omega - \omega_l(\mathbf{k})\right), \tag{96}$$

where the $F_l\left[\Xi_{nm}(\omega,\mathbf{k})\right]$ can be directly determined by the dispersion profile of the medium, but its explicit expression is not of direct concern to us here. The unit vectors $\mathbf{e}_{l,n}(\mathbf{k})$ describe the polarization of the lth mode.[8] This derivation shows that although the antihermitian part

[8] Notice that for the case of transverse modes, the degeneracy of the eigenvalue problems requires a special treatment. Indeed, in this case one has to resort to the use of polarization matrices. We omit such details here.

was not originally taken into consideration in writing up the expression of the Greens dyad (94), causality considerations *forces* us to introduce an antihermitian part. As we will show in Section 10.1, it is precisely this antihermitian part that contributes to the radiated field.[9]

It is interesting to observe again the role played by thermodynamics in the solution of Maxwell's equations. Indeed, the ultimate origin of the causality consideration introduced above can be tracked back to the thermodynamic requirement that energy decays away from the source and toward the sink. Maxwell's equations themselves are blind to the direction of power flow; they can support both (temporally) forward and backward waves. However, thermodynamics appears to fix the sign of the pole contribution around the real ω-axis and hence effectively imposes a particular form of the solution of the field equations. The reader can appreciate better the subtlety of this fact by recalling that the very concepts of source and sink are thermodynamic in nature and cannot be based ultimately on Maxwell's equations. An impulsive excitation, say an ideal Dirac delta function, can be mathematically introduced to the theory in a straightforward manner, e.g., using generalized function theory. However, the choice of the signs of the imaginary part of the pole associated with source or sink depends on energetics and dissipation, a topic that is best described macroscopically by classical thermodynamics. Since the ultimate origin of the antihermitian part of the Greens function, as shown above in equation (96), is causality, and the particular form of this depends in turn on thermodynamic consideration, and knowing that it is this part of the Greens dyad that is responsible of radiation (see Section 10.1), we can claim that the ultimate answer to the question of why an antenna can radiate appears to be purely thermodynamic in nature.

10. Applications

In this section, we provide some applications for the general Fourier approach toward the characterization of the material responses to the electromagnetic fields as sketched above.

10.1 Localization of Electromagnetic Energy Radiated by Antennas in Complex Media

In this part, we perform an explicit calculation of the electromagnetic energy radiated by an arbitrary antenna in a medium described by a nonlocal response tensor. We will show that the Fourier approach described in this paper provides a direct method to understand the structure of the near-field surrounding the antenna, and therefore the possibility of localizing energy in complex artificial media.

The method relies on calculating the total energy of the radiated field using the Fourier integral. We start from the statement of energy conservation as stated in (8). The current $\mathbf{J}$ appearing at the RHS is replaced by the current distribution on the antenna, which is taken as an external current $\mathbf{J}_{\text{ext}}$. As discussed in Section 2.1, the energy density (work) transferred to the surrounding field by this current is given by $-\mathbf{J}_{\text{ext}} \cdot \mathbf{E}$. The trick in performing general calculation is to introduce a new quantity $U_l(\mathbf{k})$, which is defined as *the density in the* $\mathbf{k}$*-space of the energy added to the surrounding field by the antenna when radiating through the* l*th mode*. That is, by energy conservation, the time-averaged total energy added to the field by all modes is given by the following equation

$$\int_{-T/2}^{T/2} dt \int d^3r \mathbf{J}_{\text{ext}}(t,\mathbf{r}) \cdot \mathbf{E}(t,\mathbf{r}) = \sum_l \int \frac{d^3k}{(2\pi)^3} U_l(\mathbf{k}). \tag{97}$$

[9] The hermitian part will contribute to the non-propagating field (near field) surrounding the source.

Expressing the electric field in terms of the vector potential in the temporal gauge as $\mathbf{E}(\omega,\mathbf{k}) = i\omega\mathbf{A}(\omega,\mathbf{k})$, using the Greens function of (93), and employing the Pareseval (power) theorem of Fourier analysis, we write the LHS as

$$\begin{aligned}\int_{-T/2}^{T/2} dt \int d^3r \mathbf{J}_{\text{ext}}(t,\mathbf{r}) \cdot \mathbf{E}(t,\mathbf{r}) &= -\int \frac{d\omega d^3k}{(2\pi)^4} \mathbf{J}^*_{\text{ext}}(\omega,\mathbf{k}) \cdot \mathbf{E}(\omega,\mathbf{k}) \\ &= \int \frac{d\omega d^3k}{(2\pi)^4} \frac{i}{\varepsilon_0\omega} J^*_{\text{ext},m}(\omega,\mathbf{k})\, G_{mn}(\omega,\mathbf{k})\, J_{\text{ext},n}(\omega,\mathbf{k}) \\ &= \int \frac{d\omega d^3k}{(2\pi)^4} \frac{i}{\varepsilon_0\omega} J^*_{\text{ext},m}(\omega,\mathbf{k}) \left[G^{\text{A}}_{mn}(\omega,\mathbf{k}) + G^{\text{H}}_{mn}(\omega,\mathbf{k})\right] J_{\text{ext},n}(\omega,\mathbf{k})\end{aligned} \tag{98}$$

Due to the presence of the factor i in the integrand, together with the fact that the integral must be real, it follows from the basic properties of hermitian and antihermitian functions (operators) that only the antihermitian part of the Greens dyad contributes to the radiation field. Now, by plugging this part as given in (96) into (98), we finally arrive to the following expression of the energy density

$$U_l(\mathbf{k}) = \frac{F_l\left[\Xi_{nm}(\omega,\mathbf{k})\right]}{\varepsilon_0} \left|\mathbf{e}^*_l(\mathbf{k}) \cdot \mathbf{J}_{\text{ext}}(\omega_l(\mathbf{k}),\mathbf{k})\right|^2. \tag{99}$$

We will propose an interpretation for the physical meaning of $U_l(\mathbf{k})$. Consider the inverse Fourier transform

$$u_{F,l}(\mathbf{r}) = \Im^{-1}\left\{\sqrt{U_l(\mathbf{k})}\right\}. \tag{100}$$

Next, we use the Parseval theorem to write

$$\int d^3r\, u_{F,l}(\mathbf{r}) \left(u_{F,l}(\mathbf{r})\right)^* = \int \frac{d^3k}{(2\pi)^3} \sqrt{U_l(\mathbf{k})}\sqrt{U_l(\mathbf{k})}. \tag{101}$$

Since the RHS is by definition the total energy radiated by the antenna in the lth mode, it follows that the integrand of the LHS, namely $u_l(\mathbf{r}) \equiv \left|u_{F,l}(\mathbf{r})\right|^2$ can be interpreted as the *spatial* distribution of the energy density radiated by the antenna through the lth mode. We have

$$\begin{aligned} u_l(\mathbf{r}) = \int\int d^3k d^3k' & \frac{\sqrt{F_l[\Xi_{nm}(\omega,\mathbf{k})] F_l[\Xi_{nm}(\omega,\mathbf{k}')]^*}}{\varepsilon_0} e^{i(\mathbf{k}-\mathbf{k}')\cdot\mathbf{r}} \\ & \times \left|\mathbf{e}^*_l(\mathbf{k}) \cdot \mathbf{J}_{\text{ext}}(\omega_l(\mathbf{k}),\mathbf{k})\, \mathbf{e}^*_l(\mathbf{k}') \cdot \mathbf{J}_{\text{ext}}(\omega_l(\mathbf{k}'),\mathbf{k}')\right|. \end{aligned} \tag{102}$$

This new quantitative measure contains information about the spatial structure of the time-averaged energy surrounding a radiator specified by its externally enforced current distribution $\mathbf{J}_{\text{ext}}(t,\mathbf{r})$. For example, it can be directly used in studying the localization of the radiated energy in an antenna inside an artificial medium described by the dispersion profile $\chi_{nm}(\omega,\mathbf{k})$.

10.2 Nonlocal Electromagnetic Media

10.2.1 General Theoretical Background from the Field of Crystal Optics

By the term *nonlocal medium* we refer to a material described by response functions similar to (47). As we we noticed previously, it follows from this definition that the material exhibits a memory-like behavior in the sense that the response to a field excreted at a particular location appears to depend on the field values at *other* locations. We will show below that this phenomenon is very general and does not just refer to a particular physical process occurring in the crystal.

First, notice that we arrived to the definition (47) through a Fourier transform approach to the electromagnetic fields. Moreover, we were able to derive a relation connecting the traditional multipole approach and the Fourier approach. We found that nonlocality or spatial dispersion arises very naturally to account for nonmagnetic media. However, it is in the nature of the Fourier approach itself to introduce the spatial spectral variable **k** into the description of the material medium, and hence one can view nonlocality as a characteristics of the formalism itself, rather than a particular label given to an exotic physical process, for example exciotons-polaritons in crystal optics.

Let us start by providing a global qualitative look at the response of material media in classical and quantum optics. This view will serve as standard theoretical background upon which we measure our understanding of how to design artificial media.

Imagine that the material is composed of a system of uncoupled (hence, independent) oscillators. Each oscillator can interact with the applied electromagnetic fields by producing a dipole moment **p**. From the basic picture of Lorentz models, we can express the functional dependence of this induced dipole moment on the temporal frequency ω in the broad Lorentzian form $\zeta/(\omega^2 - \omega_0^2)$, where ω_0 is a constant called the eigenfrequency or the resonance frequency, and ζ the oscillator strength. In general, each independent oscillator will resonate with the applied field according to its own eigenfrequency and strength, and the medium overall response will be taken as the sum of all individual resonances. In this view, it is useful to think of each oscillator as representing an 'atom', even when its actual physical dimensions are much larger than real atoms. The essential idea in the art of artificial material design is taking this conceptual framework into its extreme by assuming that one can manipulate each atom individually in order to control and tailor the resulted material responses. The assumption that the atoms are *un*coupled will be translated to the fact that the resulted eigenfrequency ω_0 and oscillator strength ζ don't depend on wavelength, or equivalently on **k**. For natural materials observed and studied through macroscopic electromagnetics, the atomic separation, for example in periodic structures like crystals, denoted here by a, is very small compared with the operating wavelength, i.e., we have $a/\lambda \ll 1$. In this case, all atoms appear to be in perfect phase synchronization and no significant coupling mechanism takes place.[10]

The situation is dramatically different in periodic structures, like photonic crystals and frequency selective surfaces, where, in this case, the operating wavelength can become appreciable compared with the characteristic spatial scale of the separation between the atoms (or unit cells), and hence interesting electromagnetic behavior can arise, like stopbands, localization, etc. It is still however possible to describe all these complex structures by employing an effective dielectric function that is *nonlocal*. Such function can contain the full information of the symmetry group of the periodic structure. Therefore, Maxwell's equations, written in terms of these equivalent responses functions, can be used to describe the electromagnetics of the medium without writing explicitly the set of the boundary conditions.[11] Aside from the economic advantage of such formulation, allowing the effective dielectric function to become nonlocal has the advantage of bringing the full power of the conceptual framework of effective medium theory right to the fore even though the artificial medium under consideration may not satisfy the natural condition of infinitesimally small atomic constituents.

---

[10] The fundamental pre-condition for this to be true is that the fields are averaged on a spatial scale much larger than this natural characteristic spatial scale, i.e., the atomic separation a.

[11] As an example for a concrete implementation of this general idea, see (6).

10.2.2 Negative Group Velocity Artificial Media

We now briefly demonstrate the above general theory by showing that a new genre of artificial (meta-) materials can be envisioned by thinking in terms of the Fourier-space formalism of this paper. Specifically, we consider an idealized isotropic and homogenous medium that exhibit both temporal and spatial dispersion.

The dispersion relation for the transverse waves is $\mathbf{k} \cdot \mathbf{k} = (\omega/c)^2 n^2(\omega, \mathbf{k})$, where we have *defined* the index of refraction as $n(\omega, \mathbf{k})^2 \equiv \varepsilon(\omega, \mathbf{k})\mu$. We will work with such general index of refraction given by $n = n(\omega, \mathbf{k})$. The group velocity is defined as $\mathbf{v}_g = \nabla_{\mathbf{k}}\omega$. It can be shown that an equation connecting the spatial and temporal dispersion such that the resulted medium supports negative group velocity propagation can be put in the following form

$$\frac{\omega}{c}\frac{\partial n(\omega, \mathbf{k})}{\partial k} - \frac{\gamma}{c}\left(1 + \omega\frac{\partial}{\partial\omega}\right) n(\omega, \mathbf{k}) = 1. \tag{103}$$

where $\gamma = -\left|\mathbf{v}_g\right|$. We call this partial differential equation the dispersion engineering equation for negative group velocity. An exact solution for this equation can be attempted for interesting special cases.

Consider the boundary-value problem consisting of the partial differential equation (103) together with

$$\frac{\partial\gamma}{\partial\omega} = 0, n(\omega, k = k_1) = \phi(\omega), \tag{104}$$

where $\omega_1 < \omega < \omega_2, \omega_1 > 0, k_1 > 0$. Here, $k_1 < k_2$ and $\omega_1 < \omega_2$ are positive real numbers and $\phi(\omega)$ is a general function representing the boundary condition of the problem. In the interesting scenario where the group velocity is constant and negative, an exact solution was found to be (4)

$$n(\omega, k) = \frac{c(k - k_1)}{\omega} + \frac{1}{\omega}\left[\omega + \gamma(k - k_1)\right]\phi(\omega + \gamma(k - k_1)). \tag{105}$$

Such solution is then theoretically feasible. It demonstrates that there is rich degrees of freedom in the material design waiting for us to discover and exploit for novel and interesting applications. A more comprehensive theory for the negative group velocity metamaterial was developed by the authors in (4).

A. Magnetic Moments in Terms of Electric Moments

A.1 The Magnetic Moment Term

Multiply the equation of continuity (7) by r_l and integrate over all space to get

$$\int d^3r \frac{\partial\rho(t, \mathbf{r})}{\partial t} r_l = -\int d^3r\, r_l \nabla \cdot \mathbf{J}(t, \mathbf{r}). \tag{106}$$

Consider first the LHS of (106). By employing the definition of the dipole moment (30), we write immediately

$$\int d^3r \frac{\partial\rho(t, \mathbf{r})}{\partial t} r_l = \frac{\partial}{\partial t}\int d^3r\, \rho(t, \mathbf{r}) r_l = \frac{\partial}{\partial t} p_l(t, \mathbf{r}). \tag{107}$$

Now we consider the RHS of (106). Write the divergence as $\nabla \cdot \mathbf{J}(t,\mathbf{r}) = (\partial/\partial r_s) J_s(t,\mathbf{r})$ and integrate by parts through the variable r_s to obtain

$$\begin{aligned} \int d^3r r_l \nabla \cdot \mathbf{J}(t,\mathbf{r}) &= \int d^3r r_l \tfrac{\partial}{\partial r_s} J_s(t,\mathbf{r}) \\ &= \int d^2r \int dr_s r_l \tfrac{\partial}{\partial r_s} J_s(t,\mathbf{r}) \\ &= \int d^2r \left[J_s(t,\mathbf{r}) r_l |_{r_s=-\infty}^{r_s=+\infty} - \int dr_s \tfrac{\partial r_l}{\partial r_s} J_s(t,\mathbf{r}) \right] \\ &= \int d^2r \left[J_s(t,\mathbf{r}) r_l |_{r_s=-\infty}^{r_s=+\infty} - \int dr_s \delta_s^l J_s(t,\mathbf{r}) \right] \\ &= \int d^2r \left[J_s(t,\mathbf{r}) r_l |_{r_s=-\infty}^{r_s=+\infty} - \int dr_s J_l(t,\mathbf{r}) \right]. \end{aligned} \tag{108}$$

Using the assumption that the surface current $J_s(t,\mathbf{r})$ vanishes on the surface of the integration volume, we obtain

$$\int d^3r r_l \nabla \cdot \mathbf{J}(t,\mathbf{r}) = -\int d^3r J_l(t,\mathbf{r}) \equiv -\mu_l(t,\mathbf{r}). \tag{109}$$

From (106), (107), and (109), we finally arrive to

$$\mu_l(t,\mathbf{r}) = \frac{\partial}{\partial t} p_l(t,\mathbf{r}). \tag{110}$$

A.2 The Magnetic Quadrable Term

Multiply the equation of continuity (7) by $r_l r_m$ and integrate over all space to get

$$\int d^3r \frac{\partial \rho(t,\mathbf{r})}{\partial t} r_n r_m = -\int d^3r r_n r_m \nabla \cdot \mathbf{J}(t,\mathbf{r}). \tag{111}$$

Consider first the LHS of (111). By employing the definition of the electric quadrable moment (31), we write immediately

$$\int d^3r \frac{\partial \rho(t,\mathbf{r})}{\partial t} r_n r_m = \frac{\partial}{\partial t} \int d^3r \rho(t,\mathbf{r}) r_n r_m = \frac{\partial}{\partial t} q_{nm}(t). \tag{112}$$

Now let us take the RHS of (111). We first decompose the magnetic moment into the sum of symmetric and anti-symmetric parts as follows

$$x_n J_m = \frac{1}{2}(x_n J_m + x_m J_n) + \frac{1}{2}(x_n J_m - x_m J_n). \tag{113}$$

Now, write again the divergence as $\nabla \cdot \mathbf{J}(t,\mathbf{r}) = (\partial/\partial r_s) J_s(t,\mathbf{r})$ and integrate by parts through the variable r_s to obtain

$$\begin{aligned} \int d^3r r_n r_m \nabla \cdot \mathbf{J}(t,\mathbf{r}) &= \int d^3r r_n r_m \tfrac{\partial}{\partial r_s} J_s(t,\mathbf{r}) \\ &= \int d^2r \int dr_s r_n r_m \tfrac{\partial}{\partial r_s} J_s(t,\mathbf{r}) \\ &= \int d^2r \left[J_s(t,\mathbf{r}) r_n r_m |_{r_s=-\infty}^{r_s=+\infty} - \int dr_s \tfrac{\partial}{\partial r_s}(r_n r_m) J_s(t,\mathbf{r}) \right] \\ &= \int d^2r \left[J_s(t,\mathbf{r}) r_n r_m |_{r_s=-\infty}^{r_s=+\infty} - \int dr_s (r_n \delta_s^n + r_m \delta_s^m) J_s(t,\mathbf{r}) \right] \\ &= \int d^2r \left[J_s(t,\mathbf{r}) r_n r_m |_{r_s=-\infty}^{r_s=+\infty} - \int dr_s \{ r_n J_s(t,\mathbf{r}) + r_m J_s(t,\mathbf{r}) \} \right] \end{aligned} \tag{114}$$

Using again the assumption that the surface current $J_s(t, \mathbf{r})$ vanishes on the surface of the integration volume, we obtain

$$\int d^3r\, r_n r_m \nabla \cdot \mathbf{J}(t, \mathbf{r}) = -\int d^3r\, [r_n J_s(t, \mathbf{r}) + r_m J_s(t, \mathbf{r})]. \tag{115}$$

From (111), (112), and (115), we reach

$$\int d^3r\, [r_n J_s(t, \mathbf{r}) + r_m J_s(t, \mathbf{r})] = \frac{\partial}{\partial t} q_{ns}(t). \tag{116}$$

The antisymmetrical part in (113) can be written readily in the form $1/2\epsilon_{lmn}\mu_{mn}(t)$, where magnetic quadrable moment μ_{mn} is defined as

$$\mu_{nm}(t) = \int d^3r\, r_m J_n(t, \mathbf{r}). \tag{117}$$

Therefore, one can express the *axial* vector as

$$\mathbf{m}(t) = \frac{1}{2} \int d^3r\, \mathbf{r} \times \mathbf{J}(t, \mathbf{r}). \tag{118}$$

It follows then

$$\begin{aligned} \epsilon_{lsn} m_n(t) &= \tfrac{1}{2} \int d^3r \epsilon_{lsn} \left(\mathbf{r} \times \mathbf{J}(t, \mathbf{r})\right)_n \\ &= \tfrac{1}{2} \int d^3r \epsilon_{lsn} \epsilon_{ns'n'} r_{s'} J_{n'}(t, \mathbf{r}) \\ &= \tfrac{1}{2} \int d^3r \left(\delta_{ss'}\delta_{nn'} - \delta_{sn'}\delta_{ns'}\right) r_{s'} J_{n'}(t, \mathbf{r}) \\ &= \tfrac{1}{2} \int d^3r \left[r_s J_n(t, \mathbf{r}) - r_n J_s(t, \mathbf{r})\right] \end{aligned} \tag{119}$$

where the the definition of the cross product (18) was used in the second equality, and the identity (22) was employed for the third equality. Thus, from (112), (113), (116), and (119) we finally arrive to

$$\mu_{ln}(t) = \frac{1}{2} \frac{\partial}{\partial t} q_{ln}(t) + \epsilon_{lns} m_s(t). \tag{120}$$

B. References

[1] V. M. Agranovich and V. L. Ginzburg, *Spatial Dispersion in Crystal Optics and the Theory of Excitons*, Interscience Pulishers (John Wiley & Sons), (translated from Russian), 1966.

[2] V. L. Ginzburg, *The Propagation of Electromagnetic Waves in Plasmas*, Pergamon Press, 1970.

[3] L. D. Landau and E. M. Lifshitz, *Electrodynamics of Continuous Media*, Pergamon Press, 1960.

[4] Said M. Mikki and Ahmed A. Ksihk, "Electromagnetic wave propagation in dispersive negative group velocity media," *Proceedings of the IEEE MTT-S International Symposium*, June 15–20, 2008.

[5] S. M. Mikki and A. A. Kishk, "Electromagnetic wave propagation in nonlocal media: negative group velocity and beyond," *Progress In Electromagnetics Research B*, Vol. 14, 149–174, 2009.

[6] S. M. Mikki and A. A. Kishk, "Derivation of the carbon nanotube susceptibility tensor using lattice dynamics formalism ," *Progress In Electromagnetics Research B*, Vol. 9, 1–26, 2008

5

Numerical Modeling of Photonic Crystal Circuits Using Fourier Series Expansion Method Based on Floquet-Modes

Koki Watanabe[1] and Kiyotoshi Yasumoto[2]
[1]*Fukuoka Institute of Technology*
[2]*Professor Emeritus of Kyushu University*
Japan

1. Introduction

Microwave waveguides are usually composed of metals and dielectrics. The high contrast of their electromagnetic parameters benefits that the electromagnetic energy is strongly confined near the waveguiding structures and propagates along them. However, metal waveguides in millimeter and sub-millimeter wave regions are often suffered from large attenuations because the conductor loss of metal rapidly increases with the operating frequency. This is a reason why optical fiber, which is a waveguide for much higher frequency ranges, is composed of dielectrics only and avoids the use of metals. Then, the dielectric waveguides are widely investigated in millimeter and sub-millimeter wave regions to get rid of the conductor loss. They are typically composed of the core and the surrounding cladding. The refractive index of the core is some larger than that of the cladding. The fields propagating along a dielectric waveguide are not totally confined as with a metallic waveguide, and they are composed of the guided modes and the radiation modes. These modes are coupled to each other when the waveguide structure is not uniform along the wave propagation. Then the bending loss tends to be large in the dielectric waveguides. This is an obstacle to realize integrated circuits. The photonic crystal waveguide attracts attention as a waveguiding structure that resolves this problem.

The photonic crystals are periodic dielectric structures that are designed to reject the propagation of electromagnetic waves at certain wavelength range. Local collapses of the periodicity supply significant advantages for field confinement, wave guiding, and directing radiation. Especially, defects introduced into the photonic crystals compose electromagnetic wave devices such as cavities, waveguides, splitter, coupler, etc. and they constitute photonic crystal circuits. The photonic crystals are composed of dielectrics only and the conductor loss is negligible in many cases. Also, the electromagnetic fields are strongly confined around the defects because any energy cannot escape through the surrounding photonic crystal. These features provide a significant progress towards reducing the size of electromagnetic wave circuits. The electromagnetic wave propagation along the photonic crystal circuits has been simulated using various numerical methods such as the beam propagation method(Koshiba et al., 2000), the finite difference time domain (FDTD) method(Taflove, 1995), and the plane wave expansion method(Benisty, 1996). These methods require adequate treatments of terminating conditions for the waves at the output ends of the circuits. However, the structure of photonic crystal waveguide is nonuniform along the wave propagation, and the Floquet-mode analysis is necessary to decompose the fields in input/output waveguides into the forward and the backward propagating components. The Floquet-modes are the eigenmodes propagating in a structure periodic along the wave propagation.

The guided Floquet-modes propagating in straight photonic crystal waveguides are sometimes analyzed by FDTD method with the help of the super-cell method(Sakoda et al., 1997) or Prony's method(Naka & Ikuno, 2002). However, since the computation errors are comparatively easy to accumulate in periodic structures, FDTD method seems to require special techniques in accurate calculations. The structure is fully periodic in the propagation direction, and several papers(Jia & Yasumoto, 2006; Tanaka et al., 1994; Yasumoto et al., 2004) introduce therefore the generalized Fourier series to expand the electromagnetic fields. Maxwell's equations and the constitutive relations yield a coupled ordinary differential-equation set in terms of the generalized Fourier coefficients. Then, the dispersion equation is derived based on the scattering-matrix (S-matrix) propagation algorithm for multilayer structures(Li, 1996a). The derived dispersion equation is written by a complex function with a complex argument and the zeros correspond to the propagation constants of the eigenmodes. These approaches make us possible to obtain the guided Floquet-modes in very high accuracy, but they are not applicable to obtain the evanescent ones.

Consideration of the evanescent modes is possible by the Fourier series expansion method (FSEM), which was originally developed to analyze the discontinuities in dielectric waveguides(Hosono et al., 1982; Yamakita et al., 1993; Yasumoto et al., 1999). This method introduces an artificial periodicity in the transverse direction and expresses the electromagnetic fields in the Fourier series expansion. The waveguiding structure is sliced into segments uniform in the propagation direction, and Maxwell's equations and the constitutive relations in each segment yield a coupled ordinary differential-equation set for the Fourier coefficients of field components. Since the coefficients of the coupled differential-equation set are constant, the general solution for each segment can be obtained by an eigenvalue/eigenvector calculation. The field coefficients are matched at the boundary between the segments, and we can obtain the propagation characteristics for composite structure. Miyamoto et al.(Miyamoto et al., 2003) calculated the Floquet-modes propagating in grating waveguides using FSEM. The Floquet-modes are obtained by the eigenvalue analysis of the transfer matrix for one periodicity cell in the propagation direction. Their formulation was applied to the Floquet-mode analyses of photonic crystal waveguides(Li & Ho, 2003; Watanabe & Yasumoto, 2009; Yasumoto & Toyama, 2001; Yasumoto & Watanabe, 2008a).

To show the basic idea of FSEM based on Floquet-mode concept, this chapter considers the two-dimensional electromagnetic fields propagating in electromagnetic wave circuits formed by defects in a photonic crystal, which consists of periodic array of rectangular cylinders. All dielectrics under consideration are linear, isotropic, lossless media, and the permeability of free space is assumed. The structure and fields are uniform in the y-direction, and two fundamental polarization are expressed by the transverse electric (TE) and the transverse magnetic (TM) polarizations, in which the electric and the magnetic fields are respectively perpendicular to the y-axis. We consider only time-harmonic fields assuming a time-dependence in $e^{-i\omega t}$, and the fields are represented by complex vectors depending only on the space variables x and z.

2. Floquet-Mode Analysis of Straight Photonic Crystal Waveguides

In this section, we consider the fields propagating in a straight photonic crystal waveguides schematically shown in Fig. 1(a). The photonic crystal consists of rectangular cylinders located parallel in rectangular lattice characterized by the periods d_x and d_z in the x- and z-directions, respectively. Each cylinder has common dimensions a_x and a_z along the x- and z-directions, and infinitely long in the y-direction. The permittivity of rectangular cylinders is denoted by ε_c and that of the surrounding media is denoted by ε_s. Choosing the appropriate parameters, the cylinder array forms the photonic crystal, which rejects the wave propagation. The waveguide is formed by straight line defects in the photonic crystal, and we assume that the fields are well confined in the x-direction.

Let $\varepsilon(x,z)$ be the permittivity distribution. Then Maxwell's curl equations and the constitutive relations yield the following relations:

$$\frac{\partial}{\partial z} H_x(x,z) - \frac{\partial}{\partial x} H_z(x,z) = -i\omega\varepsilon(x,z)\, E_y(x,z) \tag{1}$$

$$\frac{\partial}{\partial z} E_y(x,z) = -i\omega\mu_0 H_x(x,z) \tag{2}$$

$$\frac{\partial}{\partial x} E_y(x,z) = i\omega\mu_0 H_z(x,z) \tag{3}$$

$$\frac{\partial}{\partial z} E_x(x,z) - \frac{\partial}{\partial x} E_z(x,z) = i\omega\mu_0 H_y(x,z) \tag{4}$$

$$\frac{\partial}{\partial z} H_y(x,z) = i\omega\varepsilon(x,z)\, E_x(x,z) \tag{5}$$

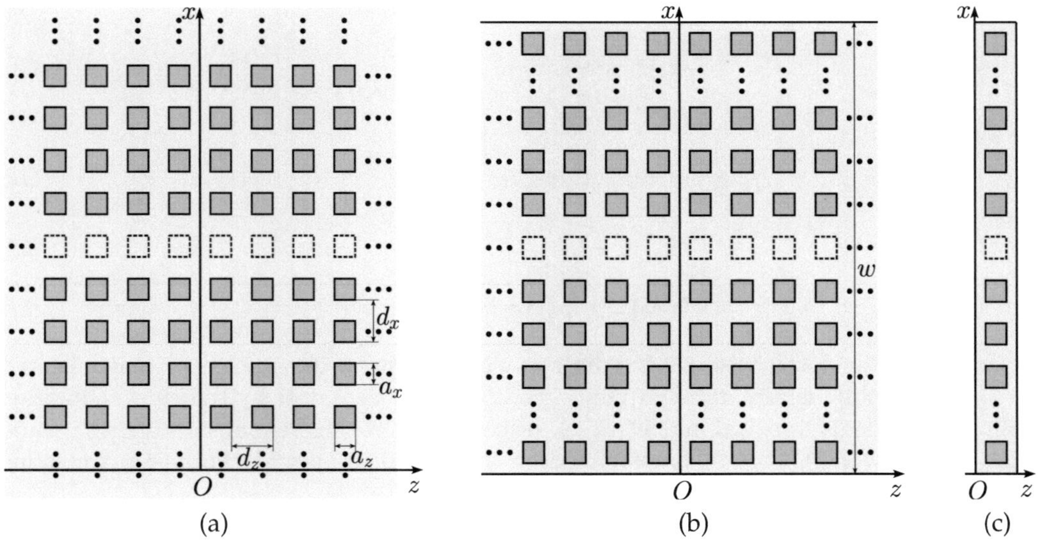

Fig. 1. Two-dimensional photonic crystal waveguide formed by rectangular cylinders, (a) original structure, (b) introduction of artificial boundaries with periodicity condition, and (c) periodicity cell for analysis.

$$\frac{\partial}{\partial x} H_y(x,z) = -i\omega\varepsilon(x,z)\, E_z(x,z). \tag{6}$$

Equations (1)–(3) correspond to the TE-polarization and Eqs. (4)–(6) correspond to the TM-polarization.

We introduce artificial boundaries at $x = 0$ and $x = w$, which are supposed to be sufficiently far from the line defect and not to make an unnecessary space near the boundaries (see Fig. 1(b)). The original electromagnetic fields in $0 < x < w$ are then approximated by periodic functions with the period w and expressed in the Fourier series expansions. For example, the y-component of electric field is approximately expressed as

$$E_y(x,z) = \sum_{n=-N}^{N} E_{y,n}(z)\, e^{i n k_w x} \tag{7}$$

where N denotes the truncation order and $k_w = 2\pi/w$. The Fourier coefficients $\{E_{y,n}(z)\}_{n=-N}^{N}$ are functions of z, and the field profile can be derived by calculating the z-dependence of the coefficients. To treat the coefficients systematically, we introduce $(2N+1) \times 1$ column matrices; for example, the coefficients of $E_y(x,z)$ are expressed by a column matrix $\boldsymbol{e}_y(z)$ that is defined by

$$\boldsymbol{e}_y(z) = \begin{pmatrix} E_{y,-N}(z) & \cdots & E_{y,N}(z) \end{pmatrix}^t \tag{8}$$

where the superscript t denotes the transpose matrix. Then, replacing all the field components by the Fourier series expansions and using the orthogonal property of the Fourier basis, Eqs. (1)–(6) yield the following relations:

$$\frac{d}{dz}\boldsymbol{h}_x(z) - i\boldsymbol{X}\boldsymbol{h}_z(z) = -i\omega\, [\![\varepsilon]\!]\, \boldsymbol{e}_y(z) \tag{9}$$

$$\frac{d}{dz}\boldsymbol{e}_y(z) = -i\omega\mu_0 \boldsymbol{h}_x(z) \tag{10}$$

$$\boldsymbol{X}\boldsymbol{e}_y(z) = \omega\mu_0 \boldsymbol{h}_z(z) \tag{11}$$

$$\frac{d}{dz}\boldsymbol{e}_x(z) - i\boldsymbol{X}\boldsymbol{e}_z(z) = i\omega\mu_0 \boldsymbol{h}_y(z) \tag{12}$$

$$\frac{d}{dz}\boldsymbol{h}_y(z) = i\omega \left[\!\!\left[\frac{1}{\varepsilon}\right]\!\!\right]^{-1} \boldsymbol{e}_x(z) \tag{13}$$

$$\boldsymbol{X}\boldsymbol{h}_y(z) = -\omega\, [\![\varepsilon]\!]\, \boldsymbol{e}_z(z) \tag{14}$$

where $\boldsymbol{X}$ denotes a diagonal matrix whose diagonal elements are $\{n k_w\}_{n=-N}^{N}$, and $[\![\varepsilon]\!]$ and $[\![1/\varepsilon]\!]$ are square Toeplitz matrices whose (n,m)-entries are given by the $(n-m)$th-order Fourier coefficients of $\varepsilon(x,z)$ and $1/\varepsilon(x,z)$, respectively. The expressions on the right-hand sides in Eqs. (9), (13), and (14) are obtained by taking into account Li's Fourier factorization rules(Li, 1996b).

The entries of $[\![\varepsilon]\!]$ and $[\![1/\varepsilon]\!]$ are functions of z. Here, we slice the analysis region into segments uniform in the z-direction, and denote the regions $l\,d_z + (d_z - a_z)/2 < z < l\,d_z + (d_z + a_z)/2$ and $l\,d_z - (d_z - a_z)/2 < z < l\,d_z + (d_z - a_z)/2$ for any integer l as segments g and s, respectively. The permittivity distribution in the segment g is also denoted by $\varepsilon_g(x)$ though that in the segment s is a constant value ε_s. The Toeplitz matrices $[\![\varepsilon]\!]$ and $[\![1/\varepsilon]\!]$ are constant matrices $[\![\varepsilon_g]\!]$ and $[\![1/\varepsilon_g]\!]$ in the segment g, and constant diagonal matrices $\varepsilon_s \boldsymbol{I}$ and $(1/\varepsilon_s)\boldsymbol{I}$ in the segment s, where $\boldsymbol{I}$ denotes the identity matrix.

Then the general solutions to the coupled differential-equation set (9)–(14) in the segments $r = g, s$ are obtained by the eigenvalue/eigenvector calculations in the following forms:

$$\begin{pmatrix} \boldsymbol{e}_y(z) \\ \boldsymbol{h}_x(z) \end{pmatrix} = \boldsymbol{Q}_r^{(e)} \begin{pmatrix} \boldsymbol{a}_r^{(e,+)}(z) \\ \boldsymbol{a}_r^{(e,-)}(z) \end{pmatrix} \tag{15}$$

$$\begin{pmatrix} \boldsymbol{h}_y(z) \\ \boldsymbol{e}_x(z) \end{pmatrix} = \boldsymbol{Q}_r^{(h)} \begin{pmatrix} \boldsymbol{a}_r^{(h,+)}(z) \\ \boldsymbol{a}_r^{(h,-)}(z) \end{pmatrix} \tag{16}$$

with

$$\boldsymbol{Q}_g^{(e)} = \begin{pmatrix} \boldsymbol{P}_g^{(e)} & \boldsymbol{P}_g^{(e)} \\ -\frac{1}{\omega \mu_0} \boldsymbol{P}_g^{(e)} \boldsymbol{Z}_g^{(e)} & \frac{1}{\omega \mu_0} \boldsymbol{P}_g^{(e)} \boldsymbol{Z}_g^{(e)} \end{pmatrix} \tag{17}$$

$$\boldsymbol{Q}_g^{(h)} = \begin{pmatrix} \boldsymbol{P}_g^{(h)} & \boldsymbol{P}_g^{(h)} \\ \frac{1}{\omega} \left[\!\left[\frac{1}{\varepsilon_g} \right]\!\right] \boldsymbol{P}_g^{(h)} \boldsymbol{Z}_g^{(h)} & -\frac{1}{\omega} \left[\!\left[\frac{1}{\varepsilon_g} \right]\!\right] \boldsymbol{P}_g^{(h)} \boldsymbol{Z}_g^{(h)} \end{pmatrix} \tag{18}$$

$$\boldsymbol{Q}_s^{(e)} = \begin{pmatrix} \boldsymbol{I} & \boldsymbol{I} \\ -\frac{1}{\omega \mu_0} \boldsymbol{Z}_s^{(e)} & \frac{1}{\omega \mu_0} \boldsymbol{Z}_s^{(e)} \end{pmatrix} \tag{19}$$

$$\boldsymbol{Q}_s^{(h)} = \begin{pmatrix} \boldsymbol{I} & \boldsymbol{I} \\ \frac{1}{\omega \varepsilon_s} \boldsymbol{Z}_s^{(h)} & -\frac{1}{\omega \varepsilon_s} \boldsymbol{Z}_s^{(h)} \end{pmatrix} \tag{20}$$

$$\boldsymbol{P}_g^{(p)} = \begin{pmatrix} \boldsymbol{p}_{g,1}^{(p)} & \cdots & \boldsymbol{p}_{g,2N+1}^{(p)} \end{pmatrix} \tag{21}$$

$$\left(\boldsymbol{Z}_r^{(p)} \right)_{n,m} = \delta_{n,m} \, \gamma_{r,n}^{(p)} \tag{22}$$

for $p = e, h$, where ${\gamma_{g,n}^{(p)}}^2$ and $\boldsymbol{p}_{g,n}^{(p)}$ denote respectively the nth-eigenvalues and the associated eigenvectors of the matrices $\boldsymbol{C}_g^{(e)} = \omega^2 \mu_0 [\![\varepsilon_g]\!] - \boldsymbol{X}^2$ for the TE-polarization and $\boldsymbol{C}_g^{(h)} = [\![1/\varepsilon_g]\!]^{-1} (\omega^2 \mu_0 \boldsymbol{I} - \boldsymbol{X} [\![\varepsilon_g]\!]^{-1} \boldsymbol{X})$ for the TM-polarization, and $\gamma_{s,n}^{(p)} = \sqrt{\omega^2 \varepsilon_s \mu_0 - (\boldsymbol{X})_{n,n}^{\;\;2}}$. The column matrices $\boldsymbol{a}_r^{(p,+)}(z)$ and $\boldsymbol{a}_r^{(p,-)}(z)$ give the amplitudes of the local normal modes of the segment $r = g, s$ propagating in the $+z$- and $-z$-directions, respectively, and the relation between the modal amplitudes at $z = z'$ and $z = z''$ is given as

$$\begin{pmatrix} \boldsymbol{a}_r^{(p,+)}(z') \\ \boldsymbol{a}_r^{(p,-)}(z') \end{pmatrix} = \boldsymbol{U}_r^{(p)}(z' - z'') \begin{pmatrix} \boldsymbol{a}_r^{(p,+)}(z'') \\ \boldsymbol{a}_r^{(p,-)}(z'') \end{pmatrix} \tag{23}$$

with

$$\boldsymbol{U}_r^{(p)}(z) = \begin{pmatrix} \boldsymbol{V}_r^{(p)}(z) & \boldsymbol{0} \\ \boldsymbol{0} & \boldsymbol{V}_r^{(p)}(-z) \end{pmatrix} \tag{24}$$

$$\left(\boldsymbol{V}_r^{(p)}(z) \right)_{n,m} = \delta_{n,m} \, e^{i \gamma_{r,n}^{(p)} z}. \tag{25}$$

The structure under consideration is periodic in the z-direction, and the Floquet theorem asserts that analysis region can be reduced in one periodicity cell to characterize the propagation property. Here, the periodicity cell is taken to be the region $0 < z < d_z$ (see Fig. 1(c)). The fields in segments g and s should be matched at the boundaries $z = (d_z \pm a_z)/2$ by the boundary conditions, which are given by continuing the coefficient column matrices for the tangential field components. Using Eqs. (15), (16), and (23), the relation between the modal amplitudes $\boldsymbol{a}_s^{(p,\pm)}(0)$ and $\boldsymbol{a}_s^{(p,\pm)}(d_z)$ are derived as

$$\begin{pmatrix} \boldsymbol{a}_s^{(p,+)}(d_z) \\ \boldsymbol{a}_s^{(p,-)}(d_z) \end{pmatrix} = \boldsymbol{F}^{(p)} \begin{pmatrix} \boldsymbol{a}_s^{(p,+)}(0) \\ \boldsymbol{a}_s^{(p,-)}(0) \end{pmatrix} \tag{26}$$

where the transfer matrix for the periodicity cell $\boldsymbol{F}^{(p)}$ is given by

$$\boldsymbol{F}^{(p)} = \boldsymbol{U}_s^{(p)}\left(\frac{d_z - a_z}{2}\right) \boldsymbol{Q}_s^{(p)^{-1}} \boldsymbol{Q}_g^{(p)} \boldsymbol{U}_g^{(p)}(a_z) \boldsymbol{Q}_g^{(p)^{-1}} \boldsymbol{Q}_s^{(p)} \boldsymbol{U}_s^{(p)}\left(\frac{d_z - a_z}{2}\right). \tag{27}$$

The Floquet-modes propagating in the photonic crystal waveguides are obtained by the eigenvalue analysis of the transfer matrix $\boldsymbol{F}^{(p)}$. Let $\beta_n^{(p)}$ and $\boldsymbol{r}_n^{(p)}$ be the nth-eigenvalues and the associated eigenvectors of $\boldsymbol{F}^{(p)}$, respectively. We define a column matrix $\boldsymbol{b}^{(p)}(z)$ by

$$\boldsymbol{b}^{(p)}(z) = \boldsymbol{R}^{(p)^{-1}} \begin{pmatrix} \boldsymbol{a}_s^{(p,+)}(z) \\ \boldsymbol{a}_s^{(p,-)}(z) \end{pmatrix} \tag{28}$$

with

$$\boldsymbol{R}^{(p)} = \begin{pmatrix} \boldsymbol{r}_1^{(p)} & \cdots & \boldsymbol{r}_{4N+2}^{(p)} \end{pmatrix} \tag{29}$$

and denote the nth-component of $\boldsymbol{b}^{(p)}(z)$ by $b_n^{(p)}(z)$. Then, from Eqs. (26) and (28), we obtain a relation:

$$b_n^{(p)}(d_z) = \beta_n^{(p)} b_n^{(p)}(0). \tag{30}$$

This implies that $\{b_n(0)\}_{n=1}^{4N+2}$ gives the amplitudes of the Floquet-modes propagating in the photonic crystal waveguide at $z = 0$, and the propagation constants are calculated by

$$\eta_n^{(p)} = -i\,\frac{\mathrm{Ln}(\beta_n^{(p)})}{d_z} \tag{31}$$

where Ln denotes the principal natural logarithm function. Also, considering Eqs. (15), (16), and (28), the Fourier coefficients of the modal profile functions corresponding to the nth-order Floquet-modes at $z = 0$ are given by

$$\begin{pmatrix} \boldsymbol{e}_y(0) \\ \boldsymbol{h}_x(0) \end{pmatrix} = \boldsymbol{Q}_s^{(e)} \boldsymbol{r}_n^{(e)} \tag{32}$$

$$\begin{pmatrix} \boldsymbol{h}_y(0) \\ \boldsymbol{e}_x(0) \end{pmatrix} = \boldsymbol{Q}_s^{(h)} \boldsymbol{r}_n^{(h)} \tag{33}$$

where Eq. (32) is for the TE-polarization and Eq. (33) is for the TM-polarization.
The propagation direction of each Floquet-mode can be judged as follows:

- if $|\beta_n^{(p)}| < 1$, the corresponding mode is the evanescent one propagating in the $+z$-direction.
- if $|\beta_n^{(p)}| > 1$, the corresponding mode is the evanescent one propagating in the $-z$-direction.
- if $|\beta_n^{(p)}| = 1$, the corresponding mode is the guided one. When the modal power carried in the z-direction is positive (negative), the corresponding mode propagates in the $+z$ ($-z$)-direction.

The z-component of the Poynting vector is given by

$$\begin{aligned} s_z(x,z) &= E_x(x,z)\, H_y(x,z)^* - E_y(x,z)\, H_x(x,z)^* \\ &= \sum_{n=-N}^{N} \sum_{m=-N}^{N} \left(E_{x,n}(z)\, H_{y,m}(z)^* - E_{y,n}(z)\, H_{x,m}(z)^*\right) e^{i(n-m)k_g x} \end{aligned} \tag{34}$$

and, integrating $s_z(x,z)$ over $0 \le x \le w$, we may obtain the following expression:

$$\int_0^w s_z(x,z)\, dx = w\left(\boldsymbol{e}_x(z) \cdot \boldsymbol{h}_y(z)^* - \boldsymbol{e}_y(z) \cdot \boldsymbol{h}_x(z)^*\right). \tag{35}$$

Therefore, the time-averaging modal power carried in the z-direction is calculated by $-w\Re(\boldsymbol{e}_y(0) \cdot \boldsymbol{h}_x(0)^*)/2$ for the TE-polarization and $w\Re(\boldsymbol{e}_x(0) \cdot \boldsymbol{h}_y(0)^*)/2$ for the TM-polarization, where $\boldsymbol{e}_y(0)$, $\boldsymbol{h}_x(0)$, $\boldsymbol{h}_y(0)$, and $\boldsymbol{e}_x(0)$ are obtained by Eqs. (32) and (33). The order of $\boldsymbol{F}^{(p)}$ $(p = e, h)$ is $4N+2$, and the eigenvalues $\{\beta_n^{(p)}\}_{n=1}^{4N+2}$, the propagation coefficients $\{\eta_n^{(p)}\}_{n=1}^{4N+2}$, and the eigenvectors $\{\boldsymbol{r}_n^{(p)}\}_{n=1}^{4N+2}$ are here supposed to be arranged in such a way that $\{\beta_n^{(p)}\}_{n=1}^{2N+1}$, $\{\eta_n^{(p)}\}_{n=1}^{2N+1}$, and $\{\boldsymbol{r}_n^{(p)}\}_{n=1}^{2N+1}$ correspond to the Floquet-modes propagating in the $+z$-direction and that $\{\beta_n^{(p)}\}_{n=2N+2}^{4N+2}$, $\{\eta_n^{(p)}\}_{n=2N+2}^{4N+2}$, and $\{\boldsymbol{r}_n^{(p)}\}_{n=2N+2}^{4N+2}$ correspond to ones propagating in the $-z$-direction.

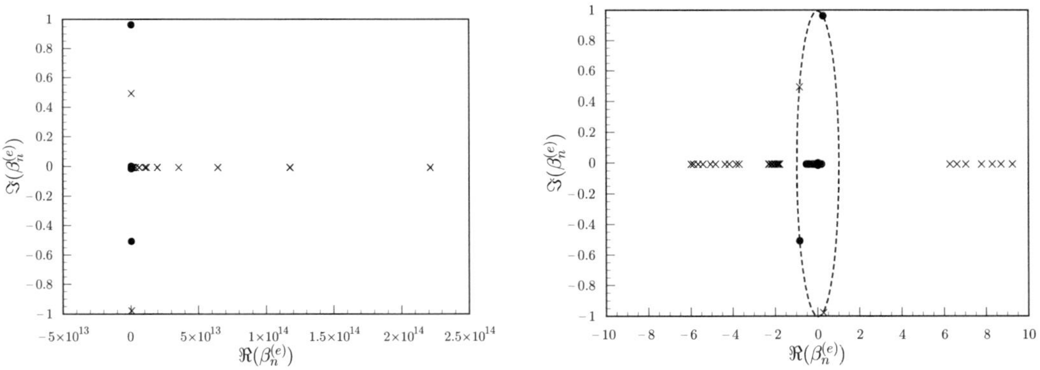

(a) (b)

Fig. 2. Distribution of the eigenvalues $\{\beta_n^{(e)}\}$ in the complex-plane, (a) whole view and (b) close view near the origin. (Reproduced from K. Watanabe and K. Yasumoto, *Progress In Electromagnetics Research*, PIER 92, 209–222, 2009, with courtesy of EMW Publishing.)

The eigenvalues $\{\beta_n^{(e)}\}$ calculated with $N = 60$ for the TE-polarization are plotted on the complex-plane in Fig. 2. The parameters of the photonic crystal are chosen as $\varepsilon_s = \varepsilon_0$, $\varepsilon_c = 12.25\,\varepsilon_0$, $d_x = d_z = 0.67\,\lambda_0$, and $a_x = a_z = \sqrt{0.41}\,d_x$. The rectangular cylinders are situated with the center at $x = (m - 1/2)\,d_x$ for positive integer m though one layer of cylinder array is removed at $x = 5.5\,d_x$ to form the waveguide structure, and $w = 11\,d_x$ is used for the periodic boundary condition. The photonic crystal waveguide with these parameters supports two guided modes. The dots and the crosses denote the eigenvalues correspond to the Floquet-mode propagating in the $+z$- and the $-z$-directions, respectively. Fig. 2(b) is a close view near the origin and the dashed curve denotes a circle with a unit radius. The values on the dashed curve correspond to the guided modes, and those corresponding to the forward and the backward propagating modes are respectively distributed inside and outside the circle. A whole view is given in Fig. 2(a) and shows that the spectral radius of the transfer matrix $\boldsymbol{F}^{(e)}$ (maximum absolute value of $\beta_n^{(e)}$) is about 2.2×10^{14}. This implies that the double precision computation leads to roundoff errors in the order of 10^{-2} and the obtained eigenvalues with $|\beta_n^{(e)}| \lesssim 10^{-2}$ may not be accurate. As a result, the calculation of the Floquet-modes propagating in the $-z$-direction are more accurate than ones propagating in the $+z$-direction, and the situation should be same for the TM-polarization.

Figure 3(a) shows the normalized propagation constants $\{\eta_n^{(e)}/k_d\}$ calculated from the eigenvalues plotted in Fig. 2, where $k_d = 2\pi/d_z$ denotes the inverse lattice constant in the z-direction. Jia and Yasumoto(Jia & Yasumoto, 2006) have calculated the normalized propagation constants of the guided Floquet-modes as 0.415946 for the even-mode and 0.219867 for the odd-mode. The normalized propagation constants of the guided Floquet-modes are plotted on the real axis in Fig. 3(a), and their values are ± 0.417 and ± 0.211. They are thought to be in a sufficiently good agreement with the reference values. If η is a propagation constant of Floquet-mode, $-\eta$ is also the propagation constant because of the structural symmetry. The

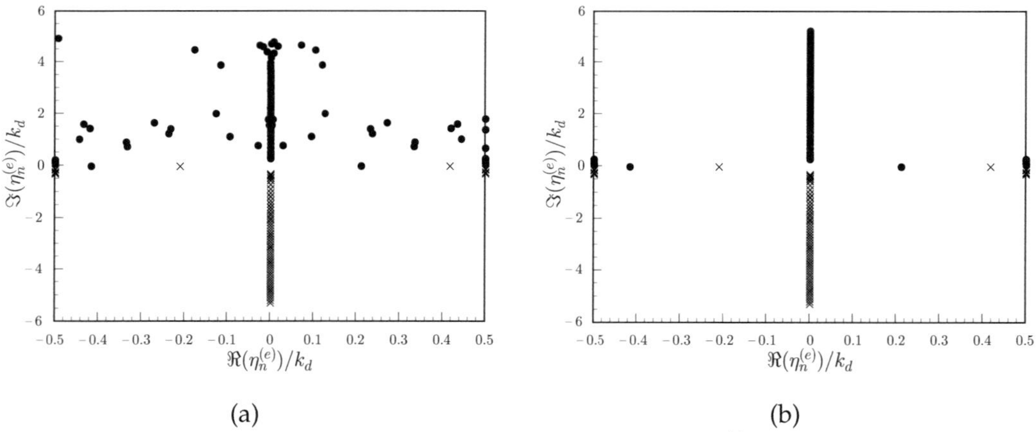

Fig. 3. Distribution of the normalized propagation constants $\{\eta_n^{(e)}/k_d\}$ with $k_d = 2\pi/d_z$ in the complex-plane, (a) direct results and (b) results with the use of the symmetric property. (Reproduced from K. Watanabe and K. Yasumoto, *Progress In Electromagnetics Research*, PIER 92, 209–222, 2009, with courtesy of EMW Publishing.)

distribution of $\{\eta_n^{(e)}\}$ should be therefore point symmetric to the origin. However, it is clearly observed in Fig. 3(a) that the symmetry for the obtained evanescent modes is broken due to the roundoff error, which is unnegligible for the propagation constants with large $\Im(\eta_n^{(e)})$.
Using the symmetric property, the accuracy of the Floquet-modes propagating in the $+z$-direction can be improved. Let $\boldsymbol{R}_1^{(p)}$ and $\boldsymbol{R}_2^{(p)}$ be $(2N+1) \times (2N+1)$ square matrices defined by

$$\begin{pmatrix} \boldsymbol{R}_1^{(p)} \\ \boldsymbol{R}_2^{(p)} \end{pmatrix} = \begin{pmatrix} \boldsymbol{r}_{2N+2}^{(p)} & \cdots & \boldsymbol{r}_{4N+2}^{(p)} \end{pmatrix}. \tag{36}$$

Then, when the error is negligible for the guided modes, more accurate results are obtained by replacing $\{\eta_n^{(p)}\}_{n=1}^{2N+1}$ and $\{\boldsymbol{r}_n^{(p)}\}_{n=1}^{2N+1}$ in the following ways:

$$\eta_n^{(p)} = -\eta_{n+2N+1}^{(p)} \tag{37}$$

$$\begin{pmatrix} \boldsymbol{r}_1^{(p)} & \cdots & \boldsymbol{r}_{2N+1}^{(p)} \end{pmatrix} = \begin{pmatrix} \boldsymbol{R}_2^{(p)} \\ \boldsymbol{R}_1^{(p)} \end{pmatrix}, \tag{38}$$

where n in Eq. (37) is any integer $1 \leq n \leq 2N+1$ and Eq. (38) is presented in Ref. (Miyamoto et al., 2003). Then, if we calculate accurately the Floquet-modes corresponding to the values located in the lower half-plane, the other Floquet-modes are obtained by the symmetry property. The normalized propagation constants of the TE polarized Floquet-modes for the same photonic crystal waveguide as in Fig. 3(a), in which the constants corresponding to the modes propagating in the $+z$-direction are obtained by Eq. (37), are shown in Fig. 3(b). This gives the typical distribution of the constants of the Floquet-modes propagating in the photonic crystal waveguides.
The present approach provides sufficiently accurate results in many applications but it has a limitation for highly accurate computation. The spectral radius of the transfer matrix $\boldsymbol{F}^{(p)}$ becomes larger with the increase of the truncation order N, and the computation with very large N may lead to significant errors for not only the evanescent modes propagating in the $+z$-direction but also the guided modes. In such case, using the symmetric property is no longer effective to improve the accuracy. For example, when the double precision computation is applied to the same waveguide as in Figs. 2 and 3, the guided modes cannot be distinguished from the evanescent ones for $N > 68$. This limitation is some relieved if the eigenvalues are computed as the Rayleigh quotients, but the validity is still limited(Watanabe & Yasumoto, 2009).

3. Floquet-Modal Expansion

Since the propagation constants $\{\eta_n^{(p)}\}_{n=1}^{4N+2}$ and the eigenvectors $\{\boldsymbol{r}_n^{(p)}\}_{n=1}^{4N+2}$ are arranged in such a way that $\{\eta_n^{(p)}\}_{n=1}^{2N+1}$ and $\{\boldsymbol{r}_n^{(p)}\}_{n=1}^{2N+1}$ correspond to the Floquet-modes propagating in the $+z$-direction and that $\{\eta_n^{(p)}\}_{n=2N+2}^{4N+2}$ and $\{\boldsymbol{r}_n^{(p)}\}_{n=2N+2}^{4N+2}$ correspond to ones propagating in the $-z$-direction, the amplitudes of the Floquet-modes $\{b_n^{(p)}(z)\}_{n=1}^{2N+1}$ and $\{b_n^{(p)}(z)\}_{n=2N+2}^{4N+2}$ correspond to the modes propagating in the $+z$- and $-z$-directions, respectively. To express clearly, we use the following notations:

$$b_m^{(p,+)}(z) = b_m^{(p)}(z) \tag{39}$$

$$b_m^{(p,-)}(z) = b_{m+2N+1}^{(p)}(z) \tag{40}$$

for $m = 1,\ldots,2N+1$. Then $b_m^{(p,\pm)}(z)$ gives the amplitude of the mth-order Floquet-mode propagating in the $\pm z$-direction. We rewrite the Fourier coefficients of the modal profile functions corresponding to the mth-order Floquet-modes propagating in the $+z$-direction as follows:

$$\begin{pmatrix} \boldsymbol{e}_{y,m} \\ \boldsymbol{h}_{x,m} \end{pmatrix} = \boldsymbol{Q}_s^{(e)} \boldsymbol{r}_m^{(e)} \tag{41}$$

$$\begin{pmatrix} \boldsymbol{h}_{y,m} \\ \boldsymbol{e}_{x,m} \end{pmatrix} = \boldsymbol{Q}_s^{(h)} \boldsymbol{r}_m^{(h)} \tag{42}$$

for $m = 1,\ldots,2N+1$. Considering Eqs. (19), (20), and (38), the coefficients of the modal profile functions corresponding to the mth-order Floquet-modes propagating in the $-z$-direction are given by

$$\begin{pmatrix} \boldsymbol{e}_{y,m} \\ -\boldsymbol{h}_{x,m} \end{pmatrix} = \boldsymbol{Q}_s^{(e)} \boldsymbol{r}_{m+2N+1}^{(e)} \tag{43}$$

$$\begin{pmatrix} \boldsymbol{h}_{y,m} \\ -\boldsymbol{e}_{x,m} \end{pmatrix} = \boldsymbol{Q}_s^{(h)} \boldsymbol{r}_{m+2N+1}^{(h)}. \tag{44}$$

The coefficients at $z = l\,d_z$ for any integer l are expressed as follows:

$$\begin{pmatrix} \boldsymbol{e}_y(l\,d_z) \\ \boldsymbol{h}_x(l\,d_z) \end{pmatrix} = \sum_{m=1}^{2N+1} b_m^{(e,+)}(l\,d_z) \begin{pmatrix} \boldsymbol{e}_{y,m} \\ \boldsymbol{h}_{x,m} \end{pmatrix} + \sum_{m=1}^{2N+1} b_m^{(e,-)}(l\,d_z) \begin{pmatrix} \boldsymbol{e}_{y,m} \\ -\boldsymbol{h}_{x,m} \end{pmatrix} \tag{45}$$

$$\begin{pmatrix} \boldsymbol{h}_y(l\,d_z) \\ \boldsymbol{e}_x(l\,d_z) \end{pmatrix} = \sum_{m=1}^{2N+1} b_m^{(h,+)}(l\,d_z) \begin{pmatrix} \boldsymbol{h}_{y,m} \\ \boldsymbol{e}_{x,m} \end{pmatrix} + \sum_{m=1}^{2N+1} b_m^{(h,-)}(l\,d_z) \begin{pmatrix} \boldsymbol{h}_{y,m} \\ -\boldsymbol{e}_{x,m} \end{pmatrix} \tag{46}$$

from Eqs. (15), (16), and (28). This gives the Floquet-mode expansion representation for the fields at $z = l\,d_z$ in the Fourier space. From Eqs. (30), (31), and (37), the amplitudes $\{b_m^{(p,\pm)}(z)\}_{m=1}^{2N+1}$ have the following dependence:

$$b_m^{(p,\pm)}(l'\,d_z) = b_m^{(p,\pm)}(l\,d_z)\,e^{\pm i(l'-l)\eta_m^{(p)}\,d_z} \tag{47}$$

for any integers l and l'.
Here, two field distributions $(\boldsymbol{E}_1, \boldsymbol{H}_1)$ and $(\boldsymbol{E}_2, \boldsymbol{H}_2)$ are known to satisfy Lorentz's reciprocal theorem:

$$\nabla \cdot (\boldsymbol{E}_1(x,z) \times \boldsymbol{H}_2(x,z) - \boldsymbol{E}_2(x,z) \times \boldsymbol{H}_1(x,z)) = 0. \tag{48}$$

Integrating over the layer between $z = l\,d_z$ and $z = l'\,d_z$ for any integers l, l' and using the Gauss theorem, we may obtain the following relation:

$$\int_0^w (\boldsymbol{E}_1(x,l\,d_z) \times \boldsymbol{H}_2(x,l\,d_z) - \boldsymbol{E}_2(x,l\,d_z) \times \boldsymbol{H}_1(x,l\,d_z)) \cdot \hat{\boldsymbol{z}}\,dx$$
$$- \int_0^w (\boldsymbol{E}_1(x,l'\,d_z) \times \boldsymbol{H}_2(x,l'\,d_z) - \boldsymbol{E}_2(x,l'\,d_z) \times \boldsymbol{H}_1(x,l'\,d_z)) \cdot \hat{\boldsymbol{z}}\,dx = 0 \tag{49}$$

where the fields are assumed to satisfy the periodic boundary condition with period w and $\hat{\boldsymbol{z}}$ denotes the unit vector along the positive z-axis. The Cartesian components of the fields are expressed in the Fourier series expansions. Then, Eq. (49) yields

$$\begin{aligned}&\langle \boldsymbol{e}_{1,y}(l\,d_z)|\boldsymbol{h}_{2,x}(l\,d_z)\rangle - \langle \boldsymbol{e}_{2,y}(l\,d_z)|\boldsymbol{h}_{1,x}(l\,d_z)\rangle \\ &\quad - \langle \boldsymbol{e}_{1,y}(l'\,d_z)|\boldsymbol{h}_{2,x}(l'\,d_z)\rangle + \langle \boldsymbol{e}_{2,y}(l'\,d_z)|\boldsymbol{h}_{1,x}(l'\,d_z)\rangle = 0\end{aligned} \tag{50}$$

for the TE-polarization, and

$$\begin{aligned}&\langle \boldsymbol{e}_{1,x}(l\,d_z)|\boldsymbol{h}_{2,y}(l\,d_z)\rangle - \langle \boldsymbol{e}_{2,x}(l\,d_z)|\boldsymbol{h}_{1,y}(l\,d_z)\rangle \\ &\quad - \langle \boldsymbol{e}_{1,x}(l'\,d_z)|\boldsymbol{h}_{2,y}(l'\,d_z)\rangle + \langle \boldsymbol{e}_{2,x}(l'\,d_z)|\boldsymbol{h}_{1,y}(l'\,d_z)\rangle = 0\end{aligned} \tag{51}$$

for the TM-polarization. For example, $\boldsymbol{e}_{1,y}(z)$ denotes the column matrix of the Fourier coefficients corresponding to the y-component of $\boldsymbol{E}_1(x,z)$. Also, the angle brackets denote an inner product defined by

$$\langle \boldsymbol{a}|\boldsymbol{b}\rangle \equiv \sum_{n=1}^{2N+1} (\boldsymbol{a})_n\,(\boldsymbol{b})_{-n+2N+2} \tag{52}$$

where $\boldsymbol{a}$ and $\boldsymbol{b}$ are vectors of length $2N+1$. Equations (50) and (51) represent Lorentz's reciprocal theorem in the Fourier space.

We choose the mth-order Floquet-mode propagating in the $+z$-direction as the first fields $(\boldsymbol{E}_1, \boldsymbol{H}_1)$ and the m'th-order Floquet-mode propagating in the $\pm z$-direction as the second fields $(\boldsymbol{E}_2, \boldsymbol{H}_2)$. The propagation constants $\eta_m^{(h)}$ and $\eta_{m'}^{(h)}$ are assumed to be equal only when $m = m'$. Then, for the TE-polarization, we use the following column matrices:

$$\begin{pmatrix} \boldsymbol{e}_{1,y}(l\,d_z) \\ \boldsymbol{h}_{1,x}(l\,d_z) \end{pmatrix} = b_m^{(e,+)}(l\,d_z)\begin{pmatrix} \boldsymbol{e}_{y,m} \\ \boldsymbol{h}_{x,m} \end{pmatrix} \tag{53}$$

$$\begin{pmatrix} \boldsymbol{e}_{2,y}(l\,d_z) \\ \boldsymbol{h}_{2,x}(l\,d_z) \end{pmatrix} = b_{m'}^{(e,\pm)}(l\,d_z)\begin{pmatrix} \boldsymbol{e}_{y,m'} \\ \pm\boldsymbol{h}_{x,m'} \end{pmatrix} \tag{54}$$

for any integer l. Substituting into Eq. (50) and using Eq. (47), we obtain

$$\left[1 - e^{i(l'-l)\left(\eta_m^{(e)} \pm \eta_{m'}^{(e)}\right)d_z}\right]\Big(\langle \boldsymbol{e}_{y,m}|\boldsymbol{h}_{x,m'}\rangle \mp \langle \boldsymbol{e}_{y,m'}|\boldsymbol{h}_{x,m}\rangle\Big) = 0. \tag{55}$$

This relation should hold for any integers l and l'. Since $\eta_m^{(h)} + \eta_{m'}^{(h)} \neq n\,k_d$ for any integer n, we have

$$\langle \boldsymbol{e}_{y,m}|\boldsymbol{h}_{x,m'}\rangle = \langle \boldsymbol{e}_{y,m'}|\boldsymbol{h}_{x,m}\rangle. \tag{56}$$

Also, since $\eta_m^{(h)} - \eta_{m'}^{(h)} = n\,k_d$ for any integer n only when $m = m'$, we derive the following relation:

$$\langle \boldsymbol{e}_{y,m}|\boldsymbol{h}_{x,m'}\rangle = \delta_{m,m'}\,\langle \boldsymbol{e}_{y,m}|\boldsymbol{h}_{x,m}\rangle. \tag{57}$$

Following the similar process, the relation for the TM-polarization is derived as

$$\langle \boldsymbol{e}_{x,m}|\boldsymbol{h}_{y,m'}\rangle = \delta_{m,m'}\,\langle \boldsymbol{e}_{x,m}|\boldsymbol{h}_{y,m}\rangle. \tag{58}$$

Equations (57) and (58) represent the orthogonal relations of the Floquet-modes.
As mentioned before, the coefficient matrices of the fields at $z = l\,d_z$ for any integer l are expressed by Eqs. (45) and (46). Calculating $\langle \boldsymbol{e}_y(l\,d_z)|\boldsymbol{h}_{x,m'}\rangle \pm \langle \boldsymbol{e}_{y,m'}|\boldsymbol{h}_x(l\,d_z)\rangle$ and $\langle \boldsymbol{e}_x(l\,d_z)|\boldsymbol{h}_{y,m'}\rangle \pm \langle \boldsymbol{e}_{x,m'}|\boldsymbol{h}_y(l\,d_z)\rangle$ for the TE- and the TM-polarizations, respectively, the orthogonal relations (57) and (58) yield the amplitudes of the Floquet-modes as

$$b_m^{(e,\pm)}(l\,d_z) = \frac{\langle \boldsymbol{e}_y(l\,d_z)|\boldsymbol{h}_{x,m}\rangle \pm \langle \boldsymbol{e}_{y,m}|\boldsymbol{h}_x(l\,d_z)\rangle}{2\,\langle \boldsymbol{e}_{y,m}|\boldsymbol{h}_{x,m}\rangle} \tag{59}$$

for the TE-polarization, and

$$b_m^{(h,\pm)}(l\,d_z) = \frac{\pm\langle \boldsymbol{e}_x(l\,d_z)|\boldsymbol{h}_{y,m}\rangle + \langle \boldsymbol{e}_{x,m}|\boldsymbol{h}_y(l\,d_z)\rangle}{2\,\langle \boldsymbol{e}_{x,m}|\boldsymbol{h}_{y,m}\rangle} \tag{60}$$

for the TM-polarization.

4. Discontinuities in Photonic Crystal Waveguides

Next, we consider nonuniform structures formed by defects in a photonic crystal as shown in Fig. 4. The fields are assumed to be well confined in $0 < x < w$ and approximated by periodic functions of x with the period w as same with in the previous sections. The structure is decomposed into three sections. The regions $z < 0$ and $z > h$ ($h = L\,d_z$) are input/output waveguide sections consisting of the straight line defects, while the region $0 < z < h$ is a transition section where the periodicity breaks. The structure is considered as a series of step-transitions of the photonic crystal waveguides and we apply the S-matrix propagation algorithm(Li, 1996a).
For $p = e,h$, the matrices $\boldsymbol{R}^{(p)}$ given by Eq. (29) and the Floquet-modal amplitudes $\{b_m^{(p,\pm)}(z)\}_{m=1}^{2N+1}$ given by Eqs. (39), (40) for the regions $(l-1)\,d_z < z < l\,d_z$ ($l = 0,\ldots,L+1$) are denoted by $\boldsymbol{R}^{(p,l)}$ and $\{b_m^{(p,l,\pm)}(z)\}_{m=1}^{2N+1}$, respectively. Also, we use the column matrix $\boldsymbol{b}^{(p,l,\pm)}(z)$ whose mth-component is $b_m^{(p,l,\pm)}(z)$. Considering $\boldsymbol{b}^{(p,l,\pm)}(z)$ gives the amplitudes

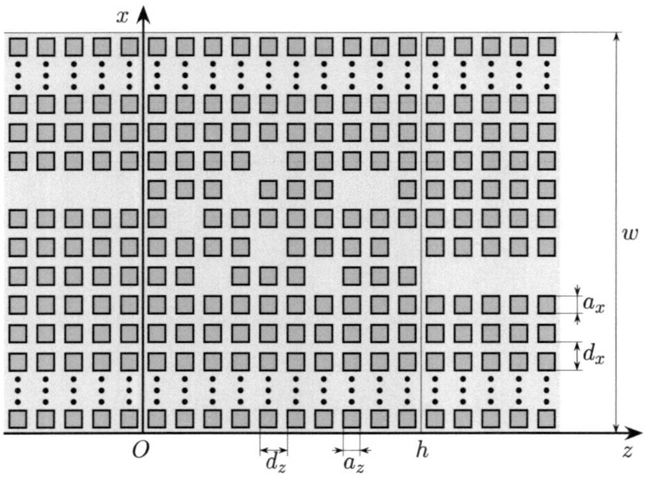

Fig. 4. Nonuniform structure formed by defects in photonic crystal.

of the Floquet-modes propagating in the $\pm z$-direction, the S-matrix of the region $0 < z < l\,d_z$ are defined by

$$\begin{pmatrix} \boldsymbol{b}^{(p,0,-)}(0) \\ \boldsymbol{b}^{(p,l+1,+)}(l\,d_z) \end{pmatrix} = \begin{pmatrix} \boldsymbol{S}_{l,11}^{(p)} & \boldsymbol{S}_{l,12}^{(p)} \\ \boldsymbol{S}_{l,21}^{(p)} & \boldsymbol{S}_{l,22}^{(p)} \end{pmatrix} \begin{pmatrix} \boldsymbol{b}^{(p,0,+)}(0) \\ \boldsymbol{b}^{(p,l+1,-)}(l\,d_z) \end{pmatrix} \tag{61}$$

where $\boldsymbol{S}_{l,11}^{(p)}$, $\boldsymbol{S}_{l,12}^{(p)}$, $\boldsymbol{S}_{l,21}^{(p)}$, and $\boldsymbol{S}_{l,22}^{(p)}$ are $(2N+1) \times (2N+1)$ square submatrices. The boundary condition at $z = l\,d_z$ $(l = 0, \ldots, L)$ is matched by equating the Fourier coefficients of tangential field components in both sides of the step-transition, and yields

$$\begin{pmatrix} \boldsymbol{b}^{(p,l+1,+)}(l\,d_z) \\ \boldsymbol{b}^{(p,l+1,-)}(l\,d_z) \end{pmatrix} = \begin{pmatrix} \boldsymbol{G}_{11}^{(p,l)} & \boldsymbol{G}_{12}^{(p,l)} \\ \boldsymbol{G}_{21}^{(p,l)} & \boldsymbol{G}_{22}^{(p,l)} \end{pmatrix} \begin{pmatrix} \boldsymbol{b}^{(p,l,+)}(l\,d_z) \\ \boldsymbol{b}^{(p,l,-)}(l\,d_z) \end{pmatrix} \tag{62}$$

with

$$\begin{pmatrix} \boldsymbol{G}_{11}^{(p,l)} & \boldsymbol{G}_{12}^{(p,l)} \\ \boldsymbol{G}_{21}^{(p,l)} & \boldsymbol{G}_{22}^{(p,l)} \end{pmatrix} = \boldsymbol{R}^{(p,l+1)^{-1}} \boldsymbol{R}^{(p,l)}. \tag{63}$$

From Eq. (62) for $l = 0$, the initial S-matrices are derived as follows:

$$\boldsymbol{S}_{0,12}^{(p)} = \boldsymbol{G}_{22}^{(p,0)^{-1}} \tag{64}$$

$$\boldsymbol{S}_{0,11}^{(p)} = -\boldsymbol{S}_{0,12}^{(p)} \boldsymbol{G}_{21}^{(p,0)} \tag{65}$$

$$\boldsymbol{S}_{0,21}^{(p)} = \boldsymbol{G}_{11}^{(p,0)} + \boldsymbol{G}_{12}^{(p,0)} \boldsymbol{S}_{0,11}^{(p)} \tag{66}$$

$$\boldsymbol{S}_{0,22}^{(p)} = \boldsymbol{G}_{12}^{(p,0)} \boldsymbol{S}_{0,12}^{(p)}. \tag{67}$$

Also, Eq. (47) gives the following relations:

$$\boldsymbol{b}^{(p,l,+)}(l\,d_z) = \boldsymbol{D}^{(p,l)} \boldsymbol{b}^{(p,l,+)}((l-1)\,d_z) \tag{68}$$

$$\boldsymbol{b}^{(p,l,-)}((l-1)\,d_z) = \boldsymbol{D}^{(p,l)} \boldsymbol{b}^{(p,l,-)}(l\,d_z) \tag{69}$$

$$\left(\boldsymbol{D}^{(p,l)}\right)_{n,m} = \delta_{n,m}\, e^{i\eta_m^{(p,l)} d_z} \tag{70}$$

for $1, \ldots, L$, where $\{\eta_m^{(p,l)}\}_{m=1}^{2N+1}$ are the propagation constants of the Floquet-modes in each region $(l-1)\,d_z < z < l\,d_z$. When the S-matrices $\boldsymbol{S}_{l-1,11}^{(p)}$, $\boldsymbol{S}_{l-1,12}^{(p)}$, $\boldsymbol{S}_{l-1,21}^{(p)}$, and $\boldsymbol{S}_{l-1,22}^{(p)}$ are given, the S-matrices $\boldsymbol{S}_{l,11}^{(p)}$, $\boldsymbol{S}_{l,12}^{(p)}$, $\boldsymbol{S}_{l,21}^{(p)}$, and $\boldsymbol{S}_{l,22}^{(p)}$ are derived from Eqs. (61), (62), (68), and (69) as follows:

$$\boldsymbol{S}_{l,12}^{(p)} = \boldsymbol{S}_{l-1,12}^{(p)} \boldsymbol{D}^{(p,l)} \boldsymbol{W}_{l,1}^{(p)^{-1}} \tag{71}$$

$$\boldsymbol{S}_{l,11}^{(p)} = \boldsymbol{S}_{l-1,11}^{(p)} - \boldsymbol{S}_{l,12}^{(p)} \boldsymbol{G}_{21}^{(p,l)} \boldsymbol{D}^{(p,l)} \boldsymbol{S}_{l-1,21}^{(p)} \tag{72}$$

$$\boldsymbol{S}_{l,22}^{(p)} = \boldsymbol{W}_{l,2}^{(p)} \boldsymbol{W}_{l,1}^{(p)^{-1}} \tag{73}$$

$$\boldsymbol{S}_{l,21}^{(p)} = \left(\boldsymbol{G}_{11}^{(p,l)} - \boldsymbol{S}_{l,22}^{(p)} \boldsymbol{G}_{21}^{(p,l)}\right) \boldsymbol{D}^{(p,l)} \boldsymbol{S}_{l-1,21}^{(p)} \tag{74}$$

with

$$\boldsymbol{W}_{l,1}^{(p)} = \boldsymbol{G}_{22}^{(p,l)} + \boldsymbol{G}_{21}^{(p,l)} \boldsymbol{D}^{(p,l)} \boldsymbol{S}_{l-1,22}^{(p)} \boldsymbol{D}^{(p,l)} \tag{75}$$

$$\boldsymbol{W}_{l,2}^{(p)} = \boldsymbol{G}_{12}^{(p,l)} + \boldsymbol{G}_{11}^{(p,l)} \boldsymbol{D}^{(p,l)} \boldsymbol{S}_{l-1,22}^{(p)} \boldsymbol{D}^{(p,l)}. \tag{76}$$

Consequently, the S-matrices $\boldsymbol{S}_{L,11}^{(p)}$, $\boldsymbol{S}_{L,12}^{(p)}$, $\boldsymbol{S}_{L,21}^{(p)}$, and $\boldsymbol{S}_{L,22}^{(p)}$ for the entire transition section are obtained by the initial matrices (64)–(67) and the recursive relations (71)–(76).

The proposed method has been used to analyze several fundamental photonic crystal circuit components(Yasumoto & Watanabe, 2008a;b). The photonic crystal consists of identical rectangular cylinders situated parallel in rectangular lattice. The permittivities of the surrounding medium and the cylinders are $\varepsilon_s = \varepsilon_0$ and $\varepsilon_c = 12.25\,\varepsilon_0$. The dimensions of the cylinders are $a_x = a_z = 0.2\sqrt{\pi}\,d_x$ and the lattice constants are $d_x = d_z = 340\,\mu$m. We consider the TE-polarized fields, and the photonic crystal waveguides formed by a straight line defect in the photonic crystal support only one guided Floquet-mode.

First, we consider a directional coupler with the coupling length with $h = 10\,d_z$ as shown in Fig. 5(a). The wavelength in free space and the artificial periodicity for FSEM are chosen as $\lambda_0 = 1$ mm and $w = 20\,d_x$. The incident wave is the guided Floquet-mode of the uniform waveguide in $z < 0$. If the guided mode is arranged to be the first-order mode, this incident condition is given as

$$b_m^{(e,0,+)}(0) = \delta_{m,1} \tag{77}$$

$$b_m^{(e,L+1,-)}(h) = 0 \tag{78}$$

for $m = 1, \ldots, 2N+1$. Let $\boldsymbol{e}_{y,m}^{(l)}$ and $\boldsymbol{h}_{x,m}^{(l)}$ denote the Fourier coefficient matrices of the modal profile functions corresponding to the mth-order Floquet modes in each region $(l-1)\,d_z < z < l\,d_z$. Then, the transmitted and the reflected powers are respectively calculated by

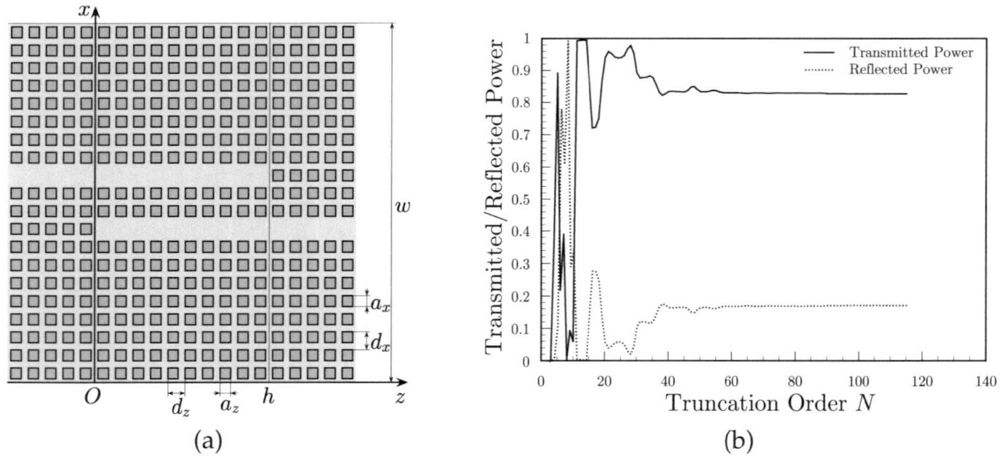

Fig. 5. Photonic crystal direction couplers: (a) structure under consideration and (b) convergence of the normalized reflection and transmitted powers as functions of the truncation order N.

$(s_1^{(e,L+1)}/s_1^{(e,0)})\,|b_1^{(e,L+1,+)}(h)|^2$ and $|b_1^{(e,0,-)}(0)|^2$ with $s_m^{(e,l)} = -w\,\Re(\boldsymbol{e}_{y,m}^{(l)} \cdot \boldsymbol{h}_{x,m}^{(l)\,*})/2$, and they are shown in Fig. 5(b) as functions of the truncation order N of the Fourier series expansions. We can see that a good convergence is obtained for $N \gtrsim 4w/d_x$.

Figures 6 and 7 are the results of the photonic crystal waveguide filters. The filters are formed with resonant cavities on the defect layers as shown in Figs. 6(a) and 7(a), and the power transmission and reflection spectra through the resonant cavities are plotted in Figs. 6(b) and 7(b) as functions of the wavelength λ_0. It is seen that the transmission spectra have resonance peaks at around $\lambda_0 = 0.91$ mm. Since the resonant cavity in Fig. 7 is weakly coupled with the feed waveguides, its resonance peak is much sharper than that shown in Fig. 6.

The present formulation is also applied to photonic crystal waveguide cranks shown in Figs. 8(a) and 9(a), in which wave propagates perpendicular partially to the z-direction and

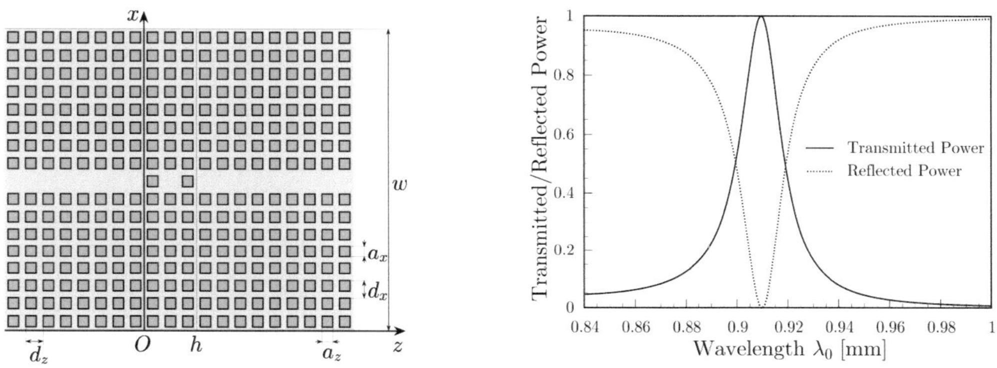

(a) (b)

Fig. 6. Photonic crystal waveguide filter with a resonant cavity strongly coupled to the feed waveguides: (a) structure under consideration and (b) power transmission spectra.

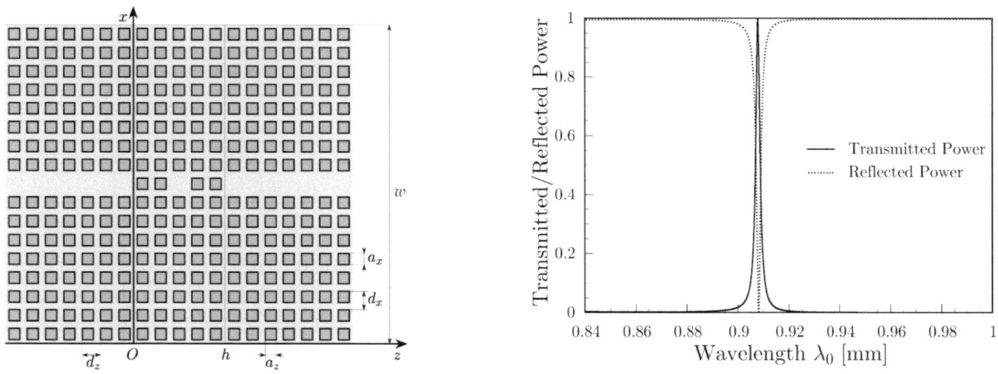

(a) (b)

Fig. 7. Photonic crystal waveguide filter with a resonant cavity weakly coupled to the feed waveguides: (a) structure under consideration and (b) power transmission spectra. (Reproduced from K. Yasumoto and K. Watanabe, *International Journal of Microwave and Optical Technology*, 3, 397–403, 2008, with courtesy of ISRAMT/IJMOT.)

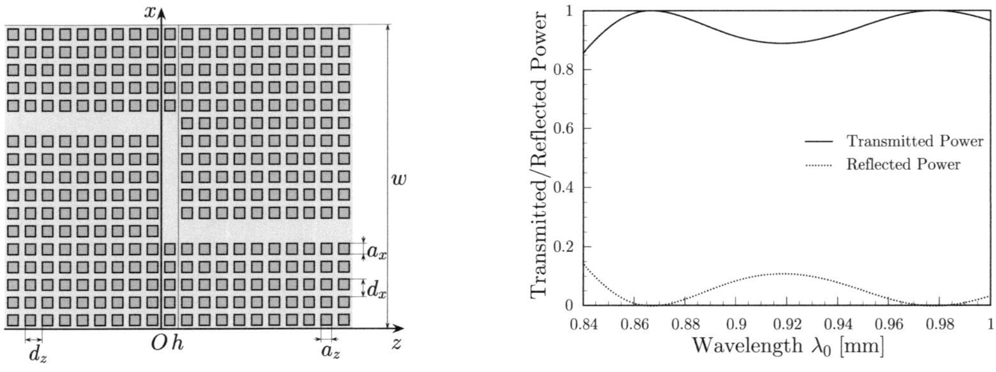

(a) (b)

Fig. 8. Conventional crank of the photonic crystal waveguide: (a) structure under consideration and (b) power transmission spectra. (Reproduced from K. Yasumoto and K. Watanabe, *Proc. of CJMW2008*, 3–8, 2008, with courtesy of IEICE.)

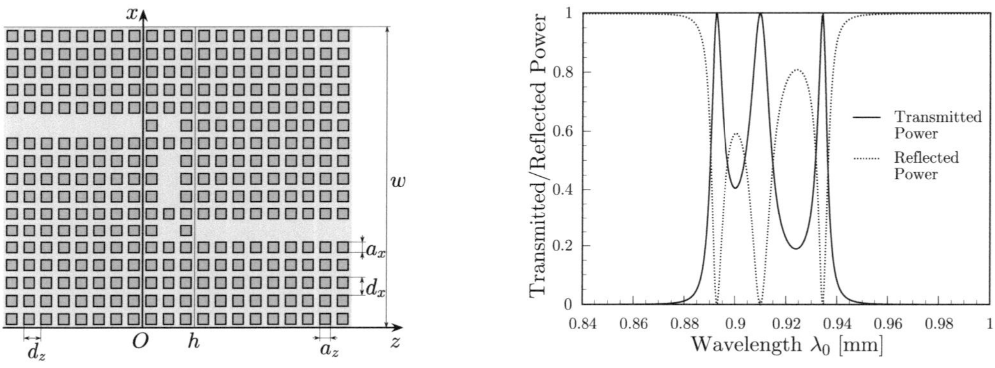

(a) (b)

Fig. 9. Photonic crystal waveguide crank with two resonant cavities: (a) structure under consideration and (b) power transmission spectra. (Reproduced from K. Yasumoto and K. Watanabe, *Proc. of CJMW2008*, 3–8, 2008, with courtesy of IEICE.)

the artificial periodicity is chosen as $w = 17d_x$. The conventional crank has no resonance in the transmission and the reflection spectra as shown in Fig. 8(b). However, when two resonant cavities are introduced in the crank, there appear three resonant peaks in the transmission and the reflection spectra as shown in Fig. 9(b). The center peak corresponds to the resonance wavelength of each isolated cavity (see Fig. 6), whereas other two peaks seem to represent the resonant coupling between two cavities.

Next, we consider the photonic crystal waveguide branches shown in Figs. 10(a), 11(a), and 12(a). The incident wave is the guided Floquet-mode of the uniform waveguide in $z < 0$ and, then, this incident condition is also given by Eqs. (77) and (78). In these structures, the section $z > h$ consists of two parallel waveguides, and we are interested in the transmitted power for each waveguide. Here we denote the upper waveguide by "1" and the lower one

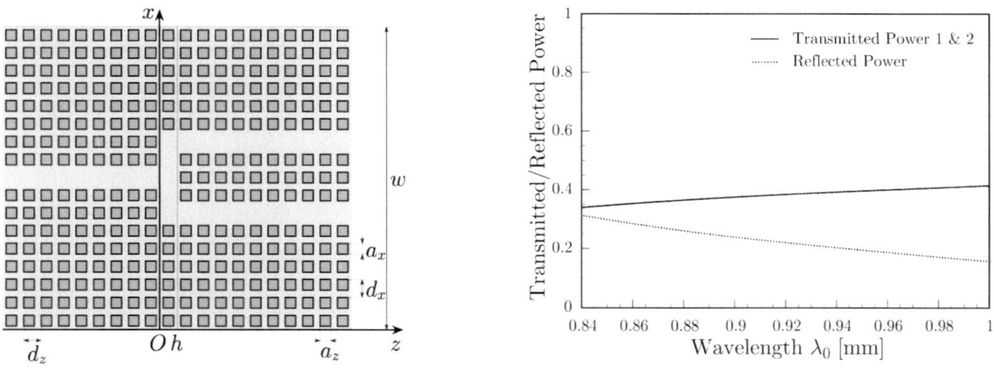

(a) (b)

Fig. 10. Symmetric branch of the photonic crystal waveguide: (a) structure under consideration and (b) power transmission spectra.

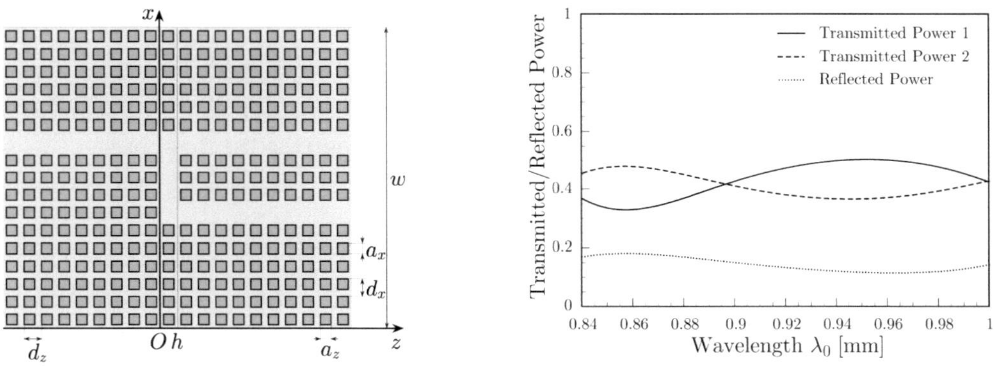

(a) (b)

Fig. 11. Asymmetric branch of the photonic crystal waveguide: (a) structure under consideration and (b) power transmission spectra.

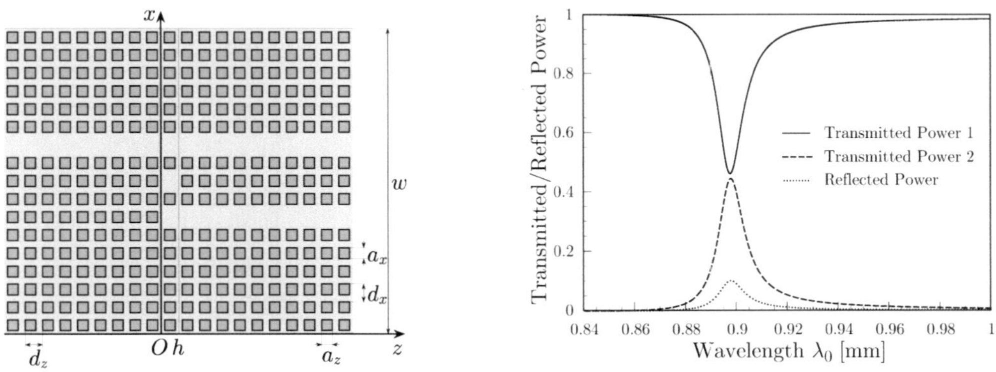

(a) (b)

Fig. 12. Asymmetric branch of the photonic crystal waveguide with a resonant cavity: (a) structure under consideration and (b) power transmission spectra. (Reproduced from K. Yasumoto and K. Watanabe, *Proc. of CJMW2008*, 3–8, 2008, with courtesy of IEICE.)

by "2." Let $\bar{\boldsymbol{e}}_{y,\nu,m}$ and $\pm\bar{\boldsymbol{h}}_{x,\nu,m}$ be the Floquet coefficient matrices of the modal profile functions corresponding to the mth-order Floquet-modes of the isolated waveguides ν ($\nu = 1,2$) propagating in the $\pm z$-direction, and $\bar{b}_{\nu,m}^{(e,l,\pm)}(z)$ be the associated modal amplitudes for the regions $(l-1)\,d_z < z < l\,d_z$ ($l = 0,\ldots,L+1$). From Eqs. (45) and (59), the relation between the amplitudes of the isolated waveguide modes and the compound modes are derived as

$$\bar{b}_{\nu,m}^{(e,l,\pm)}(l'\,d_z) = \sum_{m'=1}^{2N+1} b_{m'}^{(e,l,\pm)}(l'\,d_z)\,\frac{\langle \boldsymbol{e}_{y,m'}^{(l)} | \bar{\boldsymbol{h}}_{x,\nu,m}\rangle + \langle \bar{\boldsymbol{e}}_{y,\nu,m} | \boldsymbol{h}_{x,m'}^{(l)}\rangle}{2\,\langle \bar{\boldsymbol{e}}_{y,\nu,m} | \bar{\boldsymbol{h}}_{x,\nu,m}\rangle} \tag{79}$$

where the counter directional couplings are ignored. If the guided mode is arranged to be the first-order mode, the transmitted powers for the waveguides ν are therefore calculated by $(\bar{s}_{\nu,1}^{(e)}/s_1^{(e,0)})\,|\bar{b}_{\nu,1}^{(e,L+1,+)}(h)|^2$ with $\bar{s}_{\nu,1}^{(e)} = -w\,\Re(\bar{\boldsymbol{e}}_{y,\nu,1}\cdot\bar{\boldsymbol{h}}_{x,\nu,1}^{*})/2$. The transmitted and the reflected powers for the conventional branches are shown in Figs. 10(b), 11(b) as functions of the wavelength. It is seen that the spectra have no sharp resonant peak. On the other hand, Fig. 12(b) shows the spectra for a branch with a resonant cavity. The input power is dropped from the waveguide 1 to the waveguide 2 through the resonance, and almost equally divided into two transmission ports at the resonant wavelength with a small reflection.

The last examples are photonic crystal waveguide couplers with resonant cavities as shown in Figs. 13(a) and 14(a). The incident wave is the guided Floquet-mode of the waveguide 1 in isolation and comes from $z < 0$. If the guided mode is arranged to the first-order mode, the incident condition is given as

$$b_m^{(e,0,+)}(0) = \frac{\langle \bar{\boldsymbol{e}}_{y,1,1} | \boldsymbol{h}_{x,m}^{(0)}\rangle + \langle \boldsymbol{e}_{y,m}^{(0)} | \bar{\boldsymbol{h}}_{x,1,1}\rangle}{2\,\langle \boldsymbol{e}_{y,m}^{(0)} | \boldsymbol{h}_{x,m}^{(0)}\rangle} \tag{80}$$

$$b_m^{(e,L+1,-)}(h) = 0. \tag{81}$$

Then the transmitted and the reflected powers for the isolated waveguides ν are calculated by $(\bar{s}_{\nu,1}^{(e)}/\bar{s}_{1,1}^{(e)})\,|\bar{b}_{\nu,1}^{(e,L+1,+)}(h)|^2$ and $(\bar{s}_{\nu,1}^{(e)}/\bar{s}_{1,1}^{(e)})\,|\bar{b}_{\nu,1}^{(e,0,-)}(0)|^2$, respectively. The transmitted and

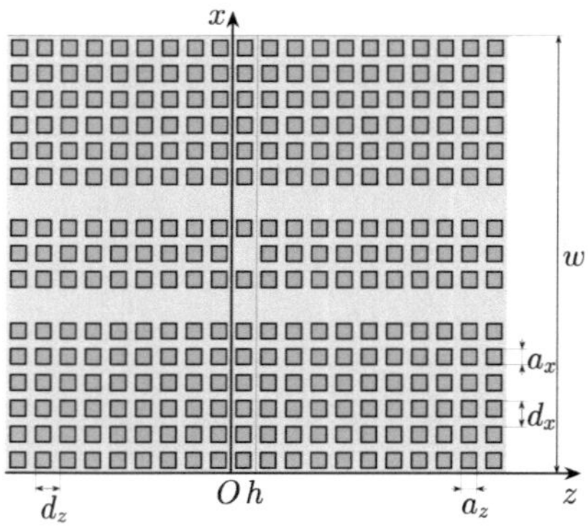

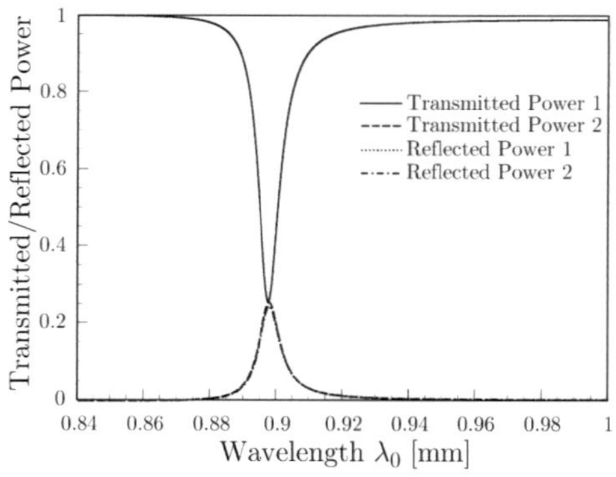

Fig. 13. Photonic crystal waveguide coupler with a resonant cavity: (a) structure under consideration and (b) power transmission spectra. (Reproduced from K. Yasumoto and K. Watanabe, *Proc. of CJMW2008*, 3–8, 2008, with courtesy of IEICE.)

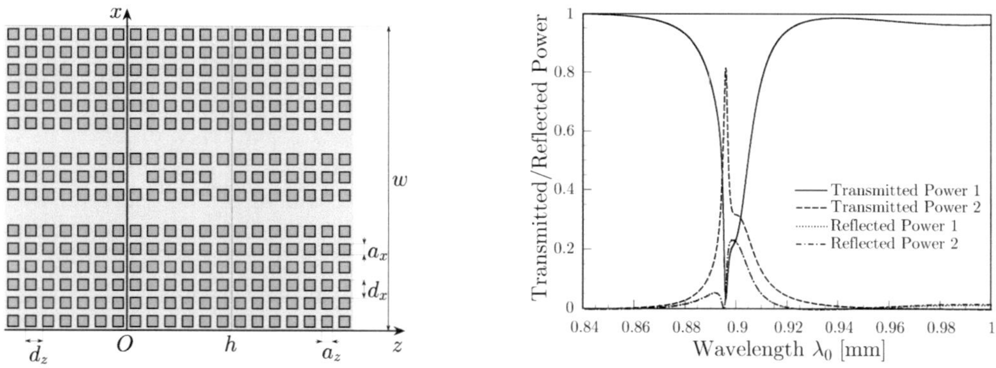

(a) (b)

Fig. 14. Photonic crystal waveguide coupler with two resonant cavities: (a) structure under consideration and (b) power transmission spectra.

the reflected powers for the photonic crystal waveguide coupler with one resonant cavity are shown in Fig. 13(b) as functions of the wavelength. For this configuration, three lines except for the solid lines overlap though it is indiscernible. The input power is equally divided into four output ports at the resonant to the cavity. Figure 14(b) shows the spectra of the coupler with two resonant cavities. Two reflection spectra still overlap but the transmission spectrum of the isolated waveguide 2 separates from them. In this case, about 80 percent of the input power is dropped from the waveguide 1 to the waveguide 2 through the resonance.

5. Conclusion

This chapter has presented the Fourier series expansion method for analyzing the photonic crystal circuit components. The method derives the Floquet-modes by the eigenvalue analysis of the transfer matrix for one periodicity cell in the propagation direction, and the input/output relations are expressed by the S-matrix for the Floquet-modes. The numerical examples for photonic crystal filters, cranks, branches, and couplers have been presented to demonstrate the effectiveness of the present method. We have dealt with the photonic crystals consisting of rectangular cylinders only. However, the method is also applied to the photonic crystals consisting of circular cylinders by introducing the staircase approximation or using a numerical integration to derive the transfer matrix for one periodicity cell. Also, when the number of rows of cylinders on each side of the waveguide is not enough, the radiation loss can be taken into account by introducing the perfectly matched layer near the artificial boundaries(Li & Ho, 2004; Yasumoto et al., 2002; Zhang & Jia, 2007).

Acknowledgments

This work was supported in part by the 2008 grant from the Japan-Indo Collaboration Project on "Infrastructural Communication Technologies Supporting Fully Ubiquitous Information Society."

6. References

Benisty, H. (1996). Modal analysis of optical guides with two-dimensional photonic band-gap boundaries, *J. Appl. Phys.* **Vol. 79**(No. 10): 7483–7492.

Hosono, T., Hinata, T. & Inoue, A. (1982). Numerical analysis of the discontinuities in slab dielectric waveguides, *Radio Sci.* **Vol. 17**(No. 1): 75–83.

Jia, H. & Yasumoto, K. (2006). Modal analysis of two-dimensional photonic-crystal waveguides formed by rectangular cylinders using an improved fourier series method, *IEEE Trans. Microwave Theory and Techniques* **Vol. 54**(No. 2): 564–571.

Koshiba, M., Tsuji, Y. & Hikari, M. (2000). Time-domain beam propagation method and its application to photonic crystal circuits, *J. Lightwave Technol.* **Vol. 18**(No. 1): 102–110.

Li, L. (1996a). Formulation and comparison of two recursive matrix algorithms for modeling layered diffraction gratings, *J. Opt. Soc. Am. A* **Vol. 13**(No. 5): 1024–1035.

Li, L. (1996b). Use of fourier series in the analysis of discontinuous periodic structures, *J. Opt. Soc. Am. A* **Vol. 13**(No. 9): 1870–1876.

Li, Z. Y. & Ho, K. M. (2003). Light propagation in semi-infinite photonic crystal and related waveguide structures, *Physical Review B* **Vol. 68**: 155101.

Li, Z. Y. & Ho, K. M. (2004). Anomalous propagation loss in photonic crystal waveguides, *Physical Review Lett.* **Vol. 92**(No. 6): 063904.

Miyamoto, T., Momoda, M. & Yasumoto, K. (2003). Numerical analysis for 3-dimensional optical waveguides with periodic structures using fourier series expansion method, *IEICE Trans. Electron.* **Vol. J86-C**(No. 6): 591–600. (in Japanese).

Naka, Y. & Ikuno, H. (2002). Analysis of characteristics of optical waveguide devices constructed by two-dimensional air-hole type photonic crystal, *Proc. Asia-Pacific Eng. Res. Forum on Microwave and Electromagnetic Theory*, pp. 209–218.

Sakoda, K., Ueta, T. & Ohtaka, K. (1997). Numerical analysis of eigenmodes localized at line defects in photonic lattices, *Phys. Rev. B* **Vol. 56**(No. 23): 14905–14908.

Taflove, A. (1995). *Computational Electrodynamics: The Finite-Difference Time-Domain Method*, Artech House, Boston.

Tanaka, H., Yamasaki, T. & Hosono, T. (1994). Propagation characteristics of dielectric waveguides with slanted grating structure, *IEICE Trans. Electron.* **Vol. E77-C**(No. 11): 1820–1827.

Watanabe, K. & Yasumoto, K. (2009). Accuracy improvement of the fourier series expansion method for floquet-mode analysis of photonic crystal waveguides, *Progress In Electromagnetics Res.* **Vol. PIER 92**: 209–222.

Yamakita, J., Matsumoto, K. & Rokushima, K. (1993). Analysis of discontinuities in anisotropic dielectric waveguides, *IEICE Technical Report.* EMT-93-87 (in Japanese).

Yasumoto, K., Jia, H. & Kai, S. (2004). Rigorous analysis of two-dimensional photonic crystal waveguides, *Proc. URSI Int. Symp. on Electromagnetic Theory*, pp. 739–741.

Yasumoto, K., Miyamoto, T. & Momoda, M. (1999). Full-wave analysis of optical waveguides using periodic boundary conditions, *Proc. SPIE* **Vol. 3666**: 170–176.

Yasumoto, K. & Toyama, H. (2001). Formulation for electromagnetic scattering and guidance by two-dimensional photonic crystals, *IEICE Technical Report.* OPE2001-93.

Yasumoto, K. & Watanabe, K. (2008a). Analysis of discontinuities in two-dimensional photonic crystal waveguides using floquet modes concept, *Int. J. Microwave and Opt. Technol.* **Vol. 3**(No. 3): 397–403.

Yasumoto, K. & Watanabe, K. (2008b). Numerical modeling of two-dimensional photonic crystal circuits using fourier modal method based on floquet modes, *Proc. China-Japan Joint Microwave Conf.*, pp. 3–8.

Yasumoto, K., Watanabe, K. & Ishihara, J. (2002). Numerical analysis of optical waveguides with the use of fourier-series expansion method combined with perfectly matched layer, *Microwave Opt. Technol. Lett.* **Vol. 34**(No. 6): 422–426.

Zhang, D. & Jia, H. (2007). Numerical analysis of leaky modes in two-dimensional photonic crystal waveguides using fourier series expansion method with perfectly matched layer, *IEICE Trans. Electron.* **Vol. E90-C**(No. 3): 613–622.

6

Computer Aided Design of Waveguide Devices by Mode-Matching Methods

Jorge A. Ruiz-Cruz
Escuela Politecnica Superior, Universidad Autonoma de Madrid
C/ Francisco Tomas y Valiente 11, 28049 Madrid
Spain

Jose R. Montejo-Garai and Jesus M. Rebollar
Dpto. de Electromagnetismo y Teora de Circuitos, Universidad Politecnica de Madrid
Ciudad Universitaria s/n, 28040 Madrid
Spain

1. Introduction

1.1 Waveguide devices and CAD tools

Technology is one of the key issues in the hardware design at the microwave and millimetre wave band. Its selection depends on many factors such as bandwidth, physical size, losses, power handling capability and cost. This has led to many different transmission media for implementing circuits and systems; the planar and the waveguide technologies are two representative examples.

The main advantages of planar structures are their small size and simple manufacturing. In addition, they can be easily integrated with Microwave Integrated Circuits (MIC) and Monolithic Microwave Integrated Circuits (MMIC). The microstrip transmission line is a common example of this technology, where circuits are made by printing a metallic strip on a dielectric substrate supported by a ground plane. On the other hand, waveguide devices (the type of components which will be discussed in this chapter) are constructed on metallic pipes that may have many forms. The electromagnetic waves are confined to the interior of the waveguiding structure. In contrast to planar devices, they are more cumbersome and bulkier. Their main advantage is their high power handling capabilities and high quality factor (Q), which leads to electric responses with lower insertion losses than planar technology components. In satellite applications, their robustness also becomes an advantage.

One aspect that has significantly modified the design of advanced waveguide components during the last decades has been the evolution of software modeling and Computer Aided Design (CAD) tools. Traditionally, the analysis of waveguide devices (such as couplers, filters or multiplexers) was based on approximate equivalent circuits made up of transmission lines to represent waveguiding regions and lumped elements (inductors, capacitors, transformers, resistors, etc.) to model dissipative effects and discontinuities between different transmission media. Most of the equivalent circuits for waveguide problems were developed at the MIT

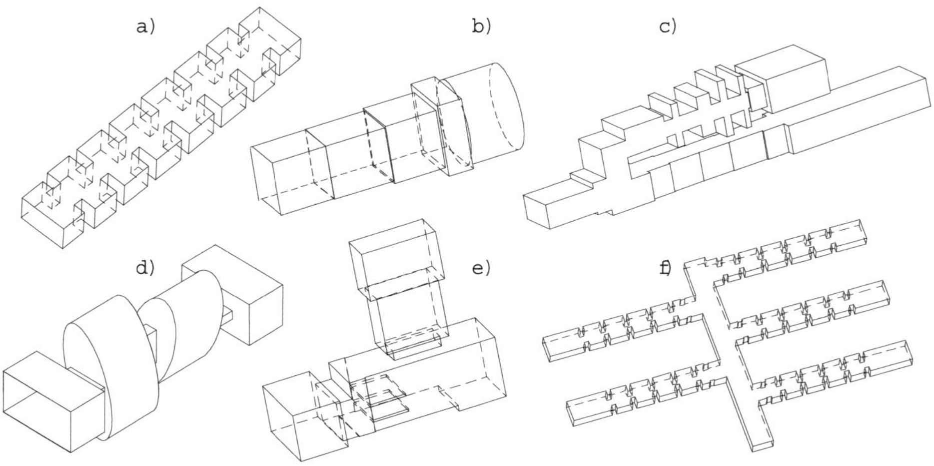

Fig. 1. Examples of waveguide devices which can be efficiently modeled by Mode-Matching: a) bandpass filter; b) square to circular waveguide transformer; c) diplexer with low-pass and high-pass filters; d) dual-Mode filter with elliptical waveguides; e) ortho-mode transducer (OMT); f) 5-channel manifold multiplexer.

Radiation Laboratory. These models, together with the advance in synthesis circuit theory, have brought about the design of many devices.

Nevertheless, the equivalent circuit approach has a number of limitations. The most significant is that equivalent circuits are only focused on modeling the fundamental mode response of the elements in the structure and the localized effect of the higher order modes. However, the interactions between the higher-order modes of the different elements and other electromagnetic effects are not taken into account. Thus, this approach leads to discrepancies between the theoretical predicted response of the device and its actual measurement. The designed prototypes following this approach hence need a relevant experimental effort and manual tuning. In a more demanding industrial sector, it seems reasonable to complement these results with other more precise methods and to evolve towards CAD tools.

New techniques to model microwave devices appear as a consequence of two facts: i) a solid comprehension of the microwave theory along with numerical methods, and ii) the development of computers. The latter allows the implementation of codes that few years ago seemed unfeasible because of their required computer resources.

There is a great variety of methods to deal with the so-called full-wave analysis of modern microwave devices (Uher et al., 1993). Among the different alternatives, two basic aspects should be addressed: i) the efficiency using the computer resources (RAM memory, disk storage and processor capabilities), and ii) the types of geometries and materials that they can handle. The methods vary from general numeric techniques (such as the Finite Element Method, Finite Differences, etc.) to quasi-analytical techniques (such as the Mode-Matching method). There also hybrid techniques combining different approaches.

Among these different CAD options, those based on Mode-Matching (MM) methods (Clarricoats & Slinn, 1966), (Drabuwitch, 1966), (Wexler, 1967) are good examples of accurate and efficient tools. The range of problems that they can tackle is more constrained in comparison with general numeric techniques. Nevertheless, when the characterization of the device under investigation can be carried out by MM (as for instance those shown in Fig. 1), the resulting codes are very efficient and facilitate the design of components with sophisticated responses. In fact, this is the objective of this chapter: to present the main concepts of the MM techniques and to show its application to several designs for common applications.

1.2 Overview of the Mode-Matching (MM) method

The starting point for MM is to segment the problem under analysis in different waveguide regions (for instance the different rectangular waveguides in Fig. 1.a), where the total electromagnetic field is represented by the superposition of modes

$$\vec{\mathbf{E}} = \sum\nolimits_n \varsigma_n^+ \vec{\mathbf{E}}_n^+ + \sum\nolimits_n \varsigma_n^- \vec{\mathbf{E}}_n^-, \quad \vec{\mathbf{H}} = \sum\nolimits_n \varsigma_n^+ \vec{\mathbf{H}}_n^+ + \sum\nolimits_n \varsigma_n^- \vec{\mathbf{H}}_n^-. \tag{1}$$

These expansions (different for each waveguide) are constructed in order to represent any possible field inside each region. The scalar complex amplitudes $\varsigma_n^\pm$ are (initially) undetermined. On the other hand, the electromagnetic field of each mode $(\vec{\mathbf{E}}_n^\pm, \vec{\mathbf{H}}_n^\pm)$ must be known in advance either analytically (as for the waveguides shown at App. A.5) or numerically (by means of a suitable numerical method).

The modal expansion (1) provides a formal solution to the Maxwell´s equations for each waveguide. However, for the complete resolution of the problem, the boundary conditions at the interface between the different segmented regions must be also fulfilled. A field-matching procedure is used to impose those boundary conditions, providing a relation between the amplitudes $\varsigma_n^\pm$ of the modes involved in all the regions. This relation is usually not simple, and requires some previous computations: the inner cross products between the modal fields.

In order to make a formal representation of the problem under investigation, the modal amplitudes in each region are usually collected in vectors. The relation of those vectors is usually expressed in terms of the Generalized Scattering Matrix (GSM). Other formulations use the Generalized Admittance or Impedance Matrix (GAM or GIM, respectively) (Conciauro et al., 1999), which are the natural option when (1) is expressed with equivalent voltages and currents instead of modal amplitudes. The term *Generalized* refers to the types of modes used in (1): propagating and evanescent modes are both required in the series to represent the field. Their amplitudes at the different regions will be related by the GSM.

As many other numerical methods, some convergence issues will appear in MM. They are related to the series in (1), which must be truncated to a finite number of modes for computational purposes. Therefore, the boundary conditions and the amplitudes obtained by MM will be an approximation whose accuracy will depend on the number of modes retained in the modal expansions. Moreover, the solution will depend not only on the number of modes used in the different regions, but also on the relation between them. This problem is known as the relative convergence problem (Mittra & Lee, 1971), (Vassallo, 1985).

These general ideas are found in the MM of very different problems. In order to see them in a more detailed form, a specific problem will be treated now: the cascade connection of different transmission systems. This is the basic structure that can be used to solve more complex problems with cubic junctions or volume-type structures with more general enclosures. The theory for the discontinuities will be developed in a unified manner, aiming to highlight the important aspects of the method not always covered in other text books. The implementation details will be given as references (as for instance the computation of the inner cross products).

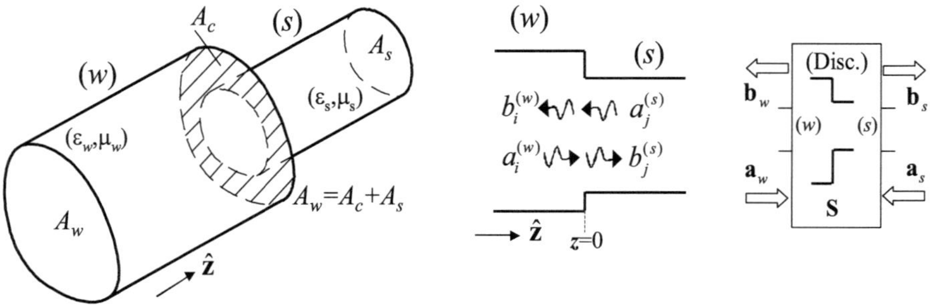

Fig. 2. Basic waveguide step with the modal amplitudes incident and scattered by the discontinuity. The Mode-Matching (MM) method provides its Generalized Scattering Matrix (GSM) as a building-block to use in more complex structures.

2. Mode-Matching method for waveguide steps

2.1 Waveguide step or discontinuity

The waveguide step or discontinuity problem arises when different transmission systems are connected in a plane transversal to the propagation direction. The type of transmission systems considered now are homogeneous cylindrical waveguides connected in the direction of their longitudinal axis z (described in App. A). The problem is detailed in Fig. 2 and it could represent, for instance, the discontinuity between two rectangular waveguides in Fig. 1.a or the discontinuity between a rectangular and a elliptical waveguide in Fig. 1.d. The goal is to have a representation of this step as a building block characterized by its GSM, which could be latter used in more complex problems.

2.2 Field-matching procedure

The electromagnetic field at both sides of the discontinuity in Fig. 2 are derived from (1). For this problem, the boundary conditions involve the fields transversal to z, evaluated at the discontinuity plane located at $z = 0$.
The transversal fields are expressed with the amplitudes of the modes incident (a) and scattered (b) by the step:

$$\vec{\mathbf{E}}_t^{(w)}\Big]_{A_w,z=0^-} = \sum_{n=1}^{N_w} (a_n^{(w)} + b_n^{(w)})\,\vec{\mathbf{e}}_n^{(w)},\ \ \vec{\mathbf{H}}_t^{(w)}\Big]_{A_w,z=0^-} = \sum_{n=1}^{N_w} (a_n^{(w)} - b_n^{(w)})\,\vec{\mathbf{h}}_n^{(w)} \tag{2a}$$

$$\vec{\mathbf{E}}_t^{(s)}\Big]_{A_s,z=0^+} = \sum_{m=1}^{N_s} (b_m^{(s)} + a_m^{(s)})\,\vec{\mathbf{e}}_m^{(s)},\ \ \vec{\mathbf{H}}_t^{(s)}\Big]_{A_s,z=0^+} = \sum_{m=1}^{N_s} (b_m^{(s)} - a_m^{(s)})\,\vec{\mathbf{h}}_m^{(s)}. \tag{2b}$$

Each term in the series[1] (already truncated to a finite number of modes N_w and N_s in each waveguide) belongs to a TEM, TE or TM mode, not necessarily propagating at the analysis frequency. The modes are orthogonal with arbitrary normalization:

$$\iint_{A_g} \vec{\mathbf{e}}_n^{(g)} \times \vec{\mathbf{h}}_m^{(g)} \cdot \hat{\mathbf{z}}dS = Q_n^{(g)}\delta_{nm},\ \ g \equiv w,s,\ \ \delta_{nm} = 1(m=n),0(m \neq n). \tag{3}$$

[1] The reversed sign for the amplitudes in the magnetic field of waveguide (s) is because the a,b modal amplitudes travel in the opoposite direction than their a,b counterpart in waveguide (w). See (30) in App. A for a more detailed description of the fields and App. A.4 for their normalization.

In this initial formulation it is considered that the cross section of waveguide (s) is completely included in the one of (w): $A_s \subseteq A_w$ (the general aperture case is studied later). The Electric and Magnetic Field Boundary Conditions (EFBC and MFBC, respectively) must be satisfied at the interface between the two waveguides:

$$\begin{cases} \text{EFBC in } A_w\text{:} \quad \hat{\mathbf{z}} \times \vec{\mathbf{E}}^{(w)} = \begin{cases} \mathbf{0}, & \text{in } A_c, z=0 \\ \hat{\mathbf{z}} \times \vec{\mathbf{E}}^{(s)} & \text{in } A_s, z=0 \end{cases} \\ \text{MFBC in } A_s\text{:} \quad \hat{\mathbf{z}} \times \vec{\mathbf{H}}^{(w)} = \hat{\mathbf{z}} \times \vec{\mathbf{H}}^{(s)}, \quad \text{in } A_s, z=0. \end{cases} \tag{4}$$

In order to impose these boundary conditions, a *Galerkin* method is used. The first step is to test the EFBC with a generic modal magnetic field $\vec{\mathbf{h}}_j^{(w)}$ of waveguide (w). The left hand side, using (2a), leads to

$$\iint_{A_w} (\hat{\mathbf{z}} \times \vec{\mathbf{E}}^{(w)}) \cdot \vec{\mathbf{h}}_j^{(w)} \hat{\mathbf{z}} dS = \sum_{n=1}^{N_w} (a_n^{(w)} + b_n^{(w)}) (\iint_{A_w} (\hat{\mathbf{z}} \times \vec{\mathbf{e}}_n^{(w)}) \cdot \vec{\mathbf{h}}_j^{(w)} \hat{\mathbf{z}} dS).$$

Since the modes are orthogonal, according to (3), the previous equation becomes

$$\iint_{A_w} (\hat{\mathbf{z}} \times \vec{\mathbf{E}}^{(w)}) \cdot \vec{\mathbf{h}}_j^{(w)} \hat{\mathbf{z}} dS = (a_j^{(w)} + b_j^{(w)}) Q_j^{(w)}. \tag{5}$$

Moreover, the integration in A_w can be divided in two terms, since $A_w = A_s \cup A_c$:

$$\iint_{A_w} (\hat{\mathbf{z}} \times \vec{\mathbf{E}}^{(w)}) \cdot \vec{\mathbf{h}}_j^{(w)} \hat{\mathbf{z}} dS = \iint_{A_c} (\hat{\mathbf{z}} \times \vec{\mathbf{E}}^{(w)}) \cdot \vec{\mathbf{h}}_j^{(w)} \hat{\mathbf{z}} dS + \iint_{A_s} (\hat{\mathbf{z}} \times \vec{\mathbf{E}}^{(w)}) \cdot \vec{\mathbf{h}}_j^{(w)} \hat{\mathbf{z}} dS.$$

Taking into account that $\hat{\mathbf{z}} \times \vec{\mathbf{E}}^{(w)} = \mathbf{0}$, in $A_c, z = 0$ and (2b), it is obtained

$$\begin{aligned} \iint_{A_w} (\hat{\mathbf{z}} \times \vec{\mathbf{E}}^{(w)}) \cdot \vec{\mathbf{h}}_j^{(w)} \hat{\mathbf{z}} dS &= \iint_{A_s} (\hat{\mathbf{z}} \times \vec{\mathbf{E}}^{(s)}) \cdot \vec{\mathbf{h}}_j^{(w)} \hat{\mathbf{z}} dS = \\ &= \sum_{m=1}^{N_s} (a_m^{(s)} + b_m^{(s)}) \{ \iint_{A_s} \vec{\mathbf{e}}_m^{(s)} \times \vec{\mathbf{h}}_j^{(w)} \cdot \hat{\mathbf{z}} dS \}. \end{aligned} \tag{6}$$

The integral with the modal fields is a complex number called the inner cross product X_{mj}. The process for obtaining equations (5) and (6) can be done for any $j = 1, \ldots, N_w$. The result can be expressed as

$$Q_j^{(w)} (a_j^{(w)} + b_j^{(w)}) = \sum_{m=1}^{N_s} X_{mj} (a_m^{(s)} + b_m^{(s)}), \; j = 1, \ldots, N_w. \tag{7}$$

The second step is to test the MFBC with the modal electric fields $\vec{\mathbf{e}}_i^{(w)}$ of waveguide (s). The left hand side provides an equation with the same previous type of inner cross product:

$$\iint_{A_s} (\hat{\mathbf{z}} \times \vec{\mathbf{H}}^{(w)}) \cdot \vec{\mathbf{e}}_i^{(s)} \hat{\mathbf{z}} dS = \sum_{n=1}^{N_w} (a_n^{(w)} - b_n^{(w)}) \{ \iint_{A_s} \vec{\mathbf{e}}_i^{(s)} \times \vec{\mathbf{h}}_n^{(w)} \cdot \hat{\mathbf{z}} dS \}. \tag{8}$$

For the right hand side, the orthogonality of the modes in waveguide (s) is applied

$$\iint_{A_s} (\hat{\mathbf{z}} \times \vec{\mathbf{H}}^{(w)}) \cdot \vec{\mathbf{e}}_i^{(s)} \hat{\mathbf{z}} dS = \iint_{A_s} (\hat{\mathbf{z}} \times \vec{\mathbf{H}}^{(s)}) \cdot \vec{\mathbf{e}}_i^{(s)} \hat{\mathbf{z}} dS = Q_i^{(s)} (b_i^{(s)} - a_i^{(s)}). \tag{9}$$

Equations (8) and (9) can be reproduced for any $i = 1,\ldots,N_s$, providing:

$$\sum_{n=1}^{N_w} X_{in}(a_n^{(w)} - b_n^{(w)}) = Q_i^{(s)}(b_i^{(s)} - a_i^{(s)}),\; i = 1,\ldots,N_s. \tag{10}$$

The resulting equations (7) and (10) are better expressed if the amplitudes are arranged in columns vectors

$$\mathbf{a}_g = \left[\cdots a_n^{(g)} \cdots\right]^t_{n=1,\ldots,N_g},\quad \mathbf{b}_g = \left[\cdots b_n^{(g)} \cdots\right]^t_{n=1,\ldots,N_g},\quad g \equiv w,s.$$

The normalization constants are collected in the diagonal matrices

$$\mathbf{Q}_g = \operatorname{diag}\left[Q_n^{(g)}\right]_{n=1,\ldots,N_g},\quad g \equiv w,s,\qquad [N_g \times N_g]. \tag{11}$$

and the inner cross product in a full rectangular matrix

$$[X_{mn}] = \iint_{A_s} \vec{\mathbf{e}}_m^{(s)} \times \vec{\mathbf{h}}_n^{(w)} \cdot \hat{\mathbf{z}} dS,\qquad [N_s \times N_w]. \tag{12}$$

With these definitions, the linear system obtained from the boundary conditions is expressed as

$$\left\{ \begin{array}{lll} \text{EFBC:} & \mathbf{Q}_w(\mathbf{a}_w + \mathbf{b}_w) = \mathbf{X}^t(\mathbf{a}_s + \mathbf{b}_s) & (N_w \text{ eqs.}) \\ \text{MFBC:} & \mathbf{X}(\mathbf{a}_w - \mathbf{b}_w) = \mathbf{Q}_s(\mathbf{b}_s - \mathbf{a}_s) & (N_s \text{ eqs.}) \end{array} \right. . \tag{13}$$

2.3 Generalized Scattering Matrix of the waveguide step

The above system (13) can be used to find the value of $\mathbf{b}_w, \mathbf{b}_s$ in terms of $\mathbf{a}_w, \mathbf{a}_s$. This relation is the Generalized Scattering Matrix (GSM) representation of the waveguide step and it can be expressed as:

$$\begin{bmatrix} \mathbf{b}_w \\ \mathbf{b}_s \end{bmatrix} = \begin{bmatrix} \mathbf{S}_{ww} & \mathbf{S}_{ws} \\ \mathbf{S}_{sw} & \mathbf{S}_{ss} \end{bmatrix} \begin{bmatrix} \mathbf{a}_w \\ \mathbf{a}_s \end{bmatrix},\; \mathbf{b} = \mathbf{S}\mathbf{a}.$$

The terms of the GSM, once (13) is used, are:

$$\mathbf{S} = \begin{bmatrix} \mathbf{Q}_w^{-1}\mathbf{X}^t\mathbf{F}\mathbf{X} - \mathbf{I}_w & \mathbf{Q}_w^{-1}\mathbf{X}^t\mathbf{F}\mathbf{Q}_s \\ \mathbf{F}\mathbf{X} & \mathbf{F}\mathbf{Q}_s - \mathbf{I}_s \end{bmatrix},\quad \mathbf{F} = 2(\mathbf{Q}_s + \mathbf{X}\mathbf{Q}_w^{-1}\mathbf{X}^t)^{-1}, \tag{14}$$

where $\mathbf{I}_g$ is the identity matrix of size N_g. Therefore, in conclusion, for obtaining the GSM is only required to select a set of modes in each waveguide, to fill the normalization and inner cross product matrices (discussed below) and to perform the previous matrix operations.

The question to address now is the effect of varying the number of selected modes on the solution of the problem. In addition, since different modal series are involved, the solution also depends on the relative truncation criterion for N_w/N_s. This problem occurs in any MM approach and is called the relative convergence problem (Leroy, 1983), (Itoh (editor), 1989, Ch. 9 and 11).

The issue to highlight is that the convergence depends not only on the number of modes, but also in the relation between the numbers used in the different expansions. If the relation is not appropriate, the results may not converge even increasing the number of modes. It has been shown that the optimum mode selection for a bifurcation in parallel-plate waveguide (Mittra & Lee, 1971) is a number of modes proportional to the aspect ratio of each waveguide cross section. The key idea is that the different expansions must reach the same maximum spatial resolution.

In practice, this result is generalized for other discontinuities and a usual criterion is to select the mode ratio the same as the aspect ratio. However, this is a starting point and does not guarantee the convergence. Other similar criterion is to take into account all the modes in the waveguides whose cutoff wavenumbers are lower than a certain $k_{c,max}$. Then, the convergence is checked looking at the results when $k_{c,max}$ is increased.

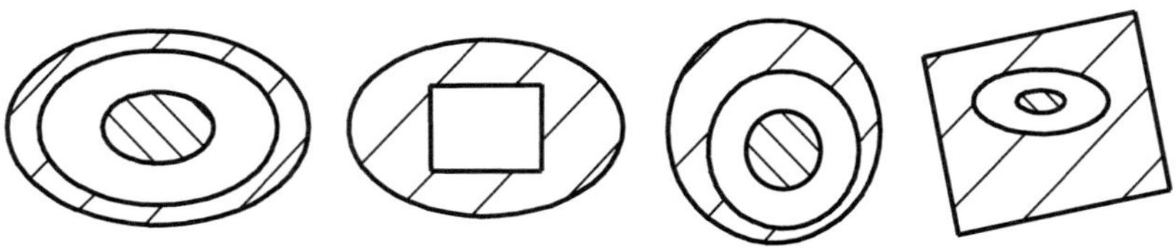

Fig. 3. Examples of cross-sections in discontinuities involving canonical waveguides (rectangular, circular, elliptical, coaxial).

2.4 Inner cross products between the modal fields

The computation of the inner cross products is better addressed if they are factorized as follows. They can be divided into a geometry factor (the normalized inner cross products) and the frequency and waveguide medium information (the modal wave impedances/admittances and, maybe, the normalization Q). According to the expressions for the modal fields in App. A (equations (33),(36),(39)), the inner products can be factorized as follows

$$X_{mn} = \iint_{A_s} \vec{\mathbf{e}}_m^{(s)} \times \vec{\mathbf{h}}_n^{(w)} \cdot \hat{\mathbf{z}} dS = \{(Q_m^{(s)})^{\frac{1}{2}} (Z_m^{(s)})^{\frac{1}{2}}\} \bar{X}_{mn} \{(Y_n^{(w)})^{\frac{1}{2}} (Q_n^{(w)})^{\frac{1}{2}}\}. \tag{15}$$

The terms $\bar{X}_{mn}$ (the normalized inner cross products) are real numbers independent of frequency and the waveguide media and exclusively dependent on the geometry of the discontinuity, since they only involve the real functions defined in App. A:

$$\bar{X}_{mn} = \iint_{A_s} \vec{\mathbf{\Phi}}_{Em}^{(s)} \times \vec{\mathbf{\Phi}}_{Hn}^{(w)} \cdot \hat{\mathbf{z}} dS = \iint_{A_s} \vec{\mathbf{\Phi}}_{Em}^{(s)} \cdot \vec{\mathbf{\Phi}}_{En}^{(w)} dS = \iint_{A_s} \vec{\mathbf{\Phi}}_{Hm}^{(s)} \cdot \vec{\mathbf{\Phi}}_{Hn}^{(w)} dS. \tag{16}$$

Thus, in a frequency sweep, they can be computed once and be multiplied by diagonal matrices to update the GSM expression (14) at any frequency point:

$$\mathbf{X} = \{\mathbf{Q}_s^{\frac{1}{2}} \mathbf{Z}_s^{\frac{1}{2}}\} \bar{\mathbf{X}} \{\mathbf{Y}_w^{\frac{1}{2}} \mathbf{Q}_w^{\frac{1}{2}}\}. \tag{17}$$

To simplify the notation, $\bar{X}_{sw}$ is used now (instead of $\bar{X}_{mn}$) for an inner product between a mode of the smaller waveguide obtained from Φ_s and one of the larger derived from Φ_w. There are two equivalent approaches (shown in Table 1) to calculate $\bar{X}_{sw}$: the surface integral formulation obtained from the definition and the contour integral formulation derived later in (Figlia & Gentili, 2002), (Bozzi et al., 2002). In addition, it can be shown that there are some

$\bar{X}_{sw}$	TEM$_w$	TE$_w$	TM$_w$
TEM$_s$	$\iint_{A_s} \nabla_t\Phi_w \cdot \nabla_t\Phi_s dS$	0	$\iint_{A_s} \nabla_t\Phi_w \cdot \nabla_t\Phi_s dS$
TE$_s$	$\iint_{A_s} \nabla_t\Phi_w \times \nabla_t\Phi_s \cdot \hat{\mathbf{z}} dS$	$\iint_{A_s} \nabla_t\Phi_w \cdot \nabla_t\Phi_s dS$	$\iint_{A_s} \nabla_t\Phi_w \times \nabla_t\Phi_s \cdot \hat{\mathbf{z}} dS$
TM$_s$	0	0	$\iint_{A_s} \nabla_t\Phi_w \cdot \nabla_t\Phi_s dS$

$$\bar{X}_{sw} = \iint_{A_s} \vec{\mathbf{\Phi}}_{Es} \times \vec{\mathbf{\Phi}}_{Hw} \cdot \hat{\mathbf{z}}\, dS$$

(s)-(w)	$\bar{X}_{sw}$
TEM-TEM	$\oint_{C_s} \Phi_w \nabla_t \Phi_s \cdot \hat{\mathbf{n}}_s dl = \oint_{C_s} \Phi_s \nabla_t \Phi_w \cdot \hat{\mathbf{n}}_s dl$
TEM-TM	$\oint_{C_s} \Phi_w \nabla_t \Phi_s \cdot \hat{\mathbf{n}}_s dl$
TE-TEM	$\oint_{C_s} \Phi_w \nabla_t \Phi_s \cdot (\hat{\mathbf{z}} \times \hat{\mathbf{n}}_s) dl = -\oint_{C_s} \Phi_s \nabla_t \Phi_w \cdot (\hat{\mathbf{z}} \times \hat{\mathbf{n}}_s) dl$
TE-TM	$\oint_{C_s} \Phi_w \nabla_t \Phi_s \cdot (\hat{\mathbf{z}} \times \hat{\mathbf{n}}_s) dl = -\oint_{C_s} \Phi_s \nabla_t \Phi_w \cdot (\hat{\mathbf{z}} \times \hat{\mathbf{n}}_s) dl$
TE-TE	$\frac{k_{cs}^2}{k_{cs}^2 - k_{cw}^2} \oint_{C_s} \Phi_s \nabla_t \Phi_w \cdot \hat{\mathbf{n}}_s dl \qquad k_{cs} \neq k_{cw}$ $\frac{-1}{2} \oint_{C_s} \Phi_s \nabla_t (\rho_{ar} \frac{\partial \Phi_w}{\partial \rho_{ar}}) \cdot \hat{\mathbf{n}}_s dl = \frac{-k_{cw}}{2} \oint_{C_s} \Phi_s \frac{\partial(\nabla_t \Phi_w \cdot \hat{\mathbf{n}}_s)}{\partial k_{cw}} dl \qquad k_{cs} = k_{cw}$
TM-TM	$\frac{k_{cw}^2}{k_{cw}^2 - k_{cs}^2} \oint_{C_s} \Phi_w \nabla_t \Phi_s \cdot \hat{\mathbf{n}}_s dl \qquad k_{cs} \neq k_{cw}$ $\frac{1}{2} \oint_{C_s} \rho_{ar} \frac{\partial \Phi_w}{\partial \rho_{ar}} \nabla_t \Phi_s \cdot \hat{\mathbf{n}}_s dl = \frac{k_{cw}}{2} \oint_{C_s} \frac{\partial \Phi_w}{\partial k_{cw}} \nabla_t \Phi_s \cdot \hat{\mathbf{n}}_s dl \qquad k_{cs} = k_{cw}$

Table 1. Normalized inner cross products: surface and contour integral formulations (Figlia & Gentili, 2002), (Bozzi et al., 2002). ($C_s = \delta A_s$ is the contour of A_s.)

type of inner cross products that are always zero (Gentili, 1991). These cases are included in the same Table 1.
The computation of the integrals in $\bar{X}_{sw}$ for a discontinuity between two rectangular waveguides can be done analytically leading to very compact expressions (in fact, this problem was one of the basis for the developing of MM (Patzelt & Arndt, 1982)). However, this is not always the case, since the modal functions can be obtained by means of other numerical methods. If a numerical evaluation of the integrals is required, the contour integral formulation provides the most efficient approach, although it is more sensitive to numerical errors than the surface integral version.
Moreover, even when the TEM, TE and TM solutions Φ_s, Φ_w are available with closed expressions (see App. A.5) the integrals may not be direct and must involve a careful study. Some discontinuity cross-sections are shown in Fig. 3 and a fast computation of $\bar{X}_{sw}$ may involve transformations among plane, circular and elliptical waves. Some examples are found in (Orfanidis et al., 2000), (Zhongxiang & MacPhie, 1995), (MacPhie & Wu, 1995), (Mongiardo & Tomassoni, 2000), (Chan & Judah, 1997).

2.5 Properties of the GSM obtained by MM

The formal properties that must satisfy the GSM (including evanescent modes) of a waveguide junction (not only for a waveguide step) can be shown by applying the classic electromagnetic theorems of Lorentz, Poynting and Self-Reaction (Haskal, 1964).

The properties are investigated now in the context of the MM formulations (Omar et al., 1994). The matter to discern is when these properties give any information about how the electromagnetic problem is being solved.

The properties will be related to the normalization of the modes. For instance, the GSM in (14) is Q-symmetric ($\mathbf{QS} = \mathbf{S}^t\mathbf{Q}$) and self-inverting ($\mathbf{S} = \mathbf{S}^{-1}$ or $\mathbf{SS} = \mathbf{I}$). This is shown exclusively by means of algebraic operations, beginning with (14) and using $\mathbf{X}\mathbf{Q}_w^{-1}\mathbf{X}^t = 2\mathbf{F}^{-1} - \mathbf{Q}_s$. Therefore, these properties hold regardless of the number of modes retained in the field expansions and the value (whether calculated correctly or not) of the inner products.

An alternative derivation of these properties is based on showing that equations (13) impose the same type of relations that the Lorentz and Self-Reaction theorems. This approach is now followed for the Poynting theorem. The proof requires to use the diagonal matrices containing the complex power carried by each mode in each waveguide (g) of the discontinuity

$$\mathbf{P}_g = \mathrm{diag}\left[P_n^{(g)}\right]_{n=1,\ldots,N_g}, \quad P_n^{(g)} = \iint_{A_g} \vec{\mathbf{e}}_n^{(g)} \times \vec{\mathbf{h}}_n^{(g)*} \cdot \hat{\mathbf{z}} dS.$$

In addition, the modal amplitudes and the matrices are arranged in matrix form, as for instance to express the relation (43) shown in the App. A.4 between $\mathbf{P}$ and $\mathbf{Q}$

$$\mathbf{P} = \mathbf{Z}^{\frac{1}{2}}(\mathbf{Y}^{\frac{1}{2}})^{\dagger}\mathbf{Q}^{\frac{1}{2}}(\mathbf{Q}^{\frac{1}{2}})^{\dagger}, \tag{18}$$

where † is the transpose and conjugate matrix operation. A further block division, exclusive for lossless ports, classifies the modes in propagating (p) and evanescent (v), regardless of its physical port:

$$\mathbf{S} = \begin{bmatrix} \mathbf{S}_{pp} & \mathbf{S}_{pv} \\ \mathbf{S}_{vp} & \mathbf{S}_{vv} \end{bmatrix}, \quad \mathbf{P} = \begin{bmatrix} \mathbf{P}_p & 0 \\ 0 & \mathbf{P}_v \end{bmatrix}, \quad \mathbf{Q} = \begin{bmatrix} \mathbf{Q}_p & 0 \\ 0 & \mathbf{Q}_v \end{bmatrix}.$$

The $\mathbf{S}_{pp}$ matrix is what is usually understood by the S-parameters of the device, and it is extracted from the GSM just taking the parameters relating propagating modes. These are the parameters (in dB) that are plotted later in Sec. 4.

The Poynting theorem is related to the complex power flowing into $+\hat{\mathbf{z}}$ at port w:

$$\Omega_{A_w} = \frac{1}{2}\iint_{A_w} \vec{\mathbf{E}}^{(w)} \times \vec{\mathbf{H}}^{(w)*} \cdot \hat{\mathbf{z}} dS = \frac{1}{2}(\mathbf{a}_w - \mathbf{b}_w)^{\dagger}\mathbf{P}_w(\mathbf{a}_w + \mathbf{b}_w). \tag{19}$$

A similar expression would hold for port s. Now, Ω_{A_w} is transformed by (13) along with the following relation

$$\mathbf{P}_w\mathbf{Q}_w^{-1}\mathbf{X}^t = (\mathbf{P}_s^{\dagger}\mathbf{Q}_s^{-1}\mathbf{X})^{\dagger}. \tag{20}$$

This last expression comes from the $\mathbf{P}$ and $\mathbf{Q}$ relation (18) and the $\mathbf{X}$ factorization (17). As a result, it can be shown that $\Omega_{A_w} = \Omega_{A_s}$, *i.e.*, the complex power flow is conserved when using (13).

This is the relation that the Poynting theorem states for this particular junction. Hence, the results from (Haskal, 1964) could be applied here. Particularly, if the ports are lossless, $\mathbf{S}_{pp}$

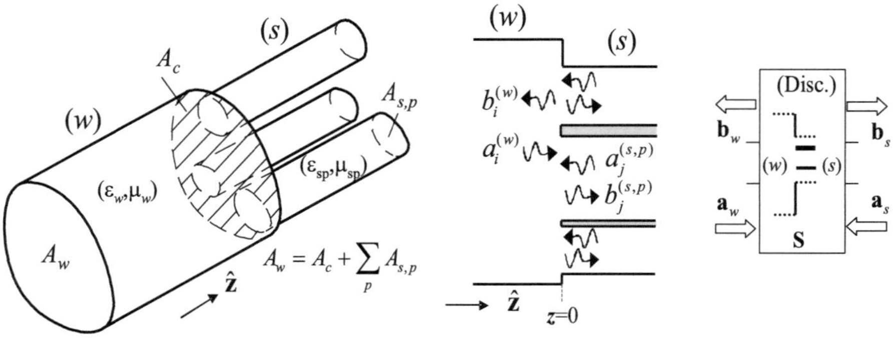

Fig. 4. Waveguide step with one main waveguide (w) and P smaller waveguides. This the multiport case (or P-furcation) whose GSM is also obtained by MM.

is P_p-unitary ($\mathbf{P}_p\mathbf{S}_{pp}^{-1} = \mathbf{S}_{pp}^{\dagger}\mathbf{P}_p$). This yields the familiar unitary property $\mathbf{S}_{pp}^{-1} = \mathbf{S}_{pp}^{\dagger}$ under a normalization with $\mathbf{P}_p = |\mathbf{Q}_p| = \mathbf{I}$.
This last result relies exclusively on the form of the linear system (13) and (20). If the real matrix $\bar{\mathbf{X}}$ is wrong and $\mathbf{X}$ is generated by (17), the P_p-unitary property continues being satisfied.
In conclusion, the GSM in (14) is self-inverting ($\mathbf{S} = \mathbf{S}^{-1}$). Upon the common modal normalization $\mathbf{Q} = \mathbf{I}$, the GSM is symmetric ($\mathbf{S} = \mathbf{S}^t$) and, for lossless ports, $\mathbf{S}_{pp}$ is unitary ($\mathbf{S}_{pp}^{-1} = \mathbf{S}_{pp}^{\dagger}$). These properties are satisfied exclusively by the form of the equations to be solved, without any relation to the boundary condition fulfillment. They are guaranteed by how the GSM is constructed, and they do not give any information about how the electromagnetic problem is being solved.

3. Extension to other type of steps and cascading

3.1 Extension to the multiport case: P-furcation

The formulation can be easily extended to the multiport case in Fig. 4, provided that the inner product matrix is constructed in blocks corresponding to the subregions:

$$\mathbf{X} = \left[\cdots\ \mathbf{X}^{(p)t}\ \cdots \right]^t_{p=1,\ldots,P}, \quad \left[\sum_{p=1}^{P} N_{s,p} \times N_w\right]. \tag{21}$$

The block matrices $\mathbf{X}^{(p)}$ are defined like (12) and contain the inner products between the electric modal fields in waveguide (s,p) and the magnetic modal fields in waveguide (w), integrating in the surface of the smaller waveguide $A_{s,p}$. In this problem, the multiport (s) collects the $N_{s,p}$ amplitudes of all the different modal series in each waveguide (s,p).

3.2 Extension to consider the losses at the discontinuity wall

An extension of the above formulation is proposed in (Shen & MacPhie, 1990) to evaluate the effect of the losses produced by real conductors in the steps of Figs. 2 and 4. The field-matching procedure relies on assuming a surface impedance boundary condition (Leontovich condition (Senior, 1960)) at the wall A_c:

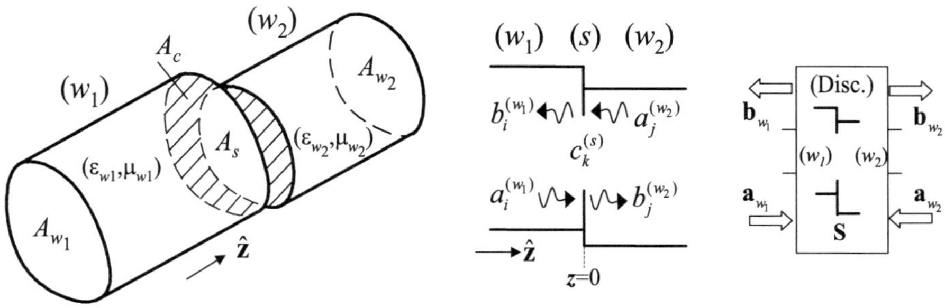

Fig. 5. Waveguide discontinuity with general aperture step (compare with Fig. 2).

$$\begin{cases} \text{EFBC in } A_w: \quad \hat{\mathbf{z}} \times \vec{\mathbf{E}}^{(w)} = \begin{cases} Z_c(\vec{\mathbf{H}}^{(w)} \times \hat{\mathbf{z}}), & \text{in } A_c, z = 0 \\ \hat{\mathbf{z}} \times \vec{\mathbf{E}}^{(s)} & \text{in } A_s, z = 0 \end{cases} \\ \text{MFBC in } A_s: \quad \hat{\mathbf{z}} \times \vec{\mathbf{H}}^{(w)} = \hat{\mathbf{z}} \times \vec{\mathbf{H}}^{(s)}, \quad \text{in } A_s, z = 0 \end{cases}$$

with surface impedance, assuming good conductor, given by:

$$Z_c = Y_c^{-1} = \frac{(1+j)}{\sigma\delta} = (1+j)\sqrt{\frac{\pi f \mu}{\sigma}}. \tag{22}$$

The boundary conditions are imposed by means of a *Galerkin* method, providing:

$$\begin{cases} \text{EFBC:} \quad \mathbf{Q}_w(\mathbf{a}_w + \mathbf{b}_w) = \mathbf{X}^t(\mathbf{a}_s + \mathbf{b}_s) + \mathbf{L}_c(\mathbf{a}_w - \mathbf{b}_w) \quad (N_w \text{ eqs.}) \\ \text{MFBC:} \quad \mathbf{X}(\mathbf{a}_w - \mathbf{b}_w) = \mathbf{Q}_s(\mathbf{b}_s - \mathbf{a}_s) \quad (N_s \text{ eqs.}). \end{cases} \tag{23}$$

The resolution of this linear system yields the GSM of the step in the following form:

$$\mathbf{S} = \begin{bmatrix} \mathbf{I} - \mathbf{F}\mathbf{Q}_w & \mathbf{F}\mathbf{X}^t \\ \mathbf{Q}_s^{-1}\mathbf{X}\mathbf{F}\mathbf{Q}_w & \mathbf{I} - \mathbf{Q}_s^{-1}\mathbf{X}\mathbf{F}\mathbf{X}^t \end{bmatrix}, \quad \mathbf{F} = 2(\mathbf{Q}_w + \mathbf{X}^t\mathbf{Q}_s^{-1}\mathbf{X} + \mathbf{L}_c)^{-1}. \tag{24}$$

The inner products in $\mathbf{X}$ remains as (12) and, in the multiport case, are constructed in blocks as (21). The new inner products are expressed as

$$[L_{c,mn}] = Z_c \iint_{A_c} \vec{\mathbf{h}}_m^{(w)} \cdot \vec{\mathbf{h}}_n^{(w)} dS, \quad [N_w \times N_w], \tag{25}$$

and are calculated by integrating the modes of the larger waveguide on the conductor wall. Coherently, when the conductivity is infinite, $\mathbf{L}_c$ is null and (24) becomes (14).

3.3 Extension for the general aperture case

This section presents another type of step shown in Fig. 5, whose main feature is that the aperture at the discontinuity does not coincide with any of the input/output waveguide cross sections. In this case, a new electric field expansion is made at the aperture surface $A_s \subseteq A_{w_1}, A_{w_2}$, expressing the boundary conditions as:

$$\vec{\mathbf{E}}^{(s)}\Big|_{A_s,z=0} = \sum_{k=1}^{N_s} c_k^{(s)} \vec{\mathbf{e}}_k^{(s)}, \begin{cases} \text{EFBC in } A_g: & \hat{\mathbf{z}} \times \vec{\mathbf{E}}^{(g)} = \begin{cases} 0, & \text{in } A_c, z=0 \\ \hat{\mathbf{z}} \times \vec{\mathbf{E}}^{(s)} & \text{in } A_s, z=0 \end{cases} \\ (g \equiv w_1, w_2) & \\ \text{MFBC in } A_s: & \hat{\mathbf{z}} \times \vec{\mathbf{H}}^{(w_1)} = \hat{\mathbf{z}} \times \vec{\mathbf{H}}^{(w_2)}, \text{ in } A_s, z=0. \end{cases}$$

The two EFBC's (in A_{w_1}, A_{w_2}) are tested by the magnetic modal fields in (w_1) and (w_2), respectively, providing $N_{w1} + N_{w2}$ equations. The MFBC is tested by $\vec{\mathbf{e}}_k^{(s)}$ (the modal electric fields of the expansion at A_s), resulting in N_s equations. This leads to the following linear system with new inner products X:

$$\begin{cases} \text{EFBC:} & \mathbf{Q}_{w_1}(\mathbf{a}_{w_1} + \mathbf{b}_{w_1}) = \mathbf{X}_{w_1}^t \mathbf{c}_s, \; \mathbf{Q}_{w_2}(\mathbf{a}_{w_2} + \mathbf{b}_{w_2}) = \mathbf{X}_{w_2}^t \mathbf{c}_s \\ \text{MFBC:} & \mathbf{X}_{w_1}(\mathbf{a}_{w_1} - \mathbf{b}_{w_1}) = \mathbf{X}_{w_2}(\mathbf{b}_{w_2} - \mathbf{a}_{w_2}) \end{cases}, \tag{26}$$
$$[X_{g,mn}] = \iint_{A_s} \vec{\mathbf{e}}_m^{(s)} \times \vec{\mathbf{h}}_n^{(g)} \cdot \hat{\mathbf{z}} dS, \; g \equiv w_1, w_2, \; [N_s \times N_g].$$

This system is solved for the GSM of the general aperture step:

$$\mathbf{S} = \begin{bmatrix} \mathbf{Q}_{w_1}^{-1} \mathbf{X}_{w_1}^t \mathbf{F} \mathbf{X}_{w_1} - \mathbf{I}_{w_1} & \mathbf{Q}_{w_1}^{-1} \mathbf{X}_{w_1}^t \mathbf{F} \mathbf{X}_{w_2} \\ \mathbf{Q}_{w_2}^{-1} \mathbf{X}_{w_2}^t \mathbf{F} \mathbf{X}_{w_1} & \mathbf{Q}_{w_2}^{-1} \mathbf{X}_{w_2}^t \mathbf{F} \mathbf{X}_{w_2} - \mathbf{I}_{w_2} \end{bmatrix} \tag{27}$$

with $\mathbf{F} = 2(\mathbf{X}_{w_1} \mathbf{Q}_{w_1}^{-1} \mathbf{X}_{w_1}^t + \mathbf{X}_{w_2} \mathbf{Q}_{w_2}^{-1} \mathbf{X}_{w_2}^t)^{-1}$.

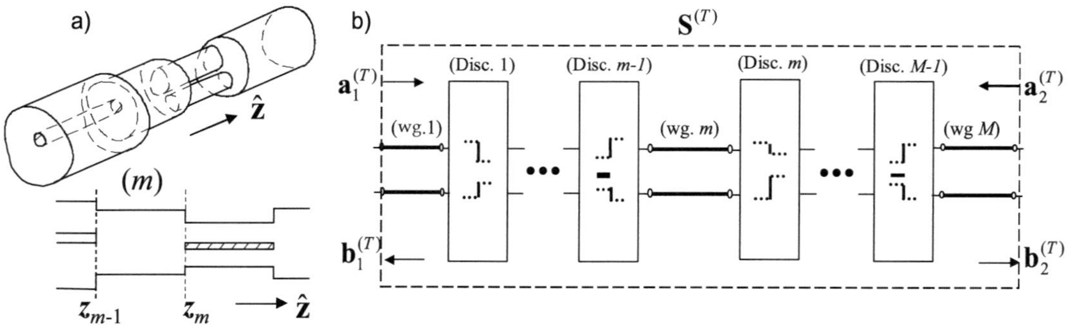

Fig. 6. a) Several waveguides with different cross section cascaded in the longitudinal direction, with its segmentation scheme in $\hat{\mathbf{z}}$; b) Cascading of GSMs. Each one could represent a waveguide step (or any other building-block in other structures).

3.4 Cascading of waveguide discontinuities

The overall characterization of a structure made up of a number of discontinuities as in Fig. 6.a is obtained by cascading the individual GSMs, using simple matrix operations. The building-blocks for the structure in Fig. 6.a belong to the waveguide steps shown in previous sections, providing the block scheme shown in Fig. 6.b.

Consider two consecutive blocks (A) and (B) in Fig. 6.b (they could be for instance the discontinuities m and $m-1$). Each block is characterized by its own GSM: $\mathbf{b}^{(A)} = \mathbf{S}^{(A)}\mathbf{a}^{(A)}$ and $\mathbf{b}^{(B)} = \mathbf{S}^{(B)}\mathbf{a}^{(B)}$. Both blocks are connected by a waveguide (which could have multiple regions) of length l, where N_{AB} modes are considered. Since the modal amplitudes vary with with $e^{-\gamma_n l}$ (see (30)), the vectors with the modal amplitudes at that region are related by:

$$\mathbf{a}_1^{(B)} = \mathbf{Y}\mathbf{b}_2^{(A)}, \ \mathbf{a}_2^{(A)} = \mathbf{Y}\mathbf{b}_1^{(B)}, \ \mathbf{Y} = \text{diag}[e^{-\gamma_n l}]_{n=1,\ldots N_{AB}}. \tag{28}$$

It is implicitly considered that $\mathbf{S}^{(A)}$ and $\mathbf{S}^{(B)}$ have computed with $N_{AB} = N_2^{(A)} = N_1^{(B)}$. After the required algebra, the new GSM $\mathbf{b}^{(C)} = \mathbf{S}^{(C)}\mathbf{a}^{(C)}$ representing the whole composite block is given by:

$$\mathbf{S}^{(C)} = \begin{bmatrix} \mathbf{S}_{11}^{(A)} + \mathbf{S}_{12}^{(A)}\mathbf{Y}\mathbf{H}\mathbf{S}_{11}^{(B)}\mathbf{Y}\mathbf{S}_{21}^{(A)} & \mathbf{S}_{12}^{(A)}\mathbf{Y}\mathbf{H}\mathbf{S}_{12}^{(B)} \\ \mathbf{S}_{21}^{(B)}\mathbf{Y}(\mathbf{I}_{AB} + \mathbf{S}_{22}^{(A)}\mathbf{Y}\mathbf{H}\mathbf{S}_{11}^{(B)}\mathbf{Y})\mathbf{S}_{21}^{(A)} & \mathbf{S}_{22}^{(B)} + \mathbf{S}_{21}^{(B)}\mathbf{Y}\mathbf{S}_{22}^{(A)}\mathbf{Y}\mathbf{H}\mathbf{S}_{12}^{(B)} \end{bmatrix}, \tag{29}$$

$$\mathbf{H} = (\mathbf{I}_{AB} - \mathbf{S}_{11}^{(B)}\mathbf{Y}\mathbf{S}_{22}^{(A)}\mathbf{Y})^{-1}, \quad [N_{AB}, N_{AB}].$$

This process can be repeated iteratively till the overall GSM representing the total structure (T) in Fig. 6.b is obtained. This procedure is not only found in the context of the Mode-Matching method. It can be applied to many other problems, where the structure is segmented in blocks. The GSM of those blocks is obtained by any suitable method and the whole response is obtained by cascading.

4. Analysis and design of waveguide devices

4.1 Design approach

The previous sections have been focused on the analysis part of the CAD for waveguide devices. This part has to be complemented with a design approach, which is outlined now. Fig. 7 illustrates the typical flow chart used in many microwave devices. From the given specifications, a circuit model (if available) is obtained to lead the synthesis of the waveguide structure. Next, an initial set of physical dimensions are determined based on simple circuits and/or simplified models. Once the complete physical model is generated, a MM simulation is performed to obtain the S-parameters of the structure.

This preliminary response of the device using the initial dimensions will usually be relatively poor, since the original design did not take into account the higher-order mode interactions among the different parts of the structure. Then, the simulated response is compared with the circuit model response, and an error or cost function is computed. Using an optimization routine, the dimensions of the device are adjusted to minimize the cost function (see Fig. 7). This process is repeated until the desired response is achieved. This approach has been followed for the structures introduced in the next subsections.

4.2 H-plane bandpass filter in rectangular waveguide

An *H*-plane filter (see Fig. 8.a) is used to illustrate the different MM formulations shown above. The filter is analyzed considering that the conductor walls are not perfect in order to evaluate the insertion loss produced by real conductors (Ruiz-Cruz et al., 2002). This subject is an important engineering task, especially for passive components in satellite communication systems.

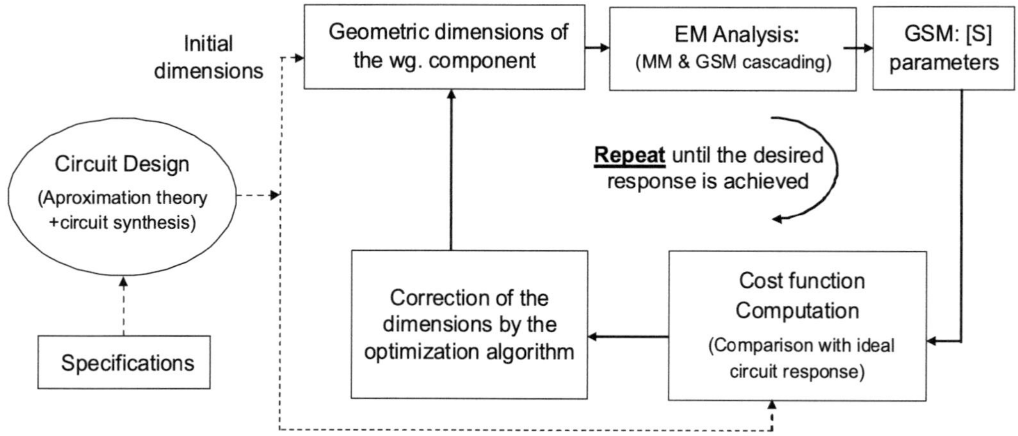

Fig. 7. Optimization cycle for designing a waveguide component.

Starting with the ideal case ($\sigma = \infty$), the different discontinuities are computed by formulation in Subsec. 2.2 and cascaded together to get the filter response. However, for the case $\sigma \neq \infty$, two models are used: i) the formulation in Subsec. 3.2, using the Leontovich condition at walls A_c (Fig. 8.b) and ii) the use of a trifurcation with lossy waveguides (Fig. 8.c). This last model simulates the solid conductors by imaginary waveguides filled with a lossy dielectric. The resultant trifurcation problem is solved as in Subsec. 3.1.
Both models take into account the losses at the lateral waveguide walls by means of modifying the propagation constant of the propagating modes (Collin, 1991), adding an attenuation constant α_c owing to the finite metal conductivity: $\gamma|_{\sigma\neq\infty} \approx \alpha_c + \gamma|_{\sigma=\infty}$. This is a common approximation for good conductors.
The results of a pass-band Chebychev filter, with 2.4% relative bandwidth in Ku frequency band, are shown in Fig. 8.d. The designed sixth order filter is implemented by means of symmetric inductive irises in WR75 waveguide. The responses that are being compared in Fig. 8.d correspond to the measurements and the simulations in the ideal and lossy cases. As expected, the three return loss responses are similar for good conductors (the simulations used $\sigma = 4.8 \cdot 10^7 S/m$).
Fig. 8.e shows the detail for the insertion loss for the Leontovich, trifurcation, FEM (HFSS, available at www.ansoft.com), circuit models and the measured results. The circuit model consists on the classic impedance inverters connected by $\lambda_g/2$ transmission lines, using the approximation for the propagation constant $\gamma|_{\sigma\neq\infty}$. It is seen that the trifurcation approach yields values slightly under the predicted by the Leontovich model. However, the agreement among the different procedures for the insertion loss and the experimental results is satisfactory.

4.3 Coaxial probe and stripline bifurcation

Other interesting problem that can been analyzed by MM is a coaxial probe inside a rectangular waveguide, shown in the inset of Fig. 9.a. From a CAD point of view, this structure can be represented as the cascade of uniform waveguides with different cross sections and, then, suitable for being solved by MM.

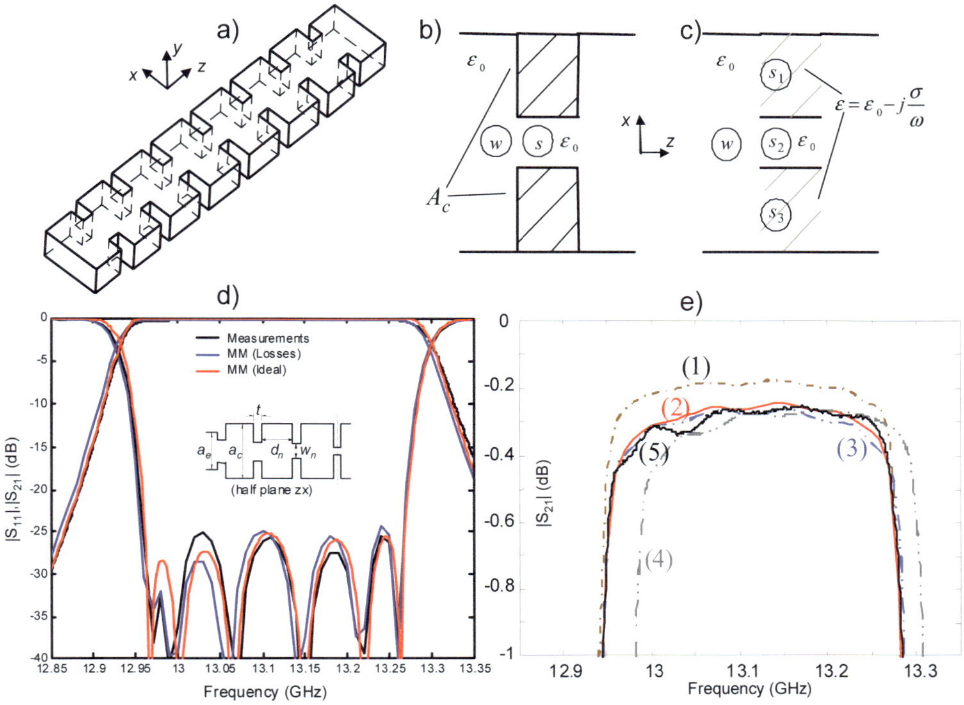

Fig. 8. a) Rectangular waveguide H-plane filter; b) Leontovich model: the cavity-iris discontinuity takes into account non perfect conductor boundary conditions at the wall A_c; c) Trifurcation model: the non perfect conductor is replaced by a virtual rectangular waveguide with lossy dielectric. d) Return and insertion loss. e) Detail of the insertion loss: 1-trifurcation model, 2-Leontovich formulation, 3-circuit model, 4-HFSS, 5-measurements. Dimensions (mm): $a_e = 19.05$, $b_e = 9.525$, $a_c = 21.9$, $t = 3.75$, $d_{1-3} = 10.51, 11.95, 12.13$, $w_{1-4} = 9.73, 6.68, 6.11, 6.03$.

In this scope, the modes of the waveguide in the probe region can be calculated in different ways. The model followed here (Ruiz-Cruz et al., 2004) consists of a generalized stripline whose inner conductor has a stepped profile in order to approximate the desired cross section of the probe. Depending on the application, the probe can have a different size than the inner conductor of the coaxial line, but it is usually very thin. Therefore, two or three steps, even one (*i.e.* square) in some cases, give accurate results. This is shown not just in one isolated discontinuity but in the response of the cascading of several waveguide steps.
The basic discontinuity to be modeled is made up of a coaxial waveguide and the generalized stripline. Other discontinuities involved in the problem are between ridge waveguides and rectangular waveguides. The modes of these waveguides have been computed using the Generalized Transverse Resonance (GTR) (Itoh (editor), 1989), but other techniques are also possible.
The coaxial-to-rectangular waveguide transition designed and measured in (Gerini & Guglielmi, 2001) is calculated in Fig. 9.a. In the structure, the inner conductor of the SMA connector is extended to contact a ridge waveguide section. The MM response

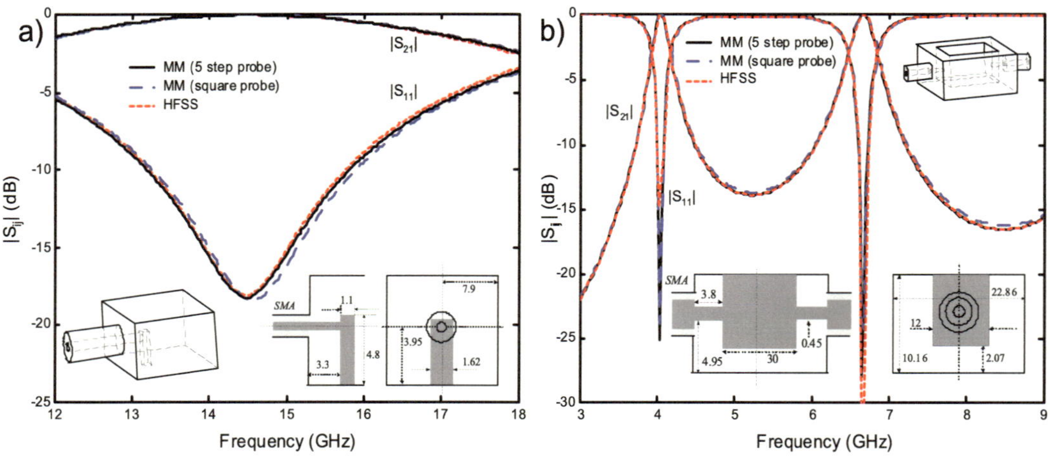

Fig. 9. a) Coaxial (50Ω SMA) to rectangular waveguide transition of (Gerini & Guglielmi, 2001). b) Back-to-back coaxial (50Ω SMA) to ridge waveguide junction. SMA dimensions (mm): $r_{in} = 0.645$, $r_{ex} = 2.045$, $\varepsilon_r = 1.97$.

is compared with the HFSS, showing good agreement but with much less computation time (as in the other examples).

To see the effect of the number of sections in the generalized stripline, the presented results are computed with a square inner conductor whose area is the same as that of the circular probe, obtaining results very close to the HFSS simulation. Since the results are very similar to those with 5 steps, the analysis can be carried out very efficiently with the simple square probe. The results for a back to back coaxial to ridge resonator junction are shown in Fig. 9.b.

Finally, another structure which can be modeled by MM is a bifurcation in stripline (or rectangular coaxial), shown in Fig. 10.a. The structure has an inner conductor and an enclosure that vary in order to divide the power into two isolated stripline ports. The response is shown in Fig. 10.b. The structure is also computed exclusively with TEM modes to illustrate the influence of the higher order modes in the response.

5. Conclusions

This chapter has introduced the main concepts of the Mode-Matching (MM) methods for the CAD of waveguide devices. The key idea is to segment the problem under analysis in different waveguide regions where the electromagnetic field is represented by the superposition of modes. At the interface between regions, those modal series have to be matched to fulfill the boundary conditions. This process leads to the Generalized Scattering Matrix (GSM) representation of the problem.

This idea has been applied to the cascading of several transmission systems with different cross-sections. This type of problem could represent many waveguide devices such as filters, transformers, N-furcations, diplexers, polarizers, etc.

Different types of basic discontinuities have been shown, with their corresponding formulations and intrinsic properties for the obtained GSM. Some examples have been introduced in

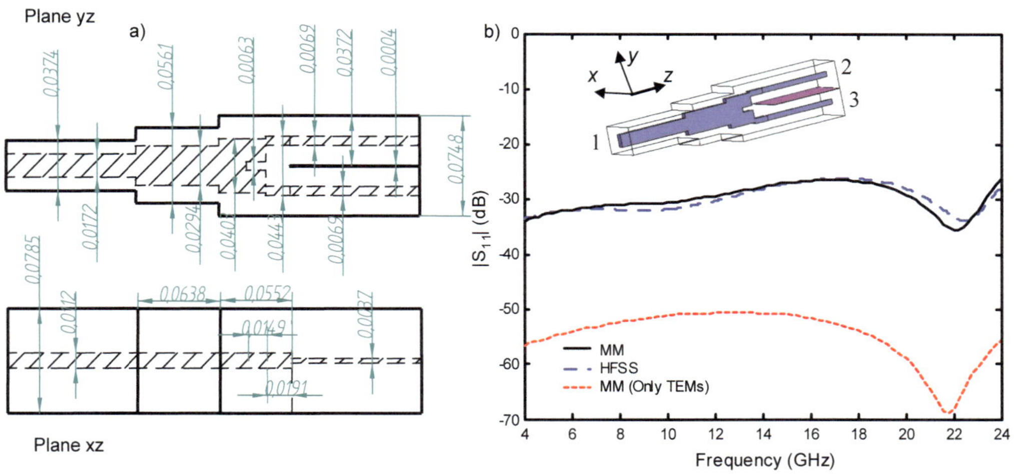

Fig. 10. Configuration of the stripline (rectangular coaxial) bifurcation ($\varepsilon_r = 5.9$). a) Actual aspect ratio and dimensions (inch). b) Return loss at port one.

order to validate the theory with measured experimental data and with the results obtained with other numerical methods.

One important idea to emphasize is the building-block concept, since it provides a simple algorithm to solve more complex problems: the discontinuities are individually characterized and then are cascaded. This idea can be extended to other problems, as long as it is known how to compute the GSM of the building blocks in the structure.

The MM technique has been used for many years and, nowadays, can be considered as an industry standard for some types of problems like the cascading of rectangular or circular waveguides in filters, couplers or horns. Since the formulation of the technique is well known, the latest developments have been oriented towards broadening the range of problems where the method can be applied. The theory shown in this chapter is the basis for more evolved formulations.

A. Appendix. Cylindrical waveguides

The total electromagnetic field in a cylindrical waveguide (see Fig. 11.a) of arbitrary cross section S, with perfect conductivity walls $\sigma = \infty$, filled with isotropic and homogeneous dielectric, can be represented by a modal series as (1) (Collin, 1991).

The homogeneous medium of electric permeability $\varepsilon = \varepsilon_r \varepsilon_0$ and magnetic permeability $\mu = \mu_r \mu_0$ may have losses $\varepsilon_r = \varepsilon_r' - j\varepsilon_r'', \mu_r = \mu_r' - j\mu_r''$, resulting in a complex wavenumber $k = \omega\sqrt{\mu\varepsilon}$ and complex intrinsic impedance of the medium $\eta = \sqrt{\mu/\varepsilon}$. For a lossless waveguide $\varepsilon_r'' = \mu_r'' = 0$.

The pairs $(\vec{\mathbf{E}}_n^{\pm}, \vec{\mathbf{H}}_n^{\pm})$, travelling along $\pm z$, respectively, are the electromagnetic fields of a TEM, TE or TM mode[2]: solutions to Maxwell's equations which satisfy the corresponding boundary conditions at the contour C. The modes are not restricted to be propagating; they have at

[2] The TEM, TE, TM modes are also called O, H, E, respectively. When it is needed to identify them, the scripts $(o), (h), (e)$ will be used.

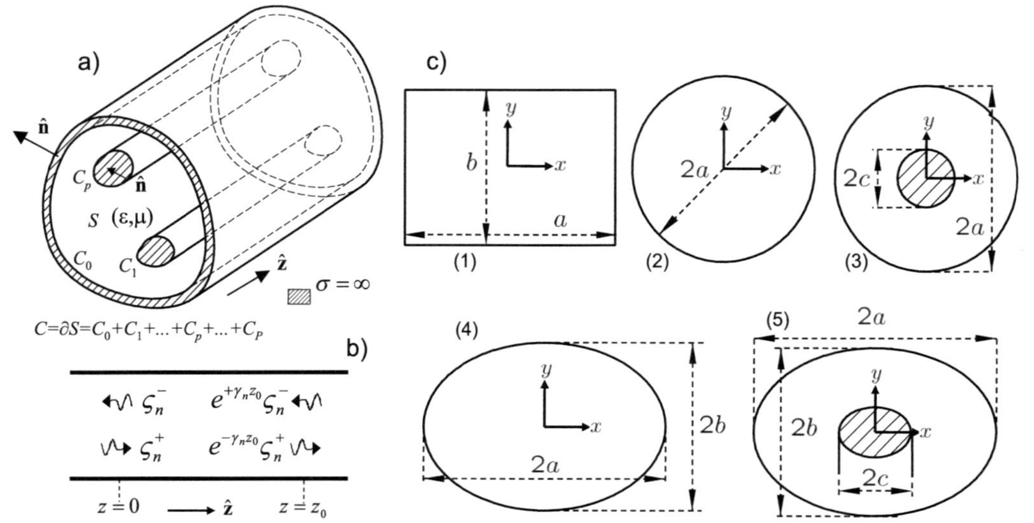

Fig. 11. Multi-conductor cylindrical waveguide closed by perfect conducting walls with homogeneous and isotropic dielectric: a) geometry, b) equivalent transmission line with $\pm\hat{\mathbf{z}}$ modal amplitudes. c) Cross section of some common canonical waveguides: rectangular, circular, circular coaxial, elliptical, confocal elliptical coaxial.

the operating frequency $f=\omega/2\pi$ a complex propagation constant that can be: pure real (evanescent mode), imaginary (propagating mode) or complex with non vanishing real and imaginary part ($\varepsilon_r''\neq 0$ and/or $\mu_r''\neq 0$). They are written as:

$$\vec{\mathbf{E}}_n^+=(\vec{\mathbf{e}}_{tn}^+ + e_{zn}^+\hat{\mathbf{z}})e^{-\gamma_n z} \qquad \vec{\mathbf{E}}_n^-=(\vec{\mathbf{e}}_{tn}^- + e_{zn}^-\hat{\mathbf{z}})e^{+\gamma_n z}$$
$$\vec{\mathbf{H}}_n^+=(\vec{\mathbf{h}}_{tn}^+ + h_{zn}^+\hat{\mathbf{z}})e^{-\gamma_n z} \quad \vec{\mathbf{H}}_n^-=(\vec{\mathbf{h}}_{tn}^- + h_{zn}^-\hat{\mathbf{z}})e^{+\gamma_n z}.$$

Since the forward and backward modal solutions are related by simple sign changes, the notation can be simplified:

$$\vec{\mathbf{e}}_n\equiv\vec{\mathbf{e}}_{tn}^+=\vec{\mathbf{e}}_{tn}^-,\ e_{zn}\equiv e_{zn}^+=-e_{zn}^-,\ \vec{\mathbf{h}}_n\equiv\vec{\mathbf{h}}_{tn}^+=-\vec{\mathbf{h}}_{tn}^-,\ h_{zn}\equiv h_{zn}^+=h_{zn}^-$$

Therefore, (1) is conveniently expressed as:

$$\vec{\mathbf{E}}=\sum_n\{\varsigma_n^+(\vec{\mathbf{e}}_n+e_{zn}\hat{\mathbf{z}})e^{-\gamma_n z}+\varsigma_n^-(+\vec{\mathbf{e}}_n-e_{zn}\hat{\mathbf{z}})e^{+\gamma_n z}\} \tag{30a}$$

$$\vec{\mathbf{H}}=\sum_n\left\{\varsigma_n^+(\vec{\mathbf{h}}_n+h_{zn}\hat{\mathbf{z}})e^{-\gamma_n z}+\varsigma_n^-(-\vec{\mathbf{h}}_n+h_{zn}\hat{\mathbf{z}})e^{+\gamma_n z}\right\}. \tag{30b}$$

The terms $\varsigma_n^{\pm}e^{\mp\gamma_n z}$ are called the modal amplitudes at an arbitrary plane z. Thus, $\varsigma_n^{\pm}$ are the modal amplitudes at $z=0$ (see Fig. 11.b). The vector functions $\vec{\mathbf{e}}_n,\vec{\mathbf{h}}_n$, and the scalars functions e_{zn},h_{zn} exclusively depend on the transversal coordinates and $\vec{\mathbf{e}}_n,\vec{\mathbf{h}}_n$ are perpendicular to $\hat{\mathbf{z}}$. They are related by the modal wave impedance/admittance:

$$\vec{\mathbf{e}}_n=Z_n\vec{\mathbf{h}}_n\times\hat{\mathbf{z}},\quad \vec{\mathbf{h}}_n=Y_n\hat{\mathbf{z}}\times\vec{\mathbf{e}}_n,\quad Z_n=Y_n^{-1}. \tag{31}$$

The formal expressions for the modal fields of any waveguide as in Fig. 11.a are given now. Later, the mode normalization used in the chapter and the specific solutions for the canonical waveguides in Fig. 11.c are introduced.

A.1 TEM modes (*o*-modes)

In a structure with $P+1$ conductors (each one with contour C_p), P linearly independent TEM (*Transversal ElectroMagnetic*) modes must be included in the modal series ($P \geq 0$). They are solution to the Laplace equation with non-homogeneous Dirichlet boundary conditions:

$$\triangle_t \Phi_n = 0, \;\; \Phi_n \Big|_{C_p} = v_{np}, \;\; p = 0, \ldots, P, \;\; n = 1, \ldots, P. \tag{32}$$

The electromagnetic field[3] is given by:

$$\begin{aligned} \vec{\mathbf{e}}_n &= Q_n^{\frac{1}{2}} Z_n^{\frac{1}{2}} \vec{\mathbf{\Phi}}_{En} = Q_n^{\frac{1}{2}} Z_n^{\frac{1}{2}} \nabla_t \Phi_n, & e_{zn} &= 0 \\ \vec{\mathbf{h}}_n &= Q_n^{\frac{1}{2}} Y_n^{\frac{1}{2}} \vec{\mathbf{\Phi}}_{Hn} = Q_n^{\frac{1}{2}} Y_n^{\frac{1}{2}} \hat{\mathbf{z}} \times \nabla_t \Phi_n, & h_{zn} &= 0. \end{aligned} \tag{33}$$

where Q_n is an arbitrary normalization constant and $\vec{\mathbf{\Phi}}_{En}, \vec{\mathbf{\Phi}}_{Hn}$ are obtained from Φ_n by simple gradient operations. The propagation constant and wave impedance for all the TEM modes are:

$$\gamma_n = j\omega\sqrt{\mu\varepsilon} = jk, \quad Z_n = Y_n^{-1} = \sqrt{\tfrac{\mu}{\varepsilon}} = \eta. \tag{34}$$

A.2 TE modes (*h*-modes)

The electromagnetic field of a TE (*Transversal Electric*) mode is obtained by solving the Helmholtz equation with homogeneous Neumann boundary conditions:

$$\triangle_t \Phi_n + k_{cn}^2 \Phi_n = 0, \quad (\nabla_t \Phi_n \cdot \hat{\mathbf{n}}) \rfloor_C = 0, \quad \iint_S |\nabla_t \Phi_n|^2 dS = k_{cn}^2 \iint_S \Phi_n^2 dS = 1. \tag{35}$$

The field, with arbitrary normalization constant Q_n, is expressed as:

$$\begin{aligned} \vec{\mathbf{e}}_n &= Q_n^{\frac{1}{2}} Z_n^{\frac{1}{2}} \vec{\mathbf{\Phi}}_{En} = Q_n^{\frac{1}{2}} Z_n^{\frac{1}{2}} \nabla_t \Phi_n \times \hat{\mathbf{z}}, & e_{zn} &= 0 \\ \vec{\mathbf{h}}_n &= Q_n^{\frac{1}{2}} Y_n^{\frac{1}{2}} \vec{\mathbf{\Phi}}_{Hn} = Q_n^{\frac{1}{2}} Y_n^{\frac{1}{2}} \nabla_t \Phi_n, & h_{zn} &= -Q_n^{\frac{1}{2}} Y_{\omega n}^{\frac{1}{2}} \frac{k_{cn}^2}{\gamma_n} \Phi_n, \end{aligned} \tag{36}$$

where the propagation constant and wave impedance are given by:

$$\gamma_n = \sqrt{k_{cn}^2 - \omega^2 \mu\varepsilon} = \sqrt{k_{cn}^2 - k^2}, \quad Z_n = Y_n^{-1} = \frac{j\omega\mu}{\gamma_n} = \frac{jk\eta}{\gamma_n}. \tag{37}$$

A.3 TM modes (*e*-modes)

The electromagnetic field of a TM (*Transversal Magnetic*) mode is obtained by solving the Helmholtz equation with homogeneous Dirichlet boundary conditions

$$\triangle_t \Phi_n + k_{cn}^2 \Phi_n = 0, \quad \Phi_n \rfloor_C = 0, \quad \iint_S |\nabla_t \Phi_n|^2 dS = k_{cn}^2 \iint_S \Phi_n^2 dS = 1. \tag{38}$$

[3] For maintaining a formal similarity with a TM mode with $k_{cn} = 0$, the electric field is calculated as $\nabla_t \Phi_n$ and not as $-\nabla_t \Phi_n$. Therefore, v_{np} is the opposite of the electrostatic voltage.

The field, with arbitrary normalization constant Q_n, is expressed as:

$$\begin{aligned} \vec{\mathbf{e}}_n &= Q_n^{\frac{1}{2}} Z_n^{\frac{1}{2}} \vec{\mathbf{\Phi}}_{En} = Q_n^{\frac{1}{2}} Z_n^{\frac{1}{2}} \nabla_t \Phi_n, & e_{zn} &= -Q_n^{\frac{1}{2}} Z_n^{\frac{1}{2}} \frac{k_{cn}^2}{\gamma_n} \Phi_n \\ \vec{\mathbf{h}}_n &= Q_n^{\frac{1}{2}} Y_n^{\frac{1}{2}} \vec{\mathbf{\Phi}}_{Hn} = Q_n^{\frac{1}{2}} Y_n^{\frac{1}{2}} \hat{\mathbf{z}} \times \nabla_t \Phi_n, & h_{zn} &= 0 \end{aligned} \tag{39}$$

where the propagation constant and wave impedance are given by[4]:

$$\gamma_n = \sqrt{k_{cn}^2 - \omega^2 \mu \varepsilon} = \sqrt{k_{cn}^2 - k^2}, \quad Z_n = Y_n^{-1} = \frac{\gamma_n}{j\omega\varepsilon} = \frac{\gamma_n \eta}{jk}. \tag{40}$$

A.4 Mode orthogonality and normalization

A very useful property of the modes is the orthogonality:

$$\iint_S \vec{\mathbf{e}}_m \times \vec{\mathbf{h}}_n \cdot \hat{\mathbf{z}} dS = Y_n \iint_S \vec{\mathbf{e}}_m \cdot \vec{\mathbf{e}}_n dS = Z_m \iint_S \vec{\mathbf{h}}_m \cdot \vec{\mathbf{h}}_n dS = Q_n \delta_{mn}, \tag{41}$$

In (Collin, 1991), this relation is shown for modes with different cutoff wave number (non degenerate). In any other case, an orthogonalization process may be carried out. Thus, in this chapter is used a set of modes (33),(36),(39) where (41) always holds.
It is stressed that the Φ solutions (32),(35),(38) and their related vectors $\vec{\mathbf{\Phi}}_E, \vec{\mathbf{\Phi}}_H$ do not depend on the waveguide dielectric and can be written as frequency independent functions that take real values. Q_n is a complex normalization factor that may vary with frequency and can be arbitrarily chosen (for instance it is common to set $Q_n = 1$).
Another important orthogonality relation is

$$\iint_S \vec{\mathbf{e}}_m \times \vec{\mathbf{h}}_n^* \cdot \hat{\mathbf{z}} dS = Y_n^* \iint_S \vec{\mathbf{e}}_m \cdot \vec{\mathbf{e}}_n^* dS = Z_m \iint_S \vec{\mathbf{h}}_m \cdot \vec{\mathbf{h}}_n^* dS = P_n \delta_{mn}, \tag{42}$$

which is related to (41) by:

$$P_n = \frac{\sqrt{Z_n}}{(\sqrt{Z_n})^*} |Q_n| \overset{\varepsilon_r''=\mu_r''=0}{=} \begin{cases} |Q_n| & n \text{ prop.} \quad (k \geq k_{cn}) \\ +j|Q_n| & n \text{ evan. TE} \quad (k < k_{cn}) \\ -j|Q_n| & n \text{ evan. TM} \quad (k < k_{cn}) \end{cases} . \tag{43}$$

Once Q_n is fixed, P_n cannot be independently defined. Moreover, for a lossless waveguide, at frequencies where $k \geq k_{cn}$ (propagating mode, operating frequency above the cutoff frequency $f \geq f_{cn} = k_{cn}/(2\pi\sqrt{\mu\varepsilon})$), γ_n is pure imaginary and Z_n is real. For $k < k_{cn}$ (*evanescent* mode, $f < f_{cn}$), γ_n is real and Z_n is pure imaginary. In terms of complex power flow, as (43) shows, P_n is real for propagating modes and imaginary for evanescent modes.

A.5 Canonical waveguides

There are several waveguide cross sections that allow analytical TEM, TE and TM solutions, as circular or coaxial sectors (elliptical sectors as well), contours described in parabolic coordinates or 30, 45 and 60 degrees triangles. The more common are the following, using standard rectangular, circular and elliptical coordinates.

[4] The TEM eqs. (33),(34) can be formally considered as a particular case of (39),(40) with $k_{cn} = 0$.

A.5.1 Rectangular waveguide

Regarding Fig. 11.c1, where the reference system is placed at the rectangular cross section center, the TE and TM solutions to (35),(38) are $((m,n) \neq (0,0))$:

$$\begin{aligned}(H_{mn})\ \Phi^{(h)}_{mn} &= (N^{(h)}_{mn})^{\frac{1}{2}} \cos\left(\tfrac{m\pi}{a}(x+\tfrac{a}{2})\right)\cos\left(\tfrac{n\pi}{b}(y+\tfrac{b}{2})\right),\ m,n=0,1,\ldots \\ (E_{mn})\ \Phi^{(e)}_{mn} &= (N^{(e)}_{mn})^{\frac{1}{2}} \sin\left(\tfrac{m\pi}{a}(x+\tfrac{a}{2})\right)\sin\left(\tfrac{n\pi}{b}(y+\tfrac{b}{2})\right),\ m,n=1,2,\ldots\end{aligned} \tag{44}$$

with cutoff wavenumber and normalization constants $(\epsilon_{mk} = 2(m=k), 1(m \neq k))$:

$$k^{\binom{h}{e}}_{c,mn} = \sqrt{(\tfrac{m\pi}{a})^2 + (\tfrac{n\pi}{b})^2},\quad N^{\binom{h}{e}}_{mn} = \left|\left((\tfrac{m\pi}{a})^2 + (\tfrac{n\pi}{b})^2\right)\tfrac{ab}{4}\epsilon_{m0}\epsilon_{n0}\right|^{-1}.$$

A.5.2 Circular waveguide

Regarding Fig. 11.c2, the TE and TM solutions to (35),(38) are:

$$\begin{aligned}&\begin{matrix}(Hc_{pr})\ \Phi^{(hc)}_{pr} \\ (Hs_{pr})\ \Phi^{(hs)}_{pr}\end{matrix} = (N^{(h)}_{pr})^{\frac{1}{2}} J_p(\tfrac{\xi'_{pr}}{a}\rho)\begin{matrix}\cos\\ \sin\end{matrix}(p\varphi),\ p = \begin{matrix}0,1,\ldots\\ 1,2,\ldots\end{matrix}\ r=1,2,\ldots \\ &\begin{matrix}(Ec_{pr})\ \Phi^{(ec)}_{pr} \\ (Es_{pr})\ \Phi^{(es)}_{pr}\end{matrix} = (N^{(e)}_{pr})^{\frac{1}{2}} J_p(\tfrac{\xi_{pr}}{a}\rho)\begin{matrix}\cos\\ \sin\end{matrix}(p\varphi),\ p = \begin{matrix}0,1,\ldots\\ 1,2,\ldots\end{matrix}\ r=1,2,\ldots\end{aligned} \tag{45}$$

with roots (Abramowitz & Stegun, 1956) and normalization constants:

$$\begin{aligned}k^{(h)}_{c,pr} &= \tfrac{\xi'_{pr}}{a},\ J'_p(\xi'_{pr}) = 0,\ N^{(h)}_{pr} = |\epsilon_{p0}\tfrac{\pi}{2}(\xi'^2_{pr} - p^2)J^2_p(\xi'_{pr})|^{-1} \\ k^{(e)}_{c,pr} &= \tfrac{\xi_{pr}}{a},\ J_p(\xi_{pr}) = 0,\ N^{(e)}_{pr} = |\epsilon_{p0}\tfrac{\pi}{2}\xi^2_{pr}J'^2_p(\xi_{pr})|^{-1}.\end{aligned} \tag{46}$$

A.5.3 Circular coaxial waveguide

The TEM mode corresponding to (33) in a circular coaxial waveguide (Fig. 11.c3) of inner radius c and outer radius a is

$$\Phi^{(o)} = (N^{(o)})^{\frac{1}{2}}\ln\tfrac{\rho}{a},\quad N^{(o)} = \left|2\pi\ln\tfrac{a}{c}\right|^{-1},\quad Z_c = \tfrac{\eta}{2\pi}\ln\tfrac{a}{c}. \tag{47}$$

The TE and TM solutions to (35),(38) are:

$$\begin{aligned}&\begin{matrix}(Hc_{pr})\ \Phi^{(hc)}_{pr} \\ (Hs_{pr})\ \Phi^{(hs)}_{pr}\end{matrix} = (N^{(h)}_{pr})^{\frac{1}{2}} R^{(h)}_p(\tfrac{\xi'_{pr}}{a}\rho)\begin{matrix}\cos\\ \sin\end{matrix}(p\varphi),\ p = \begin{matrix}0,1,\ldots\\ 1,2,\ldots\end{matrix}\ r=1,2,\ldots \\ &\begin{matrix}(Ec_{pr})\ \Phi^{(ec)}_{pr} \\ (Es_{pr})\ \Phi^{(es)}_{pr}\end{matrix} = (N^{(e)}_{pr})^{\frac{1}{2}} R^{(e)}_p(\tfrac{\xi_{pr}}{a}\rho)\begin{matrix}\cos\\ \sin\end{matrix}(p\varphi),\ p = \begin{matrix}0,1,\ldots\\ 1,2,\ldots\end{matrix}\ r=1,2,\ldots,\end{aligned} \tag{48}$$

where the radial functions with their roots (Abramowitz & Stegun, 1956) are:

$$\begin{aligned}R^{(h)}_p(u) &= Y'_p(\xi'_{pr})J_p(u) - J'_p(\xi'_{pr})Y_p(u),\quad R'^{(h)}_p(\xi'_{pr}\tfrac{c}{a}) = 0,\quad k^{(h)}_{c,pr} = \tfrac{\xi'_{pr}}{a} \\ R^{(e)}_p(u) &= Y_p(\xi_{pr})J_p(u) - J_p(\xi_{pr})Y_p(u),\quad R^{(e)}_p(\xi_{pr}\tfrac{c}{a}) = 0,\quad k^{(e)}_{c,pr} = \tfrac{\xi_{pr}}{a}.\end{aligned} \tag{49}$$

The normalization constants are

$$\begin{aligned}N^{(h)}_{pr} &= |\epsilon_{p0}\tfrac{\pi}{2}((\xi'^2_{pr} - p^2)R^{(h)2}_{pr}(\xi'_{pr}) - ((\xi'_{pr}\tfrac{c}{a})^2 - p^2)R^{(h)2}_{pr}(\xi'_{pr}\tfrac{c}{a}))|^{-1} \\ N^{(e)}_{pr} &= |\epsilon_{p0}\tfrac{\pi}{2}(\xi^2_{pr}R^{(e)\prime 2}_{pr}(\xi_{pr}) - (\xi_{pr}\tfrac{c}{a})^2R^{(e)\prime 2}_{pr}(\xi_{pr}\tfrac{c}{a}))|^{-1}.\end{aligned} \tag{50}$$

A.5.4 Elliptical and confocal elliptical coaxial waveguides

An elliptic contour (Fig. 11.c4) of axes $2a, 2b$ is described in elliptic coordinates

$$x = d_f \cosh \varrho \cos \vartheta, \quad y = d_f \sinh \varrho \sin \vartheta, \quad \varrho \geq 0, -\pi \leq \vartheta < \pi,$$

as $\varrho = \varrho_a = \mathrm{acosh}\, a/d_f$, where ${d_f}^2 = a^2 - b^2$ is half the focal distance. A second confocal ellipse with major axis $2c$ (Fig. 11.c5) is described by $\varrho = \varrho_c = \mathrm{acosh}\, c/d_f$.
The solutions to the Helmholtz equation in elliptic coordinates is written as $\Phi = R(\varrho)a(\vartheta)$, a and R being solutions to the Mathieu and modified Mathieu equation, respectively, with parameter q_0 (Mclachlan, 1964). With them, the TE and TM modes (35),(38) of the elliptical waveguide are classified as follows:

$$\begin{aligned}
(Hc_{pr})\, \Phi_{pr}^{(hc)} &= (N_{pr}^{(hc)})^{\frac{1}{2}} R_p^{(hc)}(\varrho, q'_{cpr})\, \mathrm{ce}_p(\vartheta, q'_{cpr}),\ p = 0,1,\ldots; r = 1,2,\ldots \\
(Hs_{pr})\, \Phi_{pr}^{(hs)} &= (N_{pr}^{(hs)})^{\frac{1}{2}} R_p^{(hs)}(\varrho, q'_{spr})\, \mathrm{se}_p(\vartheta, q'_{spr}),\ p = 1,2,\ldots; r = 1,2,\ldots \\
(Ec_{pr})\, \Phi_{pr}^{(ec)} &= (N_{pr}^{(ec)})^{\frac{1}{2}} R_p^{(ec)}(\varrho, q_{cpr})\, \mathrm{ce}_p(\vartheta, q_{cpr}),\ p = 0,1,\ldots; r = 1,2,\ldots \\
(Es_{pr})\, \Phi_{pr}^{(es)} &= (N_{pr}^{(es)})^{\frac{1}{2}} R_p^{(es)}(\varrho, q_{spr})\, \mathrm{se}_p(\vartheta, q_{spr}),\ p = 1,2,\ldots; r = 1,2,\ldots
\end{aligned} \tag{51}$$

Referring to both waveguides (although their radial functions are different), the roots and cutoff wavenumbers are (Alhargan & Judah, 1994), (Alhargan & Judah, 1996):

$$\begin{array}{l} R_p^{\prime(hc)}(\varrho_a, q'_{cpr}) = 0, \ \ R_p^{(ec)}(\varrho_a, q_{cpr}) = 0 \\ R_p^{\prime(hs)}(\varrho_a, q'_{spr}) = 0, \ \ R_p^{(es)}(\varrho_a, q_{spr}) = 0 \end{array}, \quad k_c = \tfrac{2}{d_f}\sqrt{q_0} = \tfrac{2}{a}\sqrt{q_0} \cosh \varrho_a,$$

The normalization constants are:

$$N^{\left(\substack{h\\e}\right)} = |2\pi q_0 \int_{\varrho_0}^{\varrho_a} R^2(\varrho)(\cosh 2\varrho - I^{\left(\substack{h\\e}\right)}) d\varrho|^{-1},\ I^{\left(\substack{h\\e}\right)} = \tfrac{1}{\pi} \int_{-\pi}^{\pi} a^2(\vartheta) \cos 2\vartheta d\vartheta, \tag{52}$$

with $\varrho_0 = 0$ for the elliptical waveguide and $\varrho_0 = \varrho_c$ for the coaxial waveguide. The angular integral can be done analytically. Then, the radial integral is computed numerically. The TEM solution (33) for the confocal elliptical coaxial waveguide is:

$$\phi^{(o)} = (N^{(o)})^{\frac{1}{2}} (\varrho - \varrho_a),\ N^{(o)} = |2\pi(\varrho_a - \varrho_c)|^{-1},\ Z_c = \tfrac{\eta}{2\pi} \ln \frac{a + \sqrt{a^2 - d_f^2}}{c + \sqrt{c^2 - d_f^2}}. \tag{53}$$

B. References

Abramowitz, M. & Stegun, I. (1956). *Handbook of Mathematical Functions*, Dover Publications Inc., New York.

Alhargan, F. & Judah, S. (1994). Tables of normalized cutoff wavenumbers of elliptic cross section resonators, *IEEE Trans. Microw. Theory Tech.* **42**(2): 333–338.

Alhargan, F. & Judah, S. (1996). Mode charts for confocal annular elliptic resonators, *IEE Proc. Microw., Antennas and Propagat.* **143**(4): 358–360.

Bozzi, M., Conciauro, G. & Perregrini, L. (2002). On the evaluation of modal coupling coefficients by contour integrals, *IEEE Trans. Microw. Theory Tech.* **50**(7): 1853–1855.

Chan, K.-L. & Judah, S. (1997). Two port scattering at an elliptical-waveguide junction, *IEEE Trans. Microw. Theory Tech.* **45**(8): 1255–1262.

Clarricoats, P. J. B. & Slinn, K. R. (1966). Numerical method for the solution of waveguide discontinuity problems, *Electronic Letters* pp. 226–227.

Collin, R. E. (1991). *Field Theory of Guided Waves*, IEEE Press, New York.

Conciauro, G., Guglielmi, M. & Sorrentino, R. (1999). *Advanced modal analysis: CAD techniques for waveguide components and filters*, John Wiley.

Drabuwitch, S. W. (1966). Multimode antennas, *Microwave Journal* **9**: 41–51.

Figlia, G. & Gentili, G. G. (2002). On the line-integral formulation of Mode-Matching technique, *IEEE Trans. Microw. Theory Tech.* **50**: 578–580.

Gentili, G. G. (1991). Properties of TE-TM Mode-Matching techniques, *IEEE Trans. Microw. Theory Tech.* **39**(9): 1669–1673.

Gerini, G. & Guglielmi, M. (2001). Full-wave CAD of a rectangular waveguide filter with integrated coaxial excitation, *IEEE Trans. Microw. Theory Tech.* **49**: 986–990.

Haskal, H. (1964). Matrix description of waveguide discontinuities in the presence of evanescent modes, *IEEE Trans. Microw. Theory Tech.* pp. 184–188.

Itoh (editor), T. (1989). *Numerical Techniques for Microwave and Millimeter-Wave Passive-Structures*, John Wiley, New York.

Leroy, M. (1983). On the convergence of numerical results in Modal Analysis, *IEEE Trans. Antennas Propagat.* **31**(4): 655–659.

MacPhie, R. H. & Wu, K. L. (1995). Scattering at the junction of a rectangular waveguide and a larger circular waveguide, *IEEE Trans. Microw. Theory Tech.* **43**(9): 2041–2045.

Mclachlan, N. (1964). *Theory and application of Mathieu functions*, Dover Pub., New York.

Mittra, R. & Lee, W. W. (1971). *Analytical techniques in the theory of guided waves*, MacMillan, New York.

Mongiardo, M. & Tomassoni, C. (2000). Modal analysis of discontinuities between elliptical waveguides, *IEEE Trans. Microw. Theory Tech.* **48**(4): 597–605.

Omar, A., Eleftheriades, G. & Katehi, L. (1994). Some important properties of waveguide junction generalized scattering matrices in the context of the mode matching technique, *IEEE Trans. Microw. Theory Tech.* **42**(10): 1896–1903.

Orfanidis, A., Kyriacou, G. & Sahalos, J. (2000). A mode-matching technique for the study of circular and coaxial waveguide discontinuities based on closed-form coupling integrals, *IEEE Trans. Microw. Theory Tech.* **48**(5): 880–883.

Patzelt, H. & Arndt, F. (1982). Double-plane steps in rectangular waveguides and their application for transformers, irises, and filters, *IEEE Trans. Microw. Theory Tech.* **82**(5): 771–776.

Ruiz-Cruz, J. A., Montejo-Garai, J. R. & Rebollar, J. M. (2002). Characterisation of waveguide discontinuities with finite wall conductivity, *Proc. 2002 IEEE Antennas and Propagation Symp. Dig., Columbus, OH* **4**: 428–431.

Ruiz-Cruz, J. A., Zaki, K. A. & Rebollar, J. M. (2004). Mode-matching analysis of a coaxial-to-stripline discontinuity applied to the modelling of a coaxial probe, *Proc. 2004 IEEE Antennas and Propagation Symp. Dig., Monterey, CA* **2**: 2139–2142.

Senior, T. B. A. (1960). Impedance boundary conditions for imperfectly conducting surfaces, *Applied Science Research, section B* **8**: 418–436.

Shen, Z. X. & MacPhie, R. H. (1990). Conservation of complex power technique for waveguide junctions with finite wall conductivity, *IEEE Trans. Microw. Theory Tech.* **38**(4): 373–378.

Uher, J., Bornemann, J. & Rosenberg, U. (1993). *Waveguide Components for Antenna Feed Systems*, Artech House, Boston.

Vassallo, C. (1985). *Thorie des guides d'ondes lectromagntiques*, Collection Technique et Scientifique des Telcommunications (CNET), Eyrolles, Paris.
Wexler, A. (1967). Solution of waveguides discontinuities by modal analysis, *IEEE Transactions on Microwave Theory and Techniques* **15**: 508–517.
Zhongxiang, S. & MacPhie, R. H. (1995). Scattering by a thick off-centered circular iris in circular waveguide, *IEEE Trans. Microw. Theory Tech.* **43**(11): 2639–2642.

7

Circuital analysis of cylindrical structures applied to the electromagnetic resolution of resonant cavities

Felipe L. Penaranda-Foix and Jose M. Catala-Civera
ITACA-UPV
Camino de Vera, s/n; 46022-Valencia
fpenaran@dcom.upv.es & jmcatala@dcom.upv.es
Spain

1. Introduction

The objective of this chapter is first to describe the generalized circuital analysis as a method to solve complex electromagnetic problems and second to apply this specific technique to the determination of the resonant frequency and Q-factor of a cylindrical cavity loaded with a dielectric material when the material is introduced inside the cavity through a hole in the upper wall.

The generalized circuital analysis as a method for solving electromagnetic problems consisting of the segmentation of the whole geometry of the microwave circuit into simpler structures which resolution can be solved in a easier way. Once the simpler structures have been solved separately, they can be joined or combined in order to give the complete solution of the complex structure.

The resolution of the resonant frequency and Quality factor of a coaxially loaded circular cavity with a dielectric material is very interesting, for instance, for the determination of the dielectric properties (complex permittivity) of materials on this type of cavities. This type of analysis on these cavities can be found in the technical literature but in all cases, the effect of the hole to introduce the dielectric material inside the cavity is neglected and in some cases, such as it will be shown in the second part of the chapter, the effect of the hole can introduce considerable errors in the determination of the resonant frequency and quality factor whose can interfere the precision of the permittivity calculations.

In next sections, the effect of the hole for the introduction of dielectric materials inside circular cavities is evaluated by solving the structure by the generalized circuital analysis. Several measurements of circular cavities with dielectric materials will confirm the effect of the hole in the precision of permittivity calculations.

2. Circuital Analysis

In this section the ability of solving complex electromagnetic problems by segmenting the whole structure in several much simpler sub-structures that can be analyzed separately is described. The *Generalized Scattering Matrix* is also introduced as the method to interconnect (combine) networks with different ports and therefore re-combine the segmented networks to characterize the original structure.

2.1 The Generalized Admittance Matrix

The circuital analysis is based on the wave impedance concept. This parameter establishes the relationship between electric and magnetic fields in each point of the space and reflects the interaction between fields and the medium. The concept of impedance (or admittance), associated to each point in a coordinates system, can be extended to the characterization of a given volume by the introduction of the *Generalized Admittance Matrix* (GAM)[1] or the *Generalized Impedance Matrix* (GIM).

As introduced by Schelkunoff in the 30ths and as indicated in (Harrington, 1961), in electromagnetic theories, the relation between the electric field components $\vec{\mathbf{E}}$ and magnetic field components $\vec{\mathbf{H}}$ is called as wave impedances. These values are punctual because they are associated to each space point. Then, if the wave impedance characterizes a unique relation between $\vec{\mathbf{E}}$ and $\vec{\mathbf{H}}$, the GAM characterizes all the volume free of sources by means of the relationship established by $\vec{\mathbf{E}}$ and $\vec{\mathbf{H}}$ on the surrounding surface.

The generalized admittance is unique as probed by the Uniqueness theorem (Harrington, 1961). This theorem establishes that the field inside a closed and lossy region can be completely determined by knowing the sources into the region and the electric and magnetic tangential field components in a fictitious surface that surrounds the region of interest.

Figure 1 shows the schematic of this situation. Region 2 is the region of interest free of sources and region 1 is the region that contains the sources. The Uniqueness theorem guaranties that it is not necessary to know the real sources of region 1 to determine in a unique way the fields $\vec{\mathbf{E}}$ and $\vec{\mathbf{H}}$ in any point within region 2 but it is enough to know the tangential electric field $\vec{\mathbf{E}}_t$ in the surface S, or the tangential magnetic field $\vec{\mathbf{H}}_t$ in the surface S or the tangential electric field $\vec{\mathbf{E}}_t$ in part of the surface S and the tangential magnetic field $\vec{\mathbf{H}}_t$ in the rest of the surface S.

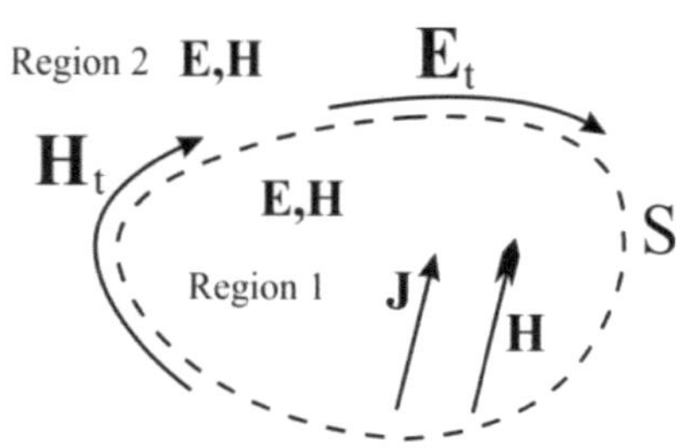

Fig. 1. Generic Surface with two types of regions: with and without sources

---

[1] In an equivalent manner we can speak of *Generalized Impedance Matrix* (GIM)

Therefore, it is clear that there exists a relationship between both tangential electric and magnetic field components, since the knowledge of one of them ensures the knowledge of all the rest electromagnetic field components. Moreover this relationship is unique due to the Uniqueness Theorem.
The relation between electric and magnetic fields can be expressed as a linear combination of base function according to the following expressions:

$$\vec{\mathbf{E}}_t = \sum_n e_n \cdot \vec{\mathbf{E}}_n \quad ; \quad \vec{\mathbf{H}}_t = \sum_n h_n \cdot \vec{\mathbf{H}}_n \tag{1}$$

where $\vec{\mathbf{E}}_n$ and $\vec{\mathbf{H}}_n$ are the base functions for the electric and magnetic fields, respectively, and the terms e_n and h_n are, respectively, the amplitudes of weight of each base function.
According to these expressions, the relation between both field components is established by a matrix. This matrix relates the weights of the magnetic field series h_n with the weights of the electric field series e_n. This matrix is known as the *Generalized Admittance Matrix* $\overline{\overline{\mathbf{Y}}}$. The inverse is called the *Generalized impedance Matrix* $\overline{\overline{\mathbf{Z}}}$:

$$\vec{\mathbf{h}} = \begin{pmatrix} h_1 \\ h_2 \\ \dots \\ h_N \end{pmatrix} = \begin{pmatrix} Y_{11} & Y_{12} & \dots & Y_{1N} \\ Y_{21} & Y_{22} & \dots & Y_{2N} \\ \dots & \dots & \dots & \dots \\ Y_{N1} & Y_{N2} & \dots & Y_{NN} \end{pmatrix} \cdot \begin{pmatrix} e_1 \\ e_2 \\ \dots \\ e_N \end{pmatrix} = \overline{\mathbf{Y}} \cdot \vec{\mathbf{e}} \quad ; \quad \vec{\mathbf{e}} = \overline{\mathbf{Z}} \cdot \vec{\mathbf{h}} \tag{2}$$

The relation between both matrices $\overline{\overline{\mathbf{Y}}}$ y $\overline{\overline{\mathbf{Z}}}$ is clearly:

$$\overline{\overline{\mathbf{Z}}} = \left(\overline{\overline{\mathbf{Y}}}\right)^{-1} \tag{3}$$

N refers to the number of terms considered in the series development of the electric and magnetic fields of equation (1).
The definition of this matrix is very important for solving complex electromagnetic problems because it permits to segment the whole structure into simpler structures by making use of circuital theories. The segmentation concept can be first attributed to Harrington en (Harrington, 1961) where diverse waveguide apertures were analyzed by dividing the structure in two parts by placing at the aperture some equivalent currents. This technique was known as *Generalized Circuital Formulation*. From this first attempts, there has been many references that made use of it. For instance, (Collin, 1966), (Collin, 1991) and (Pozar, 1990) applied this technique for solving cavities excited by slots. Another interesting example is the work of (Gentili & Melloni, 1996). Some other examples can be found in (Alessandri et al, 1994), (Gimeno & Guglielmi, 1997) y (Rebollar et al., 1994) for closed structures and in (Valero-Nogueira 1997), (Penaranda-Foix, 2001), (Penaranda-Foix & Ferrando-Bataller, 2003), (Penaranda-Foix et al., 2007a) y (Penaranda-Foix et al., 2009) for open problems.

Figure 2a shows a generic example for illustrating how the circuital theory can be applied to solve complex structures. Figure 2a distinguishes two zones: region *a* free of sources and region *b* with sources.

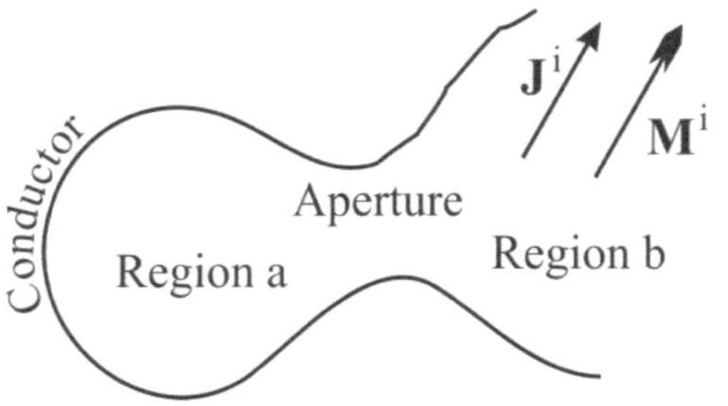

Fig. 2a. Segmentation problem: Two regions connected by the aperture

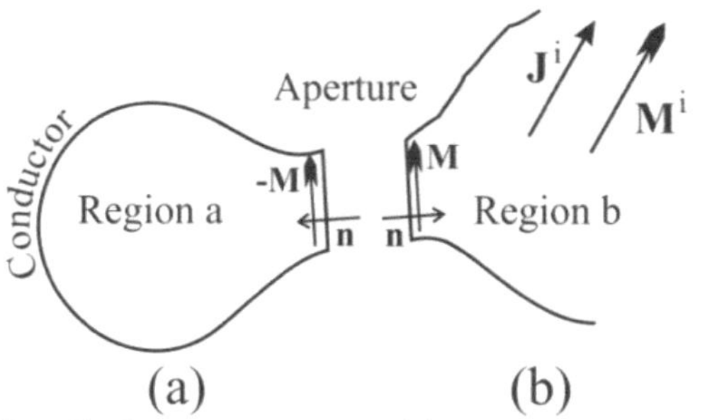

Fig. 2b. Segmentation problem: Two regions separated by the aperture

Figure 2b shows the same structure but segmented in two equivalent problems by means of the *equivalent principle*. This separation of regions is carried out by placing an electric wall in the aperture between regions *a* and *b*.

The fields generated in region *b* are the result of the sources $\mathbf{J}^i$ y $\mathbf{M}^i$ and the fields created by the equivalent magnetic current $\vec{\mathbf{M}} = \vec{\mathbf{E}} \times \hat{\mathbf{n}}$ on the aperture surface. Since the aperture is covered by an electric wall, it is not necessary the use of equivalent electric currents. On the other hand, in region *a*, the field is uniquely created by the magnetic current $-\vec{\mathbf{M}}$ at the aperture since there are no sources in this region *(a)*.

To raise this problem, continuity of tangential magnetic fields in the aperture is imposed. The total magnetic field on the aperture zone in region *b* is the summation of the field due to the impressed currents $\vec{\mathbf{H}}_t^i$ and the field due to the equivalent sources $\vec{\mathbf{M}}$ called as $\mathbf{H}_t^b(\vec{\mathbf{M}})$ from now on, then:

$$\vec{\mathbf{H}}_t^b = \vec{\mathbf{H}}_t^i + \mathbf{H}_t^b(\vec{\mathbf{M}}) \tag{4}$$

It is important to remark that both $\vec{\mathbf{H}}_t^i$ and $\mathbf{H}_t^b(\vec{\mathbf{M}})$ are fields calculated by assuming the aperture as an electric wall.

In a similar manner, the tangential magnetic field in the aperture of region *b* is due to the equivalent sources $-\vec{\mathbf{M}}$ called $\mathbf{H}_t^a(-\vec{\mathbf{M}})$ from now on, leading to:

$$\vec{\mathbf{H}}_t^a = \mathbf{H}_t^a(-\vec{\mathbf{M}}) \tag{5}$$

Where once again $\mathbf{H}_t^a(-\vec{\mathbf{M}})$ is calculated assuming the source $-\vec{\mathbf{M}}$ in the aperture where the electric wall is located.

Making both tangential components equal, we have:

$$\vec{\mathbf{H}}_t^b = \vec{\mathbf{H}}_t^i + \mathbf{H}_t^b\left(\vec{\mathbf{M}}\right) = \vec{\mathbf{H}}_t^a = \mathbf{H}_t^a\left(-\vec{\mathbf{M}}\right)$$
$$\Downarrow \tag{6}$$
$$\mathbf{H}_t^b\left(\vec{\mathbf{M}}\right) - \mathbf{H}_t^a\left(-\vec{\mathbf{M}}\right) = -\vec{\mathbf{H}}_t^i$$

Equation (6) is the basic equation for the calculation of $\vec{\mathbf{M}}$ by assuming it is known the operator that relates magnetic fields with the sources in each region.
To solve this equation two numerical methods, very similar between them, are available: The Method of Moments (MoM), see (Harrington, 1967) y (Harrington, 1993), and Modal Analysis or Mode Matching, see (Wexler, 1967).
Assuming that the unknown source $\vec{\mathbf{M}}$ can be written as a series of base functions, in a similar manner to equation (1),

$$\vec{\mathbf{M}} = \sum_n v_n \cdot \vec{\mathbf{m}}_n \tag{7}$$

where coefficients v_n need to be determined. Substituting this series development in equation (6) and making use of the operator linearity, we can write:

$$\sum_n v_n \cdot \vec{\mathbf{H}}_t^b(\vec{\mathbf{m}}_n) - \sum_n v_n \cdot \vec{\mathbf{H}}_t^a(-\vec{\mathbf{m}}_n) = -\vec{\mathbf{H}}_t^i \tag{8}$$

With the definition of the internal product:

$$\left\langle \vec{\mathbf{A}}, \vec{\mathbf{B}} \right\rangle = \iint_{Aperture} \vec{\mathbf{A}} \cdot \vec{\mathbf{B}} \cdot dS \tag{9}$$

and a set of weight functions $\{\vec{\mathbf{w}}_n\}$, that in general may differ from the base functions $\vec{\mathbf{m}}_n$, can be applied in (8), leading, thanks to the linearity of the product function, to:

$$\sum_n v_n \cdot \left\langle \vec{\mathbf{w}}_m, \vec{\mathbf{H}}_t^b(\vec{\mathbf{m}}_n) \right\rangle - \sum_n v_n \cdot \left\langle \vec{\mathbf{w}}_m, -\vec{\mathbf{H}}_t^a(\vec{\mathbf{m}}_n) \right\rangle = -\left\langle \vec{\mathbf{w}}_m, \vec{\mathbf{H}}_t^i \right\rangle \tag{10}$$

If previous equations are re-written in a matricial form, a new set of matrices $\overline{\overline{\mathbf{Y}}}^a$ and $\overline{\overline{\mathbf{Y}}}^b$ are defined as the matrices that characterizes the regions *a* and *b*, respectively.

$$Y^a\Big|_{mn} = -\left\langle \vec{\mathbf{w}}_m, -\vec{\mathbf{H}}_t^a(\vec{\mathbf{m}}_n) \right\rangle = \left\langle \vec{\mathbf{w}}_m, \vec{\mathbf{H}}_t^a(\vec{\mathbf{m}}_n) \right\rangle \quad ; \quad Y^b\Big|_{mn} = \left\langle \vec{\mathbf{w}}_m, \vec{\mathbf{H}}_t^b(\vec{\mathbf{m}}_n) \right\rangle \tag{11}$$

Matrix $\vec{\mathbf{I}}^i$ of dimensions (Mx1) is also defined as,

$$I^i\Big|_m = \left\langle \vec{\mathbf{w}}_m, \vec{\mathbf{H}}_t^i \right\rangle \tag{12}$$

and the column vector $\vec{\mathbf{V}}$, of dimensions (Nx1), as:

$$V\big|_n = v_n \tag{13}$$

On the other hand, equation (10) is now:

$$\left(\overline{\overline{\mathbf{Y}}}^{b} + \overline{\overline{\mathbf{Y}}}^{a}\right) \cdot \vec{\mathbf{V}} = \vec{\mathbf{I}}^{i} \tag{14}$$

T

hen, magnetic current $\vec{\mathbf{M}}$ is -see equations (6) and (8)-:

$$\vec{\mathbf{V}} = \left(\overline{\overline{\mathbf{Y}}}^{b} + \overline{\overline{\mathbf{Y}}}^{a}\right)^{-1} \cdot \vec{\mathbf{I}}^{i} \tag{15}$$

It is important to emphasize that equation (15) presents a clear circuital interpretation which is showed in next figure 3.

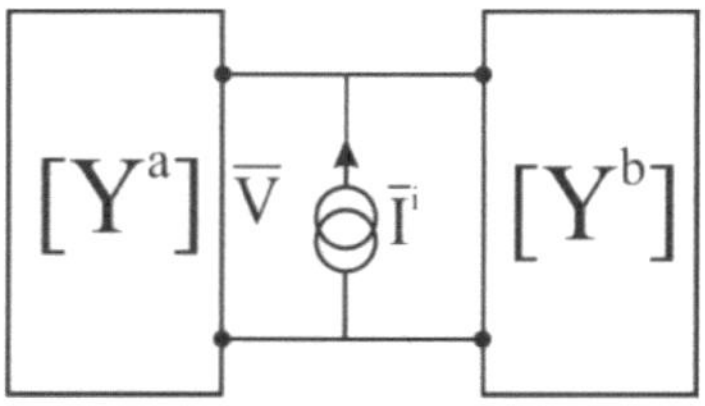

Fig. 3. Circuital interpretation of equation (15)

Figure 3 represents two networks, each one characterized by an admittance matrix ($\overline{\overline{\mathbf{Y}}}^{a}$ and $\overline{\overline{\mathbf{Y}}}^{b}$, respectively) and a current source $\vec{\mathbf{I}}^{i}$, which solution is given by equation (15). The importance of this result falls in the possibility to analyze and solve each region separately and irrespective of the other regions. Moreover, this result can be extended to networks with more port numbers, as described in (Penaranda-Foix, 2001)

2.2 The Generalized Scattering Matrix

In previous section, GAM and GIM matrices were introduced. Next section adds to these matrices the *Generalized Scattering Matrix* (GSM).

It is well known that by considering a canonic plane as the access port to the network under analysis, the electric and magnetic fields in such surface can be decomposed in forward and reflected waves in the network, respectively. This is especially evident when the wave equation is solved (Balanis, 1989), (Harrington, 1961)-.

Therefore, electric and magnetic fields in a port of the network, can be written as:

$$\begin{cases} \vec{\mathbf{E}} = \vec{\mathbf{E}}^{(i)} + \vec{\mathbf{E}}^{(r)} \\ \vec{\mathbf{H}} = \vec{\mathbf{H}}^{(i)} - \vec{\mathbf{H}}^{(r)} = \overline{\overline{\mathbf{Y}}}_0^{(i)} \cdot \vec{\mathbf{E}}^{(i)} - \overline{\overline{\mathbf{Y}}}_0^{(r)} \cdot \vec{\mathbf{E}}^{(r)} \end{cases} \tag{16}$$

By making use the known concept of characteristic admittances $\overline{\overline{\mathbf{Y}}}_0^{(i)}$ y $\overline{\overline{\mathbf{Y}}}_0^{(r)}$.

At the same time, according to equation (1), electric and magnetic fields can be written as a series development. Therefore, the *Generalized Scattering Matrix* (GSM) can be defined as the relation of the forward electric wave fields $\vec{\mathbf{E}}^{(i)}$ and the reflected wave fields $\vec{\mathbf{E}}^{(r)}$:

$$\vec{\mathbf{E}}^{(r)} = \overline{\overline{\mathbf{S}}} \cdot \vec{\mathbf{E}}^{(i)} \tag{17}$$

The relationship between the $\overline{\overline{\mathbf{S}}}$ matrix and the admittance $\overline{\overline{\mathbf{Y}}}$ and impedance $\overline{\overline{\mathbf{Z}}}$ matrices is calculated by the following equations

$$\begin{gathered}
\vec{\mathbf{H}} = \vec{\mathbf{H}}^{(i)} - \vec{\mathbf{H}}^{(r)} = \overline{\overline{\mathbf{Y}}}_0^{(i)} \cdot \vec{\mathbf{E}}^{(i)} - \overline{\overline{\mathbf{Y}}}_0^{(r)} \cdot \vec{\mathbf{E}}^{(r)} = \overline{\overline{\mathbf{Y}}} \cdot \vec{\mathbf{E}} = \overline{\overline{\mathbf{Y}}} \cdot \left(\vec{\mathbf{E}}^{(i)} + \vec{\mathbf{E}}^{(r)} \right) \\
\Downarrow \\
\left(\overline{\overline{\mathbf{Y}}} + \overline{\overline{\mathbf{Y}}}_0^{(r)} \right) \cdot \vec{\mathbf{E}}^{(r)} = \left(\overline{\overline{\mathbf{Y}}}_0^{(i)} - \overline{\overline{\mathbf{Y}}} \right) \cdot \vec{\mathbf{E}}^{(i)} \Rightarrow \vec{\mathbf{E}}^{(r)} = \left(\overline{\overline{\mathbf{Y}}} + \overline{\overline{\mathbf{Y}}}_0^{(r)} \right)^{-1} \cdot \left(\overline{\overline{\mathbf{Y}}}_0^{(i)} - \overline{\overline{\mathbf{Y}}} \right) \cdot \vec{\mathbf{E}}^{(i)} \\
\Downarrow \\
\overline{\overline{\mathbf{S}}} = \left(\overline{\overline{\mathbf{Y}}} + \overline{\overline{\mathbf{Y}}}_0^{(r)} \right)^{-1} \cdot \left(\overline{\overline{\mathbf{Y}}}_0^{(i)} - \overline{\overline{\mathbf{Y}}} \right)
\end{gathered} \tag{18}$$

In a similar manner, the admittance matrix is given by:

$$\overline{\overline{\mathbf{Y}}} = \left(\overline{\overline{\mathbf{Y}}}_0^{(i)} - \overline{\overline{\mathbf{Y}}}_0^{(r)} \cdot \overline{\overline{\mathbf{S}}} \right)^{-1} \cdot \left(\overline{\overline{\mathbf{I}}} + \overline{\overline{\mathbf{S}}} \right) = \overline{\overline{\mathbf{Z}}}^{-1} \tag{19}$$

2.3 Connecting networks

To conclude with this section, the procedure to interconnect networks of several ports and set the new $\overline{\overline{\mathbf{S}}}$, $\overline{\overline{\mathbf{Y}}}$ o $\overline{\overline{\mathbf{Z}}}$ matrices is described.

Assuming two generic networks characterized by their respective admittance matrices $\overline{\overline{\mathbf{Y}}}^{(1)}$ and $\overline{\overline{\mathbf{Y}}}^{(2)}$ where, for example, the first network presents 5 ports and the second 6 ports, respectively. (See figure 4a for more details of the networks and port numbers).

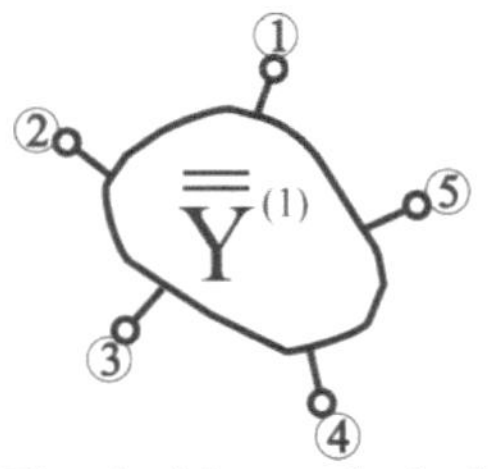

Fig. 4a. Networks before connecting

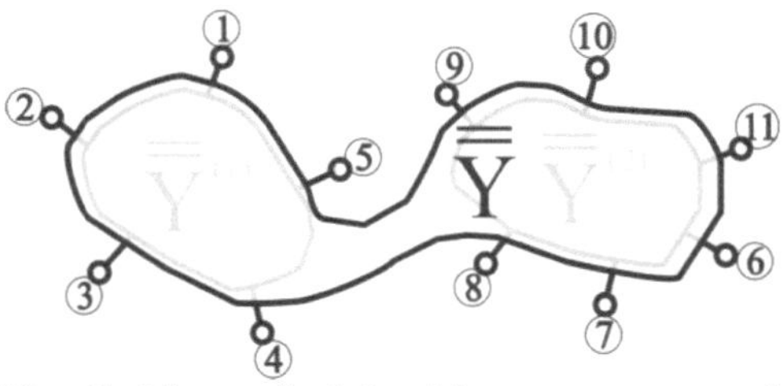

Fig. 4b. Networks joined but not connected

The first step consists of setting a global network with a number of ports equal to the sum of the number of ports of both networks and with a new numeration of ports. (It is a good practice to keep the order of the ports with the new numeration) Thus, the new GAM is witten as:

$$\overline{\overline{\mathbf{Y}}} = \begin{pmatrix} \overline{\overline{\mathbf{Y}}}^{(1)} & \overline{\overline{\mathbf{0}}} \\ \overline{\overline{\mathbf{0}}} & \overline{\overline{\mathbf{Y}}}^{(2)} \end{pmatrix} \tag{20}$$

Figure 4b shows this new network with a commoun numeration of ports. This particular example of figure 4 can be extended to a any network with N ports, as the network showed in figure 5a.

The second step consist of connecting two (or more) ports from the new network of figure 5b. By connecting these two ports, for instance ports l and k, the resulting network reduces the number of ports to N-2.

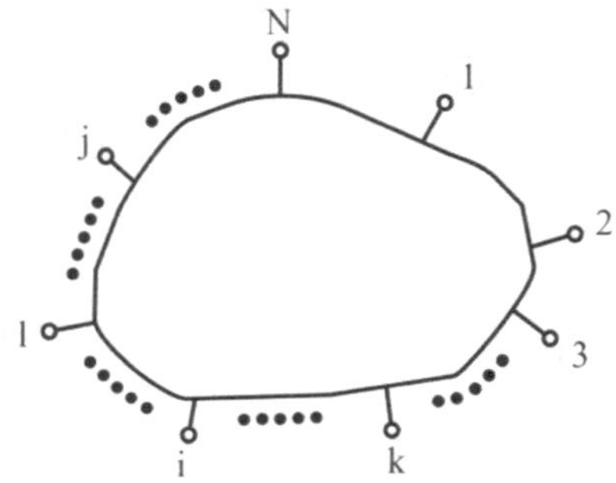

Fig. 5a. Generic network before connecting ports l and k.

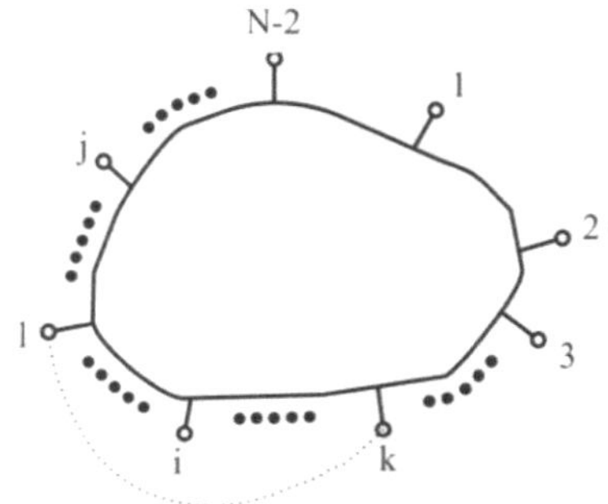

Fig. 5b. Generic network after connecting ports l and k.

The new GAM matrix with N-2 ports is calculated by imposing continuity of electric and magnetic tangential component fields between ports l and k.

$$\begin{gathered} \vec{\mathbf{h}}_k = \sum_{j=1}^{N} \overline{\overline{\mathbf{Y}}}_{kj} \cdot \vec{\mathbf{e}}_j = \vec{\mathbf{h}}_l = \sum_{j=1}^{N} \overline{\overline{\mathbf{Y}}}_{lj} \cdot \vec{\mathbf{e}}_j \\ \Downarrow \left\langle \vec{\mathbf{e}}_l = \vec{\mathbf{e}}_k \right\rangle \\ \sum_{\substack{j=1 \\ j\neq k \\ j \neq l}}^{N} \left(\overline{\overline{\mathbf{Y}}}_{kj} - \overline{\overline{\mathbf{Y}}}_{lj} \right) \cdot \vec{\mathbf{e}}_j = \left(\overline{\overline{\mathbf{Y}}}_{lk} + \overline{\overline{\mathbf{Y}}}_{ll} - \overline{\overline{\mathbf{Y}}}_{kl} - \overline{\overline{\mathbf{Y}}}_{kk} \right) \cdot \vec{\mathbf{e}}_k \end{gathered} \tag{21}$$

And operating on previous equation, the new GAM matrix with N-2 ports, called $\overline{\overline{\mathbf{Y}}}^{(T)}$, is reached:

$$\left. \overline{\overline{\mathbf{Y}}}_{pq}^{(T)} \right|_{\substack{1\le p \le (N-2) \\ 1 \le q \le (N-2)}} = \left. \overline{\overline{\mathbf{Y}}}_{pq} \right|_{\substack{1\le p \le (N-2) \\ 1 \le q \le (N-2)}} + \left(\overline{\overline{\mathbf{Y}}}_{pk} + \overline{\overline{\mathbf{Y}}}_{pl} \right) \cdot \left(\overline{\overline{\mathbf{Y}}}_{lk} + \overline{\overline{\mathbf{Y}}}_{ll} - \overline{\overline{\mathbf{Y}}}_{kl} - \overline{\overline{\mathbf{Y}}}_{kk} \right)^{-1} \cdot \left(\overline{\overline{\mathbf{Y}}}_{kq} - \overline{\overline{\mathbf{Y}}}_{lq} \right) \tag{22}$$

Operating in a similar manner, the impedance matrix GIM, is also found as:

$$\left. \overline{\overline{\mathbf{Z}}}_{pq}^{(T)} \right|_{\substack{1\le p \le (N-2) \\ 1 \le q \le (N-2)}} = \left. \overline{\overline{\mathbf{Z}}}_{pq} \right|_{\substack{1\le p \le (N-2) \\ 1 \le q \le (N-2)}} + \left(\overline{\overline{\mathbf{Z}}}_{pk} + \overline{\overline{\mathbf{Z}}}_{pl} \right) \cdot \left(\overline{\overline{\mathbf{Z}}}_{lk} + \overline{\overline{\mathbf{Z}}}_{ll} - \overline{\overline{\mathbf{Z}}}_{kl} - \overline{\overline{\mathbf{Z}}}_{kk} \right)^{-1} \cdot \left(\overline{\overline{\mathbf{Z}}}_{kq} - \overline{\overline{\mathbf{Z}}}_{lq} \right) \tag{23}$$

Also by imposing the same continuity of tangential fields, the new GSM matrix is obtained:

$$\left.\overline{\overline{\mathbf{S}}}_{pq}^{(T)}\right|_{\substack{1\le p\le(N-2)\\1\le q\le(N-2)}} = \left.\overline{\overline{\mathbf{S}}}_{pq}\right|_{\substack{1\le p\le(N-2)\\1\le q\le(N-2)}} + \overline{\overline{\mathbf{S}}}_{pl}\cdot\left(\overline{\overline{\mathbf{I}}}-\overline{\overline{\mathbf{\Delta}}}_{kl}\cdot\overline{\overline{\mathbf{S}}}_{kk}\cdot\overline{\overline{\mathbf{\Delta}}}_{lk}\cdot\overline{\overline{\mathbf{S}}}_{ll}\right)^{-1}\cdot\overline{\overline{\mathbf{\Delta}}}_{kl}\cdot\left(\overline{\overline{\mathbf{S}}}_{kq}+\overline{\overline{\mathbf{S}}}_{kk}\cdot\overline{\overline{\mathbf{\Delta}}}_{lk}\cdot\overline{\overline{\mathbf{S}}}_{lq}\right)+ \\ +\overline{\overline{\mathbf{S}}}_{pk}\cdot\left(\overline{\overline{\mathbf{I}}}-\overline{\overline{\mathbf{\Delta}}}_{lk}\cdot\overline{\overline{\mathbf{S}}}_{ll}\cdot\overline{\overline{\mathbf{\Delta}}}_{kl}\cdot\overline{\overline{\mathbf{S}}}_{kk}\right)^{-1}\cdot\overline{\overline{\mathbf{\Delta}}}_{lk}\cdot\left(\overline{\overline{\mathbf{S}}}_{lq}+\overline{\overline{\mathbf{S}}}_{ll}\cdot\overline{\overline{\mathbf{\Delta}}}_{kl}\cdot\overline{\overline{\mathbf{S}}}_{kq}\right)+ \tag{24}$$

where:

$$\overline{\overline{\mathbf{\Delta}}}_{kl}=\left(\overline{\overline{\mathbf{I}}}-\overline{\overline{\mathbf{S}}}_{kl}\right)^{-1} \quad ; \quad \overline{\overline{\mathbf{\Delta}}}_{lk}=\left(\overline{\overline{\mathbf{I}}}-\overline{\overline{\mathbf{S}}}_{lk}\right)^{-1} \tag{25}$$

Equations (22), (23) and (24) are a powerful set of equations that allow to apply the circuital method to electromagnetic problems by the segmentation procedure.
Once each simple problem or circuit has been analysed and its GAM, GIM or GSM is known, just using the previous equations the whole GAM, GIM or GSM or the orginal problem is calculated.
In the next point this general procedure will be applied to a specific problem: a cylindrical cavity with insertion hole.

3. Cylindrical cavity analysis

As an example of the circuital theory described in previous section we will analyze the circular resonant cavities used for the electromagnetic characterization of materials.
To determine accurately the dielectric and magnetic properties of materials is essential nowadays in applications such antennas, radomes, planar circuits, etc. Special attention must be dedicated to the new materials developed for special applications. But not only in those cases: in applications such as microwave heating, where losses associated to the materials determine its capability to be heated, where the measured electric variations will allow to determine how the monitorized material is changing its properties in real time. Finally, a large number of applications can be thought in the world of medicine, to get images, or even from the security view point, in order to know the electromagnetic properties of potentially dangerous materials.
One of the most widely method to determine the dielectric properties is the coaxially filled circular resonant cavity with an insertion hole in the top or in top and in the bottom, to introduce the material. These cavities are based on the use of the resonant mode TM_{010}, with no angular changes –see (Balanis, 1989), (Chen et al., 2004) and (Metaxas & Meredith, 1988)- and in the property that the insertion hole is considered as a cylindrical waveguide under cut-off so there is no propagation along it. Then the cavity, even open, can be considered as a closed cavity.
The problem geometry is shown in figure 6a, being figure 6b a section, where we can see a cavity of radius b and height h with a generic material of permittivity ε_{r2} and permeability μ_{r2} (both will be in general air: $\varepsilon_{r2}=\mu_{r2}=1$). In the center there is an insertion hole, or tube, whose radius is a and its height is larger than the cavity height. It contains the material to be measured with permittivity ε_{r1} and permeability μ_{r1}. Finally, above the measured material there is another one with permittivity ε_r and permeability μ_r (in general air again: $\varepsilon_r=\mu_r=1$). But it is important to remember that the tube is under cut-off and then there is no leakage through it and it is negligible.

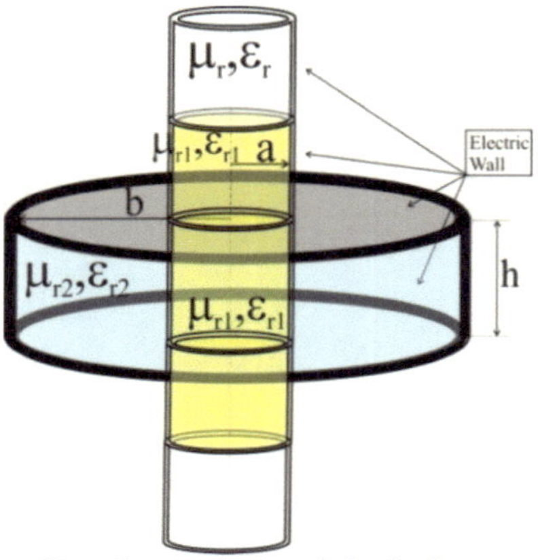

Fig. 6a. Circular cavity and the hole.

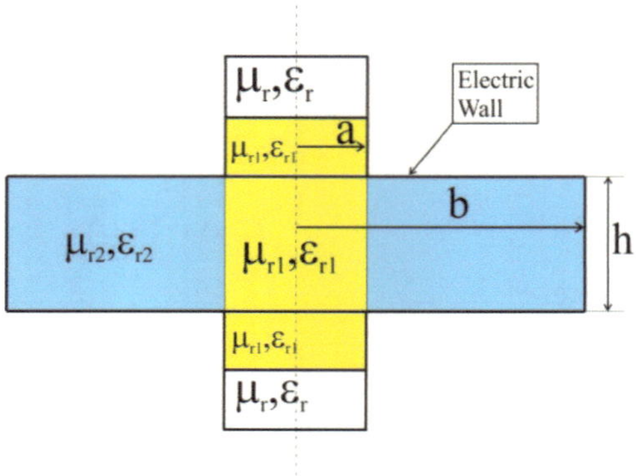

Fig. 6b. Circular cavity and the hole: section view.

As previously said, to completely analyze this structure we must divide it in simpler networks by a segmentation process. Figure 7a shows how this structure can be segmented. The segmentation consists of a division of the whole network into smaller ones, with canonical shapes in general, and then the analysis will be easier. So, the segmentation proposed in figure 7a divides the geometry into 3 different networks: one is a one-port network, in a circular ring shape with external radius *a* and external radius *b* and height *h* where the one-port in put in the inner part (r=a). This structure is shown in figure 7b. Another network is a two-port network that consists of a circular waveguide of radius *a* and height *d* whose ports are located in the top and in the bottom, as shown in figure 7c.

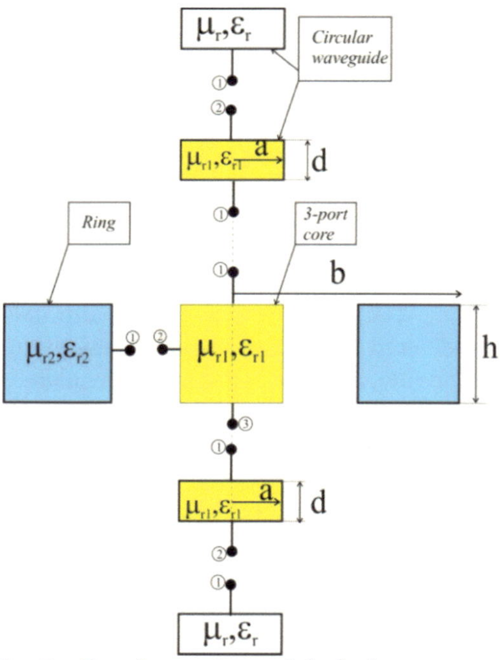

Fig. 7a. Circular cavity and the hole.

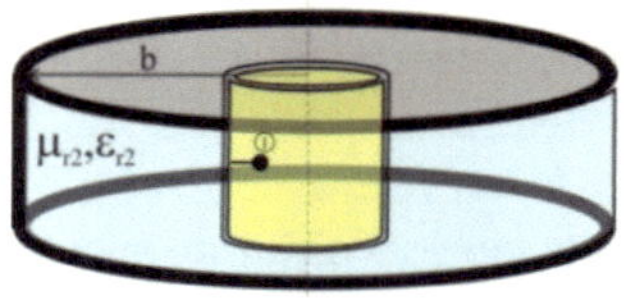

Fig. 7b. 1-port network: Ring.

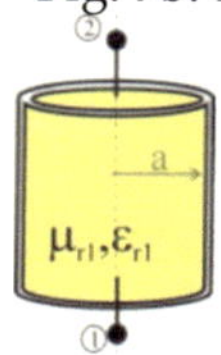

Fig. 7c. 2-port: Waveguide

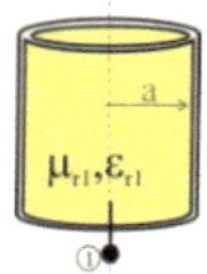

Fig. 7d. 1-port: Waveguide

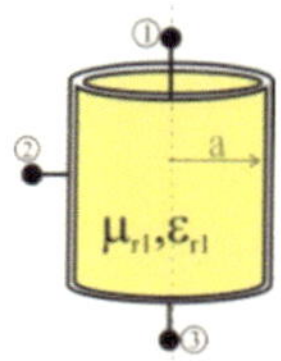

Fig. 7e. 3-port network: Core

This network can be converted in a one-port network, located in the bottom, as shown in figure 7d, just considering that there is no reflection in port 2. Finally a three-port network appears that consists of a circular tube of radius a and height h with ports in the top and in the bottom, as before, but adding a port in the circular face, in $r=a$. This geometry is shown in figure 7e.
Then the original problem is reduced to 3 simpler problems: the circular ring, a circular waveguide and the 3-port guide.

3.4 Analysis of a 1 port network

Let's start calculating the Generalized Admittance Matrix (GAM) of the 1-port network shown in figure 7d. It is the simplest one and it will allow seeing the whole procedure.
Figure 8 shows the axis to be considered for this structure.

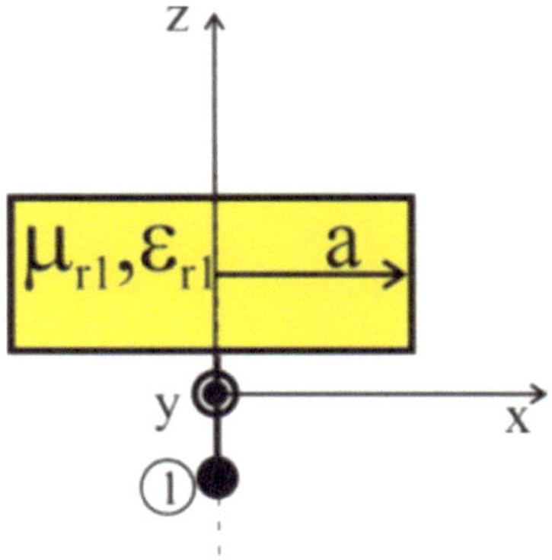

Fig. 8. One port network

The internal field in the circular waveguide is:

$$\begin{cases} E_z = \sum_{n=1}^{\infty} \frac{j \cdot k_{cn}^2}{\omega \cdot \varepsilon \cdot \mu} \cdot J_0(k_{cn} \cdot r) \cdot \left(A_n^+ \cdot e^{-\gamma_n \cdot z} + A_n^- \cdot e^{+\gamma_n \cdot z}\right) \\ \vec{\mathbf{E}}_t = E_r \cdot \hat{r} = \sum_{n=1}^{\infty} \frac{\gamma_n \cdot k_{cn}}{j \cdot \omega \cdot \varepsilon \cdot \mu} \cdot J_0'(k_{cm} \cdot r) \cdot \left(A_n^+ \cdot e^{-\gamma_n \cdot z} - A_n^- \cdot e^{+\gamma_n \cdot z}\right) \cdot \hat{r} \\ \vec{\mathbf{H}}_t = H_\varphi \cdot \hat{\varphi} = \sum_{n=1}^{\infty} \frac{k_{cn}}{\mu} \cdot J_0'(k_{cn} \cdot r) \cdot \left(A_n^+ \cdot e^{-\gamma_n \cdot z} + A_n^- \cdot e^{+\gamma_n \cdot z}\right) \cdot \hat{\varphi} \end{cases} \tag{26}$$

All the fields correspond to circular TM_{0n} modes (Balanis, 1989). As previously mentioned, only TM_{0n} modes are considered and, even there are a larger set of modes excited ,like TE_{mn} or TM_{mn} in general, these modes are not coupled with the other ones. The resonant modes associated to these TM_{0n} modes are the TM_{0np} modes. Anyway, the proposed circuital method ca be used with any set of modes, so there is no loss of generality.
The propagation constant γ_n is calculated from the cut-off wavenumbers k_{cn} as (Balanis, 1989):

$$k_{cn}^2 = \omega^2 \cdot \varepsilon \cdot \mu + \gamma_n^2 = \omega^2 \cdot \varepsilon_0 \cdot \varepsilon_{r1} \cdot \mu_0 \cdot \mu_{r1} + \gamma_n^2 = k_0^2 \cdot \varepsilon_{r1} \cdot \mu_{r1} + \gamma_n^2 = \left(\frac{p_{0n}}{a}\right)^2 \tag{27}$$

where p_{0n} are the zeros of the First Kind Bessel Functions of order 0 ($J_0(p_{0n})=0$) and its value is obtained by focing the boundary conditions in the lateral faces (Balanis, 1989). The cut-off wavenumber is always a real number (Ramo, 1994) and teh criteria to select the sign of the propagation constant is (Baker-Jarvis et al., 1994):

$$\gamma_i = \begin{cases} j \cdot \sqrt{k^2 - k_{cn}^2} = j \cdot \sqrt{k_0^2 \cdot \varepsilon_{r1} \cdot \mu_{r1} - k_{cn}^2}, \mathrm{Re}(k) \geq k_{cn} \\ \sqrt{k_{cn}^2 - k^2} = \sqrt{k_{cn}^2 - k_0^2 \cdot \varepsilon_{r1} \cdot \mu_{r1}}, \mathrm{Re}(k) < k_{cn} \end{cases} \tag{28}$$

Where the square root must be in the fist or in the forth quadrants.

Amplitudes A_n^+ y A_n^- are, respectively, the forward and backward waves in the z-axis direction, which is the propagation direction. In our particular case, because the waveguide is under cut-off, the backward waves are zero, so $A_n^- = 0$.

To get the GAM we must incide with an electric field in port 1. This electric field, as previously said, is written as a series expansions of base functions as:

$$E_i^{(1)} = \sum_{m=1}^{\infty} \alpha_m \cdot e_m^{(1)} \tag{29}$$

where the sleected base function for port 1 is:

$$e_m^{(1)} = N_m^{(1)} \cdot J_0'\left(\frac{p_{0m}}{a} \cdot r\right) \tag{30}$$

Term $N_m^{(1)}$ is a normalization term introduced in the modal analysis and we are going to use the proposed by Gentilli (Gentili, 1991). It must accomplish:

$$I_{mq} = \int_S \mathbf{e}_m^{(1)}(r) \cdot \mathbf{e}_q^{(1)}(r) \cdot dS = \delta_{mn} = \begin{cases} 0 \ , \ m \neq q \\ 1 \ , \ m = q \end{cases} \tag{31}$$

Then the normalization term is:

$$\begin{gathered} \int_{r=0}^{a} \mathbf{e}_m^{(1)}(r) \cdot \mathbf{e}_q^{(1)}(r) \cdot r \cdot dr = \delta(m-q) = \int_{r=0}^{a} \left(N_m^{(1)}\right)^2 \cdot J_1^2\left(\frac{p_{0m}}{a} \cdot r\right) \cdot r \cdot dr = \left(N_m^{(1)}\right)^2 \cdot \frac{\left(a \cdot J_1(p_{0m})\right)^2}{2} = 1 \\ \Downarrow \\ N_m^{(1)} = \frac{\sqrt{2}}{a \cdot J_1(p_{0m})} \end{gathered} \tag{32}$$

The incident field in port 1, equation (29), must equal to the electric field calculated from the inside of the structure, following equation(26). Then the forward amplitude is:

$$E_r(z=0)=\sum_{n=1}^{\infty}\frac{\gamma_n\cdot k_{cn}}{j\cdot\omega\cdot\varepsilon\cdot\mu}\cdot J_0'(k_{cm}\cdot r)\cdot A_n^+=E_i=\sum_{m=1}^{\infty}\alpha_m\cdot e_m^{(1)}=\sum_{m=1}^{\infty}\alpha_m\cdot N_m^{(1)}\cdot J_0'\left(\frac{p_{0m}}{a}\cdot r\right)$$
$$\Downarrow$$
$$A_n^+=\alpha_m\cdot N_m^{(1)}\cdot\frac{j\cdot\omega\cdot\varepsilon\cdot\mu}{\gamma_n\cdot k_{cn}}\cdot\delta(n-m) \tag{33}$$

To get the GAM, we must obtain a relation between the electric and the magnetic fields in port 1. The magnetic field in port 1 is, in the same way than the electric one:

$$H^{(1)}=\sum_{q=1}^{\infty}c_q\cdot h_q^{(1)} \tag{34}$$

where the base fucntion $h_q^{(1)}$ is the same than that used for electric field: $h_q^{(1)}=e_q^{(1)}$
And the GAM is defined as:

$$\vec{\mathbf{h}}^{(1)}=\overline{\overline{\mathbf{Y}}}_{11}\cdot\vec{\mathbf{e}}^{(1)} \tag{35}$$

where $\vec{\mathbf{e}}^{(1)}$ is a column vector with the electric field amplitudes at port 1 and $\vec{\mathbf{h}}^{(1)}$ is the column vector with the magnetic field amplitudes at port 1:

$$\vec{\mathbf{h}}^{(1)}=\begin{pmatrix}c_1 & c_2 & \dots & c_Q\end{pmatrix}^t \quad ; \quad \vec{\mathbf{e}}^{(1)}=\begin{pmatrix}\alpha_1 & \alpha_2 & \dots & \alpha_M\end{pmatrix}^t \tag{36}$$

Then identifying the inner magnetic field in port 1 with the magnetic field from equation (34) we have:

$$H_\varphi(z=0)=\sum_{n=1}^{\infty}\frac{k_{cn}}{\mu}\cdot J_0'(k_{cn}\cdot r)\cdot A_n^+=\sum_{q=1}^{\infty}c_q\cdot h_q^{(1)}=\sum_{q=1}^{\infty}c_q\cdot N_q^{(1)}\cdot J_0'\left(\frac{p_{0q}}{a}\cdot r\right)$$
$$\Downarrow$$
$$c_q=\alpha_m\cdot\frac{j\cdot\omega\cdot\varepsilon}{\gamma_m}\cdot\delta(q-m) \tag{37}$$

Then the Generalized Admittance Matrix GAM is:

$$Y_{11}\Big|_{qm}=\frac{j\cdot\omega\cdot\varepsilon}{\gamma_m}\cdot\delta(q-m)=\frac{j\cdot\omega\cdot\varepsilon_0\cdot\varepsilon_{r1}}{\gamma_m}\cdot\delta(q-m) \tag{38}$$

3.5 Analysis of a 1-port ring structure

This example will show how to calculate the GAM of the network shown in figure 7b. It is a 1-port network, as before, but the port is located in the lateral.
Figure 9 shows the axis to be considered.

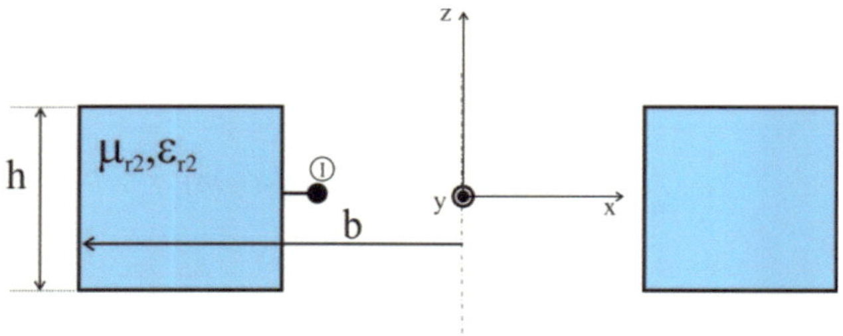

Fig. 9. One port network: ring

The fields inside are, as before:

$$\begin{cases} E_z = \sum_{n=1}^{\infty} \frac{j \cdot k_{cn}^2}{\omega \cdot \varepsilon \cdot \mu} \cdot f_{0n}(r) \cdot \left(A_n^+ \cdot e^{-\gamma_n \cdot z} + A_n^- \cdot e^{+\gamma_n \cdot z} \right) \\ \vec{\mathbf{E}}_t = E_r \cdot \hat{r} = \sum_{n=1}^{\infty} \frac{\gamma_n \cdot k_{cn}}{j \cdot \omega \cdot \varepsilon \cdot \mu} \cdot f_{0n}^{'}(r) \cdot \left(A_n^+ \cdot e^{-\gamma_n \cdot z} - A_n^- \cdot e^{+\gamma_n \cdot z} \right) \cdot \hat{r} \\ \vec{\mathbf{H}}_t = H_\varphi \cdot \hat{\varphi} = \sum_{n=1}^{\infty} \frac{k_{cn}}{\mu} \cdot f_{0n}^{'}(r) \cdot \left(A_n^+ \cdot e^{-\gamma_n \cdot z} + A_n^- \cdot e^{+\gamma_n \cdot z} \right) \cdot \hat{\varphi} \end{cases} \quad (39)$$

where functions $f_{0n}(r)$ and $f_0^{'}(r)$ are:

$$\begin{cases} f_{0n}(r) = \mathrm{J}_0(k_{cn} \cdot r) - \frac{\mathrm{J}_0(k_{cn} \cdot b)}{\mathrm{Y}_0(k_{cn} \cdot b)} \cdot \mathrm{Y}_0(k_{cn} \cdot r) \\ f_0^{'}(r) = \mathrm{J}_0^{'}(k_{cn} \cdot r) - \frac{\mathrm{J}_0(k_{cn} \cdot b)}{\mathrm{Y}_0(k_{cn} \cdot b)} \cdot \mathrm{Y}_0^{'}(k_{cn} \cdot r) \end{cases} \quad (40)$$

Again only TM_{0n} modes are considred.
The cut-off wavenumber k_{cn} is calculated, in this case, as a fucntion of the propagation constant γ_n as:

$$k_{cn}^2 = \omega^2 \cdot \varepsilon \cdot \mu + \gamma_n^2 = \omega^2 \cdot \varepsilon_0 \cdot \varepsilon_{r2} \cdot \mu_0 \cdot \mu_{r2} + \gamma_n^2 = k_0^2 \cdot \varepsilon_{r2} \cdot \mu_{r2} + \gamma_n^2 \quad (41)$$

And the propagation constant is calculated from equations (39) by applying the boundary conditions in *r=b*, that is already accomplished, and in *z=0* and *z=h*. Forcing these last two boundary conditions we have:

$$\begin{cases} E_r(z = h/2) = 0 \;\Rightarrow\; A_n^- = A_n^- \cdot e^{-2 \cdot \gamma_n \cdot h/2} = A_n^- \cdot e^{-\gamma_n \cdot h} \\ E_r(z = -h/2) = 0 \;\Rightarrow\; \sinh(\gamma_n \cdot h) = 0 \;\Rightarrow\; \gamma_n = \frac{j \cdot n}{h} \end{cases} \quad (42)$$

Then the fields inside the ring are:

$$\begin{cases} E_z = \sum_{n=1}^{\infty} \dfrac{j \cdot k_{cn}^2}{\omega \cdot \varepsilon \cdot \mu} \cdot f_{0n}(r) \cdot 2 \cdot A_n^+ \cdot e^{\gamma_n \cdot h/2} \cdot \cosh\left(\gamma_n \cdot \left(z + \dfrac{h}{2}\right)\right) \\ \vec{\mathbf{E}}_t = E_r \cdot \hat{r} = \sum_{n=1}^{\infty} \dfrac{\gamma_n \cdot k_{cn}}{j \cdot \omega \cdot \varepsilon \cdot \mu} \cdot f_{0n}'(r) \cdot (-2) \cdot A_n^+ \cdot e^{\gamma_n \cdot h/2} \cdot \sinh\left(\gamma_n \cdot \left(z + \dfrac{h}{2}\right)\right) \cdot \hat{r} \\ \vec{\mathbf{H}}_t = H_\varphi \cdot \hat{\varphi} = \sum_{n=1}^{\infty} \dfrac{k_{cn}}{\mu} \cdot f_{0n}'(r) \cdot 2 \cdot A_n^+ \cdot e^{\gamma_n \cdot h/2} \cdot \cosh\left(\gamma_n \cdot \left(z + \dfrac{h}{2}\right)\right) \cdot \hat{\varphi} \end{cases} \quad (43)$$

But the electric field in port 1 must be a series expasions on a basis functions. In this case we will use:

$$E_i^{(1)} = \sum_{m=1}^{\infty} \alpha_m \cdot \sin\left(2 \cdot \pi \cdot m \cdot \frac{z - h/2}{h}\right) + \beta_m \cdot \cos\left(2 \cdot \pi \cdot m \cdot \frac{z - h/2}{h}\right) \quad (44)$$

where the basis functions are the trigonometric functions.
This incident electric field in port 1, equation (44), must be equal to the same electric field comuted from the field inside the structure, following equation (43). Then the amplitudes inside are:

$$\left(\begin{aligned} E_z(r = a) &= \sum_{n=1}^{\infty} \frac{j \cdot k_{cn}^2}{\omega \cdot \varepsilon \cdot \mu} \cdot f_{0n}(a) \cdot 2 \cdot A_n^+ \cdot e^{\gamma_n \cdot h/2} \cdot \cosh\left(\gamma_n \cdot \left(z + \frac{h}{2}\right)\right) = E_i^{(1)} = \\ &= \sum_{m=1}^{\infty} \alpha_m \cdot \sin\left(2 \cdot \pi \cdot m \cdot \frac{z - h/2}{h}\right) + \beta_m \cdot \cos\left(2 \cdot \pi \cdot m \cdot \frac{z - h/2}{h}\right) \end{aligned} \right)$$

$$\Downarrow \quad (45)$$

$$A_n^+ = \frac{\chi_n}{h} \cdot \frac{\omega \cdot \varepsilon \cdot \mu}{j \cdot k_{cn}^2} \cdot \frac{e^{-\gamma_n \cdot h/2}}{2 \cdot f_{0n}(a)} \cdot \sum_{m=1}^{\infty} \alpha_m \cdot I_{nm}^{(s2)} + \beta_m \cdot I_{nm}^{(c2)}$$

where χ_n, $I_{nm}^{(s2)}$ y $I_{nm}^{(c2)}$ are[2]:

$$\chi_n = \begin{cases} 1 \ , \ n = 0 \\ 2 \ , \ n \neq 0 \end{cases} \quad (46a)$$

$$\begin{aligned} I_{nm}^{(s2)} &= \int_{-h/2}^{h/2} \sin\left(2 \cdot \pi \cdot m \cdot \frac{z - h/2}{h}\right) \cdot \cos\left(\pi \cdot n \cdot \frac{z + h/2}{h}\right) \cdot dz = \\ &= -2 \cdot m \cdot h \cdot \frac{\sin(n \cdot \pi/2 - m \cdot \pi)}{n + 2 \cdot m} \cdot \operatorname{sinc}(n/2 - m) \end{aligned} \quad (46b)$$

$$\begin{aligned} I_{nm}^{(c2)} &= \int_{-h/2}^{h/2} \cos\left(2 \cdot \pi \cdot m \cdot \frac{z - h/2}{h}\right) \cdot \cos\left(\pi \cdot n \cdot \frac{z + h/2}{h}\right) \cdot dz = \\ &= \frac{h}{\chi_m} \cdot \delta(n - 2 \cdot m) \end{aligned} \quad (46c)$$

[2] Function *sinc* of $I_{nm}^{(s2)}$ is defined as normal: $\operatorname{sinc}(x) = \dfrac{\sin(\pi \cdot x)}{\pi \cdot x}$

To get the GAM we must, again, obtain the relation between the electric and magnetic fields in port 1. The magnetic field in port 1 is:

$$H^{(1)} = \sum_{q=1}^{\infty} c_q \cdot \sin\left(2\cdot\pi\cdot q\cdot\frac{z-h/2}{h}\right) + d_q \cdot \cos\left(2\cdot\pi\cdot q\cdot\frac{z-h/2}{h}\right) \tag{47}$$

And the GAM is defined as:

$$\vec{\mathbf{h}}^{(1)} = \overline{\overline{\mathbf{Y}}}_{11} \cdot \vec{\mathbf{e}}^{(1)} \tag{48}$$

where $\vec{\mathbf{e}}^{(1)}$ and $\vec{\mathbf{h}}^{(1)}$ are the column vectors with the electric and magnetic amplitudes in port 1:

$$\vec{\mathbf{h}}^{(1)} = \begin{pmatrix} c_1 & c_2 & \ldots & c_Q & d_1 & d_2 & \ldots & d_Q \end{pmatrix}^t \quad ; \quad \vec{\mathbf{e}}^{(1)} = \begin{pmatrix} \alpha_1 & \alpha_2 & \ldots & \alpha_M & \beta_1 & \beta_2 & \ldots & \beta_M \end{pmatrix}^t \tag{49}$$

Then, identifying the inner magnetic field in port 1, equation (43), and the magnetic field outside, from equation (47), and substituing A_n^+ following (45):

$$\begin{pmatrix} H_\varphi(r=a) = \sum_{n=1}^{\infty} \frac{k_{cn}}{\mu} \cdot f_{0n}'(a) \cdot 2 \cdot A_n^+ \cdot e^{\gamma_n \cdot h/2} \cdot \cosh\left(\gamma_n \cdot \left(z+\frac{h}{2}\right)\right) = \\ = \sum_{q=1}^{\infty} c_q \cdot \sin\left(2\cdot\pi\cdot q\cdot\frac{z-h/2}{h}\right) + d_q \cdot \cos\left(2\cdot\pi\cdot q\cdot\frac{z-h/2}{h}\right) \end{pmatrix}$$
$$\Downarrow$$
$$\begin{cases} c_q = \dfrac{2\cdot(\chi_q - 1)}{h} \cdot \sum_{n=1}^{\infty} \dfrac{\chi_n}{h} \cdot \dfrac{\omega\cdot\varepsilon}{j\cdot k_{cn}} \cdot \dfrac{f_{0n}'(a)}{f_{0n}(a)} \cdot I_{nq}^{(s2)} \cdot \sum_{m=1}^{\infty} \alpha_m \cdot I_{nm}^{(s2)} + \beta_m \cdot I_{nm}^{(c2)} \\ d_q = \dfrac{\chi_q}{h} \cdot \sum_{n=1}^{\infty} \dfrac{\chi_n}{h} \cdot \dfrac{\omega\cdot\varepsilon}{j\cdot k_{cn}} \cdot \dfrac{f_{0n}'(a)}{f_{0n}(a)} \cdot I_{nq}^{(c2)} \cdot \sum_{m=1}^{\infty} \alpha_m \cdot I_{nm}^{(s2)} + \beta_m \cdot I_{nm}^{(c2)} \end{cases} \tag{50}$$

On the other way, the GAM is:

$$\vec{\mathbf{h}}^{(1)} = \begin{pmatrix} \vec{\mathbf{c}} \\ \vec{\mathbf{d}} \end{pmatrix} = \overline{\overline{\mathbf{Y}}}_{11} \cdot \vec{\mathbf{e}}^{(1)} = \begin{pmatrix} \overline{\overline{\mathbf{Y}}}_{11}^{(ss)} & \overline{\overline{\mathbf{Y}}}_{11}^{(sc)} \\ \overline{\overline{\mathbf{Y}}}_{11}^{(cs)} & \overline{\overline{\mathbf{Y}}}_{11}^{(cc)} \end{pmatrix} \cdot \begin{pmatrix} \vec{\boldsymbol{\alpha}} \\ \vec{\boldsymbol{\beta}} \end{pmatrix} \tag{51}$$

where each submatrix $\overline{\overline{\mathbf{Y}}}_{11}^{(ab)}$ is:

$$\left.\overline{\overline{\mathbf{Y}}}_{11}^{(ss)}\right|_{qm} = \frac{2\cdot(\chi_q - 1)}{h} \cdot \sum_{n=1}^{\infty} \frac{\chi_n}{h} \cdot \frac{\omega\cdot\varepsilon}{j\cdot k_{cn}} \cdot \frac{f_{0n}'(a)}{f_{0n}(a)} \cdot I_{nq}^{(s2)} \cdot I_{nm}^{(s2)} \tag{52a}$$

$$\left.\overline{\overline{\mathbf{Y}}}_{11}^{(sc)}\right|_{qm} = \frac{2\cdot(\chi_q - 1)}{h} \cdot \sum_{n=1}^{\infty} \frac{\chi_n}{h} \cdot \frac{\omega\cdot\varepsilon}{j\cdot k_{cn}} \cdot \frac{f_{0n}'(a)}{f_{0n}(a)} \cdot I_{nq}^{(s2)} \cdot I_{nm}^{(c2)} \tag{52b}$$

$$\left.\overline{\overline{\mathbf{Y}}}_{11}^{(cs)}\right|_{qm} = \frac{\chi_q}{h} \cdot \sum_{n=1}^{\infty} \frac{\chi_n}{h} \cdot \frac{\omega\cdot\varepsilon}{j\cdot k_{cn}} \cdot \frac{f_{0n}'(a)}{f_{0n}(a)} \cdot I_{nq}^{(c2)} \cdot I_{nm}^{(s2)} \tag{52c}$$

$$\left.\overline{\overline{\mathbf{Y}}}_{11}^{(cc)}\right|_{qm} = \frac{\chi_q}{h} \cdot \sum_{n=1}^{\infty} \frac{\chi_n}{h} \cdot \frac{\omega \cdot \varepsilon}{j \cdot k_{cn}} \cdot \frac{f_{0n}^{'}(a)}{f_{0n}(a)} \cdot I_{nq}^{(c2)} \cdot I_{nm}^{(c2)} \tag{52d}$$

3.6 Analysis of a 3-port network

Now let us calculate the GAM of the 3-port network depicted in figure 7e. This network has the two before cases, because it has ports in the top, in the bottom and in the lateral side. Figure 10 shows the axis to be used in this 3-port network.

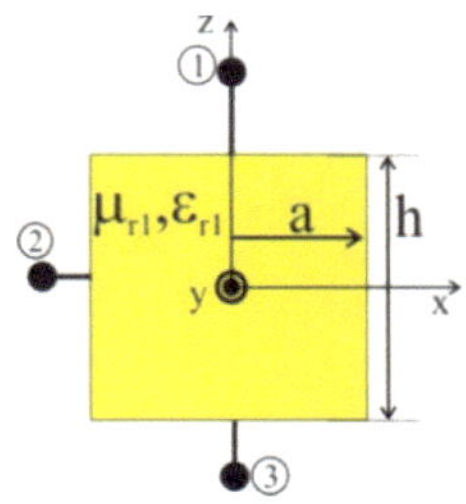

Fig. 10. Three port network: core

Note that now the GAM is 3x3 dimension matrix:

$$\vec{\mathbf{h}} = \overline{\overline{\mathbf{Y}}} \cdot \vec{\mathbf{e}} \quad \Rightarrow \quad \begin{pmatrix} \vec{\mathbf{h}}^{(1)} \\ \vec{\mathbf{h}}^{(2)} \\ \vec{\mathbf{h}}^{(3)} \end{pmatrix} = \begin{pmatrix} \overline{\overline{\mathbf{Y}}}_{11} & \overline{\overline{\mathbf{Y}}}_{12} & \overline{\overline{\mathbf{Y}}}_{13} \\ \overline{\overline{\mathbf{Y}}}_{21} & \overline{\overline{\mathbf{Y}}}_{22} & \overline{\overline{\mathbf{Y}}}_{23} \\ \overline{\overline{\mathbf{Y}}}_{31} & \overline{\overline{\mathbf{Y}}}_{32} & \overline{\overline{\mathbf{Y}}}_{33} \end{pmatrix} \cdot \begin{pmatrix} \vec{\mathbf{e}}^{(1)} \\ \vec{\mathbf{e}}^{(2)} \\ \vec{\mathbf{e}}^{(3)} \end{pmatrix} \tag{53}$$

To analyse this multiport structure we must go by columns, in such a way that all the ports are short-circuited except port *j*, where *j=(1,2,3)*. So, $\left.\vec{\mathbf{e}}^{(i)}\right|_{i \neq j} = 0$, and then the GAM elements are computed as (the *j* column in this case):

$$\vec{\mathbf{h}}^{(k)} = \left.\overline{\overline{\mathbf{Y}}}_{kj} \cdot \vec{\mathbf{e}}^{(j)}\right|_{k=1,2,3} \tag{54}$$

3.6.1 Parameters $\overline{\overline{\mathbf{Y}}}_{i1}$

To compute this column we must short-circuit ports 2 and 3. Then in $z=-h/2$ and in $r=a$ we have electric walls.

In general, the fields inside the structure will be, considering only TM_{0n} modes:

$$\begin{cases} E_z = \sum_{n=1}^{\infty} \frac{j \cdot k_{cn}^2}{\omega \cdot \varepsilon \cdot \mu} \cdot J_0(k_{cn} \cdot r) \cdot \left(A_n^+ \cdot e^{-\gamma_n \cdot z} + A_n^- \cdot e^{+\gamma_n \cdot z}\right) \\ \vec{\mathbf{E}}_t = E_r \cdot \hat{r} = \sum_{n=1}^{\infty} \frac{\gamma_n \cdot k_{cn}}{j \cdot \omega \cdot \varepsilon \cdot \mu} \cdot J_0^{'}(k_{cn} \cdot r) \cdot \left(A_n^+ \cdot e^{-\gamma_n \cdot z} - A_n^- \cdot e^{+\gamma_n \cdot z}\right) \cdot \hat{r} \\ \vec{\mathbf{H}}_t = H_\varphi \cdot \hat{\varphi} = \sum_{n=1}^{\infty} \frac{k_{cn}}{\mu} \cdot J_0^{'}(k_{cn} \cdot r) \cdot \left(A_n^+ \cdot e^{-\gamma_n \cdot z} + A_n^- \cdot e^{+\gamma_n \cdot z}\right) \cdot \hat{\varphi} \end{cases} \tag{55}$$

where the Second Kind Bessel Functions $Y_0(k_{cn}\cdot r)$ disappear because they are singular in the origin, and where the cut-off wavenumber k_{cn} and the propagation constant γ_n are related as:

$$k_{cn}^2=\omega^2\cdot\varepsilon\cdot\mu+\gamma_n^2=\omega^2\cdot\varepsilon_0\cdot\varepsilon_{r1}\cdot\mu_0\cdot\mu_{r1}+\gamma_n^2=k_0^2\cdot\varepsilon_{r1}\cdot\mu_{r1}+\gamma_n^2 \tag{56}$$

The boundary condition are then:

$$\begin{cases} E_r(z=h/2)=0 \;\Rightarrow\; A_n^-=A_n^-\cdot e^{-2\cdot\gamma_n\cdot h/2}=A_n^-\cdot e^{-\gamma_n\cdot h} \\ E_z(r=a)=0 \;\Rightarrow\; k_{cn}=\dfrac{p_{0n}}{a} \end{cases} \tag{57}$$

And the electromagnetic fields inside are:

$$\begin{cases} E_z=\sum_{n=1}^{\infty}\dfrac{j\cdot k_{cn}^2}{\omega\cdot\varepsilon\cdot\mu}\cdot \mathrm{J}_0(k_{cn}\cdot r)\cdot 2\cdot A_n^+\cdot e^{\gamma_n\cdot h/2}\cdot\cosh\left(\gamma_n\cdot\left(z+\dfrac{h}{2}\right)\right) \\ \vec{\mathbf{E}}_t=E_r\cdot\hat{r}=\sum_{n=1}^{\infty}\dfrac{-\gamma_n\cdot k_{cn}}{j\cdot\omega\cdot\varepsilon\cdot\mu}\cdot \mathrm{J}_0'(k_{cn}\cdot r)\cdot 2\cdot A_n^+\cdot e^{\gamma_n\cdot h/2}\cdot\sinh\left(\gamma_n\cdot\left(z+\dfrac{h}{2}\right)\right)\cdot\hat{r} \\ \vec{\mathbf{H}}_t=H_\varphi\cdot\hat{\varphi}=\sum_{n=1}^{\infty}\dfrac{k_{cn}}{\mu}\cdot \mathrm{J}_0'(k_{cn}\cdot r)\cdot 2\cdot A_n^+\cdot e^{\gamma_n\cdot h/2}\cdot\cosh\left(\gamma_n\cdot\left(z+\dfrac{h}{2}\right)\right)\cdot\hat{\varphi} \end{cases} \tag{58}$$

Identifying now the tangential incident field in port 1 with the ineternal field in the plane, as made in equation (33), and using the same series expansion for the incident electric field than before, as in equation (29), we have:

$$\begin{gathered} \sum_{n=1}^{\infty}\frac{\gamma_n\cdot k_{cn}\cdot 2}{j\cdot\omega\cdot\varepsilon\cdot\mu}\cdot \mathrm{J}_1(k_{cm}\cdot r)\cdot A_n^+\cdot e^{\gamma_n\cdot h/2}\cdot\sinh(\gamma_n\cdot h)=\sum_{m=1}^{\infty}\alpha_m\cdot N_m^{(1)}\cdot J_0'\left(\frac{p_{0m}}{a}\cdot r\right) \\ \Downarrow \\ A_n^+=\alpha_m\cdot N_m^{(1)}\cdot\frac{j\cdot\omega\cdot\varepsilon\cdot\mu}{\gamma_n\cdot k_{cn}}\cdot\frac{e^{\gamma_n\cdot h/2}}{-2\cdot\sinh(\gamma_n\cdot h)}\cdot\delta(n-m) \end{gathered} \tag{59}$$

Following this, parameter $\overline{\overline{\mathbf{Y}}}_{11}$ must accomplish: $\vec{\mathbf{h}}^{(1)}=\overline{\overline{\mathbf{Y}}}_{11}\cdot\vec{\mathbf{e}}^{(1)}$. The magnetic field will be, in the same way than in point 3.4 and showed in equation (34):

$$\begin{gathered} \left(\begin{aligned} H_\varphi\left(z=\frac{h}{2}\right)&=\sum_{n=1}^{\infty}\frac{k_{cn}}{\mu}\cdot \mathrm{J}_0'(k_{cn}\cdot r)\cdot 2\cdot A_n^+\cdot e^{\gamma_n\cdot h/2}\cdot\cosh(\gamma_n\cdot h)= \\ &=\sum_{q=1}^{\infty}c_q\cdot h_q^{(1)}=\sum_{q=1}^{\infty}c_q\cdot N_q^{(1)}\cdot J_0'\left(\frac{p_{0q}}{a}\cdot r\right) \end{aligned}\right) \\ \Downarrow \\ c_q=\alpha_m\cdot\frac{-j\cdot\omega\cdot\varepsilon}{\gamma_m}\cdot\frac{1}{\tanh(\gamma_n\cdot h)}\cdot\delta(q-m) \end{gathered} \tag{60}$$

Then the GAM is:

$$Y_{11}\Big|_{qm} = \frac{-j\cdot\omega\cdot\varepsilon_0\cdot\varepsilon_{r1}}{\gamma_m}\cdot\frac{1}{\tanh(\gamma_n\cdot h)}\cdot\delta(q-m) \tag{61}$$

In the same way, parameter $\overline{\overline{\mathbf{Y}}}_{31}$ is that who accomplishes: $\vec{\mathbf{h}}^{(3)} = \overline{\overline{\mathbf{Y}}}_{31}\cdot\vec{\mathbf{e}}^{(1)}$. The magnetic field is, as shown in point 3.4 and in equation (34), and equating as before:

$$Y_{31}\Big|_{qm} = \frac{-j\cdot\omega\cdot\varepsilon_0\cdot\varepsilon_{r1}}{\gamma_m}\cdot\frac{1}{\sinh(\gamma_n\cdot h)}\cdot\delta(q-m) \tag{62}$$

And parameter $\overline{\overline{\mathbf{Y}}}_{21}$, who accomplishes $\vec{\mathbf{h}}^{(2)} = \overline{\overline{\mathbf{Y}}}_{21}\cdot\vec{\mathbf{e}}^{(1)}$, is calculated in a similar way, remebering that the tangential magnetic field in port 2 is expanded in a series as shown in (47). Identifying the magnetic fields, we have, after replacing A_n^+ with the value obtained in equation (59):

$$\begin{pmatrix} H_\varphi(r=a) = \sum_{n=1}^{\infty}\frac{k_{cn}}{\mu}\cdot \mathrm{J}_0'(k_{cn}\cdot a)\cdot 2\cdot A_n^+\cdot e^{\gamma_n\cdot h/2}\cdot\cosh\left(\gamma_n\cdot\left(z+\frac{h}{2}\right)\right) = \\ = \sum_{q=1}^{\infty} c_q\cdot\sin\left(2\cdot\pi\cdot q\cdot\frac{z-h/2}{h}\right) + d_q\cdot\cos\left(2\cdot\pi\cdot q\cdot\frac{z-h/2}{h}\right)\end{pmatrix}$$
$$\Downarrow \tag{63}$$
$$\begin{cases} c_q = \frac{2\cdot(\chi_q-1)}{h}\cdot\sum_{m=1}^{\infty}\alpha_m\cdot \mathrm{J}_0'(p_{0m})\cdot\frac{\omega\cdot\varepsilon}{j\cdot\gamma_m}\cdot\frac{N_m^{(1)}}{\cosh(\gamma_m\cdot h)}\cdot I_{qm}^{(s)} \\ d_q = \frac{\chi_q}{h}\cdot\sum_{n=1}^{\infty}\alpha_m\cdot \mathrm{J}_0'(p_{0m})\cdot\frac{\omega\cdot\varepsilon}{j\cdot\gamma_m}\cdot\frac{N_m^{(1)}}{\cosh(\gamma_m\cdot h)}\cdot I_{qm}^{(c)} \end{cases}$$

where χ_n and *sinc(x)* have been defined before and $I_{qm}^{(s,c)}$ are[3] (more general than (46)):

$$\begin{aligned} I_{qm}^{(s)} &= \int_{-h/2}^{h/2}\sin\left(2\cdot\pi\cdot q\cdot\frac{z-h/2}{h}\right)\cdot\cosh(\gamma_m\cdot(z+h/2))\cdot dz = \\ &= \frac{h\cdot\pi\cdot q}{j\cdot\pi\cdot q+\gamma_m\cdot h}\cdot\sinh(\gamma_m\cdot h/2 - j\cdot\pi\cdot q)\cdot\mathrm{sinhc}(\gamma_m\cdot h/2 - j\cdot\pi\cdot q) = \\ &= -\int_{-h/2}^{h/2}\sin\left(2\cdot\pi\cdot q\cdot\frac{z-h/2}{h}\right)\cdot\cosh(\gamma_m\cdot(z-h/2))\cdot dz \end{aligned} \tag{64a}$$

[3] Function *sinhc* of $I_{qm}^{(s,c)}$ is: $\mathrm{sinhc}(x) = \frac{\sinh(x)}{x}$ (note the difference with *sinc*)

$$\begin{aligned} I_{qm}^{(c)} &= \int_{-h/2}^{h/2} \cos\left(2\cdot\pi\cdot q\cdot\frac{z-h/2}{h}\right)\cdot\cosh\left(\gamma_m\cdot(z+h/2)\right)\cdot dz = \\ &= h^2\cdot\gamma_m\cdot\frac{\operatorname{sinhc}\left(\gamma_m\cdot h-j\cdot 2\cdot\pi\cdot q\right)}{\gamma_m\cdot h+j\cdot 2\cdot\pi\cdot q} = \\ &= \int_{-h/2}^{h/2} \cos\left(2\cdot\pi\cdot q\cdot\frac{z-h/2}{h}\right)\cdot\cosh\left(\gamma_m\cdot(z-h/2)\right)\cdot dz \end{aligned} \tag{64b}$$

Then parameter $\overline{\overline{\mathbf{Y}}}_{21}$, defined as $\vec{\mathbf{h}}^{(2)} = \begin{pmatrix}\vec{\mathbf{c}}\\ \vec{\mathbf{d}}\end{pmatrix} = \overline{\overline{\mathbf{Y}}}_{21}\cdot\vec{\mathbf{e}}^{(1)} = \begin{pmatrix}\overline{\overline{\mathbf{Y}}}_{21}^{(s)}\\ \overline{\overline{\mathbf{Y}}}_{21}^{(c)}\end{pmatrix}\cdot\vec{\boldsymbol{\alpha}}$, is:

where each submatrix $\overline{\overline{\mathbf{Y}}}_{21}^{(s,c)}$ is:

$$\left.\overline{\overline{\mathbf{Y}}}_{21}^{(s)}\right|_{qm} = \frac{2\cdot(\chi_q-1)}{h}\cdot \mathrm{J}_0'(p_{0m})\cdot\frac{\omega\cdot\varepsilon}{\gamma_m}\cdot\frac{N_m^{(1)}}{\sinh(\gamma_m\cdot h)}\cdot I_{qm}^{(s)} \tag{65a}$$

$$\left.\overline{\overline{\mathbf{Y}}}_{21}^{(c)}\right|_{qm} = \frac{\chi_q}{h}\cdot \mathrm{J}_0'(p_{0m})\cdot\frac{\omega\cdot\varepsilon}{\gamma_m}\cdot\frac{N_m^{(1)}}{\sinh(\gamma_m\cdot h)}\cdot I_{qm}^{(c)} \tag{65b}$$

3.6.2 Parameters $\overline{\overline{\mathbf{Y}}}_{i2}$

For this column number 2, we must put short-circuits in ports 1 and 3. Then in z=-$h/2$ and in z=+$h/2$ we have electric walls.

In general, the inner electromagnetic fields are, as before for TM_{0n} modes:

$$\begin{cases} E_z = \displaystyle\sum_{n=1}^{\infty}\frac{j\cdot k_{cn}^2}{\omega\cdot\varepsilon\cdot\mu}\cdot \mathrm{J}_0(k_{cn}\cdot r)\cdot\left(A_n^+\cdot e^{-\gamma_n\cdot z}+A_n^-\cdot e^{+\gamma_n\cdot z}\right) \\ \vec{\mathbf{E}}_t = E_r\cdot\hat{r} = \displaystyle\sum_{n=1}^{\infty}\frac{\gamma_n\cdot k_{cn}}{j\cdot\omega\cdot\varepsilon\cdot\mu}\cdot \mathrm{J}_0'(k_{cn}\cdot r)\cdot\left(A_n^+\cdot e^{-\gamma_n\cdot z}-A_n^-\cdot e^{+\gamma_n\cdot z}\right)\cdot\hat{r} \\ \vec{\mathbf{H}}_t = H_\varphi\cdot\hat{\varphi} = \displaystyle\sum_{n=1}^{\infty}\frac{k_{cn}}{\mu}\cdot \mathrm{J}_0'(k_{cn}\cdot r)\cdot\left(A_n^+\cdot e^{-\gamma_n\cdot z}+A_n^-\cdot e^{+\gamma_n\cdot z}\right)\cdot\hat{\varphi} \end{cases} \tag{66}$$

where the cut-off wavenumber k_{cn} and the propagation constant γ_n are related as:

$$k_{cn}^2 = \omega^2\cdot\varepsilon\cdot\mu+\gamma_n^2 = \omega^2\cdot\varepsilon_0\cdot\varepsilon_{r1}\cdot\mu_0\cdot\mu_{r1}+\gamma_n^2 = k_0^2\cdot\varepsilon_{r1}\cdot\mu_{r1}+\gamma_n^2 \tag{67}$$

Imposing the boundary conditions we have:

$$\begin{cases} E_z(z=h/2)=0 \;\Rightarrow\; A_n^- = A_n^-\cdot e^{-2\cdot\gamma_n\cdot h/2} = A_n^-\cdot e^{-\gamma_n\cdot h} \\ E_z(z=-h/2)=0 \;\Rightarrow\; \gamma_n = j\cdot\beta_n = j\cdot\dfrac{n\cdot\pi}{h} \end{cases} \tag{68}$$

And the inner fields:

$$\begin{cases} E_z = \sum_{n=1}^{\infty} \frac{j \cdot k_{cn}^2}{\omega \cdot \varepsilon \cdot \mu} \cdot \mathrm{J}_0(k_{cn} \cdot r) \cdot 2 \cdot A_n^+ \cdot e^{\gamma_n \cdot h/2} \cdot \cosh\left(\gamma_n \cdot \left(z + \frac{h}{2}\right)\right) \\ \vec{\mathbf{E}}_t = E_r \cdot \hat{r} = \sum_{n=1}^{\infty} \frac{-\gamma_n \cdot k_{cn}}{j \cdot \omega \cdot \varepsilon \cdot \mu} \cdot \mathrm{J}_0'(k_{cn} \cdot r) \cdot 2 \cdot A_n^+ \cdot e^{\gamma_n \cdot h/2} \cdot \sinh\left(\gamma_n \cdot \left(z + \frac{h}{2}\right)\right) \cdot \hat{r} \\ \vec{\mathbf{H}}_t = H_\varphi \cdot \hat{\varphi} = \sum_{n=1}^{\infty} \frac{k_{cn}}{\mu} \cdot \mathrm{J}_0'(k_{cn} \cdot r) \cdot 2 \cdot A_n^+ \cdot e^{\gamma_n \cdot h/2} \cdot \cosh\left(\gamma_n \cdot \left(z + \frac{h}{2}\right)\right) \cdot \hat{\varphi} \end{cases} \quad (69)$$

Identifying the incident tangential electric field in port 2 with the inner electric field, from equation (45), and using the same series expansion in the incident field than in equation (44):

$$\left(\begin{aligned} &\sum_{n=1}^{\infty} \frac{j \cdot k_{cn}^2}{\omega \cdot \varepsilon \cdot \mu} \cdot \mathrm{J}_0(k_{cn} \cdot a) \cdot 2 \cdot A_n^+ \cdot e^{\gamma_n \cdot h/2} \cdot \cos\left(\frac{n \cdot \pi}{h} \cdot \left(z + \frac{h}{2}\right)\right) = \\ &= \sum_{m=1}^{\infty} \alpha_m \cdot \sin\left(2 \cdot \pi \cdot m \cdot \frac{z - h/2}{h}\right) + \beta_m \cdot \cos\left(2 \cdot \pi \cdot m \cdot \frac{z - h/2}{h}\right) \end{aligned} \right) \quad (70)$$

$$\Downarrow$$

$$A_n^+ = \frac{\chi_n}{h} \cdot \frac{\omega \cdot \varepsilon \cdot \mu}{j \cdot k_{cn}^2} \cdot \frac{e^{-\gamma_n \cdot h/2}}{2 \cdot \mathrm{J}_0(k_{cn} \cdot a)} \cdot \sum_{m=1}^{\infty} \alpha_m \cdot I_{nm}^{(s2)} + \beta_m \cdot I_{nm}^{(c2)}$$

Then, parameters $\overline{\overline{\mathbf{Y}}}_{12}$, $\overline{\overline{\mathbf{Y}}}_{22}$ and $\overline{\overline{\mathbf{Y}}}_{32}$ are calculated easily using the same procedure than before.

3.6.3 Parameters $\overline{\overline{\mathbf{Y}}}_{i3}$

Finally, parameters of the 3rd column are calulated exactly in the same way than parameters $\overline{\overline{\mathbf{Y}}}_{i1}$.

4. Simulations and measurements

Once all the GAM in figure 7a are obtained, it is quite easy and fast to analyse the geometries proposed in figures 6a and 6b. To do this is enough to apply the theory and equations proposed in point
2.3.

Those expressions allow interconnecting different networks just knowing the corresponding multimodal matrices GAM, GIM or GSM.

At this point is important to note that each multimodal matrix can use only one mode. Then they are called *monomode expressions*. These monomode expressions are less accurate but they are a really good seed for a more accurate result.

The monomode expressions were used and presented by the authors in (Penaranda-Foix et al., 2007b). If we look at the cavity shown in figure 7a, to get the resonant frequency is reduced to the simplest resonant condition, shown as an example in (Penaranda-Foix et al.,

2009). When only one mode is considered, the monomode analysis takes us to the expression (10) of (Penaranda-Foix et al., 2007b):

$$1+\frac{j\cdot k_0\cdot\varepsilon_{r1}}{a\cdot h\cdot\beta_{z1}^3}\cdot\frac{1}{\frac{f_1(a)}{f_0(a)}-\sqrt{\frac{\varepsilon_{r1}}{\mu_{r1}}}\cdot\frac{\mathrm{J}_1\left(k_0\cdot\sqrt{\varepsilon_{r1}\cdot\mu_{r1}}\cdot a\right)}{\mathrm{J}_0\left(k_0\cdot\sqrt{\varepsilon_{r1}\cdot\mu_{r1}}\cdot a\right)}}=0 \tag{71}$$

where all the permitivities and permeabilities but those of the central material to be measured, are air, and where:

$$f_n(a)=\frac{\mathrm{J}_n(k_0\cdot a)-\frac{\mathrm{J}_0(k_0\cdot b)}{\mathrm{Y}_0(k_0\cdot b)}\cdot\mathrm{Y}_n(k_0\cdot b)}{\mathrm{J}_0(k_0\cdot a)-\frac{\mathrm{J}_0(k_0\cdot b)}{\mathrm{Y}_0(k_0\cdot b)}\cdot\mathrm{Y}_0(k_0\cdot b)}\quad,\quad n=0,1$$
$$\beta_{z1}^3=k_0^2\cdot\varepsilon_{r1}\cdot\mu_{r1}-\left(p_{01}/a\right)^2 \tag{72}$$

To obtain equation (71) the insertion hole is only in the top, having a short-circuito in the bottom. That menas that in port 3 a short-circuited has been placed. Using the same procedure, a similar expression can be achieved with an insertion hole in the top and in the bottom. When no insertion hole is placed, the analytical analitycal procedure can be used (see Balanis, 1989)). Using the monomode equations previously determined, we arrive to the equation (8) of (Penaranda-Foix et al., 2007b):

$$\frac{f_1(a)}{f_0(a)}-\sqrt{\frac{\varepsilon_{r1}}{\mu_{r1}}}\cdot\frac{\mathrm{J}_1\left(k_0\cdot\sqrt{\varepsilon_{r1}\cdot\mu_{r1}}\cdot a\right)}{\mathrm{J}_0\left(k_0\cdot\sqrt{\varepsilon_{r1}\cdot\mu_{r1}}\cdot a\right)}=0 \tag{73}$$

Of course, this equation is the same that that obtained by Balanis. The result is normal, because the Balanis procedure consists of a monomode analysis.
The resonant frequencies obtained with and without insertion hole are different. This frequency deviation can be really important when measuring large permittivities or large aspect ratio cavities. It was 1960 when Estin (Estin & Bussey, 1960) published a first aproximate expression to estimate this error. It was a linear equation and it did not take into account the saturation effect described by the monomode procedure. It appears at large permittivities and at large aspect rates. Then in 2007 Penaranda-Foix (Penaranda-Foix et al., 2007c) proposed an alternative exponential equation obtained from the exact value and (71):

$$\frac{\delta f}{f_r}=\frac{f_r-f_0}{f_r}=\alpha\cdot\left(1-e^{-\gamma\cdot\varepsilon_{r1}^{1+\Delta}}\right) \tag{74}$$

where f_r is the resonant frequency with insertion hole and f_0 is the resonant frequency in the ideal case, without insertion hole. And the parameters are:

$$\alpha = x_1 \cdot \frac{a^{x_2} \cdot b^{x_3}}{h^{x_4}} \qquad \gamma = K_m \cdot \frac{a^3}{h \cdot b^2} \cdot \frac{1}{\alpha}$$
$$x_1 = 0.1277 \quad x_2 = 1.0464 \quad x_3 = 0.0220 \quad x_4 = 1.0462 \tag{75}$$
$$\Delta = 0.16 \quad K_m = 0.32$$

Finally, it is important to note that both equations, Estin and Penaranda-Foix, are also valid when the insertion hole goes through the cavity and it is in the top and in the bottom. In this case the relative error must be doubled. So it is enough to multiply by two equation (74).

4.1 Measurements

To probe the validity of the previous expressions, as well as the Generalised Circuital Analysis described along this chapter, some measurements were carried out. These measurements implied two diferent cavities with an insertion hole in the top and in the bottom.

Cavity number 1 has a radius *b=60 mm* and a height *h=20 mm*. The insertion hole radius is *a=6.35 mm*, and long enough to consider that is under cut-off. Cavity number 2 is exactly the same than number 1 but the external radius is *b=20 mm*.

Cavity number 1 is shown in figure 11a, where it is open. In figure 11b is shown the same cavity but closed and some samples used to be measured.

Fig. 11a. Cavity N. 1 open

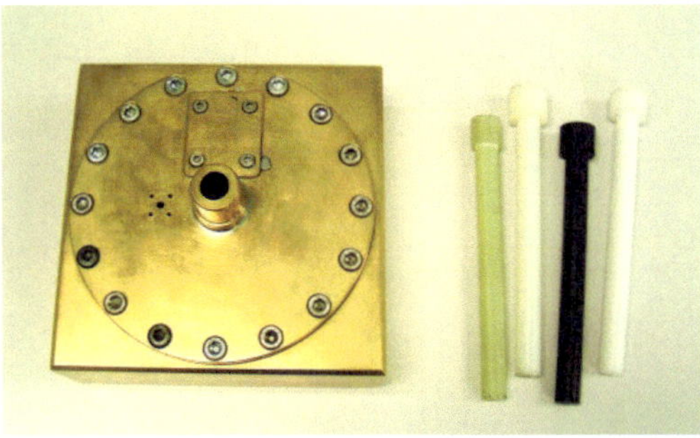

Fig. 11b. Cavity N. 1 closed and samples

The expected resonant frequency deviations is this cavities was simulated and shown in figure 12a. Cavity number 1 has a acpect ratio *a/b=0.1058* and cavity number 2 aspect ratio is *a/b=0.3175*. Two more aspect ratios have been simulated: *a/b=0.5 (b=12.7 mm)*, that is even smaller than Cavity number 2, and *a/b=0.04 (b=158.75 mm)*, larger than Cavity number 1.

Figure 12a shows the predicted frequency deviation by Estin and that predicted by (74), compared with the exact one calculated by circuital analysis following the procedure described along the chapter. And figure 12b shows the actual resonant frequency with are without insertion hole.

Table 1 shows the measurements performed with Cavity number 1 (f_u is the measured resonant frequency in GHz and Q_u is the measured quality factor) and the calculated permittivities depending the case (note that in all the cases the quality factor used Q_d has been calculated with the expression $1/Q_d = 1/Q_{meas} - 1/Q_{empty}$ to avoid the wall losses

effect, in such a way that Q_{meas} is the mesured quality factor with material and Q_{empty} is the mesured quality factor without material) for materials *M1 (Nylon 66 FG)* and *M2 (ACETAL)*:
CASE 1: This is the permittivity calculated using the circuital and multimode method, considering the up and down insertion holes. This is going to be considered as the exact and reference case.

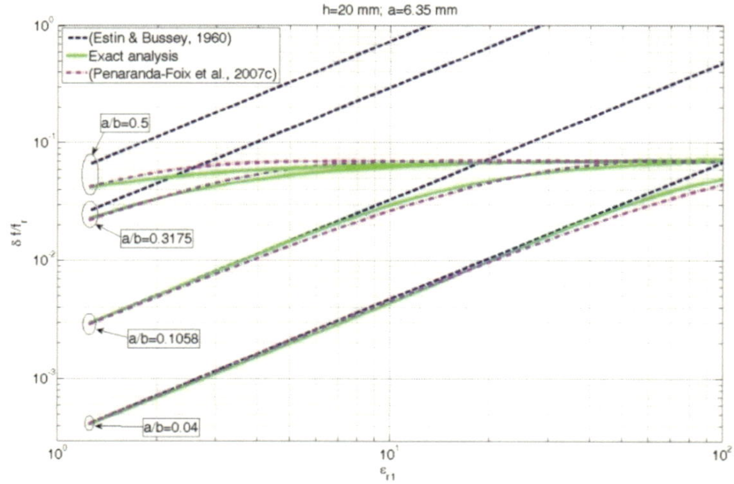

Fig. 12a. Frequency deviation for different permittivities and different aspect ratios

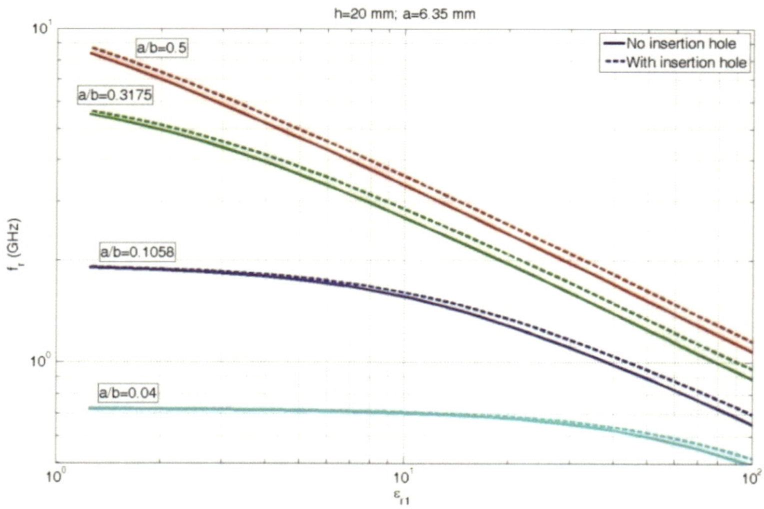

Fig. 12b. Frequency deviation for different permittivities and different aspect ratios.

CASE 2: This is the permittivity computed neglecting the effect of the insertion hole. That is using equation (73) directly.
CASE 3: This is the computed permittivity using the correction proposed by Estin (Estin & Bussey, 1960).
CASE 4: This is the permittivity using the correction proposed in (74).

In all the cases the imaginary part of the permittivity has been multiplied by 10^2 and the corrections to the resonant frequency (equations from Estin and (73)) have been applied to the complex resonant frequency, and not only to the resonant frequency- The complex resonant frequency is defined as $\Omega_r = f_r \cdot (1 + j/(2 \cdot Q))$. In Harrington (Harrington, 1961), the reader can find the origin of this concept for resonant problems.

			CASE 1		CASE 2		CASE 3		CASE 4	
	f_u	Q_u	ε'	ε''	ε'	ε''	ε'	ε''	ε'	ε''
Air	1.9157	4496	60.0274[4]	--------	--------	--------	--------	--------	--------	--------
M1	1.8517	564.8	2.8457	4.0922	2.5237	3.5545	2.8291	4.0184	2.8155	3.9760
M2	1.8471	198.8	2.9778	12.6630	2.6385	10.996	2.9588	12.4362	2.9438	12.2980

Table 1. Measurements with cavity 1.

It is important to note that the error in CASE 2 in Cavity number 2 is about 12% respect to the exact, because the insertion hole correction has not been applied. Once the correction is used, even in CASE 3 or CASE 4, errors are reduced to less than 2%.

Table 2 shows exactly the same than before, but measured in a smaller cavity: Cavity number 2.

			CASE 1		CASE 2		CASE 3		CASE 4	
	f_u	Q_u	ε'	ε''	ε'	ε''	ε'	ε''	ε'	ε''
Air	5.8350	617.7	20.0119[5]	--------	--------	--------	--------	--------	--------	--------
M1	4.6045	106.5	2.8779	3.6882	2.5410	3.1167	3.0760	4.3551	2.8903	3.7284
M2	4.5979	49.00	2.8903	8.8461	2.552	7.6725	3.0906	10.5658	2.9032	9.0356

Table 2. Measurements with cavity 2.

In this case, and due to a highest aspect ratio in Cavity number 2, the Estin formulae fails. So the error, supposing that CASE 1 is the exact one, errors are about 7% and 12% using Estin formulae or no-insertion hole approach. But it reduces again to less than 2% when using (74).

Finally, it is worth mentioning that all the above results are calculated with an uncertainty of 1.6%). This uncertainty has been calculated following the procedure described in (Baker-Jarvis et al., 1994) or (Bell, 2001). Basically it is based on the calculation of the derivative of the permittivity depending on the variable that affects the accuracy:

$$\partial \varepsilon_{r1} = \sqrt{\sum_i \left(\frac{\partial \varepsilon_{r1}}{\partial x_i} \cdot \Delta x_i \right)^2} \tag{76}$$

where x_i are the different dimensions that are involved in the accuracy of the measurement. And to compute the derivatives $\partial \varepsilon_{r1} / \partial x_i$ the three terms Lagrange polynomial interpolation formula was used:

---

4 This value is not the air permittivity but the real cavity radius calculated from the air resonant frequency.

5 Once again, this value is not the air permittivity but the real cavity radius calculated from the air resonant frequency.

$$f(x)=\sum_{i=0}^{N} a_i \cdot h_i(x), \quad h_i(x)=\prod_{\substack{j=1 \\ j\neq i}}^{N} \frac{x-x_j}{x_i-x_j} \tag{77}$$

Then three different permittivities are calculated for each source of error in order to compute the interpolation and then the derivative in (76) to calculate the final uncertainty.

5. Conclusion

Conclusions are, firstly, that an introduction and a revision of the Generalized Circuital Analysis to solve in a simplest way large electromagnetic problems have been presented.
This method has been successfully applied to a Cylindrical Cavity with insertion hole, creating four different small networks and solving each one to get the Generalized Admittance Matrix (GAM), and joining all together forcing the resonant condition to obtain the complex resonant frequency.
The origin of the problem comes from the deviation of the resonant frequency observed when the insertion hole exists. This deviation can be important if neglected, so a monomode approximation formula has been obtained as well as an optimized equation to estimate the relative error from the measured and actual resonant frequency with and without insertion hole.
The proposed expression improves the only one existing since 1960 by Estin.

6. References

Alessandri, Ferdinando; Mongiardo Mauro & Sorrentino, Roberto (1994). , Rigorous Mode Matching Analysisof Mitered E-Plane Bends in Rectangular Waveguide, *IEEE Microwave and Guided Wave Letters*, Nº 12, Vol. 4, pp. 408-410, ISSN 1531-1309

Baker-Jarvis, James; Janezic, Michael D.; Domich, Paul D. & Geyer, Richard G. (1994). Analysis of an Open-Ended Coaxial Probe with Lift-Off for Nondestructive Testing, *IEEE Trans. On Instrumentation and Measurement*, vol. 43 No 5, Oct., pp. 711-718, ISSN 0018-9456

Balanis, Constantine A., (1989). *Advanced Engineering Electromagnetics*, John Wiley and Sons, ISBN 978-0471503163

Bell, Stephanie (2001). *Beginner's Guide to Uncertainty of Measurement (Issue 2)*, National Physical Laboratory (NPL), ISSN 1368-6550. August 1999, Issue 2 with amendments March 2001

Chen, L. F.; Ong, C. K.; Neo, C. P.; Varadan, V.V. & Varadan V.K., (2004). *Microwave Electronics*, John Wiley and Sons, ISBN 0-470-84492-2

Collin, Robert E. (1991), *Field Theory of Guided Waves*, ISBN 978-0879422370, IEEE Press

Collin, Robert E. (1966), *Foundations for Microwave Engineering*, ISBN 978-0780360310, McGraw-Hill

Estin, A. J. & Bussey, H. E. (1960). Errors in Dielectric Measurements Due to a Sample Insertion Hole in a Cavity, *IEEE Trans. on MTT*, Vol. 8, No. 11, Nov., pp. 650-653, ISSN 0018-9480

Harrington, R. F. (1961). *Time-Harmonic Electromagnetic Fields*, ISBN 047120806X, New York

Harrington, Roger F. (1967). Matrix Methods for Field Problems, *Proceedings of the IEEE*, N. 2, Vol. 55, pp. 136-149, ISSN 0018-9219

Harrington, Roger F. (1993). *Field Computation by Moment Methods*, Wiley-IEEE Press, ISBN 978-0-7803-1014-8

Gentili, G.G. (1991). Properties of TE-TM mode-matching techniques, *IEEE Trans. on M.T.T.*, vol. 39, no. 9, Sept., pp. 1669-1673, ISSN 0018-9480

Gentili, G. G. & Melloni, A. (1996). Generalized Admittance Matrix Analysis of Cylindrical Cavities Coupledby Rectangular Apertures, *Journal of Electromagnetic Waves and Applications*, N° 1,Vol. 10, pp. 129-145, ISSN 0920-5071

Gimeno, Benito & Guglielmi Marco (1997). Multimode Equivalent Network Representation for Junctions between Coaxial and Circular Waveguides, *Int. Journal Microwave Millimeter-Wave CAE*,pp. 180-194, ISSN 1050-1827

Metaxas A.C. & Meredith R.J. (1988). *Industrial Microwave Heating*, The Institution of Engineering and Technology, ISBN: 978-0906048894

Penaranda-Foix, Felipe L. (2001). *Aplicación de la Teoría de Análisis Circuital Generalizado a la Resolución de Problemas de Difracción Electromagnética*, PhD in Universidad Politécnica de Valencia -Valencia (Spain).

Penaranda-Foix, Felipe L. & Ferrando-Bataller, Miguel (2003). Scattering of Inhomogeneous Cylinders by Circuital Analysis, *Microwave and Optical Technology Letters*, Vol 39, N. 2, Oct., pp. 155-159, ISSN 0895-2477

Penaranda-Foix, Felipe L.; Catala-Civera, José M.; Canos-Marin, Antoni J. & Plaza-Gonzalez, Pedro J. (2007a). Solving Cylindrically-Shaped Waveguides Partially-Filled with Isotropic Materials by Modal Techniques. *Proceedings of 11th. AMPERE 2007*, Vol. 1, pp. 67-70, ISBN 978-973-759-333-7, Oradea (Romania), 3-6 September

Penaranda-Foix, Felipe L.; Catala-Civera, Jose M.; Canos-Marin, Antoni J. & Garcia-Banos, Beatriz (2007b). Frequency Deviation Due to a Sample Insertion Hole in a Cylindrical Cavity by Circuital Analysis. Proceedings of the IMS2007, Vol. 1 pp. 1683-1686, ISBN 1-4244-0688-9/07, Honolulu (USA)

Penaranda-Foix, Felipe L.; Catala-Civera, Jose M.; Canos-Marin, Antoni J. & Garcia-Banos, Beatriz (2007c). Practical expression for frequency deviation due to insertion hole in cylindrical cavity, *Electronics Letters*, Vol 43, N. 24, 22nd Nov., pp. 1338-1340, ISSN 0013-5194

Penaranda-Foix, Felipe L.; Catala-Civera, Jose M.; Canos-Marin, Antoni J. & Garcia-Banos, Beatriz (2009). Circuital analysis of a coaxial re-entrant cavity for performing dielectric measurement. Proceedings of the IMS2009, Vol. pp. 1309-1312, ISBN 978-1-4244-2804-5, Boston (USA)

Pozar, David M. (1990), *Microwave Engineering*, ISBN 978-0471448785, Addison-Wesley Publishing Company, Inc.

Ramo, Simon; Whinnery, John R. & Duzer, Theodore van (1994). *Fields and waves in communication electronics*, John Wiley & Sons, Inc., ISBN 0471585513

Rebollar, Jesus M.; Esteban, J. & Page, J. E. (1994). Fullwave Analysis of Three and Four-Port Rectangular Waveguide Junctions, *IEEE Trans. on M.T.T.*, N° 2, Vol. 42, pp. 256-263, ISSN 0018-9480

Valero-Nogueira, Alejandro (1997). *Resolución de Problemas Electromagnéticos Complejos Mediante Análisis Circuital Generalizado*, PhD in Universidad Politécnica de Valencia-Valencia (Spain).

Wexler, A. (1967). Solution of waveguide discontinuities by modal analysis, *IEEE Trans. MTT*, Vol. 15, N. 9, pp.508–517, ISSN 0018-9480

8

Solving Inverse Scattering Problems Using Truncated Cosine Fourier Series Expansion Method

Abbas Semnani & Manoochehr Kamyab
K. N. Toosi University of Technology
Iran

1. Introduction

The aim of inverse scattering problems is to extract the unknown parameters of a medium from measured back scattered fields of an incident wave illuminating the target. The unknowns to be extracted could be any parameter affecting the propagation of waves in the medium.
Inverse scattering has found vast applications in different branches of science such as medical tomography, non-destructive testing, object detection, geophysics, and optics (Semnani & Kamyab, 2008; Cakoni & Colton, 2004).
From a mathematical point of view, inverse problems are intrinsically ill-posed and nonlinear (Colton & Paivarinta, 1992; Isakov, 1993). Generally speaking, the ill-posedness is due to the limited amount of information that can be collected. In fact, the amount of independent data achievable from the measurements of the scattered fields in some observation points is essentially limited. Hence, only a finite number of parameters can be accurately retrieved. Other reasons such as noisy data, unreachable observation data, and inexact measurement methods increase the ill-posedness of such problems. To stabilize the inverse problems against ill-posedness, usually various kinds of regularizations are used which are based on a priori information about desired parameters. (Tikhonov & Arsenin, 1977; Caorsi, et al., 1995). On the other hand, due to the multiple scattering phenomena, the inverse-scattering problem is nonlinear in nature. Therefore, when multiple scattering effects are not negligible, the use of nonlinear methodologies is mandatory.
Recently, inverse scattering problems are usually considered in global optimization-based procedures (Semnani & Kamyab, 2009; Rekanos, 2008). The unknown parameters of each cell of the medium grid would be directly considered as the optimization parameters and several types of regularizations are used to overcome the ill-posedness. All of these regularization terms commonly use a priori information to confine the range of mathematically possible solutions to a physically acceptable one. We will refer to this strategy as the direct method in this chapter.
Unfortunately, the conventional optimization-based methods suffer from two main drawbacks. The first is the huge number of the unknowns especially in 2-D and 3-D cases

which increases not only the amount of computations, but also the degree of ill-posedness. Another disadvantage is the determination of regularization factor which is not straightforward at all. Therefore, proposing an algorithm which reduces the amount of computations along with the sensitivity of the problems to the regularization term and initial guess of the optimization routine would be quite desirable.

2. Truncated cosine Fourier series expansion method

Instead of direct optimization of the unknowns, it is possible to expand them in terms of a complete set of orthogonal basis functions and optimize the coefficients of this expansion in a global optimization routine. In a general 3-D structure, for example the relative permittivity could be expressed as

$$\varepsilon_r(x,y,z) = \sum_{n=0}^{N-1} d_n f_n(x,y,z) \tag{1}$$

where f_n is the n_{th} term of the complete orthogonal basis functions.

It is clear that in order to expand any profile into this set, the basis functions must be complete. On the other hand, orthogonality is favourable because with this condition, a finite series will always represent the object with the best possible accuracy and coefficients will remain unchanged while increasing the number of expansion terms.

Because of the straightforward relation to the measured data and its simple boundary conditions, using harmonic functions over other orthogonal sets of basis functions is preferable. On the other hand, cosine basis functions have simpler mean value relation in comparison with sine basis functions which is an important condition in our algorithm.

We consider the permittivity and conductivity profiles reconstruction of lossy and inhomogeneous 1-D and 2-D media as shown in Fig. 1.

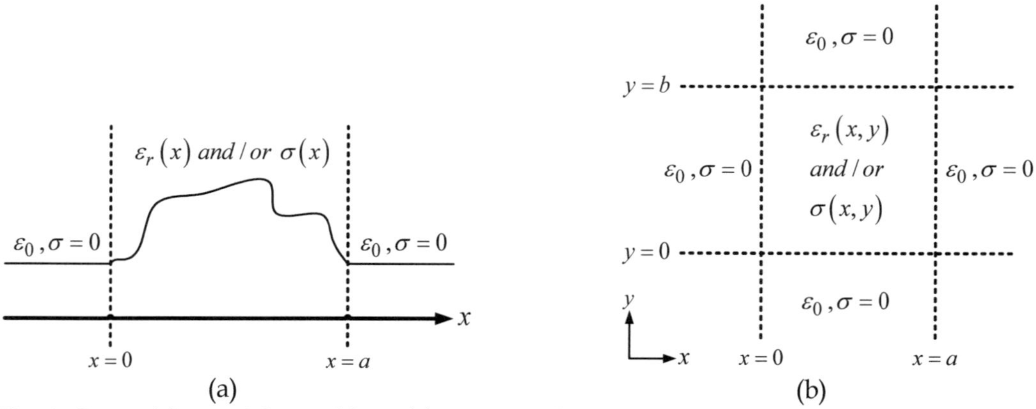

Fig. 1. General form of the problem, (a) 1-D case, (b) 2-D case

If cosine basis functions are used in one-dimensional cases, the truncated expansion of the permittivity profile along x which is homogeneous along the transverse plane could be expressed as

$$\varepsilon_r(x) = \sum_{n=0}^{N-1} d_n \cos\left(\frac{n\pi}{a}x\right) \tag{2}$$

where a is the dimension of the problem in the x direction and the coefficients, d_n, are to be optimized. In this case, the number of optimization parameters is N in comparison with conventional methods in which this number is equal to the number of discretized grid points. This results in a considerable reduction in the amount of computations. As another very important advantages of the expansion method, no additional regularization term is needed, because the smoothness of the cosine functions and the limited number of expansion terms are considered adequate to suppress the ill-posedness

In a similar manner for 2-D cases, the expansion of the relative permittivity profile in transverse x-y plane which is homogeneous along z can be written as

$$\varepsilon_r(x,y) = \sum_{n=0}^{N-1}\sum_{m=0}^{M-1} d_{nm} \cos\left(\frac{n\pi}{a}x\right)\cos\left(\frac{m\pi}{b}y\right) \tag{3}$$

where a and b are the dimensions of the problem in the x and y directions, respectively. Similar expansions could be considered for conductivity profiles in lossy cases.

The proposed expansion algorithm is shown in Fig. 2. According to this figure, based on an initial guess for a set of expansion coefficients, the permittivity and conductivity are calculated according to the expansion relations like (2) or (3). Then, an EM solver computes a trial electric and magnetic simulation fields. Afterwards, cost function which indicates the difference between the trial simulated and reference measured fields is calculated. In the next step, global optimizer is used to minimize this cost function by changing the permittivity and conductivity of each cell until the procedure leads to an acceptable predefined error.

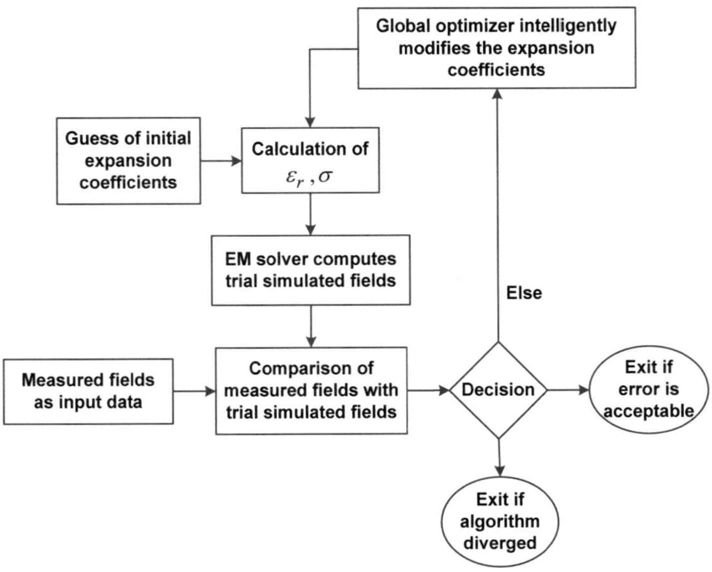

Fig. 2. Proposed algorithm for reconstruction by expansion method

3. Mathematical Considerations

As mentioned before, inverse problems are intrinsically ill-posed. Therefore, a priori information must be applied for stabilizing the algorithm as much as possible which is quite straightforward in direct optimization method. In this case, all the information can be applied directly to the medium parameters which are as the same as the optimization parameters. In the expansion algorithm, however, the optimization parameters are the Fourier series expansion coefficients and a priori information could not be considered directly. Hence, a useful indirect routine is vital to overcome this difficulty.

There are two main assumptions about the parameters of an unknown medium. For example, we may assume first that the relative permittivity and conductivity have limited ranges of variation, i.e.

$$1 \le \varepsilon_r \le \varepsilon_{r,\max} \tag{4}$$

and

$$0 \le \sigma \le \sigma_{max} \tag{5}$$

The second assumption is that the permittivity and conductivity profiles may not have severe fluctuations or oscillations. These two important conditions must be transformed in such a way to be applicable on the expansion coefficients in the initial guess and during the optimization process.

It is known that average of a function with known limited range is located within that limit, that is if

$$L_1 \le g(x) \le L_2, \quad a \le x \le b \tag{6}$$

Then

$$L_1 \le \frac{1}{b-a} \int_a^b g(x)\,dx \le L_2 \tag{7}$$

Thus, for 1-D permittivity profile expansion we have

$$1 \le d_0 \le \varepsilon_{r,\max} \tag{8}$$

For $x = 0$, (2) reduces to

$$\varepsilon_r(0) = \sum_{n=0}^{N-1} d_n \quad \Rightarrow \quad 1 \le \sum_{n=0}^{N-1} d_n \le \varepsilon_{r,\max} \tag{9}$$

and for $x = a$, we have

$$\varepsilon_r(a) = \sum_{n=0}^{N-1} (-1)^n d_n \quad \Rightarrow \quad 1 \le \sum_{n=0}^{N-1} (-1)^n d_n \le \varepsilon_{r,\max} \tag{10}$$

Using Parseval theorem, another relation between expansion coefficients and upper bound of permittivity may be written. For a periodic function $g(x)$ with period T, we have

$$\frac{1}{T}\int_T |g(x)|^2 \, dx = \sum_{n=0}^{\infty} |d_n|^2 \tag{11}$$

Based on (2), (11) may be simplified to

$$1 \le \sum_{n=0}^{N-1} |d_n|^2 \le \varepsilon_{r,\max}^2 \tag{12}$$

It is possible to achieve the similar relations for 2-D cases.

$$1 \le d_{00} \le \varepsilon_{r,\max} \tag{13}$$

$$1 \le \sum_{n=0}^{N-1}\sum_{m=0}^{M-1} d_{nm} \le \varepsilon_{r,\max} \tag{14}$$

$$1 \le \sum_{n=0}^{N-1}\sum_{m=0}^{M-1} (-1)^{n+m} d_{nm} \le \varepsilon_{r,\max} \tag{15}$$

$$1 \le \sum_{n=0}^{N-1}\sum_{m=0}^{M-1} |d_{nm}|^2 \le \varepsilon_{r,\max}^2 \tag{16}$$

By using the above supplementary equations in the initial guess of the expansion coefficients and as a boundary condition (Robinson & Rahmat-Samii, 2004) during the optimization, the routine converges in a considerable faster rate. Similar conditions can be used for conductivity profiles in lossy cases.

4. Numerical Results

Proposed method stated above is utilized for reconstruction of some different 1-D and 2-D media. In each case, reconstruction by the proposed expansion method is compared with different number of expansion functions in terms of the amount of computations and reconstruction precision.

The objective of the proposed reconstruction procedure is the estimate of the unknowns by minimizing the cost function

$$C = \frac{\sum_{i=1}^{I}\sum_{j=1}^{J}\sum_{t=1}^{T}\left|E_{ij}^{meas}(t) - E_{ij}^{sim}(t)\right|^2}{\sum_{i=1}^{I}\sum_{j=1}^{J}\sum_{t=1}^{T}(E_{ij}^{meas}(t))^2} \tag{17}$$

where $\vec{E}^{sim}$ is the simulated field in each optimization iteration. $\vec{E}^{meas}$ is measured field, I and J are the number of transmitters and receivers, respectively and T is the total time of measurement.

To quantify the reconstruction accuracy, the reconstruction errors for example for relative permittivity in 1-D case is defined as

$$e(\varepsilon) = \sqrt{\frac{\sum_{i=1}^{M_x}\left|\varepsilon_{ri} - \varepsilon_{ri}^{o}\right|^2}{\sum_{i=1}^{M_x}(\varepsilon_{ri}^{o})^2}} \times 100 \tag{18}$$

where M_x is the number of subdivisions along x axis and " ^{o} " denotes the original scatterer properties.

In all reconstructions in this chapter, FDTD (Taflove & Hagness, 2005) and DE (Storn & Price, 1997) are used as forward EM solver and global optimizer, respectively.

4.1 One-dimensional case

Reconstruction of two 1-D cases is considered in this section. The first one is inhomogeneous and lossless and the second one is considered to be lossy. In the simulations of both cases, one transmitter and two receivers are used around the medium as shown in Fig. 3.

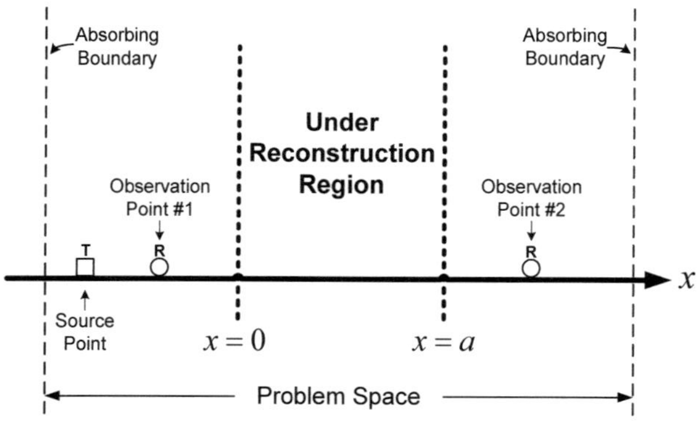

Fig. 3. Geometrical configuration of the 1-D problem

Test case #1: In the first sample case, we consider an inhomogeneous and lossless medium consisting 50 cells. Therefore, only the permittivity profile reconstruction is considered. In the expansion method, the number of expansion terms is set to 4, 5, 6 and 7 which results in a lot of reduction in the number of the unknowns in comparison with the direct method. The population in DE algorithm is chosen equal to 100 and the maximum iteration of optimization is considered to be 300. It must be noted that the initial populations in all reconstruction problems in this chapter are chosen completely random in the solution space. The exact profile and reconstructed ones by the expansion method with different number of expansion terms are shown in Fig. 4a. The variations of cost function (17) and reconstruction error (18) versus the iteration number are plotted in Figs. 4b and 4c, respectively.

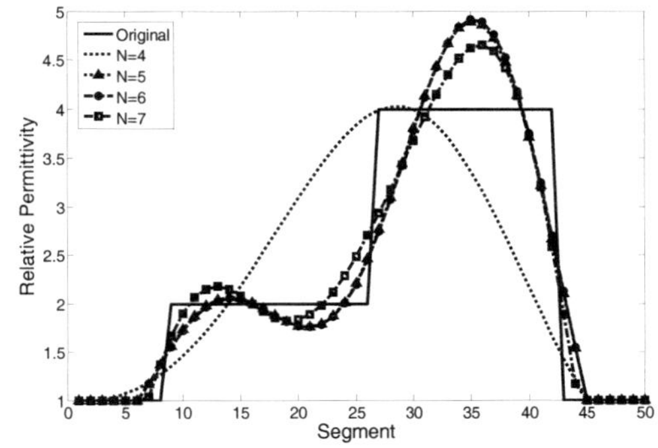

(a)

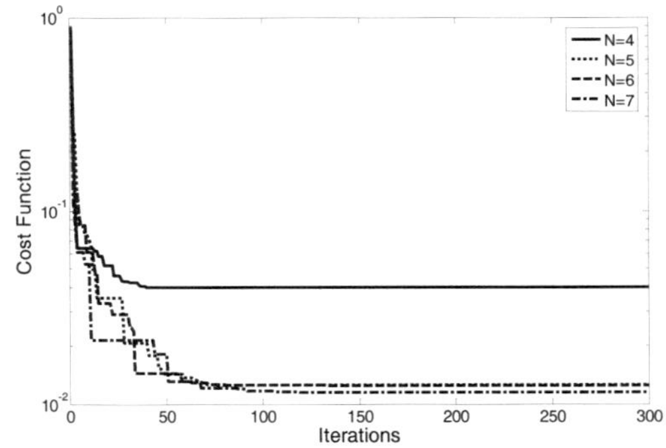

(b)

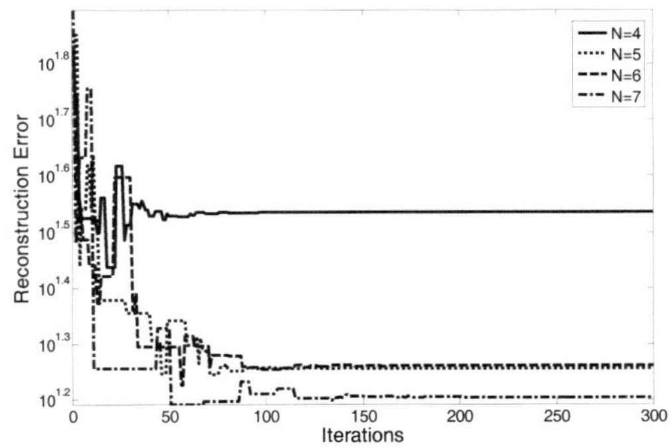

(c)

Fig. 4. Reconstruction of 1-D test case #1, (a) original and reconstructed profiles, (b) the cost function and (c) the reconstruction error

Test case #2: In this case, a lossy and inhomogeneous medium again with 50 cell length is considered. So, the number of unknowns in direct optimization method is equal to 100. In the expansion method for both permittivity and conductivity profiles expansion, N is chosen equal to 4, 5, 6 and 7. The optimization parameters are considered equal to the first sample case. The original and reconstructed profiles in addition of the variations of cost function and reconstruction error are presented in Fig. 5.

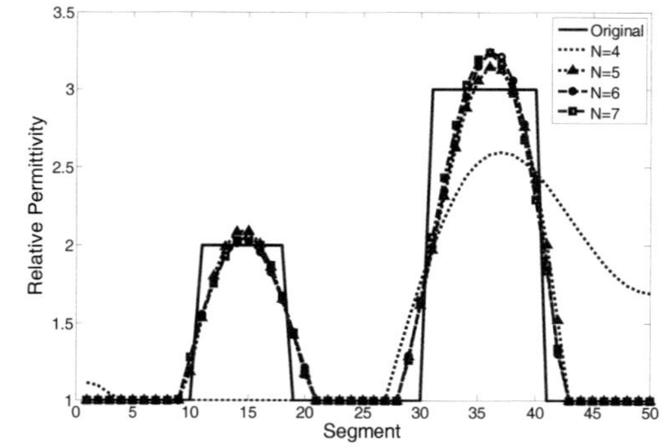

(a)

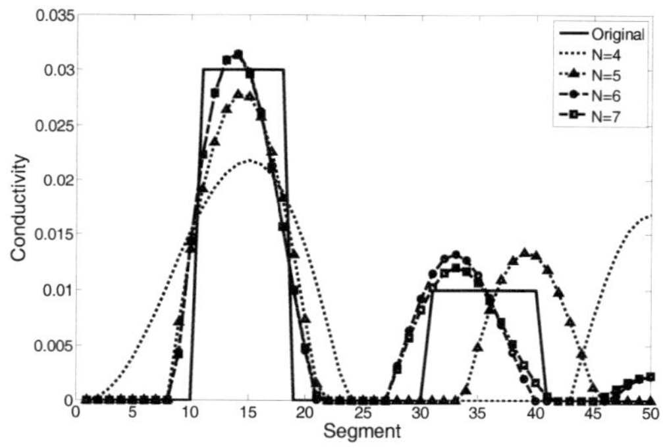

(b)

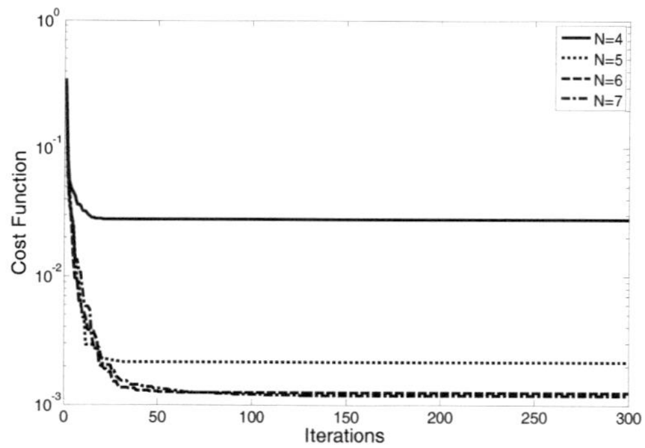

(c)

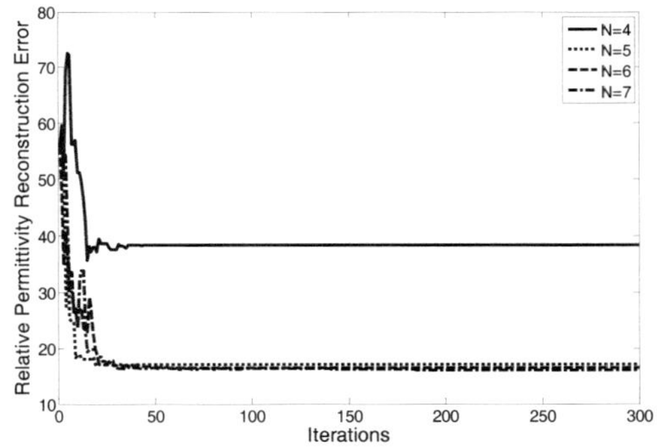

(d)

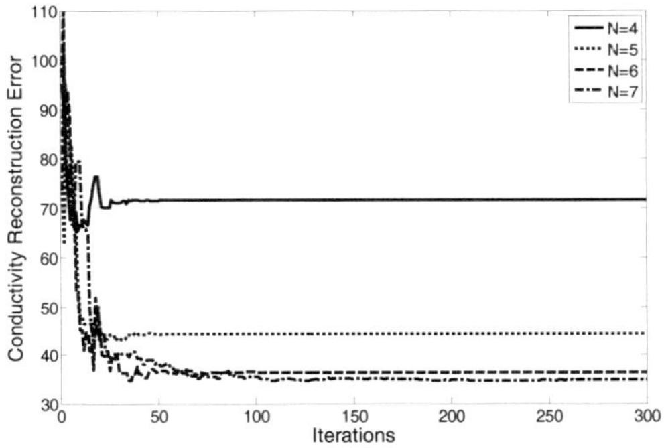

(e)

Fig. 5. Reconstruction of 1-D test case #2, (a) original and reconstructed permittivity profiles, (b) original and reconstructed conductivity profiles, (c) the cost function, (d) the permittivity reconstruction error and (e) the conductivity reconstruction error

4.2 Two-dimensional case

The proposed expansion method is also utilized for two 2-D cases. In the simulations of both cases, four transmitter and eight receivers are used as shown in Fig. 6. The population in DE algorithm is chosen equal to 100, the maximum iteration is considered to be 300.

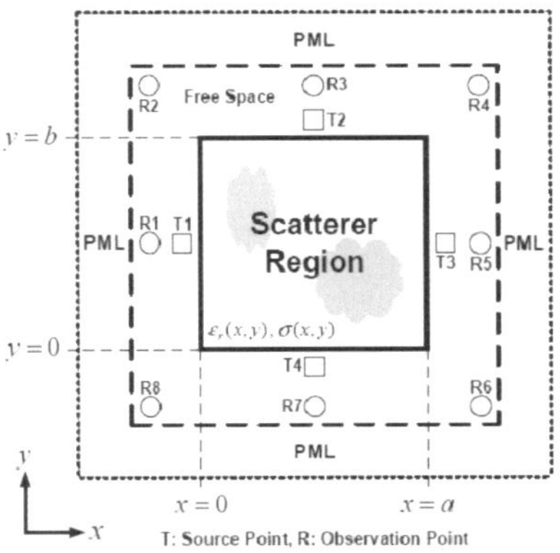

Fig. 6. Geometrical configuration of the 2-D problem

Case study #1: In the first sample case, we consider an inhomogeneous and lossless 2-D medium consisting 20*20 cells. Therefore, only the permittivity profile reconstruction is considered. In the expansion method, the number of expansion terms in both x and y directions are set to 4, 5, 6 and 7.

The original profile and reconstructed ones with the use of expansion method are shown in Fig. 7.

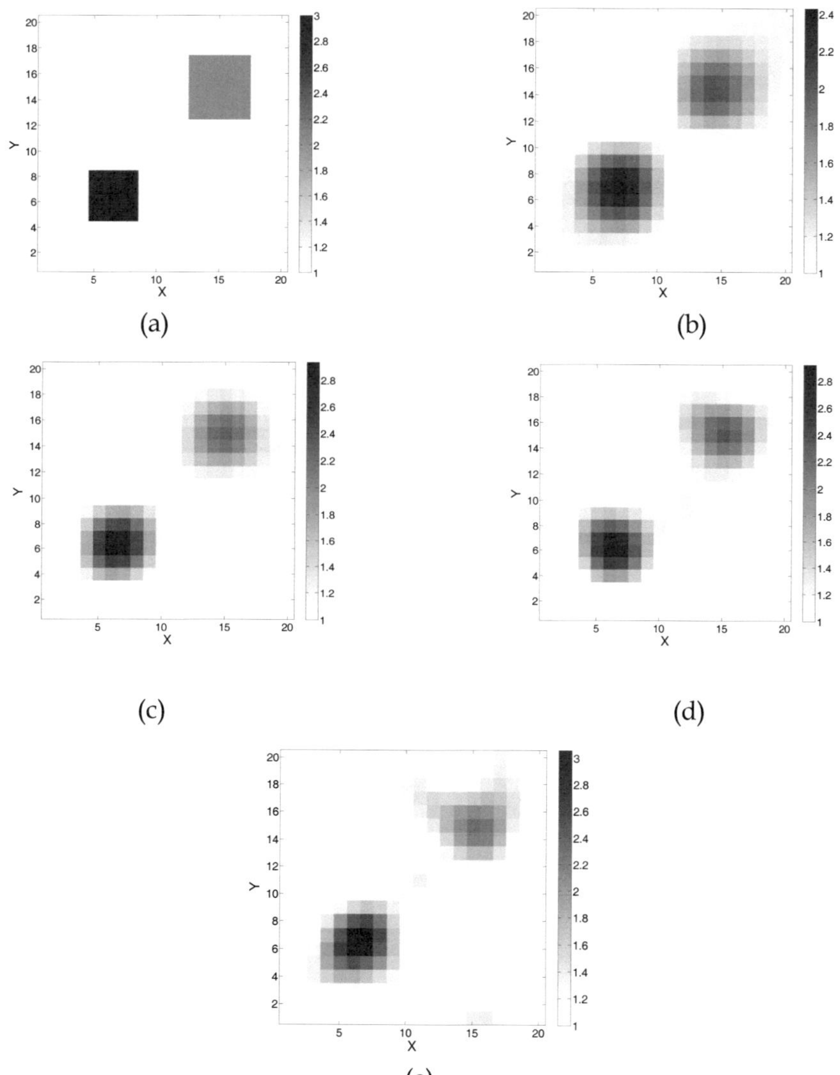

Fig. 7. Reconstruction of 2-D test case #1, (a) original profile, reconstructed profile with (b) N=M=4, (c) N=M=5, (d) N=M=6 and (e) N=M=7

The variations of cost function and reconstruction error versus the iteration number are graphed in Fig. 8.

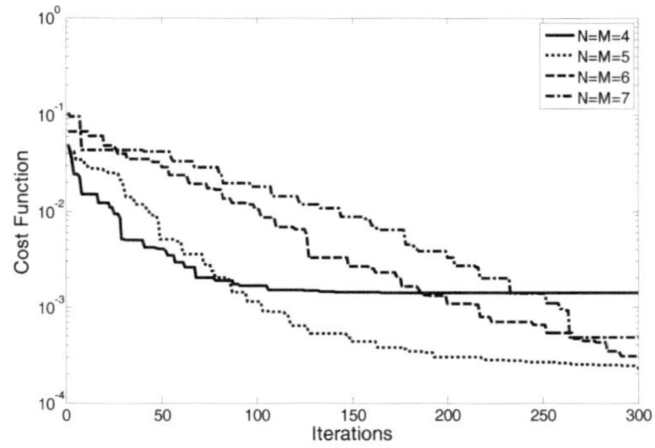

(a)

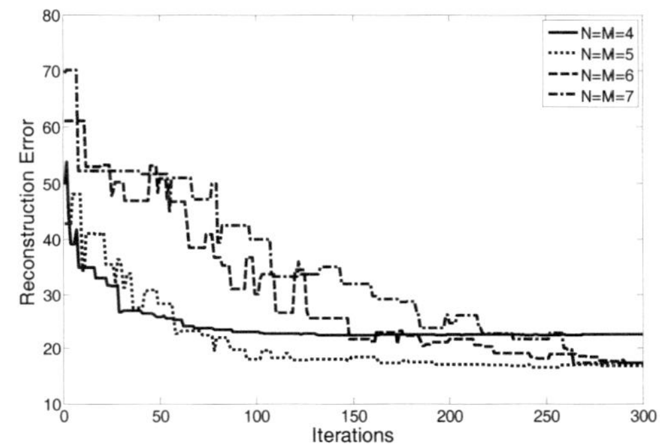

(b)

Fig. 8. Reconstruction of 2-D test case #1, (a) the cost function, (b) the reconstruction error

Case study #2: In this case, a lossy and inhomogeneous medium again with 20*20 cells is considered. Therefore, we have two expansions for relative permittivity and conductivity profiles and in both expansions, N and M are chosen equal to 4, 5, 6 and 7. It is interesting to note that the number of direct optimization unknowns in this case is equal to 800 which is really a large optimization problem. The reconstructed profiles of permittivity and conductivity are shown in Figs. 9 and 10, respectively.

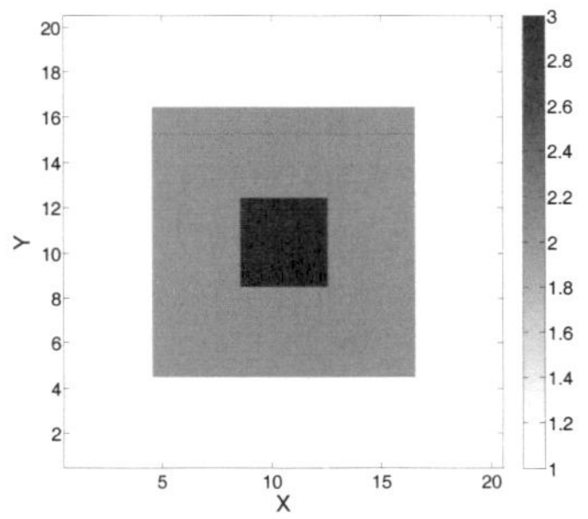

(a)

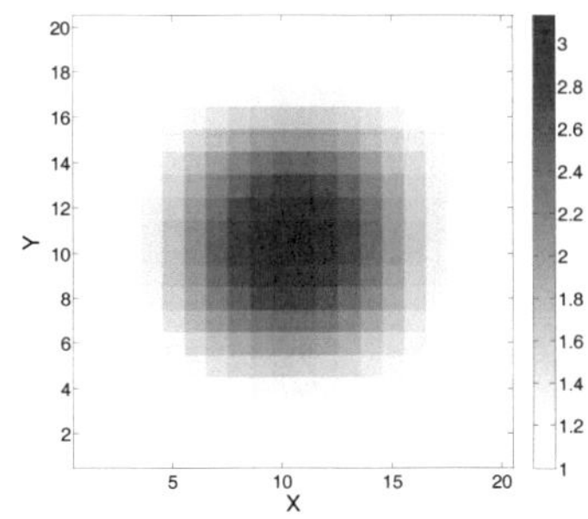

(b)

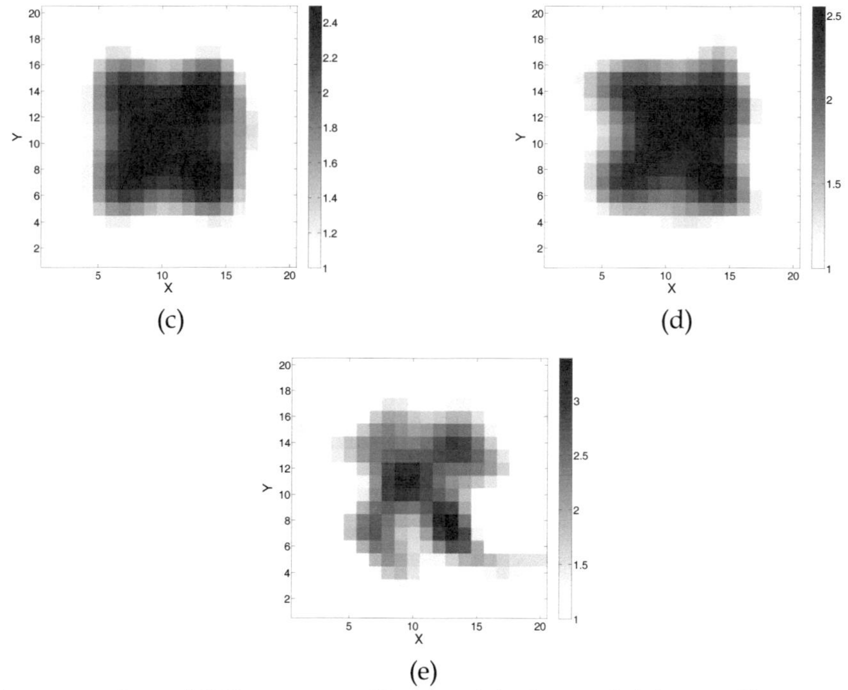

Fig. 9. Reconstruction of 2-D test case #2, (a) original permittivity profile, reconstructed permittivity profile with (b) N=M=4, (c) N=M=5, (d) N=M=6 and (e) N=M=7

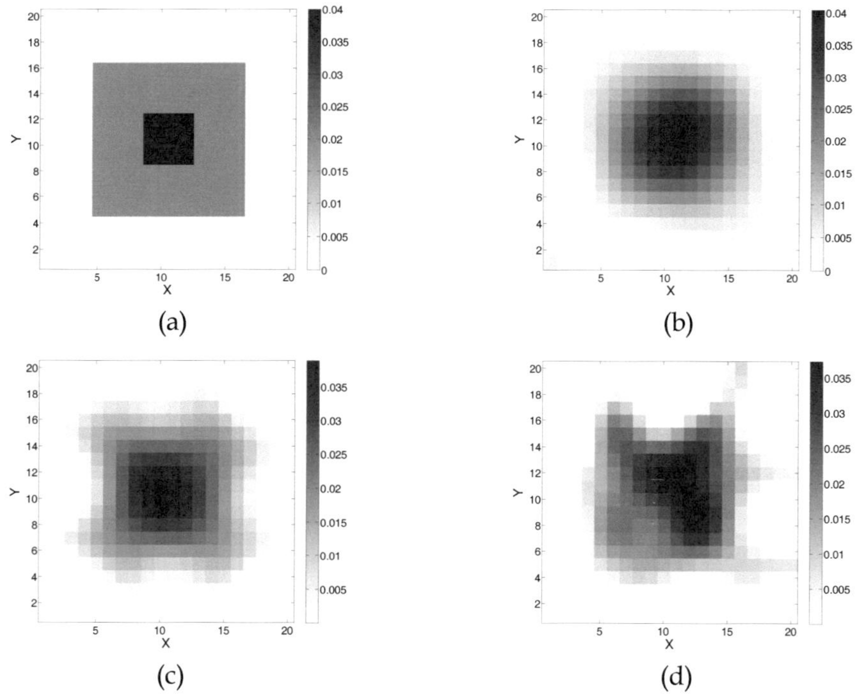

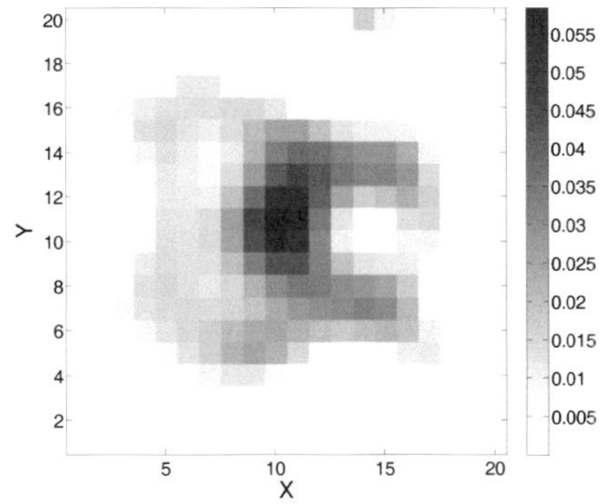

(e)

Fig. 10. Reconstruction of 2-D test case #2, (a) original conductivity profile, reconstructed conductivity profile with (b) N=M=4, (c) N=M=5, (d) N=M=6 and (e) N=M=7

The variations of cost function and reconstruction error are shown in Fig. 11.

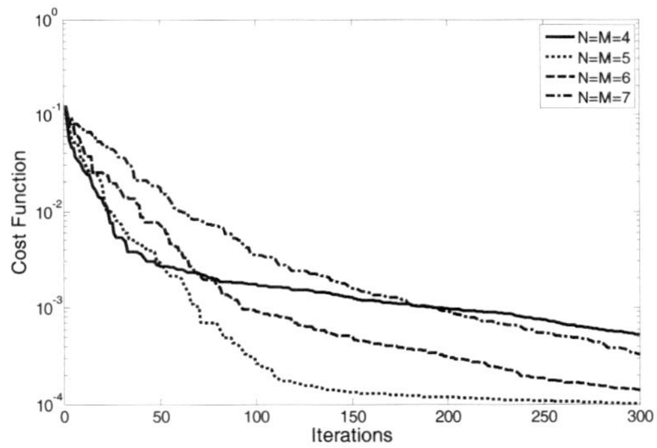

(a)

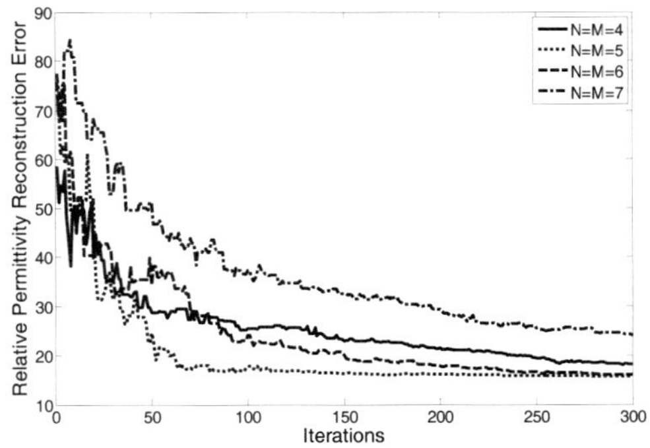

(b)

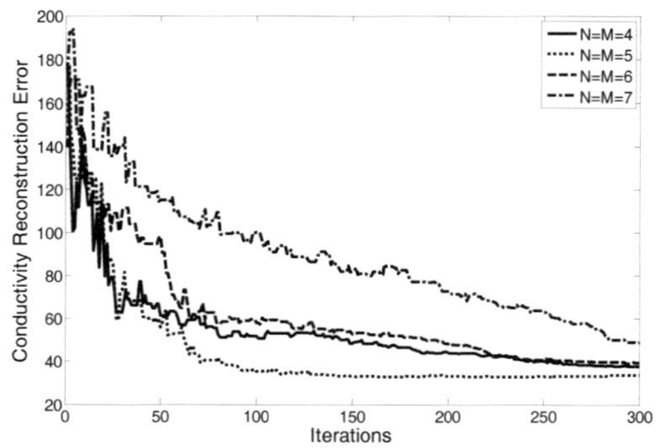

(c)

Fig. 11. Reconstruction of 2-D test case #2, (a) the cost function, (b) the permittivity reconstruction error and (c) the conductivity reconstruction error

The results of all 1-D and 2-D cases which are generally inhomogeneous and lossy or lossless media show that the proposed expansion method can tolerably reconstruct the unknown media with a considerable reduction in the amount of computations as compared to the conventional direct optimization of the unknowns.

5. Sensitivity Considerations

It is obvious that the performance of the expansion method directly depends on the number of expansion terms. Larger number of terms results in a more precise reconstruction at the expense of higher degree of ill-posedness. On the other hand, lower ones leads to a less accurate solution with higher probability of convergence of the inverse algorithm. Therefore, suitable selection of N has a notable impact on the convergence speed of the algorithm.
The reconstructed profiles of two 1-D cases with larger values of N are shown in Figs. 12 and 13 for test case #1 and #2, respectively.

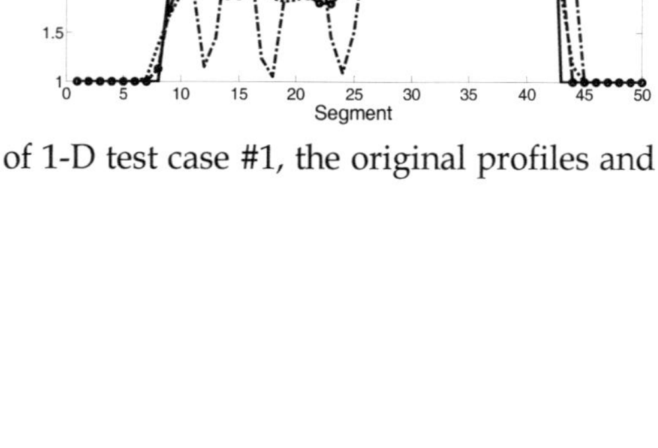

Fig. 12. Reconstruction of 1-D test case #1, the original profiles and reconstructed ones with N=7, 10 and 20

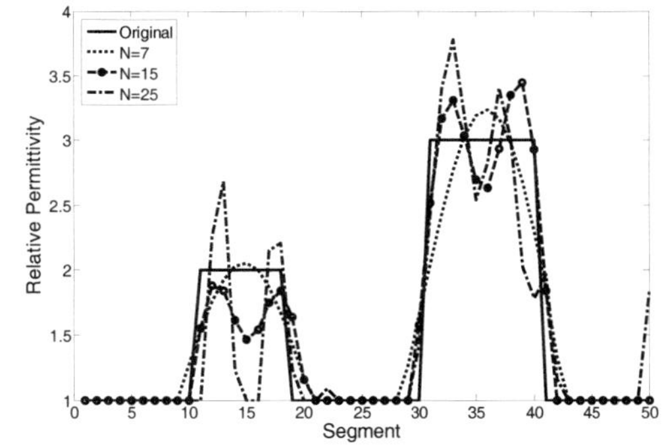

(a)

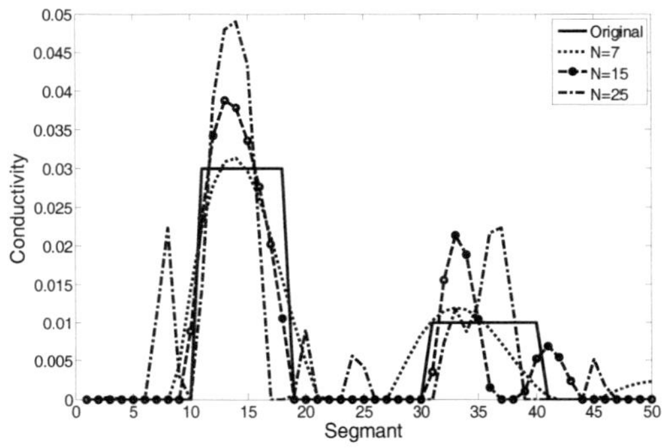

(b)

Fig. 13. Reconstruction of 1-D test case #2, the original profiles and reconstructed ones with N=7, 15 and 25, (a) permittivity profile and (b) conductivity profile

It is seen that increasing the number of expansion terms results oscillatory reconstruction because of the more ill-posedness of the problem.

We can come to similar conclusion for 2-D cases by comparing different parts of Figs. 7, 9 and 10.

Our experiences in studying various permittivity and conductivity profiles reconstruction show that choosing the number of expansion terms between 5 and 10 may be suitable for most of the reconstruction problems.

6. Conclusion

A computationally efficient method which is based on combination of the cosine Fourier series expansion, an EM solver and a global optimizer has been proposed for solving 1-D and 2-D inverse scattering problems. The mathematical formulations of the method have been derived completely and the algorithm has been examined for reconstruction of several inhomogeneous lossless and lossy cases. With a considerable reduction in the number of the unknowns and consequently the required number of populations and optimization iterations, along with no need to the regularization term, the relative permittivity and conductivity profiles have been reconstructed successfully. It has been shown by sensitivity analysis that for obtaining well-posedness as well as accurate reconstruction simultaneously, the number of expansion terms must be chosen intelligently.

7. References

Cakoni, F. & Colton, D. (2004). Open problems in the qualitative approach to inverse electromagnetic scattering theory. *Euro. Jnl. of Applied Mathematics*, Vol. 00, (1–15)

Caorsi, S.; Ciaramella, S.; Gragnani, G. L. & Pastorino, M. (1995). On the use of regularization techniques in numerical invere-scattering solutions for microwave imaging applications. *IEEE Trans. Microwave Theory Tech.*, Vol. 43, No. 3, (March). (632–640)

Colton, D. & Paivarinta, L. (1992). The uniqueness of a solution to an inverse scattering problem for electromagnetic waves. *Arc. Ration. Mech. Anal.*, Vol. 119, (59–70)

Isakov, V. (1993). Uniqueness and stability in multidimensional inverse problems," *Inverse Problems*, Vol. 9, (579–621)

Rekanos, I. T. (2008). Shape reconstruction of a perfectly conducting scatterer using differential evolution and particle swarm optimization. *IEEE Trans. Geosci. Remote Sens.*, Vol. 46, No. 7, (July). (1967-1974)

Robinson, J. & Rahmat-Samii, Y. (2004). Particle swarm optimization in electromagnetics. *IEEE Transactions on Antennas and Propagation*, Vol. 52, No. 2, (397-407)

Semnani, A. & Kamyab, M. (2008). Truncated cosine Fourier series expansion method for solving 2-D inverse scattering problems. *Progress In Electromagnetics Research*, Vol. 81, (73-97)

Semnani, A. & Kamyab, M. (2009). An enhanced hybrid method for solving inverse scattering problems. *IEEE Transaction on Magnetics*, Vol. 45, No. 3, (March). (1534-1537)

Storn, R. & Price, K. (1997). Differential evolution - A simple and efficient heuristic for global optimization over continuous space. *J. Global Optimization*, Vol. 11, No. 4, (Dec). (341–359)

Taflove, A. & Hagness, S. C. (2005). *Computational Electrodynamics: The finite-difference time-domain method*, Third Edition, Artech House

Tikhonov, A. N. & Arsenin, V. Y. (1977). *Solutions of Ill-Posed Problems*, Winston, Washington, DC

9

SLOT OPTICAL WAVEGUIDES SIMULATIONS & MODELING

Muddassir Iqbal[1], Z. Zheng[2] and J.S. Liu[2]
1. National University of Science & Technology, H-12, Islamabad, Pakistan
2. Beihang University Beijing China.

1. Introduction

In this chapter we will discuss about slot optical waveguides analysis using finite difference time domain (FDTD) algorithm. SOI slot optical waveguides invented in 2004, by Lipson at Cornell nanophotonics center [1], and experimentally demonstrated by them in forming complex nano-structures [2]. Nanophotonics group at Cornell surprised photonics researchers by discovering structure geometry; where light can be confined inside low index slot region due to electric field discontinuity.

The name slot waveguide comes from its physical shape i.e. a low index slot surrounded by two high index slabs. Slot waveguide structure has gained significant interests and importance due to its potential applications in nanophotonics especially light on chip circuits. In most basic single slot waveguide structure a high-refractive-index material is used to guide light through a low-refractive-index material. The waveguide structures are capable of guiding and confining light in such a way that very high optical intensity is obtained in a small cross-sectional area or gap filled with any material of sufficiently low refractive index, relative to the remainder of the structure. (Figure 1 is a top view of a slot waveguide structure)

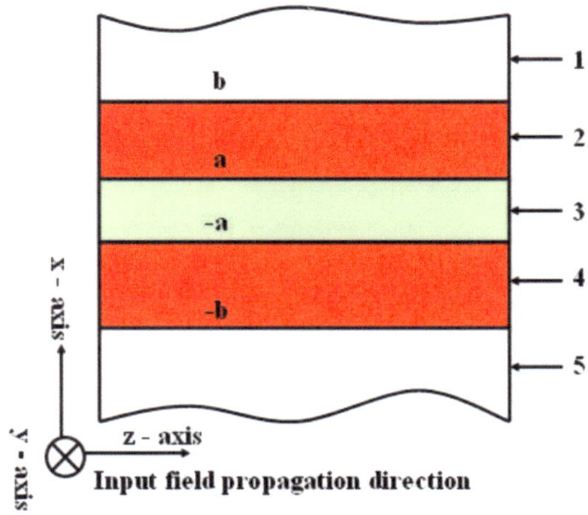

Fig. 1. Basic single slot waveguide structure.

Referring figure 1, the slot waveguide structure comprises of slabs (No.2 & 4) having high index of refraction and a slot (No.3) formed in between, which have a relatively low index of refraction. The cladding region (No.1 &5) comprises of low refractive index material or same material as used for the slot region. The analytical solution for the transverse E-field profile E_x of the fundamental TM Eigen mode of the slab-based slot waveguide is [1]:

$$E_x(x) = A\begin{cases} \frac{1}{n_S^2}\cosh(\gamma_S x); |x| < a \\ \frac{1}{n_H^2}\cosh(\gamma_S a)\cos\left[\kappa_H(|x|-a)\right] + \frac{\gamma_S}{n_S^2\kappa_H}\sinh\left[\kappa(|x|-a)\right]; a < |x| < b \\ \frac{1}{n_C^2}\left\{\cosh(\gamma_S a)\cos\left[\kappa_H(b-a)\right] + \frac{n_H^2\gamma_S}{n_S^2\kappa_H}\sinh(\gamma_S a)\sin\left[\kappa_H(b-a)\right]\right\}\exp\left[-\gamma_C(|x|-b)\right]; |x| > b \end{cases} \tag{1}$$

κ_H is the transverse wave number in high refractive index slabs. γ_χ is the field decay coefficient in the cladding. γ_σ is the field decay coefficient in the low refractive index slot waveguide. The constant A can be narrated mathematically as follows [1, 2]:

$$A = A_0 \frac{\sqrt{k_0^2 n_H^2 - k_H^2}}{k_0} \tag{2}$$

A_0 is an arbitrary constant. $\kappa_0 = 2\pi/\lambda_0$ is the vacuum wave number. Figure 2 below displays e-field distribution in a basic single slot structure.

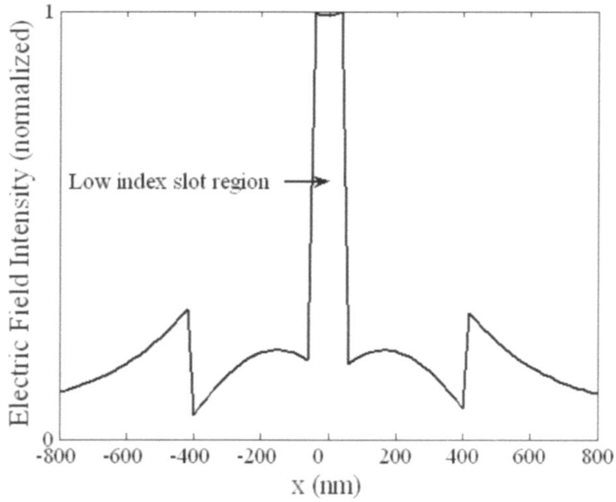

Fig. 2. High confinement of light in low index slot region.

The structure provides strong confinement in low refractive index materials, which rely on the discontinuity of the electric field perpendicular to the interface between materials with low and high refractive index. In such a case, the strongest electric field amplitude lies in the material with low refractive index. Due to the large index contrast at interfaces, the normal electric field undergoes a large discontinuity, which results in a field enhancement in the low-index region. If refractive index of slot is denoted by n_l sandwiched between two slabs (refractive index denoted by n_H), e-field enhancement in low-index region is of the order of n_H^2 / n_l^2. In order to measure the optical field confinement, the optical confinement factor can be defined as the fraction of power confined and guided in the interested regions [3]:

$$\Gamma = \frac{\int_{Regions} \mathrm{Re}\left(E \times H^*\right).\hat{z}dxdy}{\int_{total} \mathrm{Re}\left(E \times H^*\right).\hat{z}dxdy} \tag{3}$$

Where E and H are the electric and magnetic field vectors. The integrals are calculated inside the interested regions and the entire cross section regions. Optical field confinement in this chapter has been calculated based on the theory [1, 3]. Simulation of slot waveguide geometry leads to high confinement of light in low index slot region (figure 2 shows high confinement of light in low index slot).

Experimental demonstration of guiding and confining light in low index waveguides by Lipson [2], led other researchers to demonstrate slot waveguide based complex structures [4, 5]. High power confinement can also be achieved in asymmetric slot waveguides [6]. Power confined inside slot structure can be optimized by adjusting the geometry of slot and slabs [7, 8]. N. N. Feng demonstrated slot waveguide coupling capability with conventional waveguides [9]. Barrios, and Passaro proposed slot waveguides for sensing [10, 11]. Beyond single slot structure people are interested in multiple slot structure confining optical field in nanometer-thin low-index media with very high optical confinement factor [3]. Slot waveguides capable of confining high intensity electric field hence became an interesting topic of researchers to simulate and demonstrate nonlinear slot structures [12, 13]. Other than dielectrics, slot waveguides had been demonstrated by Lipson and Atwater in metals as well [14, 15]. While exploring characteristics of single and multiple slot structures, comparable light confinement in low refractive index contrast slot structures have been demonstrated [16]; a novel photonics displacement sensor based upon multiple slot waveguide structure has been proposed [17].

2. Slot Structure Analysis

As mentioned earlier, in a slot structure electric field discontinuity at the boundary between the high and low index regions results in high E-field confinement inside low index slot. Slot waveguide structure had been simulated using OptiFDTD simulation software from Optiwave Company. Finite difference time domain method (FDTD) uses a brute force discretization of Maxwell's equations. The structure is discretized using a uniform grid and the derivatives in Maxwell's equations are replaced by finite differences. The grid size used is very important, a small grid size is required to get accurate results. However using more grid points results in longer calculation times. The grid size also imposes an upper limit on the time step that can be used, because of stability requirements. FDTD can handle almost any 2-D or 3-D structure (in theory), but computation time and memory requirements limit the size of the problem that can be handled. It has been explored by us that 3dB waist of input plane and mesh size has a profound effect on the output. Hence it was decided that for the nanosize slot optical waveguide analysis using OptiFDTD, a grid size of 10 nm is to be used along x- and y-axis. However grid size in z-direction can be changed to get quicker results, it does not have profound impact on results. Sources are placed inside the waveguide to excite the waveguide mode and detectors are placed above to detect the output signal and power. Appropriate boundary conditions (e.g. PML) are used to avoid reflections at the boundaries of the computational domain and model open structures.

In a basic slot waveguide structure cladding, substrate and slot are of fused silica, whereas slabs comprising of silicon. Slot structures are ignited by placing a continuous wave (C.W.) input plane of 1.5 μm wavelength, direction of flow is in z-axis. Three observation planes had been placed at a distance of 0.5 μm, 1 μm and 1.5 μm. The observation planes were centered with the slot structure center in x-axis and y-axis. E-field distributions in spatial domain are shown in figure 3.

E-field distribution at respective propagation distances had also been checked, a combined plot at different propagation lengths is shown in figure 4 below. It was found that E-field in slabs and cladding for a propagation distance of 1 μm and 1.5 μm is at a monotonous decrease then at 0.5 μm distance and seems more stable. Hence in order to get stable results, it was decided to propagate signal for longer distance.

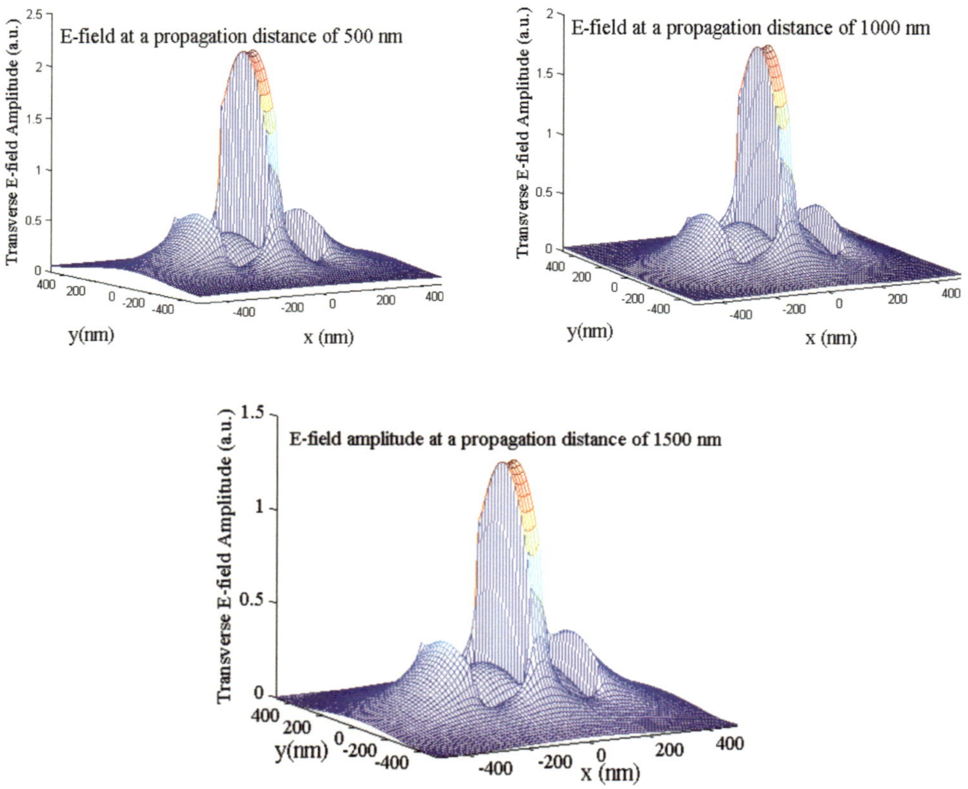

Fig. 3. E-field distribution in slot structure at different propagation distances.

However OptiFDTD ver. - 5.0 has less memory and longer simulations made the memory buffers overflow, hence most of those simulations crashed unanimously.

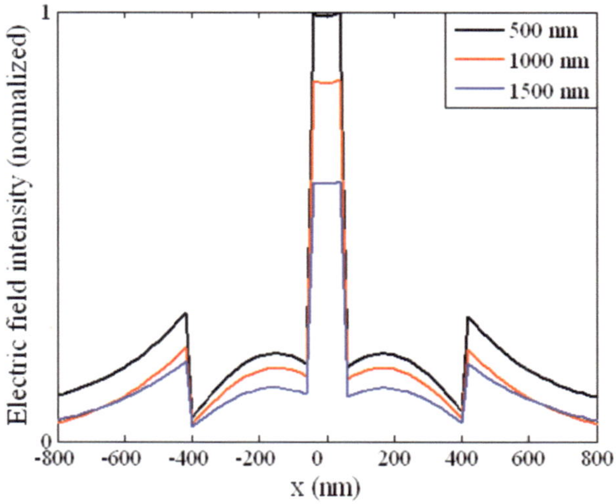

Fig. 4. Modes for single slot structure at various propagation distances.

Power confinement factor in the slot structure has been checked and found in accordance with the previous work [1].

In an extended check of power confinement factor in slot waveguide structure, the refractive index in slot region varied from 1.44 till 1.50 with a step size of 0.005. High index slabs refractive index was kept constant at 3.48, cladding and substrate refractive index kept constant at 1.44. It was found that power confinement factor varied (increased / decreased) as per the ratio, n_H^2 / n_l^2 .

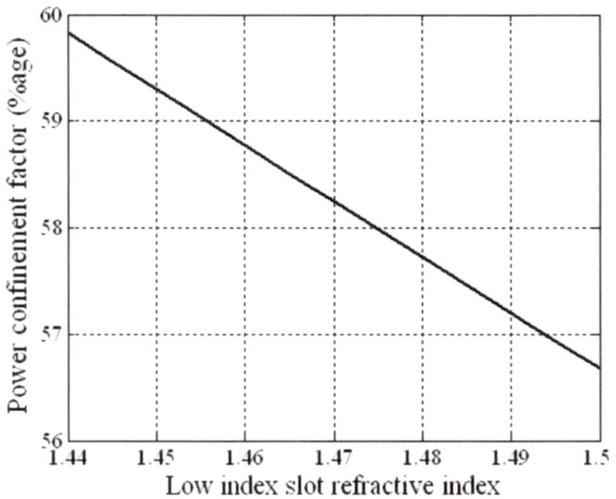

Fig. 5. Slot region power confinement dependence on refractive index.

In order to simulate complex structures, double slot structure was simulated, which was a step towards complex structures designing and simulation. A double slot structure has been simulated, where both slots of 50 nm width each are separated by a central high index slab of 100 nm width and surrounded by two 180 nm wide high index slabs. Initially it was thought that double slot structure designing is not a big issue, as it is just adding one more slot of suitable width in the waveguide structure (see figure 6). However, later it was learnt through various simulations that adding another slot of any width could not solve the problem. It requires proper width and placement of slot, as center slab width is dependent upon the placement of extra slot. Center slab width determines the amount of coupling occurring in between both the slots. The coupling effect between both slots does have an effect on power confinement factor in the low index slot region. This coupling effect determines the amount of power confined inside low index slot regions; moreover it has also been found that the center slab width can be related with the power confinement factor inside low index slot regions.

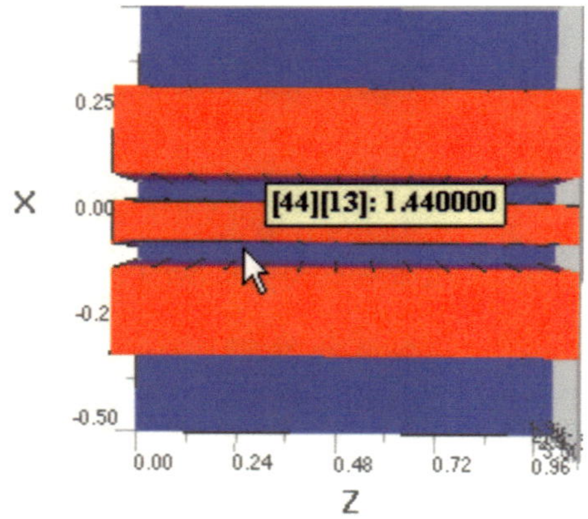

Fig. 6. Double slot structure (refractive index based view).

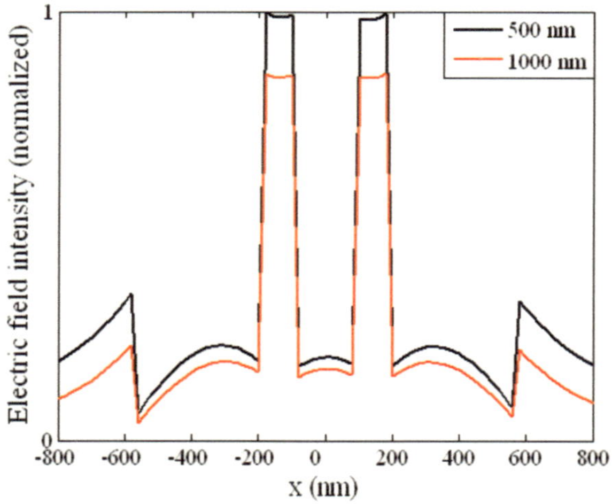

Fig. 7. E-field distribution for double slot structure at 0.5 μm & 1 μm propagation distance (low index slots R.I. = 1.44, high index slabs R.I. = 3.48, cladding & substrate R.I. = 1.44).

Further exploration of double slot structure had been done by using different refractive index materials in both low index slots. The results proved that power confinement and E-field confinement is dependent upon the contrast ratio. The more the contrast ratio between low index slot and high index slabs; the more the power confined inside low index slot region. In the current work, refractive index for left slot was 1.5, right slot was 1.44; high index slabs refractive index is 3.48. E-field distribution at a propagation distance of 0.5 μm, 0.75 μm, 1 μm, and 1.5 μm are displayed in figure 8. Referring figure 5, for a single slot structure power confined inside low index slot region follows the refractive index contrast ratio between high index slabs and low index slots. It has been found that in case of double slot structure, power confined inside low index slot regions is affected by the coupling effect as well. Coupling between low index slot regions is mainly dependent upon central high index slab width. This effect has been studied in detail, and is of use in attaining fruitful results.

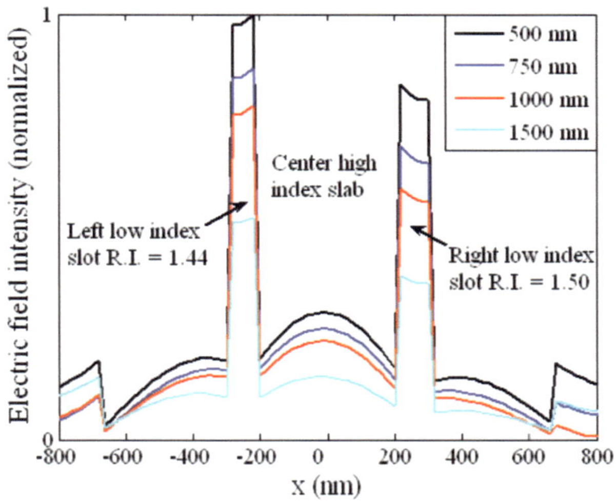

Fig. 8. E-field distribution for double slot structure with different R.I. at 500 nm, 750 nm, 1000 nm & 1500 nm propagation distance (high index slab R.I. = 3.48, cladding & substrate R.I. = 1.44).

E-field inside both slots for both distances had been analyzed for phase shift. Phase shifts had been calculated at various parametric values, minute difference in phase shift indicates its non-existence.

In another example, the double slot structure (ref. figure 6 & related details) has been repeated with cladding comprising of air. Power confined inside low index slot regions increased with a subsequent decrease of power confinement in external slabs. E-field distribution at 500 nm and 1000 nm propagation distances are shown in figure 9 below.

It was found that E-field confinement in the slot with comparatively high index of refraction is less than the other slot. This behavior of light confinement inside low index slot region is directly related to the slot and slab refractive index contrast ratio; and has already been discussed.

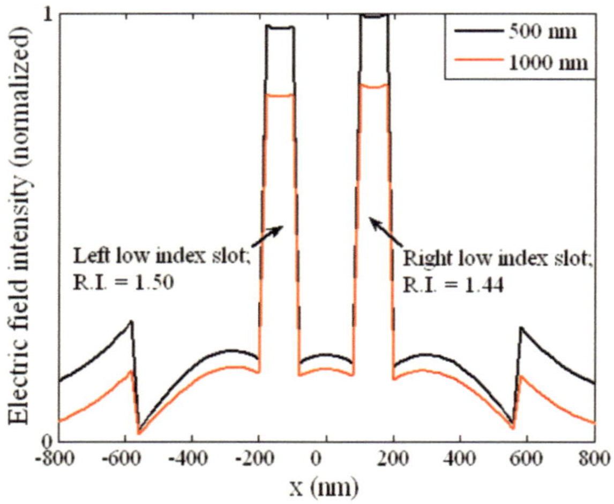

Fig. 9. Modes for double slot structure with different R.I. (cladding is air).

3. SOI Multiple Slot Structures Explored for Sensor Applications

In case of double slot structure, power confined inside low index slot regions (combined and individually) is dependent upon the center slab width. Several values have been taken at slot width 50 nm each; external slabs width 180 nm each slots and slabs height 300 nm each. Power confinement factor inside low index slot regions was found dependent upon center slab width.

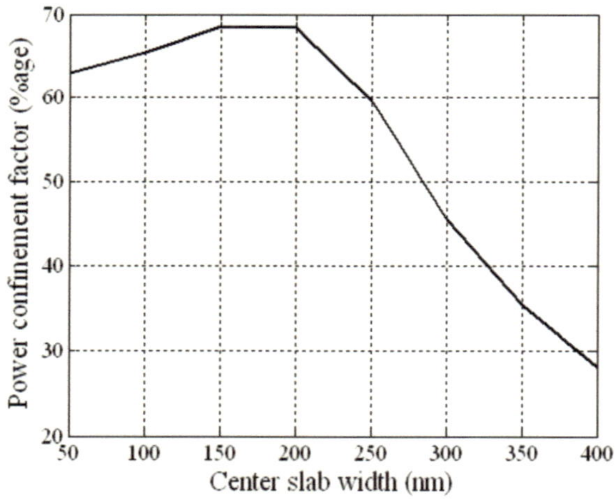

Fig. 10. Dependence of power confinement factor in double slot structure upon center slab width.

In an extended check it was found that power confined inside low index slot regions of width 50 nm each is maximum for a center slab width ranging from 150-200 nm. Center slab has a profound impact in a double slot structure on power confinement factor in low index slot region. E-field amplitude also varies for slots with different refractive index. Power confinement factor ratio between slots, normalized values of power confined in center slab and E-field amplitude ratio in slots is shown in figure 11 [17]. In order to check the variation in power confinement factor due to shift in refractive index of slots; power confinement ratio in slots is plotted for a shift in refractive index values by 0.005. Refractive index value in left slot was changed from 1.44 till 1.5 with a step of 0.005, whereas refractive index value in right slot kept constant at 1.44. Ratio in power confinement factor between both slots was calculated and it was found that the peak is shifting gradually with shift in refractive index. The results have proven the effect of shift in refractive index on power confined in low index slot regions. Numerically calculated power confinement factor ratio for shift in refractive index shows a promising behavior for probable usage in sensor systems. However the use of single slot structure in sensor systems does require a mechanism to sense the shift in modal effective index, which had been studied by others. Double slot structure showed a way to use the slot structure in sensing mechanisms; most probably by calculating and observing the shift in power confinement factor / power confinement factor ratio. The shift in power confinement factor of low index slot regions under the action of cantilever [10] type movement by central high index slab has been exploited by us in proposing double slot structures usage in forming sensor systems. This effect is discussed in detail in further half of this chapter.

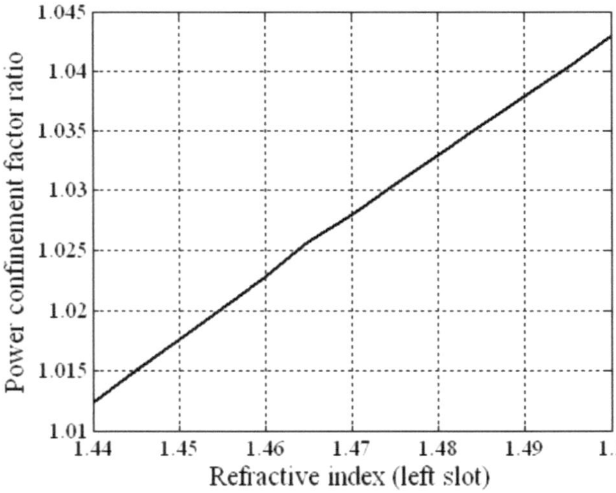

Fig. 11. Power confinement ratio in diff. refractive index slots dependence upon shift in refractive index of one slot.

Variation of refractive index only in left low index slot region only showed a profound impact upon power confinement factor ratio between both low index slot regions. Extensive research work done in this direction, so as to utilize the shift in refractive index of low index slot region (one or all in a multiple slot structure); a center slab width of 165 nm was chosen for extended extensive analysis. The reason of choosing 165 nm center slab width was that; at this center slab width coupling between low index slot regions was comparatively less. The results could help us in ascertaining the postulate that shift in refractive index of any of the low index slot region do have a profound impact on the power confinement inside low index slot regions. Refractive index in left low index slot region shifted from 1.5 till 1.48 with a step size (Δn) of 0.005. Minute shift in refractive index; which is possible due to temperature, pressure, chemical, and mechanical, or may be due to several other reasons not mentioned here; had a profound impact on power confinement factor and its ratio between both low index slot regions. Prominent shift in low index slot regions power confinement factor formed a basis of thinking to propose double slot optical waveguide structure for use in sensor applications. Power confinement ratio between both low index slots as a function of center slab width and refractive index is plotted in figure 12:

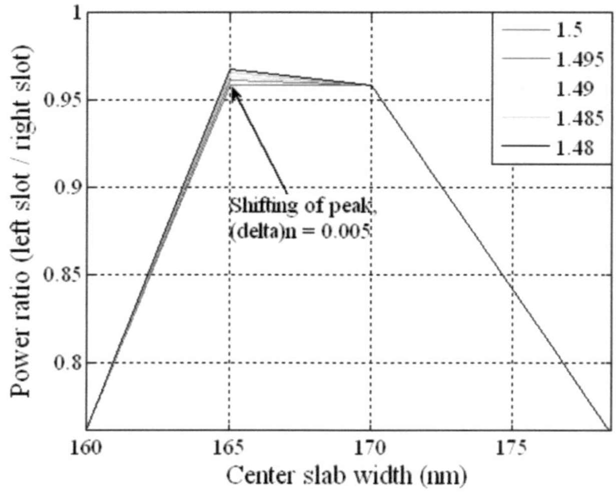

Fig. 12. Shift in power confinement ratio in low index regions due to change in refractive index.

In one of the research work related to proposing double slot structure in sensor mechanisms, numerical calculations were performed for a center slab width varying from 10 nm till 400 nm. Followed by power confinement factor ratio dependence upon shift in refractive index of low index slot region. Refractive index of right slot is kept constant at 1.44, whereas refractive index of left slot was varied from 1.44 till 1.5 with a step size of 0.005. Power confinement factor ratio was calculated by dividing the power confined in right slot with power confined in left slot for a propagation length of 1 μm. Power confinement ratio between both low index slot regions is maximum at a center slab width of 20 nm. Hence the point with center slab of 20nm width was chosen for power ratio check. For a 20 nm width of center slab, coupling between both low index slot regions is better than at center slab width of 165nm. This time power confinement factor ratio at a center slab width of 20 nm was checked, where power confinement factor inside low index slot regions is more due to increased coupling. Shift in refractive index of either of the low index slot region showed a profound impact on power confinement factor ratio. The prominent change in power confinement factor (can also be termed as confinement loss) is a contributing factor leads to proposing double slot structure for use in sensing mechanisms. Change in power confinement factor ratio indicates that a slight shift in either of the low index slot regions refractive index shows a promising change in power confinement factor ratio.

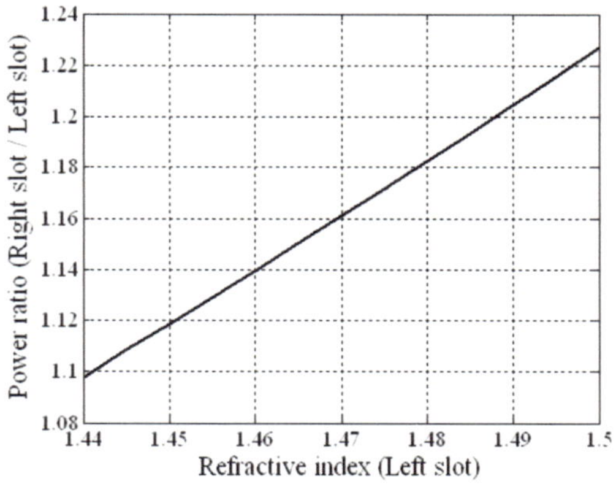

Fig. 13. Shift in power confinement ratio dependence on left low index slot R.I. (center slab width 20 nm).

Before considering power confinement / loss mechanism in double slot structure for use in sensor mechanisms, it was necessary to investigate if change in power confinement factor is dependent upon some other parametric values. Several checks have been done, in one case where refractive index in both slot regions is 1.44, detailed analysis of boundary conditions effect upon power confinement factor ratio has been done for a center slab width of 162 nm. Perfectly matched boundary conditions are used with number of layers ten and twenty. Power confinement factor ratio (right slot/left slot) at various propagation distances are plotted in figure 14 below.

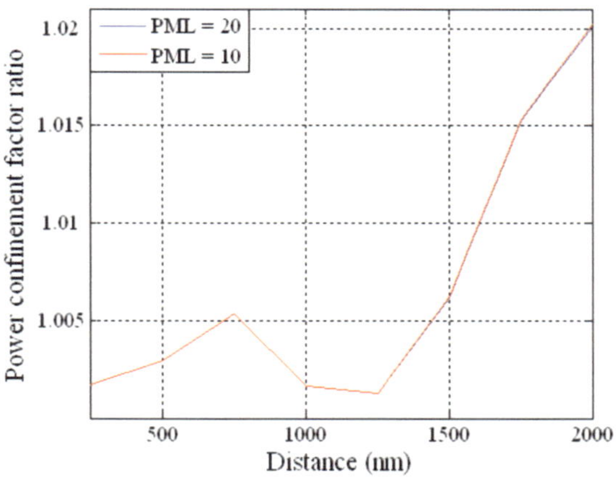

Fig. 14. Power confinement factor ratio between slots dependence on Perfectly Matched Layers boundary condition.

Later part of this subsection indicates that number of PML layers does not have profound impact on power confinement factor, hence we have kept the number of PML layers in all of our simulations to be 10. It is apparent from both the results that power confinement factor ratios are around the value of one. As the propagation conditions (boundary conditions &

input conditions) in both slot regions is same, hence ideally the ratio should be one or in close proximity of one.
Comparison between the results of both conditions, like PML=10 and PML=20 had been carried out. Power confinement factor error ratio has been plotted in figure 15. The error ratio has been taken for difference in power confinement ratio at PML-10 and PML-20 divided by actual ratio at PML-10. The error ratio is at monotonous increase hence it seems following the general conditions of error analysis.

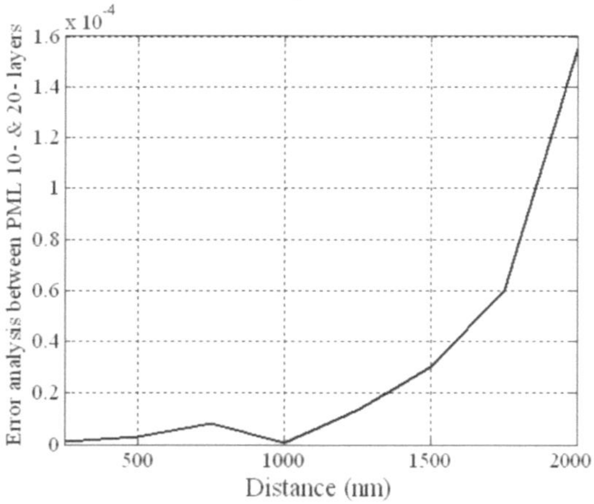

Fig. 15. Power Confinement factor ratio comparison (PML = 10 & PML = 20).

Hence it was found that power confinement factor is mainly dependent upon shift in refractive index, and shift in center slab width (double slot structure).

4. Low contrast Double Slot Structure based Optomechanical Sensor [18]

The structure is based on basic double slot structure. A head start introduction of proposed optomechanical sensor is that two low index slots of hard material dipped in a cladding of high index compressible material (suitable to term as fluid), where slots have fins on top and can move inside high index cladding under the action of an external force. Three slabs comprising of high refractive index fluid, two slots (50nm wide) of low refractive index solid material, in present work it is SiO_2 (R.I.=1.44). The cladding and substrate also comprised of SiO_2. Slabs comprising of commercially available high refractive index fluid; Gallium Halide (R.I. = 2.31) [19]. Melting point of fused silica (to be used in forming low index slots) is ~1371°C. Trihalides where Gallium is in the +3 oxidation state are Gallium Fluoride (GaF_3), Gallium Chloride ($GaCl_3$), Gallium Bromide ($GaBr_3$), and Gallium Iodide (GaI_3). Other than GaF_3 having melting point above 1000°C, all other three halides ($GaCl_3$, $GaBr_3$, and GaI_3) melting point is 78°C, 122°C, and 212°C respectively. Hence it seems technically reasonable to use fused silica based slots inside a Halide based slab.
Block diagram of proposed optomechanical sensor is shown in figure 16.
Essential assumptions in double slot structure are:

(a) Slabs comprising of high refractive index fluid.
(b) Slots body is dipped inside the slabs, there are fins on top of slots body; which are visible out of the slabs and are exposed to atmospheric conditions.

(c) Cladding is of air.

(d) Slots width remains same, the distance in between slots is varied subject to change in atmospheric conditions like temperature and pressure on the fins. As the slabs comprise of fluid, hence slots can move easily inside the slabs. Increase or decrease in central slab width (in between slots) is added into outer slabs equally

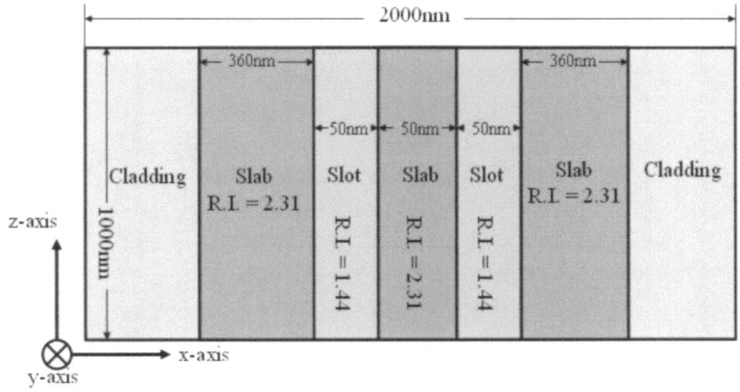

Fig. 16. Top view (z-cut) of compressible material based slot structure.

Power confinement factor in basic single slot structure (where slab refractive index is 2.31 and slot refractive index is 1.44) is investigated at various structure heights ranging from 300 nm till 500 nm with a step size of 10 nm. It has been found that for a propagation distance of 1000 nm, power confinement factor in slot waveguide shows a monotonous increase till structure height of 380nm, later on it shows a monotonous decrease till 500nm. Further analysis proved that a structure height of 410 nm is an optimized height. Hence for the current work in this section, structure height was kept at 410 nm.

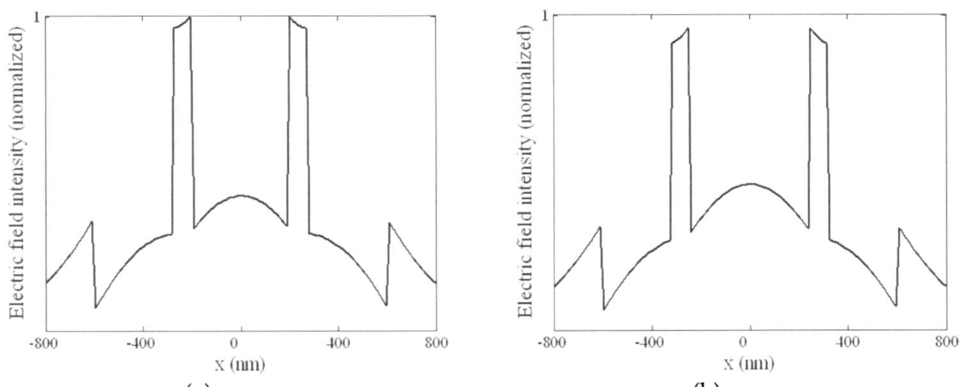

Fig.17. E-field intensity dependence upon central slab width; (a) 250 nm; (b) 300 nm.

As an example we can see that; for a case where central slab width between both slots is 300 nm and 250 nm. E-field intensity (see figure 17), E-field lines (see figure 18) and power confinement factor in low index slot region changes considerably.
The contrast ratio in present structure is considerably low, however we can see acceptable power confinement factor inside low index slot regions. E-field lines in both the cases (central slab width of 250 nm, and 300 nm) were checked. It was found that power confinement inside low index slot regions occurred due to electric lines discontinuity, plotted in figure below:

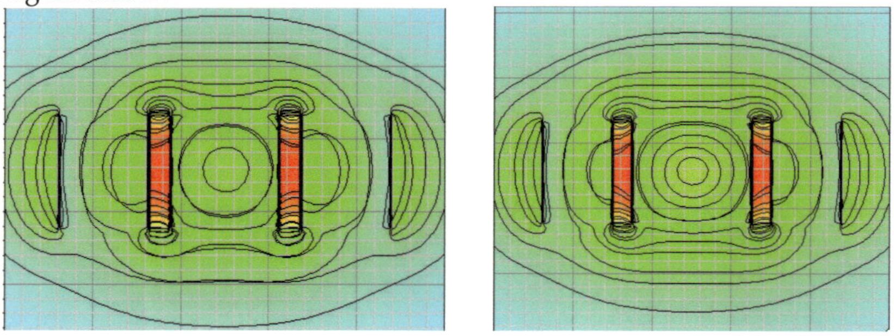

(a) (b)

Fig. 18. E-field lines at a central slab width of; (a) 250 nm; (b) 300 nm.

Power confinement factor in a low index slot has been calculated and plotted in figure 19 below. A monotonously decreasing curve was found for various widths of central slab (comprising of high index compressible material) between the slots (comprising of low index hard material). In this case power confinement factor is depending upon center slab width.

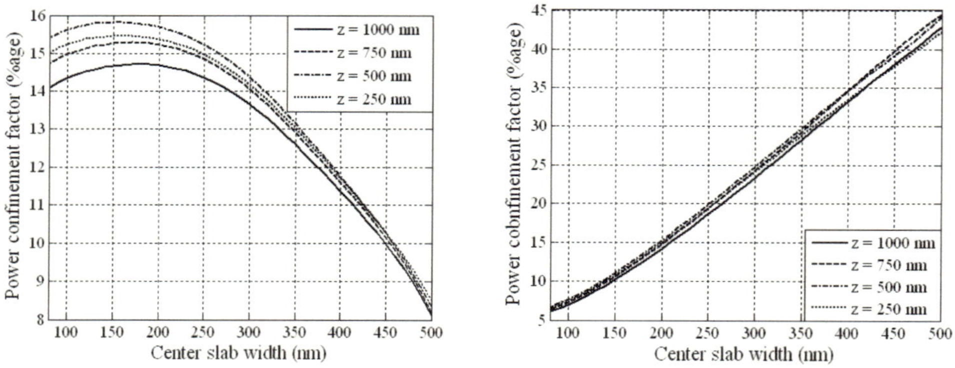

(a) (b)

Fig. 19. Power confinement factor dependence upon central slab width; (a) low index slot; (b) central high index slab.

In a careful review of the power confinement characteristics of the fluid based photonics waveguide sensor reveals that; power confinement in low index slot regions is at a monotonous decrease for a central slab width ranging from 300nm till 650nm and is as follows:

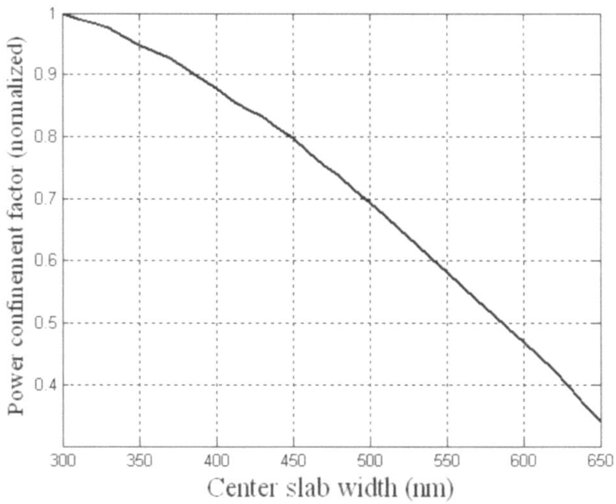

Fig.20. Normalized power confined in low index slot region.

The sensitivity of the sensor can be calculated by finding the slope of un-normalized deflection curve $((y_2-y_1)/(x_2-x_1))$, however the material sensitivity towards deflection is also required. The material sensitivity depends upon the nature of sensor requirement and its constituents. Shift in power confinement factor in this embodiment came out to be 0.435/nm.

5. Glass based Multiple Slot Structure Sensor Systems

Further to our work in last section, it was thought of that a simple and easy to realize double slot structure should be explored for sensing applications. Inspired from the work by Barrios [10], glass based double slot structure has been realized. The structure composition is that glass slabs are placed on glass substrate. Both low index slot regions and cladding is of air, for simplicity and to be practical we may say that low index slot regions and cladding is of compressible low index material. Before starting work on the proposed structure of glass and air based double slot structure, power confinement factor in glass and air based single slot structure has been investigated. Deviating from the contrast ratio of SOI slot optical waveguide, where contrast ratio is 2.42, we were thinking of realizing a structure with contrast ratio of 1.65. In a basic structure, 500 nm high, 50 nm wide single slot comprising of air, surrounded by glass slabs (R.I. = 1.65). The structure is resting on glass substrate, cladding is of air. E-field in y-cut and E-field distribution shown in figure 21 are in agreement with the basic slot structure theory [1]. High E-field confinement was found in the low index slot region.

Power confinement factor percentage inside low index slot region found out to be 9.1%. In order to enhance the power glass based double slot structure was realized. Two 500 nm high, 50 nm wide air based slots separated using a 50 nm wide glass slab, surrounded by 360 nm glass slabs. Whole structure resting on glass substrate, cladding comprising of air.

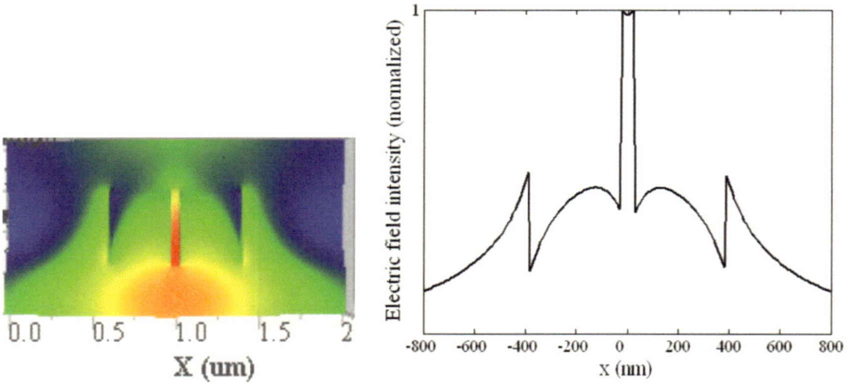

(a) (b)

Fig. 21. (a) E-field distribution; (b) E-field intensity; glass based single slot structure.

E-field distribution in y-cut and E-field intensity is shown in figure 22 (a) and (b) respectively. E-field distribution and power confinement for glass based structure had also been checked for the case where slabs comprising of glass (R.I. = 1.65); cladding, substrate & slot comprising of air. In other words for a single slot structure there are two glass slabs inserted inside an air based wafer at the requisite

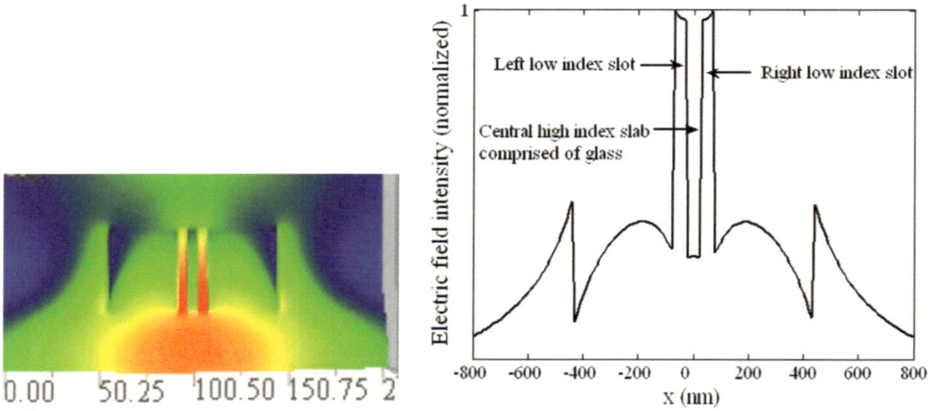

(a) (b)

Fig. 22. (a) E-field distribution; (b) E-field intensity; glass based double slot structure.

place. The placing of slabs ensures slot width, substrate depth and cladding height. Such sort of triple slot structure has been checked for power confined inside low index slot region. Power confinement factor inside low index slot region is 30.75%. The simulations supported our earlier work for the case of glass based slabs with slots comprising of air.

Moving back to the case where glass based multiple slot structure is resting on glass substrate; cladding comprising of air. Two fifty nanometer wide slots separated using a central high index glass based slab of 50nm width; surrounded by 360nm wide slabs. Whole structure resting on glass substrate, cladding is of air. Normalized E-field distribution and transverse E-field amplitude for a slot height of 1800nm is shown in figure 23 below.

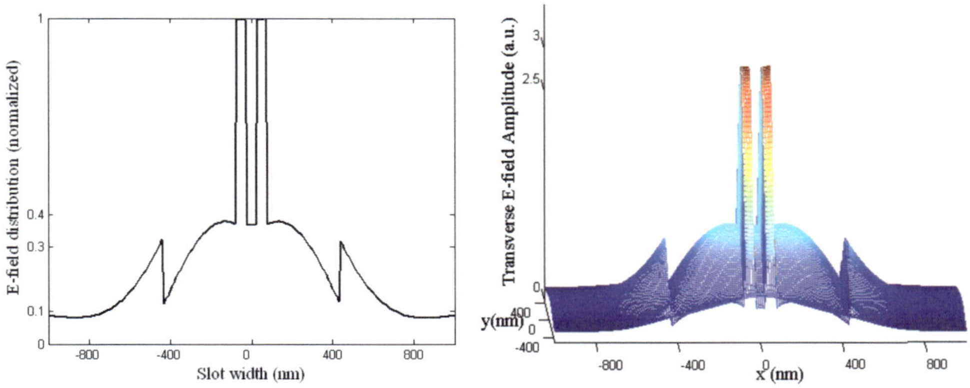

(a) (b)

Fig. 23. Normalized; (a) E-field intensity; (b) transverse E-field amplitude; glass based double slot structure.

Power confinement factor for glass based double slot structure has been investigated. For this check of power confinement factor inside low index slot regions, width of low index slot regions and central high index slab region was kept at 50 nm. Structure height was increased gradually from 300 nm till 1800 nm. Power confinement factor increases monotonously along with an increase in structure height (see figure 24 below).

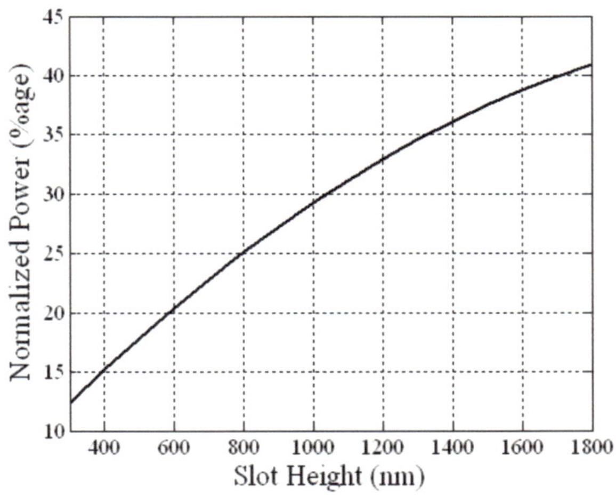

Fig. 24. Power confined in glass based double slot structure versus structure height.

Contrast ratio for SOI based slot structure is 2.42; power confined inside low index slot region for SOI based slot structure is approximately 60%. For a glass based slot structure contrast ratio is 1.65, attaining 41% power confinement factor inside low index slot regions brings the structure comparable to SOI slot structure. Glass based double slot structure is explored further for use in sensing applications.

5.1 Glass based Double Slot Structure Sensor [20]

In one of the example embodiment both 50nm wide slots are separated by a 50nm wide high index slab comprising of glass. Both slots separated by glass slab are surrounded by 360 nm wide glass slabs. The simulations scenario was created to calculate stress induced movement of center slab at one end of a long cantilever based sensor system [10]. Center slab is simulated to move (under the action of external pressure) by a step size of 5 nm each. Power confined in left slot is increasing; starting with a left slot width of 5 nm, center 50 nm glass slab is shifting right by a step of 5 nm. The left slot width is increasing by 5 nm whereas right slot width is reducing by 5 nm at each step size. Power confinement factor in both slots is same at equal width (50 nm).

It was found that power confinement factor is showing considerable change for a step size of 10 nm but for a step size of 5 nm the change is not considerable. This supports our earlier work done regarding high refractive index fluid based photonics displacement sensor.

Power confinement factor has been calculated for the case where; 50 nm slab is shifting right by a step size of 10 nm. In the start the left slot width is 5 nm, whereas right slot width is 95 nm. Results are displayed in figure 25. Power confinement factor is showing a considerable change for a shift in center slab by 10 nm; hence it can be exploited for forming a novel photonic displacement sensor.

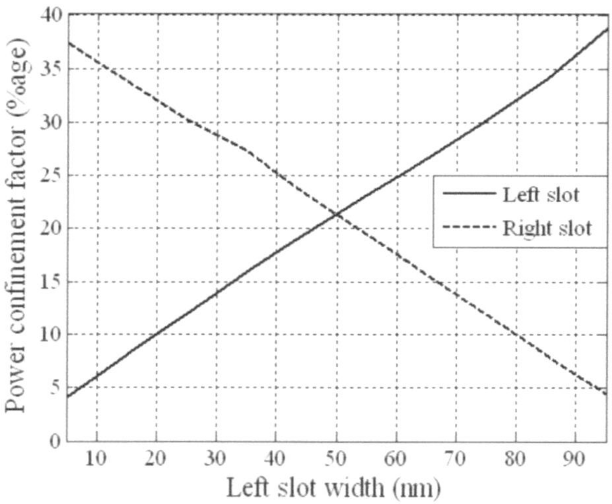

Fig. 25. Power Confinement factor in left slot region (center slab shifting right by 10nm).

5.2 Waveguide Structure

A displacement sensor based on double slot waveguide structure is designed. The structure comprising of compressible low refractive index material and hard material of relatively high index of refraction. Two low refractive index narrow slots formed between three relatively high refractive index slabs. The width of both slots and central slab is same.

The structural geometry can be changed as per requirements and designs. Figure 26 is a top view representation of double slot waveguide structure based photonics sensor. It is formed of slabs comprising of high refractive index hard material, having a width that can significantly be varied. Two low index slots comprising of low refractive index compressible material separates the high refractive index slabs by a width, which is of the nanometers range. Low refractive index slot regions may be filled with air, gas, fluid or other compressible fluids so as to allow free movement of center slab in it. High refractive index regions may be filled with relatively high refractive index glass, silicon, silicon dioxide or metal. The height of high index slab regions also defines the height of low index slot regions. The length of high index slab regions also defines the length of low index slot regions. Central high index slab has an embedded fin on it's top, force on fin's vertical surface due to physical quantities makes the center slab move like a cantilever; left or right. Cladding of the structure are of same low refractive index material as of low index slot regions; further embodiments can be comprising of air, gas, fluid or other oxides.

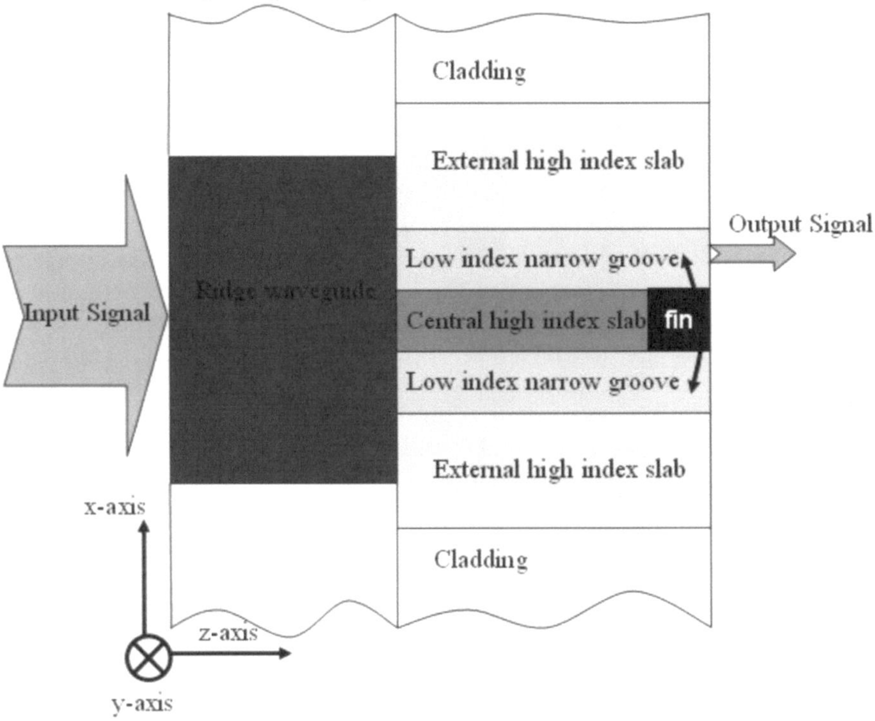

Fig. 26. Glass based photonics displacement sensor (top view; z-cut).

A channel waveguide structure is used for guiding light inside the double slot waveguide structure. Channel guide could be comprised of dielectric or any material of suitable refractive index, could be same as of high index slab regions or low index slot regions. Cladding of channel waveguide structure could be of same material as of double slot structure or could be of air, gas fluid or other metals / oxides. Another channel waveguide structure is used for directing light out of the double slot waveguide structure. The structural limitations are the same as for the earlier channel waveguide structure.

Low index slot regions may be supported by same material as used for the slabs. Other materials providing a contrast in refractive indices may also be utilized without departing from the scope of the invention. Many different structures may be used that provide a class of waveguide structures capable of guiding and confining lights in such a way that high optical intensity is obtained in both small cross sectional areas filled with any compressible material of sufficiently low refractive index, relative to the remainder core of the structure. While low index slot regions are shown as a rectangular cross section, other shapes, such as triangular or semicircular may also be used to provide suitable surfaces for defining the narrow slot regions.

5.3 Working Principle

Light is guided inside double slot structure using a conventional waveguide (channel waveguide). The height of channel waveguide structure is same as that of double slot structure. Input field was coupled with the channel waveguide and later channel waveguide forming an integral part of double slot structure is a source of guiding light in it.
Referring figure 26, the central high refractive index slab is acting as a cantilever. Surface stress on fin on the top of central slab results in its' static bending. Displacement of central slab (cantilever) under the action of stress reduces the width of one of the slot resulting in increased width of other slot. Power confined inside low index slot is directly related to the width and hence is changed accordingly. Light is directed outside the double slot waveguide structure using an embedded channel waveguide structure.
In an example embodiment of glass based photonics sensor; using glass (high refractive index slab) and air (low refractive index slot). Referring figure 26, upper slot; hereafter named as right slot and lower slot; hereafter named as left slot, are of rectangular shape with 50 nm width, 1800 nm height and 1000 nm length comprising of air (R.I.=1.00). Upper slab; hereafter named as left slab, central slab and lower slab; hereafter named as right slab surrounding both low index slot regions. Slabs comprised of commercially available high refractive index glass (R.I. = 1.65); center slab width is same as that of low index slot regions; where as outer slabs width is 360nm. Input plane is propagated along z-axis at 1.55 μm CW (optical frequency).

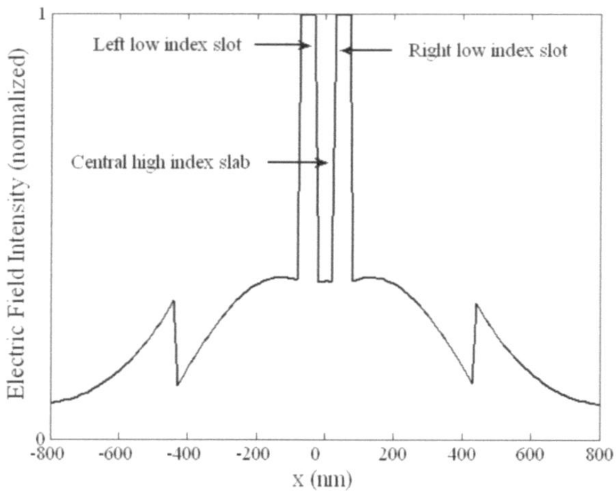

Fig. 27. (a): Normalized E-field intensity – both low index slot regions of same width.

The shift in center slab with a step size of 10 nm is causing a constant increase in left slot width starting from 5 nm till 95 nm; vice versa constant decrease in right slot width. Power

confinement factor is directly proportional to the slot width. Change in power confinement factor is being used as an indicator for shift in displacement of center slab. Normalized E-field intensity in double slot waveguide structure is displayed in figure 27 (a), (b) & (c) at three different displacement locations of center slab.
In this embodiment example (see Fig. 27(a)), E-field intensity has been calculated by keeping both slots at same width. As the slots width was same, hence power confined in both slots is same. Power confinement has been calculated using Eqn. 3 [3] and it was found same. Figure 27(a) above gives a very nice example of double slot waveguide structure, where quasi-TE mode is used. Light is totally confined inside the low index slot region. E-field was normalized with it's peak value. The E-field distribution at modal point depends upon the input conditions and geometrical shapes of slot waveguide structure. The input conditions and refractive index contrast were kept same; however geometrical conditions were changed in further embodiments. The change in geometrical conditions due to shift in central slab has a profound effect on the E-field distribution at the modal points. We will see in figures below that, for the embodiment where left low refractive index slot width is less; E-field confinement is high.
In this embodiment example (see Fig. 27(b)), numerical calculations have been done by keeping left low refractive index slot width at 5 nm. Right low refractive index slot width was 95 nm (vice versa). E-field was normalized with the E-field obtained when both the slots width is same (see fig. 27(b) below). It is evident from the figure that power was mainly confined in right low refractive index slot. Power confinement factor in both low refractive index slots has been calculated using Eqn. 3 [3].

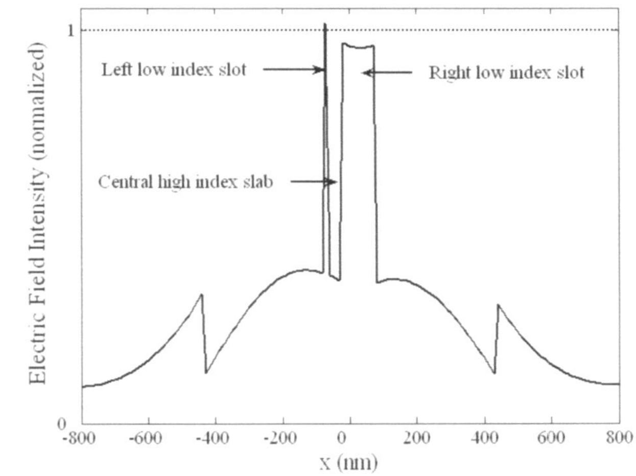

Fig. 27. (b): Normalized E-field intensity - left slot of width 5nm; right slot of width 95nm.

In this embodiment example (see Figure 27(c)), numerical calculations have been done by keeping left low refractive index slot width at 95 nm. Right low refractive index slot width was 5 nm (vice versa). E-field was normalized with the E-field obtained when both the slots width is same (see figure 27(c) below). It is evident from figure; that power was confined in left low index slot. However sharp spike of light confinement in right low refractive index groove is due to it's small cross sectional area. Power confinement factor in both low refractive index slots has been calculated [3]. The large cross sectional area of left slot attributed towards high power confinement; vice versa right low refractive index slot contains less power.

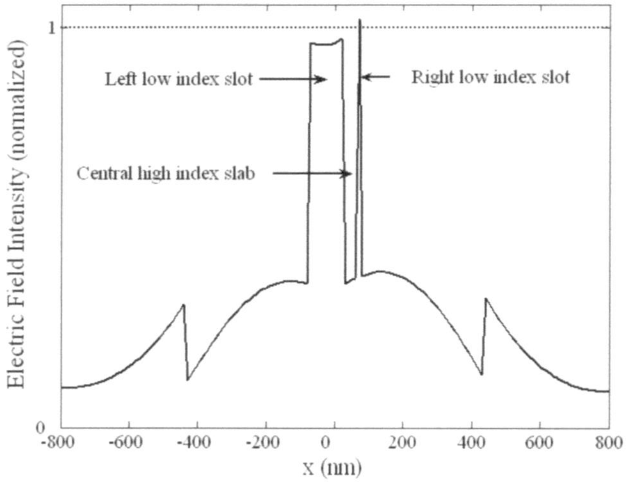

Fig. 27. (c): Normalized E-field intensity - right slot of width 5nm; left slot of width 95nm.

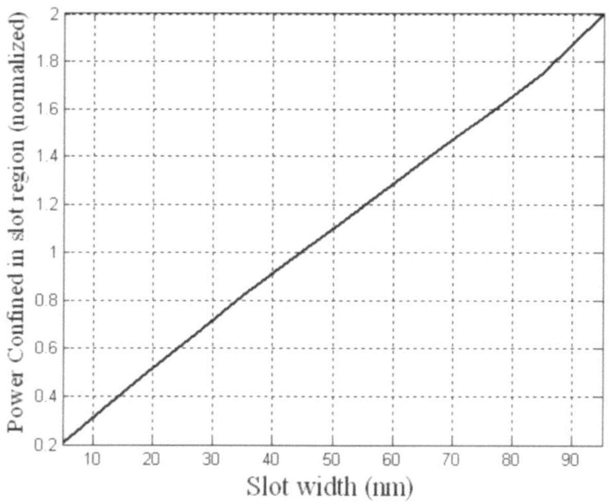

Fig. 28. Power confined (normalized) inside low refractive index slot region dependence upon slot width.

Power confined in either low index slot region is increasing along with increase in slot width. The low index slot width is dependent upon shift in center slab. Power confinement factor in both low index slots have been checked and found similar response at same displacement of central high refractive index slab. Power confined (normalized) in a low index slot region has been drawn in graphical format in figure 28. The change in slot width is due to shift in center slab under the effect of stress at fin's vertical surface. We have checked the power confinement factor at various example embodiments. Few power confinement values plotted in figure 28 are just for example. It was observed that in order to get a considerable change in power confinement factor a minimum shift of 10nm in center slab width is required. A shift in power confinement factor by 3.85 has been found for every 10nm change in center slab displacement. Hence sensitivity of proposed photonics sensor mechanism is 0.385/nm.

6. Slot Structure Coupling Structures

In order to ensure usage of glass based low contrast slot structure in forming passive optical devices, it was necessary to provide an example. Slot waveguide based single slot to double slot coupling structure (see figure 29) is realized using glass & air based single and multi slot structure.

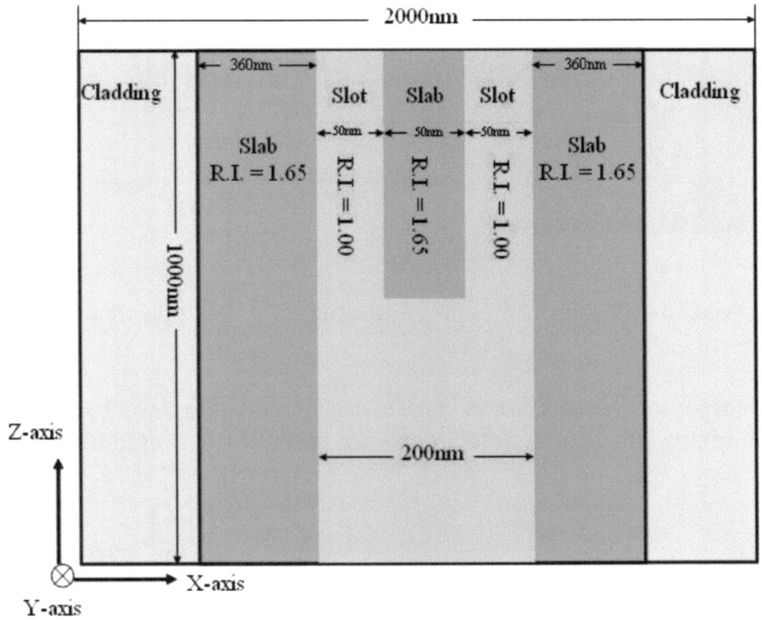

Fig. 29. Slot waveguide based Y-coupler.

In a numerical calculation using full vector finite difference mode-solver, the dimensions of double slot structure as follows:

Referring figure 29, both slots are of rectangular shape with 50 nm width, 400 nm height and 1 μm length comprising of air (R.I.=1.00). Three rectangular slabs are surrounding both the slots comprising of commercially available high refractive index glass (R.I. = 1.65). Single frequency Gaussian modulated continuous wave was used as input plane, propagated in the direction of z-axis. Numerous numerical calculations by using standard input plane were carried out at various displacement values. E-field amplitude shows that a part of optical power is also in the substrate. However due to electric field discontinuity at the boundaries of high index glass slabs and low index air slots, the e-field is confined strictly inside the low index regions. However contrast ratios do have an effect on the power confinement factor. This effect has been nullified by changing the structure geometry and changing the input plane conditions.

Referring figure 29; initially basic SOI slot optical waveguide was simulated [1]. The light was confined inside low index regions (see figure 30) and is in agreement with the existing theory.

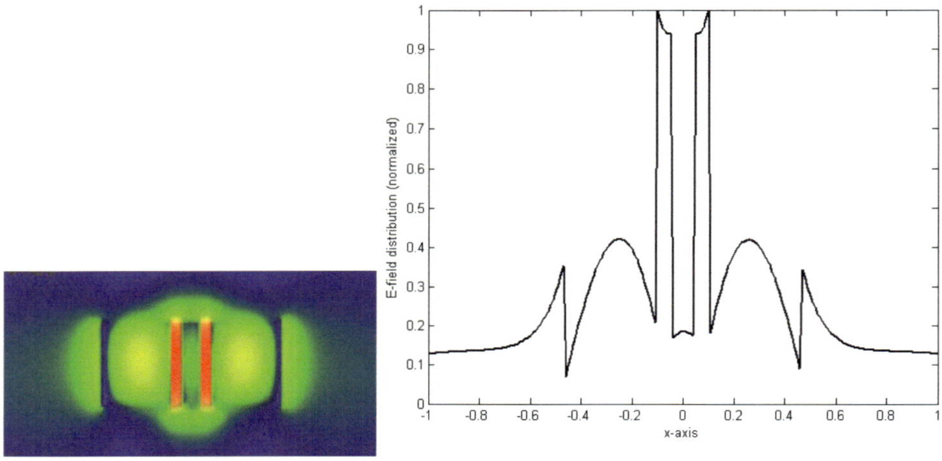

(a) (b)

Fig. 30. (a) E-field distribution; (b) E-field intensity; SOI slot optical waveguide based Y-branch coupler.

Glass based double slot waveguide structure is implemented on the Y-branch coupler model. The slots comprising of air and slabs is of glass. E-field distribution is plotted in figure below:

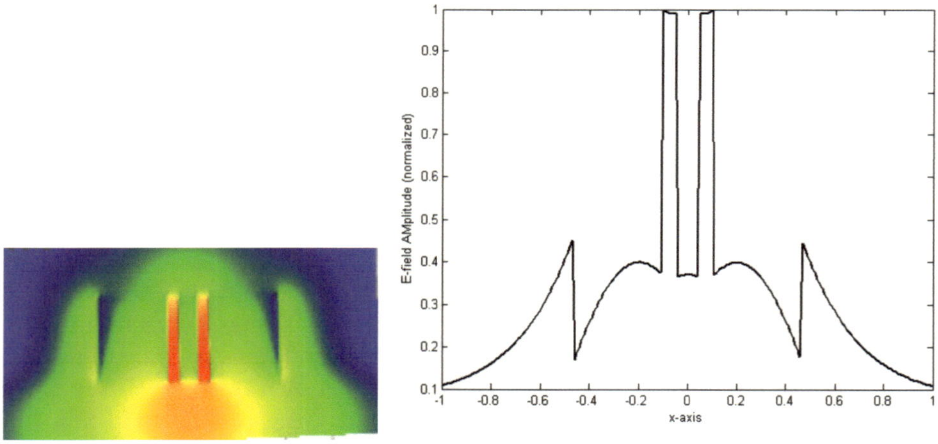

(a) (b)

Fig. 31. (a) E-field distribution; (b) E-field intensity; Double slot waveguide structure (comprising of air & glass) based Y-coupler.

E-field intensity is found splitting into two parts, hence power introduced in the y-branch coupler is splitted equally in between two ports. This is very useful in forming passive optical devices. It can be further exploited in forming complex passive optical devices.

7. Summary

In this chapter simulation and analysis of single and multiple slot waveguide structures had been carried out using FDTD algorithm. It was found that multiple slot structures not only increase the power confinement factor in low index slot regions but can also be utilized in forming sensor mechanisms. Other than traditional SOI slot optical waveguides, low contrast (glass and air based) slot optical waveguides with comparable power confinement factor had been proposed. Later the low contrast double slot optical waveguide structure had been utilized in forming cantilever based sensing mechanisms. Light confinement inside low contrast double slot structure has been explored and found comparable to SOI based slot optical waveguide structure. Based on cantilever type movement of low index slots of proposed low contrast double slot structure a simple and easy to build optomechanical sensor has been proposed. Single frequency continuous wave Gaussian pulse source has been used in simulations which is readily available. It could be most probable candidate for use in temperature, pressure and surface smoothness sensing.

8. References

[1] V. R. Almeida, Q. Xu, and M. Lipson et. al., "Guiding and confining light in void nanostructure," Optics Letters, 2004, vol.29, no.11, p.1209-1211.

[2] Q. Xu, V. R. Almeida, and M. Lipson et. al., "Experimental demonstration of guiding and confining light in nanometer-size low-refractive-index material," Optics Letters, 2004, vol. 29, no.14, p.1626-1628.

[3] N. N. Feng, J. Michel, and L. C. Kimerling, "Optical field concentration in low-index waveguides," IEEE Journal of Quantum Electronics, 2006, vol.42, no.9, p.885-890.

[4] T. Fujisawa and M. Koshiba, "All-Optical Logic Gates Based on Nonlinear Slot Waveguide Couplers," Journal of Optical Society of America B, 2006, vol.23, p.684-691.

[5] K. K. McLauchlan, S. T. Dunham, "Analysis of a Compact Modulator incorporating a hybrid silicon/electro-optic polymer waveguide," IEEE Journal of selected topics in Quantum Electronics, 2006, vol.12, no.6, p 1455-1460.

[6] P. A. Anderson, B. S. Schmidt, and M. Lipson, "High confinement in silicon slot waveguides with sharp bends," Optics Express, vol.14, no.20, 2006, p.9197-9202.

[7] R. Hainberger and P. Mullner, "Modal behavior of SOI slot waveguides," Proceedings of Symposium on Photonics Technology for 7th Framework Program Wroclaw, 2006, p.282-285.

[8] P. Mullner and R. Hainberger, "Structural optimization of silicon-on-insulator slot waveguides," IEEE Photonics Technology Letters, 2006, vol.18, no.24, p.2557-2559.

[9] N. N. Feng, R. Sun, and L. C. Kimerling et. al., "Lossless strip-to-slot waveguide transformer," Optics Letters, 2007, vol.32, no.10, p.1250-1252.

[10] C. A. Barrios, "Ultrasensitive nanomechanical photonic sensor based on horizontal slot waveguide resonator," IEEE Photonics Technology Letters, 2006, vol.18, no.22, p.2419-2421.
[11] F. D. Olio, and V. M. N. Passaro, "Optical sensing by optimized silicon slot waveguides," Optics Express, 2007, vol.15, no.8, p.4977-4993.
[12] P. Sanchis, J. Blasco, and A. Martinez et. al., "Design of silicon based slot waveguide configurations for optimum nonlinear performance," Journal of Lightwave Technology, vol. 25, no. 5, 2007, p. 1298-1305.
[13] T. Fujisawa, and M. Koshiba, "Guided modes of nonlinear slot waveguides," IEEE Photonics Technology Letters, 2006, vol.18, no.14, p.1530-1532.
[14] L. Chen, J. Shakya, M. Lipson, "Subwavelength confinement in integrated metal slot waveguide on silicon," Optics Letters, 2006, vol.31, no.14, p.2133-2135.
[15] J. A. Dionne, H. J. Lezec, H. A. Atwater, "Highly confined photon transport in subwavelength metallic slot waveguides," Nano Letters, 2006, vol.6, no.9, p.1928-1932.
[16] M. Iqbal, Z. Zheng, and J. S. Liu, "Light confinement in multiple slot structures investigated," Proceedings of IEEE, ICMMT Nanjing China, ISBN- 978-1-4244-1879-4, 2008, p.878-891.
[17] Z. Zheng, M. Iqbal, and JS Liu, patent application launched to Chinese Govt.
[18] M. Iqbal, Z. Zheng, and J. S. Liu, "Low contrast double slot structure based optomechanical sensor," Submitted for presentation in Asia Pasic Optical Conference, 09, Shanghai, China.
[19] High refractive index fluids. http://www.2spi.com/catalog/ltmic/cargille-liquid.html
[20] M. Iqbal, Z. Zheng, and J. S. Liu, "Low Refractive Index Contrast Double Slot Structure Based Cantilever Type Sensor," Submitted for presentation in Asia Pasic Optical Conference, 09, Shanghai, China.

10

Analysis and Design of SIW Components Based on H-Plane Planar Circuit Approach

Isao Ohta and Mitsuyoshi Kishihara*
University of Hyogo
Japan
**Okayama Prefectural University*
Japan

1. Introduction

A waveguide whose sidewalls are replaced with densely arranged metallic posts has been proposed. This guide enables the easy realization of circuit patterns by the arrangement of metallic posts periodically in a parallel-plate waveguide or a grounded dielectric substrate. This type of waveguide is called the post-wall waveguide (PWW) (Hirokawa & Ando, 1998;Ando et al., 1998) or the substrate integrated waveguide (SIW) (Wu, 2001; Deslandes & Wu , 2005). This SIW technology is applied to a feed waveguide for a slot array antenna, or a leakage wave antenna. In particular, recently, for the purpose of making use of merits such as low loss, low cost, and high-density integration of microwave and millimeter-wave components and subsystems, a SIW short-slot 90° hybrid coupler, and a six-port receiver consisting of the 90° couplers and/or power dividers have been developed. Since this technology is a relatively new concept, it is desired that more SIW circuit components and subsystems appear to open a new vista (Xu et al., 2005; Moldovan et al., 2006).
With regard to the analytical method of the SIW structure, the derivation of the propagation constant for the straight section of the guide has been studied on the basis of the Galerkin's method of moment (Hirokawa & Ando, 1998). Then widths of the SIW structure that is equal to the cutoff frequency of the rectangular waveguide with perfectly metalized sidewalls have been obtained. Also, empirical equations for the equivalent widths have been proposed through experiments and simulations (Xu & Wu, 2005; Cassivi et al., 2002). In (Xu & Wu, 2005), the FDTD method and the multimode calibration method are used to analyze the dispersion characteristics of the complex propagation constants of the SIW structure. However, in the case of designing and analyzing the circuit components for practical use, a full-wave em-simulator (Ansoft HFSS) has been employed (Moldovan et al., 2006). Since the simulation requires a relatively long computing time, it is desirable to develop a faster solver for the optimization requiring recursive computations.
In this study, an analysis of the SIW structure is attempted by applying the analytical technique of the H-plane waveguide discontinuities based on the planar circuit approach (Kishihara et al., 2006; Kishihara et al., 2004). This technique can reduce the computation

time considerably, because the two-dimensional structure of the SIW is well used to advantage in the analysis.

The present two-dimensional approach is about 10 times faster than the full-wave simulator (HFSS). First, a planar-circuit model of the SIW for analysis is introduced. The analytical procedure consists of 1) the derivation of the mode impedance matrices for regular-shaped planar circuits and 2) short-circuiting of fictitious ports arranged on the peripheries of the metallic posts, in accordance with the treatment of the H-plane waveguide discontinuities containing metallic obstacles. In the SIW structure, a leakage field problem outside the guide occurs due to the gaps of the arrayed metallic posts. This phenomenon should be excluded in the practical design of the passive circuit components, except in cases where it is used positively, such as in leakage wave antennas. In this paper, the leakage field is considered by connecting fictitious TEM transmission lines on the periphery of the planar circuit model and terminating them with their characteristic impedances.

Next, the *S*-parameters of the SIW straight section are calculated as a numerical example. In addition, a situation in which radiation is produced outwards from the guide is prepared intentionally by placing metallic posts at slightly broader intervals, and the validity of the above-mentioned treatment is examined. Then, the propagation constant of the SIW is calculated using the H-plane planar circuit approach along with the TRL calibration technique for a vector network analyzer (Pozar, 1998). In the analysis, the reduction of the computational time is achieved by utilizing the periodic structure of the SIW.

In this work, the present method is applied to the design of two types of SIW corners, a right-angled circular corner and a right-angled corner with a cylindrical region of air (an air-post). The corners are constructed of arrayed metallic posts similarly to an SIW straight-line section. For the corner with one air-post, a portion of the dielectric is replaced with air to obtain a matched state. The validity of the analysis and the design results are confirmed using an em-simulator (HFSS).

Moreover, a cruciform SIW quadrature hybrid (Ohta et al., 2007) is designed based on the idea similar to an H-plane crossed-waveguide quadrature 3-dB hybrid (Toda et al., 2006). The SIW has a planar structure parallel to the plane of magnetic field, and the electromagnetic field in the circuit dose not change in the direction perpendicular to the magnetic plane (H-plane). This implies that the design concept and analytical method of the SIW circuits follow those of the H-plane planar circuit (Kishihara et al., 2006; Kishihara et al., 2004). In the analysis, reduction of the computational time is successfully achieved by utilizing the periodic structure of the SIW. Finally, optimum design of crossed-SIW quadrature 3-dB hybrids is described. Good hybrid properties are obtained for some design frequencies.

2. Planar circuit model and analytical procedure

Figure 1 exhibits a portion of the SIW straight-line section. Dielectric material with relative permittivity ε_r is filled between the top and bottom metallic plates, and metallic posts of radius r are placed at width a_f and spacing s. Generally, since the height of the SIW is much less than the wavelength used in its circuit system, the electromagnetic field is constant in the height direction. Therefore, the propagation and non-propagation modes excited in the SIW are TE_{n0}-like modes, which are very similar to the TE_{n0} modes of the conventional waveguide, and hence the electric field consists of only a vertical component. In addition,

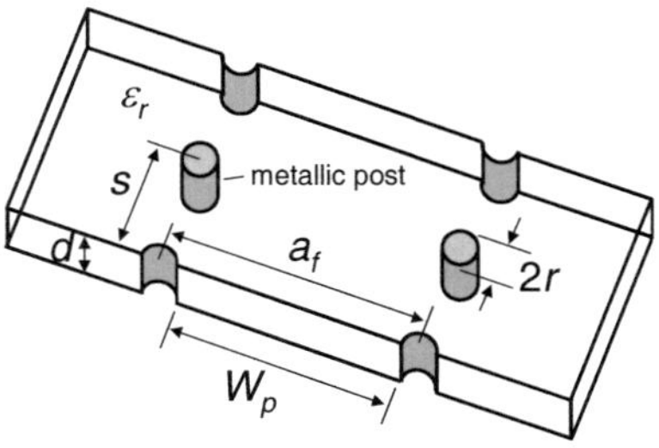

Fig. 1. Structure of SIW straight-line section

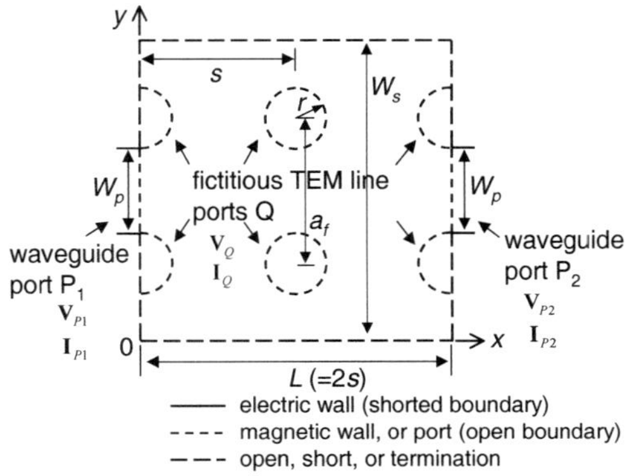

Fig. 2. Planar circuit model for SIW straight-line section

the leakage loss from the gaps between the side-wall posts is very low, because the surface current of TE_{n0}-like modes flows parallel to the metallic posts of the side-wall. In particular, if the SIW is excited by the TE_{10} (or TE_{n0}) mode of a rectangular waveguide with the same height and dielectric constant as the substrate, then the electromagnetic field in the circuit never changes in the vertical direction because the structure possesses no variation along the same direction. In other words, we can describe the circuit performance using only TE_{n0}-like modes. From this fact, in this study, we apply the H-plane planar circuit approach to analyze the SIW circuit system.

Figure 2 illustrates the planar circuit model of the SIW straight section corresponding to that in Fig.1. A rectangular area of width W_s (> $4r+W_p$) and length L (=$2s$) extended outside the posts is considered in order to treat the leakage field distribution. The solid and the dotted lines represent the electric wall (shorted boundary) and the magnetic wall (open boundary), respectively. For the periphery of the planar circuit depicted by the broken line, any boundary condition of open, shorted, or terminated with characteristic impedances can be used. It is possible to carry out the analysis with an open or shorted boundary if the influence of radiation is negligible. However, in the case that radiation losses must be taken into account, the leakage waves from the gaps of the side-wall posts should be absorbed. In this paper, we attempt to absorb the waves by fictitiously connecting many TEM parallel-plate lines of narrow width to the boundaries of the rectangular planar circuit and teminating the fictitious ports with their characteristic impedances.

The input/output waveguide ports with a width of W_p are located at the left and right sides of the circuit (ports P_1 and P_2). On the peripheries of the metallic posts, moreover, the above fictitious TEM-line ports are arranged close together without any gaps (ports Q). $\mathbf{V}_i$ and $\mathbf{I}_i$

stand for the voltage and current vectors of the ith port (i=P_1, P_2, or Q). Deriving the mode impedance matrices between the input/output ports (P_1, P_2) and the fictitious ports (Q) by the planar circuit approach, and short-circuiting only the fictitious ports arranged on the peripheries of the metallic posts, one can obtain the 2-port mode impedance matrix containing the characteristics of the arrayed posts. Therefore, the scattering parameters can be calculated.

On the basis of the planar circuit approach or the impedance Green's function approach, the voltage at any point on the planar circuit can be written as

$$V(x,y) = j\omega\mu_0 d \iint_D G(x,y \mid x_0,y_0) J(x_0,y_0) dx_0 dy_0, \tag{1}$$

where $J(x_0,y_0)$ denotes the source current normally injected into the circuit, and a Green's function $G()$ must satisfy the boundary conditions of the planar element. D denotes the two-dimensional region of the planar circuit. By expanding the fields in the input and output ports in terms of eigenmodes of the rectangular waveguide, the mode impedance matrix can be given as

$$Z_{p,q}^{i,j} = \frac{j\omega\mu_0 d}{W_i W_j} \int_0^{W_i} \int_0^{W_j} G(x_i,y_i \mid x_j,y_j) f_{i,p}(s_i) f_{i,q}(s_j) ds_i ds_j, \tag{2}$$

where W_i and W_j indicate the widths of the ith and jth coupling waveguide ports, respectively, and $f_{i,p}$ represents the eigenfunction of the waveguide with shorted or open boundaries and is given as

$$f_{i,p}(s_i) = \begin{cases} \sqrt{2}\sin\left(\dfrac{p\pi}{W_i} s_i\right) & \text{for shorted boundaries} \\ \varepsilon_p \cos\left(\dfrac{p\pi}{W_i} s_i\right) & \text{for open boundaries,} \end{cases} \tag{3}$$

where the origin of argument s_i is assigned to one side of the coupling waveguide port. s_i varies from 0 to W_i, and the integration in Eq.(2) is over the width of the port. Therefore, the xy-coordinate values must be transformed to the local coordinate value s_i or s_j of the coupling port. In Eq. (3), the sine function corresponds to the usual TE_{p0} modes of the rectangular waveguide with shorted boundaries. The cosine function is applied to the ports with open boundaries. ε_p is 1 for p=0 and $\sqrt{2}$ for $p \geq 1$. The cosine function is applied to the fictitious ports in this work, though only the fundamental TEM mode (p=0) is considered because of the sufficiently narrow port-width. The suffixes i and p denote the port and mode numbers, respectively. W_i is the width of the ith port. Green's function can be expanded in terms of eigenfunctions of the corresponding rectangular planar circuit.

$$G(x_i,y_i \mid x_j,y_j) = -\frac{1}{LW_s} \sum_{m=0}^{\infty}\sum_{n=0}^{\infty} \frac{\psi_{m,n}(x_i y_i)\psi_{m,n}(x_j,y_j)}{k^2 - \left(\frac{m\pi}{L}\right)^2 - \left(\frac{n\pi}{W_s}\right)^2} \tag{4}$$

$$\psi_{m,n} = \varepsilon_m \varepsilon_n \cos\frac{m\pi}{L} x \cos\frac{n\pi}{W_s} y \tag{5}$$

Equation (5) is applicable for the segments with open boundaries. ε_m is 1 for m=0 and $\sqrt{2}$ for $m \geq 1$.

Now, we define the mode impedance matrix of the rectangular segment between two ports, P_1 and P_2, and fictitious ports Q as

$$\begin{bmatrix} \mathbf{V}_P \\ \mathbf{V}_Q \end{bmatrix} = \begin{bmatrix} \mathbf{Z}_{PP} & \mathbf{Z}_{PQ} \\ \mathbf{Z}_{QP} & \mathbf{Z}_{QQ} \end{bmatrix} \begin{bmatrix} \mathbf{I}_P \\ \mathbf{I}_Q \end{bmatrix} \tag{6}$$

where abbreviations $\mathbf{V}_P=\{\mathbf{V}_{P1}^t, \mathbf{V}_{P2}^t\}^t$ and $\mathbf{I}_P=\{\mathbf{I}_{P1}^t, \mathbf{I}_{P2}^t\}^t$ are used. If the fictitious ports are assumed only for the metallic posts, ports Q are short-circuited by enforcing the boundary condition $\mathbf{V}_Q=0$ on the peripheries of the metallic posts. Substituting it into Eq.(6) gives the 2-port mode impedance matrix involving the discontinuity effects of the posts.

$$\mathbf{Z}_{PP}'=\mathbf{Z}_{PP}-\mathbf{Z}_{PQ}\mathbf{Z}_{QQ}^{-1}\mathbf{Z}_{QP} \tag{7}$$

Finally, by terminating the higher order modes in ports P_1 and P_2 with their characteristic impedances, the 2-port impedance matrix for the TE_{10} mode can be obtained.

Furthermore, when the leakage field outside the posts must be considered, fictitious TEM lines are connected to the periphery of the planar circuit and terminated with impedances. The fictitious TEM lines are sufficiently narrow ports with open-boundary sidewalls. In this paper, we assume a dielectric substrate that extends infinitely outside the rectangular region. Then, the leakage field from the SIW structure is considered to be a TEM mode propagating in the parallel-plate waveguide. On the basis of circuit theory, no reflection occurs at the junction between a TEM line of 1Ω and N TEM lines of NΩ. This means that the outgoing waves branch into the narrow TEM lines without reflection, if sufficiently narrow TEM lines are arranged on the periphery without gaps. Therefore, the fictitious TEM lines ought to be terminated with their characteristic impedances to suppress the reflections of the outgoing waves. Even if the transversely changing fields arrive at the periphery, their field distributions can be approximated by the sufficiently narrow TEM ports, because the field is considered constant in the narrow width. In the present treatment, only the leakage waves normal to the boundaries are absorbed, such as Mur's first-order absorbing boundary condition in the FDTD method. The leakage waves can be decomposed into the field components propagating along the x- and y-directions. At the boundaries y=0 and W_s, the leakage waves propagating along the y-direction are absorbed. The remaining components propagating along the x-direction are absorbed at the boundaries x=0 and L. In the numerical calculation, the areas that do not affect the field distribution of the main propagation mode extending slightly outside the posts and that attenuate evanescent waves should be considered. Namely, a length of about $\lambda g/4$ or more should be considered outside the posts. Then it is expected that only the TEM wave reaches the periphery.

Consequently, by deriving the mode impedance matrices among the input/output ports, the fictitious ports on the metallic posts, and the fictitious TEM lines on the periphery of the circuit, and short-circuiting and terminating the fictitious ports and TEM lines respectively, we can obtain the impedance matrix of the SIW structure containing the leakage field effects.

3. Numerical results

3.1 Straight section

The S-parameters of the SIW structure shown in Fig.1 are computed to demonstrate the usefulness of the present analytical method. The dimensions of the guide are chosen as ε_r=2.17, d=1.52 mm, and r=0.30 mm. The spacing s and the width a_f of the posts are varied in pairs as (s, a_f) = (1.00 mm, 4.92 mm), (1.50 mm, 4.85 mm), and (2.00 mm, 4.77 mm) after Ref. (Hirokawa, 1998). The SIW structure and the field distribution simulated using the HFSS at 40 GHz are depicted in Fig.3(a), and the frequency characteristics of the S-parameters are shown in Fig.3(b). In the calculation, the width of the fictitious TEM lines at the circumference of the rectangular planar circuit was chosen to be 0.1 mm, and 10 fictitious TEM lines on the peripheries of each metallic post were considered. The fictitious ports were

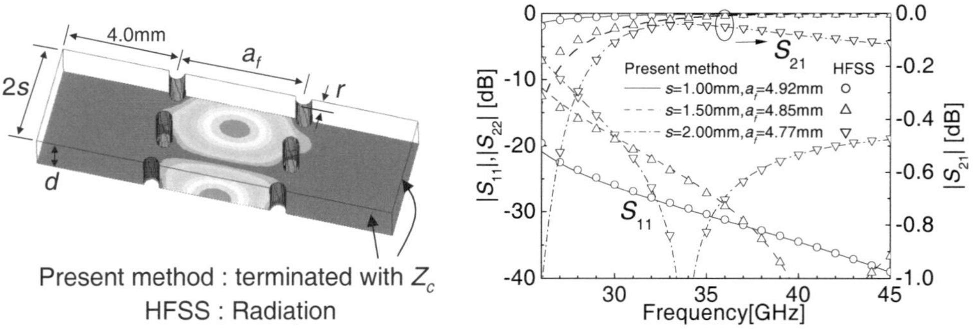

(a) (b)

Fig. 3. (a) Structure and field distribution for the SIW straight section and (b) the frequency characteristics of the *S*-parameters

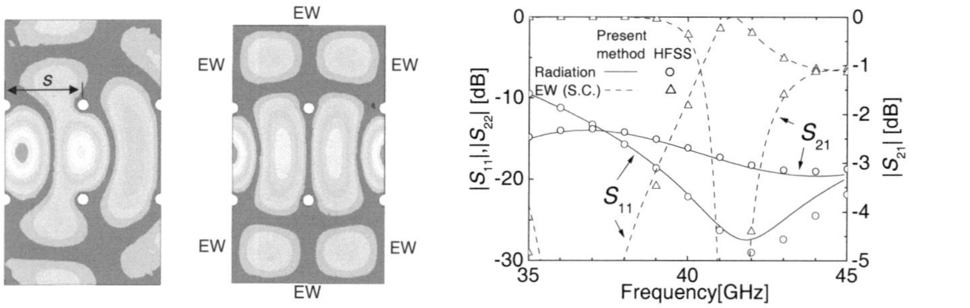

(a) (b) (c)

Fig. 4. Treatment of leakage field. (a) Structure and field distribution simulated using HFSS (boundary: radiation). (b) Structure and field distribution simulated using HFSS (boundary: electric wall). (c) Frequency characteristics of the *S*-parameters

arranged without gaps. For the doubly infinite series of $m \times n$, 100×100 modes were considered. In each exciting waveguide, 8 modes (TE_{p0}; p=1,2, ..., 8) were considered. The results simulated using the HFSS are also plotted for comparison. In the present analysis, the regions of 4.0 mm in length ($\cong$ 0.65 λ_g at 40 GHz) are considered outside the arrayed posts. The periphery of the planar circuit is terminated with the characteristic impedances of the fictitious TEM lines. The computation time at one frequency point is 11 seconds for s=2.00 mm and a_f=4.77 mm, while the HFSS requires 147 seconds (frequency=37 GHz, adaptive pass=8, Delta S<0.001) on a Pentium4 3.2 GHz PC. The present method requires about 1/13 of the computation time compared with the HFSS. The two results are in good agreement and the validity of the analysis is confirmed.

Next, the validity of the treatment of the leakage field is examined by widening the spacing s to 4.0 mm. Fig.4(a) shows the field distribution calculated using the HFSS (at 40 GHz). The TE_{10} mode is incident from the left side of the guide. It is clear that the field is leaking. When the periphery is assigned to electric walls, a resonant mode appears, as shown in Fig.4(b). The *S*-parameters obtained by the present method and the HFSS for these boundary conditions are shown in Fig.4(c). The solid and the broken lines represent the terminated and the short-circuited results for the present method, respectively. The markers are those for the HFSS.

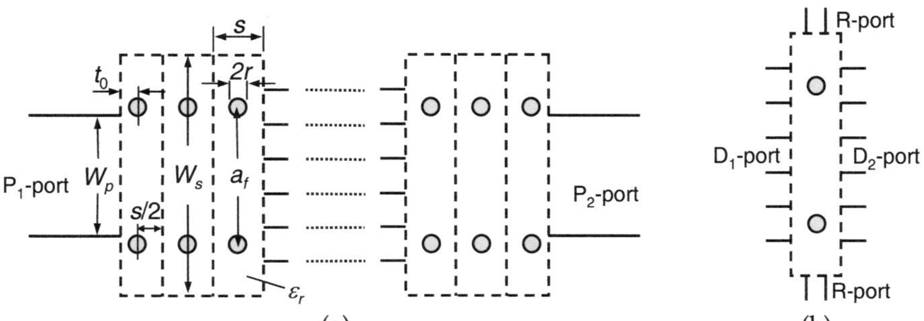

(a) (b)

Fig. 5. Analytical procedure for SIW straight-line. (a) Division of SIW line into periodic structures. (b) One-period section

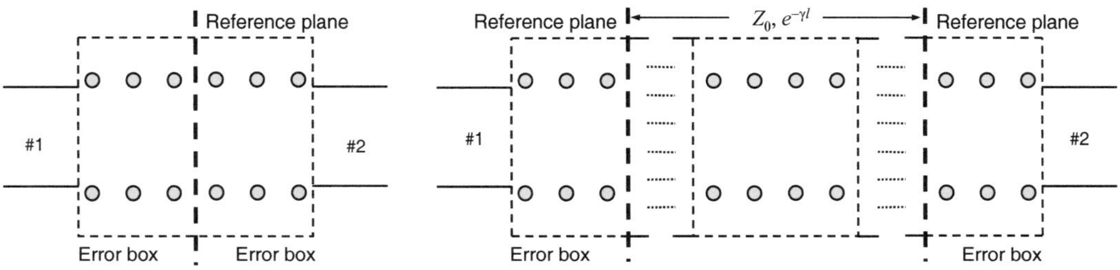

(a) (b)

Fig. 6. Thru and Line connections. (a) Thru. (b) Line

It is found that S_{21} in the solid line indicates approximately -3dB by radiation, while the results shown by the broken lines show completely different characteristics because of the short-circuited periphery. Both these results agree well with the results obtained with the HFSS. The present treatment of the leakage field is verified.

3.2 Propagation constant of SIW line

The propagation constant of the SIW line can be derived using the above technique. The straight-line has a periodical structure, except for the exciting parts at the two ends, as shown in Fig.5 (a), where a slightly modified planar circuit model is employed. The semicircular posts are eliminated and the one-period section in Fig.5 (b) is considered. In order to reduce the scale of the analysis, we first compute the mode impedance matrices of the exciting structure with the rectangular waveguide and one-period section, then use the segmentation method (Chadha & Gupta, 1981) one after the other, and finally derive the mode impedance matrix between the two exciting waveguides. Moreover, by terminating the higher-order TE_{p0} modes of the two rectangular waveguides with their characteristic impedances, we can obtain the two-port impedance matrix, and hence the scattering matrix for the TE_{10} mode. However, in the strict sense, the resultant matrix is not that of the SIW, because it contains some vagueness of the excitation region.

For that reason, the TRL calibration technique (Pozar, 1998) used in the measurement with a vector network analyzer is applied, though the "Reflect connection" is unnecessary in this case. Now, if it is assumed that the two-port scattering matrices for the "Thru" and "Line" connections shown in Figs.6 (a) and 6 (b) are given as

$$\begin{bmatrix} b_1 \\ b_2 \end{bmatrix} = \begin{bmatrix} T_{11} & T_{12} \\ T_{21} & T_{22} \end{bmatrix} \begin{bmatrix} a_1 \\ a_2 \end{bmatrix} \tag{8}$$

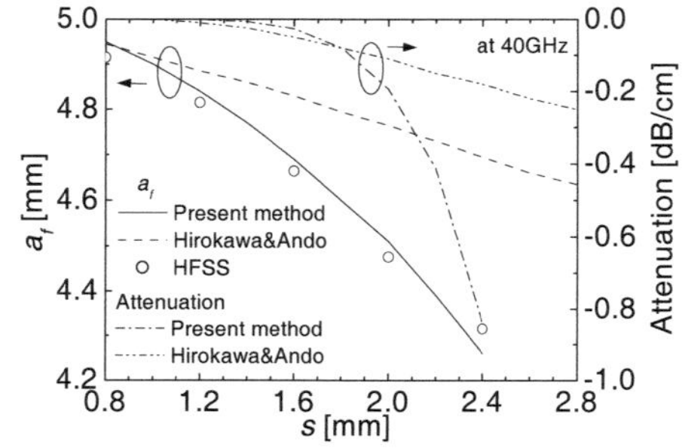

Fig. 7. Equivalent SIW width a_f and attenuation coefficient as a function of post spacing s. ($\varepsilon_r = 2.17$, $d = 1.52$ mm, $r = 0.30$ mm)

$$\begin{bmatrix} b_1 \\ b_2 \end{bmatrix} = \begin{bmatrix} L_{11} & L_{12} \\ L_{21} & L_{22} \end{bmatrix} \begin{bmatrix} a_1 \\ a_2 \end{bmatrix}, \tag{9}$$

respectively, then the propagation term of the SIW straight-line with length l can be derived as follows:

$$e^{-\gamma l} = \frac{\kappa \pm \sqrt{\kappa^2 - 4L_{12}^2 T_{12}^2}}{2L_{12}T_{12}} \tag{10}$$

$$\kappa = L_{12}^2 + T_{12}^2 - (T_{11} - L_{11})^2 . \tag{11}$$

Thus, we can derive the propagation constant γ of the SIW, excluding the error of the exciting parts, under the condition that the SIW sustains only its dominant mode (TE_{10}-like mode).

Figure 7 shows the equivalent SIW width a_f and the attenuation coefficient as a function of the post spacing s to obtain the equal phase constant (= 1.206 k_0, 1011 rad/m at 40 GHz) of the rectangular waveguide of width a_e=4.43 mm (Hirokawa, 1998) (r = 0.30 mm, d = 1.52 mm, ε_r = 2.17, $W_s = a_f + 2r + 4.00$ mm, $t_0 = r + 0.01$ mm, $W_p = a_e$). The results obtained by the present method are compared with those in Ref. (Hirokawa & Ando, 1998). It is noted that when s increases, a_f of the present method decreases rapidly. In order to verify the calculation results, the field distribution simulated using the HFSS is used. a_f can be estimated by measuring the guide wavelength. The markers in Fig.7 are those of the HFSS. The results of the present method agree well with those obtained with the HFSS. In Fig.7, the attenuation coefficient is also indicated. It is found that as the post spacing s becomes larger, the attenuation increases sharper than that in Ref. (Hirokawa & Ando, 1998). For a narrow post spacing up to about 1.8 mm, the results show fair agreement.

Figures 8(a) and 8(b) show the computed frequency dependences of the phase and the attenuation constants for s = 1.00 mm, 1.20 mm, and 1.40 mm. The phase constants (dispersion characteristics) agree well with those of the TE_{10} mode of the conventional rectangular waveguide (equivalent width a_e is 4.43 mm, which is calculated from the cutoff frequency of the SIW). The attenuation constants are less than 0.001 Np/m, 0.01 Np/m and 0.07 Np/m for s = 1.00 mm, 1.20 mm, and 1.40 mm, respectively, at operation bands over 30

GHz. It is recognized that reasonable low-loss properties are obtained for the narrow post spacing.

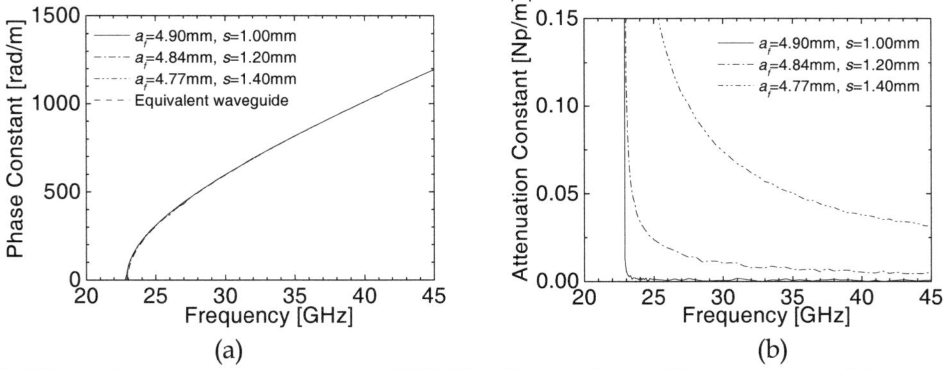

(a) (b)

Fig. 8. Dispersion characteristics of the SIW TE_{10}-like mode. (a) Phase constant. (b) Attenuation constant. (ε_r = 2.17, d = 1.52 mm, r = 0.30 mm)

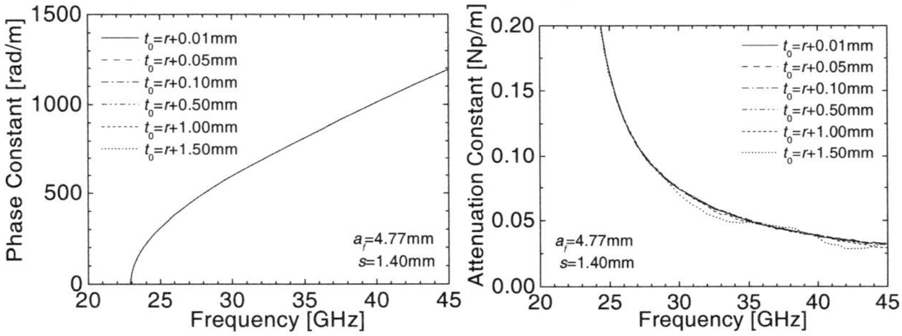

(a) (b)

Fig. 9. Dispersion characteristics with various gap t_0 in the error boxes. (a) Phase constant. (b) Attenuation constant. (ε_r = 2.17, d = 1.52 mm, r = 0.30 mm)

In order to verify that the characteristics of the exciting ports are properly eliminated, the influence of gap t_0 in the error boxes is examined. Figures 9(a) and 9(b) display the dispersion characteristics for a_f=4.77 mm and s=1.40 mm with gap t_0 varied from r+0.01 mm to r+1.50 mm. All the phase constants in Fig.9(a) are consistent with each other. The attenuation constants in Fig.9(b) are in good agreement for $t_0 \leq r$+1.00 mm. It is noted that the TRL calibration technique works well, though gap t_0 should be small to obtain accurate results. Particularly in the case of $t_0 = r + 1.50$ mm, the attenuation constant varies widely, because the attenuation at the gap is larger than that of the SIW line. When gap t_0 becomes large, excitation of the SIW results in failure.

Figures 10(a) and 10(b) also show the computed frequency dependences of the phase and attenuation constants of the SIW TE_{10}-like mode. In this calculation, the dimensions of the guide are selected to be ε_r = 2.33, d = 0.508 mm, r = 0.40 mm, s = 2.00 mm, a_f = 7.20 mm, $W_s = a_f + 2r + 4.00$ mm, $t_0 = r + 0.01$ mm, and W_p = 6.86 mm after Ref. (Xu & Wu, 2005). The cutoff frequency is estimated to be 14.3 GHz from Fig.10(a). At lower frequencies, the attenuation constant increases exponentially owing to radiation loss. The results of the present method are compared with those in Ref. (Xu & Wu, 2005) derived from the multimode calibration method. The two results are in good agreement. Thus the validity of the numerical results is confirmed.

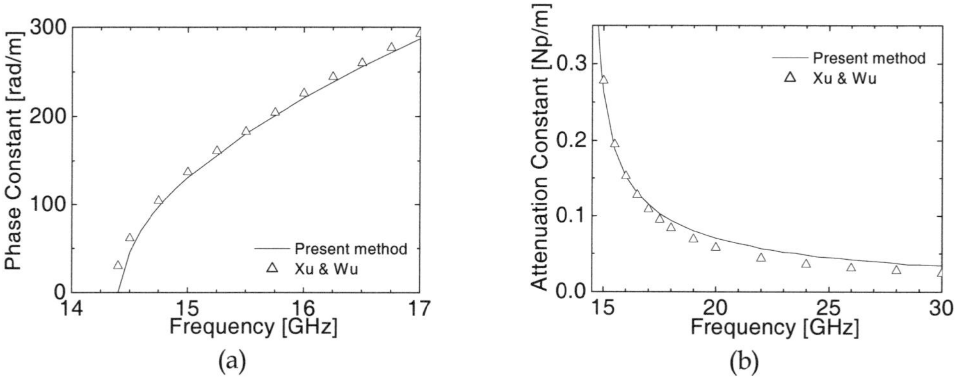

Fig. 10. Dispersion characteristics of SIW TE_{10}-like mode. (a) Phase constant. (b) Attenuation constant. (ε_r = 2.33, d = 0.508 mm, r = 0.40 mm)

In the calculation, the width of the fictitious TEM lines at the circumference of the rectangular planar circuit was chosen to be 0.1 mm (R-port), and 10 fictitious TEM lines on the peripheries of each metallic post were considered. Moreover, we employed 16 modes (TE_{q0}; q=0,1,2,...,15, TE_{00} mode: TEM mode) of the waveguide with magnetic sidewalls of width W_s, which were shown by ports D_1 and D_2 in Fig.5(b), as fictitious connecting ports for uniting each periodic segment. In each exciting waveguide, 8 modes (TE_{p0}; p=1,2,...,8) were considered.

4. Design of SIW components

4.1 SIW corners

The present method is applied to the design of SIW corners to demonstrate the usefulness and flexibility of the method. Figure 11(a) shows a right-angled circular corner constructed from the post-wall waveguide, where the metallic posts are arranged in a circular form at intervals of $\Delta\theta$. Because there is no restriction in arranging the posts, any circuit configuration can be analyzed directly by the planar circuit approach. The dimensions of the guide are chosen to be ε_r = 2.17, d = 1.52 mm, and r = 0.30 mm. Spacing s = 1.00 mm, width a_f = 4.90 mm, and angle $\Delta\theta$ = 11.25° ($a_f\,\Delta\theta$ = 0.96 mm) are used to ensure low loss. Figure 11(b) indicates the frequency characteristics of the S-parameters. It is found that small insertion losses of less than 0.01 dB as well as the relatively low reflection characteristic of less than -25dB are achieved at the operation band. The calculated results agree well with the results obtained with the HFSS.

Figure 12(a) shows the structure of a right-angled corner with one air post. The corner consists of the metallic posts arranged in a right-angle form. A portion of the dielectric material is replaced with an air region of radius R. If the dimensions and the position of the air region are optimized, a low reflection is expected. The analysis is performed by short-circuiting fictitious ports for the arrayed posts, and by the desegmentation-segmentation process (Kishihara et al., 2006; Kishihara et al., 2004) for the air post region. Namely, the analytical procedure consists of the following three steps: 1) the derivation of the mode impedance matrix of the dielectric post (the same region as the air post), 2) the extraction of the dielectric post by desegmentation, and 3) the substitution of the air post by segmentation. Figure 14 shows the frequency characteristics of the *S*-parameters obtained by optimizing the radius *R* and the position c_l of the air post. Low reflections of less than -30 dB are achieved around c_l=4.12 mm and *R*=1.16 mm, though the bandwidth is relatively narrow. It is found that this right-angled circular corner exhibits low-reflection characteristics compared with the return losses of about -10dB to 0dB of a right-angled corner without the air post. Figure 12(b) also shows the results obtained with the HFSS. The two results are in good agreement. The validity of the design results is confirmed.

4.2 SIW cruciform quadrature hybrids

In this section, we design crossed-SIW quadrature hybrids shown in Fig.13. The design frequencies are chosen in quasi-millimeter-wave regions. The dimensions of the SIW are chosen as ε_r = 2.17, *d*=0.508 mm, *r* = 0.40 mm, *s* = 1.80 mm, and a_f = 6.72 mm, because of the good propagation properties. The frequency dependences of propagation constant of the SIW TE_{10}-like mode are displayed in Fig.14. The attenuation constants are less than 0.03 Np/m in the operation band of the SIW, and reasonable low loss properties are obtained. In addition, phase constants (dispersion properties) well agree with those of the TE_{10} mode for the conventional rectangular waveguide whose width (a_e=6.32 mm) is calculated from the same cutoff frequency as the SIW. In the design, we considered the square planar circuit of $W \times W$ with open boundary, which can divide into four one-port with regard to the two symmetry planes (Toda et al., 2006). The even-odd mode analysis is applied for the analysis of the SIW cruciform hybrid. Namely, the reflection coefficients of the four one-port are derived based on the planar circuit approach, and then the scattering parameters of the entire hybrid circuit are computed. The optimization of the circuit dimensions such as the radii and positions of the metallic posts in the cross junction and at the input/output ports

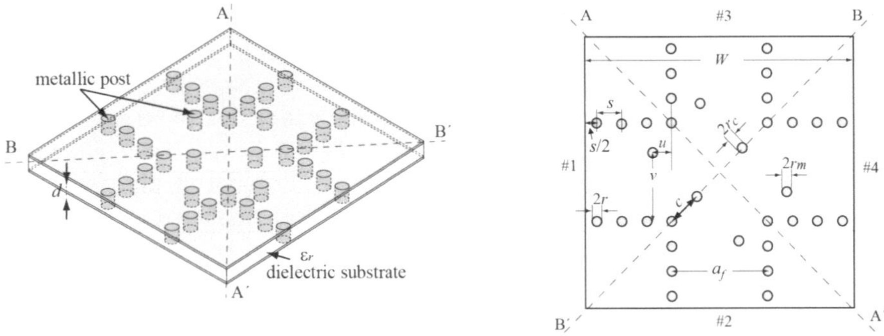

(a) (b)

Fig. 13. Structure of SIW cruciform hybrid. (a) Sketch. (b) Plane figure

were performed using Powell's method (Powell, 1964) as a mathematical technique.
Figure 15 exhibits the scattering parameters of the hybrid designed at the center frequency of 24 GHz. The dots in the figure represent the simulation results using HFSS. Both the results exhibit good agreement with each other. It is shown that the H-plane planar circuit approach is a useful design tool for the SIW circuit components. The return loss and

isolation are better than 20 dB in the bandwidth from 23.4 to 24.9 GHz, and the imbalance of the power split stays within ±0.2dB over the same bandwidth. Furthermore, we attempted to design at various center frequencies. Figure 16 (a) and (b) demonstrate the design results at 21 and 27 GHz, respectively. Although the bandwidth of the hybrid for the center frequency of 21 GHz is narrow, the hybrid designed at 27 GHz shows a relatively wide bandwidth of 26 to 32 GHz. The latter result suggests a possibility of realizing wider bandwidth by the use of widened cross-junction similarly to a crossed H-plane waveguide hybrid (Toda et al., 2006).

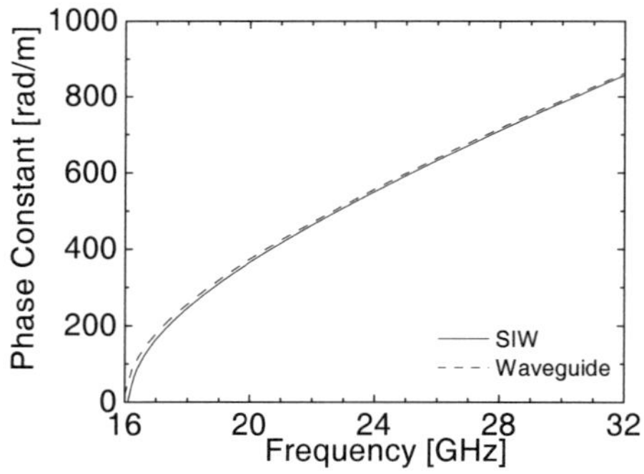

(a)

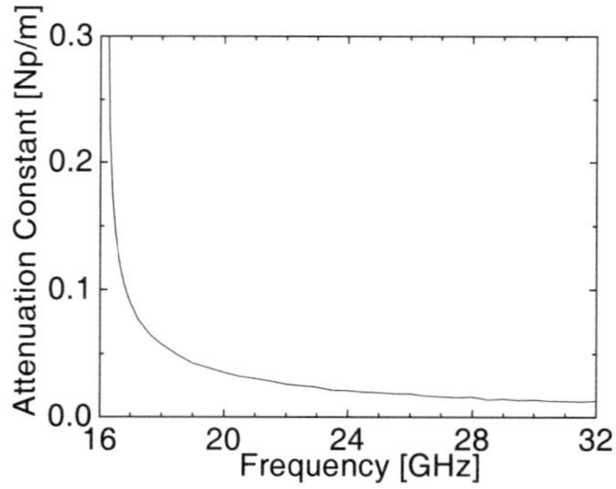

(b)

Fig. 14. Dispersion characteristics of SIW TE_{10}-like mode for the SIW cruciform hybrids. (a) Phase constant. (b) Attenuation constant. ($\varepsilon_r = 2.17$, $d = 0.508$ mm, $r = 0.40$ mm, $s = 1.80$ mm)

5. Conclusion

A method of analyzing the SIW structure was demonstrated, in which the analytical technique of the H-plane waveguide discontinuities was applied on the basis of the planar circuit approach. The leakage field was considered by connecting fictitious TEM lines and terminating them with their characteristic impedances. The propagation constants of the SIW were calculated with the use of the TRL calibration technique. Moreover, the present method was applied to the design of the SIW corners. The validity of the numerical results was verified through comparison with the results in the references and those obtained with the HFSS. In addition, a novel cruciform SIW hybrid has been proposed, and some hybrids with good hybrid properties have been designed using the H-plane planar circuit approach. The design results exhibit good agreement with the simulation results (HFSS), and the validity of the design concept is confirmed. It has been shown that the H-plane planar circuit approach is one of effective analysis techniques for the SIW circuit components, because the metallic posts can be arranged in arbitrary positions.

6. References

Ando, M.; Hirokawa, J. ; Yamamoto, T. ; Akiyama , A. ; Kimura, Y. & Goto, N. (1998). Novel single-layer waveguides for high-efficiency millimeter-wave arrays, *IEEE Transactions on Microwave Theory and Techniques*, Vol.46, No.6, pp. 792-799

Cassivi, Y., Perregrini, L., Arcioni, P., Bressan, M., Wu, K. & Conciauro, G. (2002). Dispersion characteristics of substrate integrated rectangular waveguide, *IEEE Microwave and Wireless Components Letters*, Vol.12, No.9, pp.333-335

Chadha, R. & Gupta, K.C. (1981). Segmentation method using impedance matrices for analysis of planar microwave circuits, *IEEE Transactions on Microwave Theory and Techniques*, Vol.MTT-29, No.1, pp.71-74

Deslandes, D. & Wu, K. (2005). Substrate integrated waveguide leaky-wave antenna: concept and design considerations, *Proceedings of 2005 Asia-Pacific Microwave Conference*, pp. 346-349, Dec. 2005

Hirokawa, J. & Ando, M. (1998). Single-layer feed waveguide consisting of posts for plane TEM wave excitation in parallel plates, *IEEE Transactions on Antennas and Propagation*, Vol.46, No.5, pp. 625-630

Kishihara, M., Ohta, I. & Yamane, K. (2006). Analysis of waveguide H-plane discontinuities with metallic/dielectric obstacles, *IEICE Transactions on Electronics (Japanese Edition)*, Vol.J89-C, No.7, pp. 475-484

Kishihara, M., Yamane, K. & Ohta, I. (2004). Analysis of H-plane waveguide discontinuities by segment substitution process, *Proceedings of 2004 Asia Pacific Microwave Conference*, B4.2, pp.196-197 in abstract book, Dec. 2004.

Moldovan, E., Bosisio, R.G. & Wu, K. (2006). W-band multiport substrate-integrated waveguide circuits, *IEEE Transactions on Microwave Theory and Techniques*, Vol.54, No.2, pp. 625-632

Ohta, I., Toda K., Kishihara, M. & Kawai, T. (2007). Design of cruciform substrate-integrated waveguide hybrids based on H-plane planar circuit approach, *Proceedings of 2007 Asia-Pacific Microwave Conference*, pp.683-686, Dec. 2007

Powell, M.J.D. (1964), An efficient method for finding the minimum of a function of several variables without calculating derivatives, *Computer Journal*, Vol.7, pp.155-162

Pozar, D.M. (1998). *Microwave Engineering, Second Edition*, John Wiley & Sons, 0-471-17096-8

Toda, K., Ohta, I. & Kishihara, M. (2006). H-plane crossed-waveguide hybrids, *Proceedings of 36th European Microwave Conference*, pp. 987-990, Sept. 2006

Wu, K. (2001). Integration and interconnect techniques of planar and non-planar structures for microwave and millimeter-wave circuits - Current status and future trend, *Proceedings of 2001 Asia-Pacific Microwave Conference*, pp. 411-416, Dec. 2001

Xu, X., Bosisiso, R.G. & Wu, K. (2005). A new six-port junction based on substrate integrated waveguide technology, *IEEE Transactions on Microwave Theory and Techniques*, Vol.53, No.7, pp. 2267-2273

Xu, F. & Wu, K. (2005). Guided-wave and leakage characteristics of substrate integrated waveguide, *IEEE Transactions on Microwave Theory and Techniques*, Vol.53, No.1, pp. 66-73

11

Design and Modeling of Microstrip Line to Substrate Integrated Waveguide Transitions

Ting-Yi Huang, Tze-Min Shen and Ruey-Beei Wu
National Taiwan University
Taiwan, R.O.C.

1. Introduction

Microstrip lines (MSL) are widely used in microwave systems because of its low cost, light weight, and easy integration with other components. Substrate integrated waveguides (SIW), which inherit the advantages from traditional rectangular waveguides without their bulky configuration, aroused recently in low loss and high power planar applications. This chapter proposed the design and modeling of transitions between these two common structures. Research motives will be described firstly in the next subsection, followed by a literature survey on the proposed MSL to SIW transition structures. Outlines of the following sections in this chapter will also be given in the end of this section.

1.1 Research Motives

Planar transmission lines, such as MSL and coplanar waveguides (CPW), are favorable in the integration of microwave systems. For low loss and high power applications, rectangular waveguides are often used for signal transmission between system modules. A vast of transition structures between planar circuits and rectangular waveguides have also been proposed, e.g., [Das et al., 1986; Kaneda et al., 1999; Lin & Wu, 2001]. However, systems with rectangular waveguides are often large and heavy. Transitions between rectangular waveguides and planar circuits cannot be held without extra supporting structures. Recently, SIWs were proposed as a replacement for miniaturized and light weighted applications. Nonetheless, such transitions can shed some light on the design of MSL to SIW transitions.

SIW, as its name, can be easily integrated into the substrates of planar circuits, such as printed circuit boards (PCB) and low temperature co-fired ceramics (LTCC), with their standard fabrication processes. Compared with conventional rectangular waveguides, SIW has the advantage of low-cost, compact, and easy-integration with planar circuits. Although their quality factors cannot compete with those of traditional rectangular waveguides, they are more suitable in system integration.

For system using both MSL and SIW, LTCC is often preferred for its multilayer nature and three-dimensional interconnection capability. System on package (SOP) modules can be fabricated on LTCC with various passive elements embedded. Therefore the required circuit area can be reduced significantly. This chapter aims at design and modeling of the transitions structures between MSL and SIW on LTCC substrate for better transition performance and wider bandwidth.

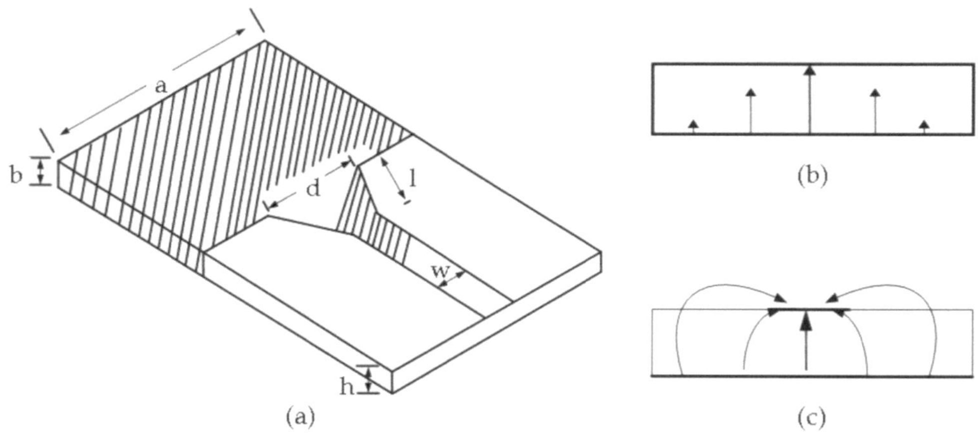

Fig. 1. MSL to SIW transition with tapered microstrip feeding. (a) Transition structure. (b) Electric field distribution in SIW cross section. (c) Electric field distribution in MSL cross section [Deslandes & Wu, 2001a].

1.2 Literature Survey

Transition structures between planar circuits and traditional rectangular waveguides have been widely studied and many high performance transitions are proposed. However, complicated structures are required to hold or support the transitions and precise fabrication processes are also needed at millimeter wave frequencies. Many of these structures can be adopted for the transitions between planar circuits and SIW directly or with little modification, but with lower cost and more reliable fabrication process. On the other hand, since planar circuits and SIW can be integrated on the same substrate, much simpler structures are able to accomplish the transition and many of them have been proposed and demonstrated with good performance.

Tapered microstrip feeding [Deslandes & Wu, 2001a] is commonly used in the transition between MSL and SIW. As shown in Fig. 1a, a MSL is connected directly to the top wall of SIW through a tapered microstrip section. The vertical components of electric field in both MSL and SIW regions are well matched, as shown in Fig. 1b and Fig. 1c, therefore the transition can be easily achieved. However, in order to reduce the discontinuity effect, a long enough tapered microstrip section is needed, up to a half wavelength in some cases. Another disadvantage of this transition structure requires the MSL on the same layer where the metal wall of SIW is located. Also, the MSL is directly connected to SIW. DC current from MSL will be shorted into ground through SIW. Transition by direct connection can also be found between CPW and SIW [Deslandes & Wu, 2001b]. As other direct connected transition structures, additional DC blocking circuit is needed because the signal line is shorted to the ground. For thick substrates, extra insertion loss may be introduced due to slot radiation.

Transition by probe feeding is often used between coaxial lines and traditional rectangular waveguides, e.g., [Collins, 1991; Liang et al., 1992; Rollins & Jarem, 1989]. MSL can also achieve probe feeding [Shih et al., 1998] by shorting its ground plane to one of the longer edge of the waveguide and stretch its signal line into the waveguide. By properly designing the stretching length, the feeding position, and the quarter wavelength transformer for impedance matching, about 40% fractional bandwidth can be achieved.

In multilayer substrates, modified probe feeding can be used for the transition between MSL and SIW. A V-band MSL to SIW transition in LTCC is proposed [Huang et al., 2003]. MSL is shorted to the bottom wall of SIW with a through hole via, which act as an excitation probe. Sizes of via pads between different layers are adjusted to achieve better impedance matching. Three different designs have also been investigated and their performances are compared [Kai et al., 2005]. However, it deserves mentioning that the thickness of SIW is much smaller than that of the metallic waveguide. Hence, the impedance matching is in general worse and will result in larger reflection. With some special compensation design, the transitions with relative bandwidth of about 10% have been reported.

A systematic design approach to enhance the bandwidth of the simple probe feeding structure has also been proposed by our group [Yau et al., 2007], as shown in Fig. 2. The feeding position is chosen to be a quarter of guided wavelength in SIW from its shorted end. Electromagnetic wave excited by via current propagates in -z direction will be in phase with that propagates in +z direction after reflection. Impedance matching can be achieved by adjusting the sizes of the via pad at MSL and the anti-pad at the top wall of SIW. Further discussions and design examples of this transition structure will be given in this chapter.

The DC shorting problem can be avoided if the transition is accomplished by coupling. Our group has also suggested the MSL to SIW transition by an open stripline stub [Yang et al., 2005]. MSL is connected to a stripline stub which stretch into the SIW for about a half wavelength. Impedance matching is done by adjusting the width of the MSL and stripline stub.

On the other hand, by opening a slot at the current return path of MSL, energy can be coupled to traditional rectangular waveguide [Grabherr et al., 1994]. By properly choosing the slot length and position, transition bandwidth can be enhanced. Transition by slot coupling can also be utilized between MSL and SIW, as shown in Fig. 3. By opening a thin slot on the common metal plane of MSL and SIW, energy can be coupled between these two structures through the slot. Transition bandwidth and frequency can be controlled by properly adjusting the lengths of the slot and the microstrip open stub. Open-type transitions often suffer from low bandwidth. In order to improve the transition bandwidth, detailed discussions and several design examples will also be presented in this chapter.

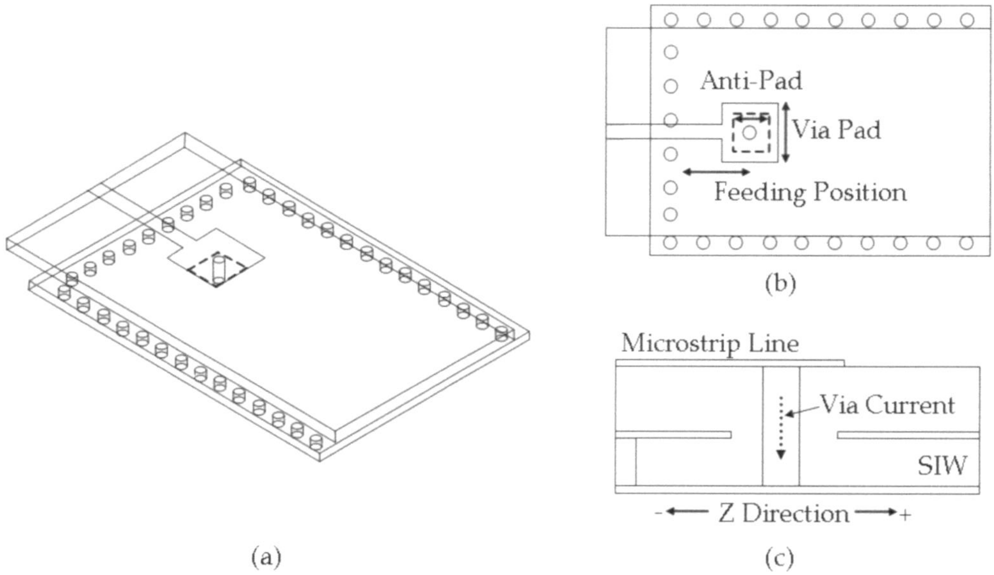

Fig. 2. MSL to SIW transition by probe feeding with via pad and anti-pad compensation. (a) 3D view. (b) Top view. (c) Side view.

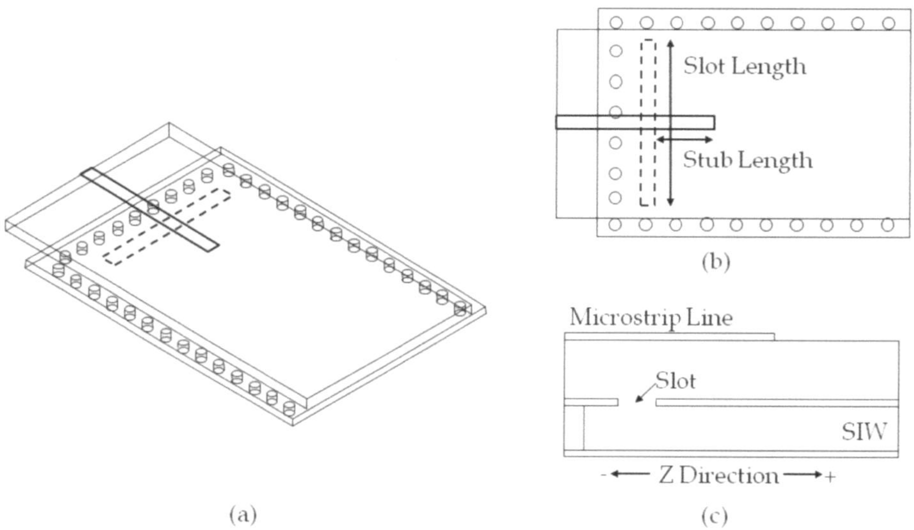

Fig. 3. MSL to SIW transition by slot coupling. (a) 3D view. (b) Top view. (c) Side view.

1.3 Section Outlines

Remaining sections of this chapter are arranged as follows. Theory of MSL to SIW transitions will be derived in section 2, including the modal analysis of SIW, calculation of input resistances for current excitation and slot coupling, and model construction with calculated input resistances. Two different kinds of transition structures, probe feeding and slot coupling, will then be presented in the following two sections. Each of these two sections begins with input resistance derivation and equivalent circuit construction by the theory proposed in section 2. Several design examples for both structures will also be presented with comparison between simulations and measurements. A brief summary and discussion will be given in the end of this chapter.

2. Theory of MSL to SIW Transitions

Rectangular waveguides are widely used in microwave systems for its high power handling ability, low radiation loss as well as low electromagnetic interference (EMI) to other circuit components. However they are also known with disadvantages such as bulky volume, heavy weight, high cost, and difficult integration with planar circuits. In addition, high precision process is required at millimeter wave frequencies. As a result, mass production is difficult for systems with rectangular waveguides.

Laminated waveguides were first proposed in 1998 [Uchimura et al., 1998], where waveguides can be embedded in multilayer printed circuit boards with their side walls replaced by via fences. In 2001, concept of substrate integrated waveguides (SIW) was also proposed [Deslandes and Wu, 2001a; Deslandes and Wu, 2001b]. Waveguides embedded in single layer substrates are demonstrated with transitions to CPWs and MSLs. These kinds of waveguides can be easily integrated with other circuit components by a standard planar circuit fabrication process. Volume and weight are also significantly reduced.

The structure and characteristic of SIWs will be introduced firstly in this section, followed by two kinds of excitation structures and calculation of their input resistances. Construction of the equivalent circuit models where the input resistances are associated will also be presented.

2.1 Modal Analysis of SIW

Figure 4a shows the structure of an SIW, which is composed of the top and bottom metal planes of a substrate and two parallel via fences in the substrate. In order to replace the vertical metal walls, via pitch must be small enough. The vias must be shorted to both metal planes to provide vertical current paths, as shown in Fig. 4b, otherwise the propagation characteristics of SIW will be significantly degraded.

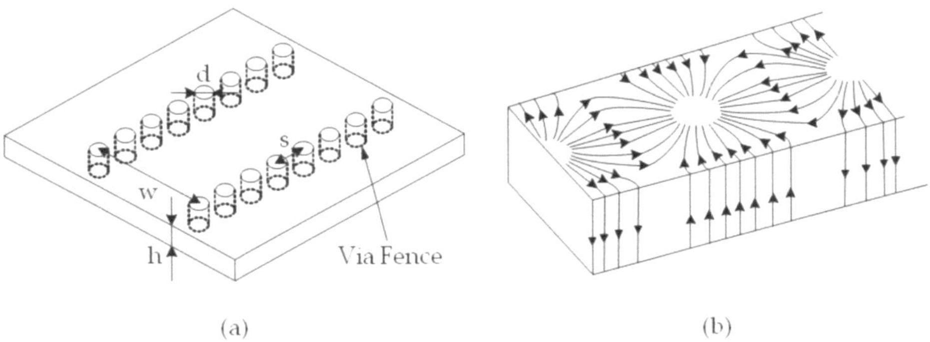

Fig. 4. (a) Structure of SIW and (b) surface current for TE_{10} mode.

Since the vertical metal walls are replaced by via fences for the SIW structures, propagating modes of SIW are very close to, but not exactly the same as, those of the rectangular waveguides. This can be verified by checking the modal surface current patterns. Only patterns with solely vertical current distributed on the side wall survive in SIWs. For example, Fig. 4b shows the TE_{10} mode surface current distribution of a rectangular waveguide. The current path will not be cut by the via fences, therefore TE_{10} mode can be supported in an SIW. This holds for all TE_{m0} modes since their current distributions on the side walls are similar. On the other hand, horizontal components of the surface current exist on the sidewalls for all TM modes and TE_{mn} modes with nonzero n's. These current paths will be cut in SIW structures, which results in radiation. Therefore we can conclude that only TE_{m0} modes exist in SIW structures. Properties of TE_{m0} modes are listed in Table 1 for later usage.

Property	**TE_{m0} modes**
Generating function	$\psi_{m0} = \cos m\pi x/a$
Cutoff wavenumber	$k_{c,m0} = m\pi/a$
Propagation constant	$\Gamma_{m0}^2 = k_{c,m0}^2 - k^2, \quad k = \omega\sqrt{\mu\varepsilon}$
Magnetic field	$\vec{H}_t = \pm\Gamma_{m0}\nabla_t\psi_{m0}e^{\pm\Gamma_{m0}z}, \quad H_{z,m0} = k_{c,m0}^2\psi_{m0}e^{\pm\Gamma_{m0}z}$
Electric field	$\vec{E}_t = \pm Z_h\hat{a}_z \times \vec{H}_t, \quad Z_h = j\omega\mu_0/\Gamma = jk_0\eta/\Gamma$
Power flow	$ab\eta k\beta_{m0}k_{c,m0}^2/4$

Table 1. Properties of TE_{m0} modes.

2.2 Input Resistance of Current Excitation

For a waveguide excited by a current source along path C with current density J in the direction τ, as shown in Fig. 5, the magnitude and patterns of different modes excited can be calculated as follows.

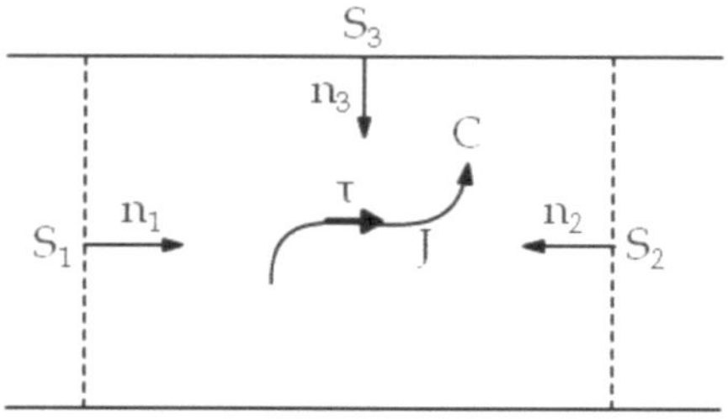

Fig. 5. Current excitation in a waveguide structure.

By the orthogonal property of waveguide modes, field propagating in the waveguide can be denoted by the linear combination of modal fields $\vec{E}_n^{\pm}$ and $\vec{H}_n^{\pm}$, which denote the electric and magnetic field of the n-th propagating mode along ±z direction, respectively. The field propagating along +z direction excited by the current can be decomposed as

$$\vec{E}^+ = \sum_n a_n \vec{E}_n^+, \quad \vec{H}^+ = \sum_n a_n \vec{H}_n^+ \cdot \tag{1}$$

In the same manner, the field propagating along -z direction excited by the current can also be decomposed as

$$\vec{E}^- = \sum_n b_n \vec{E}_n^-, \quad \vec{H}^- = \sum_n b_n \vec{H}_n^- \cdot \tag{2}$$

Consider the region V enclosed by surfaces $S = S_1 + S_2 + S_3$. By reciprocity theorem [Harrington, 2001],

$$\oint_S \hat{n} \cdot \left(\vec{E}_n^\pm \times \vec{H} - \vec{E} \times \vec{H}_n^\pm\right) dS = \int_V \vec{J} \cdot \vec{E}_n^\pm dV \cdot \tag{3}$$

For $\vec{E}_n^-$ and $\vec{H}_n^-$ modes, (3) can be reduced into

$$2a_n = -\int_C \vec{J} \cdot \vec{E}_n^- dl \cdot \tag{4a}$$

On the other hand, if $\vec{E}_n^+$ and $\vec{H}_n^+$ modes are used instead,

$$2b_n = -\int_C \vec{J} \cdot \vec{E}_n^+ dl \cdot \tag{4b}$$

Therefore the coefficients of the excited modes by the current source can be determined if only the current can be well approximated. For an SIW with equivalent width a and substrate thickness b which operates at frequencies that only TE_{10} mode propagates, the power transmitted into the SIW can be calculated by

$$P = \int_0^b \int_0^a \frac{1}{2} \mathrm{Re}\left(\vec{E}_{10} \times \vec{H}_{10}^*\right) \cdot \hat{a}_z \, dx \, dy \cdot \tag{5}$$

Here, the asterisk denotes complex conjugate operator. The resistive part of the equivalent input impedance, or simply the input resistance, seen from the impressed current excitation I can thus be given by

$$R_{in} = \frac{2P}{|I|^2} \cdot \tag{6}$$

It is worthy noting that the input resistance from (6) depends on the current distribution along the path C, but independent of the current magnitude.

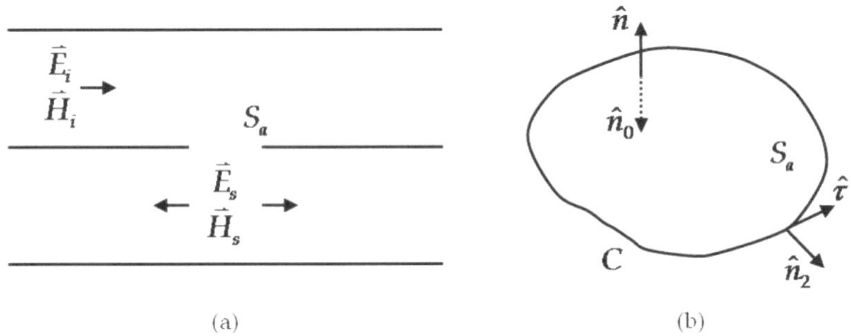

Fig. 6. (a) Waveguide excitation by slot coupling. (b) Notations on the slot aperture S_a.

2.3 Input Conductance of Slot Coupling

Waveguides can also be excited by slot coupling. Consider a small slot aperture S_a opened at the common metal wall of two waveguide structures, as shown in Fig. 6a. Incident fields $\vec{E}_i$ and $\vec{H}_i$ from the upper waveguide structure can be coupled into the lower waveguide structure through the aperture, inducing the scattered fields $\vec{E}_s$ and $\vec{H}_s$. By equivalence principle [Harrington, 2001], the aperture can be replaced by a perfect electric conductor (PEC) with properly placed magnetic current and magnetic charge.

Applying the boundary condition on the PEC, the magnetic current $\vec{J}_m$ and magnetic charge ρ_m can be determined as

$$\vec{J}_m = -\hat{n}\times\vec{E}_s, \quad \rho_m = \mu\hat{n}\cdot\vec{H}_s. \tag{7}$$

However in SIWs, magnetic field usually contains only tangential component at boundary, i.e., no magnetic charge in this case. The scattered field excitation in the lower waveguide structure is contributed solely by the magnetic current. Therefore once the electric field distribution at the slot aperture is known, the equivalent magnetic current can be calculated by (7). With a similar process in 2.2, the coefficients of excited modes can be found as

$$2a_n = \int_{S_a} \vec{J}_m \cdot \vec{H}_n^- \, dS. \tag{8}$$

The input power P can then be found by (5). If a voltage V is defined by integrating the electric field at input port, the input conductance can be determined by

$$G_{in} = \frac{2P}{|V|^2}. \tag{9}$$

Again, the value is dependent on the voltage distribution along the slot, rather than the magnitude of the voltage.

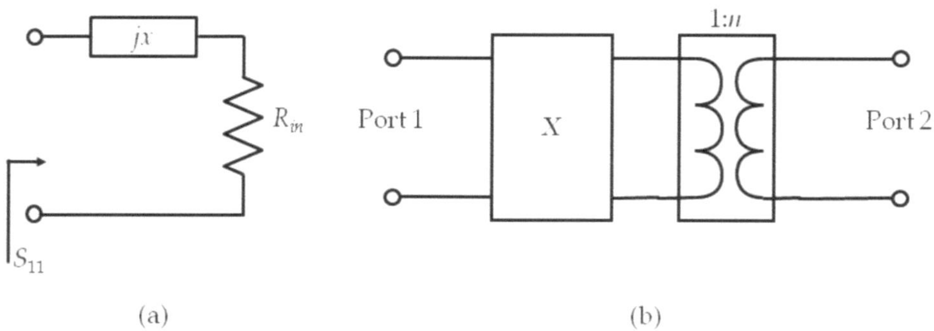

Fig. 7. (a) One port equivalent model with input resistance R_{in}. (b) Full model for the two port transition network.

2.4 Model Construction with Input Resistances

With the input resistance R is calculated in 2.2, only the reflection coefficient S_{11} at input port can be determined, as shown in Fig. 7a, where the series reactance jx is used to describe the

reactive effect which vanishes at resonance. To construct a model that can be used to determine the full scattering matrix of the two port transition network, the input resistance is replaced by a transformer with a frequency dependent ratio n determined by the calculated input resistance and the characteristic impedance of SIW. For the input conductance G_{in} calculated in 2.3, the same concept can be applied.
It should also be noted that the input port of the input resistance is defined right at the source. In practical cases additional structures are required to support the sources and usually the actual input ports are located at these structures. In general these structures are electrically small with negligible losses. Together with the series reactance jx, they can be modeled by a frequency independent reactive network $\mathbf{X}$. Elements in $\mathbf{X}$ are determined according to the physical arrangement of the structures. Since $\mathbf{X}$ is frequency independent, the values of its elements can be obtained by full-wave simulation at a single frequency. Figure 7b shows the full model of the two-port transition network. As will be demonstrated in the following two sections, with the frequency dependent transformer and the frequency independent reactive network, this model is able to explain the response of the transition structure over a wide frequency band.

3. MSL to SIW Transition with Shorting Via

This section proposes a transition structure between MSL and SIW by probe feeding with a shorting via. The transition structure is shown in Fig. 2. Input resistance will be calculated first according to 2.2. Transition model will then be constructed with the method discussed in 2.4. A Ka band and an E band transitions will then be demonstrated as examples.

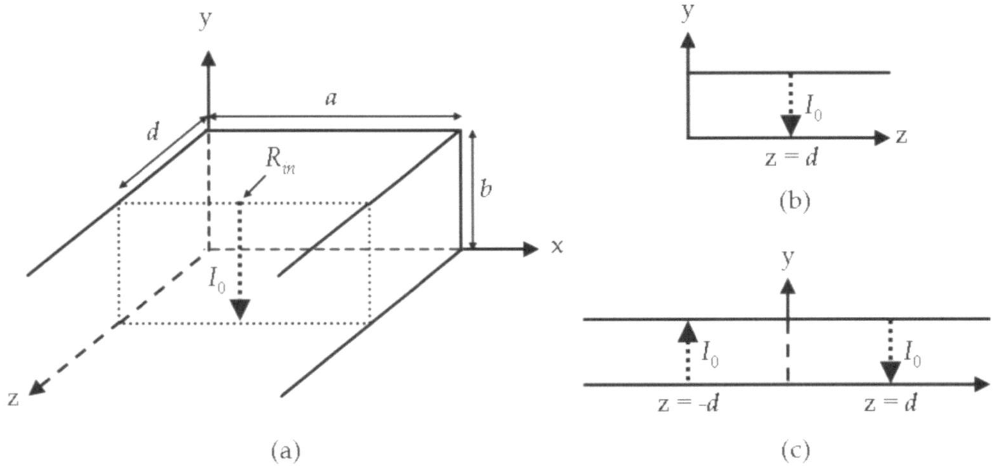

Fig. 8. Simplified problem for probe feeding. (a) Three-dimensional view. (b) Side view. (c) Side view of the equivalent problem.

3.1 Input Resistance

For a z-directed SIW with equivalent width a and substrate thickness b fed by a via probe at $z = d$, $x = a/2$, if the thickness and the via diameter are small, the current on the via can be

assumed uniformly distributed. Figure 8a shows the three-dimensional view of the simplified problem. A rectangular waveguide with one end shorted at z = 0 and the other end stretches to infinity. A linear current I_0 is used to excite the waveguide at z = *d*, as shown in Fig. 8b, the side view of the problem.
In order to use the theory developed in 2.2 for the calculation of input resistance R_{in}, image theory is applied to construct an equivalent problem, as shown in Fig. 8c. A rectangular waveguide with both ends stretching to infinity is excited by two opposite directed current sources with magnitude I_0 at z = ±*d*. Assume the only propagating mode is TE_{10}, taking a generating function from table 1 as

$$\psi_{10} = \sqrt{\frac{2}{abk\eta\beta_{10}\left(\pi/2\right)^2}} \cdot \cos\frac{\pi x}{a}, \tag{10}$$

which has been normalized to

$$\iint \vec{e}_{10} \times \vec{h}_{10}^{*} \cdot \hat{a}_z \, dx\, dy = 1 \tag{11}$$

for the convenience of power calculation. Also from table 1 the tangential field pattern of TE_{10} modes can be obtained as

$$\vec{E}_{t,10} = -jk\eta\sqrt{\frac{2}{abk\eta\beta_{10}}} \cdot \sin\frac{\pi x}{a} e^{\pm j\beta_{10}z} \hat{a}_y \tag{12a}$$

and

$$\vec{H}_{t,10} = \mp\Gamma_{10}\frac{\pi}{a}\sqrt{\frac{2a}{bk\eta\beta_{10}\pi^2}} \cdot \sin\frac{\pi x}{a} e^{\pm j\beta_{10}z} \hat{a}_x \cdot \tag{12b}$$

From (4b), coefficient of the electric field propagating in +z direction which is excited by the current I_0 at z = *d* can be calculated as

$$2b_1 = jI_0\sqrt{\frac{2k\eta b}{a\beta_{10}}} \cdot e^{-j\beta_{10}d}. \tag{13a}$$

In the same manner, the contribution of the image current can also be determined.

$$2b_1' = -jI_0\sqrt{\frac{2k\eta b}{a\beta_{10}}} \cdot e^{j\beta_{10}d} \tag{13b}$$

Since the power flows through waveguide cross section is the combination of the power excited by the current and its image, it can be easily calculate by (5), i.e.,

$$P = \left|I_0\right|^2 \frac{kb\eta}{2a\beta_{10}} \sin^2\beta_{10}d \cdot \tag{14}$$

Therefore by (6) the input resistance can be obtained.

$$R_{in} = \frac{kb\eta}{a\beta_{10}} \sin^2\beta_{10}d \cdot \tag{15}$$

It is worthy mentioning that R_{in} reaches a maximum at $\beta_{10}d = \pi/2$, i.e., the feeding is a quarter guided wavelength from the short end of the waveguide, where the -z traveling wave will be in phase with the +z traveling wave after reflection. As a result, maximum power can be transmitted into the waveguide.

3.2 Transition Model

After calculating the input resistance, transition model can be constructed by the procedures described in 2.4. As shown in Fig. 9a, R_{in} is the input resistance looking in to the shorting via in SIW while the actual input port is located at the MSL connecting to the via pad. It is obvious that the via pad can be viewed as a shunt capacitance C to the ground of the MSL. The inductance L in Fig. 9(b) includes the contribution from the input reactance jx, the shorted via inductance, and the inductance due to the extra via section from via pad to SIW. After replacing R_{in} by a transformer with frequency dependent ratio n, the full transition model is constructed as Fig. 9b.

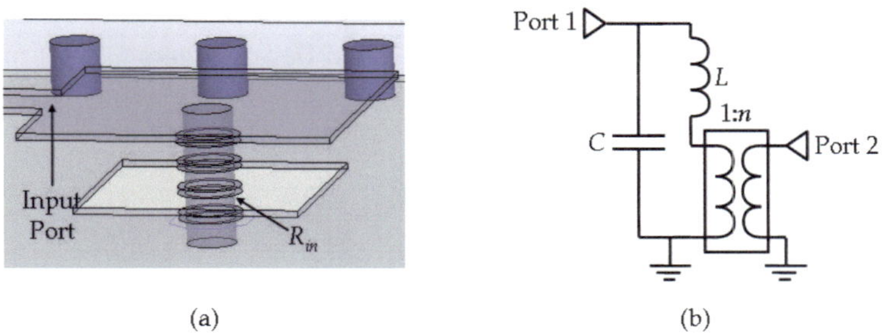

Fig. 9. Model construction according to physical structure and R_{in}. (a) Physical structure. (b) Full transition model.

In order to verify the transition model, a test transition structure was designed and simulated with Ansoft HFSS, a widely used commercial full-wave simulator. All transition structures were designed on a multilayer LTCC substrate with a relative dielectric constant 7.8 and a thickness 50.8μm for each layer. The dielectric is assumed to be lossless to simplify the investigation.

For a center frequency 30GHz, the width of SIW is chosen to be 2738μm. MSL is designed for 50Ω characteristic impedance with three layer height and 180μm width. First R_{in} is calculated separately by (15) and HFSS for different SIW heights, where in HFSS R_{in} is calculated by (6), current flows through the via is obtained by integrating the magnetic field enclosing the via and integrating the Poynting vector on SIW cross section gives power flow. Results are listed in Table 2. About 20% relative error is obtained with similar trend. This may be partly contributed to the negligence of the coupling due to the opening on the anti pad on the upper metallic wall of SIW. It can also be seen that as the height of SIW increases, the assumption of uniform current distribution in the via becomes less applicable.

SIW Layers	Formula (15)	HFSS Simulation	Relative Error
2	6.8Ω	8.1Ω	16%
4	13.6Ω	16.7Ω	18%
6	20.4Ω	25.2Ω	19%

Table 2. Input resistance with different SIW thickness by (15) and HFSS simulation.

Given the input resistance, the passive network can be adjusted for impedance matching. The main contribution of the series inductance comes from the via section above SIW, hence the adjustable range is limited. On the other hand, the value of the shunt capacitance can be easily changed by adjusting the size of the via pad and the antipad. Table 3 shows the element values tuned for the R_{in} of SIW with different layers. Size of the via pad is changed with the edge of square antipad fixed at 460μm. Also note that the closer the input resistance is to the port characteristic impedance, the wider fine tuned transition bandwidth can be obtained.

SIW Layers	R_{in} (Ω)	L (nH)	C (pF)	Viapad Size (μm)	Fractional Bandwidth
2	6.8	0.110	0.225	640	33%
4	13.6	0.118	0.17	480	43%
6	20.4	0.129	0.124	400	46%

Table 3. Element values tuned for the R_{in} of SIW with different layers and viapad sizes.

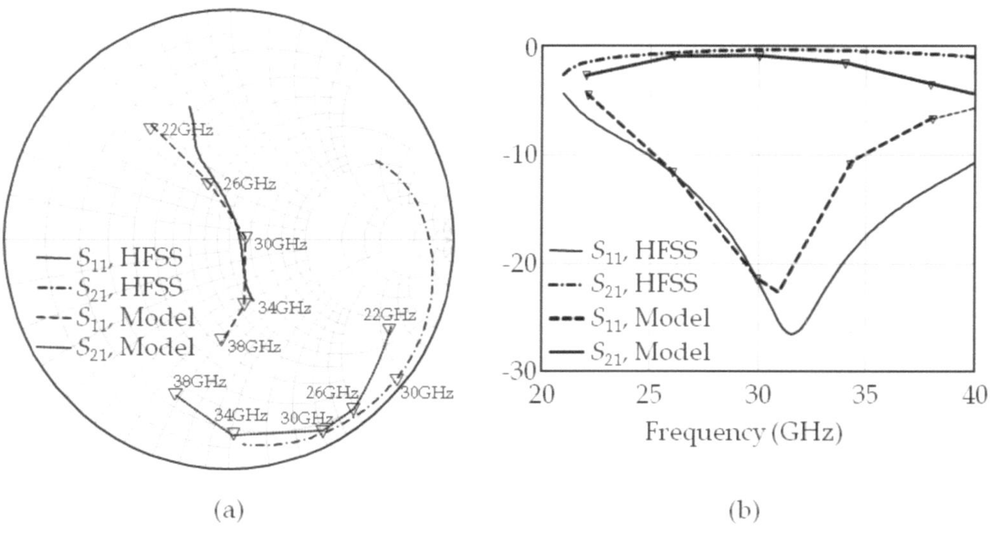

Fig. 10. Responses obtained by the shorting via transition model and HFSS. (a) Smith chart. (b) Rectangular plot.

For the test transition structure with 2-layer SIW described above, figure 10 shows the responses of the full transition model shown in Fig. 9b and the full-wave simulation results by HFSS. As can be seen, highly coherent results were obtained by the model and HFSS except a 10° phase difference in S_{21}. This may be attributed in the phase change during power transition along the via structure, which is not included in the transition model. However the model still shows its usefulness for designing MSL to SIW transitions of this type. Two practical examples were designed and fabricated on LTCC substrates, as given in the following subsections.

3.3 Ka-Band Transition Design

A transition is designed for a Ka-band transceiver module with a 31GHz center frequency, a 28-34GHz transition band for 10dB return loss and 3dB insertion loss in the back-to-back transition structure. The structure is designed on an LTCC substrate with a 7.8 relative dielectric constant and a 0.005 loss tangent at 30GHz. Thickness for each dielectric layer is 50.8μm with a 13μm silver metallization between each layer. The feeding MSL is designed for 50Ω characteristic impedance with 180μm width and 152.4μm height. The SIW is chosen to be 2738μm wide with 101.6μm thickness. Probe feeding is achieved by a shorting via located at 920μm from the shorted end of the SIW. The widths for the square viapad and antipad are 640μm and 460μm, respectively.

Figure 11a shows the simulation result by HFSS for a single transition. 9GHz bandwidth is achieved for 15dB return loss. In-band insertion loss is within 0.6dB. A back-to-back transition structure is fabricated for measurement. Figure 11b shows the comparison between simulation and measurement. As shown in the figure, 9GHz bandwidth is achieved for 10dB return loss. Insertion loss is better than 1.2dB in the entire transition band.

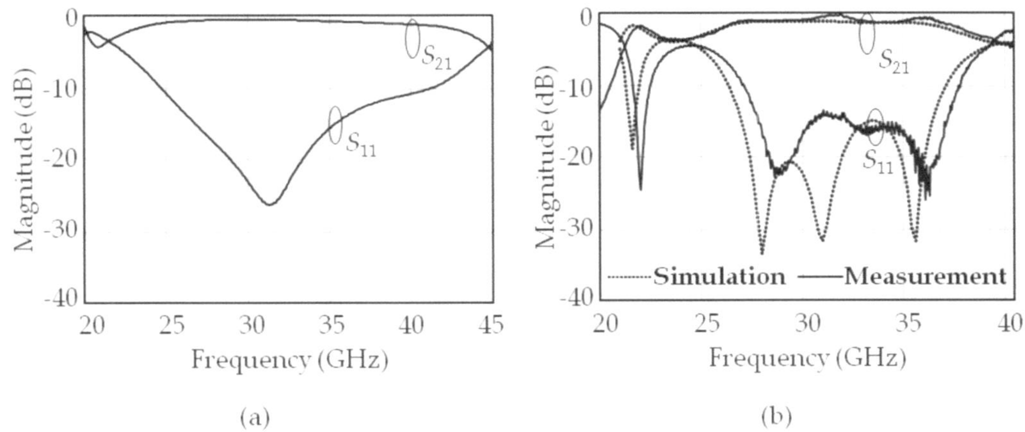

Fig. 11. Results of Ka-band MSL to SIW transition design with probe feeding by shorting via. (a) Simulation result for a single transition. (b) Comparison between simulation and measurement for the back-to-back transition.

3.4 E-Band Transition Design

Another transition is designed for an E-band transceiver module. A 73GHz center frequency is desired with a 71-76GHz transition band for 15dB return loss. The structure is designed on an LTCC substrate with a 7.8 relative dielectric constant and a 0.0078 loss tangent at 60GHz. Thickness for each dielectric layer is 50.8μm with 13μm silver metallization between each layer. The feeding MSL is designed for 50Ω characteristic impedance with 112μm width and 101.6μm height. The SIW is chosen to be 1140μm wide with 101.6μm thickness. Probe feeding is achieved by a shorting via located at 630μm from the shorted end of the SIW. The widths for the square via pad and antipad are 273μm and 373μm, respectively.

Figure 12a shows the simulation result by HFSS for a single transition. A 17% fractional bandwidth is achieved for 15dB return loss. In-band insertion loss is within 0.72dB. A back-to-back transition structure is fabricated for measurement. Figure 12b shows the comparison between simulation and measurement, where the measured result is obtained only below 75GHz owing to equipment limitation. As shown in the figure, 13GHz bandwidth is achieved for 10dB return loss. Insertion loss is better than 2.56dB in the entire transition band.

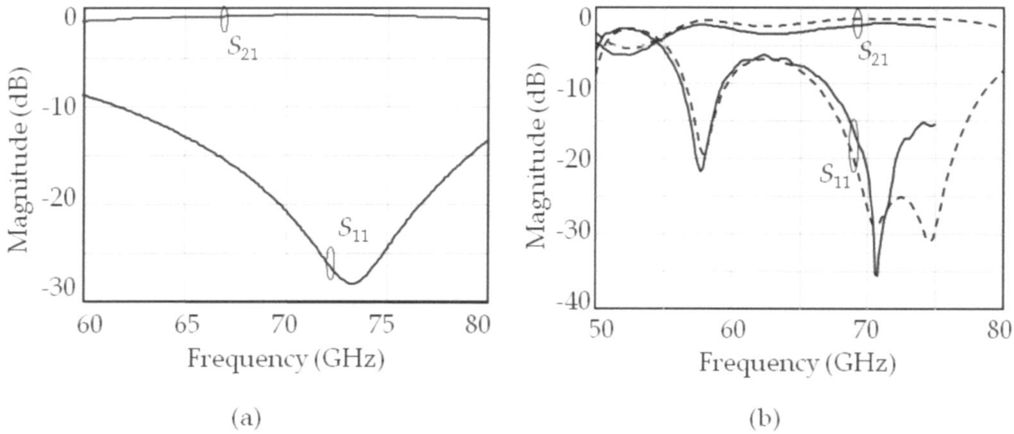

Fig. 12. Results of E-band MSL to SIW transition design with probe feeding by shorting via. (a) Simulation result for a single transition. (b) Comparison between simulation and measurement for back-to-back transition.

4. MSL to SIW Transition with Open Slot

The structure of a transition between MSL and SIW by slot coupling is shown in Fig. 3. Input conductance will be calculated first in this section followed by the construction of transition model. A Ka band and an E band transitions will also be demonstrated as examples.

4.1 Input Conductance

In contract to the probe feeding, the electric field excited by the slot is in the opposite directions for the forward and backward traveling waves. Therefore the slot should be opened as close as possible to the shorted end of the SIW to get maximum energy transfer. The transition problem can then be simplified as Fig. 13a, which shows a z-directed SIW with equivalent width a and substrate thickness b excited by a slot with length $2l$ at $z = d$. The SIW and the slot are center aligned at $x = a/2$. If the width t of the slot is small, the electric field at the slot can be assumed uniform along that direction.

In order to use the theory developed in 2.3 for the calculation of input conductance G_{in}, image theory is applied to construct an equivalent problem, as shown in Fig. 8b. A rectangular waveguide with both ends stretching to infinity is excited by two slots with the same magnetic current J_m at $z = \pm d$.

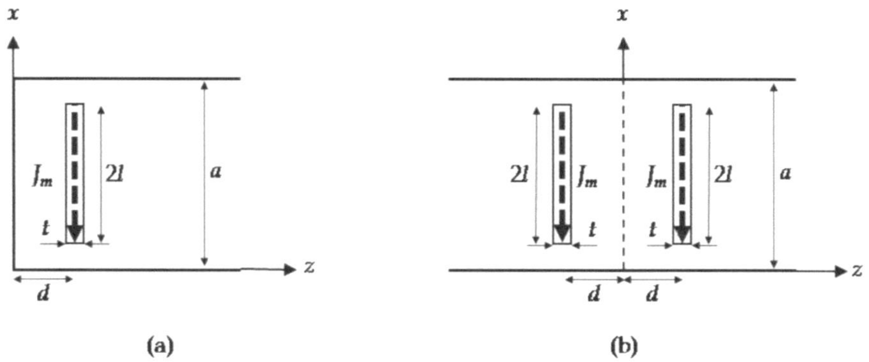

Fig. 13. (a) Top view of simplified slot coupling problem. (b) Top view of equivalent problem.

Assume the slot voltage distributes as on a slot antenna, i.e.,

$$V = V_0 \sin\left[k\left(l - \left|x - \frac{a}{2}\right|\right)\right]. \tag{16}$$

For a thin slot as described above, the electric field can be expressed as

$$\vec{E}_s = \frac{V}{t}\hat{a}_z = \frac{V_0}{t}\sin\left[k\left(l - \left|x - \frac{a}{2}\right|\right)\right]\hat{a}_z. \tag{17}$$

Therefore the magnetic current on the slot can be obtained.

$$\vec{J}_m = -\hat{n} \times \vec{E}_s = -\frac{V_0}{t}\sin\left[k\left(l - \left|x - \frac{a}{2}\right|\right)\right]\hat{a}_x. \tag{18}$$

Substitute (18) into (8) with the SIW normalized field derived in (12), the coefficient of the magnetic field propagating in +z direction which is excited by the current I_0 at z = d can be calculated as

$$2a_n = \sqrt{\frac{2}{abk\eta\beta_{10}}}\Gamma_{10}V_{10}\left(\cos k_{c,10}l - \cos kl\right)\frac{2k}{\beta_{10}^2}e^{-j\beta_{10}d}, \tag{19a}$$

On the other hand, the contribution of the image magnetic current can also be determined as

$$2a_n' = \sqrt{\frac{2}{abk\eta\beta_{10}}}\Gamma_{10}V_{10}\left(\cos k_{c,10}l - \cos kl\right)\frac{2k}{\beta_{10}^2}e^{j\beta_{10}d}. \tag{19a}$$

Power flows through waveguide cross section can then be easily calculate as

$$P = \left|V_0\right|^2 \frac{2k}{ab\eta\beta_{10}^3}\left(\cos k_{c,10}l - \cos kl\right)^2 \cos^2 \beta_{10}d. \tag{20}$$

It should be noted that maximum power occurs at $d = 0$, i.e., the slot is opened at the shorted end of the SIW, which agrees with the discussion in the beginning of this subsection. With the calculated power flow and the defined slot voltage, the input conductance can be calculated by (9) as

$$G_{in} = \frac{4k}{ab\eta\beta_{10}^3}(\cos k_{c,10}l - \cos kl)^2 \cos^2 \beta_{10}d \cdot \tag{21}$$

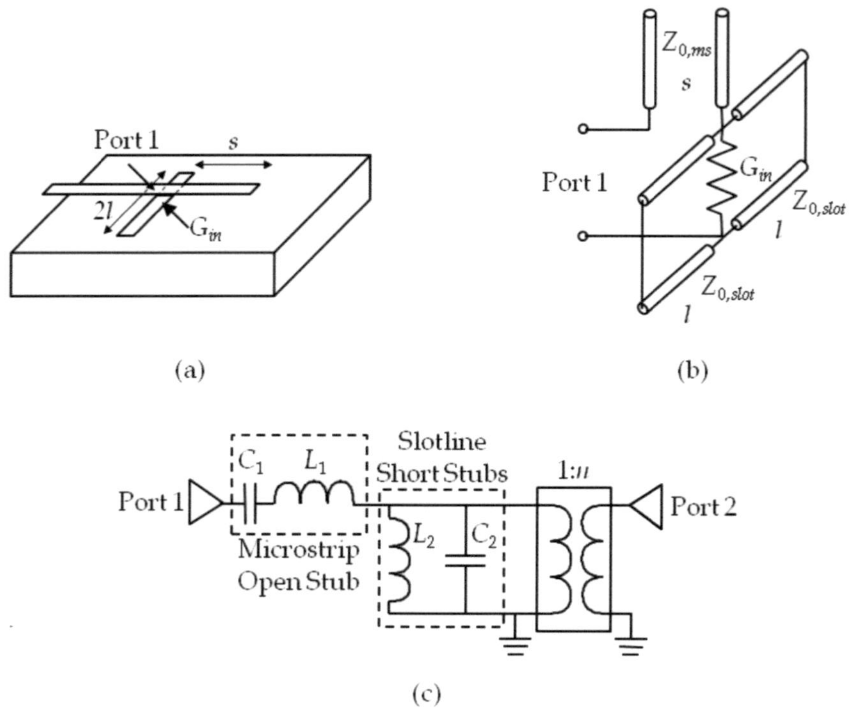

Fig. 14. Model construction for MSL to SIW transition by slot coupling. (a) Simplified transition structure. (b) Equivalent circuit using transmission line models and G_{in}. (c) Full transition model.

4.2 Transition Model

Referring to the simplified transition structure shown in Fig. 14a, G_{in} is the input conductance looking into the center of the slot on the SIW. The actual input port, i.e., Port 1, is located at the feeding MSL right above the slot. An open ended microstrip stub with length l is connected in serial with the feeding MSL. In addition, two short ended slot stubs with length $d/2$ are connected in parallel with G_{in}.

Figure 14b shows the equivalent circuit of the transition with transmission line models and G_{in}. Therefore the length for each stub can be easily determined as a quarter guided wavelength at center frequency. With these lengths, input impedance at Port 1 is equal to G_{in} at center frequency. This equivalent circuit can be further simplified around center frequency. The series microstrip open stub can be replaced by a capacitor and an inductor in serial, while the two shunt slot short stubs can be combined and replaced by a capacitor and an inductor in parallel. After replacing G_{in} by a transformer with frequency dependent ratio n, the full transition model is constructed, as shown in Fig. 14c.

A test transition structure is also designed for the verification of this model. All transition structures were designed on the same multilayer LTCC substrate with a relative dielectric constant 7.8 and a thickness 50.8μm for each layer with lossless assumption. For a center frequency 30GHz, the width of SIW a is chosen to be 2738μm. MSL is designed for 50Ω characteristic impedance with three layer height and 180μm width. Due to the limit of fabrication process, the slot cannot be opened right at the shorted end of SIW, a minimal distance satisfying the design rules from the shorting via is used, i.e., d = 200μm.

Slot Lengths	Formula (21) (mS)	HFSS Simulation (mS)
$2l = 0.7a$	9.2	11.9
$2l = 0.8a$	12.2	14.1
$2l = 0.9a$	14.6	14.9
$2l = a$	15.6	15.9

Table 4. Input conductance by (21) and HFSS simulation for a 6-layer SIW with a slot at d = 200μm and different slot lengths $2l$ in terms of SIW width a.

SIW Layers	Formula (21) (mS)	HFSS Simulation (mS)
2	46.7	31.3
3	31.5	26.3
4	23.9	21.7
5	18.8	17.9
6	15.6	15.9

Table 5. Input conductance by (21) and HFSS simulation for SIWs with different layers with a slot at d = 200μm and slot length $2l = a$.

Instead of (21), G_{in} can also be calculated directly by (9) using HFSS, where a line integration of electric field at the center of the slot gives the slot voltage and integrating the Poynting vector on SIW cross section gives the power flow. Results for different slot length on a 6-layer SIW are shown in Table 4. As slot length increases, the input resistance decreases. It is worth mentioning that as the slot length increases toward resonance, the input resistance can be calculated by (21) more accurately. On the other hand, results for SIWs with different layers and a slot length $2l = a$ are listed in Table 5. As SIW layer increases, the input conductance decreases and the agreement with simulation results becomes better.

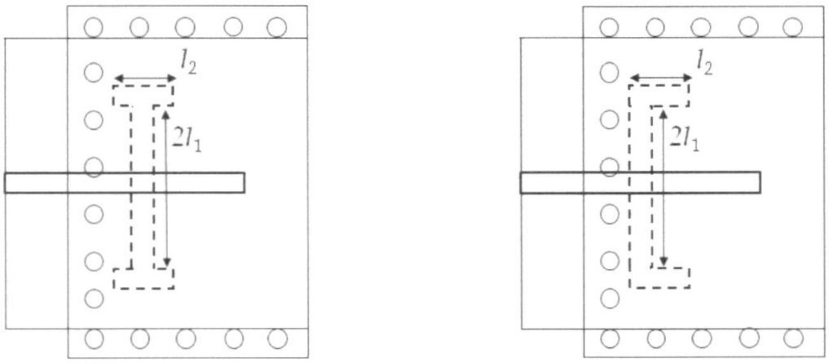

(a) (b)

Fig. 15. Modified coupling slots. (a) H-shape slot. (b) U-shape slot.

As the transition model indicated, bandwidth is maximized for G_{in} = 1/50 Siemens. This can be achieved by adjusting the slot length for SIWs with different layers. However slot length near resonance is often desired because of maximum power transition. Therefore H-shape or

U-shape slots, as shown in Fig. 15, will be used in the following designs. In these cases, the input admittances can also be found as

$$G_{in} = \frac{4k}{ab\eta\beta_{10}^3}\left[\cos k_{c,10} l_1 \cos kl_2 - \frac{k_{c,10}}{k}\sin k_{c,10} l_1 \sin kl_2 - \cos k(l_1 + l_2)\right]^2 \cos^2 \beta_{10} d \cdot \quad (22)$$

Note that the last term in the bracket vanishes when $k(l_1+l_2)=\pi/2$, i.e., total slot length equals to half wavelength. With these kinds of slots, the effective feeding lengths of the slots can be adjusted while total slot lengths are kept near resonance.

Figure 16 shows the responses for a test transition structure with design parameters mentioned above, where 6-layer SIW is used. An H-shape slot with l_1 = 2000μm, l_2 = 600μm is opened on the SIW. Phase difference between transition model and HFSS simulation can be attributed to the phase delay during power transition, which is not included in the transition model. However the model still shows its usefulness for designing MSL to SIW transitions of this type.

As shown in Fig. 17, responses of another two transition structures are obtained. By adjusting the slot shape and length, wideband transition from MSL to SIW with different layers can be obtained. Figure 17a shows the response of transition to a 10-layer SIW by an H-shape slot with l_1 = 2500μm, l_2 = 500μm. On the other hand, Fig. 17b shows the response of transition to a 4-layer SIW by a U-shape slot with l_1 = 1100μm, l_2 = 900μm. 40% fractional bandwidth is achieved in both cases. Two practical examples were designed and fabricated on LTCC substrates, as given in the following subsections.

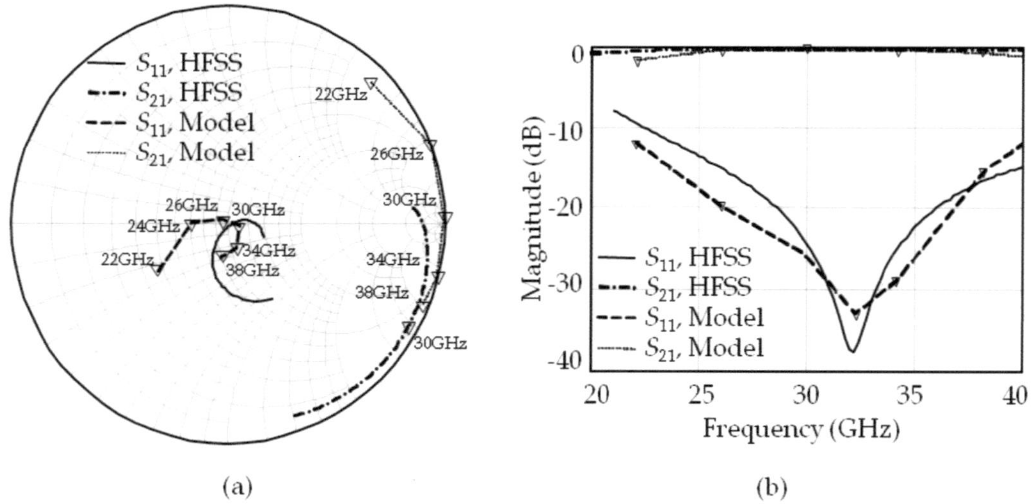

Fig. 16. Responses obtained by slot-coupled transition model and HFSS. (a) Smith chart. (b) Rectangular plot.

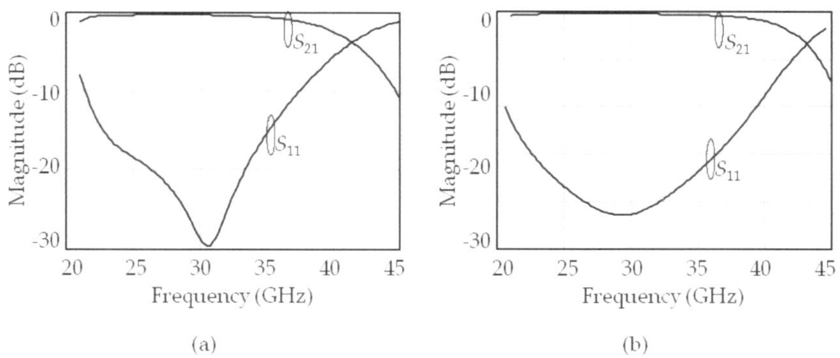

Fig. 17. Responses of slot-coupled transition with a slot at d=200μm for different slot shapes and SIW layers. (a) H-shape slot (l_1 = 2500μm, l_2 = 500μm), 10-layer SIW. (b) U-shape slot (l_1 = 1100μm, l_2 = 900μm), 4-layer SIW.

4.3 Ka-Band Transition Design

A Ka-band MSL to SIW transition by slot coupling is designed, with 31GHz center frequency, 28-34GHz transition band, 10dB return loss, and 3dB insertion loss in the back-to-back transition structure. The structure is designed on an LTCC substrate with relative dielectric constant 7.8 and loss tangent 0.005 at 30GHz. Thickness for each dielectric layer is 50.8μm with 13μm silver metallization between each layer. The feeding MSL is designed for 50Ω characteristic impedance with 180μm width and 152.4μm height. The SIW is chosen to be 2738μm wide with 304.8μm thickness. An H-shape slot with l_1 = 2000μm and l_2 = 600μm is used for the coupling.

Figure 18a shows the simulation results by HFSS for a single transition. 38% fractional bandwidth is achieved for 15dB return loss. In-band insertion loss is within 0.6dB. A back-to-back transition structure is fabricated for measurement. Figure 18b shows the comparison between simulation and measurement. As shown in the figure, more than 15GHz bandwidth is achieved for 10dB return loss. Insertion loss is better than 1.3dB in the entire transition band.

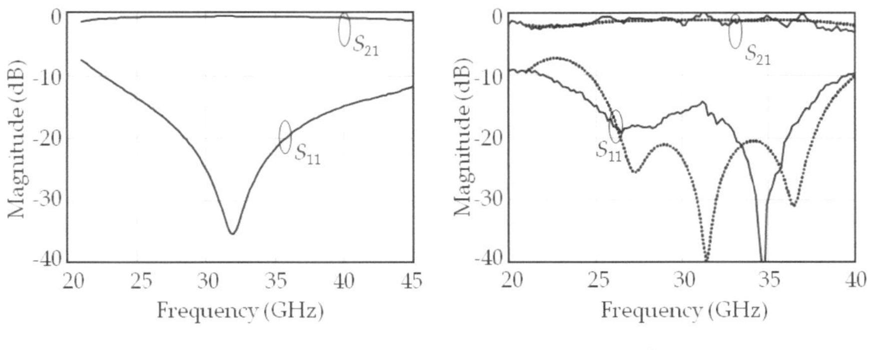

Fig. 18. Results of Ka-band MSL to SIW transition design by slot coupling. (a) Simulation result for a single transition. (b) Comparison between simulation and measurement for back-to-back transition.

4.4 E-Band Transition Design

An E-band MSL to SIW transition by slot coupling is also designed. A 73GHz center frequency is desired with 71-76GHz transition band for 15dB return loss. The structure is also designed on an LTCC substrate with relative dielectric constant 7.8 and loss tangent 0.0078 at 60GHz. Thickness for each dielectric layer is 50.8μm with a 13μm silver metallization between each layer. The feeding MSL is designed for 50Ω characteristic impedance with 112μm width and 101.6μm height. The SIW is chosen to be 1140μm wide with 203.2μm thickness. A U-shape slot with l_1 = 580μm and l_2 = 150μm is used for the coupling.

Figure 19a shows the simulation result by HFSS for a single transition. A 39% fractional bandwidth is achieved for 15dB return loss. In-band insertion loss is within 1.07dB. A back-to-back transition structure is fabricated for measurement. Figure 19b shows the comparison between simulation and measurement, where the measured result is obtained only below 75GHz owing to equipment limitation. As shown in the figure, 21GHz bandwidth is achieved for 15dB return loss. Insertion loss is better than 2.8dB in the entire transition band.

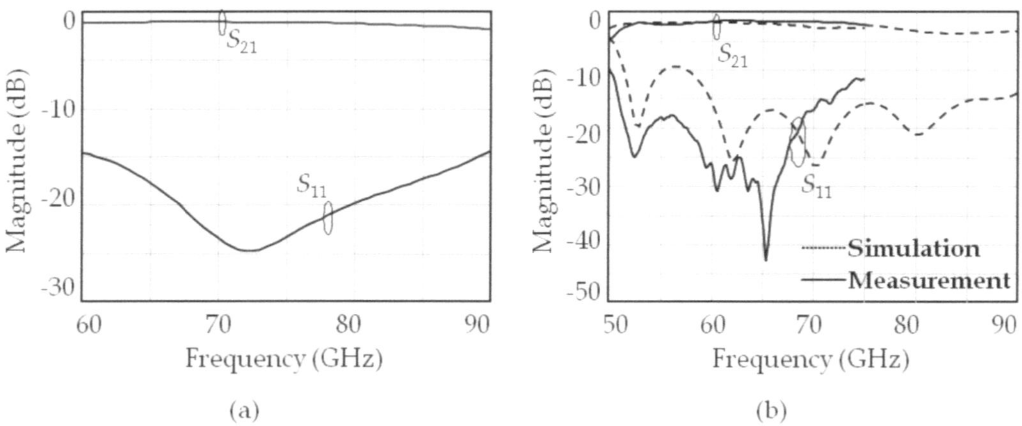

Fig. 19. Results of E-band MSL to SIW transition design by slot coupling. (a) Simulation result for a single transition. (b) Comparison between simulation and measurement for back-to-back transition.

5. Conclusions

This chapter presents systematic procedures for the design and modeling of transition structures between MSL and SIW. Input resistance or conductance of the equivalent waveguide excitation is firstly derived analytically with respect to the structural parameters of the transition. Reactive parts are then added to build the complete equivalent circuit according to the relations between voltage and current at transition discontinuities. The reactance values can be extracted by full-wave electromagnetic simulation only at center frequency. With the derived resistance and the extracted reactance values, the equivalent circuit is sufficient for wideband responses. Local compensation can then be made for maximizing the transition bandwidth. Various transition structures are designed and fabricated on LTCC substrates with center frequencies at Ka-band and E-band.

Measurements are performed on back-to-back transition structures. Good agreement between simulation and measurement results are also obtained.

For transition structures between MSL and SIW by probe feeding with a shorting via, the input resistances decrease with the thickness of SIW. Local compensation is achieved by adjusting the sizes of the via pad and the antipad. Larger transition bandwidth can be obtained for input resistances closer to the characteristic impedance of the feeding MSL. In the Ka-band designs, simulation by HFSS shows that for a single transition to SIWs with 2, 4, and 6 layers, 33%, 43%, and 46% fractional bandwidths for 15 dB return loss can be obtained, respectively. A back-to-back transition with 2-layer SIW is fabricated for measurement. Highly coherent results between simulation and measurement show that more than 30% fractional bandwidth for 10dB return loss was achieved with in band insertion loss better than 1.2dB. In the E-band design, 17% fractional bandwidth for a single transition for 15 dB return loss was obtained in HFSS simulation. Back-to-back measurement also agreed with simulation, which showed a 17% fractional bandwidth for 10dB return loss and 2.56dB in band insertion loss.

For MSL to SIW transition by slot coupling, the input conductance decreases with the thickness of SIW but increases with slot length. For different SIW layers, local compensation can be accomplished by adjusting the length and shape of the slot. Transition bandwidth can be maximized by placing the input resistances as close as to the characteristic impedance of the feeding MSL. In the Ka-band designs, simulations by HFSS show that more than 40% fractional bandwidth for 15 dB return loss can be obtained for SIWs with different layers from 4 to 10. Measurement for a fabricated back-to-back transition with 6-layer SIW agreed with simulation. More than 15GHz bandwidth for 10dB return loss was achieved with in band insertion loss better than 1.3dB. In the E-band design, 39% fractional bandwidth for 15 dB return loss was obtained in a single transition simulated by HFSS. The highly coherent results between simulation and measurement for a back-to-back structure also showed a 21GHz bandwidth for 15dB return loss. In band insertion loss is within 2.8dB. In general, the transition with slot coupling provides a wider transition bandwidth than the probe feeding.

6. References

Collins, R. E. (1991), *Field Theory of Guided Waves*, IEEE Press, ISBN: 0879422378, New York

Das, B. N.; Prasad, K. V. S. V. R.; Rao, K. V. S.; (1986). Excitation of waveguide by stripline- and mcrostrip-line-fed slots, *IEEE Trans. Microw. Theory Tech.*, Vol. 34, No. 3, Mar. 1986, pp. 321-327, ISSN: 0018-9480

Deslandes, D. & Wu, K. (2001a). Integrated microstrip and rectangular waveguide in planar form, *IEEE Microw. Wireless Compon. Lett.*, Vol. 11, No. 2, Feb. 2001, pp. 68-70, ISSN: 1531-1309

Deslandes, D. & Wu, K. (2001b). Integrated transition of coplanar to rectangular waveguide, *IEEE MTT-S Intl. Microw. Symp. Dig.*, Vol. 2, pp. 619-622, ISBN: 0780365380, Phoenix, USA, May 2001

Grabherr, W.; Huder, B. & Menzel, W. (1994). Microstrip to waveguide transition compatible with MM-wave integrated circuits, *IEEE Trans. Microw. Theory Tech.*, Vol. 42, No. 9, Sept. 1994, pp. 1842-1843, ISSN: 0018-9480

Harrington, R. F. (2001). *Time-Harmonic Electromagnetic Fields*, 2nd ed., IEEE Press, ISBN: 047120806X, New York

Huang, Y.; Wu, K. L. & Ehlert, M. (2003). An integrated LTCC laminated waveguide to microstrip line T-junction, *IEEE Microw. Wireless Compon. Lett.*, Vol. 13, No. 8, Aug. 2003, pp. 338-339, ISSN: 1531-1309

Kaneda, N.; Qian, Y. & Itoh, T. (1999). A broad-band microstrip-to-waveguide transition using quasi-Yagi antenna, *IEEE Trans. Microw. Theory Tech.*, Vol. 47, No. 12, Dec. 1999, pp. 2562-2567, ISSN: 0018-9480

Kai, T.; Hirokawa J.; Ando, M.; Nakano H. & Hirachi Y. (2005). Coaxial-line feed for post-wall waveguide in millimeter wave band, *Proc. 2005 IEEE AP-S Intl. Symp.*, pp.631-634, vol. 1A, Washington D.C., USA, July 2005.

Liang, J.-F.; Chang, H.-C. & Zaki, K. A. (1992). Coaxial probe modeling in waveguides and cavities, *IEEE Trans. Microw. Theory Tech.*, Vol. 40, No. 12, Dec. 1989, pp. 2172-2180, ISSN: 0018-9480

Lin, T.-H. & Wu, R.-B. (2001). CPW to waveguide transition with tapered slotline probe, *IEEE Microw. Wireless Compon. Lett.*, Vol. 11, No. 7, July 2001, pp. 314-316, ISSN: 1531-1309

Rollins, J. M. & Jarem, J. M. (1989). The input impedance of a hollow-probe-fed, semi-infinite rectangular waveguide, *IEEE Trans. Microw. Theory Tech.*, Vol. 37, No. 7, July 1989, pp. 1144-1146, ISSN: 0018-9480

Shih, Y. C.; Ton, T. N & Bui, L. Q. (1998). Waveguide-to-microstrip transitions for millimeter-wave applications, *IEEE MTT-S, Intl. Microw. Symp. Dig.*, Vol. 1, pp. 473-475, ISBN: 0780344715, Baltimore, USA, May 1998

Uchimura, H.; Takenoshita, T. & Fujii, M. (1998). Development of a laminated waveguide, *IEEE Trans. Microw. Theory Tech.*, Vol. 46, No 12, Dec. 1998, pp. 2438-2443, ISSN: 0018-9480

Valois, R.; Baillargeat, D.; Verdeyme, S.; Lahti, M. & Jaakola, T. (2005). High performances of shielded LTCC vertical transitions from DC up to 50 GHz, *IEEE Trans. Microw. Theory Tech.*, Vol. 53, No. 6, Jun. 2005, pp. 2026-2032, ISSN: 0018-9480

Yang, T.-H.; Chen, C.-F.; Huang, T.-Y.; Wang, C.-L. & Wu, R.-B. (2005). A 60GHz LTCC transition between microstrip line and substrate integrated waveguide, *Proc. 2005 Asia Pacific Microw. Conf.*, Vol. 1, pp. 4-7, Suzhou, China, Dec. 2005

Yau, C.-K.; Huang, T.-Y.; Shen, T.-M.; Chien, H.-Y. & Wu, R.-B. (2007). Design of 30GHz transition between microstrip line and substrate integrated waveguide, *Proc. 2007 Asia Pacific Microw. Conf.*, Bangkok, Thailand, Dec. 2007

12

Microstrip-Slot Transition and Its Applications in Multilayer Microwave Circuits

Norhudah Seman[1,2] and Marek E. Bialkowski[1]
[1]*The University of Queensland*
Australia
[2]*Universiti Teknologi Malaysia*
Malaysia

1. Introduction

The rapid development of wireless communication systems has put a lot of demand for compact RF fronts. Nowadays, these subsystems mostly operate in microwave frequency bands and include amplifiers, filters and mixers, which are accompanied by lumped capacitors, inductors, resistors and distributed circuits in the form of sections of transmission lines. With regard to the choice of technology, planar technology including microstrip, stripline, slotline or coplanar waveguide is preferred for an RF front end design because of low manufacturing cost. The demand on minimizing the occupied area by RF front end requires utilization of two sides of substrates on which passive and active devices are to be developed. A further demand on compactness calls for use of a multilayer planar approach. In such cases, the challenge is in passing a signal between two sides of a dielectric layer. A suitable transition (also named via) of wire or wireless type is required to accomplish this task. To couple DC signals wire-type transitions are unavoidable. However, when only a RF signal has to be passed, a wireless via can be a more convenient choice. The most common transition for coupling a RF signal between two sides of planar dielectric structure is the microstrip-to-slotline transition. Its convenience is that it achieves this function using wireless means, making the manufacturing process cost effective. It is important to understand the fundamental properties of this transition, as well as it is imperative to have its design guidelines. Also it is important to know its applications in basic microwave circuits and devices. This chapter provides an overview of design of a microstrip-slot transition, explains its operation and gives examples in which such transition can play a useful role.

2. Microstrip-Slot Transition

A microstrip-slot transition is a structure that uses a microstrip line on one side of a planar dielectric substrate and a slotline in the ground so that a signal is passed between the two sides. In order to perform this function with minimal power losses, the microstrip line and the slotline have to be orthogonal to each other and use suitable reactive terminations at their ends. For the microstrip line, the termination has to represent a short circuit, while for the slotline it has to be an open circuit at the two transmission lines junction. In practice, the two types of terminations can be accomplished by virtual open and virtual short circuits (open or short-circuited stubs). Different shapes can be used to realize them. They can be circular (Seman et al., 2007; Schüppert, 1998), radial (Zinieris et al., 1998), or in the form of rectangular/uniform line (Schüppert, 1988) or multi-arm lines (Soltysiak & Chramiec, 1994). One of the first microstrip-slot transitions, which was reported by Robinson and Allen in (Robinson & Allen, 1969) is shown in Fig. 1.

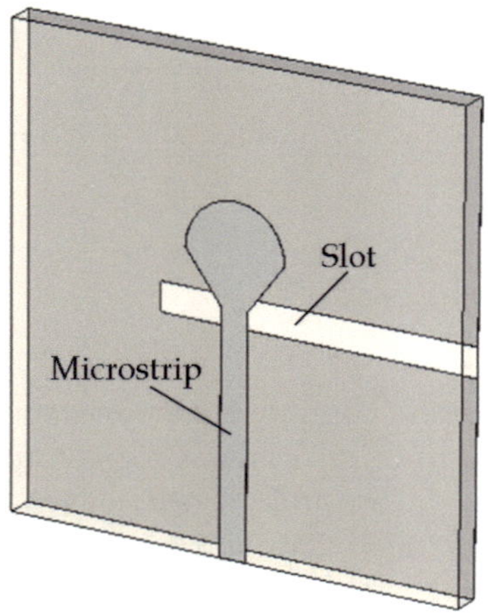

Fig. 1. Microstrip-slot transition with a radial stub and a uniform slotline (Robinson & Allen, 1969; Gupta et al., 1979).

In this configuration, the microstrip line is terminated in a radial stub, which is about one quarter wavelength at the design frequency. The slotline is terminated in a straight stub which is extended about one quarter of a wavelength beyond the microstrip line. The use of a quarter-wave length microstrip stub leads to a virtual short-circuit at the junction. In turn, a quarter-wave length slot stub makes a virtual open-circuit at the junction. In the proposed design, microstrip line characteristic impedance is assumed 50 ohm while the slotline impedance is 72 ohm. The design offers the VSWR less than 1.1 across the operational frequency band of 8 to 10 GHz which is about 20% with respect to the middle frequency of 9 GHz.

Zinieris et al. in (Zinieris et al., 1998) employed a similar approach to obtain a double slot-line-microstrip transition with a 90° crossover. Using radial stubs at the crossover between the transmission lines, the measured insertion loss of less than 1.3 dB for over the 3-15 GHz range was achieved.

Schüppert in (Schüppert, 1988) investigated four different configurations of microstrip-slot transition. These are shown in Fig. 2 and Fig. 3. The configurations in Fig. 2(a) and Fig. 3(a) use actual microstrip short-circuited stubs while the slot stub of about quarter-wave length represents a virtual open-circuit at the junction. The actual short-circuited stub is realized by using a 0.57-mm-wide copper sheet fed through the substrate and soldered at its ends to the microstrip line and ground.

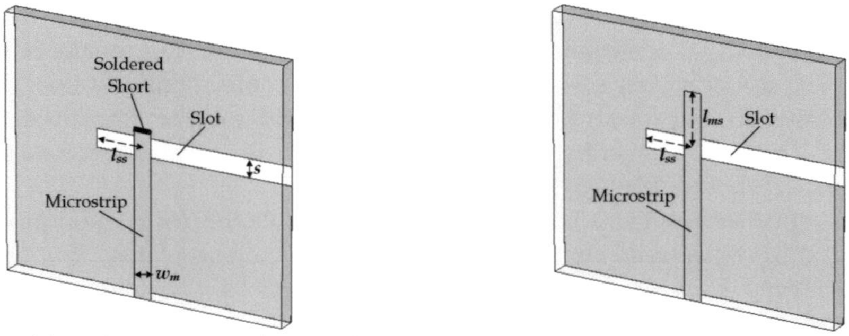

(a) soldered microstrip short and uniform λ/4 slotline

(b) virtual short with uniform λ/4 open microstrip and uniform λ/4 slotline

Fig. 2. Microstrip-slot transitions terminated by uniform line stub as investigated in (Schüppert, 1988).

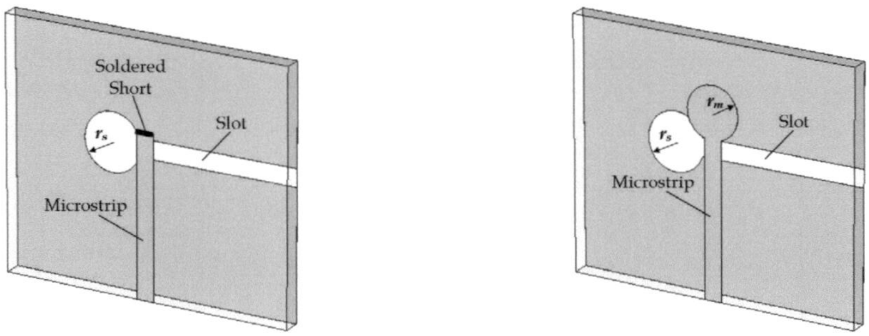

(a) soldered microstrip short and slotline terminated with virtual open circular stub

(b) microstrip and slotline terminated with short and open circular stub

Fig. 3. Microstrip-slot transitions terminated by either a soldered microstrip short or non-uniform line (circular) stub as investigated in (Schüppert, 1988).

The configurations shown in Fig. 2(b) and Fig. 3(b) replace the actual short-circuit by its virtual equivalent. The operation of these transitions is studied within the frequency range of 1 to 16 GHz. The characteristic impedances for microstrip and slotline are assumed 50 ohm. For the structures shown in Fig. 2(a) and (b), the normalized impedance factor for the slot and microstrip stub is introduced using the following expression (Schüppert, 1988):

$$v = \frac{Z_{S_S}}{50\Omega} \tag{1}$$

$$w = \frac{50\Omega}{Z_{m_S}} \tag{2}$$

where Z_{S_S} and Z_{m_S} are the characteristic impedance of slotline and microstrip uniform line stub, respectively.

It is shown that the frequency response with an improved flatness in the passband is obtained when the normalized impedance factor v is equal to w (v=w). Furthermore, the bandwidth performance is improved with increasing v=w. A wide bandwidth, referenced to 1.0 dB insertion loss, is achieved for $2 \leq w/v \leq 2.618$. The optimum result is obtained at w/v=2.618. The use of a virtual short stub accomplished with a uniform line, as shown in Fig. 2(b), offers an improved flatness in the passband and/or improved bandwidth performance. This is explained by better compensation of the junction reactance compared with the soldered short of Fig. 2(a) (Schüppert, 1988).

The positive effect of reactance compensation is also observed for non-uniform (circular) stubs of Fig. 3(b). When compared with the soldered transition of Fig. 3(a), an improved flatness is achieved. However, this is at the expense of reduced (about 10%) bandwidth. The transition with circular stubs offers a wider bandwidth compared to the one with the stub using a uniform line. The operating frequency band can be controlled through the choice of slot and microstrip stub radii. By using larger radii of circular stubs the operational frequency band is shifted downwards (Schüppert, 1988).

Soltysiak and Chramiec (Soltysiak & Chramiec, 1994) proposed multi-arm stubs to improve performance of the microstrip-slot transition. Their configuration is shown in Fig. 4.

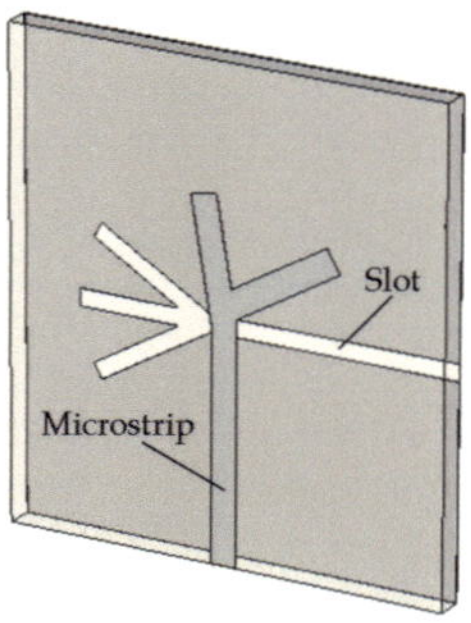

Fig. 4. Transition with multi-arm stubs proposed by P. Soltysiak and J. Chramiec (Soltysiak & Chramiec, 1994) and Schiek and Kohler (Schiek & Kohler, 1976).

They pointed out that for an improved operation, stubs of high characteristic impedance have to be used for the slotline and low characteristic impedance stubs have to be employed for the microstrip line. Also they suggested reducing the coupling between the microstrip and slotline stubs. To achieve this goal, the angle between the stubs appearing on two sides of substrate should be as large as possible. Any overlap of the microstrip and slotline parts should be avoided. Because of this requirement, the angle between the stub axes of symmetry and the input transmission lines may be different from a right angle. In the design, the input impedance of slotline composite stubs is approximately equal to the sum of single stub input impedances. The design was demonstrated experimentally for a decade-wide L-X band.

The design configuration of Fig. 4 with some modifications was included in a six-port junction design in (Schiek & Kohler, 1976). The broadband transition with an insertion loss less than 0.2 dB in the frequency band of 2 to 9 GHz was demonstrated.

3. Wireless Via for Multilayer Microstrip Structures

The microstrip-slot transition whose operation was described in the previous section can be used for an efficient power transfer between two microstrip lines located on opposite sides of a two-dielectric layer structure that is supported by a common ground plane. Such a broadband wireless via or slice interconnect using an earlier work of Schüppert (Schüppert, 1988) was proposed in (Ho et al., 1993). This *vertical* interconnect is attractive because it does not use any wires to couple the signal between the two microstrip lines. Therefore it is easy to manufacture. Its importance is that it is capable to achieve a low insertion loss coupling over an ultra wide frequency band (UWB). Therefore, it can be of significance to developing compact UWB microwave multilayer front-ends. In (Ho et al., 1993), the operation of this interconnect was demonstrated using experimental means and was limited to one example. Here, its comprehensive theoretical and experimental investigation is presented. The following sections show the design of such an UWB via aimed for operation in the 3.1 to 10.6 GHz band.

3.1 Wireless Via Design

In the undertaken design, the initial concept of (Ho et al., 1993) with some modifications is followed. The first design step includes making a rectangular slot in the ground plane which supports the two microstrip lines. In order to obtain signal transmission with minimal power losses, the two microstrip lines and the slotline are suitably terminated. Here, circular microstrip and slot stubs are chosen as terminations.

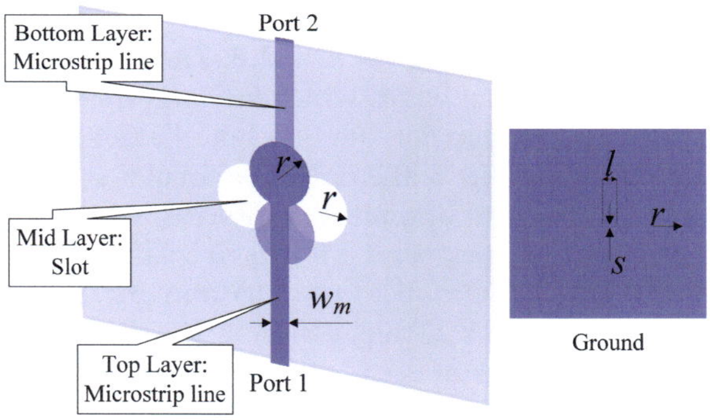

Fig. 5. Configuration of via A including details of common ground plane, with input/output port microstrip lines run in opposite directions.

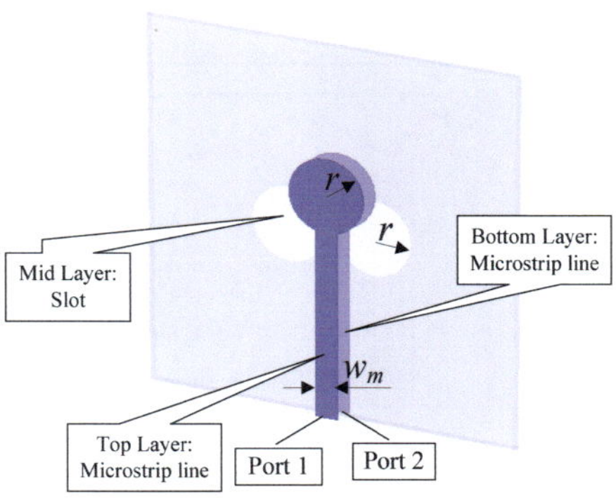

Fig. 6. Configuration of via B with input/output microstrip lines run in the same direction.

The role of virtual open circular stubs at the ends of slotline is to enforce the maximum electromagnetic power flow through the coupling slot (Ho et al., 1993). Following the recommendations in (Schüppert, 1988) and (Ho et al., 1993), the slot width is chosen such that the 50 ohm impedance is observed from the microstrip side. A better performance is expected for smaller values of impedance. However, they face manufacturing limitations because smaller characteristic impedances of slotline require a very narrow slot. Here, investigations include two types of this interconnect, which are named here as via A and B. They are shown in Fig. 5 and Fig. 6.

The difference between these two configurations is the direction of the output microstrip transmission lines. Via A has its output microstrip lines in the opposite directions while via B uses them in the same direction. Assuming that the microstrip lines have a characteristic impedance of 50 ohm, the aim is to choose suitable dimensions of a virtual microstrip short circuit and a virtual slotline open circuit so that a broadband performance in terms of low insertion losses and high return losses is obtained. The design is assisted with the full EM-wave simulation package, CST Microwave Studio (CST MWS). The design assumes a double layer of Rogers RO4003 substrate with dielectric constant of 3.38 and thickness of 0.508 mm.

3.2 Wireless Via Results

Here, the slotline characteristic impedance of 120 ohm is chosen. This gives a 0.2 mm width of the slot. The slot length is varied in order to obtain a low insertion loss and a high return loss for equivalent two-ports representing via A and B. The radius of circular stub is about twice the microstrip transmission line width which is 2 mm (Bialkowski & Abbosh, 2007). The microstrip line width is 1.18 mm for the 50 ohm characteristic impedance. For the configuration shown in Fig. 5 and Fig. 6, the radius of circular stub, *r*, the slotline length, *l* and the width, *s* is varied during the optimization process, which is performed with CST Microwave Studio across the frequency band 3.1-10.6 GHz. The optimized dimensions (w_m, *s*, *l* and *r*) of via A and B are 1.18 mm, 0.2 mm, 1.05 mm and 2.2 mm, respectively. The overall size of the structure is 28 mm x 20 mm, assuming the distance between ports 1 and 2 of 20 mm. The S-parameter responses of via A and B are compared against the reference 20 mm-long 50 ohm microstrip line in Rogers RO4003 substrate.

The CST simulated results shown in Fig. 7 reveal that the designed via A offers insertion losses less than 1 dB between 2.8 and 14 GHz while via B exhibits the 2.7 to 13.9 GHz frequency bandwidth for the same insertion loss specification. Note that Fig. 7 also shows the results for an equivalent microstrip line. This is show insight into simulations accuracy.

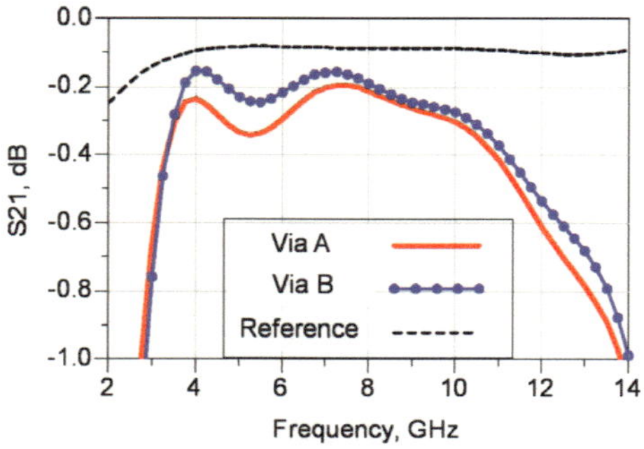

Fig. 7. Comparison of S_{21} responses of via A and B against the 20 mm-long 50 ohm reference microstrip line.

The return losses (RL) of via A and B are shown in Fig. 8. The obtained results indicate that via A has a better RL performance than via B.

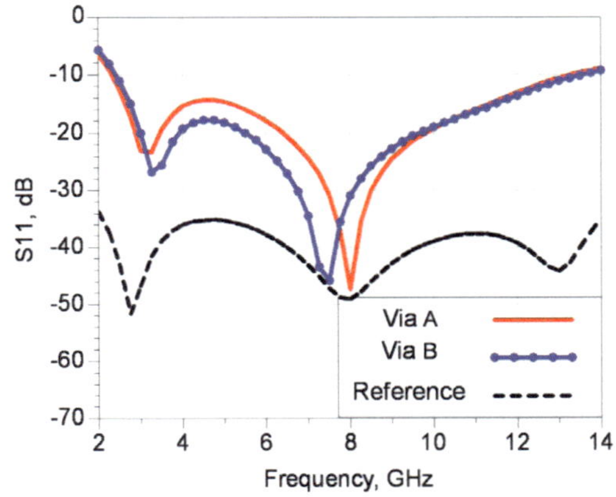

Fig. 8. Comparison of parameter S_{11} of via A and B against the 20 mm-long 50 ohm reference microstrip line.

Fig. 9 shows the results for the phase of transmission coefficient, S_{21} for both vias. The results indicate that the difference in direction of output ports in via A and B is responsible for the 180° phase difference between the two sets of results for transmission coefficient (S_{21}).

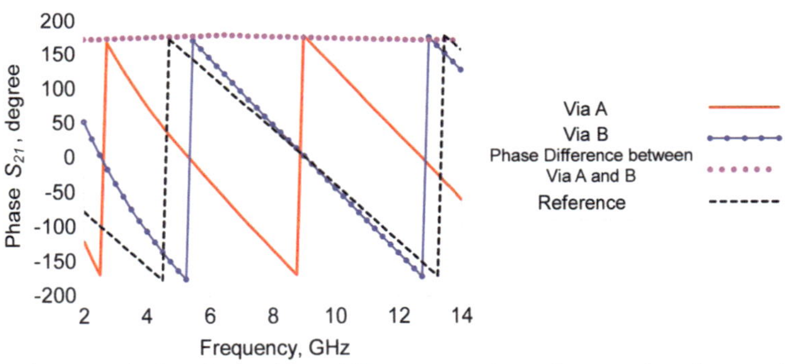

Fig. 9. Phase characteristic of via A and B compared against each other and the reference 20mm-long 50 ohm microstrip line.

The last step in the design process includes an experimental verification. To this purpose, via A with radius, r 2.2 mm is developed and tested in laboratory. The photograph of the fabricated via A is shown in Fig. 10. As seen in Fig. 10, SMA connectors are attached to the two ports for the purpose of experimental measurement. The two dielectric layers are affixed by plastic screws to minimize air gaps.

Fig. 10. Photograph of via A with a 2.2 mm radius stub and SMAs attached to the input/output ports.

Fig. 11 and Fig. 12 present a comparison between the simulated and measured results for the transmission coefficient and return loss for the designed via. The experimental S_{21} curve is slightly shifted to the lower frequency band compared to the simulated result. The measured transmission coefficient is greater than -1 dB for 2.5 to 11 GHz frequency range. The return losses at Port 1 and 2 are greater than 12 dB over the frequency band from 2.5 to 12 GHz.

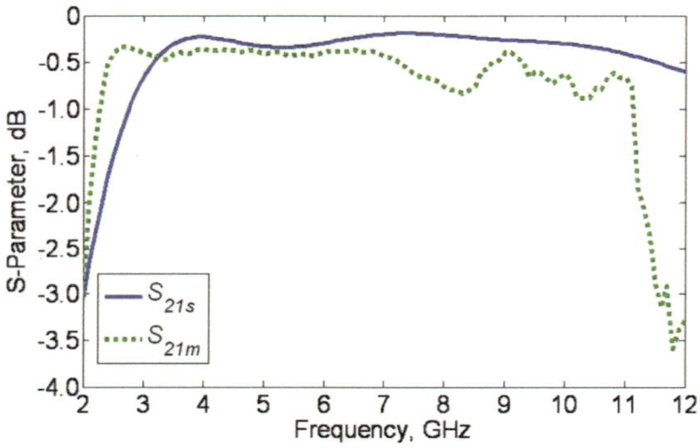

Fig. 11. Measured, *m* and simulated, *s* transmission coefficients of via A with radius 2.2 mm.

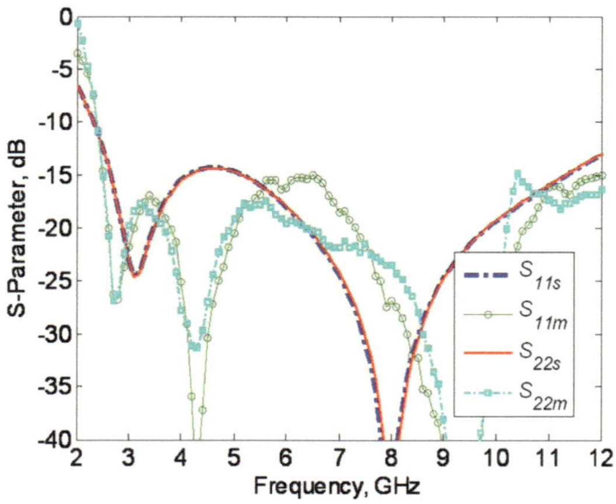

Fig. 12. Measured, *m* and simulated, *s* results for return loss at Port 1 and 2 of via A with radius, *r* of 2.2 mm.

A relatively good agreement between the simulated and measured results is achieved. The observed discrepancies can be due to the use of coaxial-to-microstrip transitions in the experiment. Nevertheless, the obtained agreement provides high confidence in using CST Microwave Studio as the design tool for the investigated via.

The next step includes the investigations into the effect of stub radii on via's performance. Fig. 13 and Fig. 14 show the simulated results for the S-parameters for via A when the circular stub radius, *r* is varied from 1.5 mm to 2.5 mm. It is observed that the smallest radius of 1.5 mm provides a 16 GHz 10-dB return loss bandwidth from 4 to 20 GHz. In turn, the radius of 2.5 mm offers a narrower bandwidth between 2 to 12 GHz. The obtained responses show that by decreasing the circular stub radius, the operating band is shifted towards higher frequencies.

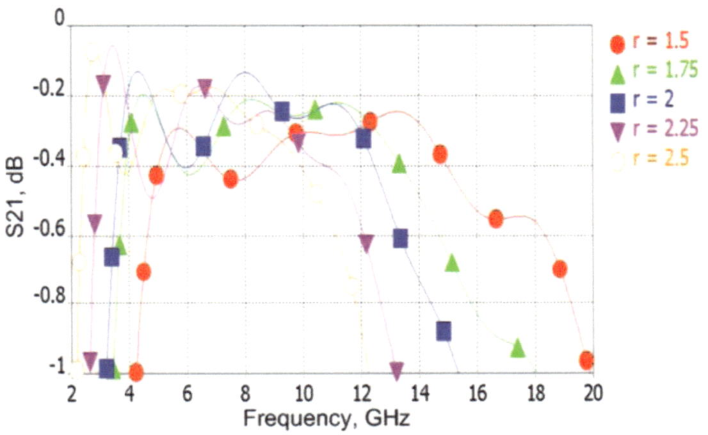

Fig. 13. Insertion loss performance of via A for the varying value of radius, *r* of circular stubs representing open or short circuits.

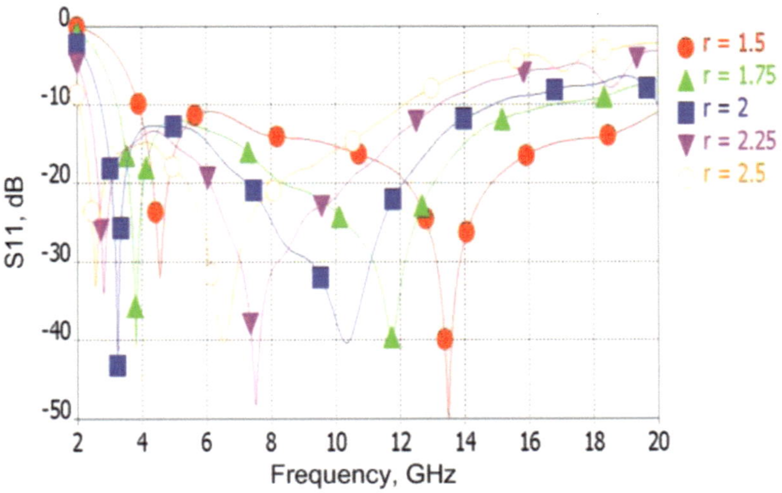

Fig. 14. Return loss performance of via A for the varying value of radius, *r* of circular stubs representing open or short circuits.

In the next step, the five vias with different stub radii are fabricated and tested experimentally. Fig. 15 shows the magnitude of measured S_{21} for the five developed vias for the frequency band from 2 to 16 GHz. The 1-dB insertion loss bandwidths are: 11.85 GHz (3.85-15.2 GHz) for radius 1.5 mm, 11.35 GHz (2.65-14 GHz) for 1.75 mm, 9 GHz (2.3-11.3 GHz) for 2 mm, 8.7 GHz (2.25-10.95 GHz) for 2.25 mm, and 8.65 (2.1-10.75 GHz) for the 2.5 mm circular stub. The return loss measured at the input port is shown in Fig. 16. The measured results confirm a UWB performance of all five manufactured vias although across slightly different frequency ranges.

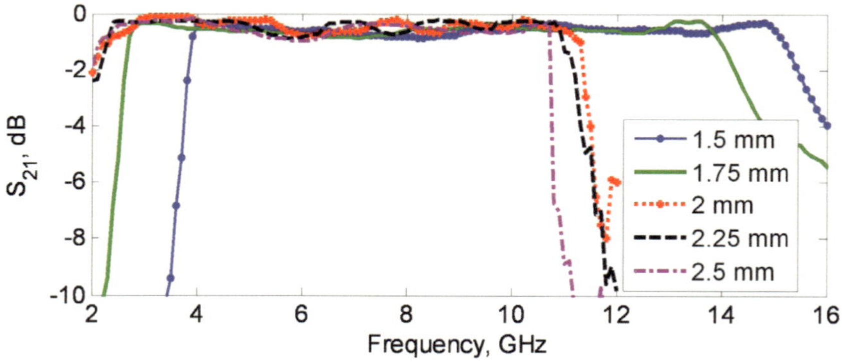

Fig. 15. Measured S_{21} performance of via A for the varying value of radius, r of circular stubs.

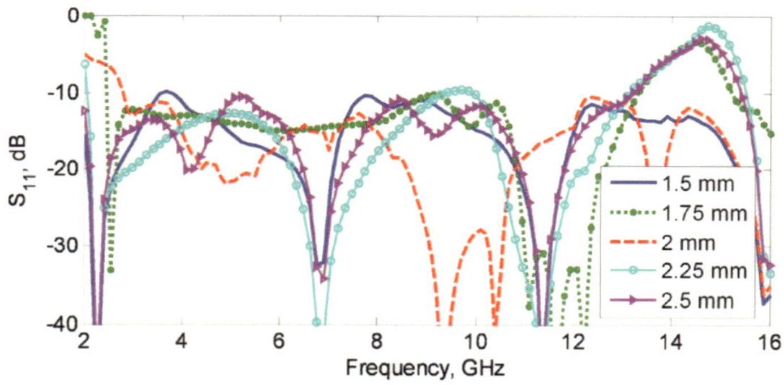

Fig. 16. Measured return loss performance of via A for the varying value of radius, r of circular stubs.

There is a slight discrepancy with the simulated results of Fig. 13 and 14. This can be due to fabrication errors, as well as due to the use of coaxial connectors and transitions which were used in the experiment but were not taken into account during CST MWS simulations.

4. Phase Shift due to Wireless Via

The presented simulation and experimental results have shown that the investigated wireless via can offer a UWB low-loss coupling between two microstrip lines located on two sides of common ground plane. This property has been demonstrated by investigating the behaviour of the wireless via magnitude of transmission and reflection coefficients as a function of frequency.

In order to complete the investigation, one has to have a close look at the phase characteristics. In this case, it is useful to compare the phase of transmission coefficient of this vertical interconnect against the one of a reference microstrip line.

Here a 50 ohm microstrip line of certain length is chosen as a reference line. The differential phase shift defined by the following equation, as adapted from (Meschanov et al., 1994), can be used to this purpose:

$$\Delta\boldsymbol{\theta}(phase) = \left\{\pm S_{21}(phase) \mp S_{21ref}(phase)\right\} \tag{3}$$

where S_{21} and S_{21ref} are the scattering matrix elements of two-ports, the transition and the reference microstrip transmission line, respectively.

Because in the vertical interconnect, the two microstrip lines are electromagnetically coupled, one can expect that the slot transition introduces a series reactance. This reactance is responsible for introducing a phase shift compared to the continuous reference microstrip line. This principle is used in so-called loaded line phase shifters (Pozar, 2005). An initial analysis of results in Fig. 9 had shown that the obtained differential phase shift was approximately constant across a wide frequency band. This property can be useful to design a wideband phase shifter, as illustrated in the following examples.

In the undertaken designs, two configurations of via A and B are utilized. An attempt is made to obtain a differential phase shift of 45°, 67.5°, 78.75° and -90° by varying the dimension of slot length, l, slot width, s, stub radius, r, and by adjusting the reference microstrip line length, L_{ref}.

Table 1. summarizes the parameters of six phase shifters, named here as T-T5, whose dimensions were worked out through manual iterations involving CST Microwave Studio.

Phase Shift ($\Delta\theta$)	Transition	Orientation	Dimension (mm)			L_{ref} (mm)
			l	s	r	
45°	T1	A	1.05	4.0	2.5	32.40
	T3				3.2	33.13
67.5°	T2	B	1.05	0.2	2.5	24.54
	T4				3.2	25.45
78.75°	T2	B	1.05	0.2	2.5	25.16
	T4				3.2	26.21
-90°	T	A	1.05	0.2	2.5	36.18
	T5				3.2	38.09

Table 1. Parameters of six phase shifters.

Table 1 shows that two orientations of via, A and B, are employed to obtain the desired phase shift. Via A is suitable to get phase shifts of 45° and -90° whilst via B is best to obtain the phase shift of 67.5° and 78.75°. It also reveals that the needed length, L_{ref} of the reference microstrip line is proportional to the required phase shift value ($\Delta\theta$). A higher value of phase shift calls for a longer reference transmission line. In the case of 45° phase, the wider slot width of 4 mm is needed compared to the rest of phase shifters which use width of 0.2 mm. The optimum slot length is the same for all of the presented phase shifters and is equal

to 1.05 mm. This length was selected to obtain the best return loss at the frequency of about 6 GHz. Table 1 also shows that with respect to the choice of microstrip and slot terminations, two circular stubs sizes were used.
The plots of magnitudes of S_{11} and S_{21} parameters versus frequency for the five wireless vias with dimensions shown in Table 1 are presented in Fig. 17.
As observed in Fig. 17, all five wireless vias feature insertion losses greater than 2 dB across the frequency band from 2 to 12 GHz. The corresponding return losses are higher than 10 dB across 2-7 GHz band and they decrease below 10 dB at higher frequencies.

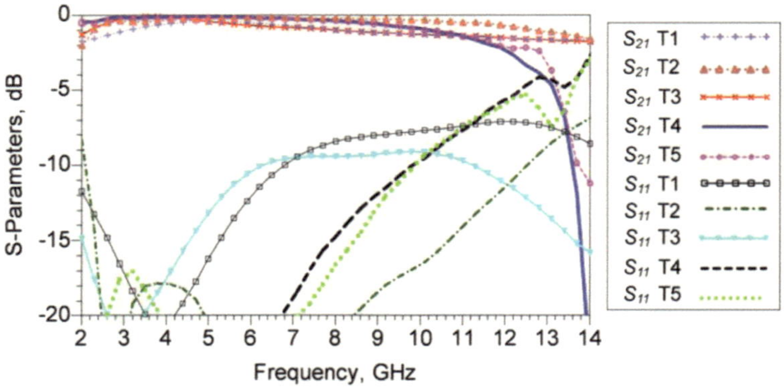

Fig. 17. Magnitudes of S-parameters of five wireless vias operating as phase shifters.

The frequency ranges over which the phase shifters of 45°, 67.5°, 78.75° and -90° with the 2.5 mm stub first radius perform well in terms of 10 dB return loss are: 6-7 GHz, 3.95-12 GHz, 3.4-12 GHz and 3.1-12.5 GHz, respectively. The 45°phase shifter with the stub radius of 3.2 mm provides a greater 10-dB return loss bandwidth of 2.8 GHz (4.2 to 7 GHz) than its counterpart with the 2.5 mm radius stub. In turn, the 67.5°and 78.75° phase shifters with the 3.2 mm stub radius offer the 10-dB RL operation in the 2.7-10.1 GHz and 2.35-10.1 GHz band respectively, which are comparable to those of the 2.5 mm counterpart. The worst results with respect to 10-dB return loss from 3.85 to 6.3 GHz are obtained for the -90° phase shifter that uses the 3.2 mm radius stub.
Fig. 18 and Fig. 19 show the simulated phase shift responses for two cases of stub radii of 2.5 mm and 3.2 mm. As observed in the graphs plotted in Fig. 18 and 19, the use of the smaller stub radius shifts the operating frequency band upwards.
By taking into account both the desired phase and the 10-dB return loss bandwidth, the best performer seems to be the 45° phase shifter. The results also show that the investigated wireless via can be used to achieve phase shifts in the range of 45°-78.5° over an ultra wide frequency band. Obtaining smaller differential phase shift may require a further increase in slot width, which is not trivial with respect to the chosen configuration of via.

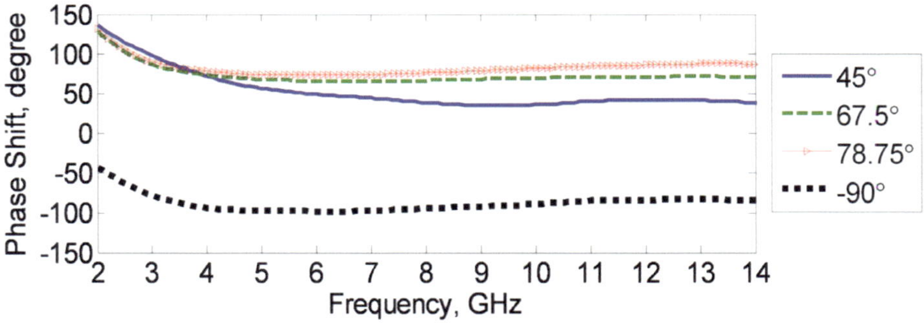

Fig. 18. Phase shift versus frequency for the phase shifters with stub radius of 2.5 mm.

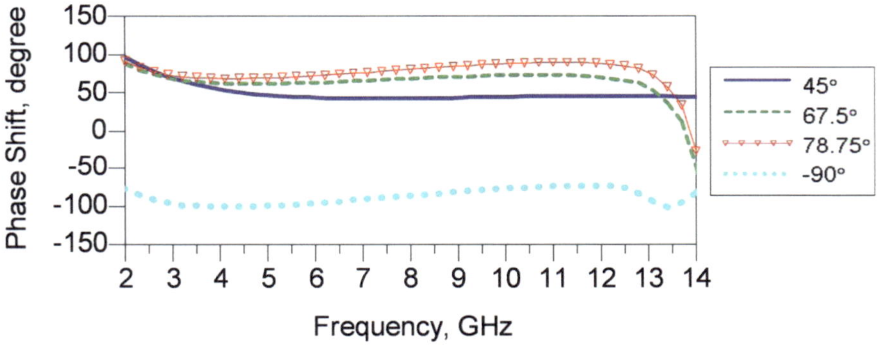

Fig. 19. Phase shift versus frequency for the phase shifters with stub radius of 3.2 mm.

This shortfall can be overcome using a different configuration of wireless via. Such an idea was explored in (Abbosh, 2007). The configuration of the microstrip-slot transition used in (Abbosh, 2007) is shown in Fig. 20.

Similarly as for the configuration of Fig. 5, this alternative vertical interconnect uses a microstrip-slot structure with two microstrip lines located on two sides of two-layer dielectric substrate coupled via a slot in the common ground to obtain a phase shifter. It was shown in (Abbosh, 2007) that the resulting phase shift is related to the coupling coefficient with the following rule: a larger coupling offers a smaller phase shift. This rule seems to be also applicable for the phase shifters based on via A and B where it was observed that a smaller phase shift requires a larger slot opening in the ground. For configuration of Fig. 20 the coupling factor and thus the phase shift can be adjusted through the choice of dimensions of elliptically shaped microstrip lines and the slot. Using this tuning mechanism, a phase shift in the range of 30° and 70° accompanied by good return losses over an ultra wide frequency band was demonstrated (Abbosh, 2007). Similarly, as for the via investigated in this article, the best results in terms of return loss and desired phase shift was offered by the 45° phase shifter.

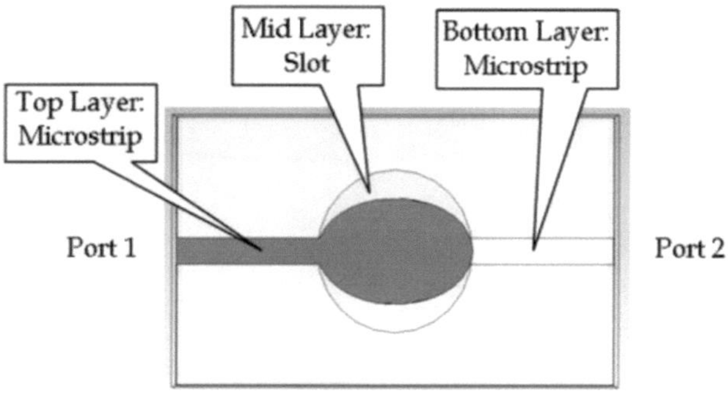

Fig. 20. Configuration of multilayer phase shifter investigated in (Abbosh, 2007).

5. Power Dividers

Besides providing an ultra wideband coupling of a microwave signal between two microstrip lines located on opposite sides of a double-layer dielectric with a common ground plane, the microstrip-slot transition can also be used to design ultra wideband in- and out-of-phase power dividers. The design of such devices is presented in this section. The resulting devices differ from commonly known uniplanar power dividers such as a lossless microstrip T-junction or Wilkinson divider (Pozar, 2005). This is because their two microstrip output ports appear on two sides of the common ground. The motivation for designing these dividers is that they can be integrated with other multilayer structures such as broadside coupled microstrip couplers (Abbosh & Bialkowski, 2007; Seman et al., 2008).

5.1 Out of Phase Power Divider

The configuration of an out-of-phase microstrip-slot power divider is shown in Fig. 21. The divider has an input port and one of the output port at the top layer and another output port at the bottom layer. This divider uses the microstrip to slot transition (Schüppert, 1988) with microstrip Port 1 to vertical slotline and a microstrip T-junction with a vertical slot as via to the microstrip output ports. The narrow slot in the common middle ground plane guides the signal from the input to the two output microstrip lines. The arrangement shown in Fig. 21 offers an equal signal division (3 dB) between two output ports with a 180° phase difference. The phase difference of 180° is due to the chosen orientation of the output ports which enable E-field lines propagate in opposite directions.

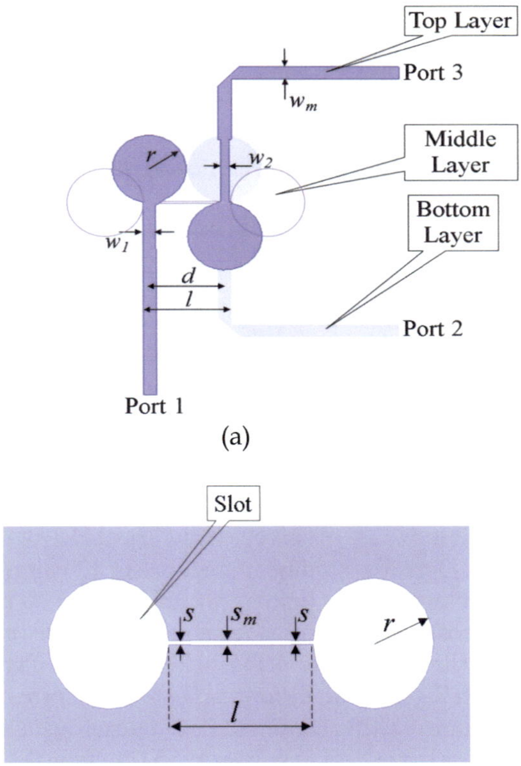

(a)

(b)

Fig. 21. The CST MWS layout of out-of-phase microstrip-slot power divider including (a) dimensions and (b) details of the slot layer in the ground plane.

As observed in Fig. 21, the design employs two mitre bends at the output ports of the divider. In order to obtain the best performance, the bend should have the optimum amount of chamfering dimension (Douville & James, 1978), as given by the following expressions:

$$D = w_m\sqrt{2} \tag{4}$$

$$x = D\left(0.52 + 0.65e^{-1.35\frac{w_m}{h}}\right) \tag{5}$$

$$A = \sqrt{2}\left(x - D/2\right) \tag{6}$$

when the condition of $0.5 \le \frac{w_m}{h} \le 2.75$ and $\varepsilon_r < 25$ is chosen. The constant w_m is the width of the transmission line and h is the thickness of substrate. For the other cases, a curve bend can be employed with an optimum bend radius to achieve minimum power loss.

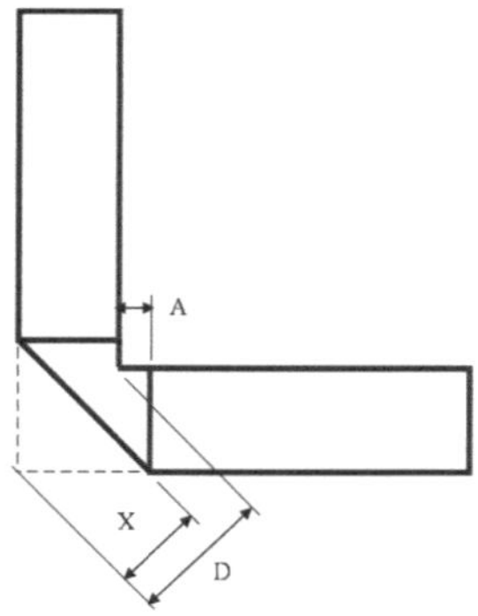

Fig. 22. Configuration of a mitre bend showing its dimensions.

In the design of power divider, input and output microstrip line characteristic impedance is assumed to be 50 ohm. The ends of microstrip lines are terminated in circular stubs representing virtual short circuits. In turn, the slotline is terminated with circular stubs representing virtual open circuits. The slot width is chosen to make a compromise between obtaining 50 ohm impedance, as observed from the side of microstrip line, and the manufacturing limitation. Here, the slotline impedance of 120 ohm is selected.
In order to improve matching of the input port, impedance steps realized by changes in width (w_1, w_2) in the microstrip lines and slotline are used. This makes the design different from the one reported in (Bialkowski & Abbosh, 2007). Slight changes in the slot width are also made, as illustrated in Fig. 21. The distance, *d* between centre of the input port and centre of the microstrip output line is chosen to be a quarter wavelength at the centre design frequency (fc = 6.85 GHz) as expressed in equation (7) (Bialkowski & Abbosh, 2007):

$$d = \frac{c}{4 f_C \sqrt{\varepsilon_{re}}} \tag{7}$$

where ε_{re} is the effective dielectric constant. It is determined using the following approximate expression:

$$\varepsilon_{re} = \frac{\varepsilon_r + 1}{2} \tag{8}$$

Length of the slot, *l* is chosen to be equal to the distance between the input port and microstrip output line plus the microstrip line width, w_m:

$$l = d + w_m \tag{9}$$

The design is accomplished with CST Microwave Studio. The dimensions of power divider are varied to obtain the best performance in the frequency band of 3.1 to 10.6 GHz. The divider's final dimensions are: wm = 1.18 mm (for 50 Ω microstrip line), $w1$ = 1.05 mm, $w2$ = 0.73 mm, s = 0.2 mm, sm = 0.22 mm, r = 3.2 mm, d = 6.51 mm and l = 7.69 mm. The prototype of the designed power divider is fabricated and tested in laboratory. Fig. 23 shows a photograph of the developed out-of-phase power divider.

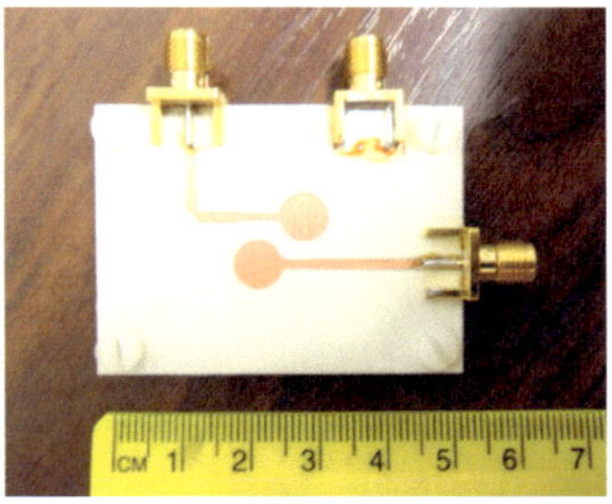

Fig. 23. Photograph of fabricated 3 dB out-of-phase power divider in microstrip-slot technology.

The simulation and measurement results for this device are presented in Fig. 24 and 25. The CST Microwave Studio's simulated responses show that the power supplied to the input port (Port 1) is equally divided between the output ports (Port 2 and Port 3) with the insertion loss less than 1.2 dB across the frequency band of 3 to 10.6 GHz. The return loss at Port 1 of the designed power divider is greater than 12.5 dB over the frequency range from 3 to 11 GHz. The phase difference between the output ports fluctuates between 178° and 180° across the same frequency band.

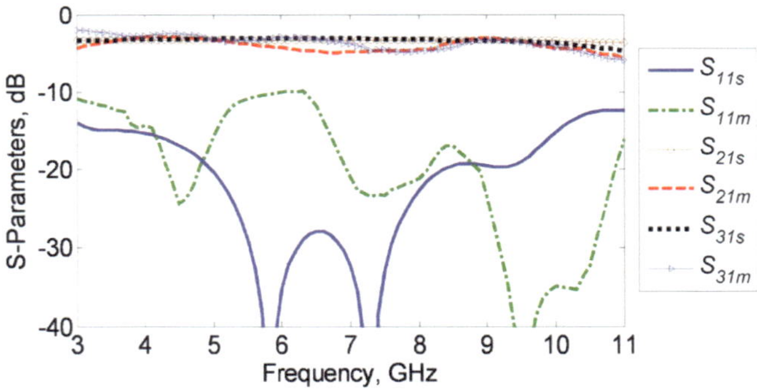

Fig. 24. The return and insertion loss performance of the designed out-of-phase microstrip-slot power divider.

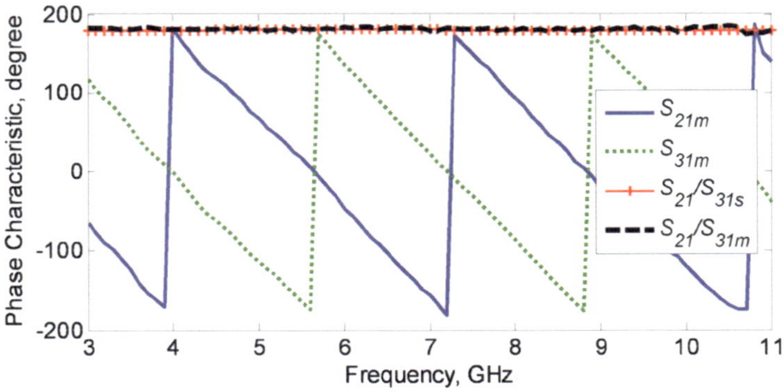

Fig. 25. Phase difference between output ports of out-of-phase power divider.

For the measured result, the average insertion losses between Port 1 and Ports 2 and 3 are 3.5 dB ± 1.43 dB from 3 to 10.4 GHz. These are slightly worse than the simulated ones. The measured return loss at Port 1 is greater than 10 dB across frequency band between 3 and 11 GHz. With respect to the phase difference between the two output ports, the fabricated prototype shows the discrepancy of ± 2° with the simulated results.

5.2 In Phase Power Divider

The design of an in-phase microstrip-slot power divider, shown in Fig. 26, is accomplished using a slight modification of the out-of-phase (180°) power divider of Fig. 21.

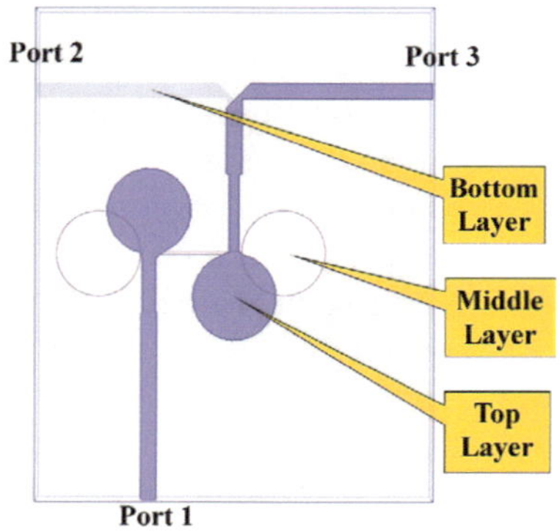

Fig. 26. Configuration of in-phase microstrip-slot power divider.

The difference is that in the in-phase power divider of Fig. 26 the microstrip lines connected to Port 2 and Port 3 are run in the same direction along the x-axis while for the out-of-phase divider they are in opposite directions. The dimensions used in the in-phase divider are the same as for the 180° power divider. The arrangement of this divider enables equal power division with same phase at the output ports. This can be explained by E-field lines being in the same direction for the two output ports.
A photograph of the fabricated in-phase power divider is shown in Fig. 27.

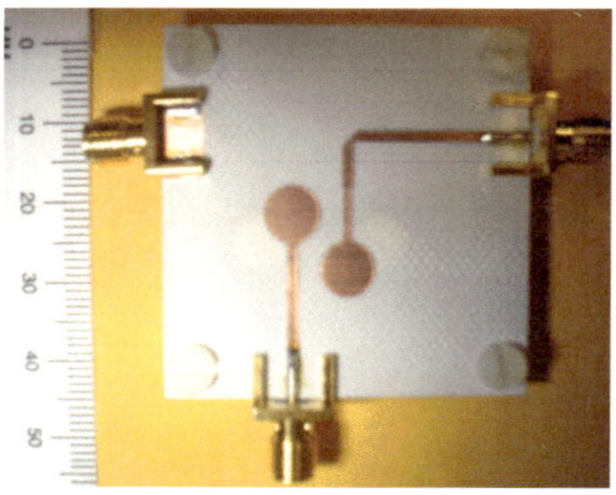

Fig. 27. Photograph of fabricated in-phase microstrip-slot power divider.

The simulated and measured results for the in-phase divider are presented in Fig. 28. The simulated power division to Port 2 and 3 is comparable to that of the out-of-phase power divider. The return loss is greater than 11 dB in the frequency band of 3 to 11 GHz, similarly as for the out-of-phase divider,. The phase difference between the output ports is approximately 0° to 3° from 3.1 to 10.6 GHz.

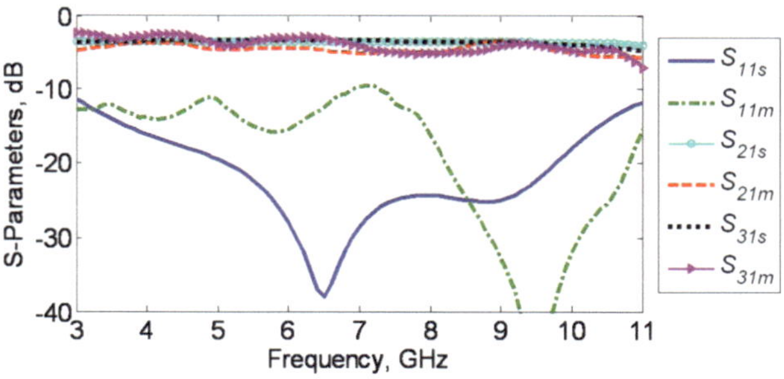

Fig. 28. Magnitudes of S-parameters of in-phase microstrip-slot power divider.

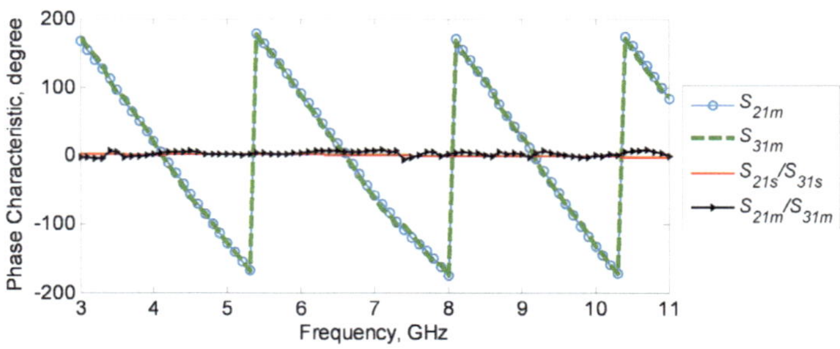

Fig. 29. Phase difference between output ports of in-phase microstrip-slot power divider.

The measured result for power division to Port 2 and 3 is 3.7 dB ± 1.3 dB and the return loss at Port 1 is greater than 9.4 dB across the frequency range between 3 and 11 GHz. The phase characteristic of the fabricated prototype shows a small deviation of ± 2° from the simulated result. The measured insertion and return losses are comparable with the simulated ones.

6. Conclusion

This chapter has reported on a microstrip-slot transition and its application to the design of wireless vertical interconnects, phase shifters and power dividers in microstrip-slot technique. The design and analysis of operation of all these components has been accomplished using CST Microwave Studio simulator. It has been shown that the use of this transition enables an ultra wideband performance of multilayer microstrip-slot vias, phase shifters and power dividers. Prototypes of these components have been fabricated and experimentally tested. The presented experimental results have demonstrated that all of these components can be manufactured using ample microwave fabrication facilities. The measured results have confirmed a very good ultra-wideband performance, as observed in simulations. Because of small size and wideband performance the investigated components should be of considerable interest to the designers of compact ultra wideband sub-systems.

7. References

Abbosh, A. M. (2007). Ultra-Wideband Phase Shifters. *IEEE Transactions on Microwave Theory and Techniques,* Vol. 55, 2007, pp. 1935-1941, ISSN. 0018-9480.

Abbosh, A. M. & Bialkowski, M. E. (2007). Design of compact directional couplers for UWB applications. *IEEE Transactions on Microwave Theory and Techniques,* Vol. 55, 2007, pp. 189-194, ISSN. 0018-9480.

Bialkowski , M. E. & Abbosh, A. (2007). Design of a compact UWB out-of-phase power divider, *IEEE Microwave and Wireless Component Letter,* Vol. 17, No. 4, April 2007, pp. 289 - 291, ISSN. 1531-1309.

Douville, R. J. P. & James, D. S. (1978). Experimental study of symmetric microstrip bends and their compensation. *IEEE Transactions on Microwave Theory and Techniques,* Vol. 26, No. 3, March 1978, pp. 175-181, ISSN. 0018-9480.

Gupta, K. C.; Garg, R. & Bahl, I. J. (1979). *Microstrip Lines and Slotlines,* Artech House, ISBN. 0-89006-074-6, USA.

Ho, C. H.; Fan, L. & Chang, K. (1993). Slot-coupled double-sided microstrip interconnects and couplers. *IEEE MTT-S International Microwave Symposium Digest,* Vol. 3, pp. 1321-1324, ISBN. 0-7803-1209-0, USA, June 1993, IEEE, Atlanta.

Meschanov, V.P.; Metelnikova, I.V.; Tupikin, V.D. & Chumaevskaya, G.G. (1994). A new structure of microwave ultrawide-band differential phase shifter. *IEEE Transactions on Microwave Theory and Techniques,* Vol. 42, No. 5, May 1994, pp. 762–765, ISSN. 0018-9480.

Pozar, D. (2005). Microwave Engineering, 3rd. Edition, John Wiley and Sons, ISBN. 978-0-471-44878-5, New Jersey.

Robinson, G. H. & Allen, J. L. (1969). Slot line application to miniature ferrite devices. *IEEE Transactions on Microwave Theory and Techniques*, Vol. 17, No. 12, December 1969, pp. 1079-1101, ISSN. 0018-9480.

Schiek, B. & Kohler, J. (1976). An improved microstrip-to-microstrip transition. *IEEE Transactions on Microwave Theory and Techniques*, Vol. 24, No. 4, April 1976, pp. 231-233, ISSN. 0018-9480.

Schüppert, B. (1988). Microstrip/slotline transitions: modeling and experimental investigation. *IEEE Transactions on Microwave Theory and Techniques*, Vol. 36, No. 8, August 1988, pp. 1272-1282, ISSN. 0018-9480.

Seman, N.; Bialkowski, M. E. & Khor, W. C. (2007). Ultra wideband vias and power dividers in microstrip-slot technology, *Proceeding of Asia-Pacific Microwave Conference 2007,* pp. 1-4, ISBN. 9781424407484, Thailand, December 2007, IEEE, Bangkok.

Seman, N.; Bialkowski, M. E. & Khor, W. C. (2008). UWB Fully Integrated Microwave Reflectometer In Multi-Layer Microstrip-Slot Technology, *Proceeding of Asia-Pacific Microwave Conference 2008,* pp. 1-4, ISBN. 9781424407484, Hong Kong, December 2008, IEEE.

Soltysiak, P. & Chramiec, J. (1994). Design of broadband transitions from microstrip to slotline. *Electronic Letters,* Vol. 30, No. 4, February 1994, pp. 328 - 329, ISSN. 0013-5194.

Zinieris, M. M.; Sloan, R. & Davis, L. E. (1998). A broadband microstrip-to-slot-line transition. *Microwave Optical and Technology Letters,* Vol. 18, No. 5, August 1998, pp. 339-342, ISSN. 0895-2477.

Vertical Transmission Lines in Multilayer Substrates and Highly-Integrated Filtering Components Based on These Transmission Lines

Taras Kushta
NEC Corporation
Japan

1. Introduction

Multilayer substrates such as interposers and printed circuit boards (PCBs) are basic interconnect technologies in modern and next-generation systems in which chip, package and board have been used as constructing elements. Consequently, multilayer substrates have been intensively studied in worldly dispersed electronics packaging research centers in which questions related to how to improve electrical, mechanical, thermal and reliable performances are on the agenda. Moreover, interconnection items affect directly on miniaturization, integration, cost-effectiveness and electrical characteristics of electronics components and, as a result, on promotion of electronics products to the market.

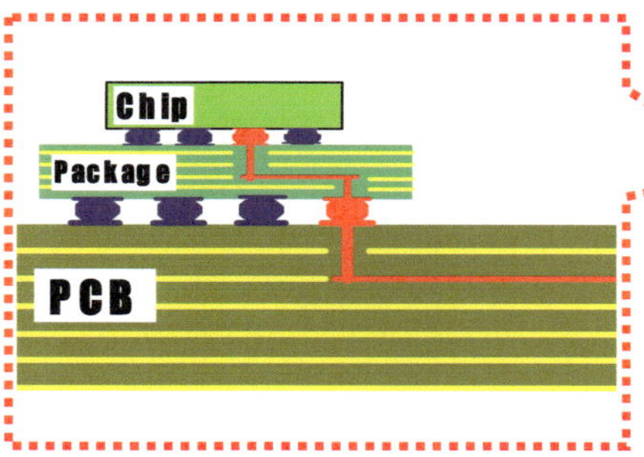

Fig. 1a. A chip-package-board part of a system

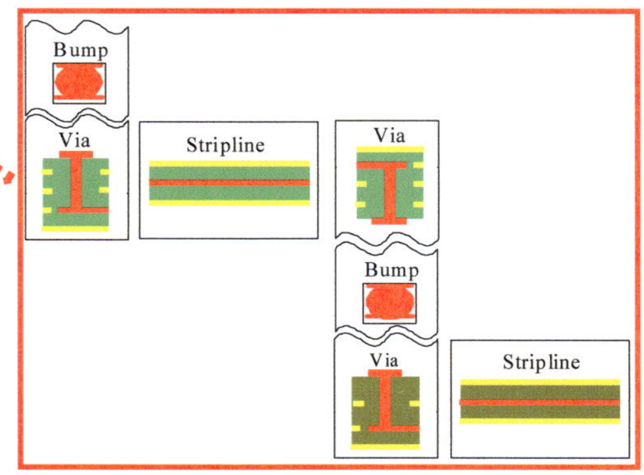

Fig. 1b. A division of an interconnection on bulding blocks

Microwave and millimeter wave areas extremely enhance difficulties in electrical design of interconnected circuits based on multilayer substrate technologies due to impedance

mismatching problems, crosstalk effects, leakage losses, unwanted resonances, dielectric and metal losses, and so on. These issues can be particularly overcome forming interconnections as well wave-guiding structures which can be also used as basic transmission lines of distributed-element passives and actives.
In Fig.1a, an example of a chip-package-board part of a system is shown. Multilayer substrate technologies are realized in the example presented by means of a package and a PCB. An interconnection in the multilayer substrates demonstrated in Fig.1a can be divided into blocks, having their specific characteristics, as shown in Fig.1b. These blocks are represented by planar transmission lines, bumps and vias for the electrical channel shown. One can generalize such building blocks by two groups - horizontal and vertical interconnections - as exhibited in Fig.2.

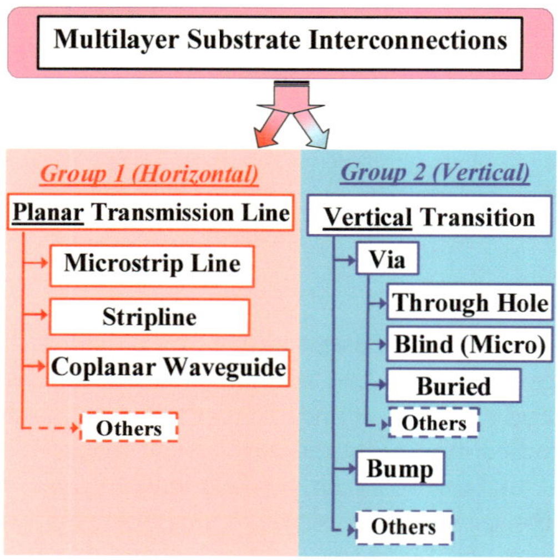

Fig. 2. A generalization of interconnections in a chip-package-board system

To design horizontal interconnections of a high electrical performance, planar transmission lines have been usually used because these structures can provide operation on one (fundamental) mode (for an example, TEM or Quasi-TEM), which has well-defined propagation constant and characteristic impedance, in a wide frequency band. That is why, short and long transmission lines have been used in high-frequency and high-speed systems. Besides that, planar transmission lines in the substrates serve not only as interconnected circuits but also as forming blocks of distributed passive and active components. Consequently, electrical study of planar transmission lines and different functional devices based on these lines has been widely and deeply presented in numerous literatures published (for an example, see comprehensive books (Hoffmann, 1987; Gupta et al., 1996), as for planar transmission lines).
In this chapter, attention will be attracted to the second group of interconnections (see Fig.2) in multilayer substrates, that is, vertical transitions.
Reasons why it will be concentrated on these structures are as following.
Firstly, it can be explained by a significant increase of the vertical transition role in achieving high electrical performance of signal interconnection paths in multilayer

substrates at microwaves and millimeter waves and a contribution of the vertical transitions to impedance mismatching, crosstalk, energy leakage, and other problems which can be excited due to these structures that can finally lead to the fault of the systems, electromagnetic interference (EMI), and other difficulties.

Secondly, it is attractive to use vertical transitions as forming elements of passives and actives (as for an example, short- or open-circuited stubs for filters) and in such way to reduce considerably their dimensions due to:

1) Three-dimensional (3-D) design;

2) Providing an approach to move a functional area for a component to a vertical transition region (see Fig.3).

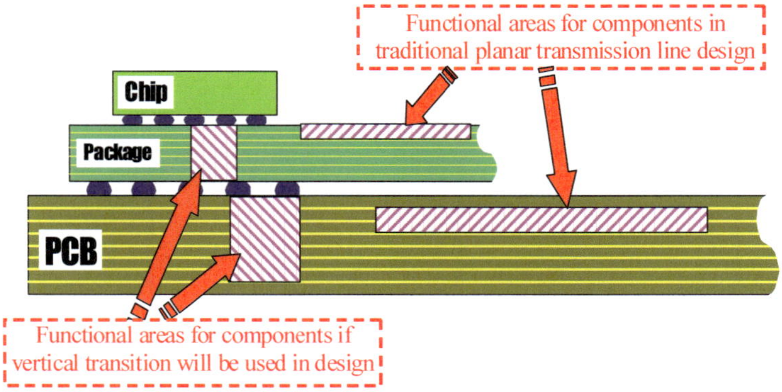

Fig. 3. Approach for miniaturization of a chip-package-board system by means of the use of vertical transitions as forming blocks of a component

2. Shield Via as Vertical Transmission Lines for Multilayer Substrates

Consider vias, as representative structures of vertical transitions, which serve usually to connect planar transmission lines disposed at different conductor layers of multilayer substrates. At microwave and millimeter wave bands, structures similar to a single signal via have poor-defined wave guiding properties and, as a result, they have increasing leakage losses with the growth of the frequency. That is why at these frequencies, propagation constant and characteristic impedance cannot be defined using traditional inductance and capacitance.

As an illustrative example, in Fig.4, the peak of the *E*-field at 10 GHz calculated by a three-dimensional full-wave technique (Weiland, 1996) in a horizontal cross-section between conductor planes of a multilayer substrate comprising the single signal via is shown. As one can see, if the single signal via is placed in the multilayer substrate, then it becomes an effective source of the parallel plate mode excitation. It acts like an antenna exciting parallel plate modes between conductor planes. As a result, such via structure leads to a dramatic reduction of the electrical performance of a whole interconnection due to in-substrate parallel plate-mode resonances and, as their consequence, signal integrity, power integrity and EMI problems. In Fig.5, an impact of the parallel plate-mode resonances on the electrical characteristics of the via is shown by means of the insertion loss. As one can see, the electrical performance of the via dramatically degrades at higher frequencies (in present example, starting from about 2GHz).

Electrical characteristics of vertical transitions can be improved by progressing from through-hole (see Fig.6a) to blind, counter-bored and buried via technologies explained respectively in Figs.6b, 6c and 6d. In these cases, stub effect (Laermans et al., 2001; Kushta et al., 2003) can be removed providing an improvement of signal transmission channel parameters, and the signal via conductor length can be shortened providing a reduction of coupling and radiating areas.
However, in spite of such advancements problems emphasized above remain at microwaves and millimeter waves.

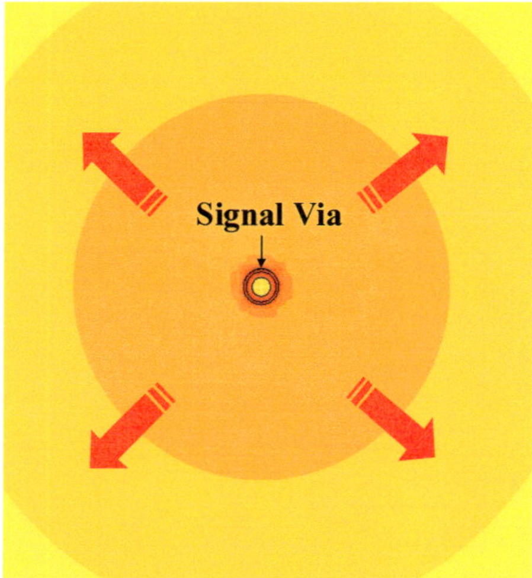

Fig. 4. Simulated peak of the *E*-field taken at 10GHz in a cross-section of a multilayer substrate comprising a single signal via

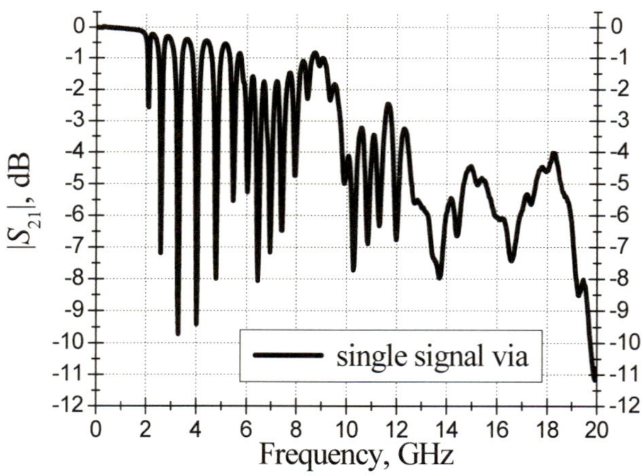

Fig.5. Experimental data for the insertion loss of the single signal via in the multilayer substrate

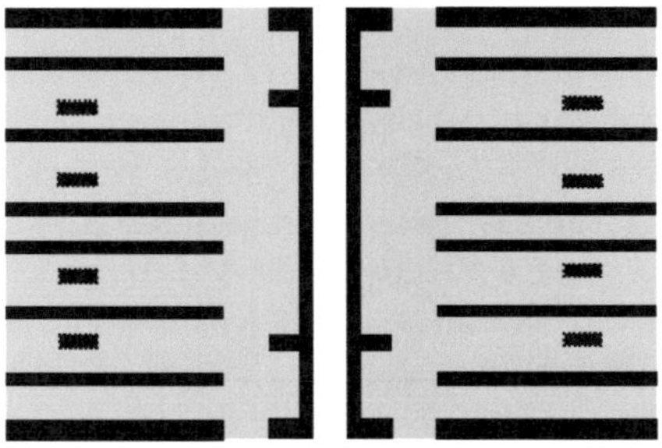

Fig. 6a. Cross-sectional view of through-hole via

Fig. 6b. Cross-sectional view of blind via

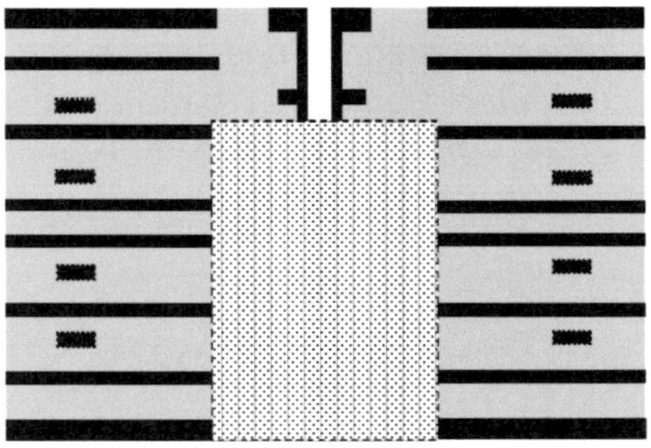

Fig. 6c. Cross-sectional view of counter-bored via

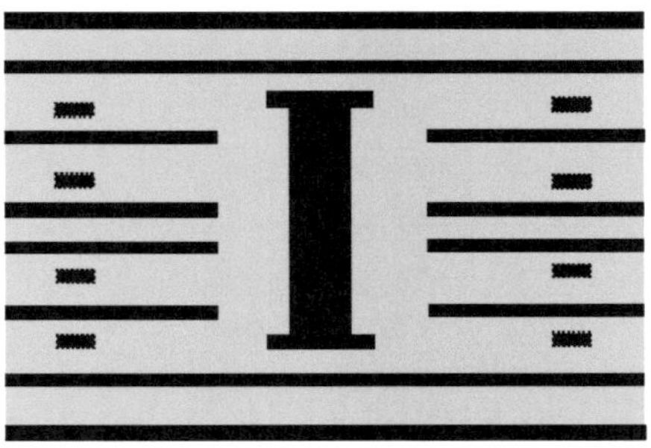

Fig. 6d. Cross-sectional view of buried via

Thus, it comes to be clear that vertical transitions including via structures become an important element in design of high-frequency and high-performance interconnections and components grounded on multilayer substrate technologies.
A solution proposed to provide a high-performance vertical transition in a multilayer substrate is based on forming a shield via as a result of the conjoint use of signal and ground vias. In this case, a specific coaxial waveguide can be formed in the vertical direction of the multilayer substrate (Pillai, 1997; Tarvainen, 2000; Kushta et al. 2002).
Following distinctive examples show advanced characteristics for the shield via compared with the single signal via case. In Fig.7, simulated peak of the *E*-field for the shield via obtained in the same way as for Fig.4 is presented for the identical dimensions of the substrate. As one can see, electromagnetic energy propagating through the shield via is disposed between signal and ground vias. This effect leads to a considerable improvement of the electrical performance for signaling as shown in Fig.8 by means of measured insertion losses (photo of the shield via experimental pattern is in Fig.9). In Fig.8 electrical characteristics of the single via are also given for comparison.
It is well known, to estimate leakage losses in a wide frequency band, *S*-parameters can be used and as for example by means of such equation:

$$Leakage\ Loss, \% = (1 - |S_{11}|^2 - |S_{21}|^2) \cdot 100\ , \quad (1)$$

where $|S_{11}|$ is the return loss and $|S_{21}|$ is the insertion loss.

In Fig.10, simulated leakage losses for single signal via and shield via with the same parameters as for Figs.4 and 7 are presented. As one can see, the application of the shield via suppresses leakage losses in considered frequency band. It also means that EMI problems can be considerably reduced by the use of such vias in electronics design (Kushta et al., 2004; Kushta & Narita, 2004).

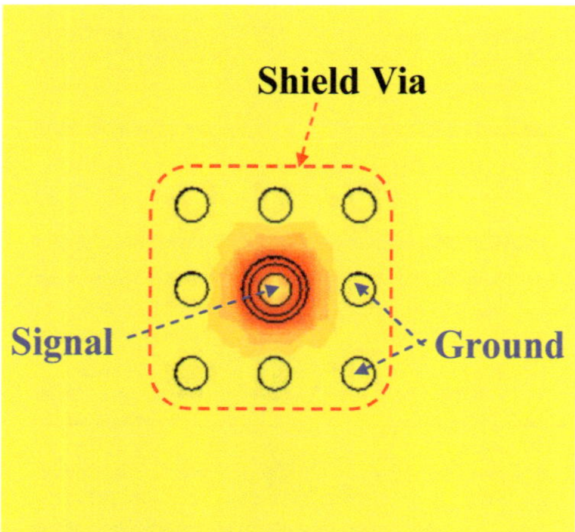

Fig. 7. Simulated peak of E-field taken at 10GHz in the cross-section of the multilayer substrate comprising a shield via

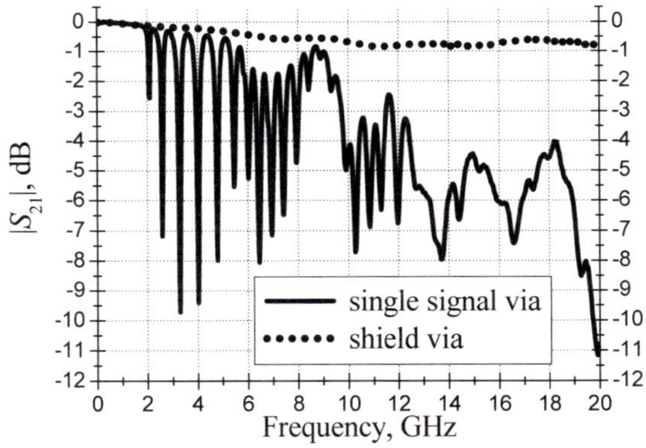

Fig. 8. Experimental data for the insertion loss of both the shield via and the single signal via in the multilayer substrate

Consider leakage effect on the electrical performance of both single and shield via structures in which a digital signal is propagating. In Fig.11, the pulse transmitted through such via structures is shown. As one can see in this figure, signal transmitted through the single signal via has not only higher insertion loss but also higher deformation of the pulse shape that is one of the most important issues in high-speed signaling because, in this case, it is necessary to apply additional techniques like pre-emphasis.

Fig. 9. Photo of the shield via formed by signal and ground vias conjointly

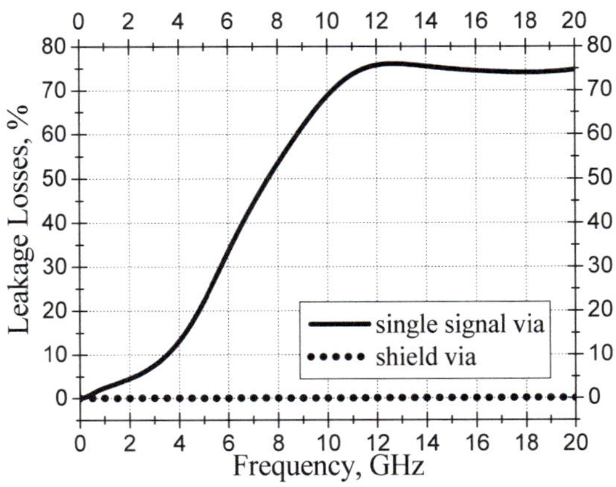

Fig. 10. Simulated leakage losses for via structures calculated according to Eq.1

On the other hand, forming the shield via in the multilayer substrate gives a possibility for a considerable improvement of the electrical performance of the vertical transitions. As follows from Fig.11, the shield via provides significantly lower loss, if it is compared with single signal via case. Moreover, the pulse shape (especially, the width for the signal transmitted) is considerably better for the shield via.

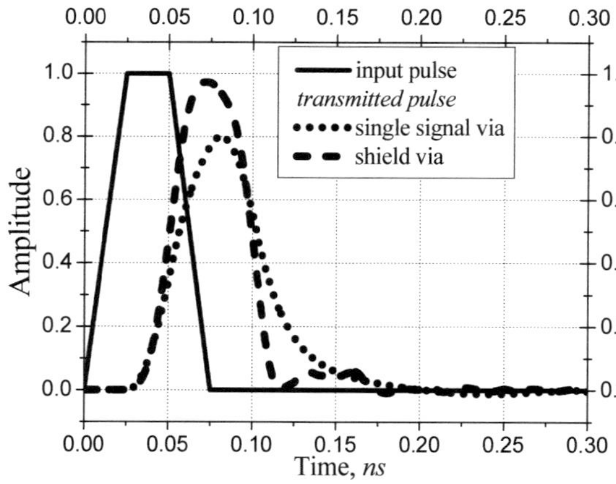

Fig. 11. Signal propagation in single signal via and shield via (transmission)

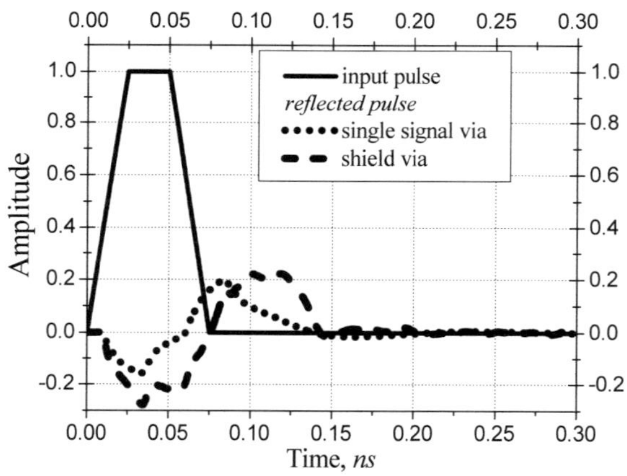

Fig. 12. Signal propagation in single signal via and shield via (reflection)

However, as follows from Fig.12, the amplitude of the reflected pulse is large enough for both via structures. That is why, providing characteristic impedance controlling in a wide frequency band is another important issue to implement the shield vias in real substrates and to achieve their electrical performance similar to that as in planar transmission lines. Therefore, an appropriate physical model showing mechanisms affecting on the electrical characteristics of such type of vertical transitions has to be defined.

Consider the shield via as in Figs.13a and 13b. This structure is formed in an 8-conductor layer substrate. Corrugated coaxial waveguide model (Kushta et al., 2002; Kushta et al., 2004) is proposed to describe physical processes in the shield via. In this model, ground vias are replaced by continuous and smooth conductive surface which acts as an outer conductive boundary and the signal via serves as an inner conductive boundary of such coaxial waveguide. Also in the model, conductive plates from conductive layers of the multilayer substrate disposed between inner and outer conductive boundaries are considered as specific corrugations of the outer conductive boundary. The corrugated coaxial waveguide model for the shield via shown in Figs. 13a and 13b is presented in Figs.14a and 14b.

In consequence, the outer conductive boundary of such corrugated coaxial waveguide model can be characterized as a surface for which the surface impedance can be approximately defined as:

$$Z_s \approx 120\pi \cdot i \cdot \sqrt{\frac{1}{\varepsilon}} \cdot \tan\left(\frac{2\pi \cdot f}{c} d \cdot \sqrt{\varepsilon}\right) \quad (2)$$

where d is the corrugation depth defined as $d = (D_r - d_{cle,r} - d_{gr})/2$, f is the frequency and c is the velocity of light in free space. Note that Eq.(2) is valid under following conditions:

$$H_{i,j} << \lambda, \quad (3)$$

where λ is the shortest wavelength in the isolation material of the multilayer substrate in considered frequency range; $H_{i,j}$ is the distance between i-th and j-th conductor planes; $j = i + 1$.

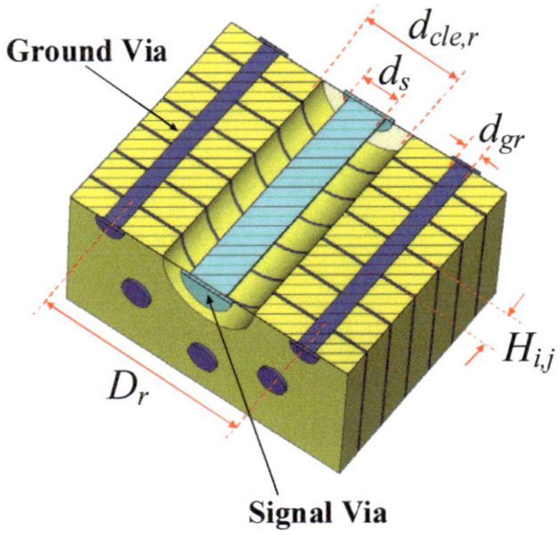

Fig. 13a. Cross-sectional view of shield via

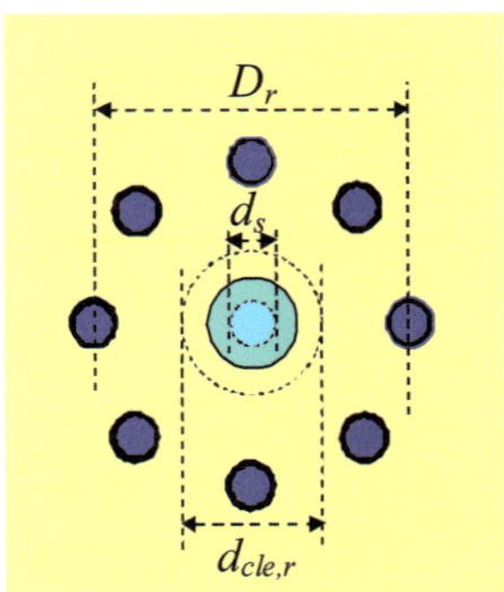

Fig. 13b. Top and bottom views of shield via

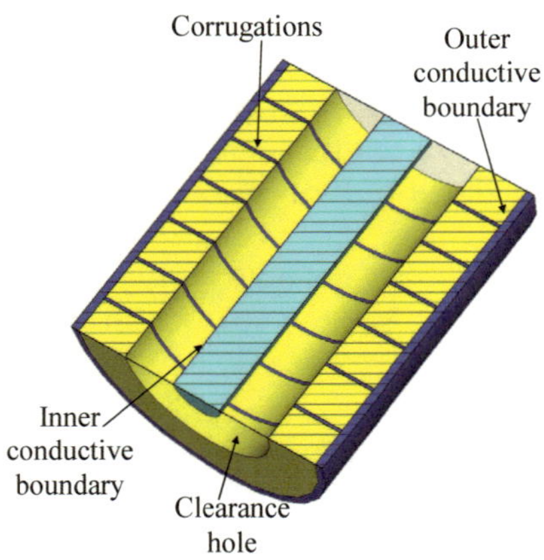

Fig. 14a. Cross-sectional view of corrugated coaxial waveguide model

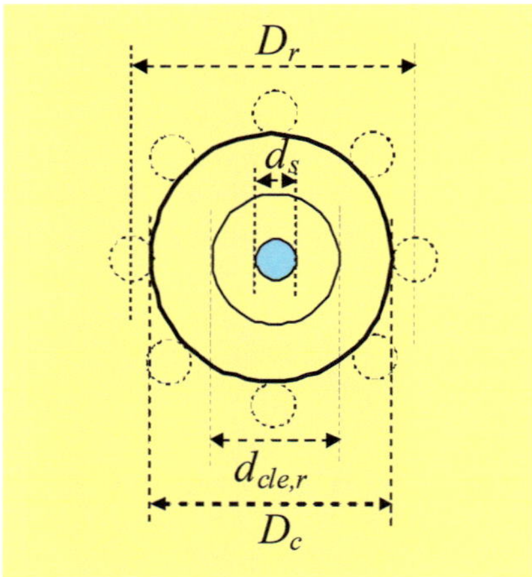

Fig. 14b. Top and bottom views of corrugated coaxial waveguide model

Eq.2 gives a simplified physical mechanism which can explain signal propagation in the shield via. In particular, if corrugations in the coaxial waveguide model are large enough, then the surface impedance of the outer conductive boundary is dependent on the frequency. It means that broadband matching of the shield via with other interconnected circuits having usually approximately constant (or weakly frequency-dependent) characteristic impedance is a difficult problem.

Thus, to provide a broadband high-performance operation of the shield via it is necessary to decrease such the corrugations as much as possible. If this condition will be satisfied, then an approximate equation for the surface impedance can be written as follows:

$$Z_s \approx 0 . \tag{4}$$

The surface impedance defined according to Eq.4 corresponds to the smooth conductive boundary and, in this case, signal propagation in the shield via can be considered as in a corresponding coaxial waveguide.

As a validation of this coaxial waveguide model, consider two types of shield vias in the multilayer substrate. The first type comprises the outer conductive boundary of a round arrangement of ground vias. The second type is consisted of ground vias with a square arrangement. From coaxial transmission line theory (Wheeler, 1979), there are known analytical formulas for the characteristic impedance of round and square coaxial waveguides. In Figs.15a and 15b, expressions for these coaxial waveguides are presented under the drawing of the corresponding structure by Equations (5) and (6), respectively.

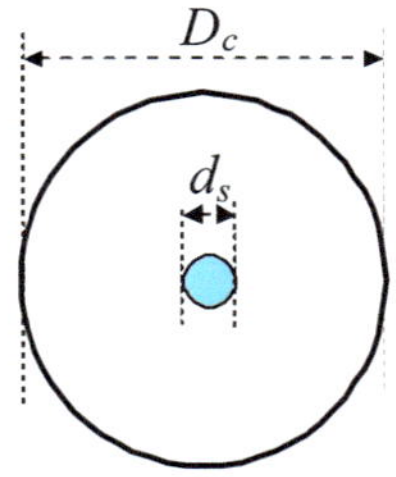

$$Z_r \approx 60 \cdot \sqrt{\frac{\mu}{\varepsilon}} \cdot \ln\left(\frac{D_c}{d_s}\right) \quad (5)$$

Fig. 15a. Cross-section view of round coaxial waveguide and its characteristic impedance

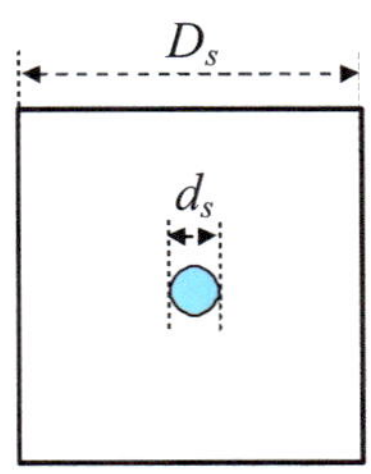

$$Z_{sq} \approx 60 \cdot \sqrt{\frac{\mu}{\varepsilon}} \cdot \ln\left(\frac{1.0787 \cdot D_s}{d_s}\right) \quad (6)$$

Fig. 15b. Cross-section view of square coaxial waveguide and its characteristic impedance

As follows from these equations, which are defined for the coaxial transmission lines with continuous and smooth inner and outer conductive boundaries, the characteristic impedance will have the same magnitude for round and square cases if the diameter of outer boundary of the round transmission line and the side of the square transmission line will satisfy the following identity:

$$D_c \approx 1.0787 \cdot D_s . \quad (7)$$

It should be noted that Eq.7 is valid if other parameters of round and square coaxial transmission lines such as the diameter of the inner conductor and constitutive parameters (such as relative permittivity, ε and relative permeability, μ) of the isolating material are the same.

So, first of all, a validation of the coaxial waveguide model will be provided in such manner. If this model is appropriate for the shield via, then identity (7) will be satisfied for shield vias with round and square arrangements of ground vias around the signal via. To verify this feature, round and square shield vias with $D_c = 3.2mm$ and $D_s = 2.967mm$ have been considered. Cross-sectional views of these via structures are presented in Figs.16a and 16b.

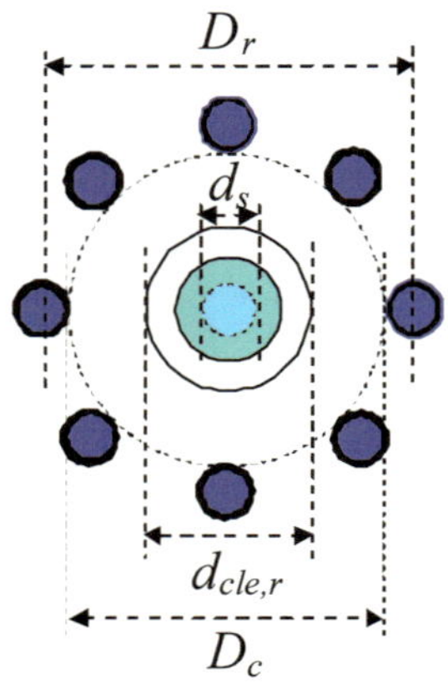

Fig. 16a. Shield via with round arrangement of ground vias

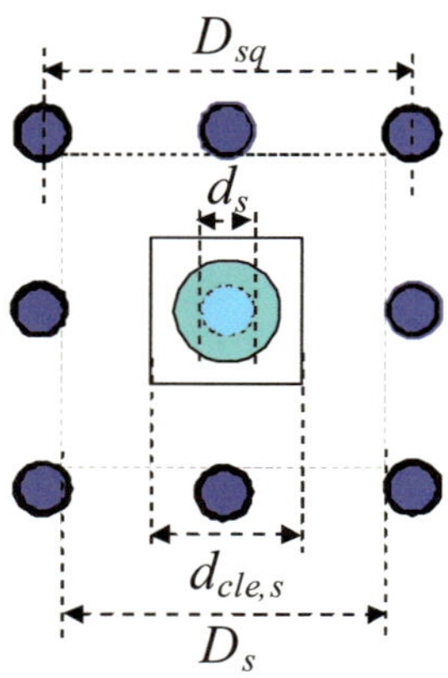

Fig. 16b. Shield via with square arrangement of ground vias

Other dimensions of aforementioned shield via structures are as following (see Fig.17): $d_{pad} = 0.95mm$, $d_{cle,r} = 1.65mm$, $d_{cle,s} = 1.53mm$ and $d_s = 0.65mm$. The shield via structures have been embedded in the substrate which consists of eight copper planar conductor layers isolated by FR-4 material with the relative permittivity of $\varepsilon = 4.17$ and loss tangent of $\tan\delta = 0.023$ as assumed in simulations. Spaces between planar conductor layers as shown in Fig.17 are: $H_1 = 0.2mm$, $H_2 = 0.385mm$ and $H_3 = 0.24mm$; the thickness of conductor planes embedded in the substrate is $t = 0.035mm$; the thickness of top and bottom conductor planes is $t_t = t_b = 0.055mm$.

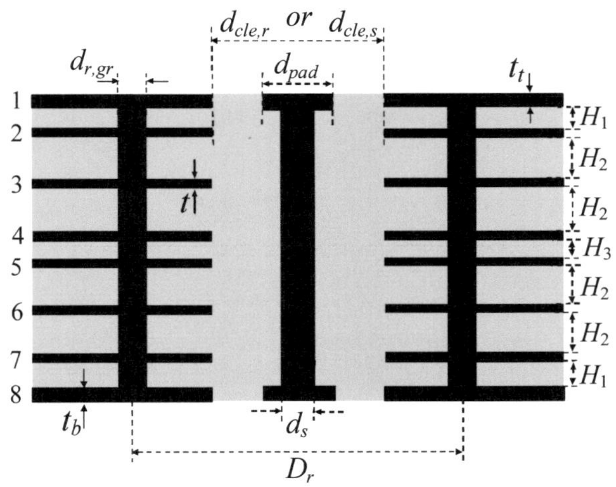

Fig. 17. Vertical cross-section view of shield via in 8-conductor-layer substrate

In Figs.18a and 18b, magnitudes of simulated *S*-parameters for two shield vias with round ($D_c = 3.2mm$) and square ($D_s = 2.967mm$) arrangements of the ground vias in the 8-conductor-layer substrate are presented. As follows from simulated *S*-parameter data shown in these figures, structures with round and square arrangements of ground vias having transverse dimensions defined according to Eq.7 demonstrate practically the same electrical performance in considered frequency band. It means also that the characteristic impedance in structures presented is the same one and, as a result, aforementioned shield vias are practically equivalent.

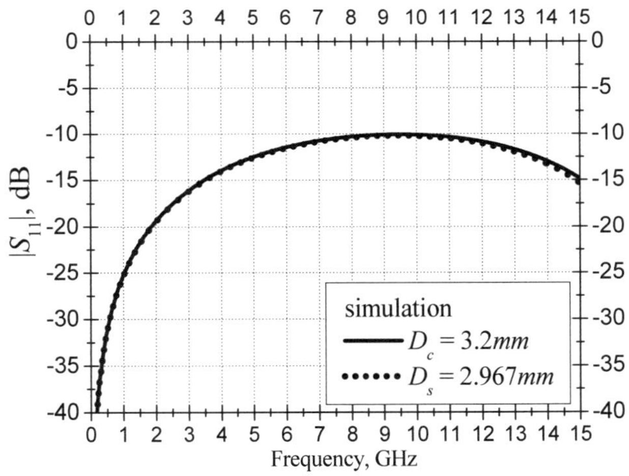

Fig. 18a. Simulated return losses for two shield vias with round and square arrangements of ground vias

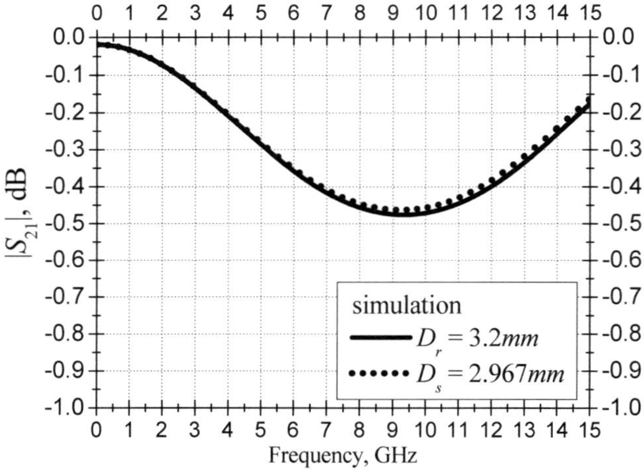

Fig. 18b. Simulated insertion losses for two shield vias with round and square arrangements of ground vias

Simulated results presented in Fig. 18a and 18b serve a proof of a simplified mechanism for signal propagations in the shield via formed by signal and ground vias conjointly as in the corresponding coaxial waveguide with smooth and continuous conductive boundaries. This consideration gives a way to define the characteristic impedance of the shield via in the multilayer substrate that is important to design well-matched interconnected circuits using multilayer substrate technologies.

Note that the corrugation depth for considered round and square coaxial waveguides is the same due to the appropriate choice of the clearance hole form and dimensions. In these cases, the round shield via has the round clearance hole, while the square shield via has the square clearance hole. Also, dimensions of the clearance holes are defined according to Eq.7.

Above-mentioned data have been obtained by three-dimensional full-wave simulations which usually give an adequate description of electromagnetic processes in a test structure. However, each theoretical model is idealized one, which does not include the frequency dependency of board isolating material, roughness and tolerances of shapes of conductive surfaces, and so on. That is why the experimental study of test structures serves not only as an evidence of their theoretical models but also gives a real wide-frequency band behavior of the structures studied.

In following Fig.19a and 19b, measured magnitudes of S-parameters for the shield vias whose simulated data are respectively presented in Figs. 18a and 18b are shown and demonstrate the electrical behavior similar to their simulation models. As follows from theoretical and experimental data, characterization of the shield vias in the multilayer substrate as specific coaxial waveguides is a vital and useful approach to design high-frequency and high-speed electrical vertical transitions.

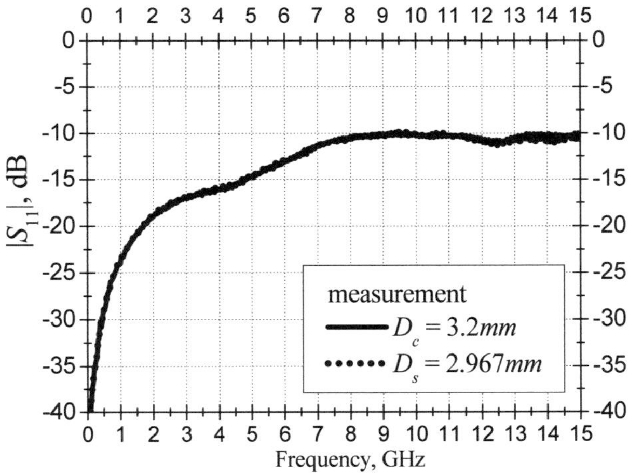

Fig. 19a. Measured return losses for two vertical transitions with round and square arrangements of ground vias

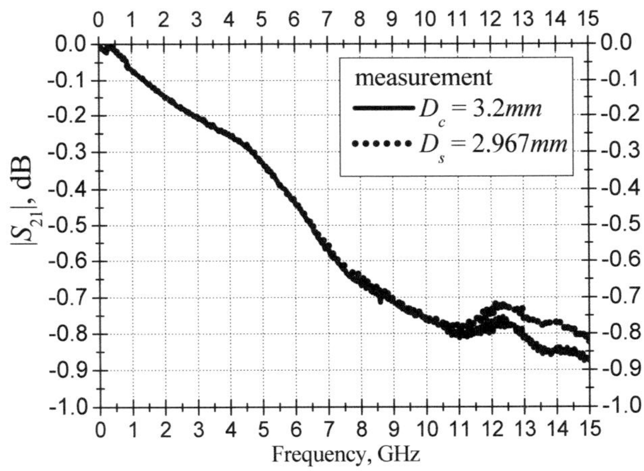

Fig. 19b. Measured insertion losses for two equivalent vertical transitions with round and square arrangements of ground vias

As another verification of the corrugated coaxial waveguide model and also as recommended design steps based on the application of this model, an effect of the distance between signal and ground vias on the electrical performance of the shield via is presented. Two square arrangements of ground vias having $D_s = 2.04mm$ and $D_s = 2.967mm$ (other parameters are the same as in aforementioned cases, except that the clearance hole has the side $d_{cle,s} = 1.16mm$) have been considered here that approximately corresponds characteristic impedances calculated according to Eq.6 as $Z_{sq} \approx 36Ohms$ and $Z_{sq} \approx 47Ohms$. Measurement data for these shield vias are shown in Figs.20a and 20b. Note that top and bottom parts of the shield vias considered were connected to 50*Ohms* coaxial cables. As follows from figures presented the highest electrical performance in all frequency band (up to 15*GHz*) is achieved for the shield via with $D_s = 2.967mm$. This shield via is better matched to $50Ohm$ cables that is an indirect validation of the coaxial waveguide model. However this is only one important point of the physical model presented because corrugations are another its key point.

Thus, as next, the clearance hole effect on the electrical performance of the shield via is shown that is associated with the corrugation depth in the physical model presented. Measurement data for two shield vias with different dimensions of the clearance hole are demonstrated in Fig.21a and 21b.

The shield vias have the same dimensions and are embedded in the same 8-conductor-layer substrate, as in above-mentioned examples. In considered shield vias, clearance holes have the square form with the side of $d_{cle,s} = 1.53mm$ and $1.16mm$ and for both shield vias $D_s = 2.967mm$. As one can see increasing the clearance hole dimensions leads to a considerable improvement of the electrical performance of the shield via in the wide frequency band.

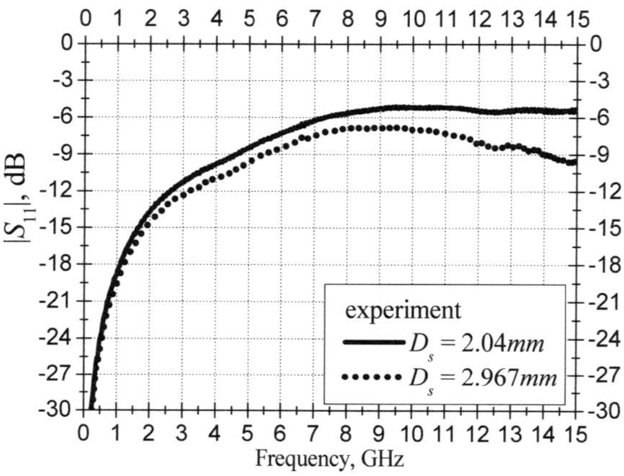

Fig. 20a. Measured return losses for shield vias with square arrangements of ground vias (effect of distance between signal and ground vias)

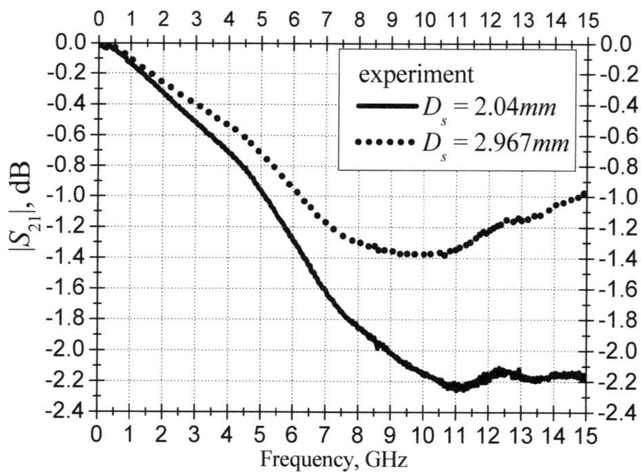

Fig. 20b. Measured insertion losses for shield vias with square arrangements of ground vias (effect of distance between signal and ground vias)

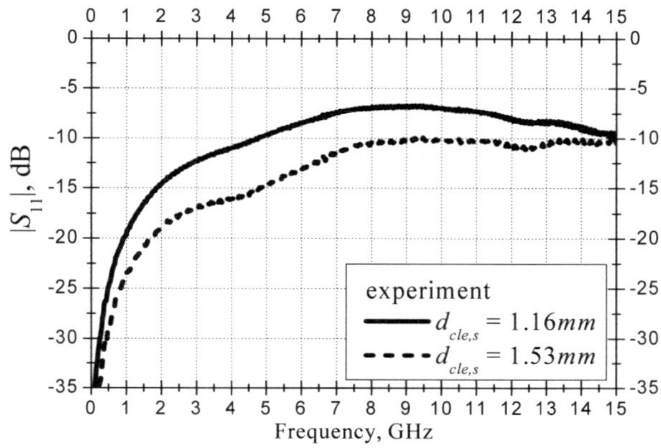

Fig. 21a. Measured return losses for shield vias with square arrangements of ground vias (clearance hole effect)

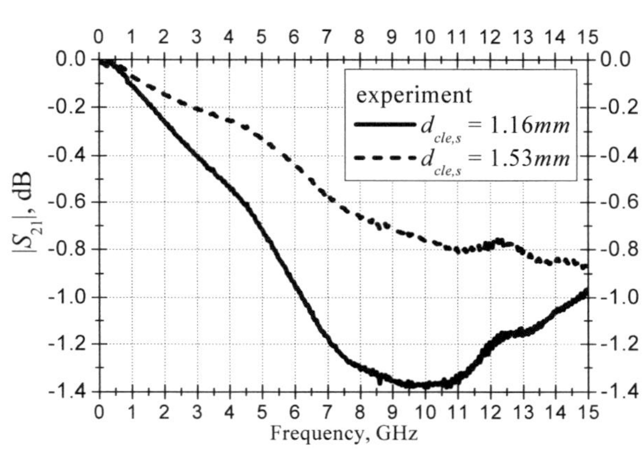

Fig. 21b. Measured insertion losses for shield vias with square arrangements of ground vias (clearance hole effect)

As a result of these considerations, two main points can be categorized as a basis which can provide a high-performance vertical transition in the form of the shield via. They are obtained as following from the corrugated coaxial waveguide model given here.

1) Signal via transversal dimensions and distance between signal and ground vias in a shield via have to be chosen in such way to provide a required characteristic impedance calculated according to an appropriate coaxial transmission line corresponding to the shield via.

2) A clearance hole has to provide minimal corrugations of the ground plates in the coaxial wave guiding channel.

As one can see, in presented examples, cases when signal is propagating from the top to the bottom of the multilayer substrate are considered. However in real applications, the shield via has to be connected to a planar transmission line disposed at a conductive layer of a multilayer substrate. And this connection can not be decided in a simple way at microwaves and millimeter waves and, that is why, it becomes an important issue. In following paragraph, a technique to provide a high-performance transition from the shield via to the planar transmission line will be shown.

3. Broadband Trasition from a Shield Via Structue to a Planar Transmission Line in a Multilayer Substrate

Thus, development of a vertical transition itself is not enough to provide a high-performance interconnection at microwaves and millimeter waves. It is important to match such vertical transition with other interconnected ciruits (Kushta & Harada, 2008), including a planar transmission line as for an example.

In Figs.22a and 22b, cross-sectional views of a shield via in a 14-conductor-layer substrate are shown. The electrical performance of the via structure is strongly-dependent on the shape and dimensions of the clearance hole as it has been shown above.

In real design, dimensions of the clearance hole can be big enough due to a large distance between the signal via and ground vias which conjointly with the radius of the signal via and constitutive parameters of an isolating material in the multilayer substrate provide controlling the characteristic impedance in the shield via. In the case of connection of the shield via to a planar transmission line such clearance hole can excite characteristic impedance mismatching problems that will be shown in following example.

Consider the model presented in Figs.22a and 22b in which the shield via is connected to a stripline disposed at the 12th conductor layer of the 14-conductor-layer substrate. The shield via has such dimensions: $d_s = 0.6mm$; $d_{pad} = 1.2mm$; $d_{cle,r} = 1.4mm$ or $d_{cle,r} = 3.4mm$; $d_{gr,r} = 0.3mm$.

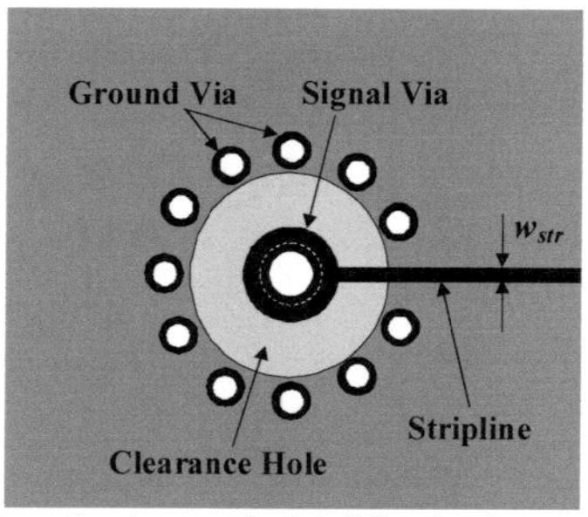

Fig. 22a. Horizontal cross-sectional view of shield via in multilayer substrate taken in the position of stripline

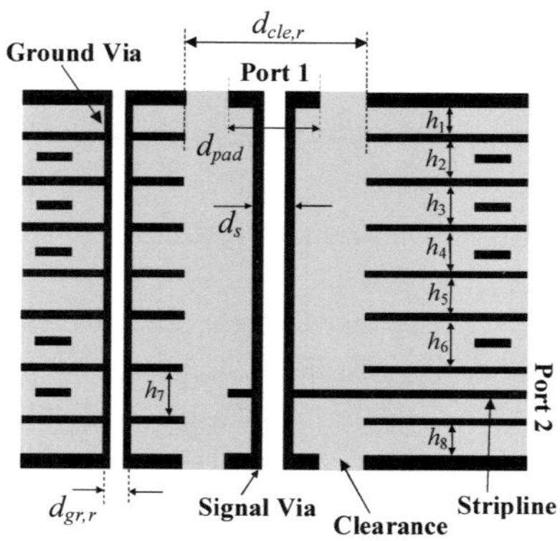

Fig.22b. Vertical cross-sectional view of shield via in multilayer substrate

Note that two dimensions of the clearance hole are considered here. The multilayer subsrate formed by PCB technologies consists of fourteen copper planar layers isolated by the FR-5 material of the relative permittivity of $\varepsilon = 3.78$ as assumed in simulations. Spaces between planar conductor layers (see Fig.22b) are: $h_1 = h_8 = 0.14mm$; $h_2 = h_3 = 0.335mm$; $h_4 = 0.56mm$; $h_5 = 0.15mm$; $h_6 = h_7 = 0.335mm$. The thickness of conductor planes embedded in the PCB is 0.035*mm*; the thickness of top and bottom conductor planes is 0.055*mm*. The signal conductor in the shield via is connected to the stripline by means of the pad having the same diameter, $d_{pad} = 1.2mm$, as via pads at top and bottom conductor layers. The width of the stripline is $w_{str} = 0.14mm$ providing the characteristic impedance of about 50*Ohms*.

Here, both TDR (Time Domain Reflectometry) and *S*-parameter data obtained by the use of the 3-D full-wave electromagnetic simulator are presented.

As input signal, the Gaussian pulse, shown in Fig.23, has been applied to stimulate a test model. Note the width of applied pulse is short (about 40*ps* at the 0.5-amplitude level). This corresponds a high-speed data transmission system.

Characteristic impedance in time domain is calculated according to following well-known equation:

$$Z(t) = \frac{1 + \rho(t)}{1 - \rho(t)} \cdot Z_0 , \tag{8}$$

where Z_0 is the characteristic impedance of input and output ports of the test structure and $\rho(t)$ is the reflection coefficient from the test model taken in time domain.

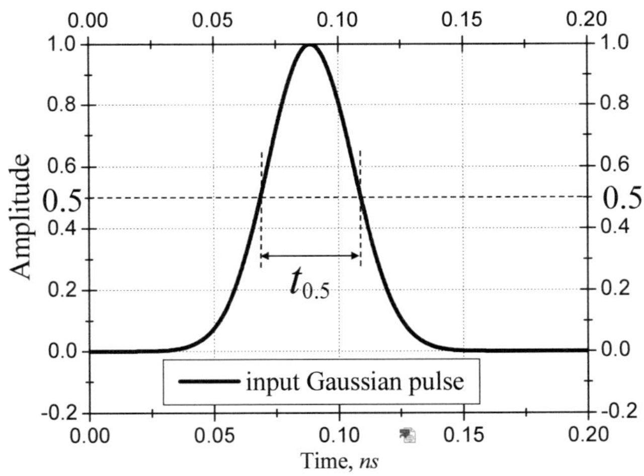

Fig. 23. Input Gaussian pulse used in simulations

In Fig.24, simulated results of the characteristic impedance are presented for models of two different clearance holes: The first is typical clearance hole defined by a technological process to provide a non-contact fabrication of the signal via and conductor layers in the PCB (for this case, $d_{cle,r} = 1.4mm$); The second is an optimized clearance hole ($d_{cle,r} = 3.4mm$) obtained according to the corrugated coaxial waveguide model presented.

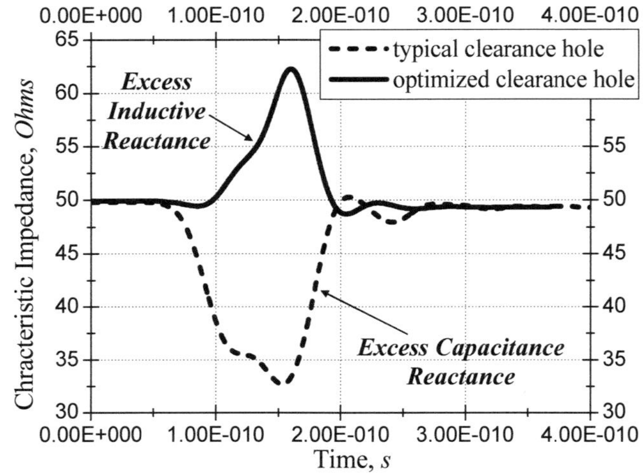

Fig. 24. Characteristic impedance in time domain for the test models

As one can see, a large impedance mismatching for the model comprising the shield via with the typical clearance hole is excited by an excess capacitance reactance due to coupling between the signal via and conductor planes of the PCB.

Optimized clearance hole removed the effect of excess capacitive reactance and considerably improved the electrical performance of the model as it can be traced from simulated *S*-parameter data presented in Figs.25a and 25b.

In spite of a considerable improvement of the interconnection comprising the shield via with the clearance hole optimized, another problem is appeared in the transition from the signal via to the stripline. This is impedance mismatching due to excess inductive reactance (see Fig.24) which is originated by the strip segment disposed between the signal via pad and the stripline. This segment acts as a flat wire inductor for which the characteristic impedance can be approximately represented by the following formula:

$$Z_w \approx iX_w = 2\pi \cdot f \cdot L, \tag{9}$$

where X_w is the inductive reactance of the strip, *f* is frequency, and *L* is the inductance of the strip segment.

To provide the characteristic impedance matching in the transition from the signal via pad to the planar transmission line in a multilayer substrate, a method to compensate the excess inductive reactance of the strip segment in the area of the clearance hole has been developed. Basis of this method can be traced by a block diagram shown in Fig.26, in which an additional capacitance reactance, $X_{add} = -1/2\pi f C_{add}$, is introduced to reduce or to suppress the effect of the excess inductive reactance of the strip segment.

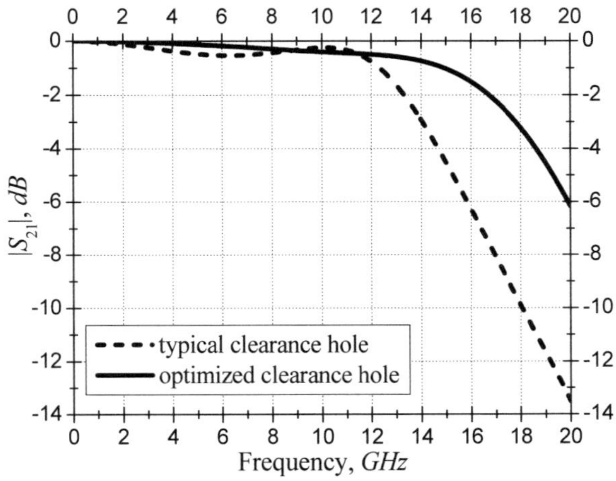

Fig.25a. Simulated insertion loss for the model shown in Figs.22a and 22b

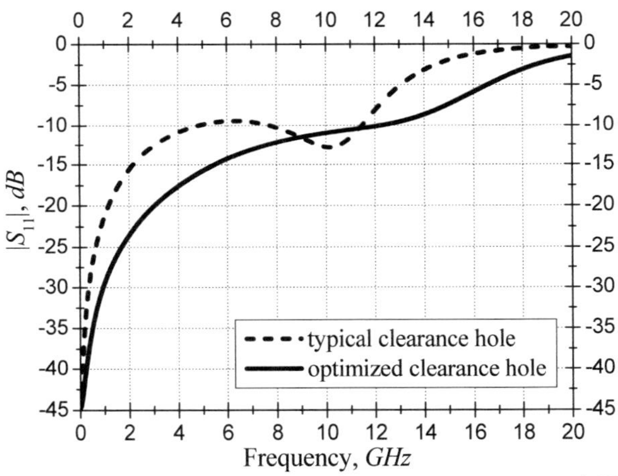

Fig.25b. Simulated return losses for the models shown in Figs.22a and 22b

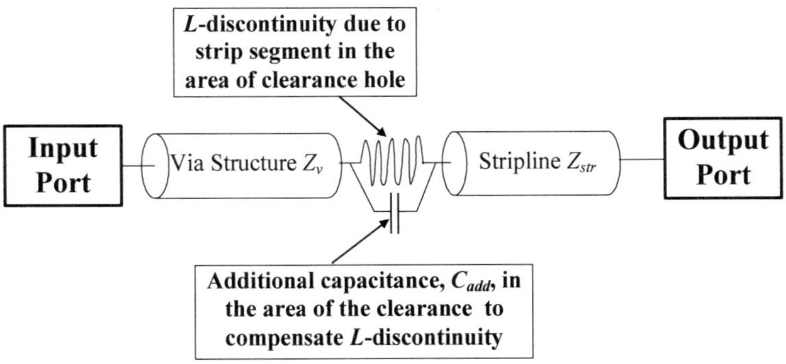

Fig.26. Block diagram of signal propagation in the model shown in Figs.22a and 22b

The additional capacitance can be obtained by forming a transition from the signal via pad to the stripline as the liner taper of the strictly-defined length. Note that the taper length, in this case, is an important parameter to control the magnitude of the additional capacitance and, as a result, the characteristic impedance matching in the considered model. In Fig.27, the horizontal cross-section view is presented for the same model as in Figs.22a and 22b, but only here a linear taper is formed as the transition from the signal via pad to the stripline. The dimensions of the via-stripline structure and parameters of the PCB are the same as for Figs.22a and 22b.

The effect of the taper length is presented in Fig.28 by means of TDR data. In this figure, four characteristic cases of via-to-stripline transitions are presented: The first is the model without application of the compensating method; The second is the transition formed as the linear taper for which the length is equal to the radius of the clearance hole; The third is the optimal taper length application; The fourth is the transition in which a long taper is used.

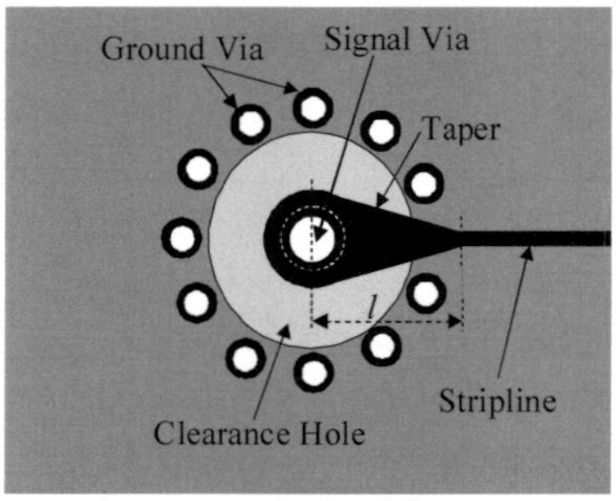

Fig. 27. Horizontal cross-sectional view of shield via in multilayer board taken in the position of stripline.

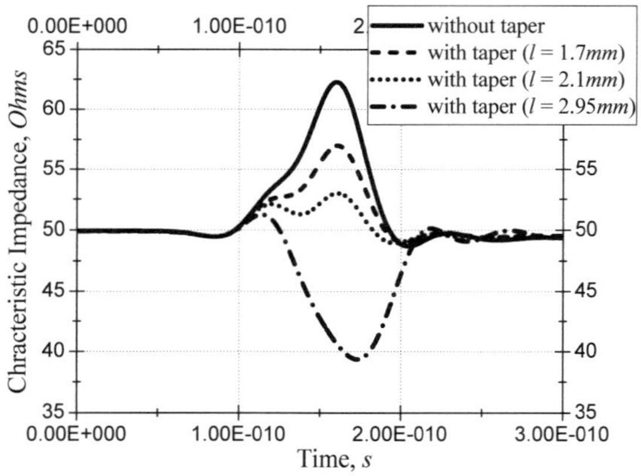

Fig. 28. Characteristic impedance in time domain

As follows from this figure, a good impedance matching (within 5*Ohms* or 10% of the nominal value) is achieved by the use of the linear taper having the length of l=2.1*mm*. Note that this length is larger than the radius of the optimal clearance hole used in the via structure.

On the one hand, the taper with the length equal to the radius of the clearance hole shows the higher excess inductive reactance than indicated nominal value (50*Ohms*). On the other hand, the long taper (2.95*mm* for presented test structure) leads to a high excess capacitance reactance.

Thus, a key point to realize a high-performance transition from the via pad to the planar transmission line is not only the use of the compensating part in the form of a linear taper but also its strictly-defined length as follows from data demonstrated in Fig.28.

Such result can be also traced by means of magnitudes of *S*-parameter data presented in Fig.29a and 29b in the frequency band up to 20*GHz*. These figures shows that the model with the linear taper of the optimal length (l=2.1*mm*) has the highest electrical performance compared with other transitions used.

To verify simulated results, experimental patterns of the same models disposed in the FR-5 multilayer board have been designed. In the measurement system, the SMA connectors have been used to provide the connection of the test model and a vector network analyzer. In Fig.30, a block diagram of the experimental patterns is presented.

In Fig.31, photo of bottom view of the test model is demonstrated. Also in Fig.32, photo of the total view of the top side of experimental patterns is shown.

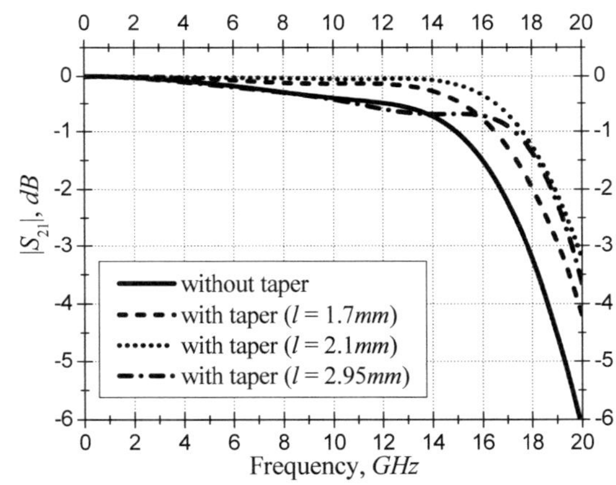

Fig. 29a Simulated insertion loss for test models

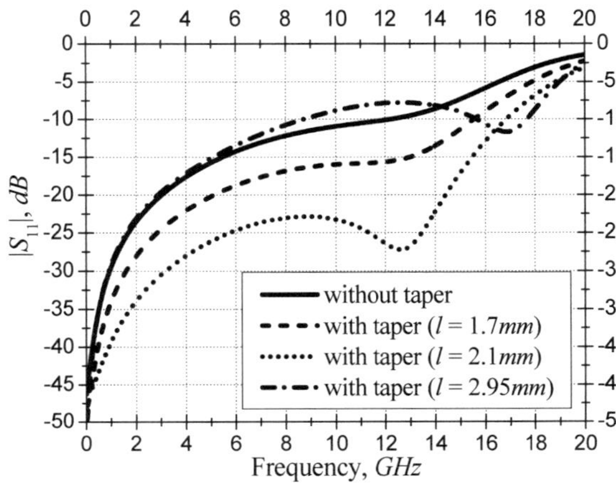

Fig. 29b Simulated return loss for test models

In Fig. 33 and 34, measured time-domain responses from the test models without the compensating technique application and with the taper optimized are respectively presented.

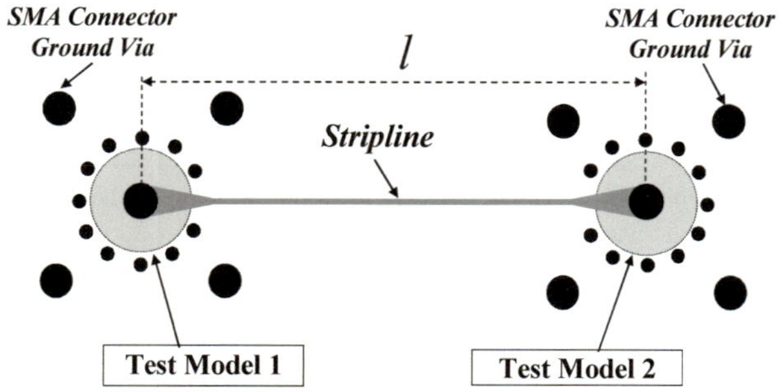

Fig. 30. Block diagram of experimental pattern

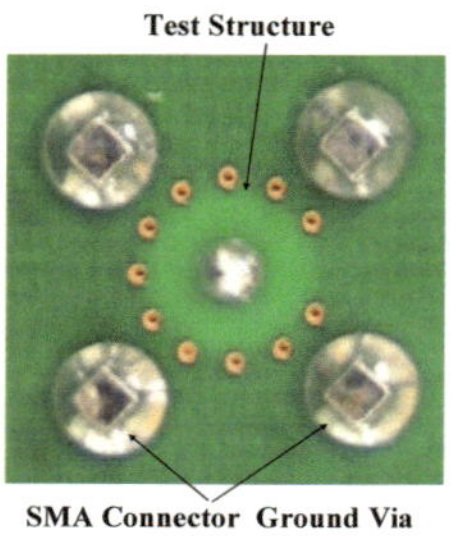

Fig. 31. Photo of experimental pattern (bottom view) of the test model

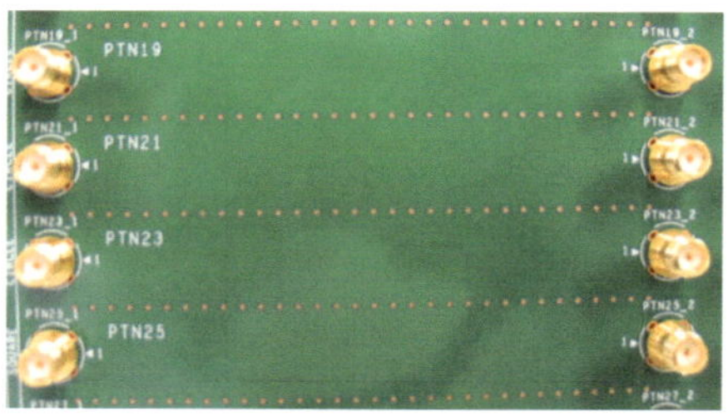

Fig. 32. Photo of experimental patterns (top view) of test models

As one can see in Fig. 33, measured TDR data demonstrate impedance mismatching up to 63*Ohms* in the area of the connection of the shield via having the characteristic impedance of about 50*Ohms* and the 50*Ohms* stripline. Also, presented measurement results are in a good agreement with the simulated data demonstrated in Fig.28.

Experimental data shown in Fig.34 make an approval of effectiveness of the use of the taper of the strictly-defined length to compensate the excess inductive reactance excited in the area of the clearance hole of the via-to-stripline transition. Thus, the use of the linear taper of l=2.1*mm* leads to impedance matching within ±10% of the nominal value 50*Ohms* (see Fig.34) that is corresponding to simulation data.

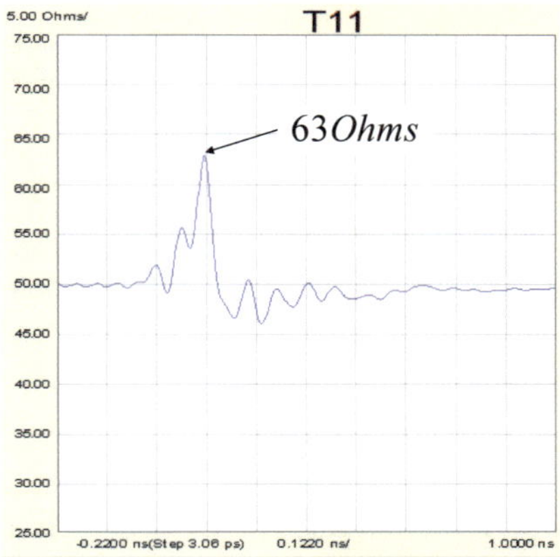

Fig. 33. Characteristic impedance in time domain for the test model without application of linear taper between signal via pad and stripline

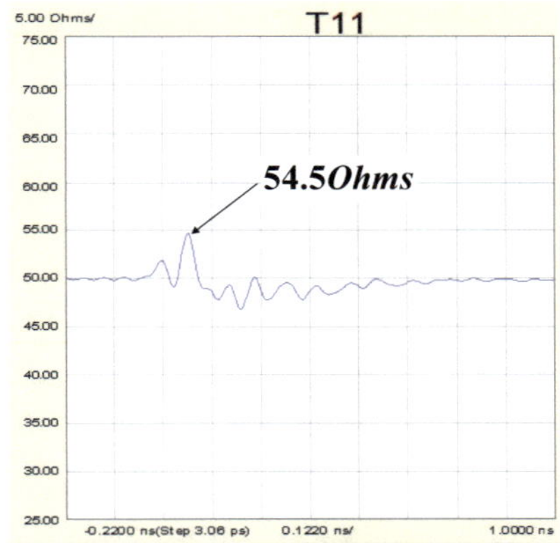

Fig. 34. Characteristic impedance in time domain for the test model with application of optimal linear taper (l=2.1mm) between signal via pad and stripline

Compensating method to improve impedance matching between a via structure and a planar transmission line disposed in a multilayer substrate is presented here. This method is a powerful technique to obtain high-performance electrical interconnections in high-speed multilayer substrates in which a liner taper of the strictly-defined length between the signal via pad and the planar transmission line is used to compensate the excess inductive reactance and to achieve required impedance matching.

4. Bandpass and Bandstop Filter Design Using Shield Via Approach

Two above-mentioned sections serve a good basis for development of high-performance interconnections and can be applied in high-frequency and high-speed multilayer substrates of present and next-generation communication and computing equipment.
Key point of this paragraph is a promotion of an approach using shield vias presented as a basis for providing miniaturization of both systems and components. Positioning of these directions is shown in Fig.35.

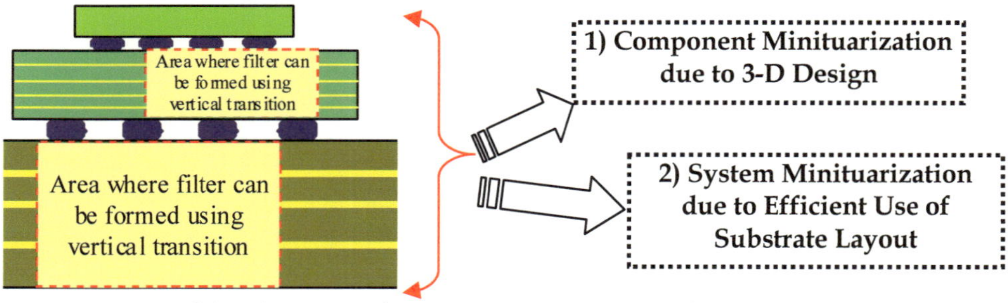

Fig. 35. Target of development of a component by means of the use of a shield via

So, we will start directly from an interconnected circuit which has been developed to realize the approach (Kushta et al., 2005; Kushta & Harada, 2008). Characteristic feature of this configuration is shown in Fig.36.

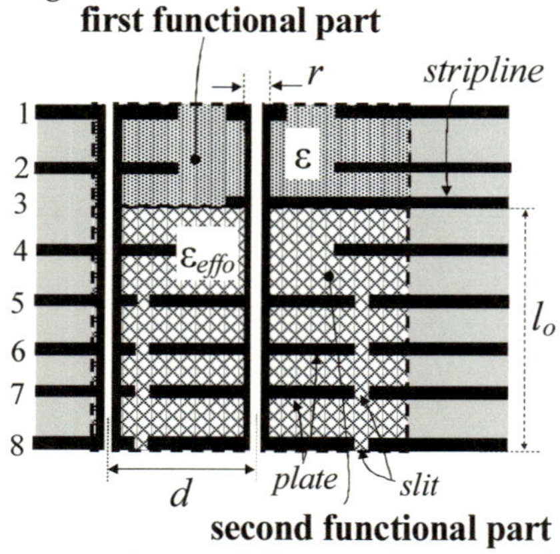

Fig. 36. A combined via structure in a multilayer substrate to form an open-circuited stub

First of all, a combined via structure, which includes two functional parts, is a distinguishing point. Consider each functional part in details.

The first functional part of the composite via structure is extended in the vertical direction from the top conductor layer to the signal conductor layer (where a stripline is disposed) of the multilayer substrate (see Fig.36). This functional part, formed as a shield via segment, provides impedance-matched low-loss signal transmission between signal pad disposed at the top conductor layer and stripline formed at signal conductor layer (third conductor layer in the structure presented). Its design can be made on the basis of the corrugated coaxial waveguide model given here.

The second functional part is extended in the vertical direction from the signal conductor layer to the bottom conductor layer. This functional part of the combined via structure serves to obtain a resonant stub. The resonant stub, besides the signal via surrounded by ground vias, comprises conductive plates connected to the signal via and separated from ground conductors by isolating slits.

Each functional part of the combined via can be characterized as a transmission line with appropriate characteristic impedance and propagation constant.

Electromagnetic properties of the first functional part can defined by means the characteristic impedance $Z_1 = f(r, d, \varepsilon)$ and propagation constant $\beta_1 = \frac{\omega}{c} \cdot \sqrt{\varepsilon}$, wherein $Z_1 = f(r, d, \varepsilon)$ shows a dependency of the characteristic impedance on transverse dimensions of the signal via, *r*, distance between the signal via and ground vias, *d*, and relative permittivity of an isolating material filling in the multilayer substrate, ε; ω is the angular frequency, *c* is the velocity of light in free space.

For the second functional part, the characteristic impedance and propagation constant can be represented as following: $Z_2 = f(r, d, \varepsilon_{eff})$ and $\beta_2 = \frac{\omega}{c} \cdot \sqrt{\varepsilon_{eff}}$, wherein, besides dimensional dependency for the characteristic impedance similar to the first functional part, the second functional parameters are a function of ε_{eff} which is dependent on dimensions and form of the conductive plates connected to the signal via, the distance between these conductive plates in the vertical direction, and the relative permittivity of the isolating material filling in the multilayer substrate.

Thus, by means of the combined via structure, we have obtained a cost-effective approach to design a miniaturized stub in which the resonant frequency is dependent on the stub length, l_0, and ε_{eff} , which can be much higher (!) than the relative permittivity of the substrate isolating material and in such way providing compactness of a filter using such stub.

Consider a representative example of the combined via structure used to form the open-circuited stub for which top, bottom and cross-sectional views are shown in Figs.37a, 37b, 37c and 37d.

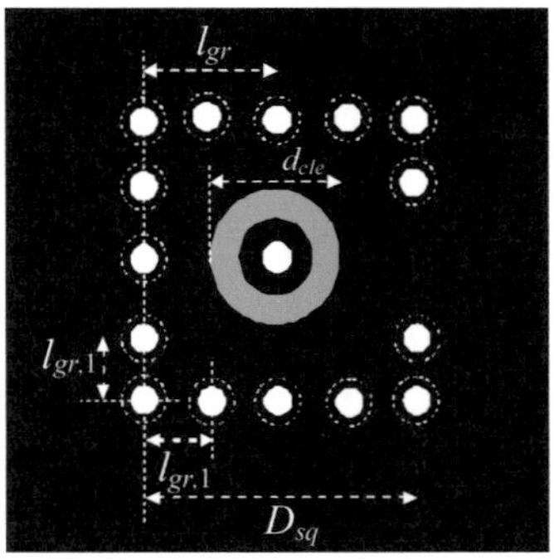

a) Top view

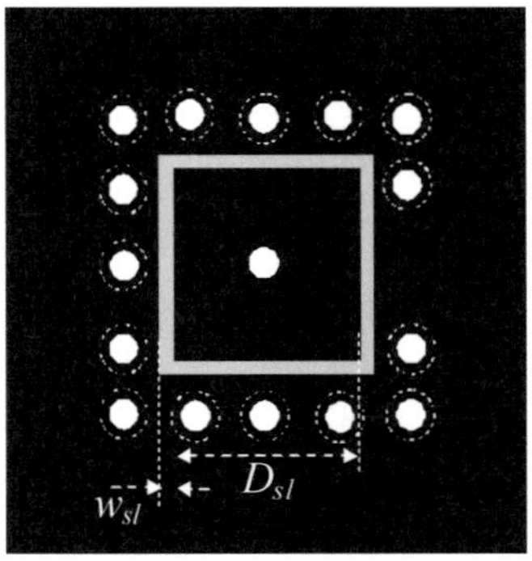

b) Bottom view

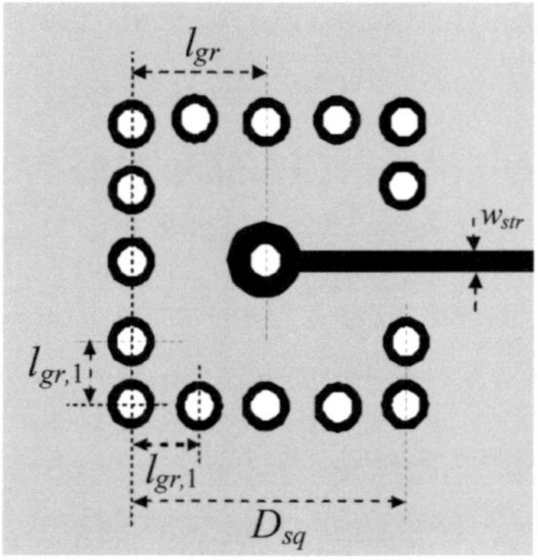

c) Horizontal cross-sectional view at signal layer

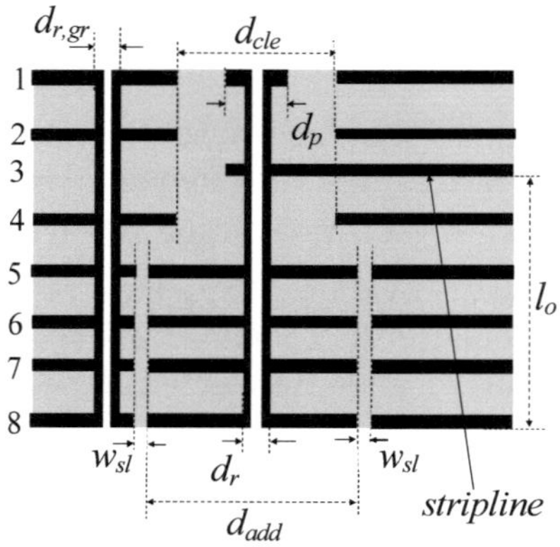

d) Vertical cross-sectional view

Fig. 37. Open-circuited stub formed by combined via structure in 8-conductor-layer PCB

Here, the stub is formed as a part of the shield via between the stripline disposed at the third conductor layer and pad formed at the bottom conductor layer of an 8-conductor-layer PCB Copper planar conductor layers of this PCB are isolated by the FR-4 material with the relative permittivity of $\varepsilon = 4.17$ as assumed in simulations. Arrangement of ground vias in the shield via has a square contour. Dimensions of the shield via (see Figs.37a, 37b, 37c, and 37d) are as following: $d_r = 0.65mm$, $d_{cle} = 1.65mm$, $d_p = 0.95mm$, $d_{r,gr} = 0.25mm$, $l_{gr} = 1.66mm$, $l_{gr,1} = 0.83mm$, $D_{sq} = 3.32mm$.

Conductive plates connected to the signal via to control the resonance frequency of the stub by means ε_{eff} have a square form with side of $d_{add} = 1.1mm$. The plates are separated from ground plates at the conductor layers by isolating slits of $w_{sl} = 0.2mm$.

Total thickness of the PCB including all conductor layers is 1.847*mm*. The thickness of the conductor layers is as follows: 0.062*mm* for layer 1; 0.069*mm* for layer 8; 0.035*mm* for layers 2, 3, 4, 5, and 6. The distance between conductor layers is as follows: 0.146*mm* is between layers 1 and 2; 0.123*mm* is between layers 2 and 3; 0.138*mm* is between layers 3 and 4; 0.677*mm* is between layers 4 and 5; 0.138*mm* is between layers 5 and 6; 0.138*mm* is between layers 6 and

7; and 0.146*mm* is between layers 7 and 8. Width of the stripline, w_{st}, at the third layer is 0.19*mm*.

In simulations, the same 3-D full-wave electromagnetic field solver has been used. Electromagnetic properties of such open-circuited stub are studied by means of *S*-parameter matrices in the frequency band up to 20*GHz* and are presented in Fig.38. Note that in these simulations signal is propagating between the pad of the combined via structure at the top conductor layer and the end of the stripline embedded in the PCB. From presented results the bandstop effect of the open-circuited stub at the central frequency of about 8*GHz* is clearly traced. This example demonstrates a possibility to form a filter using such open-circuited stub as a building block. The resonance frequency, f_{ro}, of the open-circuited stub can be defined as following:

$$f_{ro} = \frac{c}{4 \cdot \sqrt{\varepsilon_{effo}} \cdot l_o}, \tag{10}$$

where l_o is the length of the stub (or the second functional part in the vertical direction) as shown in Fig.37d.

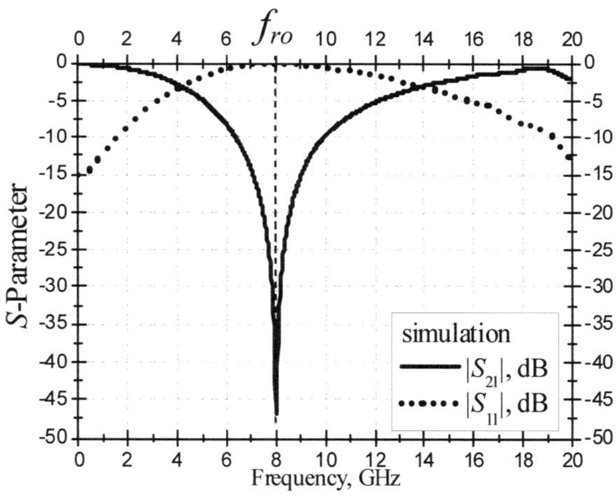

Fig. 38. Simulated insertion ($|S_{21}|$-parameter) and return ($|S_{11}|$-parameter) losses for open-circuited stub in 8-conductor-layer PCB

In the similar manner, a short-circuited stub can be obtained by means of another combined via structure in a transition from this via structure to a stripline as presented in Fig.39. In this case, the pad of the second functional part disposed at the bottom conductor layer is connected to the ground plane. Magnitudes of *S*-parameters for such configuration are demonstrated in Fig.40. Dimensions of the via structures, conductor plates connected to the signal via, stripline, and PCB are the same as for the aforementioned open-circuited stub case. Shown data demonstrate the bandpass effect with the central resonance frequency about 13*GHz*. This resonance frequency can be defined for this structure as following:

$$f_{rsh} = \frac{c}{4 \cdot \sqrt{\varepsilon_{effs}} \cdot l_{sh}}, \tag{11}$$

It is important to note that $\varepsilon_{effo} \neq \varepsilon_{effs}$, because different number of conductive plates is used to form the effective medium in short-circuited and open-circuited stubs. As one can see from Figs. 38 and 40, conductive plates connected to the signal via are powerful parameter to control resonance characteristics of both open- and short-circuited stubs.

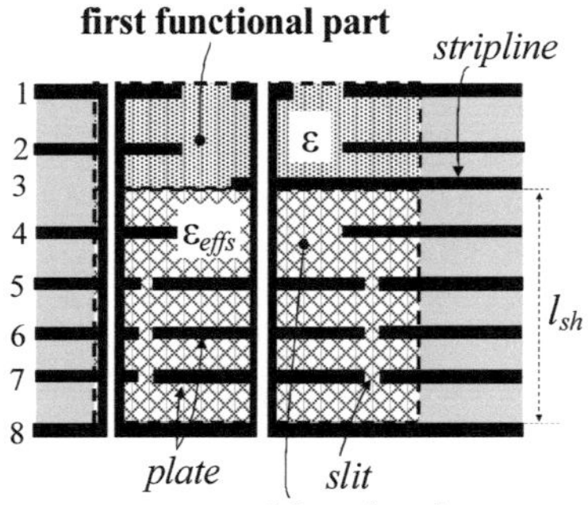

Fig. 39. A combined via structure in a multilayer substrate to form a short-circuited stub

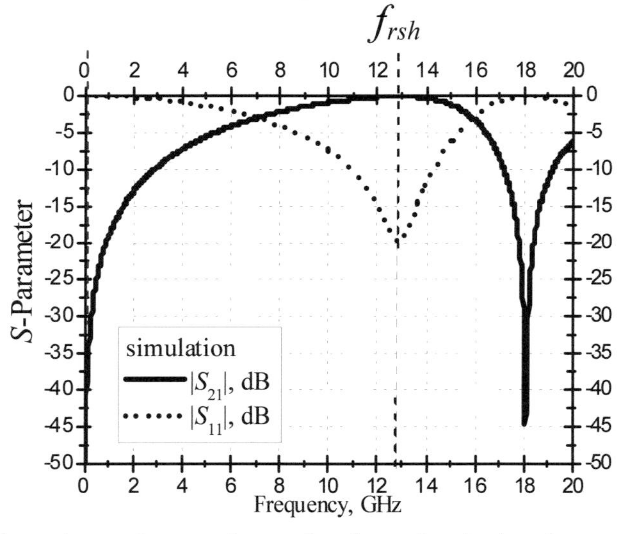

Fig. 40. Simulated insertion and return losses for short-circuited stub

As the next step, it will be shown here that the combined via structures can be effectively used to form high-performance filtering components. In Figs.41a and 41b, 41c and 41d, top and bottom views, horizontal and vertical cross-sectional views of a bandpass filter are shown, respectively. This filter is designed by the use of two identical combined via structures (discussed above) forming short-circuited stubs in the 8-conductor-layer PCB. Photographs of an experimental pattern of such bandpass filter are presented in Fig.42. Dimensions of the shield via, conductive plates connected to the signal vias, isolating slits, stripline connecting two signal vias, and PCB are the same as for the combined via structure forming aforementioned short-circuited stub. The distance between the centers of signal via conductors, L_f, is 3.32*mm*. Input and output ports of the filter are at the pads of the signal vias as shown in Fig.41d.

Simulated data of magnitudes of *S*-parameters for the considered bandpass filter in the 8-conductor-layer PCB are shown in Fig.43. In this figure one can define clearly-expressed bandpass properties of the filter in the frequency band from about 9*GHz* to 15*GHz*.

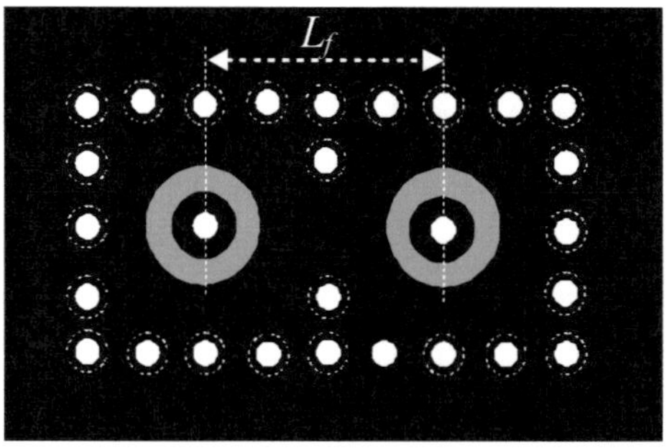

a) Top view

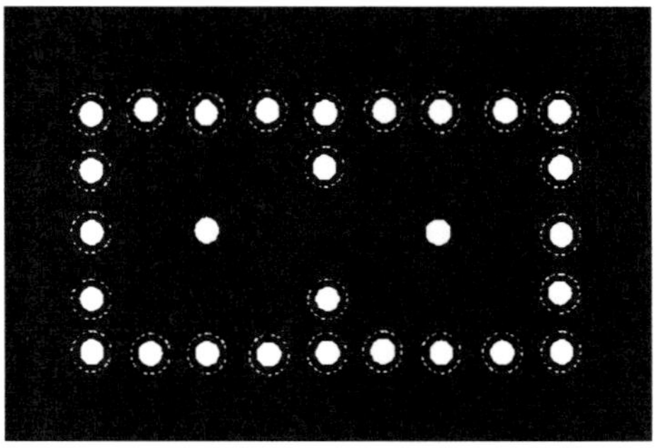

b) Bottom view

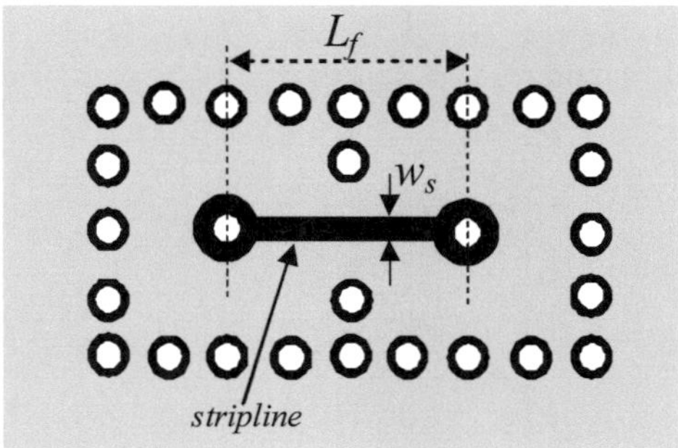

c) Cross-sectional view at signal layer

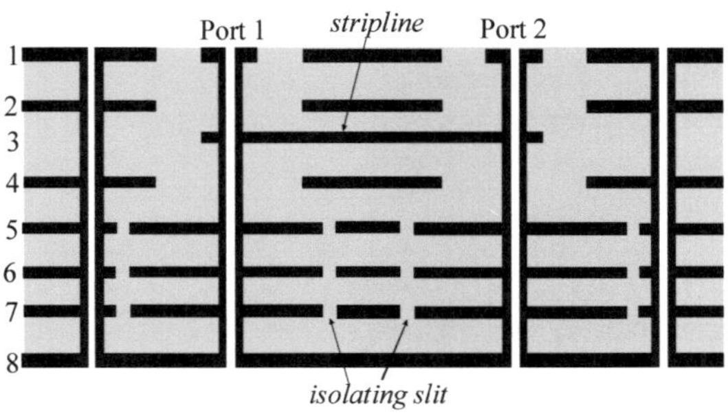

d) Cross-sectional view

Fig. 41. Bandpass filter in 8-conductor-layer PCB

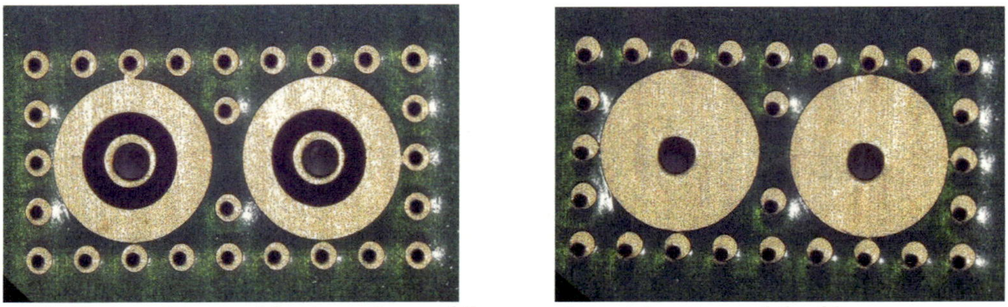

Fig. 42. Photos of top and bottom views of the experimental pattern of the filter

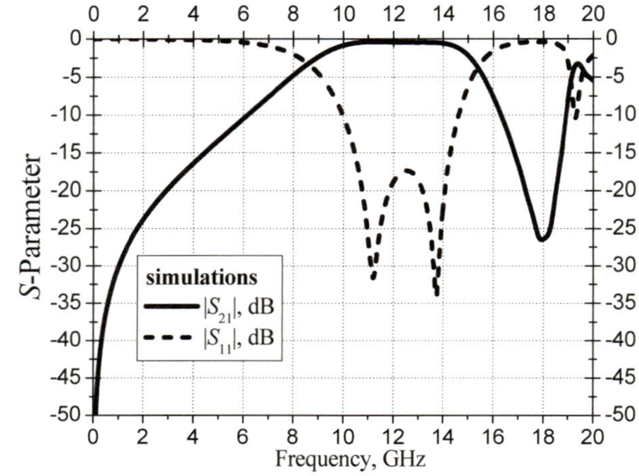

Fig. 43. Simulated insertion and return losses for bandpass filter formed by two short-circuited stubs

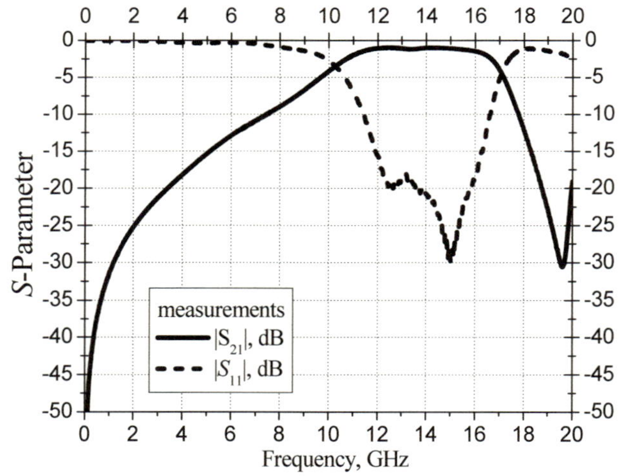

Fig. 44. Measured insertion and return losses for bandpass filter formed by two short-circuited stubs

Experimental verification of the bandpass filter presented and proposed 3-D approach are shown in Fig.44. In this figure, measured data for the filter having the same dimensions as for Fig.43 are presented. These data demonstrate similar bandpass characteristics and are in a good agreement with the simulation results.

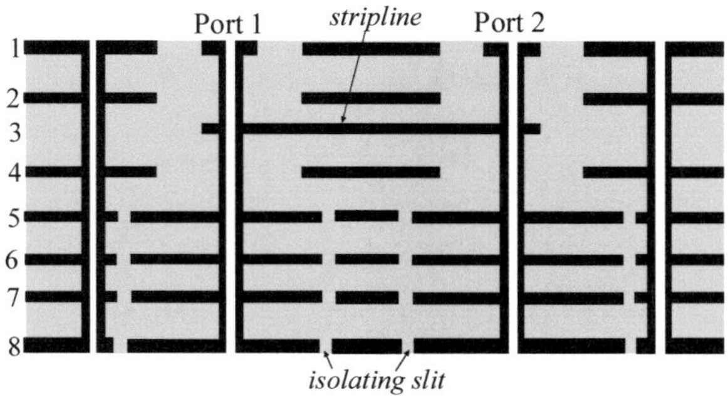

Fig. 45. Cross-sectional view of filter formed by two open-circuited stubs

Another type of filters can be obtained if open-circuited stubs will be used instead of short-circuited ones. Cross-section view of such filter is presented in Fig.45.

In Fig.46, simulated magnitudes of *S*-parameters are shown. This filter demonstrates clearly-expressed bandstop properties with the central frequency of about 8*GHz*.

As a validation of simulated results, in Fig.47, measured data for the filter are presented. As one can see, these data are corresponding to simulation with a good accuracy, including the central frequency.

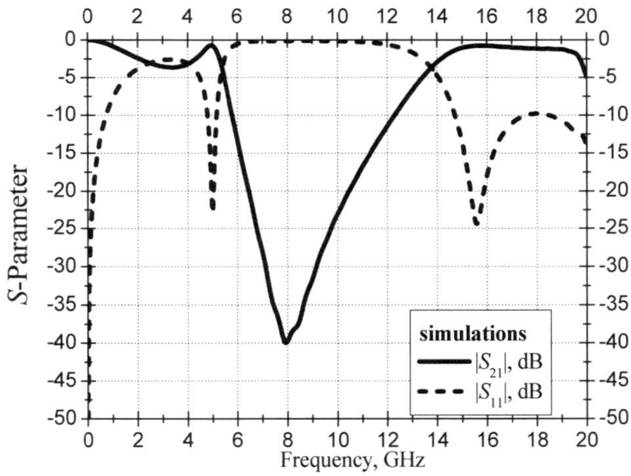

Fig. 46. Simulated insertion and return losses for bandpass filter formed by two short-circuited stubs in 8-conductor-layer PCB

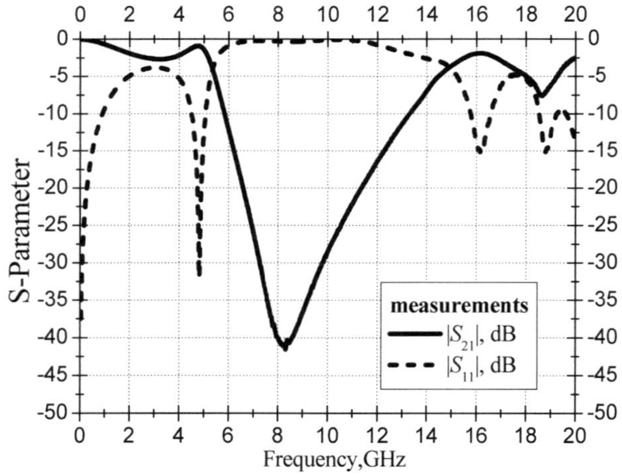

Fig. 47. Measured insertion and return losses for bandpass filter formed by two short-circuited stubs in 8-conductor-layer PCB

Developed approach can be applied to obtain a multipole filter. In Fig.48, a compact four-pole bandpass filter is presented. This filter consists of four identical short-circuited stub elements (as in two-pole filter) connected by the stripline. In Fig.49, simulated *S*-parameters for the filter are shown. As one can see, the increase of quantity of stub elements can control bandpass characteristics. In present example, it leads to sharpening the pass band compared with the two-pole filter.

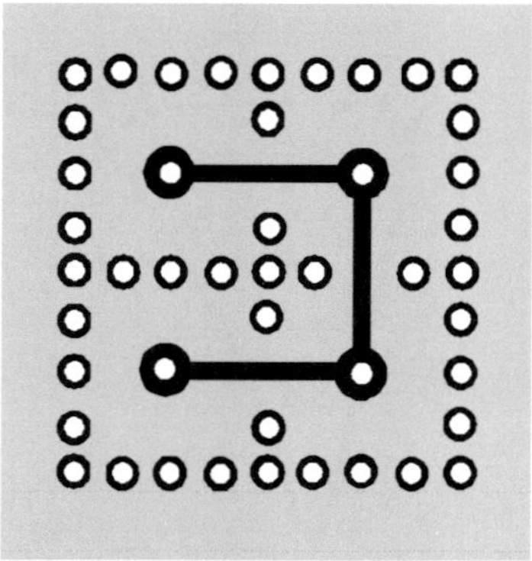

Fig. 48. Cross-sectional view at the signal layer of the bandpass filter formed by four short-circuited stubs in 8-conductor-layer PCB

Thus, these examples of the bandpass and bandstop filters demonstrate applicability of presented approach to design compact and cost-effective filtering components using shield vias in a multilayer substrate.

Required frequency band and desired characteristics of the filter can be achieved by an appropriate choice of dimensions and quantity of the short-circuited or open-circuited stubs. Also, an optimization of the parameters of the filter can be provided by the determination of an appropriate distance between the stubs. Such design techniques will be studied and presented in the future in details.

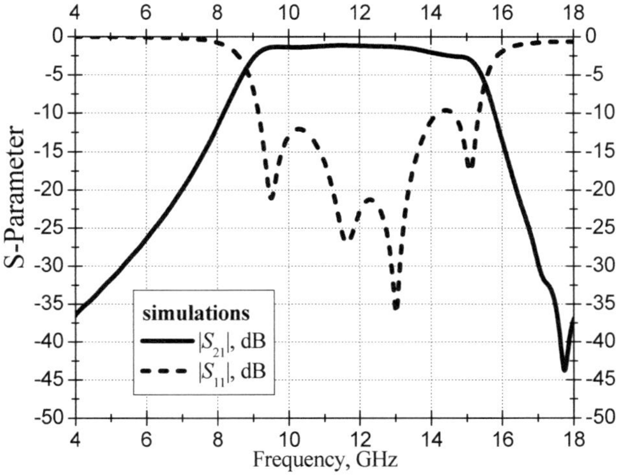

Fig. 49. Simulated insertion and return losses for bandpass filter formed by four short-circuited stubs in 8-conductor-layer PCB

5. Conclusion

In this chapter, advantages of shield vias disposed in a multilayer substrate at the higher frequencies have been shown. These structures have low leakage losses, well-controlled characteristic impedance and definite propagation constant. Presented physical model for the electromagnetic behavior of the shield via is consistent with results obtained by simulations and measurements and can be used for optimum design of vertical transitions in multilayer substrates for high-speed and high-frequency applications. Moreover, vertical transitions demonstrating the electrical performance similar to planar transmission lines can be obtained in multilayer substrate structures.

Also, a powerful 3-D approach to design filtering components using shield vias is presented here. A distinguishing point of this approach is a combined via structure comprising two functional parts. The first functional part is responsible for low-loss signal transmission in a wide frequency band and can be design by means of the corrugated coaxial waveguide model presented here for shield via. The second functional part serves to form a compact short-circuited or open-circuited resonant stub acting as a building block of a filter and comprises conductor plates connected to the signal via to control characteristic impedance and propagation constant in this functional part. Simulation and measurement data for bandpass and bandstop filters presented demonstrate applicability of the approach to design compact and cost-effective filters for highly integrated systems at microwave and millimeter waves with an improved EMI performance.

6. References

Gupta, K.C.; Garg R. & Bahl I. (1996). *Microstrip Lines and Slotlines*, 2nd Ed., Artech House, ISBN 0-89006766-X.

Hoffmann R, (1987). *Handbook of Microwave Integrated Circuits*, Artech House, ISBN 0-89006163-7.

Kushta, T.; Narita K.; Kinoshita Ya.; Tohya H.; Saeki T.; Kaneko T.; Kamiya H. & Kitao K. (2002). Characterization of via structures in multilayer printed circuit boards, *in Proceedings (CD-ROM) of International Union of Radio Science, XXVIIth General Assembly*, no.p1627, pp. 1-4, Maastricht, the Netherlands, August 2002.

Kushta, T.; Narita K.; Kaneko T.; Saeki T. & Tohya H. (2003). Resonance stub effect in a transition from a through via hole to a stripline in multiplayer PCB's, *IEEE Microwave and Wireless Components Letters*, Vol.13, No.5, May 2003, pp.169-171.

Kushta, T.; Narita K.; Saeki T. & Tohya H. (2004). High-isolated and high-performance vertical transitions for multilayer printed circuit boards, *Procedings of International Symposium on Electromagnetic Compatibility*, Vol.1, pp.57-60, Sendai, Japan, June 2004.

Kushta, T.; Narita, K.; Saeki, T. & Tohya H. (2004). Effect of ground via shielding on the electrical performance of via interconnections embedded in a multilayer PCB," *Journal of Japan Institute of Electronics Packaging*, Vol.7, No.4, July 2004, pp.314-321.

Kushta T. & Narita K. (2004). Shielded via structures for high-speed printed circuit boards, *IEICE Technical Report*, Vol.104, No.328 (EMCJ2004-55~69), October 2004, pp.29-34.

Kushta, T.; Narita K. & Kuriyama T. (2005). Novel filtering components based on multilayer substrate technologies, *Proceedings of International Symposium on Microwave and Optical Technology*, pp.827-830, Fukuoka, Japan, August 2005.

Kushta, T. & Harada, T. (2008). A broadband transition from a via structure to a planar transmission line in a high-speed multilayer board, *Proceedings of International Conference on Electronics Packaging*, pp.134-139, Tokyo, Japan, June 2008.

Kushta, T. & Harada, T. (2008). New 3-D design of filtering components using multilayer board technologies, *Proceedings of the 38th European Microwave Conference*, pp.218-221, Amsterdam, the Netherlands, October 2008.

Laermans, E.; Geest, J. D.; De Zutter, D.; Olyslager, F.; Sercu S. & Morlion, D. (2001). Modeling differential via holes, *IEEE Transactions on Advanced Packaging*, Vol. 24, August 2001, pp. 357–363.

Pillai, E.R. (1997). Coax via – a technique to reduce crosstalk and enhance impedance match at vias in high-frequency multilayer packages verified by FDTD and MoM modeling, *IEEE Transactions Microwave Theory and Techniques*, Vol.45, October 1997, pp.1981-1985.

Tarvainen, T. (2000). Simplified Modeling of Parallel Plate Resonances on Multilayer Printed Circuit Boards, *IEEE Transactions on Electromagnetic Compatibility*, Vol. 42, August 2000, pp. 284-289.

Weiland, T. (1996). Time domain electromagnetic field computation with finite difference methods, *International Journal of Numerical Modeling: Electronic Networks, Devices and Fields*, Vol.9, July-August 1996, pp.295-319.

Wheeler,H. (1979). Transmission-line properties of a round wire in a polygon shield, *IEEE Transactions Microwave Theory and Techniques*, Vol.27, August 1979, pp.717-721.

14

Impedance Transformers

Vitaliy Zhurbenko, Viktor Krozer and Tonny Rubæk
Technical University of Denmark
Denmark

1. Introduction

Impedance matching is an important aspect in the design of microwave and millimeter wave circuitry since impedance mismatches may severely deteriorate the overall performance of electronic systems.

In high-power applications, the standing electromagnetic wave resulting from mismatch in a transmission line is highly undesirable as it leads to amplitudes of voltage and current which might be several times higher than those in a matched line. This can lead to disruption or even damage of the dielectric in the transmission line. A reflected electromagnetic wave can also result in frequency pulling of signal generators connected to the mismatched transmission line, thereby shifting the oscillation frequency from the desired.

In transceiver applications, antenna mismatch leads to signal power loss and lower signal-to-noise ratio, thereby deteriorating the overall transmit or receive performance.

When designing low-noise amplifiers, it is often required to control the input network mismatch. Generally, it is not possible to design an amplifier which has the optimum input impedance for minimum noise figure equal in value to the optimum impedance for maximum gain. The input network is then should be mismatched in order to provide a low-noise operation.

Impedance transformers can also be effectively used to improve selectivity of resonant circuits and are very useful in filter design. Low values of source and load impedance decrease the loaded quality factor Q and increase the bandwidth of a given resonant circuit. This makes it very difficult to design even a basic LC high-Q resonant circuit for use between two very low values of source and load resistances. A common method to overcome this problem is to use impedance transforming circuits to present the resonant circuit with a source or load resistance that is much larger than what is actually present. Consequently, by utilizing impedance transformers, both the Q of the resonator and its selectivity can be increased.

Matching a complex impedance in a wide frequency range is most commonly achieved by using one of the following techniques:

- passive two-port networks consisting of reactive components;
- passive two-port networks consisting of resistive components;

Wideband matching can also be achieved by means of ferrite circulators in which the reflected wave is guided to an absorbing load, and ferrite isolators in which the transmission losses are different for the incident and reflected waves.
For a wideband matching, it is preferable to place the matching network as close as possible to the load, as it is illustrated in Fig. 1.

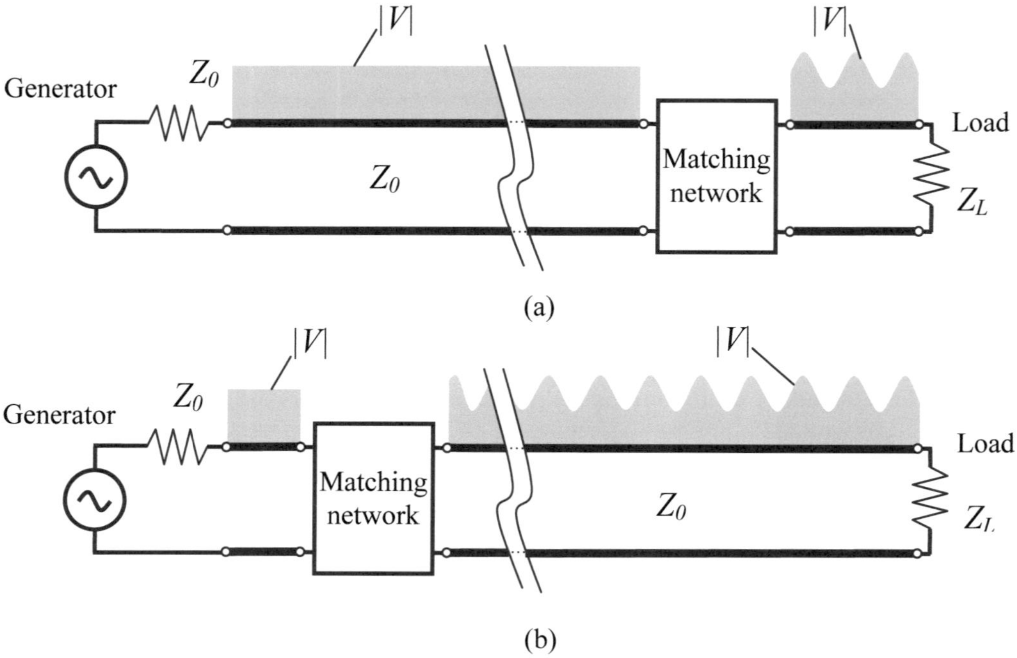

Fig. 1. Voltage standing wave patterns. Placing of the matching network with regard to the generator for wideband (a), and narrowband (b) matching.

This concept will be demonstrated in the later section 3.1 by considering an example of matching a complex load using shunt stubs.
In this chapter, different techniques for wideband matching are presented. Sections 2 thru 4 briefly present some of the well-known matching techniques while the use of coupled transmission lines for wideband matching is treated in depth in Section 5. The first part of the chapter includes a discussion of resistive and reactive lumped elements in Section 2, different types of stub matching in Section 3, and the use of series of transmission lines in Section 4. Since these techniques are all thoroughly treated in the literature, only the design-considerations relevant for applying the techniques for wideband matching are treated here while the reader is referred to the literature for specifics, such as the relevant formulas for calculating the values of the different components.
The use of coupled transmission lines for wideband impedance matching is not as widely used as the techniques described in sections 2 thru 4. Hence, in Section 5, a detailed presentation of this technique is given.

2. Matching Using Resistive and Reactive Lumped Elements

Resistive elements or attenuators can be effectively used to lower the level of the reflected signal from the load in a very wide frequency range. It should be noted, though, that the efficiency of such matching networks is low because they attenuate not only the reflected but also the incident wave.

Another type of matching network uses lumped reactive components to match a complex load impedance to a desired complex impedance. For moderate bandwidths, the component values of two-element matching networks can be found relatively easy by first choosing a pair of initial values on the basis of the Smith Chart and then applying computational optimization. To increase the bandwidth, more than two reactive elements are required. The synthesis and optimization of multi-element wideband matching circuits can be accomplished by means of software tools, which are currently available in a wide variety. The implementation of this type of transformers in microwave and millimeter wave range is limited due to the low Q-factor of lumped components. Therefore, lumped element matching is usually employed only at low frequencies, or in applications where compact size is a major requirement, e.g., in monolithic microwave integrated circuits design (Kinayman & Aksun, 2005).

3. Stub matching

This section is dedicated to matching circuits that use open-circuited or short-circuited transmission line sections, connected in parallel with the load or transmission feed line. This is a well developed matching technique which is often used in microwave and millimeter wave circuits. In this section, some of the important operational principles and properties of shunt stub matching circuits are discussed. More detailed analysis of this type of matching technique is available in the literature (Pozar, 1998), (Kinayman & Aksun, 2005).

3.1 Single-Stub Matching

This is one of the most simple and convenient ways of matching a transmission line with a load which has real or complex impedance. This method was developed by Tatarinov V. V. in 1931 and is widely used for narrow-band matching in microwave and millimeter wave applications. It consists of a short circuited or open circuited stub and a piece of transmission line between the load and the stub. An example of the single-stub matching circuit is shown in Fig. 2.

There are several choices of electric distance θ_d from the load to the matching stub. In the first case (Fig. 2 (a)), the distance between the load and matching stub is chosen as short as possible while this distance is chosen to be several times longer in the second case (Fig. 2 (b)). The responses of these two matching circuits are shown in Fig. 3. The 10 dB reflection loss bandwidth of the circuit in Fig. 2 (a) is 10.3 % while the same parameter for the circuit in Fig. 2 (b) is equal to 1.9 %. Thus, by using $\theta_d = 56.85°$ instead of $\theta_d = 282.05°$, the bandwidth is increased by more than a factor of 5.

There is also a difference in the wideband response of the matching circuits. The circuit with long distance between the load and the matching stub demonstrates more passbands in the same frequency range.

3.2 Double-Stub Matching

Single-stub matching can match any load impedance, but it requires a variable electric length of the transmission line between the load and the stub. This poses practical difficulties for adjustable tuners.

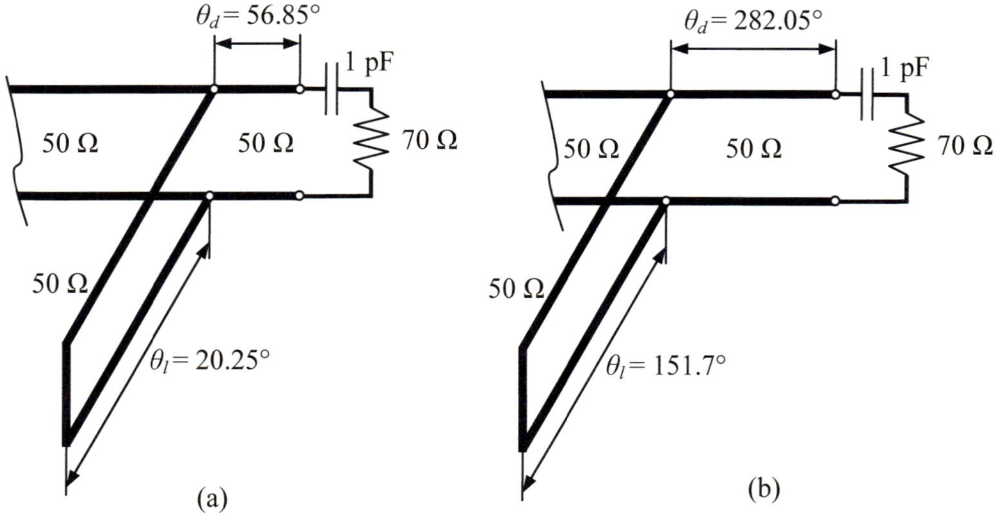

Fig. 2. Two single-stub matching solutions. (a) wideband, (b) narrowband. The load is matched at f_0 = 1 GHz.

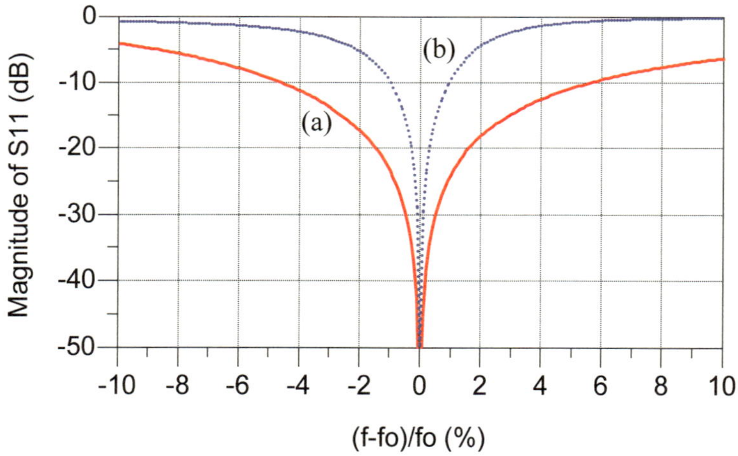

Fig. 3. Magnitude of S_{11} versus offset frequency for the matching circuits in Fig. 2. Here, f_0 = 1 GHz is the center frequency of operation.

Therefore, it would be more useful to have the length fixed and still be able to match a wide range of load impedances. This can be achieved with double-stub matching, as shown in Fig. 4, which allows for an arbitrary electric distance between the load and the stub.

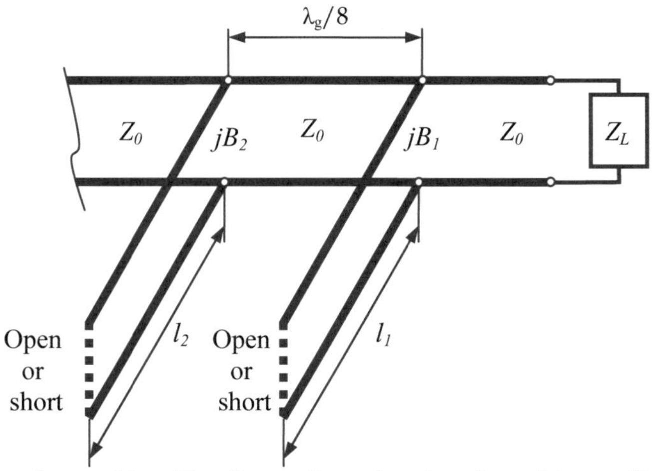

Fig. 4. Double-stub matching. The first stub can be placed at arbitrary distance from the load.

It should be noted that stub spacings near 0 or $\lambda_g/2$ (where λ_g is the guided wavelength) lead to matching networks that are very frequency sensitive (Pozar, 1998), and consequently, very narrowband. In practice, stub spacings are usually chosen as odd number of $\lambda_g/8$, for example $\lambda_g/8$, $3\lambda_g/8$ or $5\lambda_g/8$.

3.3 Triple-Stub Matching

The double-stub matching circuit can not match all load impedances. For a specified distance between two stubs, the matching is possible only for limited values of loads, which depend on amplitude and phase of the standing wave. This limitation arises from the fact, that the stub itself can not change the real part of the impedance at the point of connection to the transmission line.

This limitation can be overcomed by using a triple-stub matching as the one shown in Fig. 5.

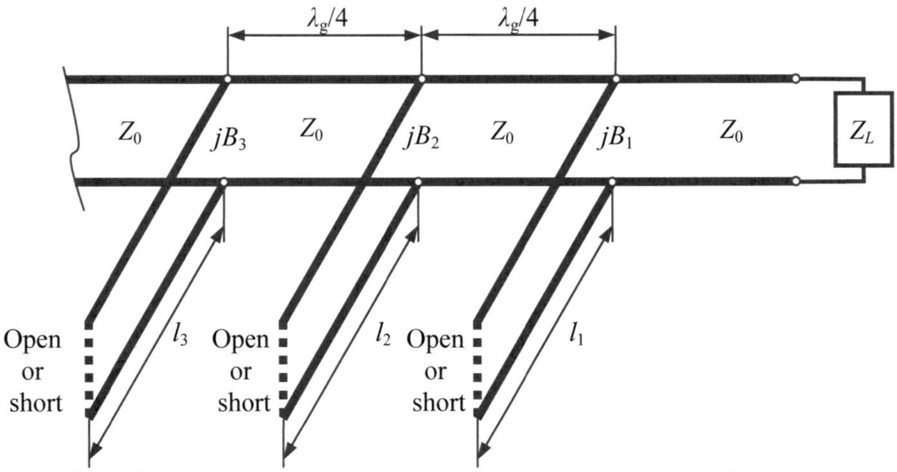

Fig. 5. Triple-stub matching. The first stub can be placed at arbitrary distance from the load.

It allows for an arbitrary distance between the load and the stub and also allows to match arbitrary load impedance. The operation of triple-stub matching circuit can be treated as a combination of two double-stub matching circuits and stub spacings are usually chosen as $\lambda_g/4$.

4. Series Transmission Line Matching

This section is dedicated to matching circuits that use series transmission lines, such as single section quarter-wave transformer, multisection transformers, and tapered transmission lines.

4.1 The Quarter-Wave Transformer

The quarter-wave transformer is one of the most simple and practical circuits for impedance matching, especially for matching of real load impedances. It is also possible to match a complex load using the quarter-wave transformer, but this requires an additional length of transmission line between the load and the quarter-wave transformer to transform the complex load impedance into a real impedance. A circuit employing a quarter-wave transformer is shown in Fig. 6.

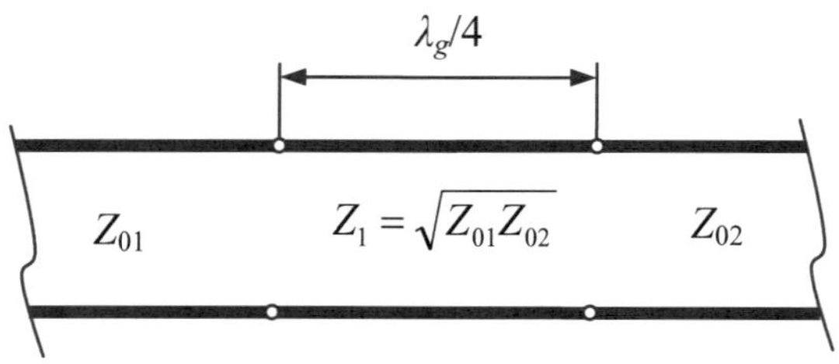

Fig. 6. A single section quarter-wave matching transformer.

One of the main drawbacks of this transformer is the requirement to have available a transmission line with an impedance of $Z_1 = \sqrt{Z_{01}Z_{02}}$. In some cases, e.g., matching with coaxial cable, the required quarter wave transmission line calls for a nonstandard value of the characteristic impedance.

4.2. Transformers with Fixed Values of Characteristic Impedance

Another useful type of series transformers are those which are based on transmission lines with the same characteristic impedances as the lines which should be matched. Such transformers are convenient for interconnection of standard lines as well as transmission lines with different geometry, where realization of transmission lines of arbitrary characteristic impedance involve difficulties.

The simplest realization of such transformer is shown in Fig. 7 (a) and described in detail by (Aizenberg et al., 1985).

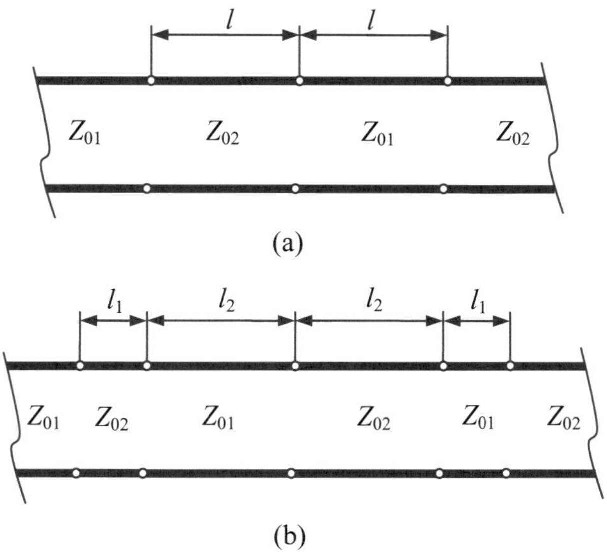

Fig. 7. Transformers with fixed values of characteristic impedance consisting of (a) two sections and (b) four sections.

This transformer consists of two transmission line sections. The characteristic impedances of these lines are the same as impedances of lines to be matched. The length of one section is

$$l = \frac{\lambda_g}{2\pi}\theta = \frac{\lambda_g}{2\pi}\operatorname{atan}\left(\frac{1}{\sqrt{n + \frac{1}{n} + 1}}\right), \tag{1}$$

where $n = Z_{02}/Z_{01}$ is the transformation ratio.

For small values of n, the value of l approaches $\lambda_g/12$, implying that the total length of the transformer approaches $\lambda_g/6$. For increasing n, the value of l approaches 0.

The operating frequency band of the described transformer is about 5 % narrower in comparison to the quarter-wave transformer, and its length for practical values of n is 1.5 - 2 times shorter.

The response of the transformer in Fig. 7 (a) is shown in Fig. 8 (curve (a)) and compared to the response of the conventional quarter-wave transformer (Fig. 8 curve (b)). For transformation ratio 2:1 the electrical length of the section in Fig. 7 (a) is equal to 28.1°. The achieved for this ratio bandwidth at 20 dB return loss level is 31 %.

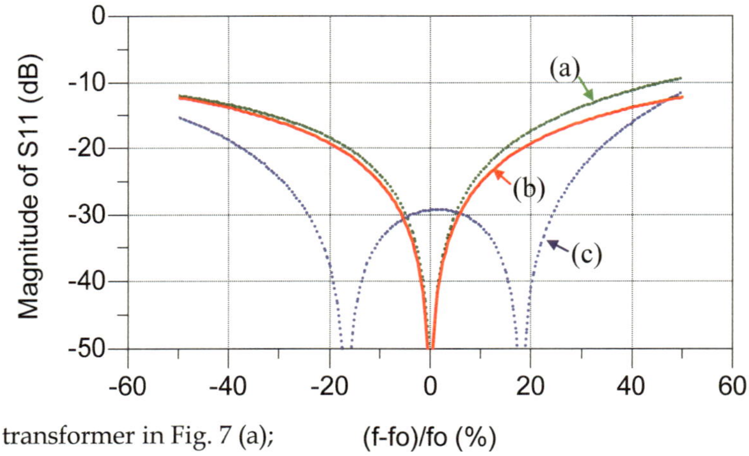

(a) $\lambda_g/6$ transformer in Fig. 7 (a);

(b) quarter-wave transformer;

(c)transformer in Fig. 7(b).

Fig. 8. Comparison of matching characteristic of quarter-wave transformer (Fig. 6) and transformers with fixed values of characteristic impedance (Fig. 7). The transformation ratio is 2:1. Here, f_0 is the center frequency of operation.

A more broadband stepped impedance transformer is shown in Fig. 7 (b)). It consists of four sections with the length of the outermost sections being shorter than the length of sections in the middle. Fig. 8 (curve (c)) shows the magnitude of S_{11} for a transformer with the fallowing parameters: the transformation ratio is 2:1; $\theta_1/\theta_2 = l_1/l_2 = 0.35$.
Here

$$\theta_1 = 2\pi\, l_1/\ \lambda_g, \quad \text{and} \quad \theta_1 = 2\pi\, l_1/\ \lambda_g \tag{2}$$

are the electrical lengths of the sections in Fig. 7 (b).
The total length of the transformer is $2l_1 + 2l_2 = 0.346\lambda_g$. The achieved bandwidth at 20 dB return loss level is 71 %. The bandwidth and inband reflection level of this type of transformer depend on length of the sections (Aizenberg et al., 1985).

4.3 Tapered Transmission Lines

As described above, the bandwidth of the quarter-wave transformer is limited. In order to extend its operating frequency band, multisection transformers, with different characteristic impedance in each section, may be used. In contrast to the transformers described in the previous section, the lengths of the sections used in the multisection transformer can be chosen equal to each other. The desired reflection coefficient response as a function of frequency can be achieved by properly choosing the characteristic impedance of the transmission line sections. In the limit of an infinite number of sections, the multisection transformer becomes a continuously tapered line. There are many ways to choose the taper profile. By changing the type of taper, one can obtain different passband characteristics. Several taper profiles may be considered: linear, exponential, triangular, and so on.

For a given taper length, the Klopfenstein taper has been shown to be optimum in the sense that the reflection coefficient is minimum over the passband (Pozar, 1998). Alternatively, for a specified level of reflection coefficient, the Klopfenstein taper yields the shortest matching section. However, it should be noted that the response of this taper has equal level of ripples in its passband.

In many cases, the relation between the physical dimensions and the characteristic impedance of a guiding structure is complicated and the generation of an optimal tapering configuration is thus not a trivial task. This implies that a linear or exponential tapering of the physical dimensions of the transmission line is often chosen for practical implementations.

5. Coupled Line Transformers

In recent years, coupled transmission lines have been suggested as a matching element due to greater flexibility and compactness in comparison to quarter wavelength transmission lines (Jensen et al., 2007). It has been demonstrated that matching real and complex loads with coupled lines leads to more compact realizations and could therefore become important at millimeter-wave frequencies for on-chip matching solutions. Another area where coupled line structures are useful is matching of antenna array structures, as successfully demonstrated by (Jaworski & Krozer, 2004).

As it was shown above, the quarter-wave transformer is simple and easy to use, but it has no flexibility beyond the ability to provide a perfect match at the center frequency for a real-valued load, although a complex load of course can be matched by increasing the overall length of the transformer. The coupled line section provides a number of variables which can be utilized for matching purposes. These variables are the even and odd mode impedances and loads of the through and coupled ports. This loading can be done in form of a feedback connection which provides additional zeros for broadband matching.

These variables can be chosen to provide a perfect match or any desired value of the reflection coefficient at the operating frequency. The bandwidth of the coupled line transformer can be further increased in case of mismatch. In addition, it is also possible to match a complex load.

In the lower GHz range the loading of the through and coupled ports can be done with lumped elements which allows for easy matching of both real and imaginary impedance values. At higher frequencies it is not possible to use lumped elements, but the difference between the even and odd mode impedances is a parameter which makes it possible to turn a mixed real and imaginary control load at the through port into a purely imaginary one, which can be implemented with a transmission line stub.

5.1 Symmetric Coupled Line Section

Coupled line impedance transformers are very useful at millimetre wave frequencies where they successfully perform direct current blocking and can handle large impedance transformation avoiding transverse resonances which occur in a conventional low impedance quarter-wave transformer. The most common configuration of the transformer is shown in Fig. 9.

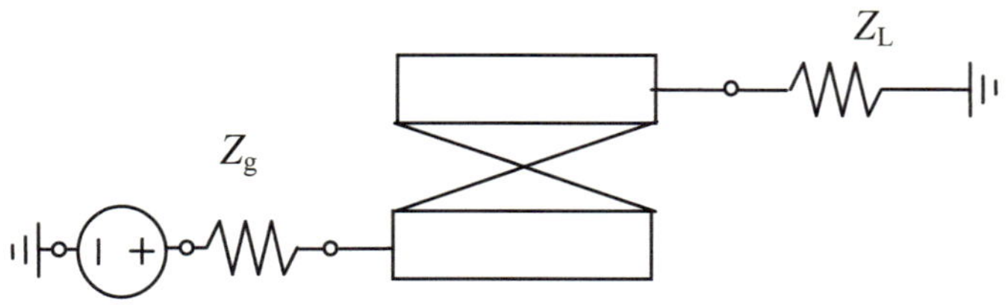

Fig. 9. Symmetric coupled transmission line transformer.

In this configuration, the diagonal terminals of the coupled line section are loaded with generator (Z_g) and load (Z_L) impedances. The opposite terminals are open circuited. In this standard configuration however, the electrical performance of the coupled lines transformer in terms of insertion loss and bandwidth can not compete with performance of the corresponding quarter-wave transformer (Mongia et al., 1999).

5.2 Asymmetric Coupled Line Section

Symmetric coupled lines represent a restricted configuration of the more general class of coupled lines. They allow for a simpler analysis, however, for wideband applications asymmetric coupled lines are preferable. For example, the bandwidth of a forward-wave directional coupler using asymmetric coupled transmission lines is greater than the one formed using symmetric ones (Jones & Bolljahn, 1956).
In this section the design of a wideband impedance transformer based on asymmetric coupled lines is described.
The considered wideband impedance transformer is based on asymmetric, uniform coupled lines in nonhomogeneous medium. A microstrip line is one of the most commonly used classes of transmission lines in nonhomogeneous medium. Edge-coupled microstrip lines are shown in Fig. 10.

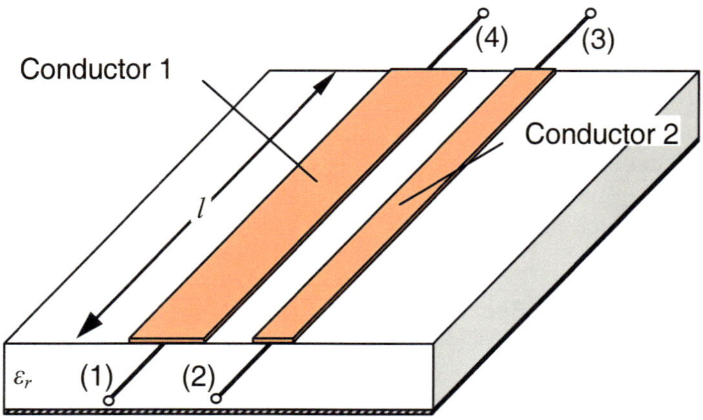

Fig. 10. A coupled microstrip line four-port.

For the purpose of analysis, this coupled line four-port is transformed to a two-port network with arbitrary load using impedance matrix representation.
The investigations presented in this book are only for the most commonly used configuration, when diagonal terminals of the coupled lines are loaded with generator and load impedances. Thus, the entire circuit can be represented as a two-port network, which performs impedance transformation between a generator impedance Zg connected to a port 1 and a load impedance Z_L connected to a port 3, as shown in Fig. 11.

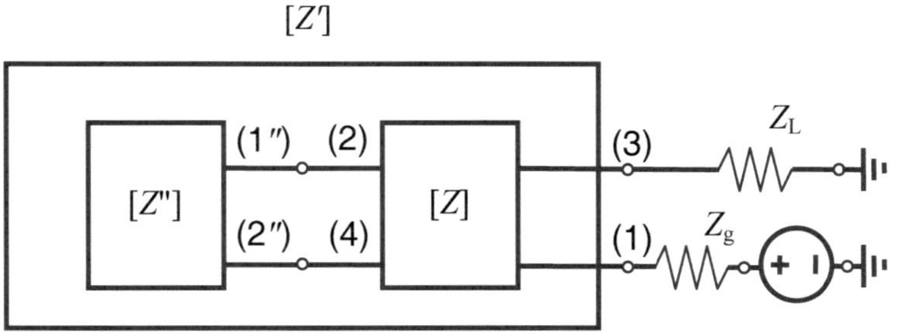

Fig. 11. Two-port network representation for the coupled line impedance transformer.

As it can be seen in Fig. 11, the network consists of the coupled line four-port network described by an impedance matrix [Z] and arbitrary load matrix at opposite terminals described by matrix [Z"]. In practice, ports 2 and 4 are in general either short-circuited or open-circuited with a corresponding representation of the two-port network [Z"].
The magnitude of S_{11} is equal to

$$\left|S_{11}\right|_{\mathrm{dB}} = 20\log\left(\left|\frac{Z_{\mathrm{IN}}\left(Z_{ij}, Z''_{ij}, Z_L\right) - Z_g}{Z_{\mathrm{IN}}\left(Z_{ij}, Z''_{ij}, Z_L\right) + Z_g}\right|\right), \tag{3}$$

where Z_{IN} is the input impedance of the transformer, which is a function of the load impedance Z_L, impedance matrix elements of coupled lines Z_{ij} and the arbitrary load Z''_{ij} (*i* and *j* are the indexes of the matrix elements). Using the general impedance matrix representation for coupled lines (Tripathi, 1975) and boundary conditions at ports 2 and 4 the input impedance is expressed by

$$Z_{\mathrm{IN}} = Z_{11} + Z_{12}\cdot a_1 + Z_{14}\cdot b_1 - \frac{\left(Z_{13} + Z_{12}\cdot a_2 + Z_{14}\cdot b_2\right)^2}{Z_{33} + Z_{32}\cdot a_2 + Z_{34}\cdot b_2 + Z_{\mathrm{L}}}, \tag{4}$$

where

$$a_1 = \frac{Z_{41}\left(Z_{24} + Z''_{12}\right) - Z_{21}\left(Z_{44} + Z''_{22}\right)}{\left(Z_{22} + Z''_{11}\right)\left(Z_{44} + Z''_{22}\right) - \left(Z_{24} + Z''_{12}\right)\left(Z_{42} + Z''_{21}\right)}, \tag{5a}$$

$$a_2 = \frac{Z_{43}\left(Z_{24} + Z''_{12}\right) - Z_{23}\left(Z_{44} + Z''_{22}\right)}{\left(Z_{22} + Z''_{11}\right)\left(Z_{44} + Z''_{22}\right) - \left(Z_{24} + Z''_{12}\right)\left(Z_{42} + Z''_{21}\right)} , \tag{5b}$$

$$b_1 = -\frac{Z_{41}}{\left(Z_{44} + Z''_{22}\right)} - \frac{\left(Z_{42} + Z''_{21}\right)}{\left(Z_{44} + Z''_{22}\right)} \cdot a_1 , \tag{5c}$$

$$b_2 = -\frac{Z_{43}}{\left(Z_{44} + Z''_{22}\right)} - \frac{\left(Z_{42} + Z''_{21}\right)}{\left(Z_{44} + Z''_{22}\right)} \cdot a_2 . \tag{5d}$$

A total number of six quantities is required to describe asymmetric coupled lines (Mongia et al., 1999), being: Z_{c1} and $Z_{\pi 1}$, which are, respectively, the characteristic impedances of line 1 for c and π modes of propagation; γ_c and γ_π, the propagation constants of c and π modes; R_c and R_π, the ratios of the voltages on the two lines for c and π modes. Thus, the elements of the impedance matrix are given by

$$Z_{11} = Z_{44} = \frac{Z_{c1}\coth(\gamma_c l)}{1 - R_c/R_\pi} + \frac{Z_{\pi 1}\coth(\gamma_\pi l)}{1 - R_\pi/R_c} , \tag{6a}$$

$$Z_{12} = Z_{21} = Z_{34} = Z_{43} = \frac{Z_{c1}R_c\coth(\gamma_c l)}{1 - R_c/R_\pi} + \frac{Z_{\pi 1}R_\pi\coth(\gamma_\pi l)}{1 - R_\pi/R_c} , \tag{6b}$$

$$Z_{13} = Z_{31} = Z_{24} = Z_{42} = \frac{Z_{c1}R_c\operatorname{csch}(\gamma_c l)}{\left(1 - R_c/R_\pi\right)} + \frac{Z_{\pi 1}R_\pi\operatorname{csch}(\gamma_\pi l)}{\left(1 - R_\pi/R_c\right)} , \tag{6c}$$

$$Z_{14} = Z_{41} = \frac{Z_{c1}\operatorname{csch}(\gamma_c l)}{\left(1 - R_c/R_\pi\right)} + \frac{Z_{\pi 1}\operatorname{csch}(\gamma_\pi l)}{\left(1 - R_\pi/R_c\right)} , \tag{6d}$$

$$Z_{22} = Z_{33} = \frac{Z_{c1}R_c^2\coth(\gamma_c l)}{1 - R_c/R_\pi} + \frac{Z_{\pi 1}R_\pi^2\coth(\gamma_\pi l)}{1 - R_\pi/R_c} , \tag{6e}$$

$$Z_{23} = Z_{32} = \frac{Z_{c1} R_c^2 \operatorname{csch}(\gamma_c l)}{\left(1 - \frac{R_c}{R_\pi}\right)} + \frac{Z_{\pi 1} R_\pi^2 \operatorname{csch}(\gamma_\pi l)}{\left(1 - \frac{R_\pi}{R_c}\right)} , \tag{6f}$$

where l is the length of the coupled line section, as it is shown in Fig. 10. These relations are substituted into (5) and (4) to calculate the input impedance and finally the reflection coefficient of the transformer.

From relation (3) it can be seen that the matching properties of the transformer depend not only on coupled line parameters, but also on load of ports 2 and 4, which are described by elements Z''_{ij}. This dependence introduces additional degree of freedom during design procedure and can be used to expand the bandwidth of the impedance transformer, as shown below.

Loading With Transmission Line

As an example, terminals 2 and 4 can be loaded with a microstrip transmission line. The impedance matrix of the transmission line with characteristic impedance Z_0, length l, and propagation constant γ is given by

$$[Z''] = \begin{bmatrix} Z_0 \coth(\gamma l) & \frac{Z_0}{\sinh(\gamma l)} \\ \frac{Z_0}{\sinh(\gamma l)} & Z_0 \coth(\gamma l) \end{bmatrix} . \tag{7}$$

The transformer configuration is shown in Fig. 12.

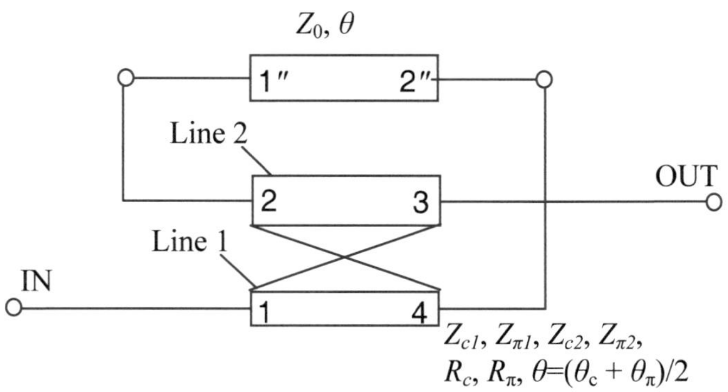

Fig. 12. Schematic illustration of the transformer based on coupled line section and a transmission line load.

In order to simplify further calculations, the transmission lines are considered to be lossless, and electrical lengths of the coupled line section $(\theta_c + \theta_\Pi)/2$ and the microstrip transmission line θ are assumed equal, resulting in

$$\gamma l = j\beta l = j\theta, \quad \gamma_c l = j\theta_c, \quad \gamma_\Pi l = j\theta_\Pi \ , \tag{8}$$

$$\theta = (\theta_c + \theta_\Pi)/2 \ , \tag{9}$$

where θ_c and θ_Π are the electrical lengths of the coupled line section for c and π mode respectively. θ is a function of frequency and can be used for the analysis of the spectrum of the transformer reflection coefficient. The response (3) for the transformer of Fig. 12 is shown in Fig. 13.

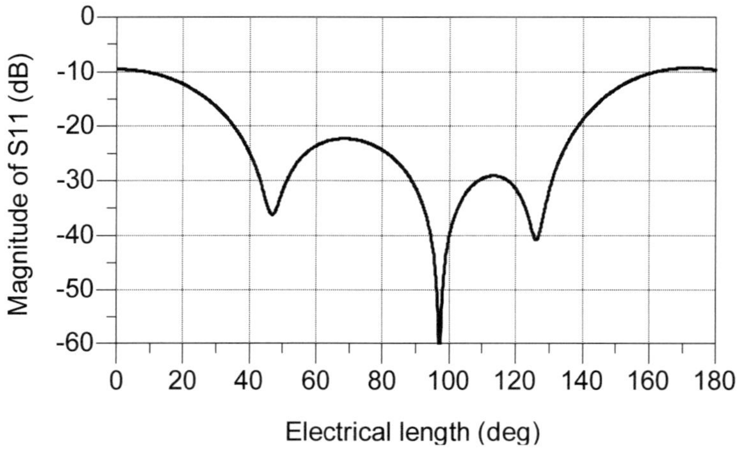

Fig. 13. Response of transformer shown in Fig. 12. The transformation ratio is 1:2.

As it can be seen in Fig. 13, this transformer configuration exhibits an additional minimum in the magnitude of S_{11} in comparison to the traditional impedance transformer based on coupled line section with open-circuited terminals (Kajfez, 1981). These minima are non-uniformly distributed in the frequency domain. This is due to the differences in electrical lengths θ_c and θ_Π for two coupled line modes in nonhomogeneous medium.

For the case of homogeneous medium the propagation constants for the two modes are equal, $\gamma_c = \gamma_\Pi$, and hence the electrical lengths for the two propagating modes are also equal. It is therefore possible to obtain three equidistant reflection zeros in the spectrum of the reflection coefficient. Because transmission lines in a homogeneous medium are a special case of transmission lines in a nonhomogeneous medium the expressions given above are also valid for response calculations.

It can be depicted from the data in Fig. 14 that the transformer provides wideband operation with uniformly distributed reflection zeros in the frequency domain. In addition, the distance between the zero locations can be varied by adjusting the parameters of the structure.

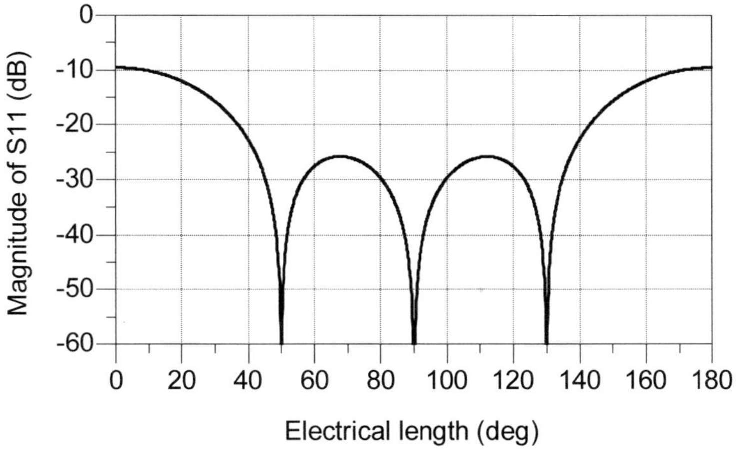

Fig. 14. Response of the transformer in homogeneous medium case.

The electrical length of the transformer is equal to a quarter wavelength at the center frequency. Comparing the results in Fig. 13 and Fig. 14 it can be deduced that the impedance transformer in nonhomogeneous medium has approximately the same bandwidth as the one in homogenous medium. However, in many cases, like for example in surface mount technology, it is more useful to deal with microstrip structures.

Loading With Stepped Impedance Transmission Line

The differences in electrical lengths of the coupled lines in nonhomogeneous medium can be compensated by introducing a stepped impedance transmission line, as it is shown in Fig. 15.

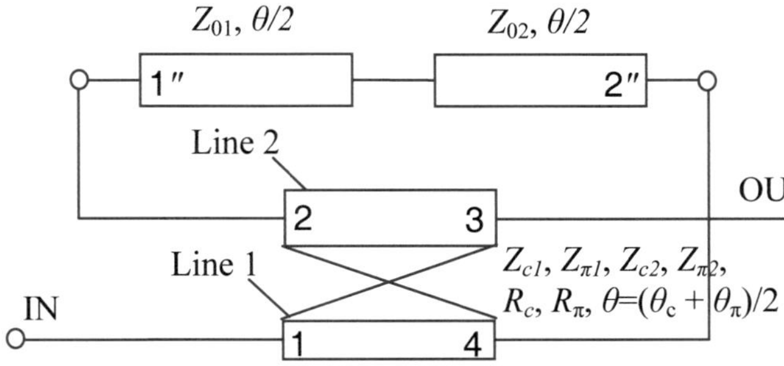

Fig. 15. Schematic illustration of the wideband impedance transformer.

The transformer consists of asymmetric coupled lines described by the electrical parameters Z_{c1}, Z_{c2}, $Z_{\pi 1}$, $Z_{\pi 2}$, which are, respectively, the characteristic impedances of line 1 and 2 for the c and π modes of propagation; θ_c and θ_π, the electrical lengths for the c and π modes; R_c and R_π, the ratios of the voltages on the two lines for the c and π modes. The stepped impedance transmission line consists of two equal length transmission lines with characteristic impedances Z_{01} and Z_{02}, as shown in Fig. 15. The electrical length of each transmission line is set to be half the electrical length of the coupled line section to reduce the number of design parameters.

For the purpose of analysis, this structure is transformed into a two-port network with arbitrary load using an impedance matrix representation. Thus, the entire circuit can be represented as a two-port network, which performs impedance transformation between a generator impedance Z_g connected to the port 1 and a load impedance Z_L connected to the port 3, as shown in Fig. 11. The magnitude of S_{11} at the port 1 is defined by (3). The input impedance Z_{IN} in (3) is calculated using relations (4)-(6) together with the corresponding elements of the impedance matrix [Z"] for the stepped impedance transmission line. A series connection of two transmission lines shown in Fig. 16 can be described as a connection of two two-port networks.

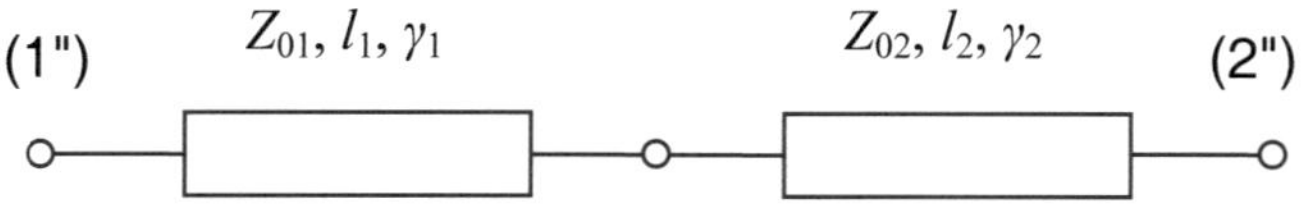

Fig. 16. Series connection of transmission lines.

The impedance matrices of the transmission lines with characteristic impedances Z_{01}, Z_{02}, lengths l_1, l_2, and propagation constants γ_1, γ_2 are given by

$$\left[Z^{(1)}\right]=\begin{bmatrix} Z_{11}^{(1)} & Z_{12}^{(1)} \\ Z_{21}^{(1)} & Z_{22}^{(1)} \end{bmatrix}=\begin{bmatrix} Z_{01}\coth(\gamma_1 l_1) & \dfrac{Z_{01}}{\sinh(\gamma_1 l_1)} \\ \dfrac{Z_{01}}{\sinh(\gamma_1 l_1)} & Z_{01}\coth(\gamma_1 l_1) \end{bmatrix}, \tag{10}$$

$$\left[Z^{(2)}\right]=\begin{bmatrix} Z_{11}^{(2)} & Z_{12}^{(2)} \\ Z_{21}^{(2)} & Z_{22}^{(2)} \end{bmatrix}=\begin{bmatrix} Z_{02}\coth(\gamma_2 l_2) & \dfrac{Z_{02}}{\sinh(\gamma_2 l_2)} \\ \dfrac{Z_{0}}{\sinh(\gamma_2 l_2)} & Z_{02}\coth(\gamma_2 l_2) \end{bmatrix}. \tag{11}$$

Impedance matrix for the overall circuit in Fig. 16 is derived using boundary conditions at the common terminal. At this terminal the voltages of two two-ports are equal, and currents are equal and oppositely directed.

Thus, impedance matrix elements are found to be:

$$\begin{aligned} Z_{11}'' &= Z_{11}^{(1)}-\frac{\left(Z_{12}^{(1)}\right)^2}{Z_{11}^{(2)}+Z_{11}^{(1)}} \\ &= Z_{01}\coth(\gamma_1 l_1)-\frac{Z_{01}^2}{\left(Z_{01}\coth(\gamma_1 l_1)+Z_{02}\coth(\gamma_2 l_2)\right)\cdot\sinh^2(\gamma_1 l_1)}, \end{aligned} \tag{12a}$$

$$Z''_{12} = Z''_{21} = \frac{Z^{(1)}_{12} Z^{(2)}_{12}}{Z^{(2)}_{11} + Z^{(1)}_{11}}$$
$$= \frac{Z_{01} Z_{02}}{\left(Z_{01} \coth(\gamma_1 l_1) + Z_{02} \coth(\gamma_2 l_2)\right) \cdot \sinh(\gamma_1 l_1) \cdot \sinh(\gamma_2 l_2)}, \tag{12b}$$

$$Z''_{22} = Z^{(2)}_{11} - \frac{\left(Z^{(2)}_{12}\right)^2}{Z^{(2)}_{11} + Z^{(1)}_{11}}$$
$$= Z_{02} \coth(\gamma_2 l_2) - \frac{Z^2_{02}}{\left(Z_{01} \coth(\gamma_1 l_1) + Z_{02} \coth(\gamma_2 l_2)\right) \cdot \sinh^2(\gamma_2 l_2)}. \tag{12c}$$

In case of transmission lines with equal electrical length $\theta/2$ (12) can be rewritten as

$$Z''_{11} = Z_{01} \coth\left(\frac{\theta}{2}\right) - \frac{2 Z^2_{01}}{(Z_{01} + Z_{02}) \cdot \sinh(\theta)}, \tag{13a}$$

$$Z''_{12} = Z''_{21} = \frac{2 Z_{01} Z_{02}}{(Z_{01} + Z_{02}) \cdot \sinh(\theta)}, \tag{13b}$$

$$Z''_{22} = Z_{02} \coth\left(\frac{\theta}{2}\right) - \frac{2 Z^2_{02}}{(Z_{01} + Z_{02}) \cdot \sinh(\theta)}. \tag{13c}$$

These equations are used for the calculation of elements of the matrix [Z"] in Fig. 11.
Thus, the analysis of the structure now can be performed using (3).
It can be depicted from the data in Fig. 17 that the transformer provides wideband operation, and the electrical length of the transformer is equal to a quarter wavelength at the center frequency.

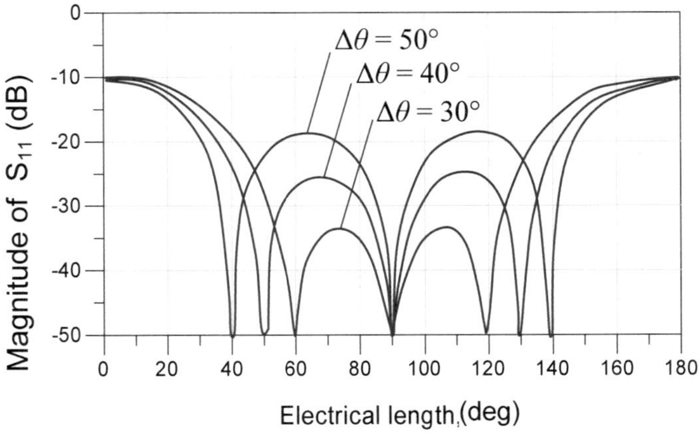

Fig. 17. Response of the 50-100 Ω impedance transformer shown in Fig. 15.

In addition, the distance between the minima locations $\Delta\theta$ can be varied by adjusting the parameters of the structure. This distance $\Delta\theta$ characterizes operating frequency bandwidth of the transformer. The characteristics of the transformer for three different values of $\Delta\theta$ are shown in Fig. 17. As it can be seen, the in-band level of the reflection coefficient depends on parameter $\Delta\theta$. The estimation of the maximum level of the return loss between minima for different transformation ratios can be found using the data shown in Fig. 18.

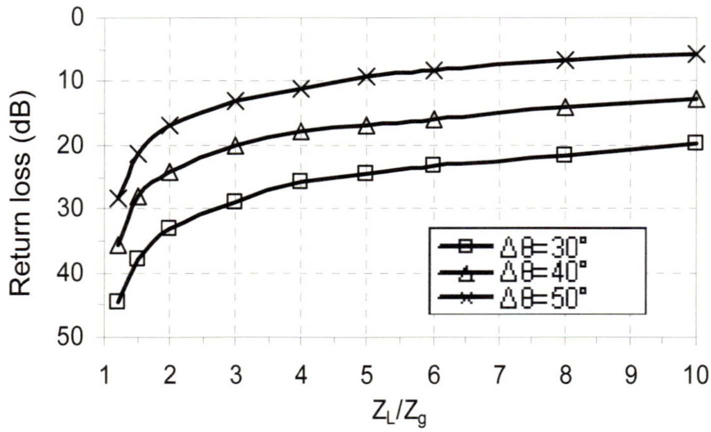

Fig. 18. The minimum level of the return loss between minima in Fig. 17.

As expected, the level of in-band return loss for the transformer increases with reducing of transformation ratio, and reaches the absolute maximum at $Z_l/Z_g = 1$.

5.3 Multisection Coupled Line Transformers

To further increase the bandwidth, it is possible to create an impedance transformer using more coupled line sections connected in series. The example of a microstrip two section impedance transformer is shown in Fig. 19.

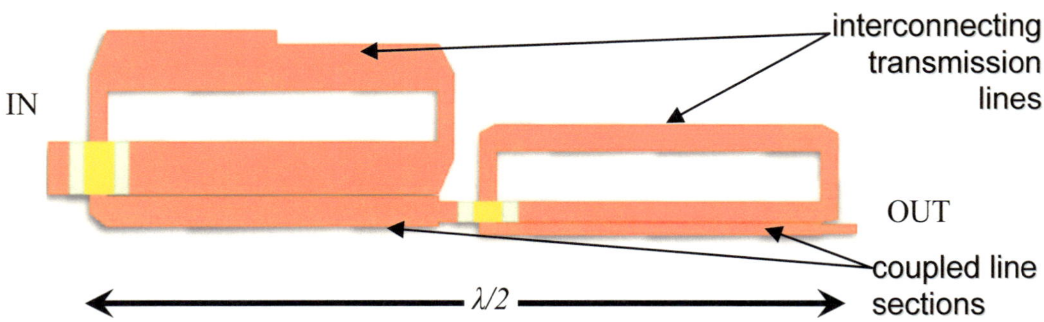

Fig. 19. Layout of the 12.5-50 Ω multi section impedance transformer.

The total electrical length of the transformer is equal to half a wavelength at the center frequency. The response of the transformer is shown in Fig. 20.

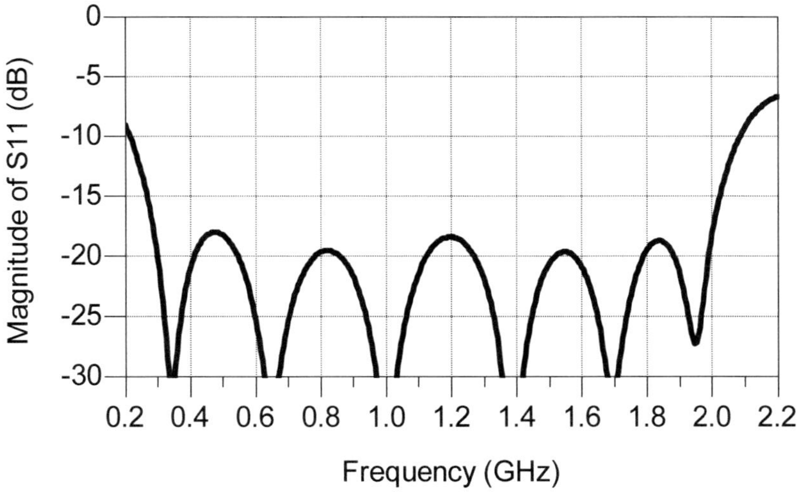

Fig. 20. Response of the 12.5-50 Ω impedance transformer shown in Fig. 19.

The transformer exhibits six minima in the spectrum of reflection coefficient. The achieved fractional matching bandwidth is beyond a decade at -10 dB reflection coefficient level. The distance between the minima locations can be varied by adjusting the parameters of the structure.

6. References

Aizenberg G. Z., Belousov S. P., Zhurbenko E. M., Kliger G. A. & Kurashov A. G. (1985). *Korotkovolnovye antenny*, 2nd ed., Moscow, Radio i Svaz (in Russian).

Jaworski G. & Krozer V. (2004). Broadband matching of dual-linear polarization stacked probe-fed microstrip patch antenna, *Electronics Letters*, vol. 40, no. 4, pp. 221-222.

Jensen T., Zhurbenko V., Krozer V. & Meincke P. (2007). Coupled Transmission Lines as Impedance Transformer, *IEEE Transactions On Microwave Theory And Techniques*, vol. 55, no. 12, pp. 2957-2965.

Jones E. M. T., & Bolljahn J. T. (1956). Coupled Strip Transmission Line Filters and Directional Couplers, *IRE Trans. Microwave Theory & Tech.*, vol. MTT-4, pp. 78-81.

Kajfez D., Bokka S. & Smith C. E. (1981). Asymmetric microstrip dc blocks with rippled response, *IEEE MTT-S Int. Microwave Symp. Dig.*, pp. 301–303.

Kinayman N. & Aksun M. I. (2005). *Modern Microwave Circuits*. Artech House, Inc.

Mongia R., Bahl I., Bhartia P.(1999). *RF and microwave coupled line circuits.* Norwood: Artech House microwave library.

Pozar D. M. (1998). *Microwave Engineering*. Wiley.

Tatarinov V. V. (1931). O pitanii beguschei volnoi korotkovolnovyh antenn i ob opredelenii ih soprotivlenia, *Vestnik elektrotehniki*, no. 1 (in Russian).

Tripathi V. K. (1975). Asymmetric coupled transmission lines in an inhomogeneous medium, *IEEE Trans. Microwave Theory & Tech.*, vol. 23, no. 9, pp. 734-739.

15

Design of Compact Planar Ultra-Wideband Bandpass Filters

Yasushi Horii
Kansai University
Japan

1. Introduction

Since the Federal Communication Commission (FCC) adopted a First Report and Order in 2002, a wide frequency range from 3.1 GHz to 10.6 GHz has been released for the marketing and operation of new types of wireless communication systems incorporating ultra-wideband (UWB) technology. Because of the high-speed data handling capability, the UWB system is expected to be used for delivering real-time HDTV video streaming, transmitting of non-compressible audio/visual signals, and replacing USB cables with wireless connections. In addition, due to the low-power consumption, new short-range wireless applications have attracted considerable attentions in the fields of home electronics, home entertainment, security sensors, and health care devices. To meet today's huge demands, the research on UWB devices has been greatly accelerated.

One of the difficulties in developing the UWB system is the bandwidth utilized for communication. Since the system uses very short impulse signals to transmit bit-data trains, it inherently requires an extremely wide frequency range. This unique feature always presses us to develop new technologies. In an RF front-end design, for instance, high-performance bandpass filters and antennas have been too large when we engaged in the conventional design methodology. An initial UWB filter was realized by a combination of low-pass and high-pass filters. Since then, tremendous efforts have been devoted to this subject, and a way to use a multi-mode resonator (MMR) was proposed in 2005. In this design, the first three resonant frequencies of the MMR are placed equally within the UWB band so as to create the huge passband response. By applying this technology, overall filter dimension was drastically reduced to less than 10 mm x 15 mm. After that, a variety of UWB filters have been proposed based on the MMR configuration. For example, some had a ground defected structure (DGS) or a periodic band gap (PBG) structure, and some used plural stages of MMRs to obtain the high selectivity performance. However, such approaches can sometimes cause an increase of the filter dimensions, leading to fabrication difficulties in a practical system design. Again, it should be noted that the short-range UWB systems are extremely low-power consumption and need to be pocket-sized or smaller.

My goal is to develop a super-compact planar UWB filter based on the microstrip line configuration. For this purpose, the following requirements are assumed;

(1) For frequency response

- To meet the FCC spectrum musk regulation
- Low insertion loss (less than 0.5 dB)
- Low ripples (less than 0.5 dB)
- Mild group delay variation (less than 0.2 ns)
- Transmission zeros above and below the passband

(2) For configuration
- Super-compact design (less than 5.0 mm x 5.0 mm)
- To use a microstrip line configuration
- No defections on a ground plane
- Compatible with the conventional PCB technology
- Low cost and mass productivity

In Sec. 2, a useful equivalent circuit is developed for design of super-compact UWB bandpass filters. In Sec. 3, actual filter designs are presented together with convenient design procedure to optimize the filter response.

2. Equivalent circuits

UWB bandpass filters need to be designed in an extremely wide frequency range from DC to around 20 GHz to meet the FCC spectrum musk regulation. The free-space wavelength at both passband edges are 96.8 mm at 3.1 GHz and 28.3 mm at 10.6 GHz. The wavelength will be compressed much more when a high permittivity substrate is applied to the filter. Therefore, even a small metal patch with 5 mm length can easily act as a resonator. This indicates that the UWB filters having a few millimeter dimension, work as a concentrated constant circuit at the lower frequency and as a distributed constant circuit at the higher frequency. Therefore, when the filter is designed, we should take these effects into consideration.

An equivalent circuit of the UWB bandpass filter, which is demonstrated in this chapter, is shown in Fig. 1. The notations, CL1, CL2, CL3 and CL4 in the figure, represent a coupled-line, each of which consists of tightly coupled three transmission lines placed in parallel with a small spacing. Among the coupled lines, a center line is short-circuited (SC) for grounding and the lines on both sides are open-circuited (OC) at their top, while these lines are connected to each other at their bottom. The coupled-line on the left-hand side (CL1 and CL2) is also coupled to that on the right-hand side (CL3 and CL4) with a coupling capacitance C_0. In addition, a coupled-inductor (CI) with a self-inductance L_0 and a mutual-inductance M_0 is presented in parallel with the capacitor C_0.

The typical circuit parameters of the filter are assumed as shown in Table 1, and the corresponding filter response is calculated by using the commercial circuit simulator ANSOFT Designer SV2. The result, presented in Fig. 2, shows that the filter has a huge flat passband together with deep transmission zeros just below and above the passband. In addition, well-suppressed out-of-passbands are obtained in a wide frequency range.

This UWB response can be explained by looking at the circuit from two points of view; one way is to look at the circuit as a *concentrated constant circuit*, which is available at the lower frequency, while the other way is to look at it as a *distributed constant circuit* in the higher frequency. According to such points of view, the equivalent circuits can also be expressed in a different way as shown in Fig. 3 and Fig. 5, respectively.

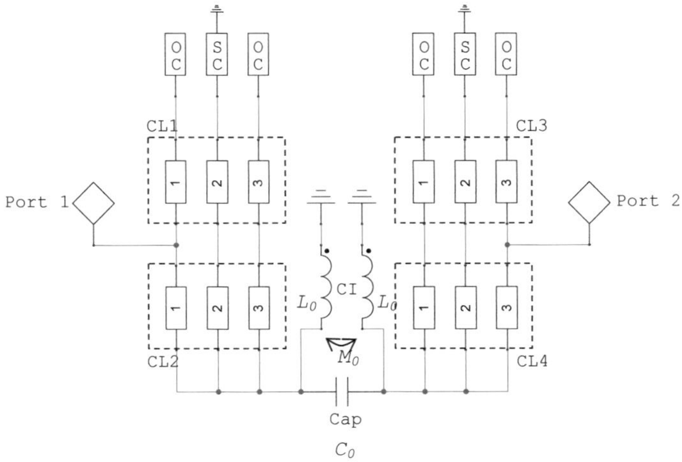

Fig. 1. An basic equivalent circuit of the super-compact planar UWB bandpass filters studied in this chapter. The circuit is configured by some lumped elements and distributed constant lines.

Coupled-lines	CL1	CL2	CL3	CL4
length	2.90 mm	1.60 mm	2.90 mm	1.60 mm
line width 1	0.50 mm	0.50 mm	0.50 mm	0.50 mm
spacing	0.35 mm	0.35 mm	0.35 mm	0.35 mm
line width 2	0.30 mm	0.30 mm	0.30 mm	0.30 mm
spacing	0.35 mm	0.35 mm	0.35 mm	0.35 mm
line width 3	0.50 mm	0.50 mm	0.50 mm	0.50 mm
Open circuit (OC)		Short circuit (SC)		
line width	0.50 mm	line width	0.30 mm	
line length	0.00 mm	line length	0.50 mm	
Coupled-inductor (CL)		Substrate		
self-inductance L_0	7.0 nH	thickness		0.80 mm
Mutual-inductance M_0	3.5 nH	relative permittivity		2.62
Capacitance C_0	0.70 pF			

Table 1. Circuit parameters of the equivalent circuit shown in Figs. 1, 3, and 5. Scattering characteristics of these circuits are presented in Fig. 2, 4, and 6. These parameters are tuned so as to obtain the UWB response.

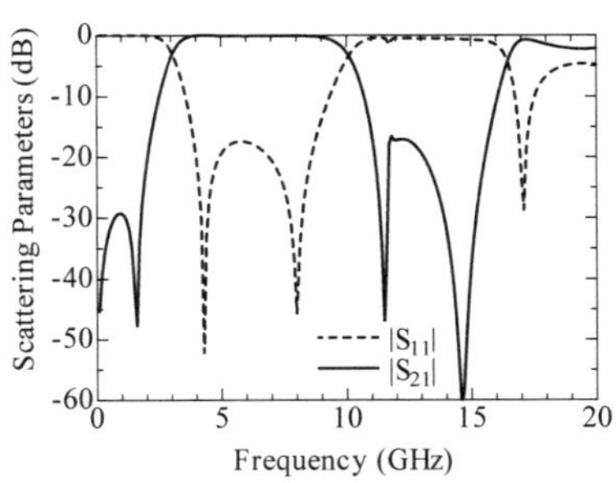

Fig. 2. Scattering characteristic of a UWB bandpass filter, calculated by an equivalent circuit shown in Fig. 1. Circuit parameters are listed in Table 1.

Fig. 3a, derived partially from the original equivalent circuit in Fig. 1, presents the circuit which creates the lower frequency response. The coupled-inductance (L_0, L_0 and M_0) inserted at the center of the figure can be transformed into three inductors (L_1, L_2 and L_3) as shown Fig. 3b, and a new series inductor L_2 forms an LC-tank circuit with the original series capacitor C_0. As a result, corresponding to the resonance of the LC-tank, a transmission zero is created below the passband. In addition, two shunt inductors (L_1 and L_3) and a series capacitor (C_0) stop the DC and the lower frequency signals from passing through the circuit. Contrarily, the circuit components related to create the higher frequency response are derived from the original circuit in the same manner. The resultant equivalent circuit and its scattering characteristic are shown in Figs. 5 and 6, respectively. The length of the coupled-lines are chosen so that the open-circuited lines in the coupled-line work as a quoter-wavelength open-stub in a frequency range from 11 GHz to 15 GHz. Consequently, two stopbands are created above 11 GHz in accordance with the lines.
This study clearly indicates that the lower and the higher frequency responses in Fig. 2 are produced by the combination of the concentrated constant circuit and the distributed constant circuit, and each response can be controlled independently by tuning corresponding circuit parameters. In other words, the lower and the higher cutoff frequencies of the passband are created by different mechanisms. Thus, the equivalent circuit presented in Fig. 1 gives us a great insight of a new filter topology.

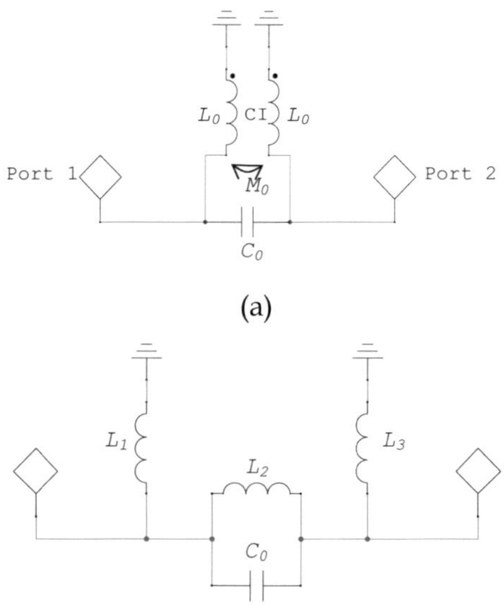

Fig. 3. An equivalent circuit, describing the lower frequency response of the UWB filter. (a) The circuit configured by lumped elements, which is derived from the original equivalent circuit presented in Fig. 1. (b) The coupled-inductor in Fig. 3a can be transformed equivalently into three inductors. An LC-tank creates a transmission zero below the UWB passband.

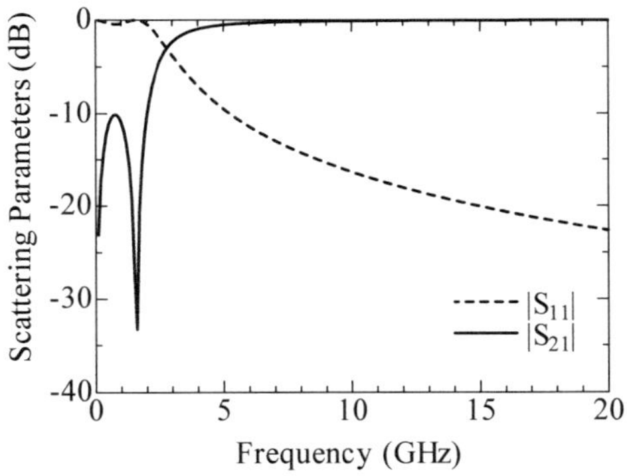

Fig. 4. Scattering characteristic of the equivalent circuit shown in Fig. 3a with circuit parameters listed in Table 1.

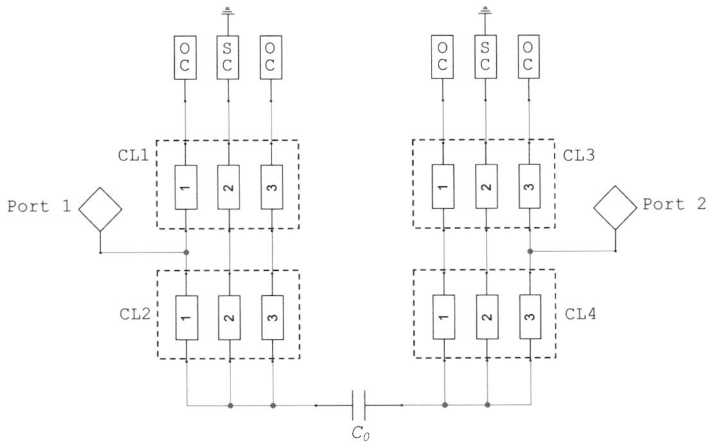

Fig. 5. An equivalent circuit, describing the higher frequency response of the UWB filter. The circuit is mainly composed of distributed constant transmission lines, which work as a quoter-wavelength open-stub to create a stopband above the UWB passband.

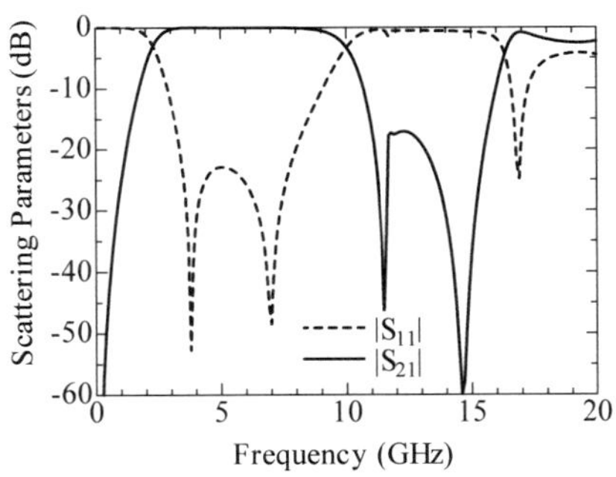

Fig. 6. Scattering characteristic of the equivalent circuit shown in Fig. 5 with circuit parameters listed in Table 1.

3. Planar UWB bandpass filters

3.1 UWB bandpass filters with tightly coupled E-shaped electrodes

Fig. 7 shows a circuit layout of a planar-type UWB bandpass filter designed in accordance with the equivalent circuit presented in Fig 1. All the circuit elements, designed using microstrip line configuration, are fabricated on a dielectric substrate with thickness H_{SUB} and relative permittivity ε_r. The coupled-lines in Fig. 1, CL1-CL2 and CL3-CL4, are created by tightly coupled three transmission lines. Length and width of the lines are designated by (L_1, W_1), (L_2, W_2), (L_3, W_3), (L_4, W_4), (L_5, W_5), and (L_6, W_6) starting from the leftmost line. The spacing between the lines in the coupled-line is shown by S. The thin line with width W_B is given to the bottom of each coupled-line to connect them together. The lines on both sides in the coupled-line are open-circuited at their top, while the center line is short-circuited for grounding. Because the shape of the coupled-line resembles to the letter "E", the coupled-line is called an "E-shaped electrode". The input and output microstrip lines with width W_{MSL} are connected to the leftmost and rightmost lines with a small offset D from the bottom of the electrode. By applying such an architecture, the distributed constant circuit presented in Fig. 5 is composed.

Next, we need to consider how to create the concentrated constant circuit parameters in the actual model. If the circuit layout was develped as a precise copy of the equivalent circuit, the overall filter dimension would be large. For significant size reduction, the function of the concentrated constant circuit should be realized by utilizing the architecture of the distributed constant circuit. Fortunately, the series capacitance C_0 is yielded by placing two E-shaped electrodes very closely with a small gap G. In addition, it is also expected that two short-circuited lines in the electrodes yield a required self-inductance L_0 and a mutual-inductance M_0 by a magnetic coupling between them. Thus, the required concentrated circuit behaviour observed at the lower frequency can be created by fine-tuning of the filter layout of the distributed constant circuit. These structural parameters, summarized in Table 2, were determined by using a full-wave EM-simulator Ansoft HFSS based on the finite element method. Design methodology is introduced in the latter subsection 3.4.

Fig. 8 presents the frequency dependence of the transmission coefficient $|S_{21}|$ and the reflection coefficient $|S_{11}|$. It can be read from the figure that the filter has a very wide passband from 3.8 GHz to 10.4 GHz (-3 dB criteria), which corresponds to the relative bandwidth of 93.6 %. The minimum insertion loss in the passband is about 0.16 dB at 9.5 GHz, and ripples are less than 0.5 dB from 5.3 GHz to 10.0 GHz. Furthermore, the $|S_{11}|$ is less than -10 dB from 5.1 GHz to 10.0 GHz. In addition, this filter has attenuation poles at 1.5 GHz and 11.3 GHz, and two reflection poles at 7.3 GHz and 9.5 GHz. The location of these poles is deeply related to the selectivity of the passband and the group delay of the filter.

Fig. 9 shows a photograph of the prototype model fabricated on a Rexolite 2200 substrate with a relative permittivity of 2.62 and a thickness of 0.8 mm. A circuit board plotter LPKF Protomat C30s was used to cut the metallization on the substrate along the contour of the electrodes and microstrip lines, and then unnecessary metallizations were removed by hand. To create an input and an output ports, two SMA connectors were soldered to the microstrip lines. Outer dimension of the prototype model was 10 mm x 15 mm x 0.8 mm.

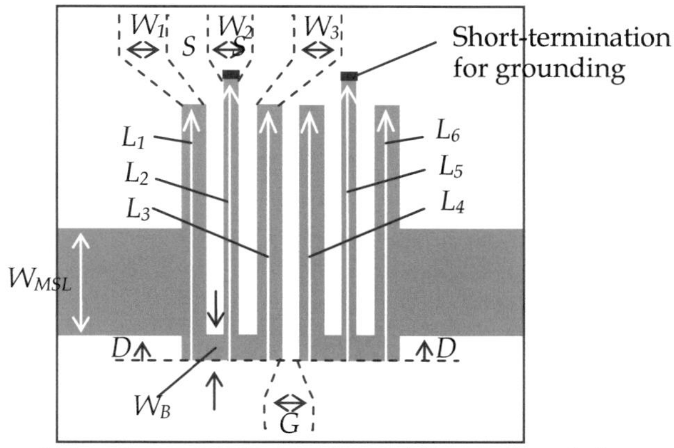

Fig. 7. A circuit layout of the planer UWB bandpass filter with tightly coupled E-shaped electrodes. Structural parameters are listed in Table 2.

Coupled-lines							
L_1	L_2	L_3	L_4	L_5	L_6	*Spacing bw lines*	*Spacing bw ELs**
5.0	5.5	5.0	5.0	5.5	5.0		
W_1	W_2	W_3	W_4	W_5	W_6	S	G
0.5	0.3	0.5	0.5	0.3	0.5	0.35	0.05

	I/O microstip lines			Substrate	
W_B	D	W_{MSL}	Z_0	H_{SUB}	ε_r
0.5	0.5	2.0	50 ohm	0.8	2.62

Table 2. Structural parameters of the planar UWB bandpass filter with tightly coupled E-shaped electrodes shown in Fig. 7. Unit for dimensional parameters is in *mm*.

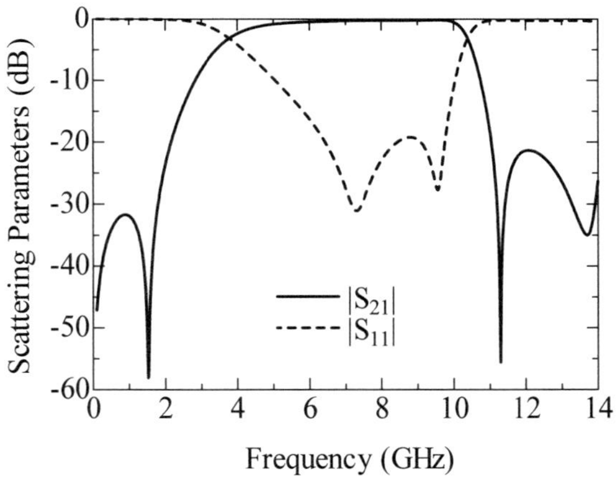

Fig. 8. Simulated scattering characteristic of the planar UWB bandpass filter with tightly coupled E-shaped electrodes shown in Fig. 7 with circuit parameters listed in Table 2.

Fig. 9. A prototype of the planar UWB bandpass filter with tightly coupled E-shaped electrodes, illustrated in Fig. 7 with circuit parameters listed in Table 2. The filter is fabricated on a Rexolite 2200 substrate with a relative permittivity of 2.62 and a thickness of 0.8 mm.

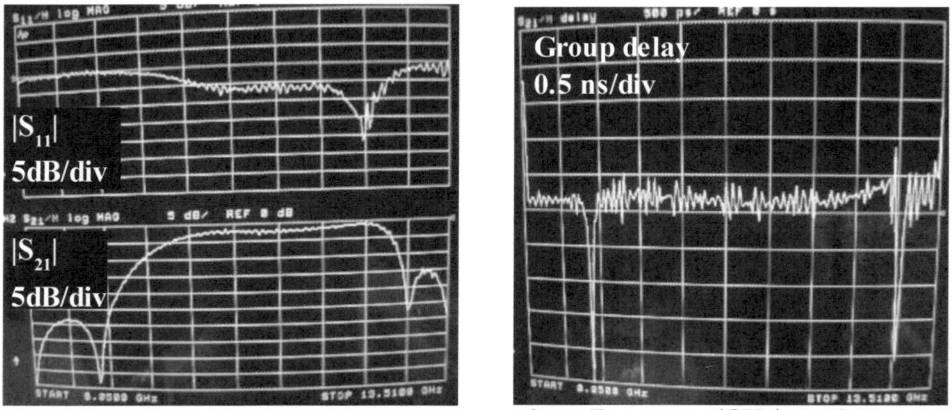

0 Frequency (GHz) 13.5   0 Frequency (GHz) 13.5

(a) (b)

Fig. 10. Measured scattering characteristics of the planar UWB bandpass filter with tightly coupled E-shaped electrodes. (a) The reflection coefficient $|S_{11}|$ (top) and the transmission coefficient $|S_{21}|$ (bottom). Unit in vertical axis is 5 dB/div. (b) Group delay. Unit is 0.5 ns/div.

Fig. 10a shows a magnitude of the scattering coefficients $|S_{11}|$ and $|S_{21}|$ of the filter. It can be read from the curve $|S_{21}|$ that the wide passband from 5.0 GHz to 11.3 GHz and deep attenuation poles at 2.3 GHz and 12.2 GHz are obtained. Although the passband is a little bit higher than the theory due to too much milling of the substrate in the pattern making process, the shape of the measured $|S_{21}|$ shows good agreement with the theory as a whole. The minimum insertion loss of less than 0.8 dB is attained at 10.8 GHz. In general, the insertion loss is caused by conductor loss, dielectric loss and radiation loss. However, the insertion loss of the passband is mainly caused by the wider gap than the initial design, leading to a weakened coupling between the E-shaped electrodes.

Fig. 10b presents a group delay measured between SMA connectors. The result includes small vibrations because the network analyzer used for this measurement was quite old and it lacked stability. However, the real group delay can be estimated by reading the average value from the measured result. The variation of the group delay is estimated around 0.2 ns in the passband and it gets slightly higher at up to 0.4 ns at the higher edge of the passband. A mild group delay is quite useful to convey the short pulse signal from the input to the output without serious distortion.

For size reduction of the filter, using the higher dielectric permittivity substrate is the well-known standard approach. As for the design examples, two filters fabricated on the substrate with ε_r=5 or ε_r=10 are designed with the structural parameters summarized in Tables 3. The simulated scattering characteristics for these models are shown in Fig. 11. Though the dielectric and conductor losses are not included in the simulation, it can be confirmed that these filters have a flat UWB passband with the reflection coefficient of less than -15 dB. In addition, compared with the conventional UWB filters engaged in multiple-mode resonators, significant size reduction is attained.

Coupled-lines							
L_1	L_2	L_3	L_4	L_5	L_6	*Spacing bw lines*	*Spacing bw ELs**
3.7	3.7	3.7	3.7	3.7	3.7		
W_1	W_2	W_3	W_4	W_5	W_6	S	G
0.3	0.3	0.3	0.3	0.3	0.3	0.35	0.01
Input / Output microstrip lines				Substrate			

	I/O microstip lines			Substrate	
W_B	D	W_{MSL}	Z_0	H_{SUB}	ε_r
0.3		0.83	50 ohm	0.5	5.0

(a) * between Electrodes

Coupled-lines							
L_1	L_2	L_3	L_4	L_5	L_6	*Spacing bw lines*	*Spacing bw ELs**
2.7	2.7	2.7	2.7	2.7	2.7		
W_1	W_2	W_3	W_4	W_5	W_6	S	G
0.2	0.2	0.2	0.2	0.2	0.2	0.2	0.008
Input / Output microstrip lines				Substrate			

	I/O microstip lines			Substrate	
W_B	D	W_{MSL}	Z_0	H_{SUB}	ε_r
0.2		0.44	50 ohm	0.5	10.0

(b) * between Electrodes

Table 3. Structural parameters of the miniaturized UWB bandpass filter with tightly coupled E-shaped electrodes fabricated on the higher permittivity substrate. (a) Substrate with ε_r=5 is used. (b) Substrate with ε_r=10 is used.

The total area of the filter body (area of the E-shaped electrodes and the gap between them) is 3.21 mm x 3.7 mm x 0.5 mm for ε_r=5 model and 2.008 mm x 2.7 mm x 0.5 mm for ε_r=10 model, respectively.

Contrary to the super-compactness, it should be noted that the filters fabricated on the higher permittivity substrate always require an extremely small gap between the electrodes in order to create the necessary series capacitance C_0. For instance, when the substrate with ε_r=10 is used, the gap width of 0.008 mm is required. However, 0.05 mm gap is the recent fabrication limit for the standard PCB technology. If the narrower gap needs to be fabricated, the high-cost laser-based process would be indispensable for structuring the circuit.

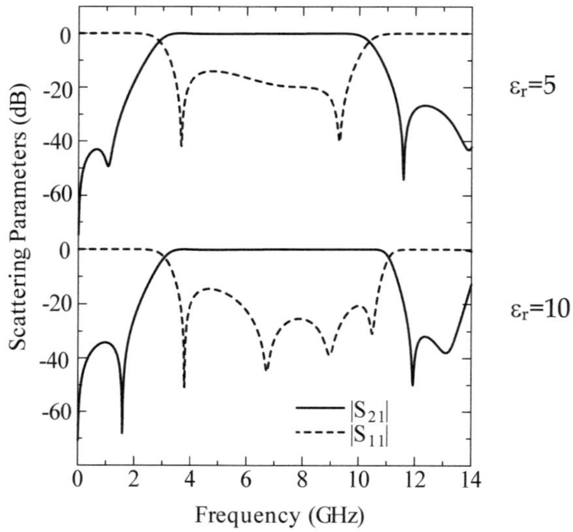

Fig. 11. Simulated scattering characteristics of the miniaturized UWB bandpass filter with tightly coupled E-shaped electrodes. Filters are fabricated on the high permittivity substrate with $\varepsilon_r=5$ (top) or $\varepsilon_r=10$ (bottom). Structural parameters are summarized in Tables 3a and 3b, respectively.

3.2 UWB bandpass filters with capacitor-loaded E-shaped electrodes

The model introduced in the former subsection would be useful if the small gap between the E-shaped electrodes could be fabricated accurately using the conventional PCB technology. However, as mentioned above, when we design the smaller filter using the higher permittivity substrate, the gap width will be significantly narrower.

The easier way to reduce such fabrication difficulties is to introduce a chip capacitor onto the gap to compensate for the insufficient gap capacitance. A circuit layout of a new planer UWB bandpass filter with capacitor-loaded E-shaped electrodes is shown in Fig. 12. The small chip capacitor with capacitance C_P is loaded between the E-shaped electrodes with an offset P from the bottom of the electrodes.

To begin with, the relation between the gap width G and the transmission coefficient $|S_{21}|$ is shown in Fig. 13a when the chip capacitor is not yet installed on the gap. Except for the value of the gap width G, structural parameters are the same as the model shown in Fig. 7. Apparently, when the gap becomes wider, the passband response is significantly damaged and reduced. To study the usefulness of the chip capacitor, let's start the filter design with G=0.3 mm.

Fig. 13b presents the $|S_{21}|$ when the chip capacitance C_P is varied from 0.0 pF (without a capacitor) to 2.0 pF. The capacitor is assumed to be fabricated at the bottom of the electrodes (P=0.0 mm), and the gap width is fixed as G=0.3 mm. In this structure, the required series capacitance C_0 is mostly supplied by the chip capacitor C_P. It is confirmed from the graph

that the response around 3 GHz is improved drastically by choosing an appropriate capacitance C_P. In this study, when C_P is 0.5 pF, the flatter passband response is obtained.
In addition, the stopband response at the higher frequency above the passband is also important. The offset of the chip capacitor P is related to the position of the transmission zeros in the higher frequency region, because the effective length of the third and forth lines of the electrodes, which work as an open-stub band elimination filter, are varied by changing the chip offset P. Fig 13c shows the simulated $|S_{21}|$ when the P is varied from 0.0 mm to 1.5 mm. Calculation is carried out for the fixed gap width G=0.3 mm and the chip capacitance C_P=0.5 pF. It can be read from the graph that the transmission zero obtained above the passband shifts slightly higher with the increase of P, and a pulse-like response around 13 GHz disappears from there. As a result, the insertion loss of more than 30 dB is attained above 11.5 GHz by choosing as P=1.5 mm. According to these theoretical considerations, the structural parameters of the final model are derived as listed in Table 4.

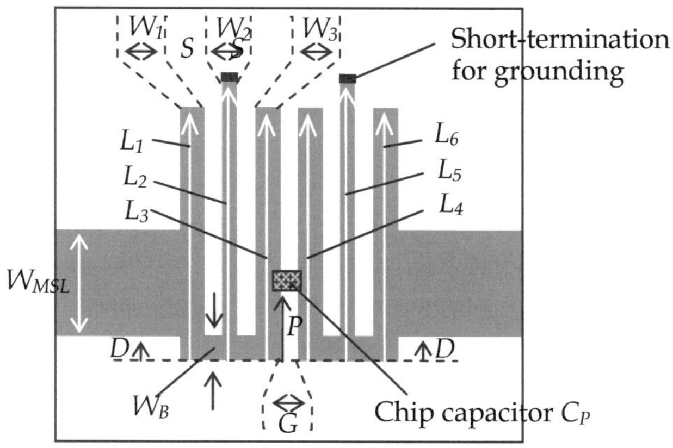

Fig. 12. A circuit layout of a planer UWB bandpass filter with capacitor-loaded E-shaped electrodes. Structural parameters are listed in Table 4.

As shown in Fig. 14, a prototype model with structural parameters listed in Table 4, is fabricated to confirm the theoretical predictions. As the chip capacitor, a high frequency multi-layered ceramic capacitor VK105 from TAIYO YUDEN was used and soldered between the electrodes with the offset P=1.5 mm.
For comparison, two types of filters are demonstrated; one is the filter before installing the chip capacitor (corresponding to the top graph in Fig. 15a), and the other is the filter after installing the chip capacitor (corresponding to the bottom graph in Fig. 15a). As predicted by the simulation in Fig.13a, the measured transmission coefficient $|S_{21}|$ is badly damaged due to the shortage of the series capacitance. However, by supplying the additional capacitance with C_P=0.5 pF, an extremely wide and flat passband is created from 3.1 GHz to 11.4 GHz, corresponding to the fractional bandwidth of 114 %. The return loss of more than 10 dB is attained from 3.7 GHz to 11.0 GHz. Furthermore, as shown in Fig. 15b, the measured group delay of the filter is quite mild, and the in-band group delay variation is less than 0.2 ns. Thus, the filter response is improved quite easily and drastically by supplying the small chip capacitor onto the gap.

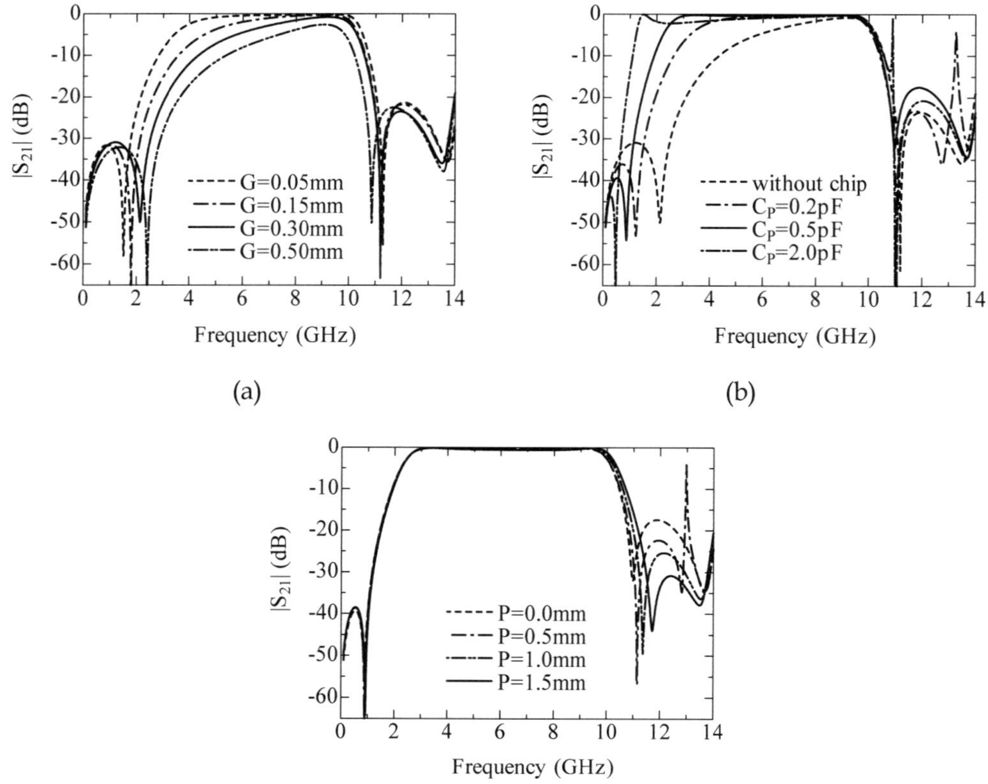

(c)

Fig. 13. Calculated transmission coefficients $|S_{21}|$ of the planer UWB bandpass filter with capacitor-loaded E-shaped electrodes shown in Fig. 12. Design starts with the initial parameters listed in Table 2. (a) The effect of the gap width G when the chip capacitor is not installed. (b) The effect of the chip capacitor C_P, when G=0.3 mm and P=0.0 mm are assummed. (c) Tuning of the higher frequency response by changing the chip offset P. The parameters G=0.3 mm and C_P=0.5 pF are assummed.

Coupled-lines							
L_1	L_2	L_3	L_4	L_5	L_6	*Spacing bw lines*	*Spacing bw ELs**
5.0	5.5	5.0	5.0	5.5	5.0		
W_1	W_2	W_3	W_4	W_5	W_6	S	G
0.5	0.3	0.5	0.5	0.3	0.5	0.35	0.3
	I/O microstrip lines			Substrate		Chip capacitor	
W_B	D	W_{MSL}	Z_0	H_{SUB}	ε_r	C_P	P
0.5	0.5	2.0	50 ohm	0.8	2.62	0.5 pF	1.5

* between Electrodes

Table 4. Structural parameters of the planer UWB bandpass filter with capacitor-loaded E-shaped electrodes shown in Fig. 12. Unit for dimensional parameters is in *mm*.

Fig. 14 A prototype of the UWB bandpass filter with capacitor-loaded E-shaped electrodes, shown in Fig. 12 with circuit parameters listed in Table 4. The filter is fabricated on a Rexolite 2200 substrate with a relative permittivity of 2.62 and a thickness of 0.8 mm. A 0.5 pF chip capacitor is loaded between the electrodes.

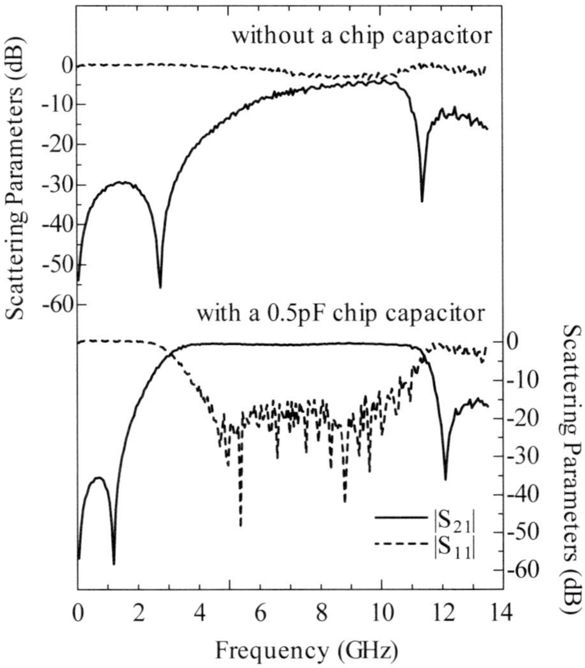

(a)

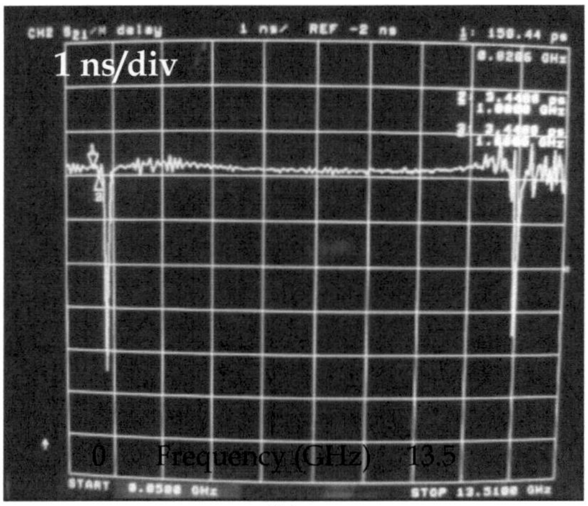

(b)

Fig. 15. Measured scattering characteristics of the UWB bandpass filter with capacitor-loaded E-shaped electrodes. (a) The reflection coefficient $|S_{11}|$ and the transmission coefficient $|S_{21}|$ of the filters without a chip capacitor (top half of the graph) and with a 0.5 pF chip capacitor (bottom half of the graph). Basic structural parameters are listed in Table 5. (b) The group delay between SMA connectors. Unit is 1 ns/div.

3.3 Asymmetric UWB bandpass filters with improved out-of-passband response

The UWB bandpass filters with tightly-coupled or capacitor-loaded E-shaped electrodes were both attractive in respects to the compactness, low-cost fabrication, and mass-productivity. In addition, the superior passband selectivity was realized by creating the transmission zeros just below and above the passband. However, the FCC specification also requires a high rejection level in the stopband. In order to improve the stopband response, an asymmetric UWB bandpass filter with capacitor-loaded E-shaped electrodes was newly developed. This technology is introduced in this subsection.

As mentioned above, the transmission zeros above the passband are created by the quarter-wavelength open-stubs in the electrodes. This means that the stopband response can be improved by tuning the length of the stubs so that the transmission zeros are located at regular intervals in the stopband region. This fine-tuning process can be done simply by using an optimizer installed in the commercial EM simulators, or by changing the length of the stubs manually.

In the actual design, the fine-tuning process, started with the symmetric UWB bandpass filter with capacitor-loaded E-shaped electrodes presented in Fig. 12. Since the lines in the electrodes couple to each other, this process will be time-consuming and tedious if the optimization is done manually.

Fig. 16 shows the circuit layout of the optimized asymmetric bandpass filter, the structural parameters of which are summarized in Table 5. For comparison, the simulated scattering characteristics are shown in Fig. 17, together with the original symmetric filter shown in Fig. 12 with the structural parameters given in Table 4. The optimized asymmetric bandpass filter has the wide stopband from 11.5 GHz to 18.5 GHz with the rejection level of more than 10 dB, while the original flat and wide passband response is maintained.

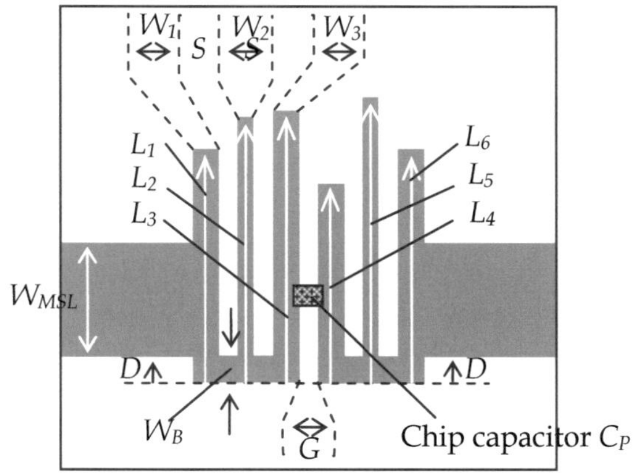

Fig. 16. A circuit layout of the asymmetric UWB bandpass filter with capacitor-loaded E-shaped electrodes. Structural parameters are listed in Table 6 and FEM-simulated scattering characteristics are presented in Fig. 17.

Coupled-lines							
L_1	L_2	L_3	L_4	L_5	L_6	*Spacing bw lines*	*Spacing bw ELs**
4.5	4.8	4.9	3.7	5.1	4.2		
W_1	W_2	W_3	W_4	W_5	W_6	S	G
0.5	0.3	0.5	0.5	0.3	0.5	0.3	0.5
W_B	I/O microstrip lines			Substrate		Chip capacitor	
	D	W_{MSL}	Z_0	H_{SUB}	ε_r	C_P	P
0.5	0.5	2.0	50 ohm	0.8	2.62	0.5	1.0

* between Electrodes

Table 5. Structural parameters of the asymmetric UWB bandpass filter with capacitor-loaded E-shaped electrodes shown in Fig. 16. Unit for dimensional parameters is in *mm*.

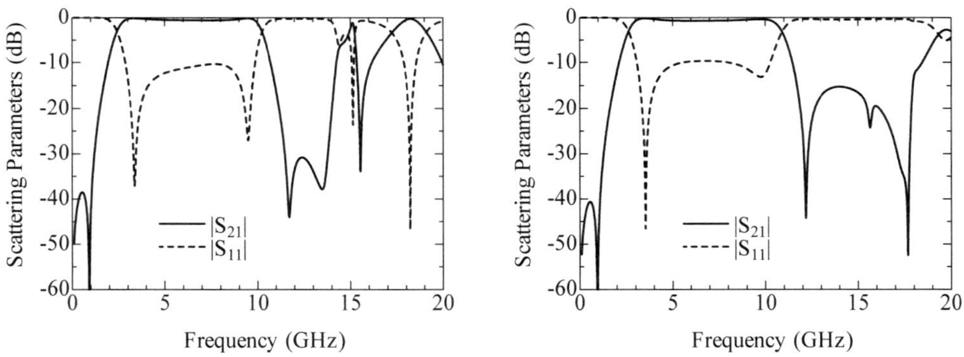

(a) (b)

Fig. 17. FEM-simulated scattering characteristics. (a) The symmetric UWB bandpass filter with capacitor-loaded E-shaped electrodes (original model). (b) The asymmetric UWB bandpass filter with capacitor-loaded E-shaped electrodes (optimized model).

3.4 Design procedure of UWB bandpass filters

Different from other UWB bandpass filters proposed so far, the couplings between the filter components play an important role in this filter to obtain a superior UWB response in a limited filter dimension. For example, the filter uses the electric coupling between the electrodes, the electric coupling between the thin lines of the electrodes, and the magnetic coupling between the short-circuited stubs of the electrodes. However, this makes it difficult to use the conventional filter theory for this kind of filter design. Instead, the filter can be designed easily in accordance with a simple design procedure as follows;

1) The length of the open-stubs in each E-shaped electrodes (L_1, L_3, L_4, and L_6) needs to be chosen as a quarter-wavelength so as to obtain some transmission zeros above the UWB passband. This process is usually done under L_1=L_3=L_4=L_6. The higher cutoff frequency is roughly determined by this.
2) The length of the short-stubs in each electrode (L_2 and L_5) needs to be chosen so as to obtain a transmission zero below the UWB passband. The lower cutoff frequency is roughly given by this step.
3) The coupling between the E-shaped electrodes should be tuned by changing the gap width between the electrodes, G, so as to create a flat UWB passband. It should be noted that the larger gap can easily damege the response at the lower edge of the passband.
4) If a chip capacitor with an appropriate capacitance is loaded between the gap, the fabrication difficulty will be reduced significantly.
5) Fine-tuning of other structural parameters are needed to obtain the better filter response. The center frequency of the UWB passband can be adjusted in some extent by the width of the line W_B at the bottom of the E-shaped electrode. The in-band ripples in the transmission coefficient can be reduced in this process.
6) The higher stopband response can be improved slightly by giving an appropriate offset D to the input and output microstrip lines connecting to the electrodes.

By this, an excellent UWB passband will be attained. To create the better stopband response, fine-tuning of the open-/short-stubs in the electrodes is quite useful as demonstrated in the former discussion.

7) To create the better stopband, the line length in the electrodes should be tuned independently by using an optimizer installed in the commercial EM simulators or by changing the length manually.

According to these instructions, all the filters presented in this chapter were designed.

4. Conclusions

In this chapter, three types of compact UWB bandpass filters with plural transmission zeros below and above the passband are introduced together with the useful design procedure. Compared with other conventional filters, the proposed filter can be made drastically small in size with the help of a tiny chip capacitor. In addition, the circuit pattern of the filter is given only at the top of the substrate and a perfect ground plane is remained without any defections. This will be a great advantage when the filter is fabricated on a double-layered substrate. Furthermore, the microstrip-line-based filter patterns can be printed out together with other circuit petterns at the same time using the common PCB process. Thus, the proposed filters have tremendous attractive features in the engineering of the latest UWB technology, in respects to the super-compactness, easy fabrication, excellent compatibility to other circuits, low-cost and mass productivity.

5. References

Federal Communications Commission (2002), "Revision of Part 15 of the Commission's Rules Regarding UWB Transmission Systems, First report and order", *Federal Communication Commissions (FCC)*, 02.V48.

Ishida, H. & Araki, K. (2004) "Design and analysis of UWB bandpass filter with ring filter", *IEEE International Microwave Symp.*, Proceedings.

Zhu, L.; Sun S. & Menzel W. (2005), "Ultra-wideband (UWB) bandpass filters using multiple-mode resonator", *IEEE Microwave Wireless Compon. Lett.*, vol.15, no.11, pp.796-798.

Horii, Y. (2007). "A compact planar ultra-wideband bandpass filter composed of coupled E-shaped electrodes", *European Microwave Conf.*, Proceedings.

Horii, Y.; Ishikawa, K.; Kaneko, T. & Azuma, Y. (2007). "Development of design procedure for compact planar ultra-wideband (UWB) bandpass filters", *Asia-Pacific Microwave Conf.*, Proceedings.

Horii, Y. (2008). "Super-compact planar ultra-wideband (UWB) bandpass filter composed of capacitor-loaded E-shaped electrodes", *European Microwave Conf.*, Proceedings, pp.361-364.

Kaneko, T.; Azuma, Y.; Ishikawa, K. & Horii, Y. (2008). "Ultra-wideband (UWB) bandpass filter composed of asymmetric E-shaped electrodes for improved out-of-passband", *Asia-Pacific Microwave Conf.*, Proceedings.

16

A Dual-Mode Wide-Band Bandpass Filter Using the Microstrip Loop Resonator with Tuning Stubs

Jessada Konpang
Rajamangala University of Technology Krungthep
Thailand

1. Introduction

The broadband wireless access (BWA) is an important issue in current developments of the modern wireless communication system. To meet this trend, the bandpass filters with relatively wide bandwidth are frequently required in the RF front ends. In microwave communication systems, the bandpass filter is an essential component, which is usually used in both receivers and transmitters. Thus, the quality of bandpass filers is extremely important. Planar filters are currently a popular structure because they can be fabricated using printed circuit technology and are suitable for commercial applications due to their small size and lower fabrication cost (D.M. Pozar, 1998). Therefore, how to design a bandpass filter at low cost and with high performance is currently of great interest. Microstrip bandpass filters can be easily mounted on a dielectric substrate and can provide a more flexible design of the circuit layout. The dual-mode resonators filter have been known for years. The compact high performance microwave bandpass filters are highly desirable in the wireless communications systems. Consequently, the dual-mode filters have been used widely for the system because of their advantages such as small size, light weight, low loss and high selectivity. Many authors (Hsieh & Chang, 2001, 2003), (Konpang, 2003) and (Chen et al., 2005) have proposed the wide-band bandpass filters using dual-mode ring resonators with tuning stubs but the configurations still occupy a large circuit area, which is not suitable for wireless communication systems where the miniaturization is an important factor. Therefore, it is desirable to develop new types of dual-mode microstrip resonators not only for offering alternative designs, but also for miniaturizing filters. On the other hand, the modern wireless communication systems require the bandpass filters having effective out-of-band spurious rejection and good in-band performance. The resonators with reasonable spurious are required to meet the out-of-band requirements. The microstrip open-loop resonators have a wide stopband resulting from the dispersion effect and the slow-wave effect (Hong & Lancaster, 1997) and (Görür, 2002).

In this book, a dual-mode wide-band bandpass filter using the microstrip loop resonators with tuning stubs is proposed. Basic concepts and design equations for microstrip lines introduced in section 2. The bandpass filter is based on the bandstop filter employing direct-connected feed lines on the orthogonal of the microstrip loop resonators. The introduction of two tuning open stubs connecting opposite to the ports widens the passband and sharpens the stopbands. Then, a dual-mode can be used to improve the narrow stopbands for lower side band and higher sideband. The design descriptions dual-mode wide-band bandpass filter are discussed in section 3.

2. Transmission lines

Basic concepts and design equations for microstrip lines, dual-mode wide-band bandpass filter using the microstrip loop resonator with tuning stubs are briefly described.

2.1 Microstrip lines

2.1.1 Microstrip structure

The general structure of a microstrip is illustrated in Figure 1. A conducting strip (microstrip line) with a width w and a thickness t is on the top of a dielectric substrate that has a relative dielectric constant ε_r and a thickness h, and the bottom of the substrate is a ground (conducting) plane.

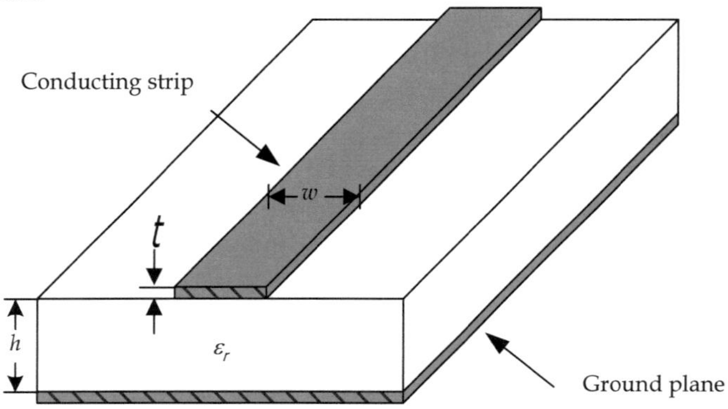

Fig. 1. General microstrip structure

2.1.2 Waves in microstrips

The fields in the microstrip extend within two media-air above and dielectric below so that the structure is inhomogeneous. Due to this inhomogeneous nature, the microstrip does not support a pure TEM wave. This is because that a pure TEM wave has only transverse components, and its propagation velocity depends only on the material properties, namely the permittivity ε and the permeability μ. However, with the presence of the two guided-wave media (the dielectric substrate and the air), the waves in a microstrip line will have no vanished longitudinal components of electric and magnetic fields, and their propagation velocities will depend not only on the material properties, but also on the physical dimensions of the microstrip.

2.1.3 Quasi-TEM approximation

When the longitudinal components of the fields for the dominant mode of a microstrip line remain very much smaller than the transverse components, they may be neglected. In this case, the dominant mode then behaves like a TEM mode, and the TEM transmission line theory is applicable for the microstrip line as well. This is called the quasi-TEM approximation and it is valid over most of the operating frequency ranges of microstrip.

2.1.4 Effective dielectric constant and characteristic impedance

In the quasi-TEM approximation, a homogeneous dielectric material with an effective dielectric permittivity replaces the inhomogeneous dielectric-air media of microstrip. Transmission characteristics of microstrips are described by two parameters, namely the effective dielectric constant ε_{re} and characteristic impedance Z_c, which may then be obtained by quasistatic analysis. In quasi-static analysis, the fundamental mode of wave propagation in a microstrip is assumed to be pure TEM. The above two parameters of microstrips are then determined from the values of two capacitances as follows

$$\varepsilon_{re} = \frac{C_d}{C_a}$$

$$Z_c = \frac{1}{c\sqrt{C_a C_d}} \tag{1}$$

in which C_d is the capacitance per unit length with the dielectric substrate present, C_a is the capacitance per unit length with the dielectric substrate replaced by air, and c is the velocity of electromagnetic waves in free space ($c \approx 3.0 \times 10^8 m/s$).

For very thin conductors (i.e., $t \to 0$), the closed-form expressions that provide an accuracy better than one percent are given as follows (Hong & Lancaster, 2001).

For $w/h \leq 1$:

$$\varepsilon_{re} = \frac{\varepsilon_r + 1}{2} + \frac{\varepsilon_r - 1}{2}\left\{\left(1 + 12\frac{h}{w}\right)^{-0.5} + 0.04\left(1 - \frac{w}{h}\right)^2\right\} \tag{2a}$$

$$Z_c = \frac{\eta}{2\pi\sqrt{\varepsilon_{re}}}\ln\left(\frac{8h}{w} + 0.25\frac{w}{h}\right) \tag{2b}$$

where $\eta = 120\pi$ ohms is the wave impedance in free space.

For $w/h \geq 1$:

$$\varepsilon_{re} = \frac{\varepsilon_r + 1}{2} + \frac{\varepsilon_r - 1}{2}\left(1 + 12\frac{h}{w}\right)^{-0.5} \tag{3a}$$

$$Z_c = \frac{\eta}{\sqrt{\varepsilon_{re}}}\left\{\frac{w}{h} + 1.393 + 0.677\ln\left(\frac{w}{h} + 1.444\right)\right\}^{-1} \tag{3b}$$

Accurate expression for the effective dielectric constant is

$$\varepsilon_{re} = \frac{\varepsilon_r + 1}{2} + \frac{\varepsilon_r - 1}{2}\left(1 + \frac{10}{u}\right)^{-ab} \tag{4}$$

Where $u = w/h$, and

$$a = 1 + \frac{1}{49}\ln\left(\frac{u^4 + \left(\frac{u}{52}\right)^2}{u^4 + 0.432}\right) \frac{1}{18.7}\ln\left(1 + \left(\frac{u}{18.1}\right)^3\right)$$

$$b = 0.564\left(\frac{\varepsilon_r - 0.9}{\varepsilon_r + 3}\right)^{0.053}$$

The accuracy of this model is better than 0.2% for $\varepsilon_r \le 128$ and $0.01 \le u \le 100$.
The more accurate expression for the characteristic impedance is

$$Z_c = \frac{\eta}{2\pi\sqrt{\varepsilon_{re}}}\ln\left[\frac{F}{u} + \sqrt{1 + \left(\frac{2}{u}\right)^2}\right] \tag{5}$$

where $u = w/h$, $\eta = 120\pi$ ohms, and

$$F = 6 + (2\pi - 6)\exp\left[-\left(\frac{30.666}{u}\right)^{0.7528}\right]$$

The accuracy for $Z_c\sqrt{\varepsilon_{re}}$ is better than 0.01% for $u \le 1$ and 0.03% for $u \le 1000$.

2.1.5 Guided wavelength, propagation constant, phase velocity, and electrical length

Once the effective dielectric constant of a microstrip is determined, the guided wavelength of the quasi-TEM mode of microstrip is given by

$$\lambda_g = \frac{\lambda_0}{\sqrt{\varepsilon_{re}}} \tag{6a}$$

where λ_0 is the free space wavelength at operation frequency *f*. More conveniently, where the frequency is given in gigahertz (GHz), the guided wavelength can be evaluated directly in millimeters as follows:

$$\lambda_g = \frac{300}{f(GHz)\sqrt{\varepsilon_{re}}} mm \tag{6b}$$

The associated propagation constant (β) and phase velocity (v_p) can be determined by

$$\beta = \frac{2\pi}{\lambda_g} \tag{7}$$

$$v_p = \frac{w}{\beta} = \frac{c}{\sqrt{\varepsilon_{re}}} \tag{8}$$

where c is the velocity of light ($c \approx 3.0 \times 10^8 m/s$) in free space.
The electrical length (θ) for a given physical length (l) of the microstrip is defined by

$$\theta = \beta l \tag{9}$$

Therefore, $\theta = \pi/2$ when $l = \lambda_g/4$, and $\theta = \pi$ when $l = \lambda_g/2$. These so-called quarterwavelength and half-wavelength microstrip lines are important for design of microstrip filters.

2.1.6 Synthesis of *w/h*

Approximate expressions for w/h in terms of Z_c and ε_r are available.
For $w/h \le 2$

$$\frac{w}{h} = \frac{8\exp(A)}{\exp(2A) - 2} \tag{10}$$

with

$$A = \frac{Z_c}{60}\left\{\frac{\varepsilon_r + 1}{2}\right\}^{0.5} + \frac{\varepsilon_r - 1}{\varepsilon_r + 1}\left\{0.23 + \frac{0.11}{\varepsilon_r}\right\} \tag{11}$$

and for $w/h \ge 2$

$$\frac{w}{h} = \frac{2}{\pi}\left\{(B-1) - \ln(2B-1) + \frac{\varepsilon_r - 1}{2\varepsilon_r}\left[\ln(B-1) + 0.39 - \frac{0.61}{\varepsilon_r}\right]\right\} \tag{11}$$

with

$$B = \frac{60\pi^2}{Z_c\sqrt{\varepsilon_r}}$$

These expressions also provide accuracy better than one percent. If more accurate values are needed, an iterative or optimization process based on the more accurate analysis models described previously can be employed.

2.1.7 Effect of strip thickness

So far we have not considered the effect of conducting strip thickness t (as referring to Figure 1). The thickness t is usually very small when the microstrip line is realized by conducting thin films; therefore, its effect may quite often be neglected. Nevertheless, its effect on the characteristic impedance and effective dielectric constant may be included.
For $w/h \le 1$:

$$Z_c(t) = \frac{\eta}{2\pi\sqrt{\varepsilon_{re}}}\ln\left\{\frac{8}{w_e(t)/h} + 0.25\frac{w_e(t)}{h}\right\} \tag{12a}$$

For $w/h \ge 1$:

$$Z_c(t) = \frac{\eta}{\sqrt{\varepsilon_{re}}}\left\{\frac{w_e(t)}{h} + 1.393 + 0.667\ln\left(\frac{w_e(t)}{h} + 1.444\right)\right\}^{-1} \tag{12b}$$

where

$$\frac{w_e(t)}{h} = \begin{cases} \dfrac{w}{h} + \dfrac{1.25}{\pi}\dfrac{t}{h}\left(1 + \ln\dfrac{4\pi w}{t}\right)(w/h \le 0.5\pi) \\ \dfrac{w}{h} + \dfrac{1.25}{\pi}\dfrac{t}{h}\left(1 + \ln\dfrac{2h}{t}\right)(w/h \ge 0.5\pi) \end{cases} \tag{13a}$$

$$\varepsilon_{re}(t) = \varepsilon_{re} - \frac{\varepsilon_r - 1}{4.6}\frac{t/h}{\sqrt{w/h}} \tag{13b}$$

In the above expressions, ε_{re} is the effective dielectric constant for $t = 0$. It can be observed that the effect of strip thickness on both the characteristic impedance and effective dielectric constant is insignificant for small values of t/h. However, the effect of strip thickness is significant for conductor loss of the microstrip line.

2.1.8 Dispersion in microstrip

Generally speaking, there is dispersion in microstrips; namely, its phase velocity is not a constant but depends on frequency. It follows that its effective dielectric constant ε_{re} is a function of frequency and can in general be defined as the frequencydependent effective dielectric constant $\varepsilon_{re}(f)$. The previous expressions for ε_{re} are obtained based on the quasi-TEM or quasistatic approximation, and therefore are rigorous only with DC. At low microwave frequencies, these expressions provide a good approximation. To take into account the effect of dispersion, the formula of $\varepsilon_{re}(f)$ is given as follows (Hong & Lancaster, 2001).

$$\varepsilon_{re}(f) = \varepsilon_r - \frac{\varepsilon_r - \varepsilon_{re}}{1 + (f/f_{50})^m} \tag{14}$$

where

$$f_{50} = \frac{f_{TM_0}}{0.75 + \left(0.75 - 0.332\varepsilon_r^{-1.73}\right)w/h} \tag{15a}$$

$$f_{TM_0} = \frac{c}{2\pi h\sqrt{\varepsilon_r - \varepsilon_{re}}}\tan^{-1}\left(\varepsilon_r\frac{\sqrt{\varepsilon_{re} - 1}}{\varepsilon_r - \varepsilon_{re}}\right) \tag{15b}$$

$$m = m_0 m_c \le 2.32 \tag{16a}$$

$$m_0 = 1 + \frac{1}{1 + \sqrt{w/h}} + 0.32\left(\frac{1}{1 + \sqrt{w/h}}\right)^3 \tag{16b}$$

$$m_c = \begin{cases} 1 + \dfrac{1.4}{1 + w/h}\left\{0.15 - 0.235\exp\left(\dfrac{-0.45f}{f_{50}}\right)\right\} & \text{for } w/h \le 0.7 \\ 1 & \text{for } w/h \ge 0.7 \end{cases} \tag{16c}$$

where c is the velocity of light in free space, and whenever the product $m_0 m_c$ is greater than 2.32, the parameter m is chosen equal to 2.32. The dispersion model shows that the $\varepsilon_{re}(f)$ increases with frequency, and $\varepsilon_{re}(f) \rightarrow \varepsilon_r$ as $f \rightarrow \infty$. The accuracy is estimated to be within 0.6% for $0.1 \le w/h \le 10$, $1 \le \varepsilon_r \le 128$ and for any value of h/λ_0.

The effect of dispersion on the characteristic impedance may be estimated by

$$Z_c(f) = Z_c \frac{\varepsilon_{re}(f) - 1}{\varepsilon_{re} - 1}\sqrt{\frac{\varepsilon_{re}}{\varepsilon_{re}(f)}} \tag{17}$$

where Z_c is the quasi-static value of characteristic impedance obtained earlier.

2.2 Microstrip discontinuities

Microstrip discontinuities commonly encountered in the layout of practical filters include junctions, bends and open stubs. Generally speaking, the effects of discontinuities can be more accurately modeled and taken into account in the filter designs with full-wave electromagnetic (EM) simulations.

2.2.1 Junction

The junction is used when we wish to split a signal into another paths. The asymmetrical microstrip line T junction is indicated in Figure 2.

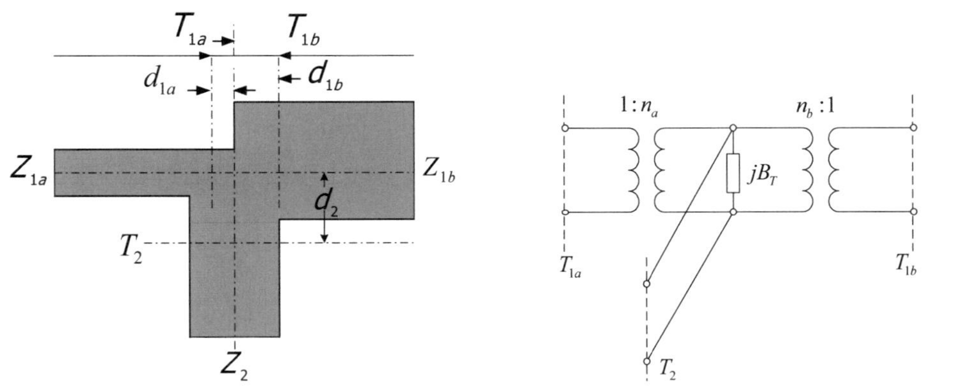

Fig. 2. Asymmetric microstrip line T junction and Model

The equations for a symmetric T junction to model the asymmetric junction. The equations are

$$\frac{d_{1a}}{D_2} = 0.055\left[1 - 2\frac{Z_{1a}}{Z_2}\left(\frac{f}{f_{p1}}\right)^2\right]\frac{Z_{1a}}{Z_2} \tag{18}$$

$$\frac{d_{1b}}{D_2} = 0.055\left[1 - 2\frac{Z_{1b}}{Z_2}\left(\frac{f}{f_{p1}}\right)^2\right]\frac{Z_{1b}}{Z_2} \tag{19}$$

$$\frac{d_2}{D_1} = 0.5 - \left[0.05 + 0.7\exp\left(-1.6\frac{\sqrt{Z_{1a}Z_{1b}}}{Z_2}\right) + 0.25\frac{\sqrt{Z_{1a}Z_{1b}}}{Z_2}\left(\frac{f}{f_{p1}}\right)^2 - 0.17\ln\frac{\sqrt{Z_{1a}Z_{1b}}}{Z_2}\right]\frac{\sqrt{Z_{1a}Z_{1b}}}{Z_2} \tag{20}$$

$$n_a^{\;2} = 1 - \pi\left(\frac{f}{f_{p1}}\right)^2\left[\frac{1}{12}\left(\frac{Z_{1a}}{Z_2}^2\right) + \left(0.5 - \frac{d_2}{D_1}\right)^2\right] \tag{21}$$

$$n_b^{\;2} = 1 - \pi\left(\frac{f}{f_{p1}}\right)^2\left[\frac{1}{12}\left(\frac{Z_{1b}}{Z_2}^2\right) + \left(0.5 - \frac{d_2}{D_1}\right)^2\right] \tag{22}$$

$$\frac{B_T}{Y_2}\frac{\lambda_1}{D_1} = 5.5\frac{\varepsilon_r + 2}{\varepsilon_r}\left[1 + 0.9\ln\frac{\sqrt{Z_{1a}Z_{1b}}}{Z_2} + 4.5\frac{\sqrt{Z_{1a}Z_{1b}}}{Z_2}\left(\frac{f}{f_{p1}}\right)^2 - 4.4\exp\left(-1.3\frac{\sqrt{Z_{1a}Z_{1b}}}{Z_2}\right) - 20\left(\frac{Z_2}{\eta_0}\right)^2\right]n^{-2}\frac{d_1}{D_2} \tag{23}$$

where

$$D = \frac{\eta_0 h}{\left(\sqrt{\varepsilon_{re}}Z_c\right)} \tag{24}$$

$$f_p(GHz) = \frac{0.4Z_c}{h} \tag{25}$$

2.2.2 Bends

Right-angle bend and mitered bend of microstrips may be modeled by and equivalent T-network, as shown in Figure 3. (Kupta et al., 1996) have given closed-form expressions for evaluation of capacitance and inductance:

$$C = 0.001h\left[(10.35\varepsilon_r + 2.5)\left(\frac{w}{h}\right)^2 + (2.6\varepsilon_r + 5.64)\left(\frac{w}{h}\right)\right]pF \tag{26}$$

$$L = 0.22h\left\{1 - 1.35\exp\left[-0.18\left(\frac{w}{h}\right)^{1.39}\right]\right\} \text{ nH} \tag{27}$$

for the microstrip mitered bend, and as

$$C = 0.001h\left[(3.93\varepsilon_r + 0.62)\left(\frac{w}{h}\right)^2 + (7.6\varepsilon_r + 3.8)\left(\frac{w}{h}\right)\right] pF \tag{28}$$

$$L = 0.44h\left\{1 - 1.062\exp\left[-0.177\left(\frac{w}{h}\right)^{0.947}\right]\right\} \text{ nH} \tag{29}$$

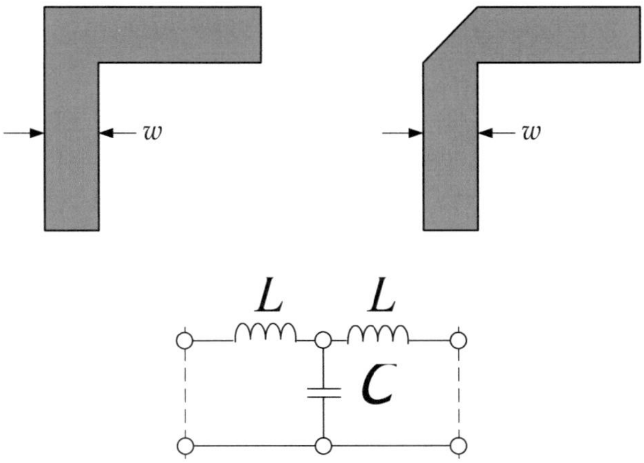

Fig. 3. Right-angle bend, mitered bend and model

2.2.3 Open stub

According to the transmission line theory, the input admittance of an open circuited transmission line having a characteristic admittance $Y_c = 1/Z_c$ and propagation constant $\beta = 2\pi/\lambda_g$ is give by

$$Y_t = jY_c \tan\left(\frac{2\pi}{\lambda_g}l\right) \tag{30}$$

Where l is the length of the stub. If $l < \lambda_g/4$ this input admittance is capacitive. The input admittance may be approximated by

$$Y_t \approx jY_c\left(\frac{2\pi}{\lambda_g}l\right) = j\omega\left(\frac{Y_c l}{v_p}\right) \tag{31}$$

Where v_p is the phase velocity of propagation in the stub. It is now clearer that such a open circuited stub is equivalent to a shunt capacitance $C = Y_c l / v_p$, as indicated in Figure 4.

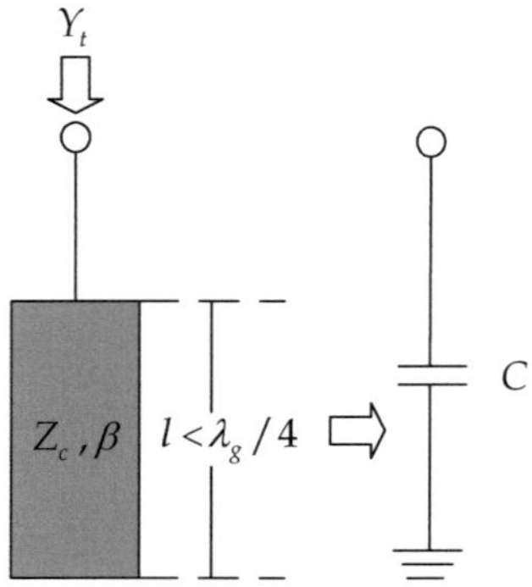

Fig. 4. Open circuit stub and Model

At the open end of a microstrip line with a width of w , the fields do not stop abruptly but extend slightly further due to the effect of the fringing field. This effect can be modeled either with an equivalent shunt capacitance C_p or with an equivalent length of transmission line

$$l_{open} = \frac{cZ_c C_p}{\sqrt{\varepsilon_{re}}} \tag{32}$$

Where c is the light velocity in free space. A closed-form expression for l_{open} / h is given by

$$\frac{l_{open}}{h} = \frac{\xi_1 \xi_3 \xi_5}{\xi_4} \tag{33}$$

where

$$\xi_1 = 0.434907 \frac{\varepsilon_{re}^{0.81} + 0.26(w/h)^{0.8544} + 0.236}{\varepsilon_{re}^{0.81} - 0.189(w/h)^{0.8544} + 0.87} \tag{34}$$

$$\xi_2 = 1 + \frac{(w/h)^{0.371}}{2.35\varepsilon_r + 1} \tag{35}$$

$$\xi_3 = 1 + \frac{0.5274 \arctan\left[0.084(w/h)^{1.9413/\xi_2}\right]}{\varepsilon_{re}^{0.9236}} \tag{36}$$

$$\xi_4 = 1 + 0.0377 \arctan\left[0.067(w/h)^{1.456}\right]\left\{6 - 5\exp\left[0.036(1-\varepsilon_r)\right]\right\} \tag{37}$$

$$\xi_5 = 1 - 0.218 \exp(-7.5w/h) \tag{38}$$

The accuracy is better than 0.2% for the range of $0.01 \le w/h \le 100$ and $\varepsilon_r \le 128$.

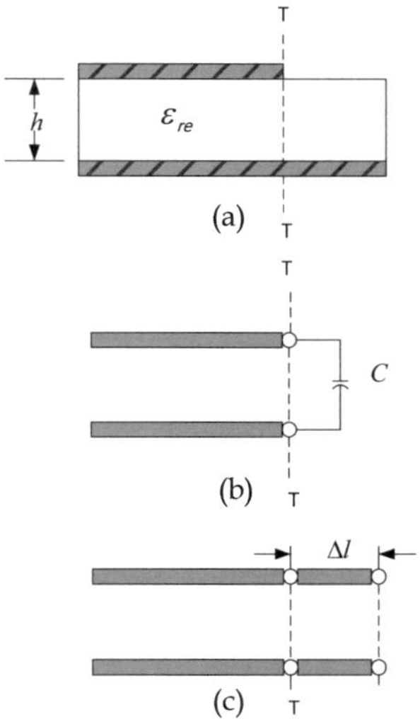

Fig. 5. (a) Microstrip open-end discontinuity (b) equivalent capacitance representation, and (c) equivalent line length representation

3. Dual-mode wide-band bandpass filter design

The bandpass filter is based on the bandstop filter employing direct-connected feed lines on the orthogonal of the microstrip loop resonators. The introduction of two tuning open stubs connecting opposite to the ports widens the passband and sharpens the stopbands. Then, a dual-mode can be used to improve the narrow stopbands for lower side band and higher sideband.

3.1 Bandstop characteristics

3.1.1 Bandstop filter (Type A)

The first bandpass filter is based on the bandstop filter employing direct-connected feed lines on the orthogonal of the microstrip loop resonator (Konpang et al., (2007). The microstrip loop resonator with direct-connected feed lines on the orthogonal depicted in Fig. 6 is a bandstop configuration. The resonator consists of four identical branches with attached to an outer corner of the square loop. The bandstop filter is designed at fundamental resonant frequency f_0 = 2.45 GHz and fabricated on a RT/Duroid substrate having a thickness h = 1.27 mm with relative dielectric constant ε_r = 6.15. The filter was designed and simulated by IE3D program. The dimensions of the loop are l_f = 8 mm, s = 0.715 mm, w_1 = 1.85 mm, w_2 = 0.75 mm, w_3 = 1.35 mm and a = 9.3 mm.

The equivalent microstrip loop circuit as shown in Fig. 7 is divided into input and output ports forming a shunt circuit denoted by the upper and lower parts, respectively. The capacitance jB_{T1} is the T-junction effect between the feed line and the microstrip loop resonator (Hsieh & Chang, 2003). The capacitance jB_{T2} is the junction effect between the loop resonator with each branch. The analysis of the characteristic of the microstrip loop resonators is performed by IE3D program. Fig.8 presents the simulation results of the microstrip loop using direct-connect orthogonal feed lines, the frequency response exhibits bandstop behaviours.

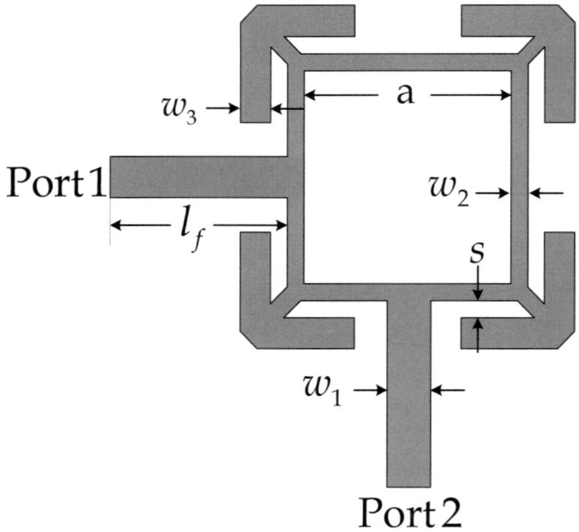

Fig. 6. Microstrip loop resonator using direct-connected orthogonal feeders (Type A)

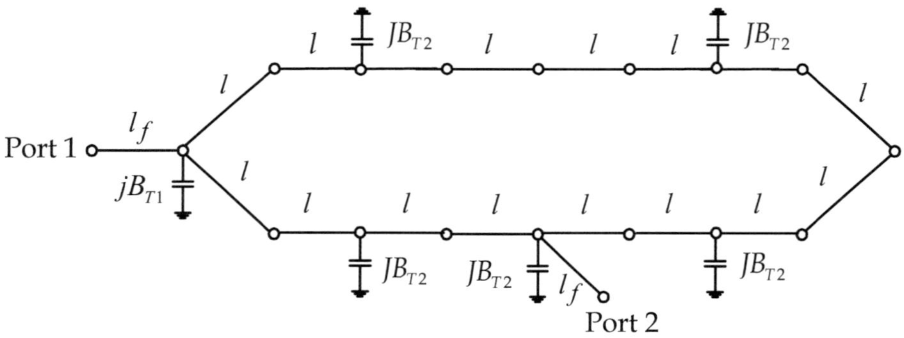

Fig. 7. Equivalent circuit of the microstrip loop resonator using direct-connected orthogonal feed lines

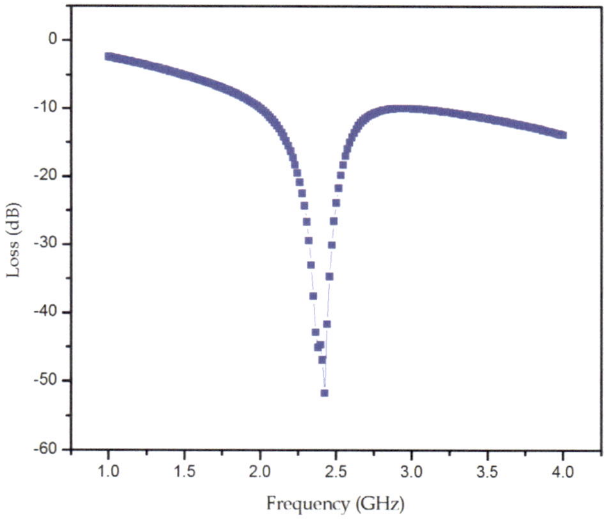

Fig. 8. Simulation results of the microstrip loop resonator using direct-connected orthogonal feed lines

3.1.2 Bandstop filter (Type B)

The second bandpass filter is based on the bandstop filter employing direct-connected feed lines on the orthogonal of the microstrip loop resonator (J. Konpang, 2008). The microstrip loop resonator with direct-connected feed lines on the orthogonal depicted in Fig.9 is a bandstop configuration. The resonator consists of four identical branches with a small square patch attached to an inner corner of the square loop.

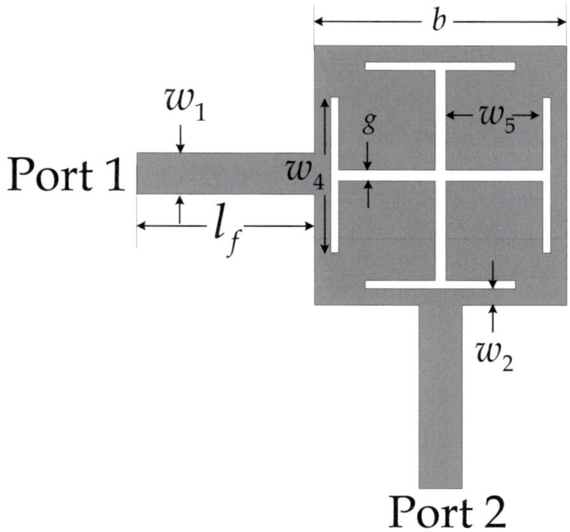

Fig. 9. Microstrip loop resonator using direct-connected orthogonal feeders (Type B)

The bandstop filters is designed at fundamental resonant frequency f_0 = 2.45 GHz and fabricated on a RT/Duroid substrate having a thickness h = 1.27 mm with relative dielectric constant ε_r = 6.15. The filter was designed and simulated by IE3D program. The dimensions of the loop are l_f = 8 mm, w_1 = 1.85 mm, w_2 = 0.75 mm, w_4 = 6.8 mm, w_5 = 4.4 mm, g = 0.4 mm and b = 11.5 mm.

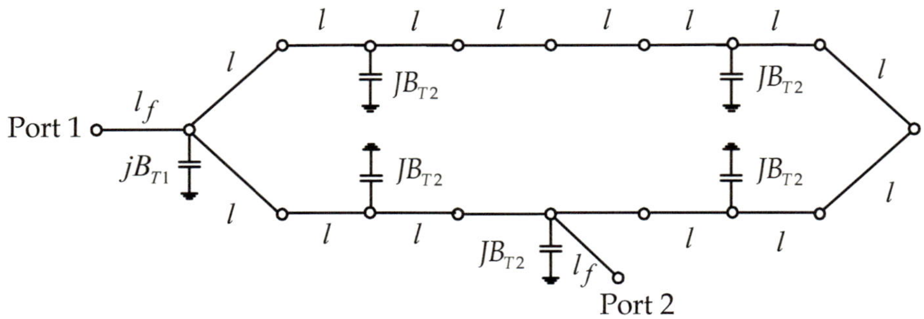

Fig. 10. Equivalent circuit of the microstrip loop resonator using direct-connected orthogonal feed lines

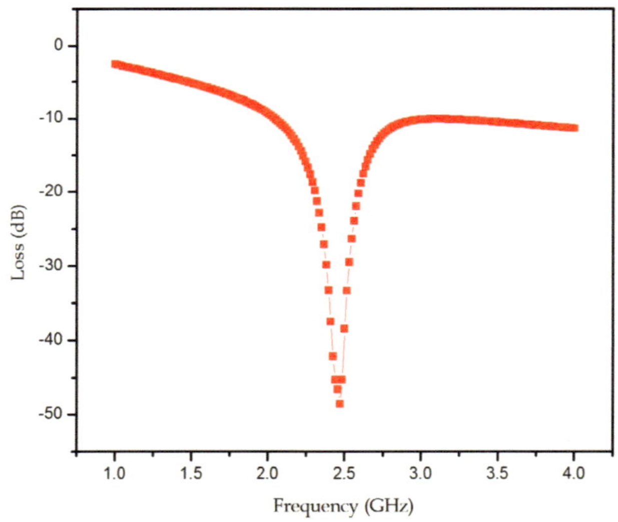

Fig. 11. Simulation results of the microstrip loop resonator using direct-connected orthogonal feed lines

The equivalent microstrip loop circuit is shown in Fig.10 is divided into input and output ports forming a shunt circuit denoted by the upper and lower parts, respectively. The capacitance jB_{T1} is the T-junction effect between the feed line and the microstrip loop resonator (Hsieh & Chang, 2003). The capacitance jB_{T2} is the junction effect between the loop resonator with each branch. The analysis of the characteristic of the microstrip loop resonators is performed by IE3D program. Fig.11 presents the simulation results of the microstrip loop using direct-connect orthogonal feed lines. The frequency response exhibits bandstop behaviours.

3.2 Two tuning stubs for a single-mode bandpass filter

3.2.1 Single-mode bandpass filter (Type A)

Based on bandstop filter, The first resonator is modified by adding two tuning stubs connecting opposite to the ports. The resonator (Type A) with tuning stubs is shown in Fig. 12. The length of tuning opened-stub is $l_t = \lambda_g / 4$.

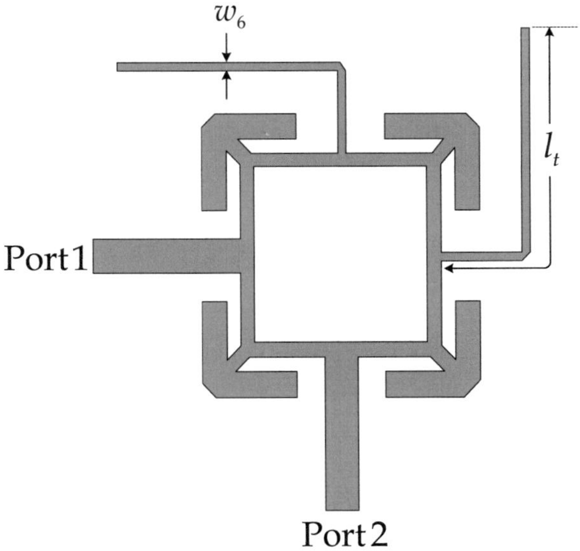

Fig. 12. Structure of two tuning stubs for single-mode (Type A)

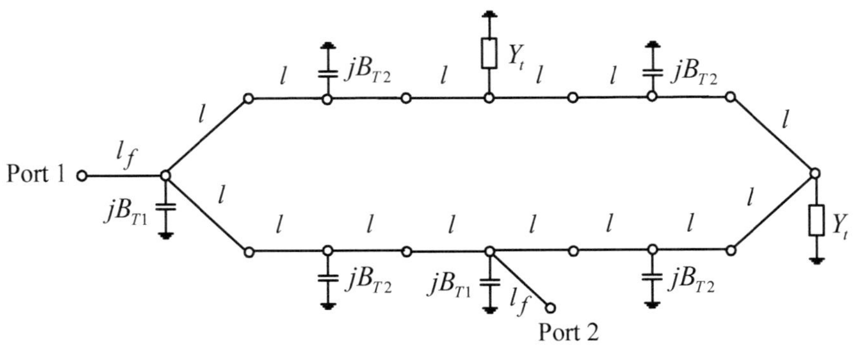

Fig. 13. Equivalent circuit of a microstrip loop resonator with tuning stubs

The equivalent circuit of resonator with tuning stubs is shown in Fig. 13. Y_t is the admittance reflecting into the stubs Y_t can be expressed by

$$Y_t = y_o \tanh(\gamma lt + l_{open}) + jB_{T3} \tag{39}$$

where y_o is the characteristic admittance of the stub, γ is the complex propagation constant, l_{open} is the equivalent open effect length and jB_{T3} is the capacitance of the T-junction between the microstrip loop with stubs l_t.

Fig. 14. Photograph of a single-mode bandpass filter (Type A)

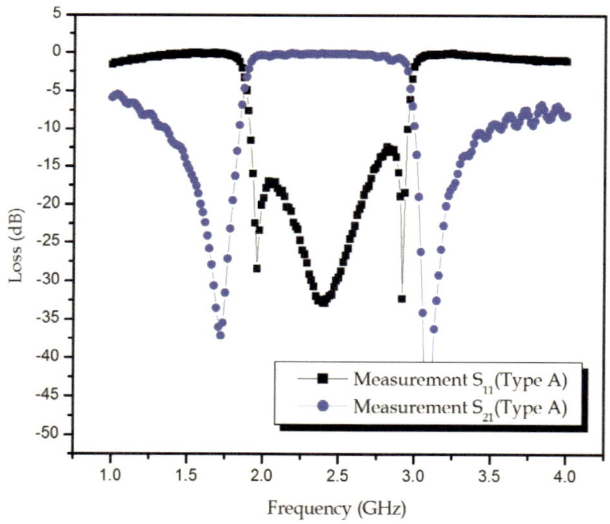

Fig. 15. Measurement for single-mode bandpass filter

The implemented of single-mode resonators filter is pictured in Fig. 14. The measurement results of the microstrip loop with tuning stubs of l_t = 16.25 mm, $w_6 = 0.4$ mm. The frequency response of the filter is portrayed in Fig. 15. The introduction of two tuning stubs connecting opposite to the ports widens the passband and sharpens the stopbands. The

single-mode filter exhibits the 3-dB fractional bandwidth of the filter is 37%, the insertion loss better than 0.26 dB and return loss greater than 12.6 dB in the passband.
In fact, this approach can be interpreted as using two stopbands induced by two tuning stubs in conjunction with the wide passband. In some cases, an undesired passband below the main passband may require a high passband section to be employed in conjunction with this proposing approach.

3.2.2 Single-mode bandpass filter (Type B)

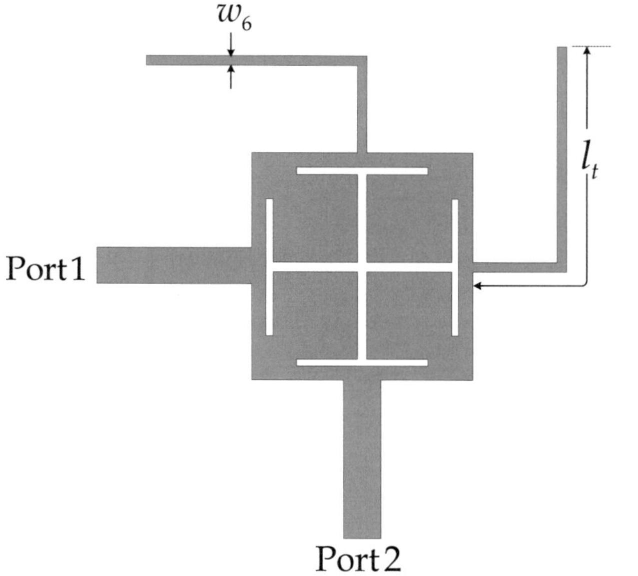

Fig. 16. Structure of two tuning stubs for single-mode (Type B)

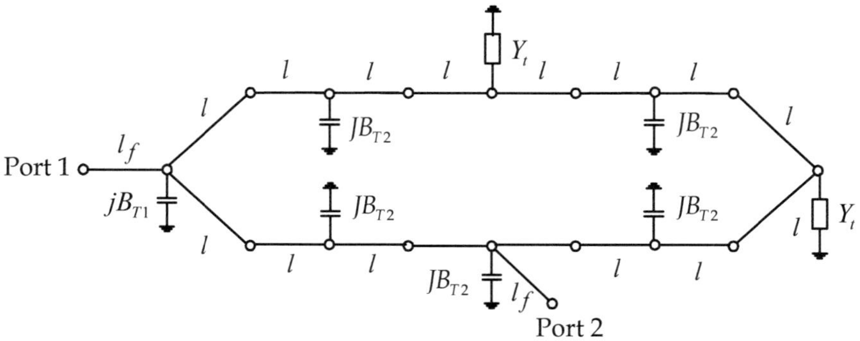

Fig. 17. Equivalent circuit of a microstrip loop resonator with tuning stubs

Fig. 18. Photograph of a single-mode bandpass filter

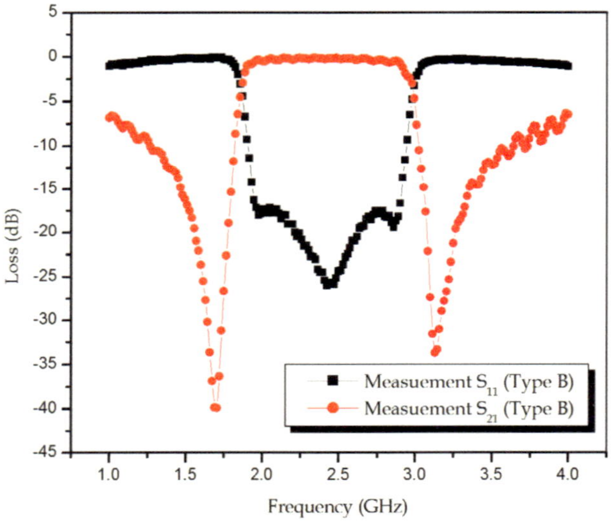

Fig. 19. Measurement for single-mode bandpass filter

Based on bandstop filter. The second resonator is modified by adding two tuning stubs connecting opposite to the ports. The resonator (Type B) with tuning stubs is shown in Fig 16. The length of tuning opened-stub is $l_t = \lambda_g / 4$. The equivalent circuit of resonator with tuning stubs is shown in Fig 17.
The implemented of single-mode resonator filter is pictured in Fig. 18. The measurement results of the microstrip loop with tuning stubs of l_t = 15.35 mm, $w_6 = 0.4$ mm . The frequency response of the filter is portrayed in Fig. 19. The introduction of two tuning stubs connecting opposite to the ports widens the passband and sharpens the stopbands. The

single-mode filter exhibits the 3-dB fractional bandwidth of the filter is 36%, the insertion loss better than 0.19 dB and return loss greater than 17 dB in the passband.

3.3 Dual-mode bandpass filter

3.3.1 Dual-mode bandpass filter (Type A)

By observing the frequency response in Fig. 15, the two stopbands for lower sideband and higher sideband of the filter propose a narrow bandstop. Based on a dual-mode can be used to improve the narrow stopbands.

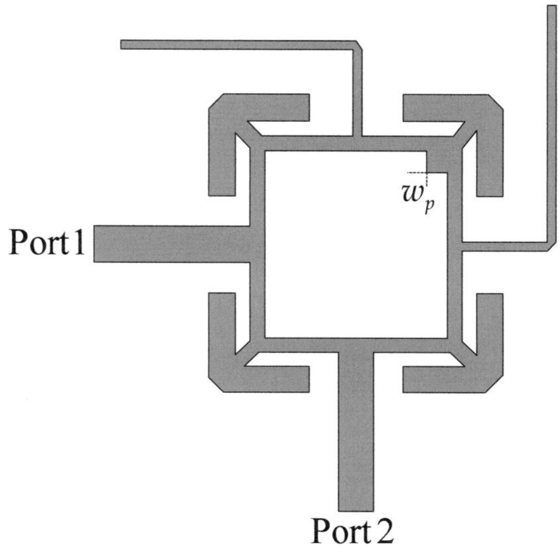

Fig. 20. Structure of dual-mode bandpass filter (Type A)

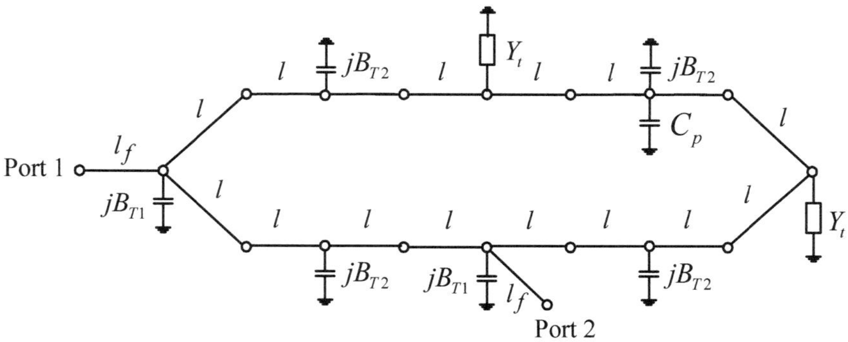

Fig. 21. Equivalent circuit of a dual-mode bandpass filter

Fig. 22. Photograph of a dual-mode bandpass filter

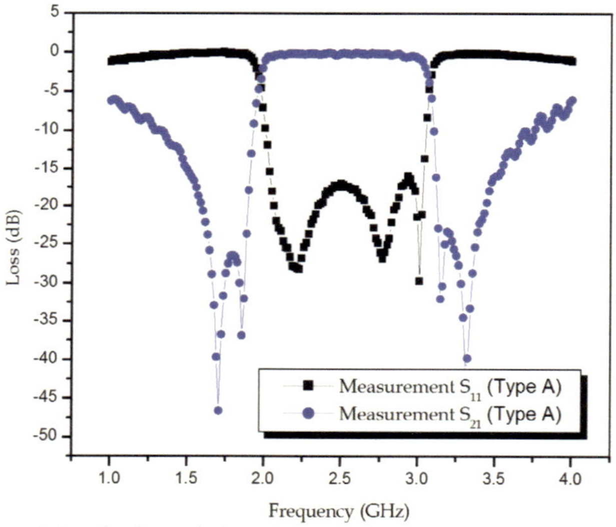

Fig. 23. Measurement of the dual-mode bandpass filter

A square perturbation stub inner corner the loop resonator in Fig. 20. The square stub perturbs the fields of the loop resonator so that the resonator can excite a dual-mode around the stopbands in order to improve the narrow stopbands. By increasing the size of the pertubation stub, the stopband bandwidth between two modes is increaseded. The length of the pertubation stub is $w_p = 1$ mm.

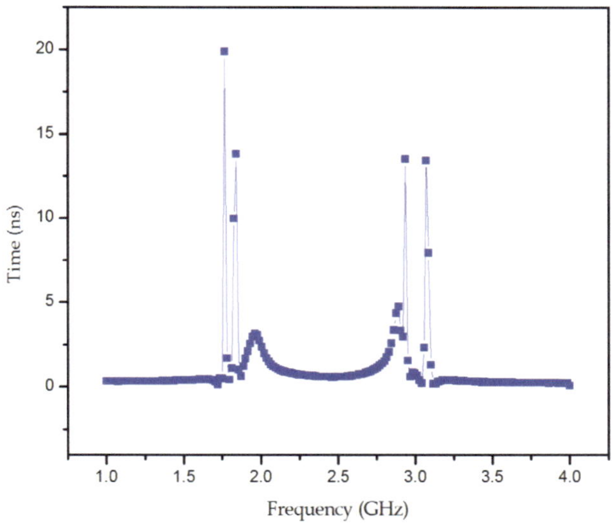

Fig. 24. Measurement group delay of the dual-mode bandpass filter

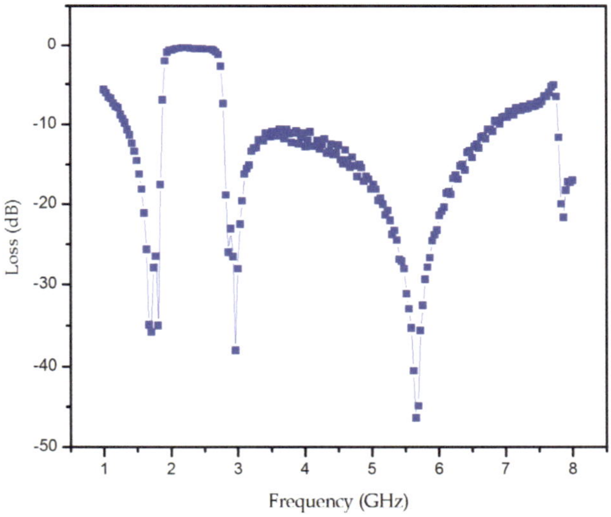

Fig. 25. Measurement wide-band of the dual-mode bandpass filter

Fig.21 delineates the equivalent circuit of the dual-mode bandpass filter (Type A). The asymetric step capacitance C_p can be calculated by

$$C_p = w_p(0.012 + 0.0039\varepsilon_r)\, pF \tag{40}$$

The implemented of dual-mode resonators filter (Type A) is pictured in Fig. 22. The frequency response of the filter is portrayed in Fig. 23. The 3-dB fractional bandwidth of the filter is 36%, the insertion loss is better than 0.34 dB and the return loss is greater than 17 dB in the passband. The group delay of the dual-mode filter can be calculated by

$$\tau = -\partial \frac{\angle S_{21}}{\partial \omega} \tag{41}$$

where $\angle S_{21}$ is the insertion-loss phase and ω is the frequency in radians per second. Fig. 24 shows the group delay of the filter. Within the passband, the group delay is below 2 ns.

The measurement of wide-band response is shown in Fig. 25. Unlike the conventional structure of the wide-band filters using dual-mode ring resonators with tuning stubs, the filter exhibits a wide stopband due to four identical branches at the outer corner of the square loop and proposes the first spurious resonance frequency of the dispersion effect.

3.3.1 Dual-mode bandpass filter (Type B)

By observing the frequency response in Fig. 19, the two stopbands for lower sideband and higher sideband of the filter propose a narrow bandstop. Based on a dual-mode can be used to improve the narrow stopbands.

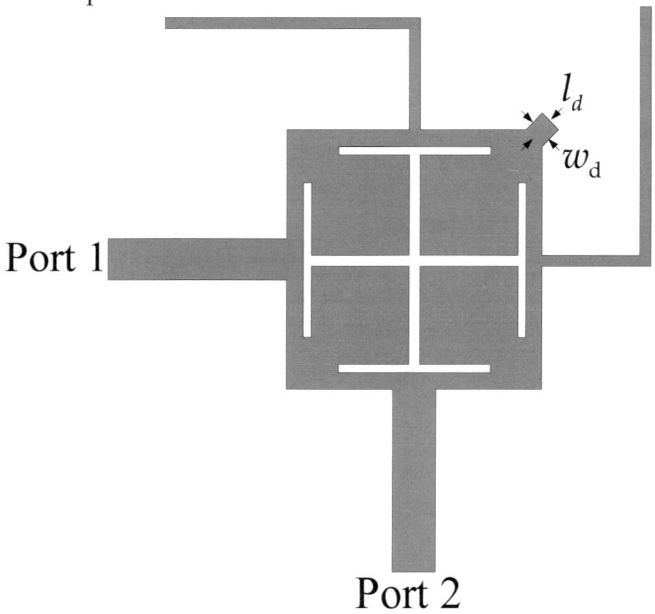

Fig. 26. Structure of dual-mode bandpass filter (Type B)

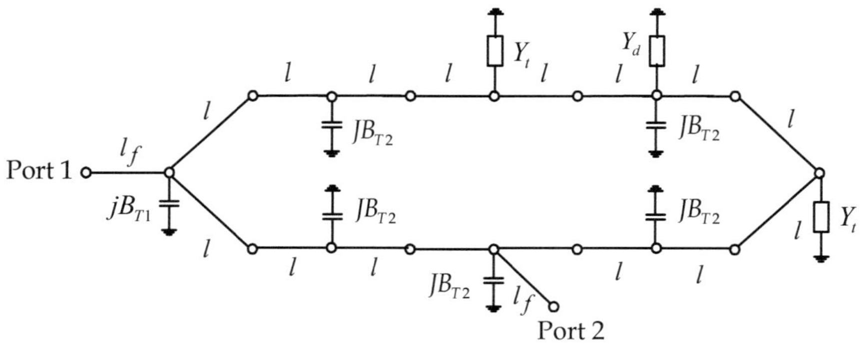

Fig. 27. Equivalent circuit of a dual-mode bandpass filter

Fig. 28. Photograph of a dual-mode bandpass filter

A square perturbation stub outward corner the loop resonator in Fig. 26. The square stub perturbs the fields of the loop resonator so that the resonator can excite a dual-mode around the stopbands in order to improve the narrow stopbands. By increasing the size of the pertubation stub, the stopband bandwidth between two modes is increaseded. The length of the pertubation stub are w_d = 0.7 mm and l_d = 0.7 mm. Fig.27 delineates the equivalent circuit of the dual-mode bandpass filter. Y_d is the admittance reflecting into the pertubation stub. Y_d can be expressed by

$$Y_d = y_o \tanh(\gamma l_d + l_{open}) + jB_{T3} \tag{42}$$

where y_o is the characteristic admittance of the stub, γ is the complex propagation constant, l_{open} is the equivalent open effect length and jB_{T3} is the capacitance of the junction between the microstrip loop with pertubation stub l_d.

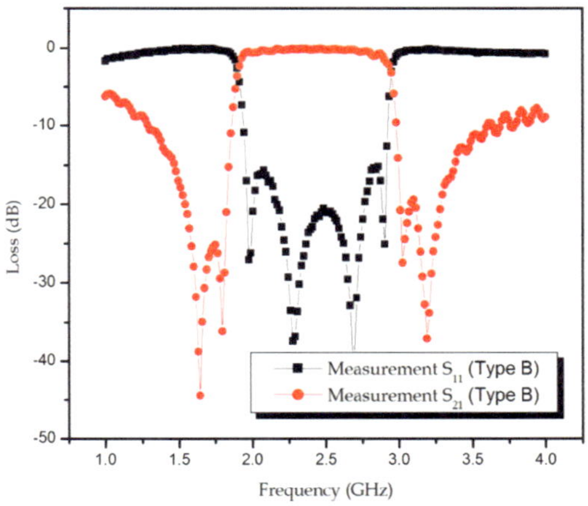

Fig. 29. Measurement of the dual-mode bandpass filter

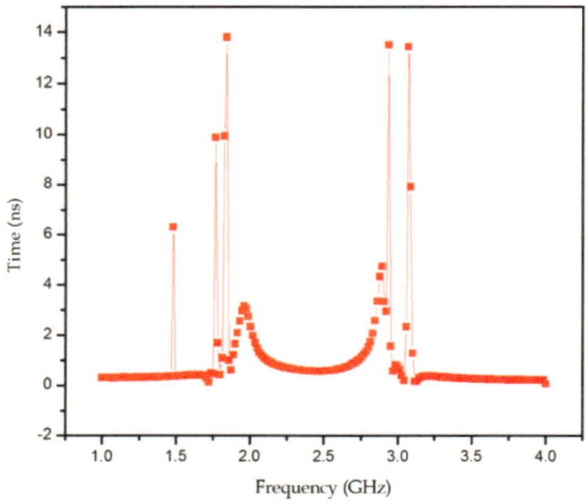

Fig. 30. Measurement group delay of the dual-mode bandpass filter

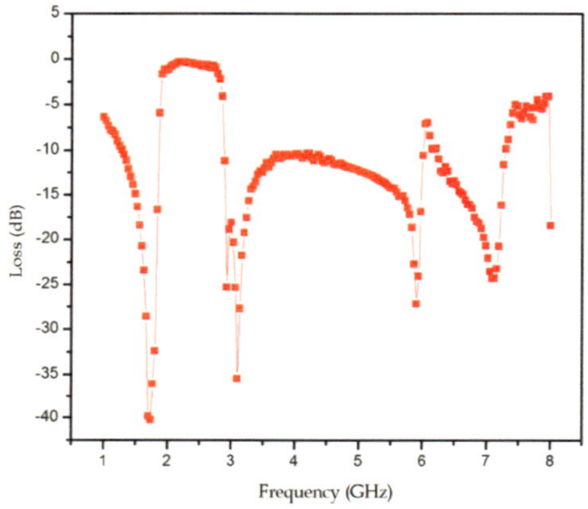

Fig. 31. Measurement wide-band of the dual-mode bandpass filter

The implemented of dual-mode resonators filter is pictured in Fig. 28. The frequency response of the filter is portrayed in Fig. 29. The 3-dB fractional bandwidth of the filter is 36%, the insertion loss is better than 0.15 dB and the return loss is greater than 15 dB in the passband. Fig. 30 shows the group delay of the filter. Within the passband, the group delay is below 2 ns. The measurement of wide-band response is shown in Fig. 31. The filter have a wide stopband resulting from the dispersion effect and the slow-wave effect.

4. Conclusions

In this book, A dual-mode wide-band bandpass filter using the microstrip loop resonators with tuning stubs is proposed here.

4.1 Conclusion

The dual-mode bandpass filter is based on the bandstop filter employing direct-connected feed lines on the orthogonal of the microstrip loop resonators. The introduction of two tuning open stubs connecting opposite to the ports widens the passband and sharpens the stopbands. Then, a dual-mode can be used to improve the narrow stopbands for lower side

band and higher sideband. The filters are designed 2.45 GHz. The 3-dB fractional bandwith of the filter are more than 36%. The group delay of the filter within the passband are below 2 ns. The filters can suppress unwanted passband to below -10 dB.
The first resonator (Type A) consists of four identical branches with attached to an outer corner of the square loop with outer tuning stubs. The filters proposes 0.34 dB insertion loss and return loss greater than 17 dB.
The second resonator (Type B) consists of four identical branches with a small square patch attached to an inner corner of the square loop with outer tuning stubs.The filter proposes 0.15 dB insertion loss and return loss greater than 15 dB.
The both filters (Type A) and (Type B) have a wide stopband resulting from the dispersion effect and the slow-wave effect.

4.2 Problem and suggestion for furure work

There are some problems. The filters can suppress unwanted passband to below -10 dB. On the other hand, the modern wireless communication systems require the bandpass filters having effective out-of-band spurious rejection and good in-band performance. This problem can be overcome by bandstop filter. Microstrip bandstop filter using shunt open stubs and spurlines are presented (Tu & Chang 2005). Basically, by cascading more identical open-stub and spurelines filter, a deeper rejection and a wider rejection bandwidth can be achieved at the expense of increasing circuit size and insertion loss.

5. References

D.M. Pozar (1998). *Microwave Engineering,* 2nd ed. New York: Wiley

Lung-Hwa Hsieh & Kai Chang (2001). Compact, low insertion-loss, sharp-rejection wideband bandpass filters using dual-mode ring resonators with tuning stubs, *Microwave and Optical Technology Letters*, Vol.45, No.4, pp.312-315

Lung-Hwa Hsieh & Kai Chang (2003). Compact,Low Insertion-Loss,Sharp-Rejection and Wide-Band Microstrip Bandpass Filters, *IEEE Transactions on Microwave theory and Techniques*, Vol.51, No.4, pp.1241-1246

Konpang, J. (2003). A Wideband Bandpass Filter Using Square-Loop Resonators with Tuning Stubs, *Master Thesis*, Department of Electrical Engineering, King Mongkut's Institute of Technology North Bangkok

Chu-Yu Chen, Cheng-Ying Hsu & Sung- Fong Lin (2005). A Novel Compact Miniaturiaturized Wideband Microstrip Bandpass Filters with Dual-Mode Ring Resonators, *Microwave and Optical Technology Letters*, Vol.45, No.4, pp.312-315

Jia-sheng Hong & Michael J. Lancaster (1997). Theory and Experiment of Novel Microstrip Slow-Wave Open-Loop Resonator Filters, *IEEE Transactions on Microwave theory and Techniques*, Vol.45, No.12, pp.2358-2365

Adnan Görür (2002). A Novel Dual-Mode Bandpass Filter With Wide Stopband Using the Properties of Microstip Open-Loop Resonator, *IEEE Microwave and Wireless Components Letters*, Vol.12, No.10, pp.386-388

Jia-Sheng Hong & M. J. Lancaster (2001). *Microstrip Filters for RF/Microwave Applications*, New York: John Wiley & Sons, Inc.

K.C. Gupta, R.Garg, I.Bahl & P.Bhartia (1996). *Microstrip Line and Slotlines*, 2nd ed., Artch House

Konpang, J.; Jumneansri, C.; Anunvrapong, P.; & Wongmethanukroah, J.; (2007). A dual-mode wide-band bandpass filter using the microstrip loop resonator with tuning stubs, *Microwave Conference, 2007. European*, pp. 791-794

Konpang, J. (2008). A dual-mode wide-band bandpass filter using slotted patch resonator with tuning stubs, *Microwave Conference, 2008. APMC 2008. Asia-Pacific*, pp.1-4

Wen-Hua Tu & Kai Chang (2005). Compact Microstip Filter Using Open Stub and Spurline, *IEEE Microwave and Wireless Components Letters*, Vol.15, No.4, pp.268-270

IE3D Version *8*, (2001). *Zeland Software*, Inc., Fremont, CA

17

Tunable Dielectric Microwave Devices with Electromechanical Control

Yuriy Poplavlo, Yuriy Prokopenko and Vitaliy Molchanov
National Technical University of Ukraine "Kiev Polytechnic Institute"
Ukraine

1. Introduction

One of the trends of modern telecommunication systems development is use of high-tunable passive components, such as tunable resonators, phase shifters, etc. These components are the key elements of smart antennas, phased-array antennas, tunable oscillators, filters and so on. Many ways are known to design tunable microwave system:

- $\mu(H)$: tuning of ferrite material permeability by magnetic field;
- $\varepsilon(E)$: tuning of ferroelectric material permittivity by electric field;
- $\sigma(E)$: tuning of semiconductor material conductivity by electric field;
- $\sigma(\Phi)$: optical impact tuning of semiconductor material conductivity under light beam Φ;
- Δ: tuning by the mechanical reconfiguration of resonant (or transmission) part of microwave subsystem.

Components with magnetic and electric tuning, such as $\mu(H)$, $\sigma(E)$ (Campbell & Brown, 2000; Ellinger et al., 2001; Lucyszyn & Robertson, 1992) and also $\varepsilon(E)$ (Rao et al., 1999; Deleniv et al., 2003; Kim et al., 2005) have frequency limitation of about 30-40 GHz due to the increased loss at higher frequencies. Optical tuning that exploits conductivity change $\sigma(\Phi)$ (Lee et al., 1999; Ling et al., 2005) under the light beam Φ also introduces considerable loss at the millimeter waves. Therefore, usual tunable components that control material's intrinsic properties $\mu(H)$, $\sigma(E)$, or $\varepsilon(E)$ have fundamental limitations at millimeter waves. The main reason is that microwaves interact with "active" material (ferrite, semiconductor, or ferroelectric) which is a part of microwave line, and transmitted energy is partially absorbed in this material.

On the contrary, the mechanical system of control is not a part of microwave propagation route so it does not contribute to the microwave loss. One but important disadvantage of mechanical control is a relatively low tuning speed.

Recent achievements in the piezoelectric actuator and MEMS technologies open an opportunity to combine advantages of mechanical and electrical tuning techniques. However, for such applications the tuning system should be highly sensitive to rather small displacement of device's components. The key question is how to achieve such high sensitivity of the system characteristics to small displacement of device's parts. This could be achieved if parts displacement provides strong perturbation of the electromagnetic field distribution. For that a variable dielectric discontinuity (the air gap) is created on the way of

the electric field lines. This air gap is placed between the dielectric parts or the dielectric and an electrode. An alteration of the air gap dimension leads to substantial transformation in electromagnetic field distribution and changes such components characteristics as resonant frequency, phase of propagating wave and so on.
This transformation could be described in terms of medium's effective dielectric permittivity (ε_{eff}). Effective dielectric permittivity of inhomogeneous medium is dielectric permittivity, which brings numerically same macro parameters to the system of the same geometrical configuration. Effective permittivity is convenient parameter to describe devices with TEM wave propagating, where propagation constant is proportional to $\sqrt{\varepsilon_{eff}}$. But it can be used to describe other devices as well. For example, effective permittivity of partially loaded waveguide can be stated as such permittivity of fully loaded waveguide, which gives numerically the same propagation constant as in partially loaded waveguide.

2. Tunable effective dielectric constant

The simplest example of a tunable effective dielectric constant is a waveguide partially filled with dielectric, Fig. 1. The air gap between dielectric material and broad wall of waveguide dramatically reduces measured value of dielectric permittivity, and it is the main component of measurement uncertainty. This fact is well known for waveguide technique of dielectric permittivity measurement. However, this phenomenon is applicable for tunable devices design as well (Jeong et al, 2002).

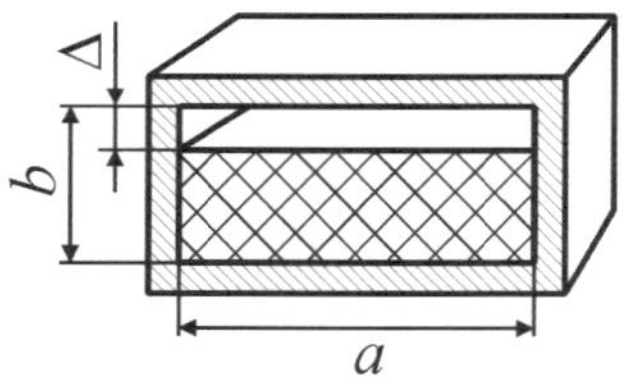

Fig. 1. Partially filled waveguide

Effective permittivity for the basic mode of rectangular waveguide can be found as:

$$\varepsilon_{eff} = \frac{\left(\pi/a\right)^2 + \gamma^2}{k^2},$$

where γ is the propagation constant, k is the wave number in free space, and a is the width of waveguide.
The results of effective permittivity simulation are presented in Fig. 2. As one can see, there is a strong influence of air gap onto effective parameters, especially for high-ε materials. The main reason of such high sensitivity is the location of dielectric discontinuity. The air gap is located across electrical field of waveguide's basic mode and acts as strong perturbation of electromagnetic field, which value depends on air gap's size Δ.
One of possible uses of effective permittivity transformation is tuning of phase shifters. The nature of the phase shift can be explained with Fig. 3. Wavelength in dielectric filled part of

waveguide is shortened proportionally to $\sqrt{\varepsilon_{eff}}$. Because of partial loading of waveguide, there is the nonzero component $\mathbf{E}_z$ of electric field in the direction of propagation. In combination with the component $\mathbf{E}_y$, which is orthogonal to media boundary, it gives resultant vector **E**, which crosses media boundary at certain slope. Refraction at the dielectric media boundary changes slope of resultant vector **E**. So, traveling wave makes its path of two ways: one inside of dielectric, and another one in the air. Because of refraction, the ratio of the way in dielectric and air respectively changes as air gap changes. Simply speaking, the control over the traveling wave phase shift is obtained by the varying part of the way, which wave travels outside of dielectric.

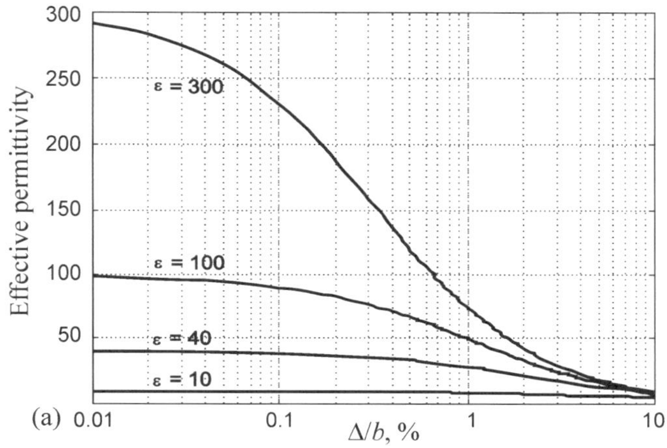

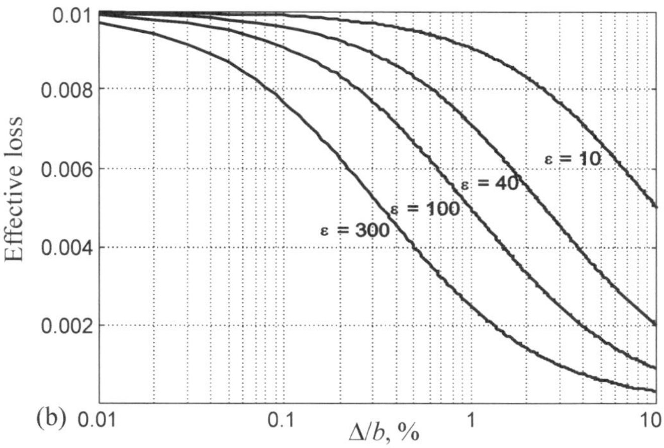

Fig. 2. Effective parameters of partially loaded waveguide: a) effective permittivity; b) effective loss; Δ is the air gap, *b* is the waveguide height

This idea was verified experimentally (Jeong et al., 2002). Phase shifter was made inside of rectangular waveguide section. It can be made either in symmetric or asymmetric fashion (the last is shown in Fig. 4). Controlled element consists of dielectric slab supported by the metal plate. This plate is rigidly attached to the piezoelectric actuator. Under applied control voltage the air gap Δ is controlled via actuators variable extension. Parameters of used dielectric materials are listed in Table I.

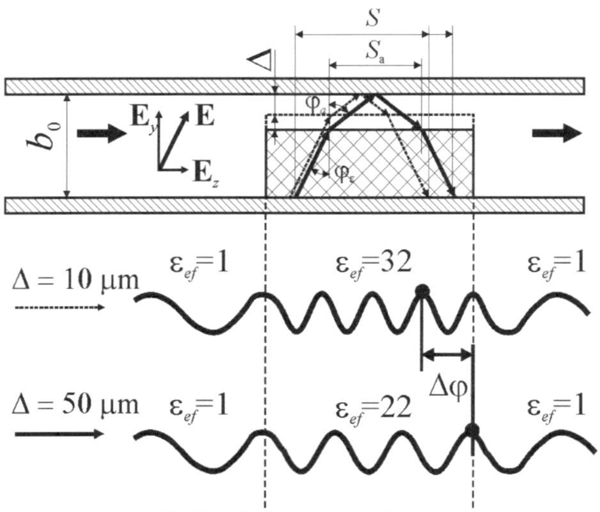

Fig. 3. Phase shift nature in partially loaded waveguide

Material	ε	tanδ @ 10 GHz	tanδ @ 40 GHz
Al_2O_3	11.6	$0.7 \cdot 10^{-4}$	$4 \cdot 10^{-4}$
$(Mg,Ca)TiO_3$	21	$2 \cdot 10^{-4}$	$8 \cdot 10^{-4}$
$BaTi_4O_9$	37	$3 \cdot 10^{-4}$	10^{-3}
BLT	85	$2 \cdot 10^{-3}$	–

Table I. Parameters of used dielectric materials

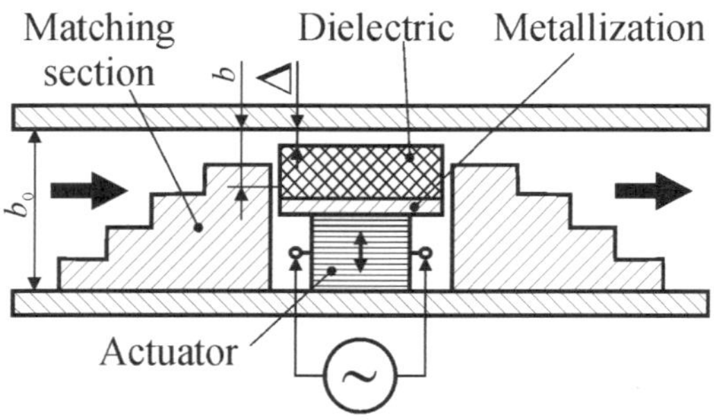

Fig. 4. Waveguide phase shifter experimentally studied design

Fig. 5 illustrates measured control curves. They have almost linear character and promising values. Fig. 6 demonstrates measured *S*-parameters of the phase shifter. It is expected that increase of operation frequency can make this design competitive with solid state devices.

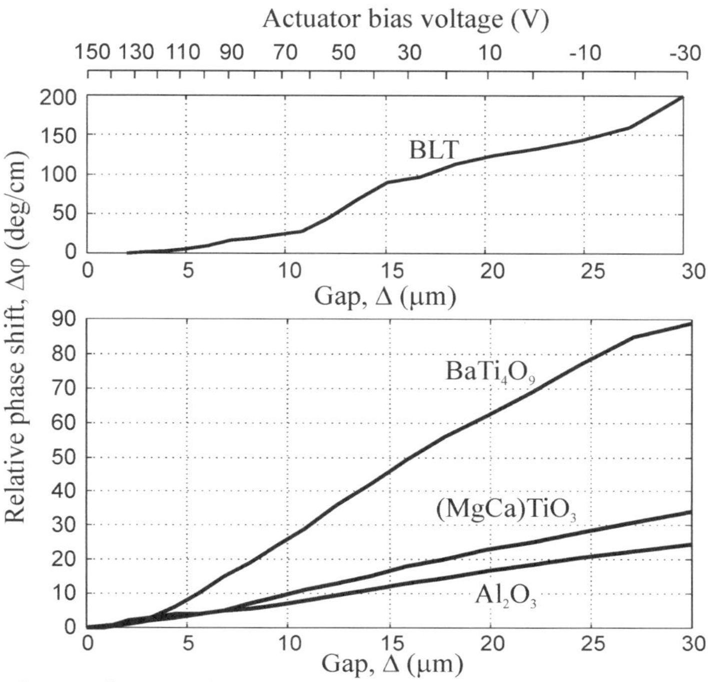

Fig. 5. Measured control curves @10.5 GHz

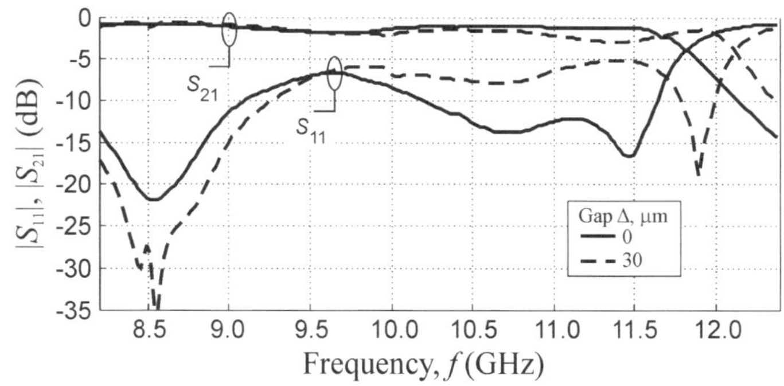

Fig. 6. Measured S-parameters of the device. Dielectric plate of 1mm height and 10 mm length made of material with $\varepsilon = 21$

3. Tunable dielectric resonators

Electromechanical control of high quality dielectric resonator frequency is known for a long time. One of the examples is two cylindrical dielectric resonators with the $H_{01\delta}$ mode separated by the air slot (Δ), constituting a binary dielectric resonator, Fig. 7, a (Wakino et al., 1987). Electric field components in the binary dielectric resonator are located in its basic plane. In contrast, a split dielectric resonator (Poplavko et al., 2001), also of $H_{01\delta}$ type, has a slot located athwart to the electric field components for the lowest resonant mode (Fig. 7, b). This split dielectric resonator shows much larger tunability than binary dielectric resonator, as it is shown in Fig. 7, c.

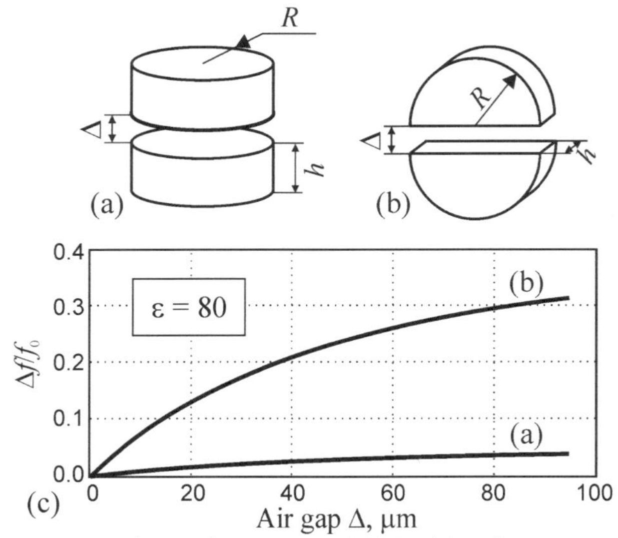

Fig. 7. dielectric resonator mechanical tuning at 10 GHz: (a) ordinary manner; (b) proposed manner; (c) characteristics comparison

One example of split dielectric resonator testing is shown in Fig. 8. No change in a quality factor Q is observed during air slot alteration. Conformable split dielectric resonators are used in the high-Q tunable filter (about 20%) in a waveguide near the central frequency of 10 GHz (Poplavko et al., 2001).

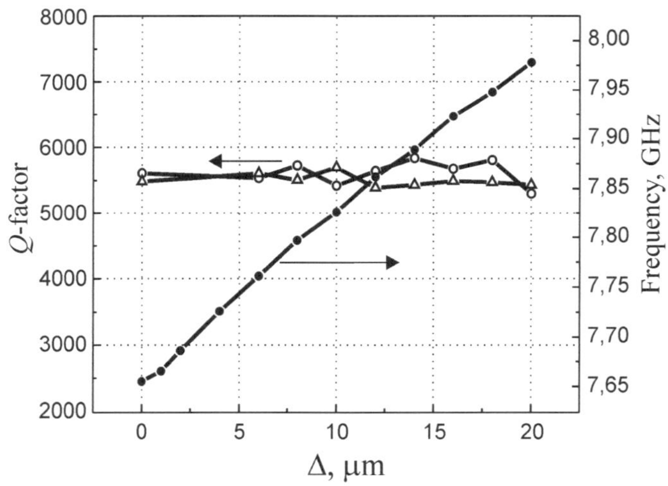

Fig. 8. Resonant frequency f_0 and Q-factor of split dielectric resonator vs. slot

Tunability of the split dielectric resonator can be explained as the alteration in the split dielectric resonator's effective permittivity (ε_{eff}), Fig. 9.

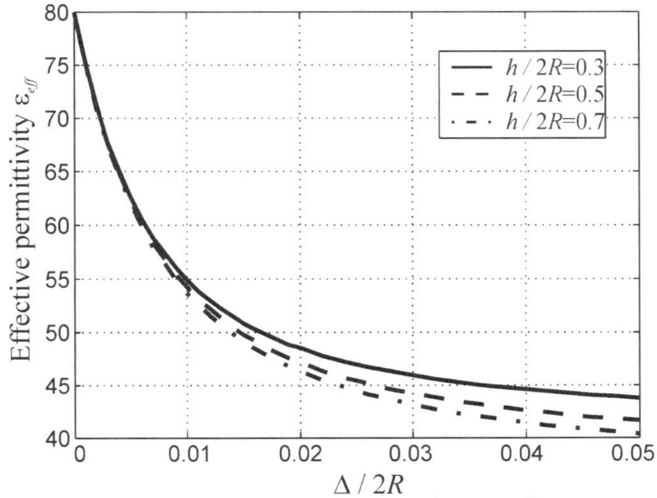

Fig. 9. Effective permittivity versus normalized value of air gap between two parts of disk dielectric resonator shown in Fig. 1*b*. Permittivity of dielectric material is equal to 80; *D* is dielectric resonator diameter while *h* is dielectric resonator thickness.

In the considered case, the value of ε_{eff} decreases about 2 times; correspondingly, split dielectric resonator resonant frequency increases up to 30%. Tunability slightly rises with the ratio of h / $2R$ where R is split dielectric resonator radius, and h is its thickness.
An advantage of such method of frequency control is high Q-factor preservation. The unloaded quality factor can be expressed as $Q_0^{-1} = T \tan\delta$, where T is the energy filling factor, which depends only on dielectric constant and domain size (tan δ is the loss tangent of dielectric material). Due to electromagnetic energy accumulation in a slot the factor T shows a trend to decrease, Fig. 10. As a result, intrinsic Q-factor of the split dielectric resonator can even rise with resonant frequency increase.

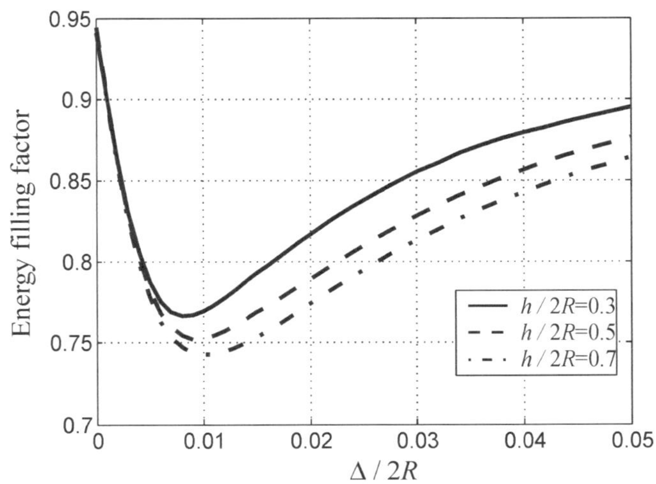

Fig. 10. Energy filling factor versus normalized value of air gap between two parts of cylindrical dielectric resonator. Dielectric constant of material is 80

This frequency control method could be applied to split dielectric resonator of different shapes, including rectangular, ring or sphere. Rectangular and spherical split dielectric resonator are shown in Fig. 11, and their effective permittivity dependences are shown in Fig. 12.

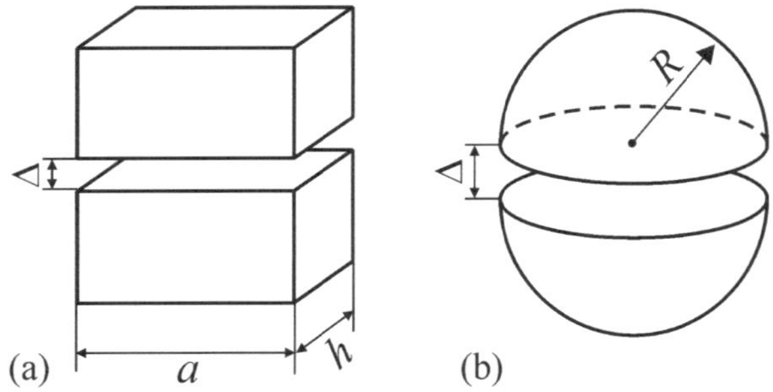

Fig. 11. Rectangular and spherical split dielectric resonator

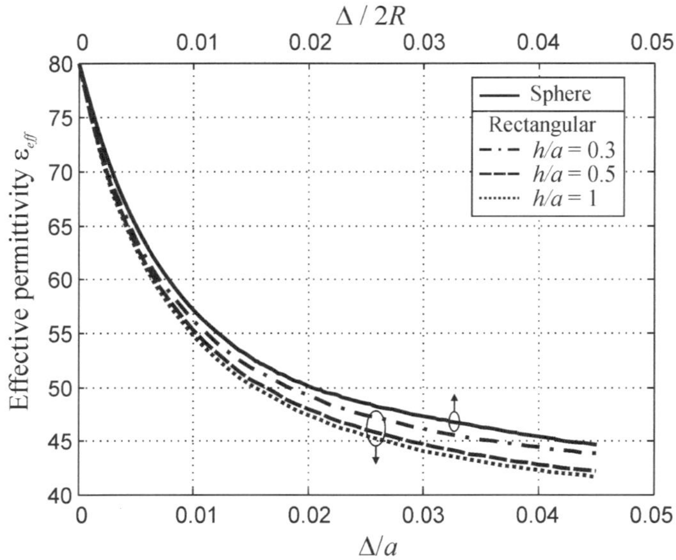

Fig. 12. Effective permittivity of rectangular and spherical split dielectric resonators

4. Electromechanically tunable microstrip phase shifter

Principal designs of piezo-driven phase shifter based on the microstrip line are shown in Fig. 13. Experiments and calculations show that their phase shift is strongly dependent on design architecture.

Only one of designs (shown in Fig. 13, a) was published previously (Yun & Chang, 2002).

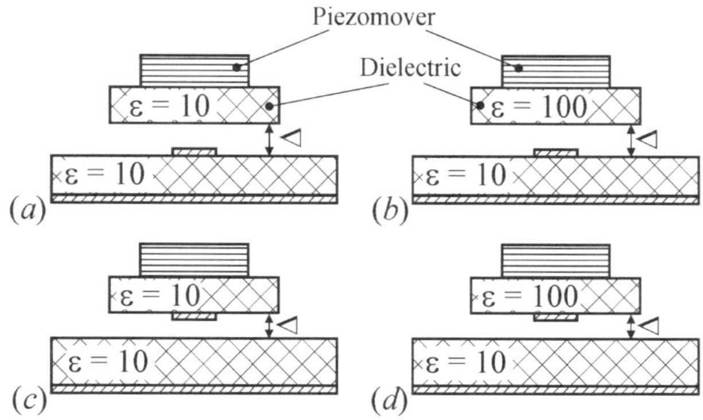

Fig. 13. Mechanically tuned microstrip phase shifters.

However, it is obvious that other designs shown in Fig. 13, b, c, d shows higher effect because dielectric discontinuity is created in the plane perpendicular to electrical field of the microstrip line. The effectiveness was verified and proved experimentally. The best result is obtained with the new idea of "detached" upper electrode, Fig. 13, c, d that is electrode disconnected from substrate and attached to the moveable dielectric plate. Close to these cases phase shift would be obtained if the bottom electrode would be disconnected. Simulation in Fig. 14 confirms that stronger perturbation of electromagnetic field distribution results in higher differential phase shift.

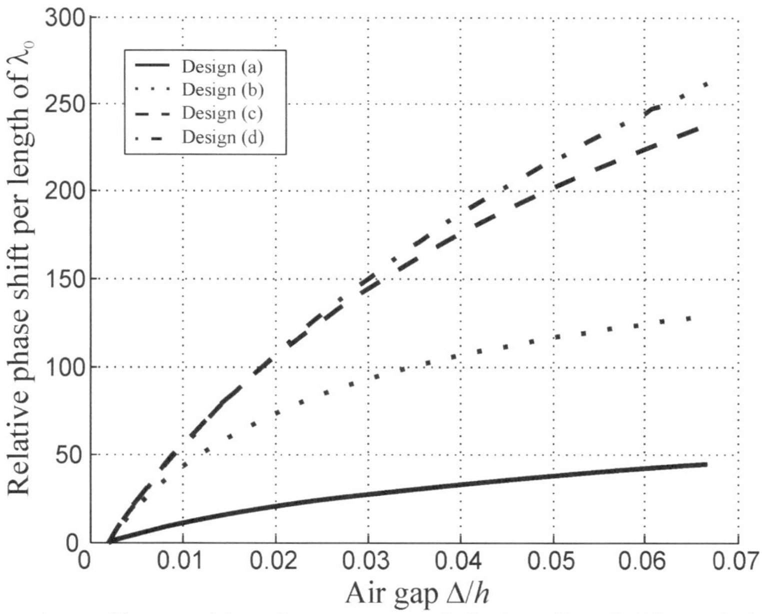

Fig. 14. Comparison of known (*a*) and new proposed devices (*b, c, d*). Phase shift (standardized on wavelength) is shown as function of tunable air gap

This effect also could be explained in terms of effective permittivity change. As one can see in Fig. 15, designs with detachable electrode exhibit larger change in effective dielectric constant, which in turn is observed as larger phase shift.

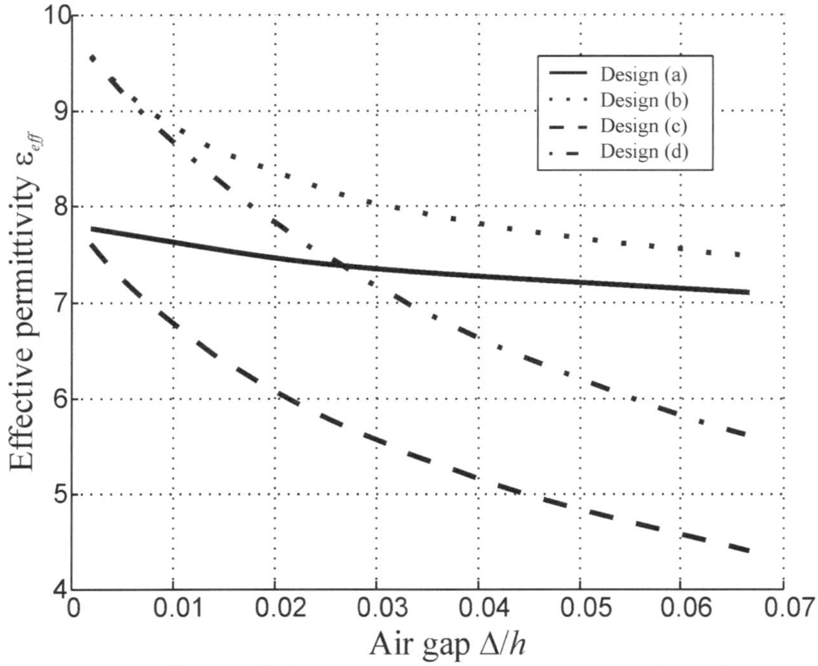

Fig. 15. Comparison of effective dielectric permittivity ε_{eff} in known (*a*) and new proposed devices (*b, c, d*)

In other words, propagation constant at a given frequency f can be estimated as $\gamma = \frac{2\pi f}{c}\sqrt{\varepsilon_{eff}}$ where c is the light velocity. So the main task of device analysis is to determine effective permittivity for prescribed geometrical configuration. This problem is solved numerically using finite element method.

5. Two resonators impedance-step filter controlled from bottom

Principal design and characteristics of a band-transmitting filter (that can be used as a phase shifter) is shown in Fig. 16 together with filter's characteristics. Experimental result was obtained with the network analyzer. Filter is arranged on the right-angled alumina substrate where two impedance steps resonators are deposited (the length of resonator is 22 mm, the ratio between high and low impedance parts ~10, substrate thickness 0,65 mm, substrate dielectric constant ε = 9,2).

With the purpose of tuning, the substrate, located under the filter, imitates a "tunable dielectric". Namely, the part of ground electrode (just under the coupling part of filter) is removed and substituted by the piezoelectric actuator, which is closely adjacent to the substrate, Fig. 17.

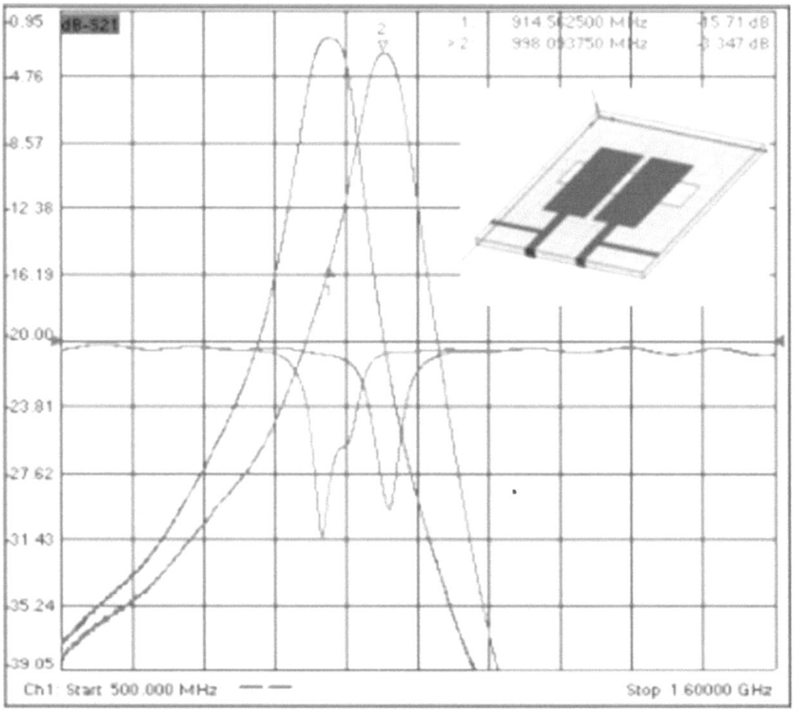

Fig. 16. Two-resonator tunable filter design and characterization.

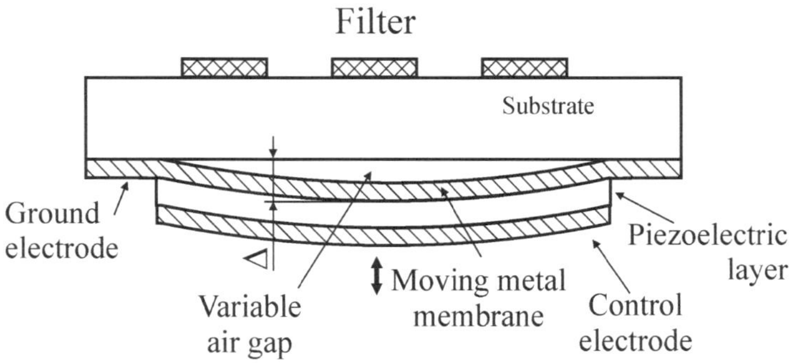

Fig. 17. The concept of filter tuning: side view of moving ground electrode under the substrate.

Actuator's upper electrode is simultaneously a ground electrode of the substrate. Due to the actuator, the thickness of the narrow air gap (Δ) is electrically controlled. Such a "tunable substrate" can be described as dielectric in which effective permittivity is controlled. The scope of the ε_{eff} change depends on the substrate ε and relationship Δ/h where h is substrate thickness. In our experiments the effective permittivity of the layered dielectric "alumina - air" decreases from $\varepsilon_{eff} \approx 7$ till $\varepsilon_{eff} \approx 3$ while the range of a gap change was from Δ ~ 10μm till Δ ~ 100μm under the voltage of about 300 V.

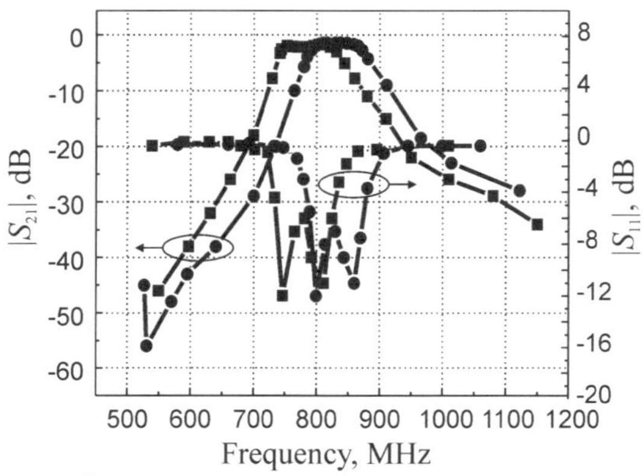

Fig. 18. Filter characteristics for two positions of dielectric plate: central resonant frequency shifts of about 10% while filter attenuation remains less than 1 dB.

Calculation and tuning of studied structures was made by the method of the FEM simulation. The results of calculation show good agreement with the experiment.

Any tunable band-pass filter can be used as a phase shifter but only at the frequency range of its bandwidth. In a given experiment this bandwidth looks rather narrow, and controlling voltage seems too big for many applications. That is why another design and different way of filter controlling is proposed below.

Experimental prototype of studied "*tunable filter – phase shifter*" is shown on photograph.

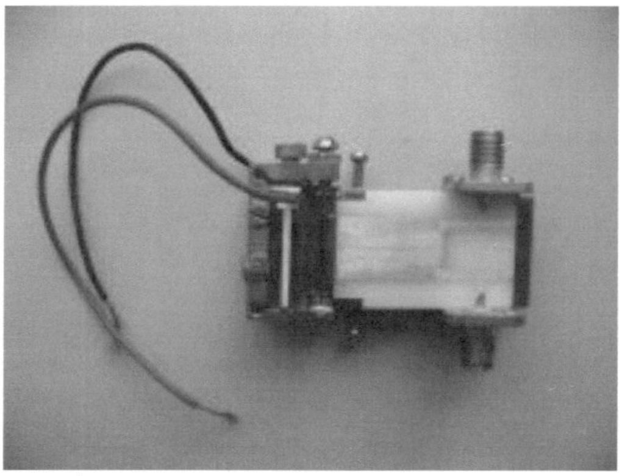

Fig. 19. Photo of experimental prototype.

6. Electromechanically tunable coplanar line

Electromagnetic filed of microwave transmission lines deposited onto substrate is mainly confined in the substrate right under electrodes and in the inter-electrode space to certain degree. Because of that the dielectric body is moved up and down above the line's surface, as it is shown in Fig. 20, a, makes small perturbation of electromagnetic field distribution. To improve device's controllability it is necessary to arrange tighter dependence of electromagnetic field on moving dielectric body position. For that it is proposed to situate a signal strip of coplanar waveguide on moving dielectric body and let them lift together (Prokopenko et al., 2007), Fig. 20, b.

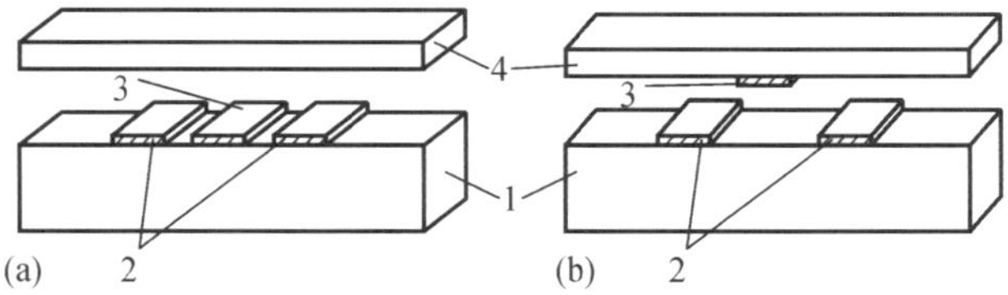

Fig. 20. Coplanar line based phase shifters with signal line: a - on the substrate, b - on the moving dielectric body. 1 - substrate, 2 - ground electrodes, 3 - signal line, 4 - moving dielectric body

Fig. 21 shows simulations of near 50 Ω coplanar lines with dielectric permittivity of both substrate and movable dielectric body ε=12 for presented in Fig. 20 designs. Here and after relative phase shift is calculated for the device of length equal to wavelength in vacuum. Qualitative conclusion is that under other same conditions the device with detaching electrode exhibits greater relative effective permittivity change, and thus its relative phase shift more than 1.5 times exceeds one from counterpart.
Obviously, strong perturbation of electromagnetic filed improves device's controllability. But quantitatively it depends on a number of design factors, such as line geometry, impedance and ratio of the substrate's and movable dielectric permittivity. Generally low impedance lines tend to exhibit higher controllability. This can be achieved not only by use of high-permittivity materials, but with proper layout as well.

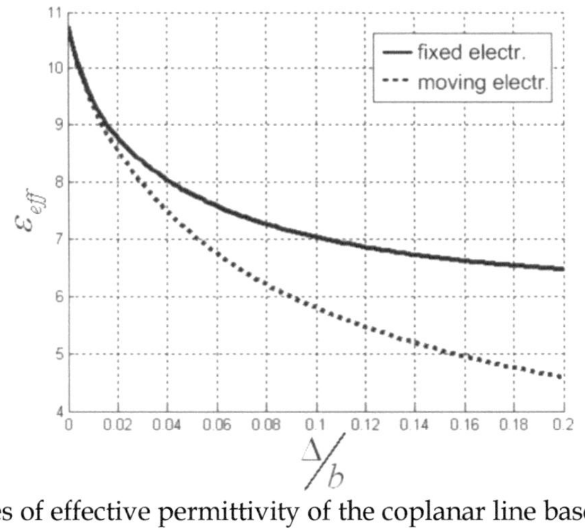

Fig. 21. Dependencies of effective permittivity of the coplanar line based phase shifters

To prove presented ideas, one scaled up experiment was performed. Experimental setup consists of coplanar dielectric (ε = 4.3) substrate in the aluminium fixture. The signal line is soldered to the bonding pads at the sides of the substrate (Fig. 22, a), whereas being glued to the moveable dielectric, which in turn is attached to micrometer screw (see the photo in Fig. 22, b).

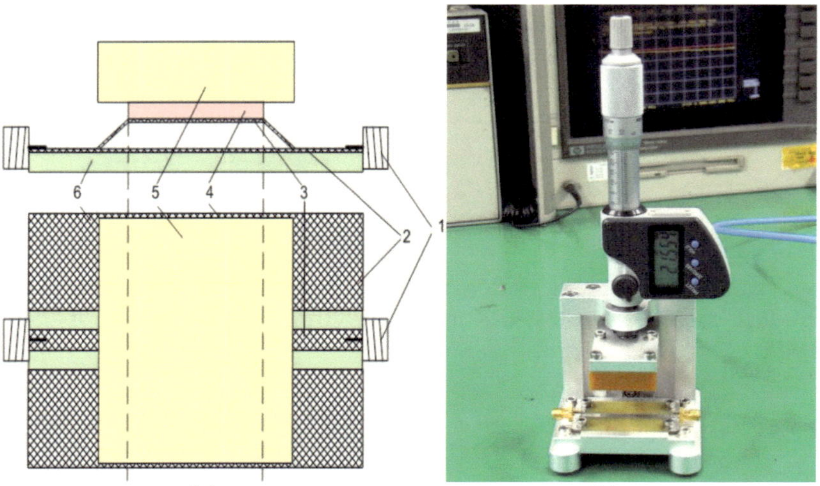

(a) (b)

Fig. 22. Experimental setup: (a) - schematic, (b) - photo. 1 - connector; 2 - ground electrodes; 3 - movable electrode; 4 - movable dielectric; 5 - low-ε support; 6 - substrate

Fig. 23 and Fig. 24 present results of experimental investigation of the proposed phase shifters.

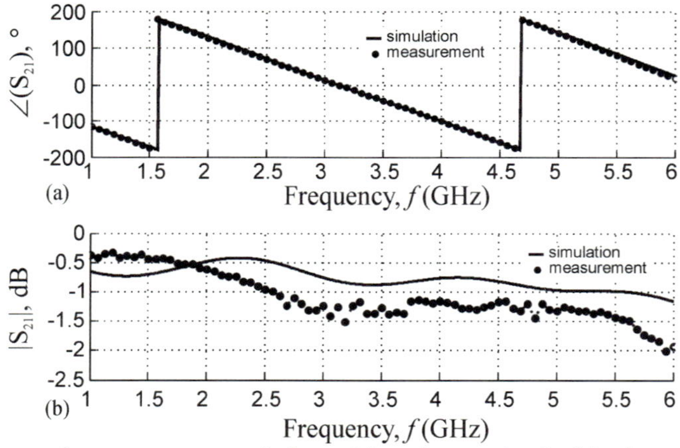

Fig. 23. Simulation and measurement of phase (a) and magnitude (b) of transmission coefficient S_{21} for the coplanar line based phase shifter with moving signal electrode

Fig. 25 presents simulation and measurement of control curve for both discussed designs. There is a good agreement between simulation and measurement result.

7. Conclusion

Main mechanisms of piezoelectric control by the ε_{eff} of some devices based on dielectric layers are discussed. It is supposed that the most effective way is to use a composition "microwave dielectric - air gap", controlled by the fast actuator. At that, a minimal loss is inserted in tunable component.

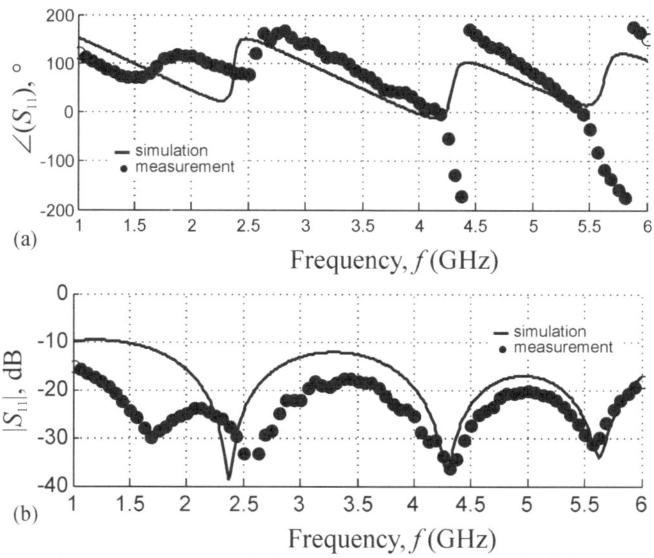

Fig. 24. Simulation and measurement of phase (a) and magnitude (b) of reflection coefficient S_{11} for the coplanar line based phase shifter with moving signal electrode

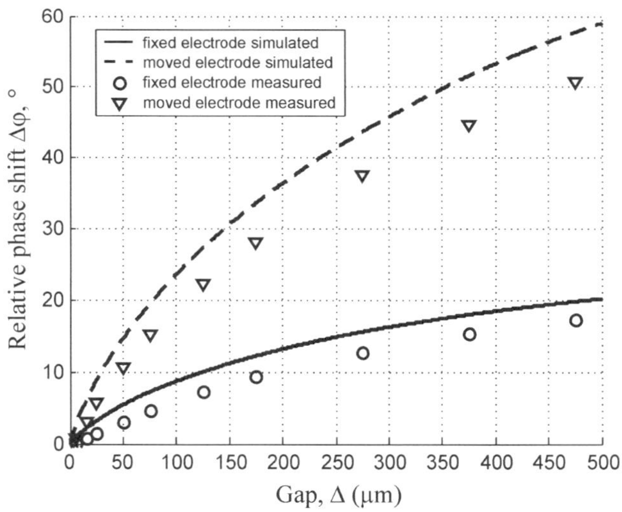

Fig. 25. Coplanar line controlled with teflon slab. f = 5 GHz, Substrate: 40 × 30 × 1.6 mm, ε= 4,3, Movable dielectric: 30 × 20 × 1 mm, ε = 2.08 Signal line width is 3 mm

Using high quality microwave dielectrics, it is possible to realize low loss filters and phase shifters as in the microwaves so as in the millimeter waves. Proposed structures are studied as in the rectangular waveguide, so in some microstrip designs.

Proposed way of control allows to increase device's controllability while maintain low loss. Simulations are proved by the experiment. With scaling down and move to the higher frequencies, the amount of required displacements could be reduced to tens micrometers, thus allowing an application of small size and fast piezo-actuators or MEMS.

8. References

Campbell, C.F. & Brown, S.A. (2000). A compact 5-bit phase-shifter MMIC for K-band satellite communication systems. *IEEE Trans. Microwave Theory Tech.,* Vol. 48, (Dec. 2000), P. 2652–2656.

Deleniv, A.; Abadei, S. & Gevorgian, S. (2003). Tunable ferroelectric filter-phase shifter. *IEEE MTT-S International Microwave Symposium Digest*.– June 2003.– Vol. 2, P. 1267–1270.

Ellinger, F.; Vogt, R. &, Bachtold, W. (2001). Compact reflective-type phase-shifter MMIC for C-band using a lumped-element coupler. *IEEE Trans. Microwave Theory Tech.*, Vol. 49, (May 2001), P. 913–917.

Jeong, M.; Kazmirenko, V.; Poplavko, Y.; Kim, B. & Baik, S. (2002). Electrically Tunable Phase Shifters With Air-Dielectric Sandwich Structure. *Proceedings of Material Research Society Symp.*, pp. H3.12.1–H3.12.6, Vol. 720.

Kim, Ki-Byoung; Yun, Tae-Soon; Kim, Hyun-Suk; Kim, Il-Doo; Kim, Ho-Gi & Lee, Jong-Chul. (2005). Coplanar ferroelectric phase shifter on silicon substrate with TiO2 buffer layer. *Proceedings of European Microwave Conference*, pp. 649-652, ISBN 2-9600551-0-1, Paris France, Sep. 2005.

Lee, S.-S.; Udupa, A. H.; Erlig, H.; Zhang, H.; Chang, Y.; Chang, D.H.; Bhttacharya, D.; Tsap, B.; Steier, W.H.; Dalton, L.R. & Fetterman, H.R. (1999). Demonstration of a photonically controlled RF phase shifter. *IEEE Microwave Guided Wave Lett.*, Vol. 9, (Sept. 1999), P. 357–359.

Ling, Liao; Ansheng, Liu; Jones, R.; Rubin, D.; Samara-Rubio, D.; Cohen, O.; Salib, M. & Paniccia, M. (2005). Phase modulation efficiency and transmission loss of silicon optical phase shifters. *IEEE Journal of Quantum Electronics*, Vol. 41, (Feb. 2005), P. 250–257.

Lucyszyn, S. & Robertson, I. D. (1992). Synthesis techniques for high performance octave bandwidth 180° analog phase shifters. *IEEE Trans. Microwave Theory Tech.*, Vol. 40, (Apr. 1992), P. 731–740.

Poplavko, Yu.M.; Prokopenko, Yu.V.; Molchanov, V.I. & Dogan, A. (2001). Frequency-tunable microwave dielectric resonator. *IEEE Trans. Microwave Theory Tech.*, Vol. 49, (June 2001), P. 1020 - 1026.

Poplavko, Y. ; Golubeva, I.; Kazmirenko, V.; Jeong, M. & Baik, S. (2005). Piezo-controlled Dielectric Phase Shifter with Microstrip and Coplanar Lines. *Proceedings of European Microwave Conference*, pp. 1335 - 1338, ISBN 2-9600551-0-1, Paris France, Sep. 2005.

Prokopenko, Y.; Golubeva, I.; Kazmirenko, V. & Poplavko, Y. (2007). Coplanar line based low loss microwave phase shifters with electromechanical control. *Proceedings of European Microwave Conference*, pp. 1582-1585, Munich Germany, Oct. 2007.

Rao, J.B.L.; Patel, D.P. & Krichevsky, V. (1999). Voltage-controlled ferroelectric lens phased arrays. *IEEE Trans. Antennas Propagat.*, Vol. 47, (Mar. 1999), P. 458–468.

Wakino, K.; Tamura, H. & Ishikava, Y. (1987). Dielectric resonator device, *USA Patent #4, 692, 727*, Sep.8, 1987.

Yun, T.-Y. & Chang, K. (2002). Analysis and optimization of a phase shifter controlled by a piezoelectric transducer. *IEEE Trans. Microwave Theory Tech.*, Vol. 50, (January 2002), P. 105-111.

Dual Mode Microstrip Ring Resonator with Composite-Right/Left-handed Line

M.K.Haldar, Hieng Tiong Su and Kian Kiong Fong
Swinburne University of Technology (Sarawak Campus)
Sarawak, Malaysia

1. Introduction

Microwave and RF filters play an important role in various electronic systems, including cellular radio, satellite communications and radar. Filters are used in these systems in order to discriminate between wanted and unwanted signal frequencies. High performance filters are desirable for good signal reception and therefore for a better system performance. The demands for high performance filters are mainly due to the stringent frequency spectrum requirements following the emerging of new applications for modern communication systems. High performance filters are filters with low insertion loss, high frequency selectivity, phase linearity and potentially no harmonic response. Following the advancement of modern technologies, design considerations have been extended to achieve compact size and light-weight, making the filter design a more challenging task. Although enormous amount of literature on various filter theories is available, new filters are continually developed and reported in major journal and conference publications, to suit severe design specifications.

In general, microwave filters are divided into two broad classes, they are distributed type and lumped-element type. At microwave frequencies the use of distributed circuit elements in implementing passive microwave devices is widespread. They differ from lumped circuits as one or more dimensions are a significant fraction of the operating wavelength. Design formulae are available in many texts. Distributed filters can take the form of planar structures or waveguide cavity and they are preferable for high Q filter design. However, the latter has the advantage of low or no spurious harmonic responses.

In this chapter, we will give a new design perspective for a potentially high performance filter namely a dual-mode microstrip ring resonator with composite-right/left-handed (CRLH) line, for suppression of first harmonic. In section 2, we will first describe the terminologies of Left-Handed (LH) and Right-Handed (RH) transmission lines and show how their wave propagation properties are different using their transmission line models. In Section 3, we will give an overview of ring resonator's research, how ring resonator can be used in a single mode or a dual-mode resonator design.

In section 4, we will discuss the principle of operation of a composite-right/left-handed line ring resonator, and explain how harmonic supprssion can be achieved.

In sections 5, we will present the implementation of CRLH ring resonator. We will give the formulation use in the design and show how the left-handed line is incorporated into a ring resonator.
In section 6, we will take the circuit-modeling approach to analyse the CRLH ring resonator in greater details. This serves as initial design guidelines to quickly determine the filter layout dimensions, given a filter specification. The final design can then be simulated using commercial electromagnetic simulator. Some measurement results will be presented.
In section 7 and 8, we express other design considerations for the CRLH ring resonator and give suggestions for future developments.

2. Left and Right Hand Transmission Lines

The term left-handed transmission line is relatively new although such lines have long been known. The term comes from a speculative paper (Veselago, 1968), which considered the electromagnetic properties of a material with negative permittivity and permeability. Among other interesting properties, such a material will have a negative refractive index. However, it was many years later that experimentalists (Shelby et al, 2001) demonstrated such a material. These materials were called left-handed metamaterials.
Consider a uniform plane wave in a right-handed rectangular Cartesian coordinate system. The direction of the Poynting vector is always given by the direction of motion of a right-handed cork screw as it is rotated from the electric field vector to the magnetic field vector. Assuming the fields to vary as $e^{j\left(\omega t-\vec{k}\bullet\vec{r}\right)}$, where ω is the angular frequency (radians/s), t is time, $\vec{k}$ is the wave vector, $\vec{r}$ is the position vector, and $j=\sqrt{-1}$. Maxwell's curl equations can be written in the SI units as

$$\vec{k}\times\vec{E}=\omega\mu\vec{H} \tag{1}$$

$$\vec{k}\times\vec{H}=-\omega\varepsilon\vec{E} \tag{2}$$

where $\vec{E}$ and $\vec{H}$ represent the electric and magnetic field vectors and ε and μ are the permittivity and permeability of the medium considered to be isotropic.
These equations show that the wave vector, $\vec{k}$ is perpendicular to the electric and magnetic field vectors. If both ε and μ are positive, the direction of $\vec{k}$ is given by the direction of $\vec{E}\times\vec{H}$, i.e., the Poynting vector. The direction of the wave vector is therefore given by the direction of motion of a right-handed cork screw as it is rotated from the electric field vector to the magnetic field vector. Hence, such a medium is called right-handed. Most materials occurring in nature are right-handed. On the other hand, if both ε and μ are negative the direction of $\vec{k}$ is given by the direction of $-\vec{E}\times\vec{H}$, i.e., opposite to the direction of the Poynting vector. The direction of the wave vector is therefore given by the direction of motion of a left-handed cork screw as it is rotated from the electric field vector to the magnetic field vector. Hence, such a medium is called left-handed. The Poynting vector is associated with the direction of energy flow, while the wave vector represents the direction of motion of the wave fronts. The former therefore represents the direction of group velocity while the latter represents the direction of phase velocity. Hence the phase and group

velocities are in the same direction in a right-handed material and in opposite directions in a left-handed material. Waves with opposite directions of phase and group velocities have been known for a long time and have been used in backward wave oscillators. These waves travel in periodic structures and are called slow waves (Beck, 1958) because their phase velocities are less than the phase velocity of light in the medium in which these periodic structures are embedded.

It is a common practice in electrical engineering to model wave propagation by transmission line theory which can represent both slow and fast waves. Thus it is expected that a left-handed material can be represented by a transmission line. Fig.1 shows the transmission line model represented by a distributed series impedance, Z Ohms/m and a distributed shunt admittance, Y Siemens/m.

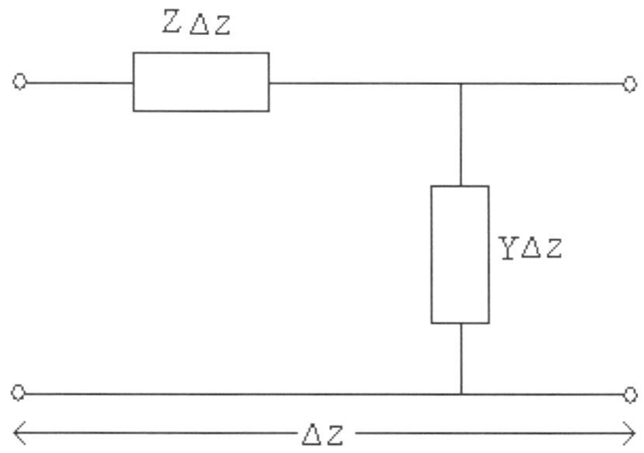

Fig. 1. Transmission line model for an infinitesimal length Δz

For a lossless transmission line the characteristic impedance, Z_0 and the phase constant, β are given by

$$Z_0 = \sqrt{Z/Y} \tag{3}$$

$$j\beta = \sqrt{ZY} \tag{4}$$

Consider now the common transmission line of Fig.2(a), in which the series impedance is an inductance, L Henrys/m and the shunt admittance is a capacitance, C Farads/m .

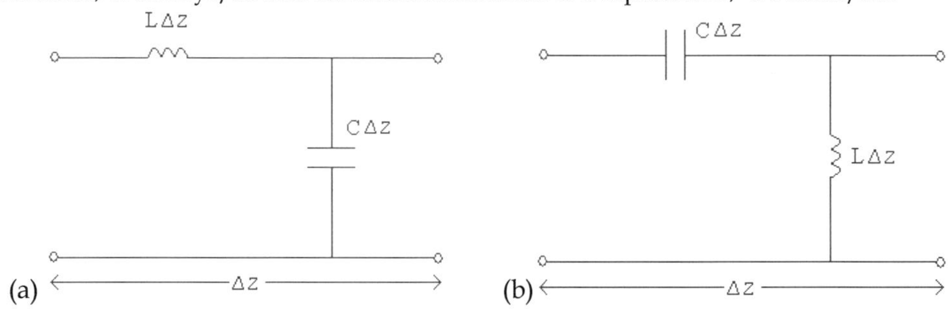

Fig. 2. (a) Model of right-handed line, (b) Model of left-handed line

Noting that $Z = j\omega L$ and $Y = j\omega C$,

$$Z_0 = \sqrt{L/C} \tag{5}$$

$$\beta = \omega\sqrt{LC} \tag{6}$$

The phase velocity is $\omega/\beta = 1/\sqrt{LC}$ and the group velocity is $d\omega/d\beta = 1/\sqrt{LC}$. Both have the same sign and hence are in the same direction. Thus this type of line can represent a right-handed material and is called a right-handed line.

Now consider the transmission line of Fig.2(b), in which the series impedance is a capacitance, C Farads/m and the shunt admittance is an inductance, L Henrys/m. Noting that $Z = 1/j\omega C_\ell$ and $Y = 1/j\omega L_\ell$,

$$Z_0 = \sqrt{L/C} \tag{7}$$

$$\beta = 1/[\omega\sqrt{LC}] \tag{8}$$

The phase velocity is $\omega/\beta = \omega^2\sqrt{LC}$ and the group velocity is $d\omega/d\beta = -\omega^2\sqrt{LC}$. They are of opposite signs and are therefore in the opposite directions. Hence this type of line can represent a left-handed material and is called a left-handed line.

Two points are to be noted. Firstly, the phase velocity is low at low frequencies when the wave can be regarded as a slow wave. At very high frequencies both phase and group velocities can be arbitrarily large which just indicates that the model cannot be right at high frequencies. The second and most important point for this chapter is that the phase of a right-handed line, proportional to β increases with frequency (in this case linearly). On the other hand, the phase of a left-handed line decreases with frequency (in this case inversely). Although the exact form of variation of β with frequency may not be the same as in the transmission lines considered, the nature of the phase variation is always correct, because $d\omega/d\beta$ is positive for the right-handed line and negative for the left-handed line.

One must now be careful about the word composite in the context of this chapter. For the transmission line, the series impedance may be a series combination of an inductance and capacitance and the shunt admittance can be a parallel combination of an inductance and capacitance. This has the characteristic of a band pass filter. For a certain range of frequencies, the phase has the characteristic of a right-handed line, i.e., it increases with frequency. For another range of frequencies, the phase has the characteristic of a left-handed line, i.e., it decreases with frequency. Such lines showing both types of behaviour have been termed composite-right/left-handed lines (Lai et al, 2004). The composite- right/left-handed line considered in this chapter is a combination of a right-handed line and a left-handed line.

3. Ring Resonators – Single and Dual mode

The microstrip ring resonator was first introduced for measuring dispersion in microstrip lines (Wolff & Knoppik, 1971). However, because of its compact nature, and simplicity of operation, it has been widely used as a resonator in bandpass RF filters. Ring resonators of various shapes - rectangular, square, circular, meander - as well as different types of coupling have been reported. Fig.3 shows a circular microstrip ring resonator.

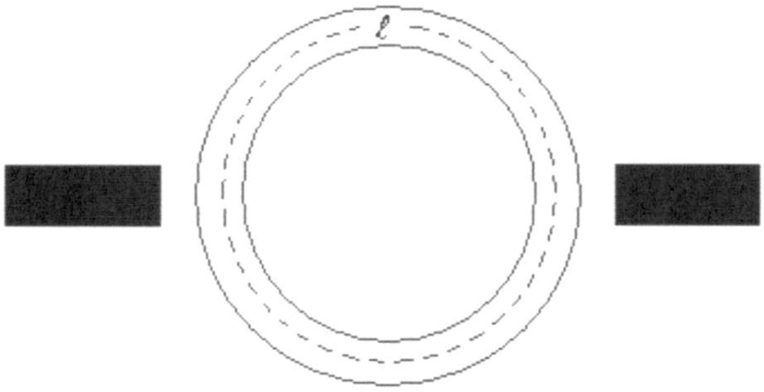

Fig. 3. A single mode microstrip ring resonator with simple microstrip line feeds

The basic principle of operation is that at the resonant frequency, a standing wave exists in the ring. For this to happen, the total phase shift around the ring must be an integer multiple of 2π. If the mean length of the resonator is ℓ, the condition is

$$\beta\ell = 2N\pi \tag{9}$$

where N is an integer and β is the phase constant.
The resonant frequencies are obtained from (9) as

$$f = Nv_p / \ell \tag{10}$$

where, $v_p = 1/\sqrt{LC}$ is the phase velocity of the microstrip line.

For a band pass filter of order n and a symmetric response about the centre frequency, one requires n such resonators each resonant at the centre frequency. The desired frequency response is obtained by the choice of coupling between the resonators. These resonators are called single mode ring resonators, because there is a single resonance at the fundamental frequency. One can on the other hand have two closely spaced resonances near the fundamental. Such ring resonators are called dual mode and were first reported by Wolff (Wolff, 1972). A single dual mode resonator with two close resonant frequencies, f_1 and f_2 is equivalent to two coupled single mode resonators of resonance frequency f_0 and a coupling coefficient

$$k = (f_1 - f_2)/f_0 \tag{11}$$

$$f_0 = \sqrt{f_1 f_2} \tag{12}$$

If one uses dual mode ring resonators, a bandpass filter of order n will require n/2 dual mode resonators as compared to n single mode resonators. This results in a far more compact filter.

A dual mode resonator can be obtained from a single mode resonator design by various ways, such as by having unequal length arms between the feeds, a perturbation in the form of a notch or patch in one of the arms or by unequal characteristic impedance of the two arms. These are shown in Fig.4.

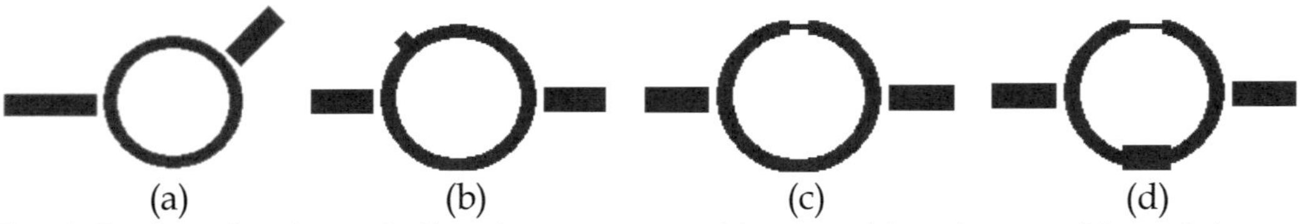

Fig. 4. Dual mode microstrip line ring resonators: (a) unequal length arms, (b) patch in one arm, (c) notch in one arm, and (d) different characteristic impedances of the arms.

From filter design (Chebyshev, Elliptic etc.) one knows the centre frequency, f_0 as well as the various coupling coefficients. On can then consider replacing a pair of single mode resonator by first calculating f_1 and f_2 using (11) and (12) and the known values of f_0 and the coupling coefficient, k. The mean length of the single mode ring resonator is already known from (10) for the given f_0. Thus one needs to determine the perturbation etc. to obtain the dual mode resonant frequencies f_1 and f_2. Very often the design is obtained by trial and error simulation. However, circuit methods can often provide a good initial design which can then be refined by simulation. Cicuit methods have been extensively discussed by Chang and Hsieh (Chang & Hsieh, 2004).

4. Harmonic Suppression in Bandpass Filters and the Use of Right/Left-handed lines

Unfortunately, many bandpass filters have passbands at the harmonics. This is easy to see in the context of ring resonator filters. Equation (10) shows that the ring resonates at the harmonics. Thus if the coupling between the resonators is constant with frequency, the filter will also have passbands at the harmonics. In practice, the couplings are not constant, but whatever they are, it is expected that the filter will have poor return loss at the harmonics. This is also true of bandpass filters employing dual mode ring resonators. In some applications low attenuation at the harmonics is undesirable.

Several papers report the reduction of harmonic response of ring resonators. One technique (Carroll & Chang, 1994, Karacaoglu et al, 1996, Chang & Hsieh, 2004) is to incorporate a low pass filter in the ring. This filter is built from stepped impedance lines. In Carroll and Chang's resonator, the first harmonic (N=2) was suppressed but with additional loss at the fundamental. In the resonator of Karacaoglu et al, the suppression is 9 dB at the first harmonic and about 6 dB at the second harmonic.

A left-handed line can be incorporated as part of the ring to suppress the first harmonic. As the resonator consists of a right-handed line and a left-handed line, it has been called a composite-right/left-handed line ring resonator (Allen et al, 2006). The principle of operation is completely different from the low pass filter technique and is illustrated in Fig.5.

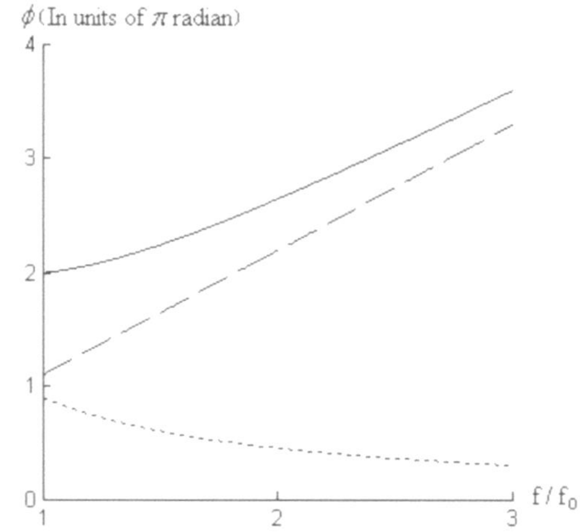

Fig. 5. Phase shifts in the ring resonator. Dashed line: Right-handed line. Dotted line: Left-handed line. Solid line: Total phase shift in the ring

At the resonant frequency, f_0 the phase shift of the right -handed line is φ_R and the phase shift of the left-handed line is φ_L and $\varphi_R + \varphi_L = 2\pi$ as required by ring resonance at the fundamental. According to (6) and (8), the phase shifts vary with frequency as $\frac{f}{f_0}\varphi_R$ for the right-handed line and as $\frac{f_0}{f}\varphi_L$ for the left-handed line. The variation of the phase shifts as well as their sum with the normalized frequency f / f_0 are shown in Fig.5. It is seen that the sum of the phase shifts (solid line) is not 4π at the first harmonic (N=2) as required by ring resonance. This is because the phase of the left-handed line decreases with frequency. Hence the first harmonic is suppressed. However the ring resonance condition can be satisfied at higher frequencies, because the phase shift of the right-handed line increases linearly with frequency while the phase shift of the left-handed line reduces slowly as it is inversely proportional to frequency. However for the suppression of the first harmonic, the precise form of the phase variation of the left-handed line is not important.

5. Implementation of Composite-Right/Left-handed ring resonator

5.1 Left-handed line as an iterative network

Unfortunately, transmission lines with series capacitance are not available. Slow wave structures can be used as left-handed lines only within a range of frequencies. It appears that left-handed metamaterials made with slow wave structures have little to do with filters (Lai et al, 2004). In any case, ring resonators incorporating slow wave structures have not been reported – this may be the subject of future research. Thus a practical way to implement the left-handed line considered here is to use iterative networks made up of lumped series capacitances and lumped shunt inductances. The theory of such networks using the image impedance method is well known (Matthaei et al, 1980). We will consider symmetric networks for which the two image impedances are equal and are called characteristic impedance. Two types of networks using π and T sections are shown in Fig.6.

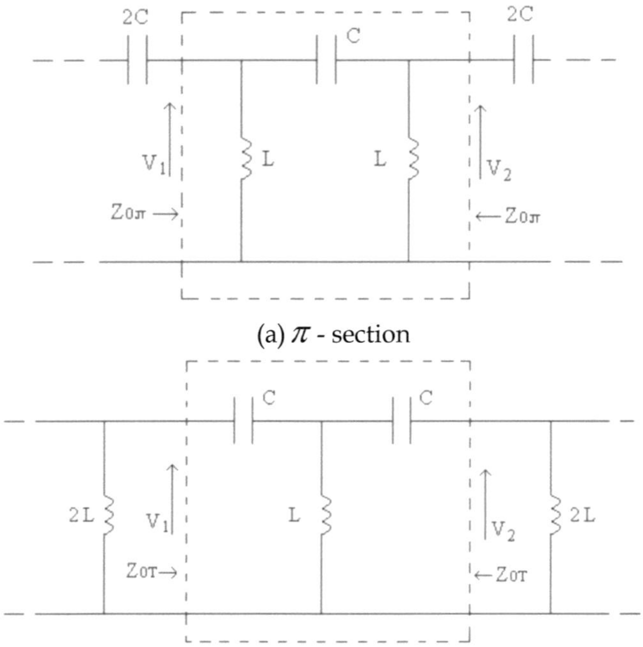

(a) π - section

(b) T – section

Fig. 6. Left-handed lines represented by iterative networks : (a) π - section, (b) T - section Boxes represent the unit cell of the infinite iterative network

The unit cell (enclosed in the boxes of Fig.6) is a high pass filter. Propagation through a cell is given by

$$V_2 = V_1 \exp(-\Gamma) \tag{13}$$

$$\Gamma = \alpha + j\beta$$

and in the passband, $\cos\beta = \left(1 - \omega^2 LC/2\right)$ (14)

For N cells, the phase shift is $N\beta$. The cut-off occurs at $\cos\beta = -1$. The cut-off frequency obtained from (14) is

$$f_c = \frac{1}{\pi\sqrt{LC}} \tag{15}$$

From (14) , it can be shown that

$$\beta = 2\sin^{-1}\frac{f_c}{f} \tag{16}$$

The variation of β against the normalized frequency f/f_c is shown in Fig.7. The phase shift decreases with frequency which is the characteristic of a left-handed line.
The characteristic impedances, $Z_{0\Pi}$ and Z_{0T} are given by

$$Z_{0\Pi} = \sqrt{\frac{L}{C}}\left(1-\frac{f_c^2}{f^2}\right)^{-1/2} \tag{17}$$

$$\text{and } Z_{0T} = \sqrt{\frac{L}{C}}\left(1-\frac{f_c^2}{f^2}\right)^{1/2} \tag{18}$$

The characteristic impedances are imaginary below cut-off and vary widely in the passband.

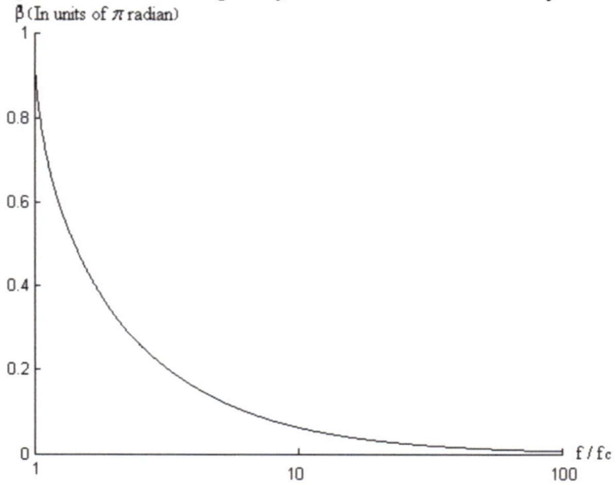

Fig. 7. Variation of the phase, β of a unit cell with frequency

5.2 Incorporating the left-handed line in a ring resonator

The first design is reported by Allen et al (Allen et al, 2006). The schematic diagram of the resonator is shown in Fig.8.

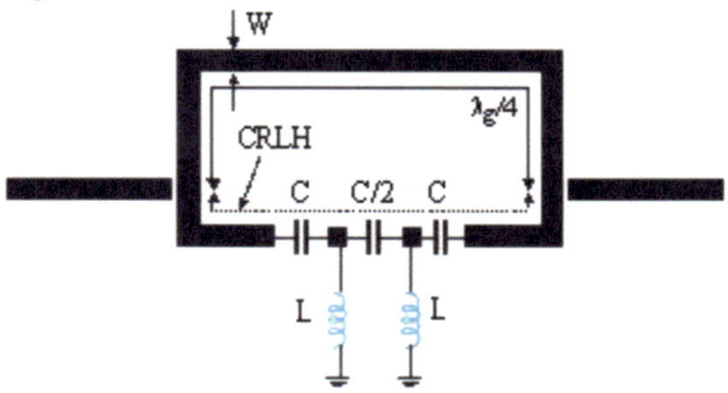

Fig. 8. Ring resonator of Allen et al (From Su & Haldar, 2007, © 2007 IEEE)

The upper part of the resonator is formed by a right-handed (microstrip) line quarter wave long at the centre frequency and of characteristic impedance $50\,\Omega$. For this characteristic impedance, the width of the microstrip line is W = 1.2 mm for the microstrip substrate of dielectric constant 10.2 and thickness 50 mil. The lower part of the ring is formed by a combination of the microstrip line and the left-handed line. The microstrip line has a total length of 12.3 mm. The left-handed line is formed by two T section unit cells (see Fig.6) with C = 7 pF, L = 8.7 nH. The gap between the input and output lines and the ring resonator is 0.2 mm. The design frequency is 0.88 GHz. The dual mode is said to occur due to the unequal values of the characteristic impedance of the T-section and the 50 Ω characteristic impedance of the microstrip line.

f_c is calculated to be 0.456 GHz. The total phase shift of the lower part of the ring, (the composite-right/left-handed line) is given by

$$\varphi = 2\pi f\sqrt{\varepsilon_{eff}}\,12.3\times 10^{-3}\ell / c + N\times 2\sin^{-1}(f_c / f) \tag{19}$$

where ε_{eff} is the effective dielectric constant of the microstrip line.

Table 1 shows the phase shift calculated for the upper and lower parts of the resonator and the total phase shift in the ring at the fundamental and harmonics. None of the total phase shifts are integer multiples of 2π . Clearly, (19) can not be employed for resonator design.

Frequency	Phase shift in upper arm (radians)	Phase shift in lower arm eqn. (5) (radians)	Total phase shift around the ring (radians)
f(= 0.88GHz)	0.5π	0.864π	1.364π
2f	π	0.675π	1.675π
3f	1.5π	0.733π	2.233π

Table 1. Phase shifts along the ring (From Su & Haldar, 2007, © 2007 IEEE)

Fig.9 shows the simulated variation of the magnitude of S_{21} with frequency. There is a peak occuring close to the cut-off frequency.

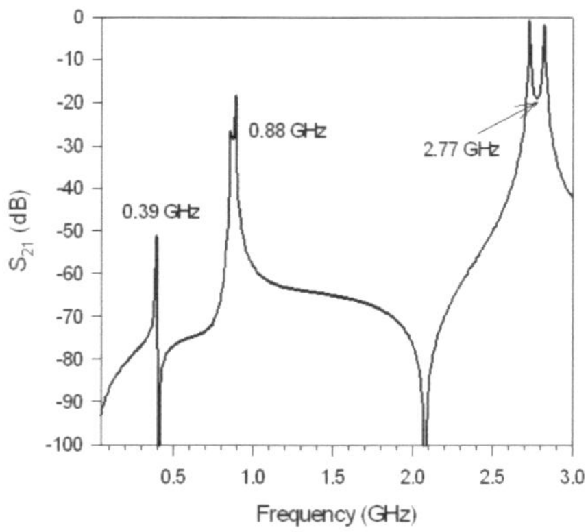

Fig. 9. Variation of the magnitude of S_{21} with frequency for the resonator of Fig.8 (From Su & Haldar, 2007, © 2007 IEEE)

It has been verified by simulation (Su & Haldar, 2007) that an increase in the mismatch of the characteristic impedance of the lines increases the frequency difference of the dual mode and appears to increase the insertion loss between the fundamental and the second harmonic (N=3).

Fig. 10 shows a variation of the resonator. A lower resonant frequency was used to get higher value capacitors to reduce the lower cut-off frequency. The value of the inductor was unchanged. To reduce the number of discrete components, the inductors were replaced by short circuited lines. The length of the line, ℓ is given by

$$\omega_0 L = Z_0 \tan \beta \ell \tag{20}$$

where ω_0 is the resonant frequency, Z_0 and β are the characteristic impedance and propagation constant of the microstrip line implementing the inductor. A large value of Z_0 is preferred to keep the line length short to avoid resonance of the line till about the third harmonic.

Fig.11 shows the simulated variation of the magnitude of S_{21} with frequency. It is interesting to note that harmonics upto the second harmonic (N=3) have been suppressed. The composite ring resonator can suppress several harmonics if the phase shift in the ring at the fundamental is produced mainly by the left-handed line.

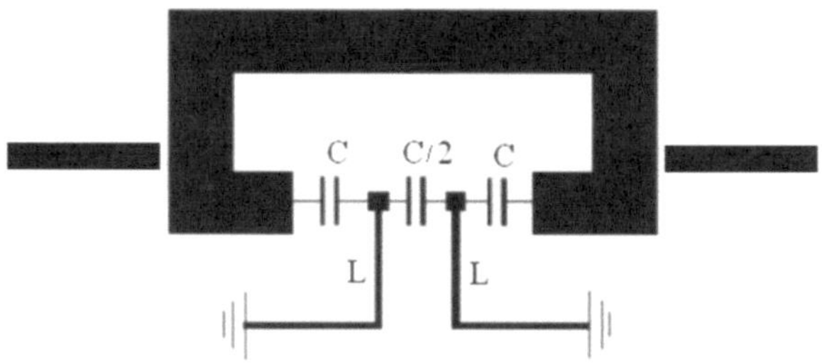

Fig. 10. Redesigned resonator at lower frequency (From Su et al, 2008, © 2008 IEEE)

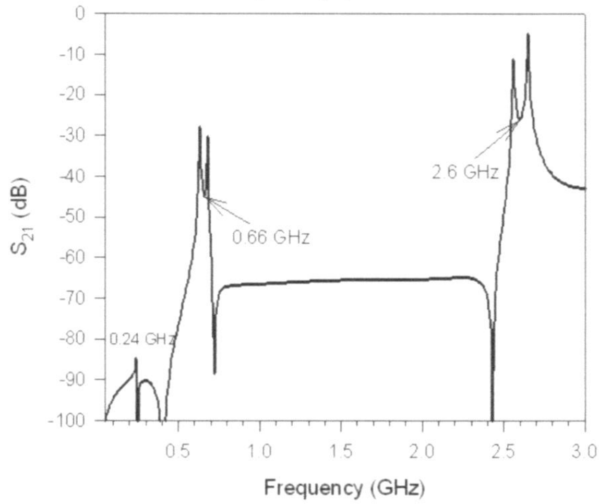

Fig. 11. Variation of the magnitude of S_{21} with frequency for the resonator of Fig.10 (From Su et al, 2008, © 2008 IEEE)

6. Filter design using circuit modelling and simulation of composite-right/left-handed line ring resonators

Table 1 in section 5.2 shows that although the principle of operation is valid, the infinitely extended iterative T- sections theory cannot model the resonator operation well. The size of the resonator limits the number of T and п sections. Many of the reported designs appear to have been carried out by simulation and multi-resonator filters are rarely reported. Circuit models (Chang, 2004) can provide an initial design, which can be refined by simulation.
The authors' intention is to reduce the number of lumped elements. Hence a п -section is used. Fig.12(a) shows the ring resonator with a п -section. The corresponding microstrip version is shown in Fig.12(b). The inductors are implemented with short-circuited transmission lines. Thus the circuit uses only one discrete element.

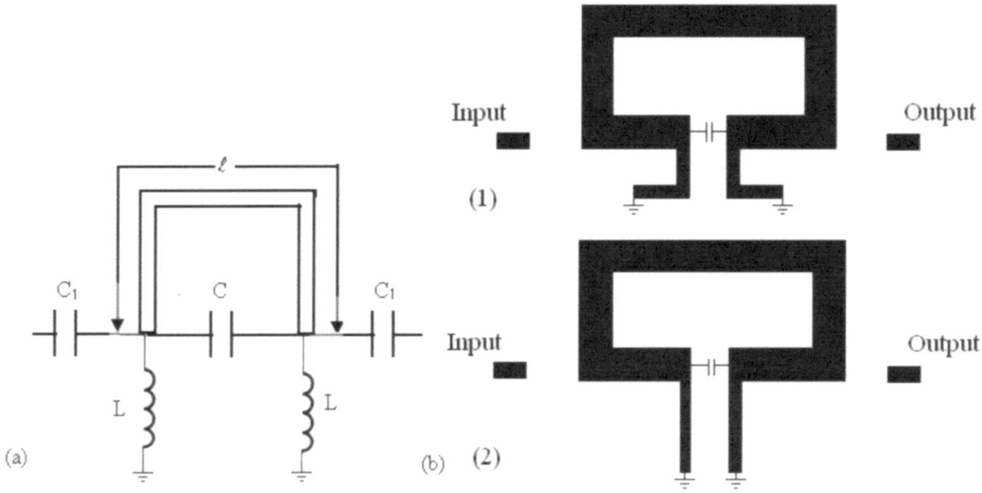

Fig. 12. (a) Circuit model of a ring resonator incorporating a п -section with weak coupling to RF input and output (b) Microstrip layout (From Fong et al, 2009)

Consider the microstrip transmission line of length ℓ, characteristic admittance $Y_{0\ell}$ and propagation constant β, in parallel with the п -section. The Y-parameter of the transmission line considered as a 2-port network is given by

$$Y = \begin{bmatrix} -jY_{0\ell}\cot(\beta\ell) & \dfrac{jY_{0\ell}}{\sin(\beta\ell)} \\ \dfrac{jY_{0\ell}}{\sin(\beta\ell)} & -jY_{0\ell}\cot(\beta\ell) \end{bmatrix} \tag{21}$$

The Y-parameter of the п - section is given by

$$Y = \begin{bmatrix} \frac{1}{j\omega L} + j\omega C & -j\omega C \\ -j\omega C & \frac{1}{j\omega L} + j\omega C \end{bmatrix} \tag{22}$$

For the two networks in parallel, the overall Y-matrix is given by the sum of the Y matrices as follows

$$Y = \begin{bmatrix} -Y_{0\ell}\cot(\beta\ell) + \frac{1}{j\omega L} + j\omega C & \frac{jY_{0\ell}}{\sin(\beta\ell)} - j\omega C \\ \frac{jY_{0\ell}}{\sin(\beta\ell)} - j\omega C & -Y_{0\ell}\cot(\beta\ell)\frac{1}{j\omega L} + j\omega C \end{bmatrix} \tag{23}$$

The resonant frequencies are obtained by setting the determinant of (23) to zero. From the resulting quadratic equation, one can show that the resonant frequencies, ω_1 and ω_2 satisfy the equations

$$\frac{2\omega_1^2 LC - 1}{\omega_1\omega_2 L^2} = Y_{0\ell}^2 \tag{24}$$

$$\frac{1}{\omega_2 L} = Y_{0\ell}\tan(\beta\ell/2) \tag{25}$$

These equations allow one to design dual mode resonators required by a filter design. For example consider the design of a fourth order Chebyshev filter with a centre frequency, f_0 of 0.6 GHz, 10 % fractional bandwidth and a passband ripple of 0.5 dB. This would require four single mode resonators with resonant frequency f_0 with coupling coefficients between resonators 1 and 2 and between 3 and 4, $K_{12} = K_{34} = 0.07$. Resonators 1 and 2 can be replaced by a single dual mode resonator and resonators 3 and 4 can be replaced by an identical dual mode resonator. One now has to design a dual mode resonator with the required coupling coefficient and then couple two such resonators for the required coupling coefficient of $K_{23} = 0.06$. Finally one designs the load and source coupling to get the required external quality factor.

The resonant frequencies of the dual modes are calculated from (11) and (12) using the values of f_0 and K_{12}. Then choosing a standard value of 10 pF for the capacitor, C, the required value of L is calculated from (24) to be 12.9 nH and 4.4 nH. The smaller value is chosen as both lumped inductors and short-circuited transmission lines have higher self-resonant frequencies for lower inductance values. Using (20), this value is implemented by a short-circuited transmission line of characteristic impedance 50 Ω and length 9.8 mm (width is 1.2 mm). The characteristic impedance of the right-handed transmission line is chosen as 28 Ω. RT/Duroid 6010.2 from Rogers Corporation is chosen as the substrate. The width of

the line is calculated to be 3.2 mm. After determining the effective dielectric constant, the length ℓ of the line is then calculated from (24) to be 61.5 mm.
To check the circuit model against simulated results (SONNET, 2008) weak coupling to input and output are used to obtain sharp resonance peaks. For the circuit model of Fig.12(a), this coupling is produced by the small coupling capacitor, C_1 = 0.05 pF. For the microstrip implementations of Fig.12(b) the weak coupling is provided by the gaps. To obtain the response of the circuit, the Y-matrix of the resonator is converted to Z-matrix, added to the Z-matrix of the coupling capacitors and then reconverted to Y-matrix. S_{21} for reference impedance of 50 Ω ($Y_0 = 1/50\,S$) is calculated from this Y-matrix using

$$S_{21} = \frac{-Y_{21}Y_0}{(Y_{11}+Y_0)(Y_{22}+Y_0)-Y_{12}Y_{21}} \tag{26}$$

Fig.13. compares the variation of the magnitude of S_{21} with frequency for the circuit model and simulations of Fig.12 (b) (1) and (2).

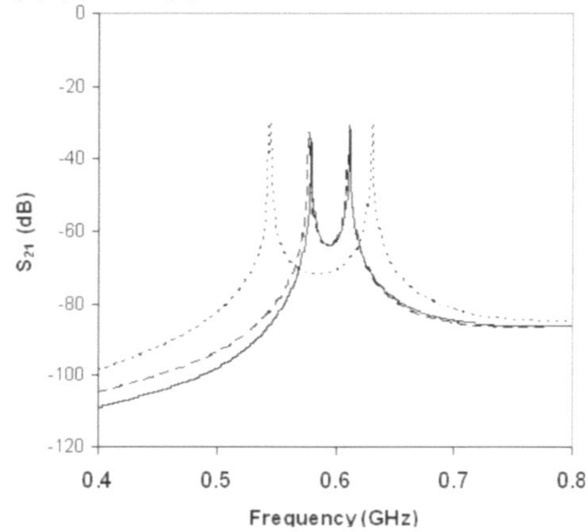

Fig. 13. Comparison of circuit model and simulation. Solid line: circuit model. Dashed line: Fig.12(b)(1). Dotted line: Fig12(b)(2) (From Fong et al, 2009)

Good agreement is obtained between the circuit model and the simulation of Fig.12 (b) (1). The difference with the simulation of Fig.12(b) (2) is ascribed to capacitive coupling between the short circuited lines. To offset this, the inductances of the lines have to be increased by increasing their lengths. Good agreement is obtained when the lengths are increased to 10.8 mm.
For the Chebyshev filter, the required external quality factor, $Q_{ext} = 16.7$. The coupling is obtained by tapping one of short-circuited lines at the source end and at the load end. The resistance in parallel with the inductance is obtained from

$$Q_{ext} = \frac{R}{\omega_0 L} \tag{27}$$

The calculated value of R is 276.3 Ω. An inductor tap transforms the resistance R to 50 Ω load/source impedance. The approximate formula (for high quality factor) is

$$R = 50\left(\frac{L}{L_T}\right)^2 \tag{28}$$

where, L_T is the inductance value at the tap.

The tap position is then given by

$$R = 50\left(\frac{\ell}{\ell_T}\right)^2 \tag{29}$$

where the lengths ℓ and ℓ_T are measured from the short circuit.

Equation (29) is approximate because inductance of a short-circuited line does not vary linearly with length (see equation 20). However, this is adequate because the circuit model establishes an initial design, which is fine tuned by simulation. The gap between two resonators is adjusted to get the filter. Fig. 14 shows a picture of the 4-pole Chebyshev filter. Fig. 15 and 16 show measured and simulated results for the filter.

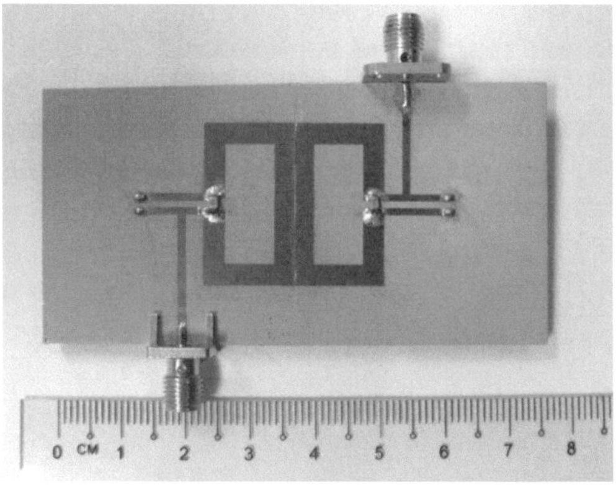

Fig. 14 A picture of the 4-pole Chebyshev filter.

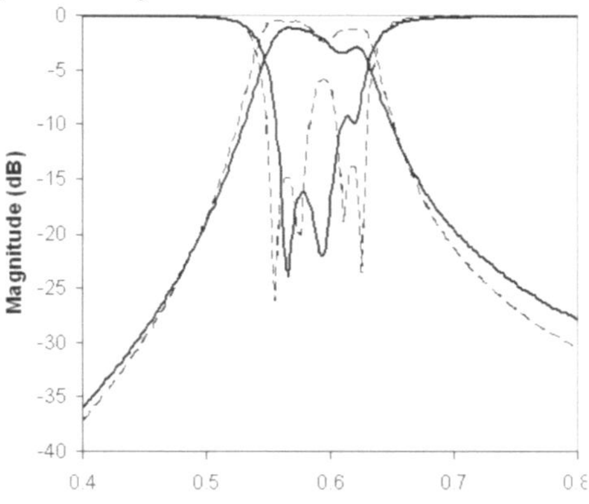

Fig. 15. Variation of magnitudes of S_{21} and S_{11} with frequency. Measured: Solid lines. Simulated (with losses): Dashed lines (From Fong et al, 2009)

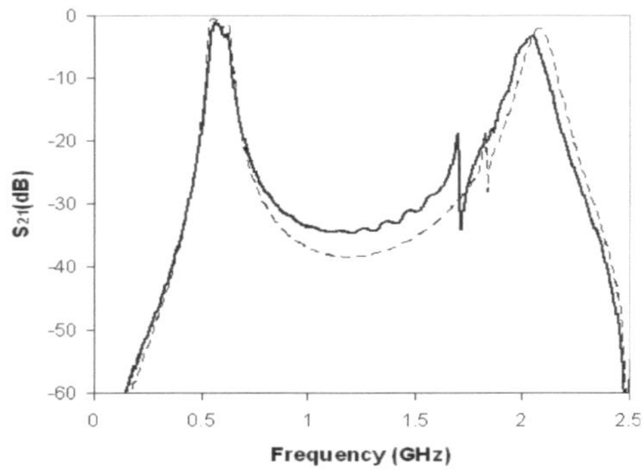

Fig. 16. Wideband frequency response showing harmonic reduction. Measured: Solid line. Simulated: Dashed line (From Fong et al, 2009)

7. Other considerations for composite-right/left-handed line ring resonators

7.1 Practical considerations for lumped capacitors

The reader might have noticed that the centre frequency of the filter designed is quite low. This is a very conservative design to avoid the self resonances of the lumped capacitor. However, it is possible to design for higher frequencies by using the circuit model of the capacitor. Such a circuit model is shown in Fig. 17.

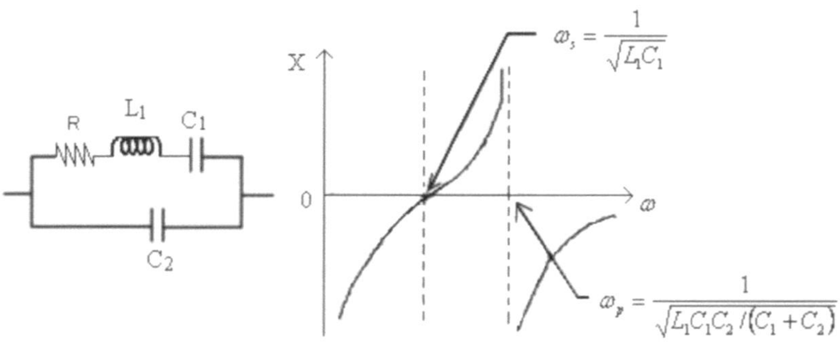

Fig. 17. Capacitor model: (a) Equivalent Circuit (b) Variation of reactance with frequence (ignoring R) showing the series and parallel resonant frequencies. (From Su et al, 2008, © 2008 IEEE)

The model parameters can be obtained by measuring the variation of the S-parameters with frequency and then adjusting the element values of the model to get best agreement with the calculated and measured results. For example, for the 10 pF capacitor considered here, C_1 = 4.22 pF, C_2 = 5.78 pF, R = 0.06 Ω and L_1 = 0.42 nH. The capacitor behaves as a capacitor below the series resonant frequency. So it appears that resonator design using lumped capacitors is limited to frequencies below the series resonant frequency.

Lumped inductors are rarely as good as lumped capacitors both in terms of self resonance and quality factor. Thus when lumped inductors are used, the maximum frequency of operation may be limited further. Fortunately, they can be replaced by short-circuited transmission lines, but one needs to be careful about the effect resonance of the line on harmonic suppression.

7.2 Tunable composite-right/left-handed line ring resonators

Tunable resonators have been reported by Allen et al (Allen et al, 2007). A typical design is shown in Fig.18. As in their earlier work (Allen et al, 2006), two T sections are used with the two capacitors of each section are replaced by varactor diodes. The dc bias is provided through high characteristic impedance lines with shunt radial stubs.

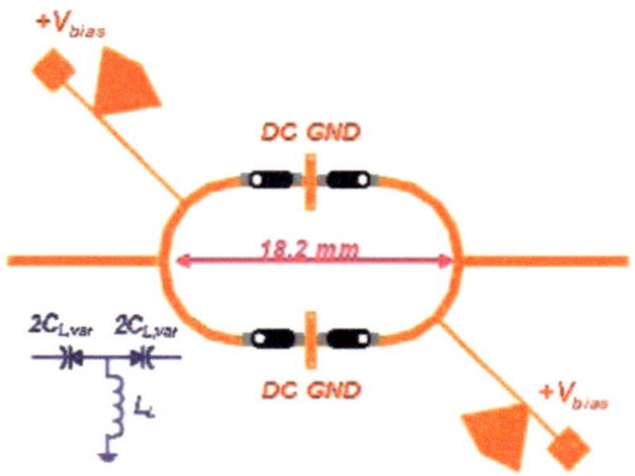

(a)

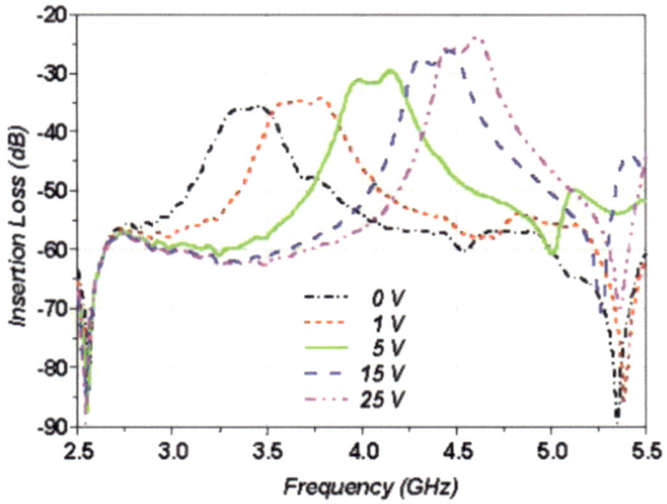

(b)

Fig. 18. Tunable resonator (a) Layout (b) Variation of insertion loss with frequency for different bias voltages (From Allen et al, 2007 © 2007 IEEE)

From Fig.18(b), one can see that the centre frequency of the resonator can be shifted by using varactor diodes.

8. Conclusion and future developments

Starting with the explanation of right/left-handed lines, the authors have described recent developments in Dual Mode Microstrip Ring Resonator with Composite-Right/Left-handed Lines. The authors have discussed a circuit technique for the analysis and design. It is shown that a large number of T or π sections may not be needed - the authors describe a resonator with only one π section requiring just one capacitor. However such a section has DC short circuits at both ends of the capacitor. Hence the section is not suitable for designing tunable filters in which dc biased varactor diodes replace capacitors. The design of filters using ring resonators with composite-right/left-handed lines has been explained and demonstrated.

What are the future developments?
Possible developments are as follow:

(1) Can one design compact higher order filters?
(2) Can one use lumped capacitors and inductors near their resonant frequencies to design ring resonators with Composite-Right/Left-handed Lines? It may be possible to use the parallel resonance of a capacitor.
(3) Can one use slow wave structures to replace lumped capacitors for ring resonator with Composite-Right/Left-handed Lines.
(4) Can one design ring resonators with composite-Right/Left-handed Lines using coplanar transmission lines?

9. References

Allen, C.A.; Leong, K.M.K.H. & Itoh, T. (2006). Dual-mode composite-right/left-handed transmission line ring resonator. *Electronics Letters,* 42, 2, (2006) pp. 96-97, ISSN: 0013-5194.

Allen, C.A.; Leong, K.M.K.H. & Itoh, T. (2007). Fixed frequency and tunable metamaterial based ring resonators with narrowly spaced resonances , *Microwave Symposium Digest, IEEE MTT-S International,* pp. 1623-1626, ISBN:1-4244-0688-9, Honolulu, USA, June, 2007.

Beck, A.H.W (1958). *Space Charge Waves,* Pergamon Press, New York

Carroll, J.M. & Chang K. (1994) Microstrip mode suppression ring resonator. *Electronics Letters,* 30, 22, (1994) pp.1861-1862, ISSN: 0013-5194

Chang, K. & Hsieh, L.H. (2004). *Microwave Ring Circuits and Related Structures,* John Wiley & Sons. Inc, ISBN: 978-0-471-44474-9, New York

Fong, K.K. ; Su, H.T. & Haldar, M.K. (2009). Circuit modelling for the design of composite-right/left-handed ring resonator, Submitted to *2009 Asia Pacific Microwave Conference,* Singapore, December, 2009.

Karacaoglu, U.; Sanchez-Hernandes, D. ; Robertson, I.D. & Guglielmi, M. (1996). Harmonic suppression in microstrip dual-mode ring-resonator bandpass filters, *Microwave Symposium Digest, IEEE MTT-S International,* pp. 1635-1638, ISBN:0-7803-3246-6, conference location, June, 1996.

Lai, A.; Itoh, T. & Caloz, C. (2004). Composite Right/Left-Handed Transmission Line Metamaterials. *IEEE Microwave Magazine,* Vol., No., (2004) pp.34-50), ISSN:1527-3342.

Matthaei, G.; Young, L. & Jones, E.M.T. (1980). *Microwave Filters, Impedance-Matching Networks and Coupling Structures,* Artech House, ISBN: 0-89006-099-1, Norwood, MA (USA).

Shelby, R.A.; Smith, D.R. & Schultz, S. (2001). Experimental verification of negative index of refraction. *Science,* Vol.292, No.5514, (2001) pp.77-79, ISSN 0036-8075 (print), 1095-9203 (online).

SONNET 11.52 Sonnet Software Inc., Pittsburgh, PA, 2008.

Su, H.T. & Haldar, M.K. (2007). Redesign of a dual-mode ring resonator using composite-right/left-handed line, *2007 Asia Pacific Microwave Conference*, pp. 2269-2272, ISBN: 978-1-4244-0748-4, Bangkok, December, 2007.

Su, H.T. ; Fong, K.K. ; Haldar, M.K. & Wong, M.L.D. (2008). New 4-pole dual-mode resonator using composite-right/left-handed line, *2008 Asia Pacific Microwave Conference*, pp. 1-4, ISBN: 978-1-4244-2641-6, Hong Kong, China, December, 2008.

Vaselago, V.G. (1968). The electrodynamics of substances with simultaneously negative values of ε and μ . *Soviet Physics Uspekhi,* 10, 4, (1968) ppp.509-514, ISSN: 0038-5670

Wolff, I. & Knoppik, N (1971). Microstrip bandpass filter using degenerate modes of a microstrip ring resonator. *Electronics Letters,* 7, 26, (1971) pp. 779-781, ISSN: 0013-5194.

Wolff, I. (1972). Microstrip ring resonator and dispersion measurements on microstrip lines. *Electronics Letters,* 8, 12, (1972) pp.302-303, ISSN: 0013-5194.

19

Electrically small resonators for metamaterial and microwave circuit design

Marta Gil, Francisco Aznar, Adolfo Vélez,
Miguel Durán-Sindreu, Jordi Selga, Gerard Sisó,
Jordi Bonache and Ferran Martín
Universitat Autònoma de Barcelona (GEMMA/CIMITEC Group)
Spain

1. Introduction

A new branch in microwave engineering arose just few years ago with the emergence of metamaterials in 2000 (Smith et al., 2000). The implementation of the first artificial medium with negative effective dielectric permittivity and magnetic permeability opened the door to the experimental study of a new kind of media: left-handed media. The possibility of the artificial implementation of such media allowed the corroboration of many of their electromagnetic properties, predicted years before by Viktor Veselago (Veselago, 1968). Since the year 2000, the interest stirred up by these new materials has given rise to numerous works in a wide range of scientific branches. The possibilities that metamaterials offer to create artificial media with controllable characteristics has permitted the creation of a growing number of completely new applications. Undoubtedly, the most innovative and spectacular application of such artificial media is their use in the implementation of cloaking structures to achieve invisibility, which can be accomplished thanks to the engineering of the refraction index of the different layers of the cloaking shield (Schurig et al., 2006). Within the vast number of new applications of metamaterials, one of the most productive ones is the implementation of microwave devices by means of artificial transmission lines. The following sections will deal with one of the approaches devoted to this purpose: the resonant-type approach. Different subwavelength resonators employed in the design of metamaterial transmission lines based on the resonant-type approach will be studied. The equivalent circuit models of different kinds of metamaterial transmission lines, as well as the parameter extraction methods employed as design and corroboration tools will be also presented. In closing, a selection of application examples of resonant-type metamaterial transmission lines in the design of microwave devices will be presented.

2. Sub-wavelength resonators

The implementation of the first effective medium with left handed properties (Smith et al., 2000) was possible thanks to the employment of small metallic resonators known as split-ring resonators (SRRs). These resonators had been previously presented (Pendry et al., 1999) as the first non-magnetic resonator capable of exhibiting negative values of the magnetic permeability around its resonance frequency. This was one of the characteristics which made the SRR suitable for the synthesis of such a medium; the second one was its small electrical size. At the resonance frequency, the SRR perimeter is smaller than half the wavelength of the exciting wave. These small dimensions allow the use of SRRs in the implementation of effective media, which requires small unit cell sizes (smaller than the wavelength). By this means, the incident radiation does not detect the internal configuration of the medium, but the effective properties of the whole medium.
The split-ring resonator is formed by two concentric metallic open rings (see Fig.1). The resonator can be excited by an axial (z direction in the figure) time-varying external magnetic field, which induces currents in the rings. The splits present in the rings force the current to flow as displacement current between them. The current loop is thus closed through the distributed capacitance that appears between the inner and the outer ring.

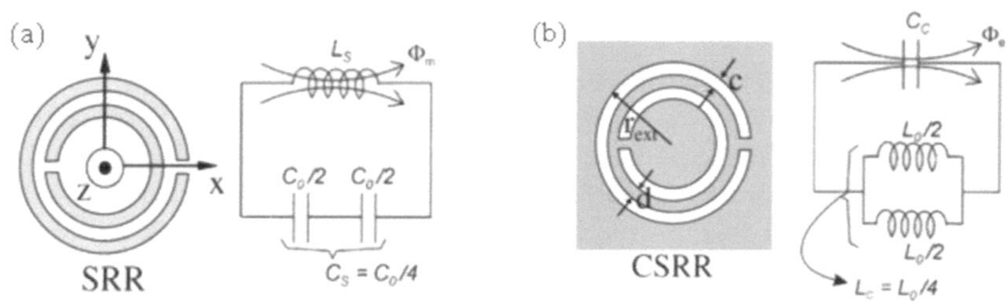

Fig. 1. Scheme of the split-ring resonator (SRR) (a) and the complementary split-ring resonator (CSRR) (b) and their equivalent circuit models. Metallic parts are depicted in grey, whereas etched parts are depicted in white.

The resonator can be modelled as is shown in Fig. 1(a) (Baena et al., 2005). $C_0/2$ is the capacitance related with each of the two SRR halves, whereas L_s is the resonator self-inductance. C_0 can be obtained as $C_0=2\pi r C_{pul}$, where C_{pul} represents the per unit length capacitance between de rings forming the resonator. As for L_s, it can be approximated to the inductance of a single ring with the average radius of the resonator and the width of the rings, *c*. Taking into account the circuit model of the resonator, its resonance frequency can be calculated as:

$$\omega_0 = \frac{1}{\sqrt{L_s C_s}} \tag{1}$$

As long as the inductance and the capacitance of the resonator can be increased (within the technology limits), the resonance frequency of the SRR can be decreased, reducing its electrical size.
The application of the Babinet principle to the structure of the SRR leads to its complementary counterpart: the complementary split-ring resonator (CSRR) (Falcone et al.,

2004), depicted in Fig. 2(b). In the CSRR the rings are etched on a metallic surface and its electric and magnetic properties are interchanged with respect to the SRR: the CSRR can be excited by an axial time-varying electric field and exhibits negative values of the dielectric permittivity. The equivalent circuit model of the CSRR is shown in Fig. 2(b). The resonance frequency of the CSRR is almost the same of the frequency of a SRR with the same dimensions.

Both resonators, the SRR and the CSRR can be employed in the synthesis of effective media (Smith et al., 2000; Shurig et al., 2008) and metasurfaces (Falcone et al., 2004), as well as, of course, artificial transmission lines (Martín et al. 2003; Falcone et al., 2004). Next sections will deal with the implementation of metamaterial transmission lines based on the resonant-type approach, as well as their application to microwave device design.

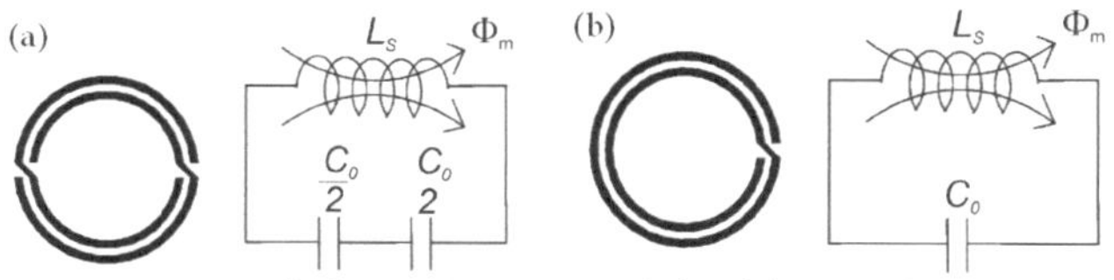

Fig. 2. SRR-based resonators and their circuit models. (a) Non-bianisotropic SRR. (b) Spiral resonator. Metallic parts are depicted in black.

Taking SRRs and CSRRs as starting points, numerous resonators have been proposed. The new resonators are obtained by modifying the topology of the original ones in order to decrease their electrical size or obtain certain symmetry properties. Figures 2, 3 and 4 show several examples of different resonators obtained following different strategies. In Fig. 2, two different resonators can be found and two more could be obtained as their complementary structures. The first one is known as non-bianisotropic split-ring resonator. It has the same electrical size as the SRR, but it has been designed to avoid the cross-polarisation (bianisotropic) effects that the original SRR exhibits (Baena et al, 2005). The electrical size can be reduced enhancing either the capacitance or the inductance of the resonator. This is achieved in the spiral resonator (Fig. 2(b)) thanks to the increase of the total capacitance of the resonator in such a way that the resonance frequency is half the one of the SRR.

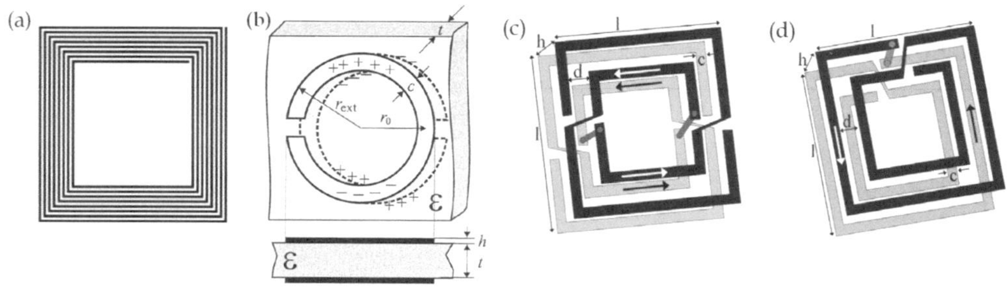

Fig. 3. SRR-based compact resonators. (a) Spiral resonator with 8 loops. (b) Broadside-coupled split-ring resonator (BC-SRR). (c) Two-layer multi spiral resonator (TL-MSR). (d) Broad-side coupled spiral resonator with four turns (BC SR (4)).

There are other strategies devoted to the miniaturisation of the resonators. One of them is the addition or enlargement of the metallic strips in order to increase the inductance of the whole structure. One example of the application of this strategy is the spiral shown in Fig. 3(a) (Alici et al., 2007; Bilotti et al., 2007). On the other hand, the capacitance can be enhanced designing the resonator so that their strips are broad-side coupled (see broad-side coupled SRR in Fig. 3(b)) (Marqués et al., 2003). This strategy requires the use of two metal layers on which lay the different parts of the resonator. By this means the strips are broad-side instead of edge-side coupled and the capacitance is enhanced, especially if thin substrates are employed. Both strategies are combined in the examples shown in Fig. 3(c) and (d) (Aznar et al., 2008b). In these two resonators the strips are elongated connecting the two metal layers by means of vias, increasing the inductance of the resonator. Additionally, the capacitance is enhanced thanks to the broadside coupling. Following these strategies, the electrical size of the resonators can be drastically reduced.

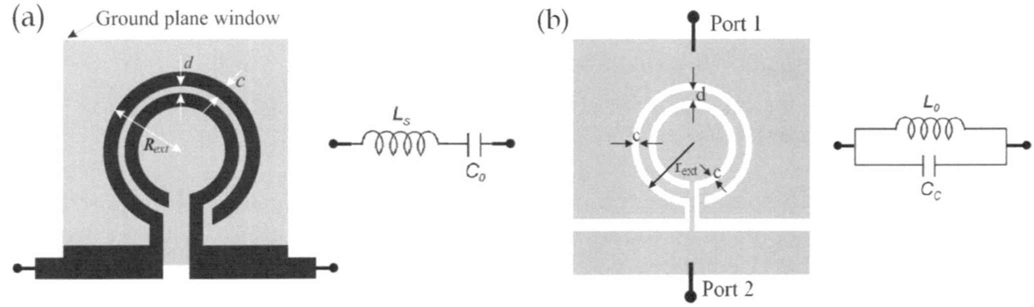

Fig. 4. (a) Open split-ring resonator (OSRR) and its equivalent circuit model. (b) Open complementary split-ring resonator (OCSRR) and its equivalent circuit model.

A different kind of SRR-based structures are open resonators. Figure 4 shows the layouts and equivalent circuit models of the open SRR (Martel et al., 2004) and the open complementary SRR (Vélez et al., 2009a). As can be seen in the layout, the OSRR is based on the SRR and is obtained by truncating the rings forming the resonator and elongating them outwards. The OCSRR can be obtained as the complementary particle of the OSRR, in a similar way as the CSRR is obtained from the SRR. The resonators shown in Fig. 4 can be implemented either in microstrip or in coplanar technology (Durán-Sindreu et al., 2009). The equivalent circuit models of the resonators are also shown in Fig. 4. The equivalent circuit model of the OSRR is a series LC resonator (Martel et al., 2004). The inductance L_s can be obtained as the inductance of a ring with the average radius of the resonator and the same width, c, of the rings forming the OSRR. The capacitance C_0 is the distributed edge capacitance that appears between the two concentric rings. In a similar way, the OCSRR can be modelled by means of a parallel LC resonant tank (Vélez et al., 2009), where the inductance L_0 is the inductance of the metallic strip between the slot hooks and the capacitance is that of a disk with radius r_0-$c/2$ surrounded by a metallic plane separated by a distance c.

The small size of all these resonant particles makes them suitable for the implementation of microwave devices based on resonant-type metamaterial transmission lines with small dimensions and even new functionalities. This will be illustrated in the following sections.

3 Resonant-type Metamaterial Transmission Lines

The previously presented resonators can be employed in many different applications and, as has already been pointed out, one of them is the implementation of metamaterial transmission lines. Soon after the implementation of the first negative refraction index medium (Smith et al., 2000), the same concepts were applied to the synthesis of planar one dimensional media, giving rise to the first left-handed transmission lines (Eleftheriades et al., 2002; Caloz & Itoh, 2002). This first approach devoted to the implementation of metamaterial transmission lines consisted on periodically loading a conventional transmission line with series capacitive and shunt-connected inductive elements. Some examples of such lines can be found in Fig. 5 together with the equivalent circuit model of the unit cell. The loading elements are modelled by C_L and L_L, whereas L_R and C_R represent the line elements.

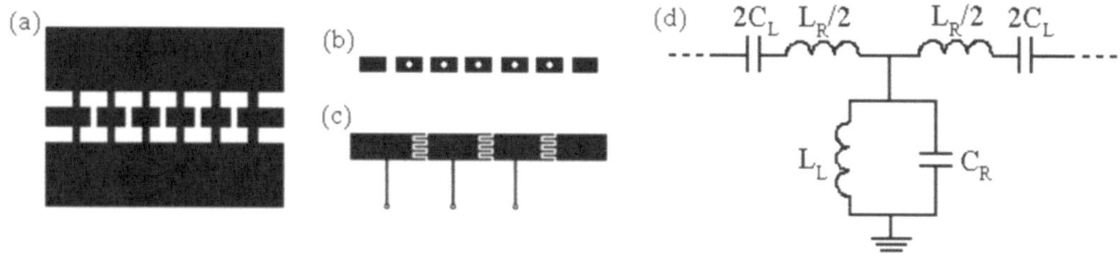

Fig. 5. Examples of CRLH transmission lines based on the LC-loaded transmission line approach and their equivalent circuit model. (a) Conventional coplanar waveguide transmission line loaded with capacitive gaps and metallic connections to the ground planes. (b) Conventional microstrip transmission line loaded with capacitive gaps and vias. (c) Conventional microstrip transmission line loaded with interdigital capacitors and shunt connected inductive stubs. (d) Equivalent circuit model for the LC-loaded transmission line unit cell.

In such a medium, we can express the dispersion relation of such lines as:

$$\cos(\phi) = 1 + \frac{Z_s(\omega)}{Z_p(\omega)} \tag{2}$$

where Z_s and Z_p are the series and shunt impedance of the circuit model, respectively. This expression gives rise to the generic dispersion diagram shown in Fig. 6 (a) (regions corresponding with positive group velocity have been chosen). Two transmission bands can be identified: the first one corresponds to the left-handed propagation region, in which the propagation constant is negative, whereas in the second one the propagation is right-handed. Moreover, the effective dielectric permittivity and magnetic permeability can be obtained from the values of the series impedance and the shunt admittance, respectively, revealing the frequency dependence of the sign of ε_{eff} and μ_{eff} that gives rise to the different right and left-handed bands (Marqués et al., 2008). At those frequencies in which the propagation is dominated by the loading elements, the propagation is left-handed (β<0 region), whereas the parasitic line elements give rise to the right-handed transmission band (β>0 region). This composite behaviour gives the name composite right/left-handed (CRLH) to these transmission lines. Both transmission bands are usually separated by an intermediate frequency gap in which transmission is forbidden (See Fig. 6(a)). However, it is possible to force both limits of the gap (ω_{G1} and ω_{G2}) to coincide ($\omega_{G1}=\omega_{G2}=\omega_0$) in order to make it disappear (Caloz & Itoh, 2005). In this case, known as balanced case, a continuous transition between the left- and the right-handed transmission bands is obtained giving rise to a wide band exhibiting backward and forward propagation at different frequencies (Fig. 6(b)).

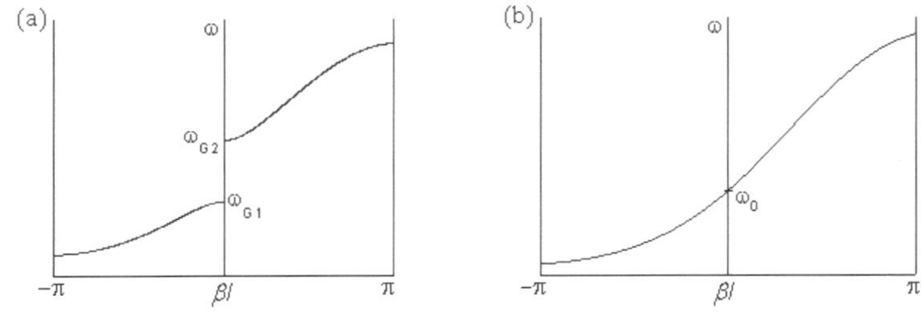

Fig. 6. Dispersion diagrams of a CRLH transmission line (a) for the unbalanced case (b) for the balanced case.

A second approach devoted to the implementation of composite right/left-handed transmission lines was proposed by some of the authors soon after the works presenting the LC-loaded transmission line: the resonant-type approach (Martín et al., 2003). In this approach, subwavelength resonators like the ones presented in the previous section are used to (in combination with other elements) load conventional transmission lines and obtain CRLH artificial lines.

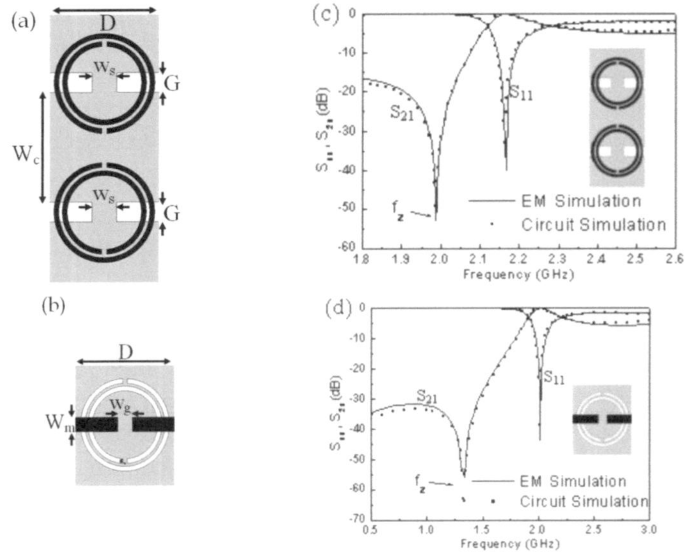

Fig. 7. Layouts and frequency responses of two resonant-type metamaterial transmission line unit cells. Metallic parts are depicted in black for the top and in gray for the bottom layer of the substrate. (a) SRR-based CRLH transmission line implemented in coplanar technology. (b) CSRR-based CRLH transmission line implemented in microstrip technology. (c) Frequency response of the structure shown in (a). (d) Frequency response of the structure shown in (b).

This approach allows the use of both, metallic and complementary resonators in coplanar and microstrip technology (see Fig. 7). Thanks to the small size of the loading resonators, the resulting transmission lines are compact and can be employed in the design of microwave devices with reduced dimensions employing commercial substrates. The location of SRRs on the bottom layer of a coplanar waveguide allows the excitation of the resonators by the magnetic field and provides negative values of the magnetic permeability. In order to obtain left-handed propagation, negative dielectric permittivity is also required, what is obtained by means of metallic junctions between the line and the ground planes (see Fig. 7 (a)) (Martín et al., 2003). It is also possible to use SRRs to load a microstrip transmission line locating them on the top layer of the substrate, close to the signal strip. By this means, the SRRs are excited by the magnetic field and, if they are combined with vias, provide a left-handed transmission band in a certain frequency range in the vicinity of the resonance frequency of the rings. Of course, the effect of the parasitic line elements provides a second transmission band with right-handed characteristics; in other words: resonant-type metamaterial transmission lines do also exhibit a composite behaviour. This allows the design of balanced transmission lines, which are interesting for broadband applications (Gil et al., 2007a).

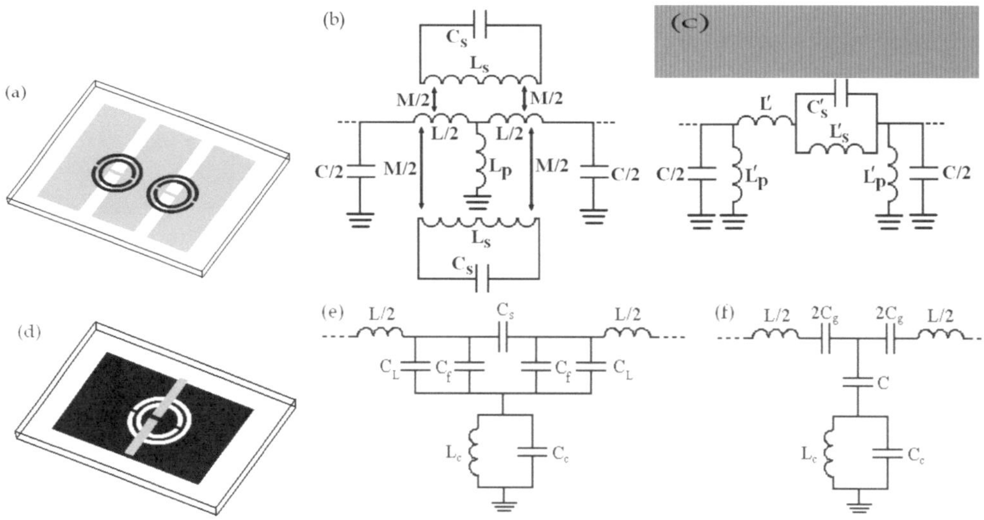

Fig. 8. Schemes of two resonant-type metamaterial transmission lines and their equivalent circuit models. Metallic parts are depicted in black for the bottom and in gray for the top layer of the substrate. (a) SRR-based transmission unit cell. (b) Equivalent circuit model for the structure shown in (a). (c) Modified circuit model for the structure shown in (a). (d) CSRR-based unit cell. (e) Equivalent circuit model for the structure shown in (d). (f) Modified circuit model for the structure shown in (d).

Other resonators can be employed as well in the design of this kind of metamaterial transmission lines, as is the case of CSRRs. In this case, the most usual configuration is the one in which the resonators are etched on the ground plane of a microstrip transmission line and combined with capacitive gaps etched on the signal strip (see Fig. 7(b)) (Falcone et al., 2004), although other configurations are possible (Gil et al., 2008a). The response of both structures (SRR- and CSRR-based) (performed with the *ADS Momentum* commercial simulation software), which are very similar, can be observed in Fig. 7. They both exhibit a left-handed transmission band preceded by a transmission zero, which represents the main difference with respect to the LC-loaded transmission line approach given that in that case the transmission zero is found at the origin. The right-handed band appearing above the left-handed one is not shown in the graphs. In Fig. 7 the electromagnetic simulations of the structures are compared with the responses given by their equivalent circuit models, which can be found in Fig. 8. As can be seen, the models perfectly describe the behaviour of the structures at the frequency range of interest.

For the SRR-based unit cell (Fig. 8(a)) the circuit model is the one shown in Fig. 8 (b). The resonators are modelled by the resonant tanks formed by L_s and C_s, which are coupled to the line by means of the mutual inductance M. The line parameters are L and C, whereas L_p represents the metallic strips. This model is the improvement of a previously existing one (Martín et al, 2003) and the parameters have a more realistic physical meaning and provide a better description of the behaviour of the structure (Aznar et al., 2008a). Nevertheless, both, the former and the new proposed model can be transformed into the circuit shown in Fig. 8(c) which, as is shown in Fig. 7, reproduces the response of the structure in a very proper way. The transformation equations are the following ones:

$$L_s' = 2M^2 C_s \omega_o^2 \frac{\left(1+\dfrac{L}{4L_p}\right)^2}{1+\dfrac{M^2}{2L_p L_s}} \tag{3}$$

$$C_s' = \frac{L_s}{2M^2\omega_o^2}\left(\frac{1+\dfrac{M^2}{2L_p L_s}}{1+\dfrac{L}{4L_p}}\right)^2 \tag{4}$$

$$L' = \left(2+\frac{L}{2L_p}\right)\frac{L}{2} - L_s' \tag{5}$$

$$L_p' = 2L_p + \frac{L}{2} \tag{6}$$

and they allow the calculation of the parameters of the circuit (c) in terms of the parameters of the circuit (b). The inversion of these equations provides the opposite transformation. Circuit (c), much simpler, is usually the one used to study the behaviour of the structure and perform parameter extractions.

Regarding the CSRR-based structure (Fig. 8(d)), a new and improved circuit model has also been recently proposed. It is the circuit shown in Fig. 8(e), which provides a more accurate description of the behaviour of the structure and is able to explain certain discrepancies that the previous one, which is shown in Fig. 8(f), presented. In the circuit (e), the resonator is modelled by the resonant tank formed by L_c and C_c, the line parameters are L and C_L and the gap is modelled by the π-structure formed by C_s and C_f, which take into account the series and the fringing capacitances due to the presence of the capacitive gap. The circuit (f) is perfectly able to reproduce the behaviour of the structure, as can be corroborated in Fig. 7(b), where its response is compared with the electromagnetic simulation of the structure. Nevertheless, circuit (e) can be transformed into circuit (f), by means of the following equations:

$$C_{par} = C_f + C_L \tag{7}$$

$$2C_g = 2C_s + C_{par} \tag{8}$$

$$C = \frac{C_{par}(2C_s + C_{par})}{C_s} \tag{9}$$

so that the circuit (f), much simpler can substitute circuit (e) for a more straightforward work. Circuit (f), in which C was formerly interpreted merely as the coupling capacitance between the line and the resonator and C_g was the gap capacitance, didn't predict neither the important change that C experiences nor the change on the position of the transmission zero (given by expression (10)) that occurs when the capacitive gap is eliminated. However, expressions (7) to (9) predict perfectly this behaviour (Aznar et al, 2008c).

$$\omega_z = \frac{1}{\sqrt{L_c(C + C_c)}} \tag{10}$$

As has already been pointed out, circuits shown in Fig. 8(c) and (f) are able to reproduce the response of the structures (a) and (d). This is illustrated in Fig. 7, where the full wave simulations of both structures are compared with the responses of their equivalent circuit models. The values of the circuit elements employed to obtain the compared responses have been obtained applying a parameter extraction method suitable for each of the structures. The parameter extraction method consists on the imposition of several conditions obtained either from a simulated or a measured response of one of the structures in order to obtain the necessary conditions to obtain the values of all the parameters of the circuit. In both structures, the model is formed by five circuit elements, so the number of impositions must also be five. The first of the parameter extraction methods was the corresponding to the CSRR-based structure (Bonache et al., 2006b), whereas the method for the SRR-based structure was proposed later (Aznar et al., 2008d). The main differences between both methods are due to the fact that they are based on a T- and a π-model, respectively. In both methods, two structures are employed: the first one contains all elements and, in the second one, the capacitive gap (or the metallic strips in the case of the SRR-based unit cell) is eliminated (see Fig. 9). This allows the determination of the five circuit elements in both cases. In the example shown in Fig. 9 the extraction is carried out from the measurement of the fabricated structure and losses have been taken into account by means of R. Its value has been adjusted by tuning until matching of the insertion loss level is achieved.
In the first of the methods, corresponding with the CSRR-based structure modelled by a T-circuit model, the imposed conditions are the transmission zero frequency (given by expression (13)), the frequency at which the phase is 90° and the resonance frequency of the resonator. The two first frequencies can be directly identified from the representation of the

S_{21} parameter (from the magnitude and the phase), whereas the third one can be identified from the representation of the S_{11} parameter in the Smith Chart (see Fig.10).

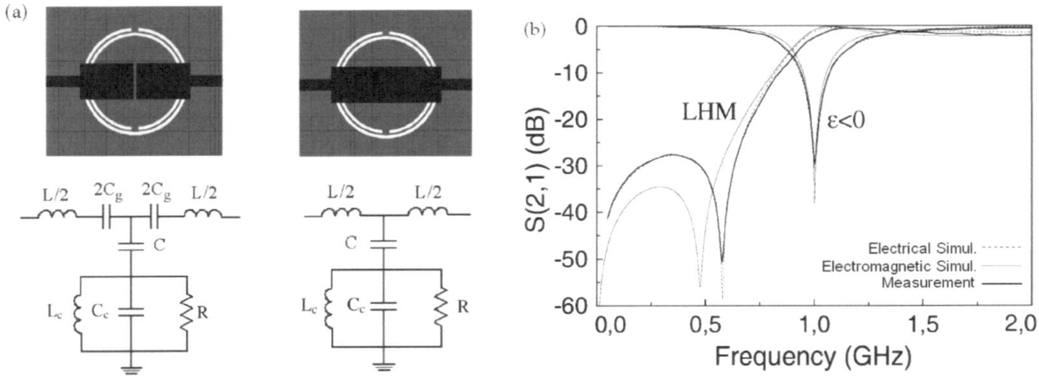

Fig. 9. (a) Layouts and circuit models of the structures employed in the parameter extraction method for the CSRR-based unit cell. (b) Frequency responses of the measurement, the electromagnetic simulation and the electric simulation employing the extracted parameters.

The resonance frequency of the CSRR has a particularity: at that frequency, the parallel branch opens and the input impedance is formed just by the impedance of the output port and the series branch impedance. This means that the real part of this impedance is the output port impedance (usually 50 ohm) and the imaginary part is given by the series branch impedance which consists of an inductance in the case of the structure without gap and an inductance and a capacitance in the case with gap.

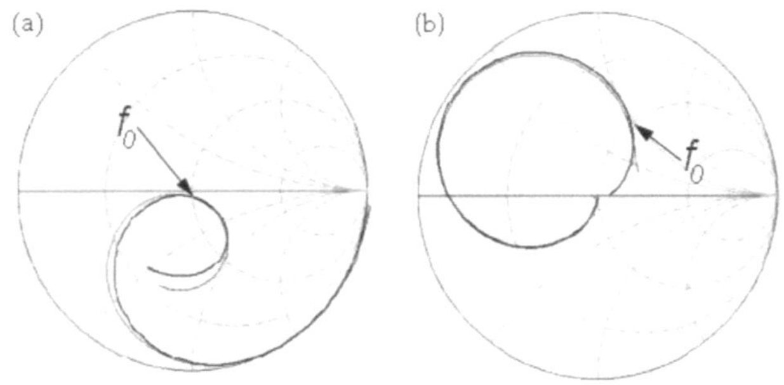

Fig. 10. Representation of the S_{11} parameter for the identification of the CSRR resonance frequency (a) for the complete (LH) structure (b) for the structure without gap (ε<0). Measurement and electric simulation.

The resonance frequency can be found at the point where the S_{11} curve meets the unit resistance circle in the Smith Chart. Given that the reactance of the input impedance at this point can be read from the chart, the calculation of the series branch elements can be easily carried out from the curves of both structures. In the case of the complete structure, at the resonance frequency the series branch is capacitive, whereas in the structure without gap it is inductive (see Fig. 10). By this means, we obtain three of the five necessary conditions, that is, the resonance frequency of the rings, which is $\omega_0 = 1/(L_cC_c)^{1/2}$, and the values of the reactance for both structures, which allow us to obtain the values L and C_g. As has been previously mentioned, the two other necessary conditions will be the transmission zero frequency and the 90º-phase frequency, at which the series and shunt impedances of the circuit have opposite signs (Z_s=-Z_p) (Bonache et al., 2006b). Once the parameters have been extracted, the electric simulation presents a very good fitting with the original curve (simulation or measurement), as figures 7, 9 and 10 corroborate.

In the case of the SRR-based structures, the parameter extraction method is very similar, but with the corresponding modifications for a π-circuit. As in the previous case, two different structures can be employed: a complete structure and a second one without metallic strips connecting the line with the ground planes. The 90º-phase and the transmission zero frequencies are imposed as well. However, the expression for the transmission zero frequency is in this case the following:

$$\omega_z = \frac{1}{\sqrt{L'_s C'_s}} \tag{11}$$

which is the frequency at which the series branch opens. One more useful frequency is that at which the series impedance nulls. When that happens, the input impedance is formed just by the impedance of the output port and the shunt impedance. This will correspond with the point in the Smith Chart in which the S_{11} curve crosses the unit conductance circle and the subsceptance of the input impedance can be directly read from the Smith Chart. This subsceptance corresponds with the shunt admitance of the structure, which is formed by an inductance and a capacitance in the case of the complete structure and only a capacitance in the structure without metallic strips. This allows the determination of the frequency at which the series impedance nulls, w_s, which can be obtained as:

$$\omega_s = \sqrt{\frac{1}{L'_s C'_s} + \frac{1}{L' C'_s}} \tag{12}$$

as well as the determination of the elements of the shunt impedance, L'_p and C. As occurred in the previous one, this parameter extraction method provides a very good fitting between the electrical response given by the extracted parameters and the original curve.

Similar parameter extraction methods have been recently applied to structures based on open split-ring resonators (OSRRs) and open complementary split-ring resonators (OCSRRs) (Fig. 11 (a) and (d), respectively). The followed strategy is very similar to the previous ones and it is even simpler, given that the equivalent circuit models consist in just three elements (see Fig. 11 (c) and (f)). In order to provide a good description of the structure, in both cases a small but not negligible phase shift must be taken into account at both sides of the resonator (Fig. 11 (b) and (e)). Taking this into account, the resulting equivalent circuit models are the ones shown in Fig. 11 (c) and (f). For the parameter extraction of the OSRR circuit, the imposed conditions are the following ones. As in the previous case, the frequency at which the series impedance nulls is found in the Smith Chart as the point in which the curve crosses the unit conductance circle, what occurs at:

$$\omega_{Zs=0} = \frac{1}{\sqrt{L'_s C_s}} \tag{13}$$

This gives us also the value of the shunt capacitance, which can be obtained from the value of the susceptance of the input impedance at that frequency. In addition, the reflection zero frequency, at which the characteristic impedance of the structure is matched to the ports (usually 50 Ω), is also identified and imposed, being the expression of the characteristic impedance:

$$Z_O = \sqrt{\frac{Z_s(\omega)Z_p(\omega)}{2Z_s(\omega)+Z_p(\omega)}} \tag{14}$$

and from it, the third of the parameters of the circuit can be obtained. The method employed for open complementary split-ring resonators is very similar. One of the imposed frequencies is the one at which the shunt impedance opens:

$$\omega_{Zp\to\infty} = \frac{1}{\sqrt{L'_p C'_p}} \tag{15}$$

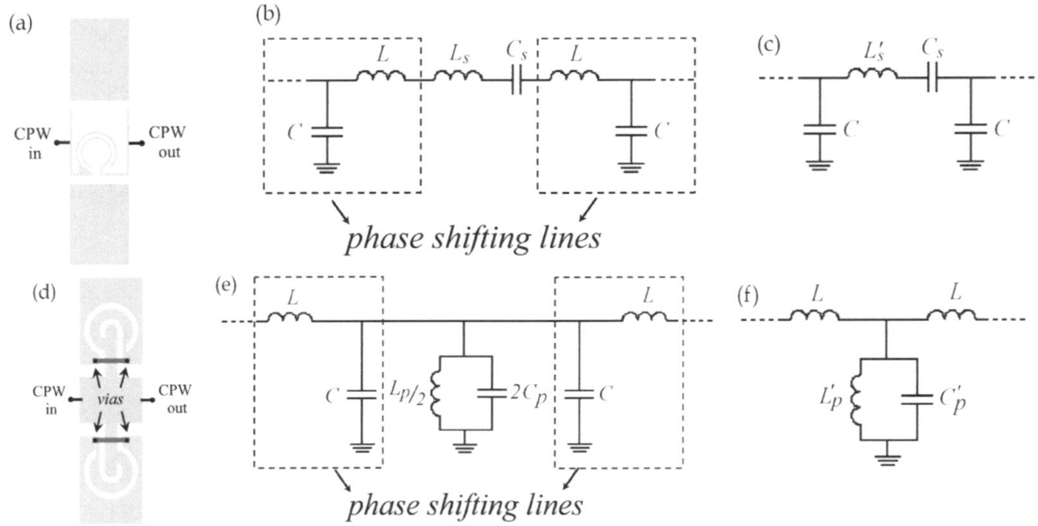

Fig. 11. (a) Layout of an OSRR implemented in coplanar waveguide technology. (b) Equivalent circuit model taking into account the phase shifting lines. (c) Simplified equivalent circuit model. (d) Layout of an OCSRR implemented in coplanar waveguide technology. (e) Equivalent circuit model taking into account the phase shifting lines. (f) Simplified equivalent circuit model.

This frequency can be found as the point where the S_{11} curve crosses the unit resistance circle in the Smith Chart and the value of the corresponding reactance of the input impedance gives us also the value of the series inductance, *L*. Furthermore, the characteristic impedance of the structure is forced to be matched to the ports at the reflection zero frequency, giving rise to the following expression:

$$Z_O = \sqrt{Z_s(\omega)(Z_s(\omega) + 2Z_p(\omega))} \tag{16}$$

With these three conditions, all parameters of the equivalent circuit model can be determined. For both structures, the frequency response of the circuit model properly fits the electromagnetic simulation of the structure (see Fig. 12).

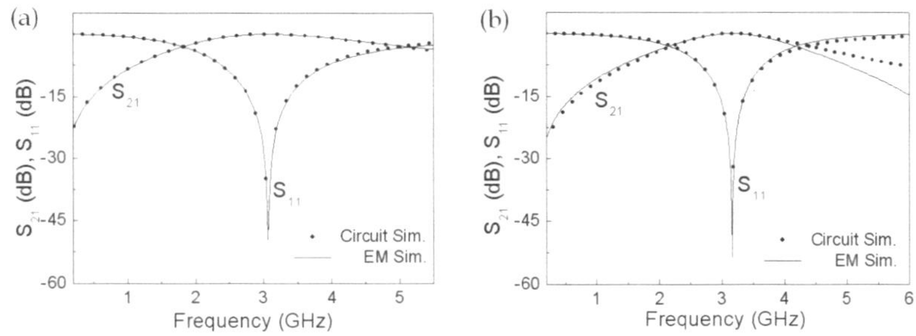

Fig. 12. Electromagnetic and electric simulations of the structures shown in Fig.11 (a) and (d). The electric parameters have been obtained by means of the described parameter extraction method. (a) Simulations of the OSRR structure. (b) Simulations of the OCSRR structure.

OSRR and OCSRR structures can be cascaded to implement composite right/left-handed transmission lines with controllable characteristics for their use in the design of microwave devices (Durán-Sindreu et al., 2009).
These results prove the validity of the proposed circuit models, as well as the diverse developed parameter extraction methods, which are an important tool in the design of this kind of artificial transmission lines.

4. Applications

Metamaterial transmission lines have two main characteristics which make them very interesting for the design of microwave devices. One of them is their small size, which allows device miniaturisation. The second one is the controllability of their electrical characteristics, that is, the characteristic (or Bloch) impedance Z_0 and the electrical length, βl. Such controllability is higher than in conventional transmission lines, where these magnitudes strongly depend on the line dimensions, determining the size of the final device requiring specific values of the line characteristics. Artificial transmission lines, however, offer the possibility of tailoring these properties to some extent (Gil et al., 2006), allowing the design of very competitive devices, even with new functionalities.
Simple examples of microwave devices including transmission lines with specific values are power dividers. They can be implemented by means of resonant-type metamaterial transmission lines and an important size reduction can be achieved (Gil et al., 2007b, Aznar et al., 2009b). Figure 13 shows one example of power dividers implemented by means of metamaterial impedance inverters, in which the impedance and the electrical length have been tailored to exhibit the required values at the design frequency (f_0=1.5 GHz). Figure 13(a) compares the layouts of two power dividers implemented in microstrip technology. The first one employs two resonant-type metamaterial transmission lines based on CSRRs as

impedance inverters with Z_0=70.71Ω and βl=90°. In the second one, the inverters are conventional transmission lines.

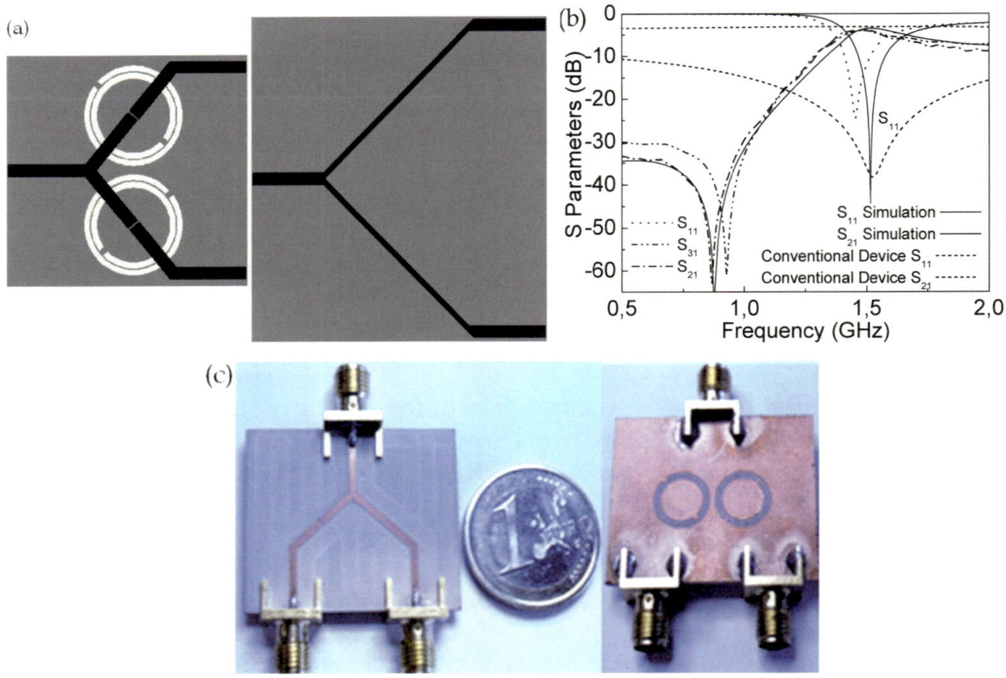

Fig. 13. (a) Comparison between the layouts of a conventional and a "metadivider" formed by two impedance inverters. (b) Frequency response of the power dividers shown in (a). (c) Photograph of the fabricated device. The device has been implemented employing a commercial substrate with thickness h=1.27mm and ε_r=10.2. The dimensions of the metamaterial inverters are: inverter length l=12.5mm, gap width g=0.16mm, CSRR external radius r_{ext}= 4.76mm, ring with c=0.5mm.

The metamaterial inverters have approximately half the length of the conventional ones so, thanks to their use, a 50% of size reduction can be achieved. Taking into account the frequency response of both devices (Fig. 13(b)), it can be seen that they exhibit similar loss levels (close to the ideal value, -3dB) at the design frequency, although the bandwidth of the metamaterial divider is narrower. This kind of dividers is, therefore, suitable for narrow band applications in which the size reduction is an important aspect. Other resonators and kinds of metamaterial transmission lines can be employed to implement similar power dividers (Aznar et al., 2009b, Gil et al., 2008b).

This kind of transmission lines can also be applied for bandwidth enhancement purposes (Sisó et al., 2007). Figure 14 shows a rat-race hybrid employing four metamaterial inverters. Three of them are right-handed inverters (βl=+90°), whereas a left-handed inverter with βl=-90° substitutes for the 270° inverter present in conventional hybrids.

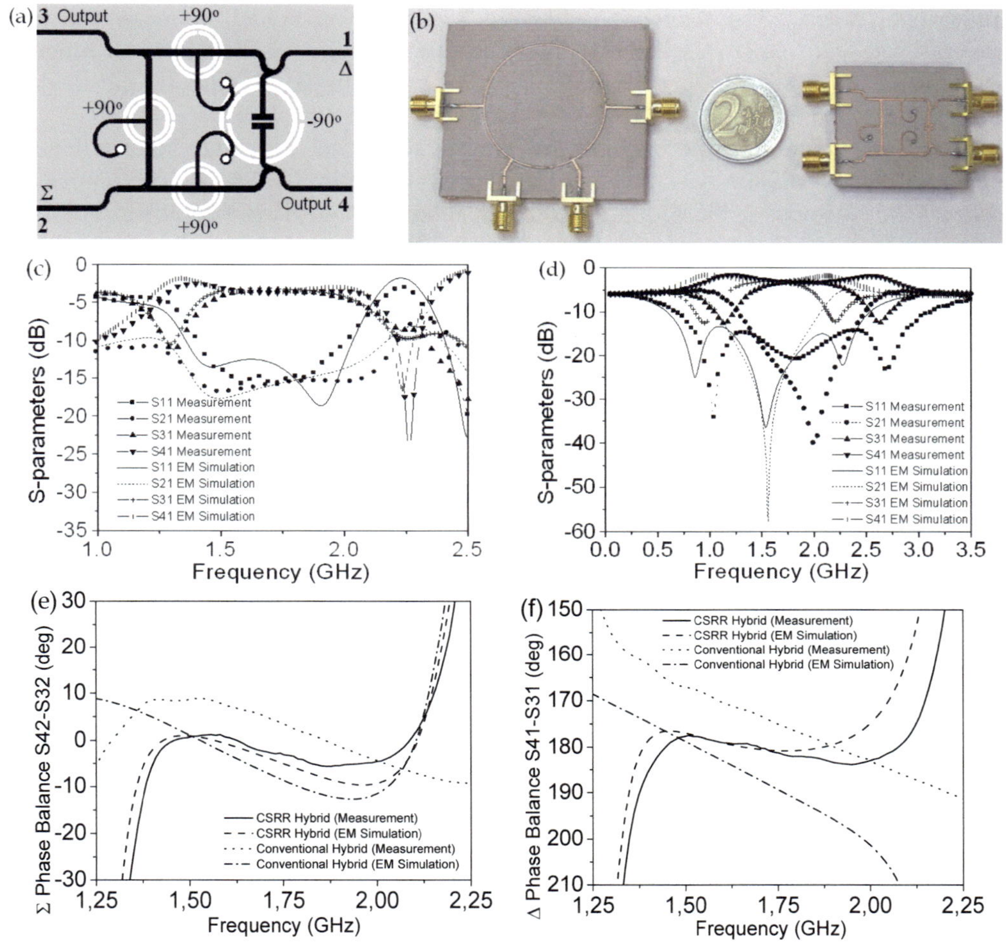

Fig. 14. (a) Layout of a rat-race hybrid coupler implemented by means of metamaterial transmission lines. The device has been implemented employing a commercial substrate with thickness h=635μm and ε_r=10.2. The relevant dimensions of the metamaterial inverters are, for the +90° inverters: stub length l=11.50mm, CSRR external radius r_{ext}=3.14mm, ring width c=0.34mm; for the -90° inverter l=10.2mm, gap separation g=0.37mm, gap width w=2.57, external resonator radius r_{ext}=5.1mm, ring width c=0.44mm. (b) Comparison between the sizes of a conventional and a metamaterial coupler. (c) S-parameters for the metamaterial coupler. (b) S-parameters for the conventional coupler. (e) Comparison of the phase balance of the Σ-port for the conventional and the metamaterial coupler. (f) Comparison of the phase balance of the Δ-port for the conventional and the metamaterial coupler.

As a result, the total size of the device is about 3 times smaller than a conventional one (see comparison in Fig. 14 (b)). Power splitting, isolation and matching of the metamaterial hybrid are similar to the ones of a conventional one (Fig. 14 (c) and (d)). Nevertheless, as can be seen in Fig. 14 (e) and (f), the phase balance presents a broader bandwidth than a conventional device. The dispersion diagram has been controlled to make the phase difference between the corresponding lines keep almost constant over a wider range. By these means, not only the size is minimized, but also the bandwidth of the device is improved.

As has been previously mentioned, the controllability of the line characteristics of metamaterial transmission lines opens the door to new application and functionalities (Sisó et al., 2009). Nowadays, there is an increasing interest in devices exhibiting multi-band operation. However, conventional transmission lines do not offer the possibility of designing such components working at arbitrary chosen frequencies. The manipulation of the dispersion diagram of metamaterial transmission lines allows the design of multi-band devices in which the operation frequencies can be chosen within certain margins. This can be achieved thanks to the composite behaviour of such transmission lines, given that the different operation frequencies can be chosen at the different transmission bands that these lines exhibit (Bonache et al., 2008; Sisó et al., 2008b).

If we consider the CSRR-based transmission line (Fig. 8(d)) and its equivalent circuit model (Fig. 8(f)), we can express the dispersion relation and the characteristic impedance of the structure as:

$$\cos(\phi) = 1 + \frac{Z_s(\omega)}{Z_p(\omega)} \tag{17}$$

$$Z_0 = \sqrt{Z_s(\omega)(Z_s(\omega) + 2Z_p(\omega))} \tag{18}$$

Impedance inverters are widely employed in microwave devices, so we will consider the design of a dual-band impedance inverter. It involves the imposition of the required phase (ϕ_1=-90°, ϕ_2=+90°) and impedance (Z_1, Z_2) values at the two frequencies of interest (f_1, f_2). The elements of the circuit model can be expressed in terms of these values. Given that the imposition of the phase and impedance represent four conditions, one more condition must be imposed in order to univocally determine the five parameters of the circuit model (Fig. 8(f)). This additional imposition can be, for example, the balance condition. In this case, the circuit elements can be expressed as:

$$L = \frac{2(Z_1\omega_1 + Z_2\omega_2)}{\omega_2^2 - \omega_1^2} \tag{19}$$

$$C_g = \frac{\omega_2^2 - \omega_1^2}{2\omega_1\omega_2(Z_1\omega_2 + Z_2\omega_1)} \tag{20}$$

$$C = \frac{2(Z_1\omega_2 + Z_2\omega_1)}{(Z_2^2 - Z_1^2)\omega_1\omega_2} \tag{21}$$

$$C_c = \frac{Z_1\omega_2 + Z_2\omega_1}{(\omega_2^2 - \omega_1^2)Z_1 Z_2} \tag{22}$$

$$C_g = \frac{(Z_1\omega_1 + Z_2\omega_2)(\omega_2^2 - \omega_1^2)Z_1 Z_2}{\omega_1\omega_2(Z_1\omega_2 + Z_2\omega_1)^2} \tag{23}$$

It can be observed that all parameters, except C, are positive as long as $\omega_2 > \omega_1$. The capacitance C, however, can be negative if $Z_2 < Z_1$, or even infinity if $Z_2 = Z_1$ (as most applications require). From this, it can be concluded that, employing the considered structure, the synthesis of impedance inverters with the same impedance value at two operating frequencies requires the use of non-balanced structures (Bonache et al., 2008).
An application example a of dual-band impedance inverter designed by means of CSRRs can be seen in Fig. 15 (Sisó et al., 2008a). The device has been designed to work at the mobile GSM bands (f_1=0.9GHz and f_2=1.8GHz). At these frequencies, the CSRR-based impedance inverter exhibits the required values of impedance (Z_0=35.35Ω) and phase (βl_1=-90° at f_1 and βl_2=+90° at f_2). Figure 15 (b) shows these two magnitudes, whereas Fig. 15 (c) shows the performance of the whole power divider. It can be seen that both, the inverter and the divider behave as expected at the design frequencies. The use of a metamaterial transmission line as impedance inverter allows the reduction of the final device size as well as the dual-band operation. The design of devices working at more than two bands is under study. Several generalized models for multiband operation have been proposed (Eleftheriades, 2007; Sisó et al., 2008c) and implementations by means of the CL-loaded approach have been already carried out (Papanastasiou et al., 2008).

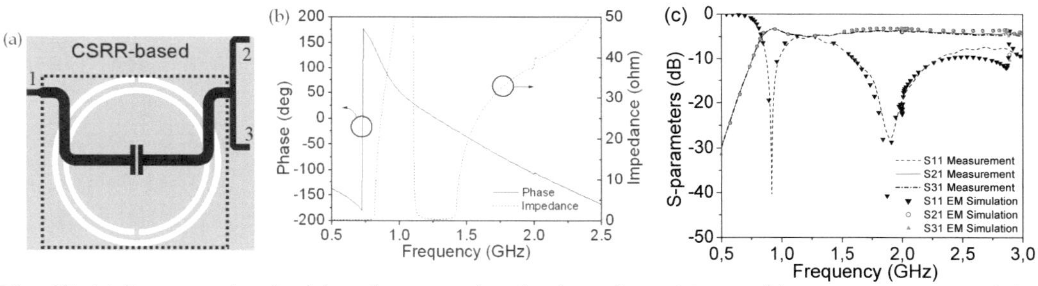

Fig. 15. (a) Layout of a dual-band power divider based on CSRRs. (b) Representation of the phase and the impedance of the inverter used in the implementation of the power divider shown in (a). (c) Simulated and measured frequency response of the power divider. The device has been implemented employing a commercial substrate with thickness h=635μm and ε_r=10.2. The metamaterial inverter dimensions are: marked active area A= 17.7mm x 15.8mm, resonator external radius r_{ext}=7.9mm, ring width c=0.5mm, line width w=1.0mm, gap separation g=0.27mm.

The fact that resonant-type metamaterial transmission lines exhibit a frequency selective behaviour suggests their application in filter design. This group has developed during the last years a wide work in this field, designing several kinds of filters based on subwavelength resonators (Gil et al., 2008b). The different responses that these structures provide allow the design of several kinds of filters, like low pass, high pass and band pass filters. Broadband filters, for example, can be designed making use of balanced transmission lines, whereas non balanced lines can be used for narrow band pass filter design. Hybrid structures, for example, can be used for both, narrow and broadband filters. The hybrid approach is based on the structure combining complementary resonators and capacitive gaps and includes shunt connected inductive stubs (see Fig.16a), which contribute to obtain inductive shunt impedance and provide more design flexibility. The addition of the stubs also creates an additional transmission zero above the left handed transmission band, which can be used in the design of filters with important frequency selectivity and very symmetric responses.

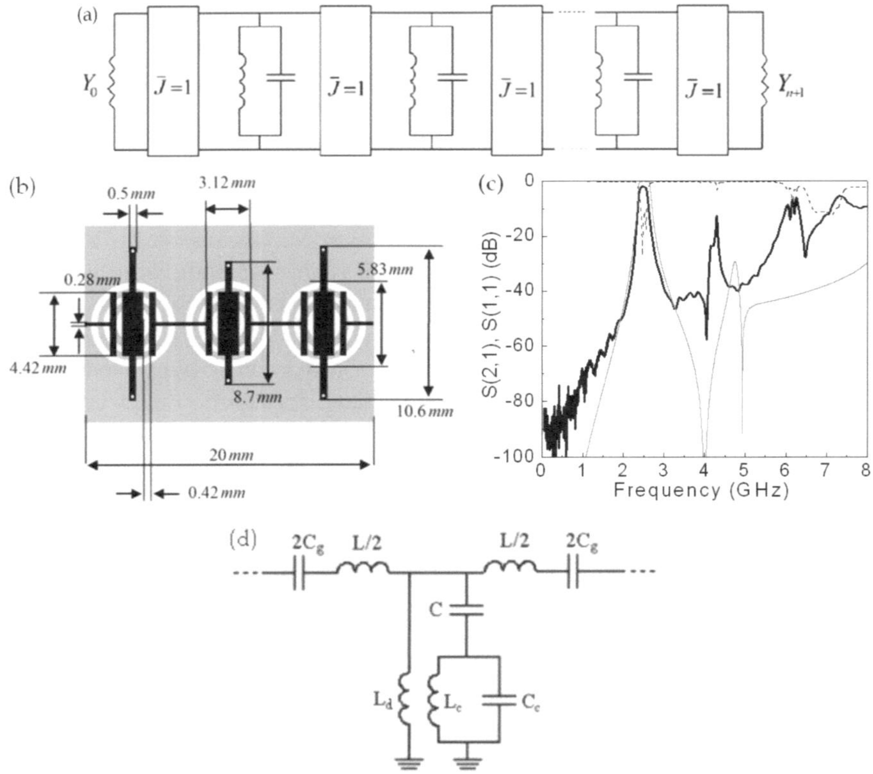

Fig. 16. (a) Band pass filter network with impedance inverters and LC resonant tanks as shunt resonators (b) Layout of a Chebyshev filter based on a hybrid structure and relevant dimensions. The device has been implemented employing a commercial substrate with thickness h=1.27mm and ε_r=10.2. (c) Electrically simulated and measured frequency response of the filter shown in (b). (d) Electrical circuit model of the hybrid unit cell.

The hybrid structure allows, for example, the design of band pass filters under standard (Chebyshev, for example) approximations with controllable bandwidth and compact dimensions (Bonache et al., 2006a). The design flexibility of these structures allows to control the position of the transmission zeros, the filter bandwidth, the ripple, etc so that a complete design methodology can be applied. The model shown in Fig. 16(a) can be applied to this structure as a basis for standard moderate and narrow band-pass filter design. When designing Chebyshev filters, the element values of the low pass filter prototype, g_i can be determined by the order and ripple level. Additionally, the bandwidth of each resonator (Δ_i) is set by the filter fractional bandwidth. These parameters, together with the filter central frequency and the position of the transmission zero allow the determination of all necessary circuit parameters. Each of the unit cells forming the filter is designed to, on the whole, exhibit characteristic impedance equal to the reference impedance of the ports (Z_0=50Ω) and phase ϕ=90° at the considered central frequency of the pass band. By this means, one unique unit cell acts as a resonator and exhibits the required phase without needing the addition of different stages acting as inverters and resonators, what involves a considerable size reduction. Furthermore, the series and shunt impedances have to be set to Z_s=-jZ_0 and Z_p=jZ_0, respectively, at f_0. The 3dB bandwidth:

$$\Delta = \frac{\omega_2 - \omega_1}{\omega_0} \tag{24}$$

and the transmission zero position:

$$f_z = \frac{1}{2\pi\sqrt{L_c(C_c + C)}} \tag{25}$$

are also imposed to determine all required elements. The bandwidth for a parallel resonant tank can be expressed in terms of its capacitance C_{eq} and inductance L_{eq} as:

$$\Delta = \frac{2}{Z_0}\sqrt{\frac{L_{eq}}{C_{eq}}} \tag{26}$$

These parameters can be related with the low pass prototype element values by the expressions

$$C_{eq} = \left[\frac{1}{FBW\omega_0 Z_0}\right] g_i \tag{27}$$

and

$$L_{eq} = \frac{1}{\omega_0^2 C_{eq}} \tag{28}$$

what allows us to obtain the following expression for the resonator bandwidth:

$$\Delta_i = \frac{2FBW}{g_i} \tag{29}$$

The two 3dB frequencies can be chosen to be equidistant to the central frequency, f_0. Considering that the series impedance is roughly constant within the pass band, the two 3dB frequencies can be considered to be those at which the shunt impedance is $Z_0/2$ and infinity, respectively, whereas it is Z_0 at f_0. These conditions can be expressed as:

$$\frac{L_d L_c \omega_1^3 (C + C_c) - L_d \omega_1}{L_c \omega_1^2 (C + C_c) - C\omega_1^2 L_d (L_c C_c \omega_1^2 - 1) - 1} = \frac{Z_o}{2} \tag{30}$$

$$L_c\omega_2^{\ 2}(C+C_c)-C\omega_2^{\ 2}L_d(L_cC_c\omega_2^{\ 2}-1)-1=0 \tag{31}$$

$$\frac{L_dL_c\omega_o^{\ 3}(C+C_c)-L_d\omega_o}{L_c\omega_o^{\ 2}(C+C_c)-C\omega_o^{\ 2}L_d(L_cC_c\omega_o^{\ 2}-1)-1}=Z_o \tag{32}$$

what, together with expression (25) leads us to the values of the shunt impedance elements. Finally, if the series inductance is neglected, the series capacitance can be expressed as:

$$C_g=\frac{1}{2Z_o\omega_o} \tag{33}$$

This methodology has given rise to very competitive filters. Figure 16 shows the layout and frequency response of an order-3 band pass filter designed following this strategy. As can be seen, very symmetric and selective responses can be obtained and the resulting devices are very compact. The layout of the device shown in Fig. 16a has a length of 0.4λ, λ being the signal wavelength at the central filter frequency.

Purely resonant and hybrid structures can be balanced to implement broadband filters (Selga et al., 2009; Gil et al., 2007c; Gil et al., 2007d). With such lines, broad responses are obtained as a result of forcing the frequency band gap that usually separates the left- and the right-handed transmission bands to disappear. As Fig. 17 shows, high pass and band pass filters can be implemented by means of balanced transmission lines. Purely resonant structures (Fig. 17(a)) exhibit a transmission zero below the pass band, which provides a sharp cut-off. Additionally, the rejection level can be improved increasing the number of unit cells. Figure 17(b) shows the frequency response of a high pass filter formed by 3 unit cells like the one shown in Fig. 17(a) in which the achieved rejection level is 40dB at the stop band (Selga et al., 2009). In such kind of filters, the upper limit of the band is not controlled. However, if it is necessary, additional resonators can be included in the design in order to reject the signal at the desired frequency or even to create attenuation poles within the transmission band (Gil et al., 2007d).

Hybrid structures, on the other hand, exhibit a pass band response and there is no need to include additional resonators. Figure 17(c) and (d) show the layout and performance of an UWB band pass filter based on hybrid unit cells including CSRRs. The filter was designed to satisfy quite restrictive specifications, like -80dB rejection level at 2GHz, or a total size smaller than 1cm^2. The resulting device was formed by four identical unit cells implemented on a thin (0.127mm thickness) substrate with ε_r=10.2 and satisfied all imposed specifications, including the size limitation (Gil et al., 2007c). In Fig. 17(c), the dashed rectangle marks 1cm^2 area. As can be seen in Fig. 17(d), the filter is very selective, exhibits low insertion loss level at the pass band and a wide rejection band above the transmission band.

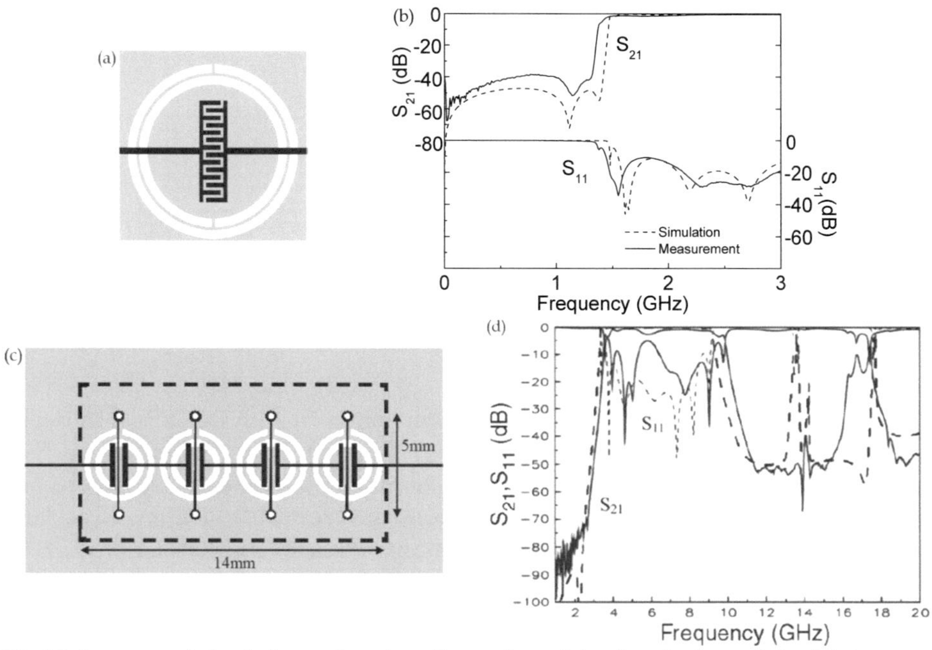

Fig. 17. (a) Layout of the balanced unit cell employed in the design of the high pass filter implemented in commercial substrate with ε_r=11 and thickness h=1.27mm. External radious of the resonators r_{ext}=4.17mm, ring width c=0.37mm, line width w=0.30mm, interdigital gap separation g=0.16mm. (b) Frequency response of high pass filter implemented by means of 3 unit cells like the one shown in (a). (c) Layout of a UWB band pass filter based on a balanced hybrid structure. Relevant dimensions are: line width w=0.126 mm, external radius of the resonator r_{ext}=1.68mm, ring width c=0.32mm and ring separation d=0.19 mm, inductor width is w_l=0.10mm and the gap distance g=0.4 mm (d) Simulated and measured frequency response of the filter shown in (c).

Other resonators can be employed to implement different kinds of filters, as is the case of OCSRRs. As has been previously mentioned, OCSRRs can be implemented in microstrip or coplanar technology. Figure 18 shows two examples of filters designed using these resonators. The filter in Fig. 18(a) is simply a low pass filter formed by 5 identical unit cells implemented with OCSRRs in microstrip technology (Aznar et al., 2009a). The design does not follow any standard approximation and a spurious band is present close to the cut-off frequency. In order to eliminate the first spurious band, a second stage formed by four unit cells is added to the low pass filter (Fig. 18(b)). As a result, the stop band is spurious free over a wider frequency range. Figure 18(c) includes the measured frequency responses of the two fabricated prototypes. As can be seen, a very sharp cut-off is obtained, as well as an important rejection level in the stop band, whereas insertion losses are very low in the pass band. Their frequency responses have been compared with the simulated response of a 0.5dB ripple elliptic filter with similar specifications, which must be an order 7 filter.

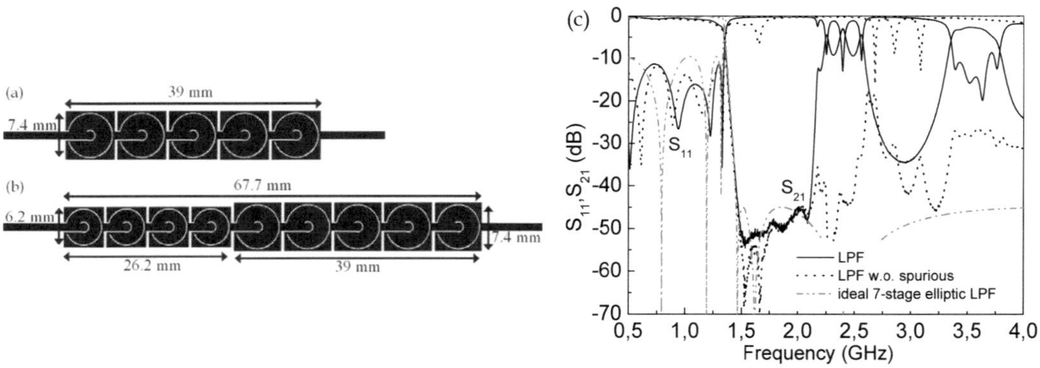

Fig. 18. (a) Layout of a microstrip low pass filter implemented with OCSRRs. The device was implemented in commercial substrate with ε_r=11 and h=1.27mm, the width of the gaps forming the OCSRRs is c=0.2mm, whereas d=2.46mm and r_{ext}=3.5mm. (b) Layout of a low pass filter including 4 OCSRRs for spurious with c=0.2mm, d=1.83mm, r_{ext}=2.9mm. (c) Measured responses of the filters (a) and (b) compared with the simulation of a 7-order elliptic filter.

Chebyshev filters can also be implemented by means of OCSRRs applying a similar methodology as the employed in the design of the filter shown in Fig. 16 (Vélez et al., 2009). This technique has been applied to the design of an order 5 filter with 0.1dB ripple, fractional bandwidth FBW=80% and central frequency f_0=5GHz. The resulting device can be observed in Fig. 19. Each section has been designed to satisfy the requirements to obtain the desired filter performance. This requires the implementation of the central unit cell by means of two resonators given that a smaller shunt inductance is required.

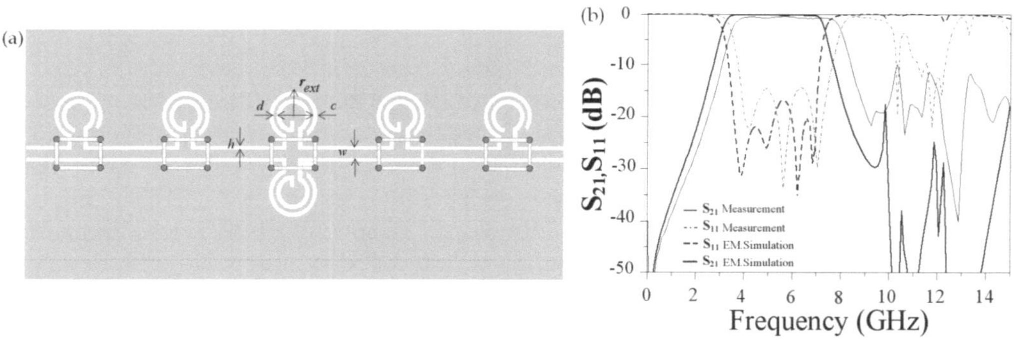

Fig. 19. (a) Layout of Chebyshev band pass filter based on OCSRRs implemented in coplanar waveguide technology. The device has been implemented in commercial substrate with ε_r=10.2 and 635μm thickness and c=0.22mm, d=0.15mm, r_{ext}=1.45mm, w=0.38mm, h=0.17mm.(b) Measured and simulated frequency response of the filter shown in (a).

The total length of the final device is 27mm, what corresponds to 0.9 times the wavelength of the signal at the central filter frequency. As can be seen, the filter exhibits good values of the insertion and return losses within the pass band. These two filters are some of the application examples of OCSRRs for the implementation of microwave devices, which can be designed in coplanar and microstrip technologies.

5. Conclusion

In this chapter, different kinds of resonant-type metamaterial transmission lines based on subwavelength resonators have been presented and studied. There are several types of resonators which allow their use in the implementation of this kind of artificial transmission lines and their small size is exploited in order to achieve device miniaturisation. Besides their small size, metamaterial transmission lines allow the control of their electrical characteristics, opening the door to very competitive and innovative application possibilities. Among these applications, some microwave devices based on resonant-type metamaterial transmission lines have been shown. Power dividers, hybrid couplers and filters are some of the components which can be implemented by means of these transmission lines. The designed devices have compact dimensions and good performances, which are comparable to those of conventional devices. In some cases, there are some restrictions, like bandwidth, making such devices only suitable for narrow band applications. However, in other cases, the performance of conventional devices is even beaten by metamaterial devices. The manipulation of the dispersion diagram (also known as "dispersion engineering") is one of the tools that allows the improvement of certain performances or even to achieve new functionalities, like is the case of multi-band operation. Contrary to what occurs with conventional transmission lines, the use of metamaterial lines allows to choose arbitrary frequencies in multi-band applications. The manipulation and design of resonant-type metamaterial transmission lines requires equivalent circuit models providing a good description of the structures. These models have been presented, together with the parameter extraction methods that provide the circuit model parameters from the frequency response of the structure. These parameter extraction methods, besides being a very useful design tool, allow the corroboration of the proposed circuits as correct models for the corresponding structures. Work is in progress in the design of new devices based on the presented or new structures. New and more efficient design tools are being developed, as well as devices with new functionalities are being studied.

6. References

Alici, K. B.; Bilotti, F.; Vegni L. & Ozbay, E. Miniaturized negative permeability materials, *Applied Physics Letters*, vol. 91, August 2007, pp. 071121-3, ISSN 0003-6951.

Aznar, F.; Bonache, J. & Martín, F., Improved circuit model for left-handed lines loaded with split ring resonators, *Applied Physics Letters*, vol. 92, February 2008, pp. 043512-3, ISSN 0003-6951.

Aznar, F.; García-García, J.; Gil, M.; Bonache J. & Martín, F. Strategies for the miniaturization of metamaterial resonators, *Microwave and Optical Technology Letters*, vol. 50, May 2008, pp. 1263-1270, ISSN 1098-2760.

Aznar, F.; Gil, M.; Bonache, J. & Martín, Modelling metamaterial transmission lines: a review and recent developments, *Opto-Electronics Review*, vol. 16, September 2008, pp. 226-236, ISSN 1230-3402.

Aznar, F.; Gil, M.; Bonache, J.; Jelinek, L.; Baena, J. D.; Marqués, R. & F. Martín, Characterization of miniaturized metamaterial resonators coupled to planar transmission lines through parameter extraction, *Journal of Applied Physics*, Vol 104, December 2008, 114501, ISSN 0021-4922.

Aznar, F.; Vélez, A.; Bonache, J.; Menés J. & Martín, F. Compact lowpass filters with very sharp transition bands based on open complementary split ring resonators, *Electronics Letters*, Vol. 45, No. 6, March 2009, pp. 316-317, ISSN 0013-5194.

Aznar, F.; Gil, M.; Bonache, J. Valcarcel, A. & Martín, Miniaturization of narrow-band power dividers by using CPW metamaterial transmission lines, *Microwave and Optical Technology Letters*, Vol. 51, No. 4, April 2009, pp. 926-929, ISSN 0895-2477.

Baena, J. D.; Bonache, J.; Martín F.; Sillero, R. M.; Falcone, F.; Lopetegi, T.; Laso, M. A. G.; García-García, J.; Gil, I.; Portillo, M. F. & Sorolla, M. Equivalent-circuit models for split-ring resonators and complementary split-ring resonators coupled to planar transmission lines, *IEEE Transactions on Microwave Theory and Techniques*, vol. 53, April 2005, pp. 1451-1461, ISSN 0018-9480.

Bilotti, F.; Toscano, A.; Vegni, L.; Aydin, K.; Alici K. B. & Ozbay, E. Equivalent-circuit models for the design of metamaterials based on artificial magnetic inclusions, *IEEE Transactions on Microwave Theory and Techniques*, vol. 55, December 2007, pp. 2865-2873, ISSN 0018-9480.

Bonache, J.; Gil, I.; García-García, J. & Martín, F. Novel microstrip band pass filters based on complementary split rings resonators, *IEEE Transactions on Microwave Theory and Techniques*, vol. 54, January 2006, pp. 265-271, ISSN 0018-9480.

Bonache, J.; Gil, M.; Gil, I.; Garcia-García, J. & Martín, F., On the electrical characteristics of complementary metamaterial resonators, *IEEE Microwave and Wireless Components Letters*, vol. 16, October 2006, pp. 543.545, ISSN 1531-1309.

Bonache, J.; Sisó, G.; Gil, M.; Iniesta, A.; García-Rincón J. & Martín, F. Application of Composite Right/Left Handed (CRLH) Transmission Lines based on Complementary Split Ring Resonators (CSRRs) to the Design of Dual-Band Microwave Components, *IEEE Microwave and Wireless Components Letters*, Vol. 18 , Issue 8, August 2008, pp. 524-526, ISSN 1531-1309.

Caloz C. & Itoh, T. Application of the transmission line theory of left-handed (LH) materials to the realization of a microstrip "LH line", *Proceedings of IEEE Antennas and Propagation Society International Symposium*, vol. 2,pp. 412-415, ISBN 0-7803-7330-8, San Antonio, Texas, (USA), June 2002.

Caloz, C. & Itoh, T. (2005). *Electromagnetic Metamaterials: Transmission Line Theory and Microwave Applications*, Wiley Interscience, ISBN 978-0-471-66985-2.

Durán-Sindreu, M.; Aznar, F.; Vélez, A.; Bonache, J. & Martín, F. New composite, Right/Left handed transmission lines based on electrically small, open resonators, *Proceedings of IEEE MTT-S International Microwave Symposium*, pp. 45-48, ISBN 978-1-4244-2804-5, Boston (MA), USA, June 2009.

Eleftheriades, G. V.; Iyer A. K. & Kremer, P. C. Planar negative refractive index media using periodically L-C loaded transmission lines, *IEEE Transactions on Microwave Theory and Techniques*, vol. 50, December 2002, pp. 2702-2712, ISSN 0018-9480.

Eleftheriades, G. V. A Generalized Negative-Refractive-Index Transmission-Line (NRI-TL) Metamaterial for Dual-Band and Quad-Band Applications, *IEEE Microwave and Wireless Components Letters*, vol. 17, 2007, pp. 415-417, ISSN 1531-1309.

Falcone, F.; Lopetegi, T.; Laso, M. A. G.; Baena, J. D.; Bonache, J.; Beruete, M.; Marqués, R.; Martín, F. & Sorolla, M. Babinet principle applied to the design of metasurfaces and metamaterials, *Physical Review Letters*, vol. 93, November 2004, 197401, ISSN 1079-7114.

Gil, M.; Bonache, J.; Gil, I.; García-García, J. & Martín, F., On the transmission properties of left handed microstrip lines implemented by complementary split rings resonators, *Int. Journal Numerical Modelling: Electronic Networks, Devices and Fields*, vol. 19, March 2006 , pp 87-103, ISSN 0894-3370.

Gil, M.; Bonache, J.; Selga, J.; García-García, J. & Martín, F. Broadband resonant type metamaterial transmission lines, *IEEE Microwave and Wireless Components Letters*, vol. 17, February 2007, pp. 97-99, ISSN 1531-1309.

Gil, M.; Bonache, J.; Gil, I.; García-García, J. & Martín, F., Miniaturization of planar microwave circuits by using resonant-type left handed transmission lines, *IET Microwave Antennas and Propagation*, Vol.1, February 2007, pp. 73-79, ISSN 1751-8725.

Gil, M.; Bonache, J.; García-García, J.; Martel, J. & Martín, F. Composite right/left-handed metamaterial transmission lines based on complementary split-rings resonators and their applications to very wideband and compact filter design", *IEEE Transactions on Microwave Theory and Techniques*, Vol. 55, No. 6, June 2007, pp. 1296-1304, ISSN 0018-9480.

Gil, M.; Bonache, J. & Martín, F. Metamaterial filters with attenuation poles in the pass band for ultra wide band applications, *Microwave and Optical Technology Letters*, Vol. 49, Issue 12, December 2007, pp 2909-2913, ISSN 0895-2477. .

Gil, M.; Bonache, J. & Martín, F. Synthesis and applications of new left handed microstrip lines with complementary split-ring resonators etched on the signal strip, *IET Microwaves, Antennas & Propagation*, Vol. 2, Issue 4, June 2008 pp. 324 - 330, ISSN 1751-8725.

Gil, M.; Bonache, J. & Martín, F. Metamaterial Filters: A Review, *Metamaterials*, Vol. 2, Issue 4, December 2008, pp 186-197, ISSN 1873-1988.

Martel, J.; Marqués, R.; Falcone, F.; Baena, J. D.; Medina, F.; Martín F. & Sorolla, M. A new LC series element for compact bandpass filter design, *IEEE Microwave and Wireless Components Letters*, vol. 14, May 2004, pp. 210-212, ISSN 1531-1309.

Marqués, R.; Mesa, F.; Martel, J. & Medina, F. Comparative analysis of edge- and broadside-coupled split ring resonators for metamaterial design - Theory and experiments, *IEEE Transactions on Antennas and Propagation*, vol. 51, October 2003, pp. 2572-2581, ISSN 0018-926X.

Marqués, R.; Martín, F. & Sorolla, M. (2008). *Metamaterials with Negative Parameters*, John Wiley & Sons, Inc., ISBN 978-0-471-74582-2.

Martín, F.; Bonache, J.; Falcone, F.; Sorolla M. & Marqués, R. Split ring resonator-based left-handed coplanar waveguide, *Applied Physics Letters*, vol. 83, December 2003, pp. 4652-4654., ISSN 0003-6951.

Papanastasiou, A. C.; Georghiou, G. E. & Eleftheriades, G. V. A quad-band Wilkinson power divider using generalized NRI transmission lines, *IEEE Microwave and Wireless Components Letters*, vol. 18, August 2008, pp. 521-523, ISSN 1531-1309.

Schurig, D.; Mock, J. J.; Justice, B. J.; Cummer, S. A.; Pendry, J. B.; Starr A. F. & Smith, D. R. Metamaterial electromagnetic cloak at microwave frequencies, *Science*, vol. 314, 2006, pp. 977-980, ISSN 0036-8075.

Selga, J.; Aznar, F.; Vélez, A.; Gil, M.; Bonache, J. & Martín, F. Low-pass and high-pass microwave filters with transmission zero based on metamaterial concepts, *Proceedings of IEEE International Workshop on Antenna Technology (iWAT2009): Small Antennas and Novel Metamaterials*, DOI 10.1109/IWAT.2009.4906914, March 2009, Santa Monica (CA), USA.

Sisó, G.; Bonache, J.; Gil, M.; García-García, J. & Martín, F, Compact rat-race hybrid coupler implemented through artificial left handed and right handed lines, *Proceedings of the IEEE MTT-S Int'l Microwave Symposium*, pp. 25-28, ISBN: 1-4244-0688-9, Honolulu, Hawaii (USA), June 2007.

Sisó, G.; Bonache, J. & Martín, F., Dual-Band Y-Junction Power Dividers Implemented Through Artificial Lines Based on Complementary Resonators, *Proceedings of the IEEE MTT-S Int'l Microwave Symposium*, pp. 663-666, ISBN: 978-1-4244-1780-3 Atlanta (USA), June 2008.

Sisó, G.; Bonache, J.; Gil, M. & Martín, F, Enhanced bandwidth and dual-band microwave components based on resonant-type metamaterial transmission lines, *International Journal of Microwave and Optical Technology*, Vol. 3, Issue 3, July 2008, pp. 345-352, ISSN 1553-0396.

Sisó, G.; Gil, M.; Aznar, F.; Bonache, J. & Martín, F. Generalized Model for Multiband Metamaterial Transmission Lines, *IEEE Microwave and Wireless Components Letters*, Vol. 18, Num. 11, November 2008, pp 728-730, ISSN 1531-1309.

Sisó, G.; Gil, M.; Aznar, F.; Bonache, J. & Martín, F. Dispersion engineering with resonant-type metamaterial transmission lines, *Laser & Photonics Review*, vol. 3, Issue 1-2, February 2009, pp. 12-29, ISSN 1863-8880.

Smith, D. R.; Padilla, W. J.; Vier, D. C.; Nemat-Nasser S. C. & Schultz, S. Composite medium with simultaneously negative permeability and permittivity, *Physical Review Letters*, vol. 84, May 2000, pp. 4184-4187, ISSN 0031-9007.

Vélez, A.; Aznar, F.; Bonache, J.; Velázquez-Ahumada, M. C.; Martel J. & Martín, F. Open complementary split ring resonators (OCSRRS) and their application to wideband cpw band pass filters, *IEEE Microwave and Wireless Components Letters*, vol. 19, April 2009, pp. 197-199, ISSN 1531-1309.

Vélez, A.; Bonache, J. & Martín, F. Metamaterial transmission lines with tunable phase and characteristic impedance based on complementary split ring resonators, *Microwave and Optical Technology Letters*, Vol. 51 Issue 8, August 2009, pp. 1966-1970, ISSN 0895-2477.

Veselago, V. G. The electrodynamics of substances with simultaneously negative values of ε and μ, *Soviet Physics Uspekhi*, January 1968, pp. 509-514, ISSN 0038-5670.

20

Compact CPW Metamaterial Resonators for High Performance Filters

Ibraheem A. I. Al-Naib[1], Christian Jansen[1] and Martin Koch[2]
[1]*Technische Universität Braunschweig*
[2]*Philipps-Universität Marburg*
Germany

1. Introduction

Since the origin of physics, scientists were concerned with the interaction of electromagnetic waves with matter. The electromagnetic properties of a material are commonly described by the electric permittivity epsilon (ε) and the magnetic permeability (μ). Soon it occurred that these dielectric parameters would have to be complex quantities to account for propagation losses and phenomena such as birefringence showed that they would have to be considered tensors in order to describe anisotropic behaviour.

Recently, the invention of artificial materials, which consist of periodically arranged, resonant, metallic sub-wavelength elements, led to a new class of materials offering a custom tailored dielectric response in certain frequency bands of interest. Today, many applications benefit from the unique electromagnetic properties that such artificial materials, also called metamaterials (MTM), offer. Especially planar metamaterials, which are easily facilitated into existing microwave circuitry, are of high practical interest, e.g. for high performance filters, antennas, and other microwave devices (Caloz & Itoh, 2005, Eleftheriades & Balmain, 2005, Marques et al., 2008). Aside from the device performance, miniaturization is a key issue in the design of metamaterial resonators, as a high integration density is a mandatory prerequisite to compete in mass-markets such as wireless communications.

In this book chapter, we will review some recently proposed planar metamaterial resonator concepts, illuminating their strengths and weaknesses in comparison to existing approaches, e.g. the ones discussed in (Marques et al., 2008). We will focus on structures integrated on coplanar waveguides (CPW) as this technology offers some distinct design advantages compared to conventional microstrip lines, e.g. the easy realization of shunts and the possibility of mounting active and passive lumped components (Simons, 2001, Wolff, 2006). The remainder of this chapter is structured as follows. We will start with a short introduction to metamaterials and discuss some of the most prominent applications. After this general section, we will focus on CPW based metamaterial filter concepts.

The first concept which falls under this category is the complementary split ring resonator (CSRR) with and without bandwidth modifying slots, as introduced in (Ibraheem & Koch, 2007). These resonators provide a distinct stopband characteristic, which can be adjusted by

modifying the slot lengths, offering great design flexibility. However, the stopband response suffers from a spurious rejection band close to the main resonance, which has its origin in the differing electrical lengths of the inner and the outer resonator arm. This issue leads us to the complementary u-shaped split resonator (CUSR) (Al-Naib & Koch, 2008a), which cancels the spurious resonance by equalizing the electrical length of both resonator arms. Furthermore we will show that by combining a CSRR and a split ring resonator with strip lines in series, compact bandpass filters can be obtained, which could be very useful, e.g. in front end filter designs (Al-Naib et al., 2008).
Apart from the CSRR based structures, we will also discuss circular multiple turn complementary spiral resonators (CSRs) (Al-Naib & Koch, 2008b), which enable extremely small electrical footprint filters as each turn elongates the effective resonator length, thus lowering the resonance frequency.

1.1 The basics of Metamaterials

To better understand what sets metamaterials apart from ordinary media, Fig. 1 illustrates a schematic with μ and epsilon on the x- and the y-axis, respectively.

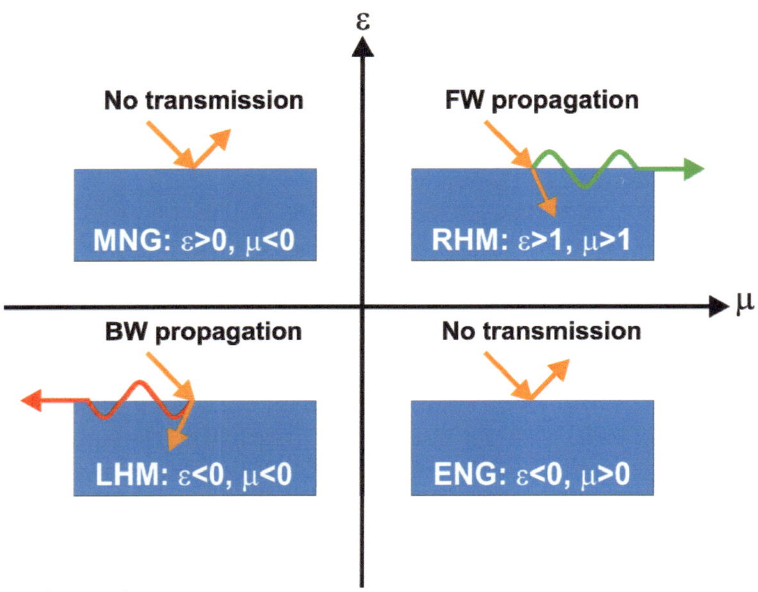

Fig. 1. Material classification.

Four regimes can be identified: The best known is the one where both ε and μ assume values larger than one. Most materials encountered in nature fall inside this upper right quadrant and are also referred to as right-handed media (RHM) or double positive material (DPS). Exceptions are ferrites, which are found in the upper left quadrant. In these materials μ becomes negative (MNG) while epsilon remains at positive values. The inverse scenario, where epsilon is negative (ENG) while μ remains positive is found in plasmas, which are grouped in the lower right corner of Fig. 1. The untapped lower-left corner contains left-handed materials, not yet discovered in nature.

Metamaterials can access all above mentioned regimes, at least in a limited spectral region, enabling new exciting applications such as superlenses or cloaking devices but also improving existing ones such as filter or antenna structures. The basic idea behind metamaterials lies in the combination of electric and/or magnetic resonances in such a way that ε and μ take the desired values in a certain frequency band. To illustrate this approach, we will now discuss a negative ε and a negative μ metamaterial and then combine both to obtain a left-handed medium.

1.1.1 ENG Metamaterials

Practically, ENG metamaterials consists of thin metallic wires. In the late nineties of the last century, concepts similar to the one shown in Fig. 2a have been explored (Pendry et al., 1996). Pendry et al. showed that its behaviour can be explained by the plasma resonance inside the metallic rods (Pendry et al., 1996, Pendry et al., 1998). As illustrated in Fig. 2b, epsilon starts with negative values in the lower frequency range (and hence, only evanescent modes are allowed to propagate) and then transits to positive values in the higher frequency region, through the plasma frequency (f_{pe}). ENG structures with plasmonic response have been suggested for the realization of sub-wavelength antennas with enhanced radiation properties and waveguide miniaturization (Engheta & Ziolkowski, 2006, Erentok & Ziolkowski, 2005, Gay-Balmaz et al., 2002, Hrabar et al., 2005).

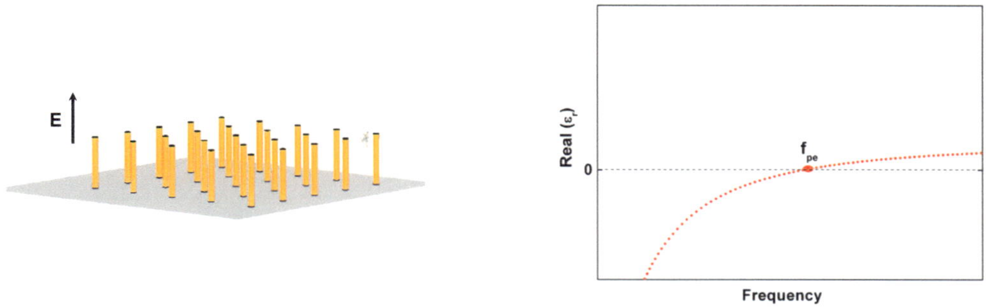

(a) (b)

Fig. 2. Wire medium with an applied electric field along the axes of the wires (a) and its effective electric permittivity value (b).

1.1.2 MNG Metamaterials

Aside from his work on metallic wire media, Pendry also proposed a novel type of magnetically excited resonator, the so called Split Ring Resonator (SRR), as shown in Fig. 3a (Pendry et al., 1999). The resonator consists of a pair of concentric rings, with slits etched in two opposing sides. By adequately exciting the SRR with a time varying magnetic field in the axial direction, a strongly resonant magnetic response can be observed as depicted in Fig. 3b.

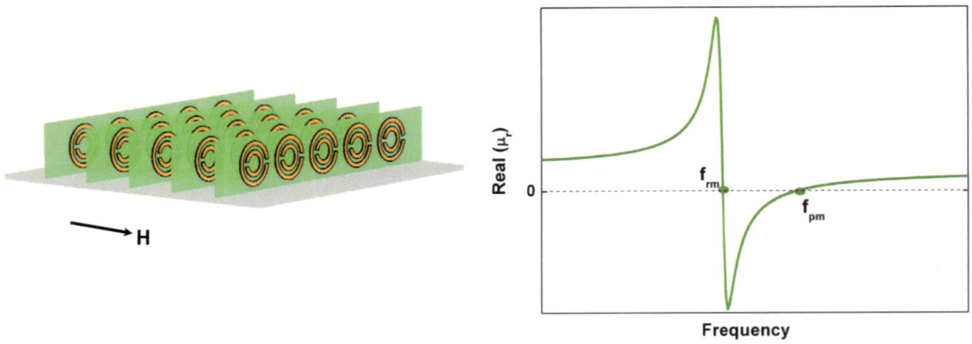

(a) (b)

Fig. 3. SRRs medium with magnetic field along their axis (a) and its analytical calculation of the effective magnetic permeability value (b).

1.1.3 LHM Metamaterial

In 2000, Smith et al. combined both the metallic wire ENG and the split ring resonator MNG media and were the first to experimentally observe a LHM (Shelby et al., 2001, Smith et al., 2000). Fig. 4a depicts a schematic of the prominent structure employed for these initial experiments. Fig. 4b shows the effective real part of ε and μ for the wire array and SRRs, respectively. f_{pe} and f_{pm} are the electric and magnetic plasma frequency while f_{rm} is the magnetic resonance frequency of the SRRs. The shaded area marks the resulting bandwidth between f_{rm} and f_{pe} for which a left handed behaviour with a negative effective refractive index is obtained.

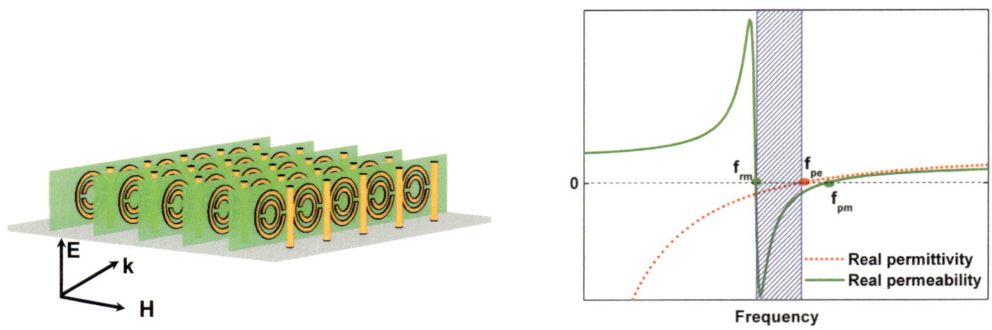

(a) (b)

Fig. 4. (a) Schematic of the LHM, resulting from the combination of an array of SRR particles with an array of thin wires (b) real part of effective permittivity (dotted) and permeability (solid) versus frequency.

1.2 Applications for metamaterials

In this section, we will briefly review the most prominent of the manifold suggested applications for metamaterials, in each case giving a short explanation of the benefit metamaterials can offer.

1. **Superlens -** Pendry proposed that a slab of LHM can be used as a lens which is free from all aberrations observed in a lens made with positive refractive index (Pendry, 2000). However, it was shown that very small deviations of the material parameters from the ideal conditions could lead to the excitation of resonances that cause deterioration of the performance of the lens. Nevertheless, scientists have been working to overcome different difficulties to improve the resolution (Ramakrishna & Grzegorczyk, 2008).
2. **Cloaking -** Another natural application for metamaterials is the development of gradient index media (Smith et al., 2005) because the value of the permittivity and permeability can be engineered at any point within the structure by adjusting the scattering properties of each unit cell (Driscoll et al., 2006, Greegor et al., 2005). By implementing complex gradients independently in the permittivity and permeability tensor components, it has been shown that an entirely new class of materials can be realized by the process of transformation optics (Leonhardt, 2006, Pendry & Smith, 2006). A recent example utilized metamaterials to form an "invisibility cloak" that was demonstrated to render an object invisible to a narrow band of microwave frequencies (Schurig et al., 2006).
3. **Scattering reduction** - metamaterials can be used for the reduction of electromagnetic wave scattering (Lagarkov & Kisel, 2001, Pacheco et al., 2002, Alu & Engheta, 2005). Recently a theoretical analysis based on Mie scattering was presented in (Alu et al., 2005) which indicated that metal, coated with metamaterials, has a drastically reduced scattering coefficient.
4. **Novel microwave components -** metamaterials can be employed as sub-wavelength resonators and zero phase delay lines. The advantage over that of RHM materials is the very small dimension of the resonator (Eleftheriades et al., 2004, Engheta, 2002). Moreover, low ε metamaterials can be employed to build high-gain antennas (Feresidis et al., 2005, Wang et al., 2006). Furthermore, compact antennas are realizable utilizing artificial magnetic conductors.

In the remainder of this chapter we will focus on planar metamaterials for planar microwave devices. The two main concepts for such metamaterials will be discussed in the following section.

2. Planar Microwave Metamaterials – A Brief Review

In order to bring metamaterial technology to microwave components, compatibility with planar circuit technology is mandatory. Two approaches have been introduced to meet this challenge: The first one employs a transmission line with integrated capacitive and inductive elements while the second one relies on planar metamaterial resonators loaded to a coplanar waveguide.

2.1 Transmission Line approach

A transmission line, loaded with reactive elements, such as capacitors and inductors, can be designed to exhibit capacitive effective series impedance while the effective shunt impedance remains inductive. The main advantage of such concept is its compatibility with conventional planar circuits. Fig. 5a illustrates the transmission line model of a TL-based metamaterials and Fig. 5b shows one of the proposed implementations (Caloz & Itoh, 2002). Several applications of TL-based metamaterials have been proposed and experimentally validated. For more information, the inclined reader is referred to (Caloz et al., 2002, Iyer & Eleftheriades, 2002, Oliner, 2002).

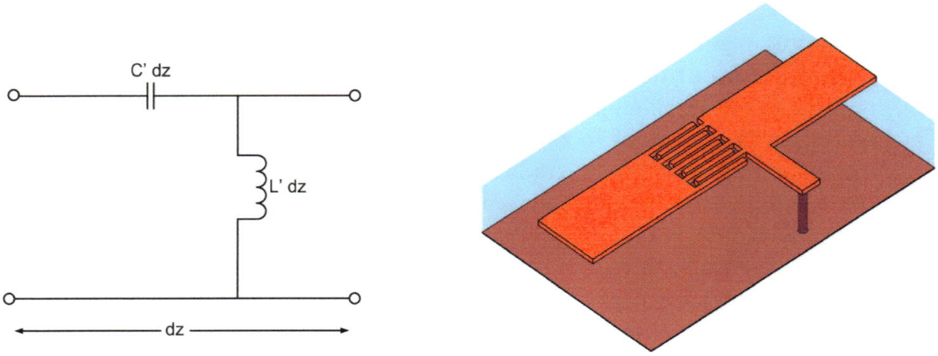

Fig. 5. Transmission line model for a TL-based metamaterials (a) and the proposed implementation (b).

2.2 CPW Approach

Coplanar waveguide technology offers a variety of advantages over the conventionally employed microstrip line. Among the benefits is the easy facilitation of shunts as well as the series surface mounting of active and passive devices. Furthermore, due to the absence of a ground plane a single metallization layer suffices which leads to reduced fabrication costs compared to microstrip technology.

Magnetic resonator structures, such as the SRR, can be employed as metamaterials if the exciting H-field is normal to the plane containing the resonator. In this operation mode, the induced currents will lead to the desired resonance of the magnetic permeability. Fig. 6 depicts the electric (E-) field and magnetic (H-) field in a CPW line. In the vicinity of the gaps lies an area where the H-field is normal to the CPW plane. Integrating magnetic resonators in this location should yield an efficient excitation, enabling a planar metamaterial structure. In the following we will discuss different implementations of this basic concept.

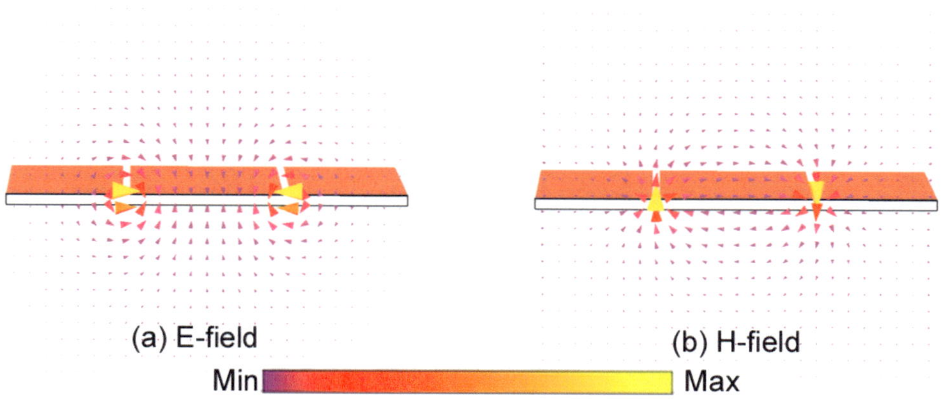

Fig. 6. The transverse electric- (a) and magnetic- (b) field in CPW.

A first implementation of a single metallization layer, CPW-based planar design introduced the SRRs directly on the slots of the CPW as depicted in the inset of Fig. 7a (Falcone et al., 2004). The gaps of the CPW are broadened to provide enough space to hold the SRRs. The SRRs are excited magnetically because the magnetic field is confined to the gaps. Fig. 7b shows the numerically obtained transmission parameters. The transmission depicts bandstop behaviour due to the magnetic resonance. Unfortunately, with regards to the return loss, the performance is quite poor because the structure is highly mismatched since the CPW line impedance is the ratio between the central conductor width and the air gap separation. Therefore, having the SRRs inside the slots of CPW puts tight restrictions to the line impedance.

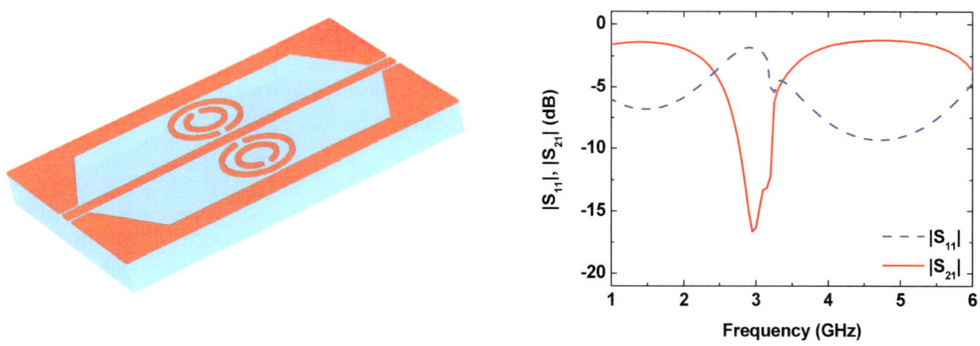

(a) (b)

Fig. 7. Single metallized CPW with SRRs inside the slots (a) and its S-parameters (b).

To overcome the mismatch problem, another approach has been proposed. Here, the SRRs are placed on the bottom side of the dielectric layer as shown in Fig. 8a (Martin et al., 2003a). In this case, the CPW can be designed to have almost the same impedance as if no SRRs were present. Four unit cells are needed to achieve good behaviour as shown in Fig. 8b.

However, the advantage of having a single metal layer is no longer maintained, reducing the applicability of this concept.

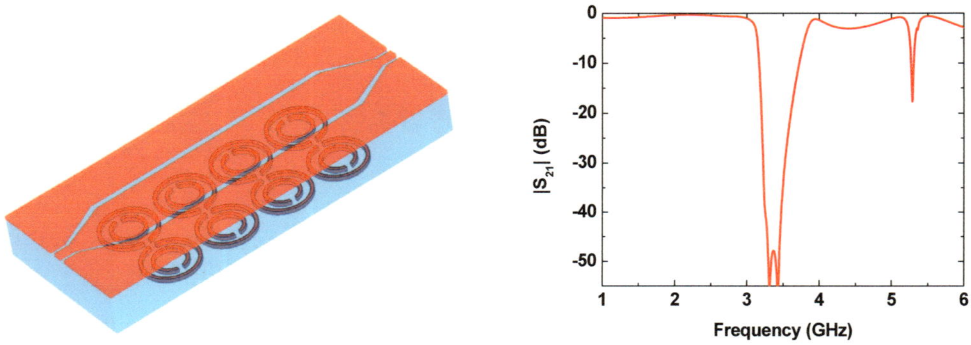

(a) (b)

Fig. 8. CPW with backside loaded SRRs (a) with its transmission characteristic (b).

3. Planar Microwave Metamaterials – Recent Advances

This section reviews novel planar metamaterial resonator concepts which have recently been proposed. The first concept is the complementary split ring resonator (CSRR). Unlike the structures introduced in sec. 3.1, which consist of two metal layers, the CSRRs have been integrated into the CPW layer to maintain the advantages single layered structures offer. However, the stopband response suffers from a spurious rejection band close to the main resonance. To resolve this issue, the complementary u-shaped split resonator (CUSR) discussed in section 3.2, which cancels the spurious resonance, has been introduced. Apart from the CSRR based structures, circular multiple turn complementary spiral resonators (CSRs) are discussed in section 3.4, which feature an extremely small electrical footprint as each turn elongates the effective resonator length. Finally, combining a CSRR and an SRR with strip lines in series leads to compact bandpass filters introduced in section 3.5. Such devices could be very useful, e.g. in front-end filter designs.

3.1 The Novel CSRR/CPW

As neither of the designs discussed in section 2.2 could satisfy the demand for an impedance matched, single layered, planar metamaterial, a new concept had to be developed. At the core of this concept stands the integration of complementary SRRs (CSRRs) into a CPW making use of the Babinet principle which leads to the structures shown in Fig. 9a & 9b. For clarity's sake, we will limit the discussion in this work to a single CSRR pair. However, cascading of the CSRRs is possible, resulting in even better performance. At the edges of the structure we added the CPW tapers to exclude measurement errors due to soldering connectors to the devices. The taper function was verified through simulation and experiment to provide maximum matching between the two sides of the CPW. A standard mask/photoetching technique is used to fabricate the structures using an FR-4 substrate (dielectric constant $\varepsilon_r = 4$, loss tangent $\tan\delta = 0.02$, thickness $h = 0.5$ mm). With a single pair of CSRRs we obtain a considerably higher suppression than reported in (Martin et al., 2003a).

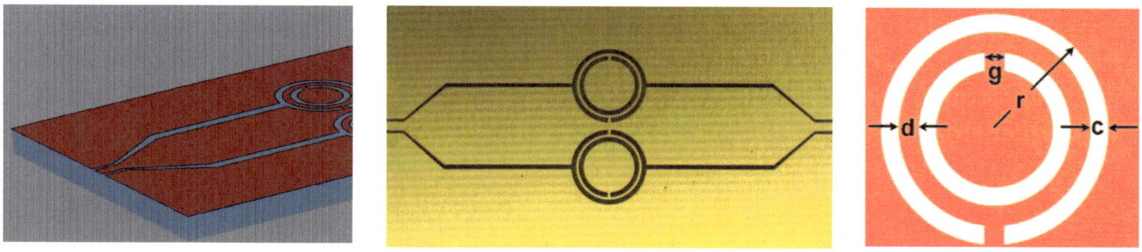

(a) (b) (c)

Fig. 9. (a) 3D layout of CPW/CSRR structure (b) top-view of the fabricated structure (c) schematic of CSRR with its dimensions.

First, a simulation of a single unit cell with periodic boundary conditions in propagation direction is performed using the Eigenmode solver of CST Microwave studio (CST). The dimensions of the unit cell are illustrated in the inset of Fig. 9c. An external radius of r = 3.6 mm, a width of c = 0.27 mm, a separation of d = 0.43 mm, and a length of the "metallic bridge" g = 0.43 mm are chosen so that the structures operate in the C-band. Calculating the dispersion by varying the phase shift in propagation direction between 0° and 180° reveals a photonic band gap between 4.2 GHz and 5.6 GHz (c.f. Fig. 10a). This bandgap is produced by the complementary rings which exhibit an effective negative dielectric permittivity. Hence, a stopband behaviour in the transmission magnitude with a centre frequency around 4.9 GHz is expected.

We use Ansoft HFSS software (HFSS) (a 3D full-wave solver based on the finite element method with adaptive iterative meshing) to simulate the transmission through the structures with a very high mesh resolution by specifying the maximum change in the S-parameters between two successive iterations to be 0.3%. Furthermore, a fine discrete sweep with a 0.02 GHz step size for the band between 4.5-5.5 GHz is performed, where sharp spectral transients are expected. The simulated magnitude response (dotted line) for the insertion loss is shown Fig. 10b. The insertion loss for the CSRR structure shows the expected stopband behaviour close to 5 GHz. Between 4.6 GHz and 5.5 GHz, an attenuation higher than 10 dB is achieved, which is in good agreement with the dispersion analysis.

An HP E8361A vector network analyzer (VNA) with a microstrip test fixture (Wiltron 3680) is employed to identify the S-parameters of the fabricated structure in the frequency band between 2 and 10 GHz. A thru-short-line (TRL) kit was used to calibrate the system. For the whole band of interest, the return loss is better than 28 dB. The measurements of the fabricated structure are shown in Fig. 10b (solid line). A good agreement between the measured S-parameters and the simulated HFSS results, confirming the dispersion analysis carried out with CST MWS. The small shifts in the resonance frequency can be ascribed to inhomogeneities in the ring dimensions.

Beneath the low-frequency limit of the artificial band gap, the structure exhibits excellent matching without any significant insertion loss. In the upper passband a good performance up to 7 GHz is achieved, but for higher frequencies (shaded area in Fig. 10) a spurious

resonance leads to an undesired dip in the transmission response at 9.1 THz. The origin of this dip and countermeasures to remove it are discussed in the following section.
Another remarkable aspect about the CSRR is the high rejection level in the forbidden band of nearly 30 dB, which is at least twice as high as the suppression of the SRR-based structures introduced in (Martin et al., 2003a). Thus, CSRRs provide an effective way to eliminate frequency parasitics in CPW structures. By cascading multiple CSRR unit cells even higher suppression levels can be achieved.

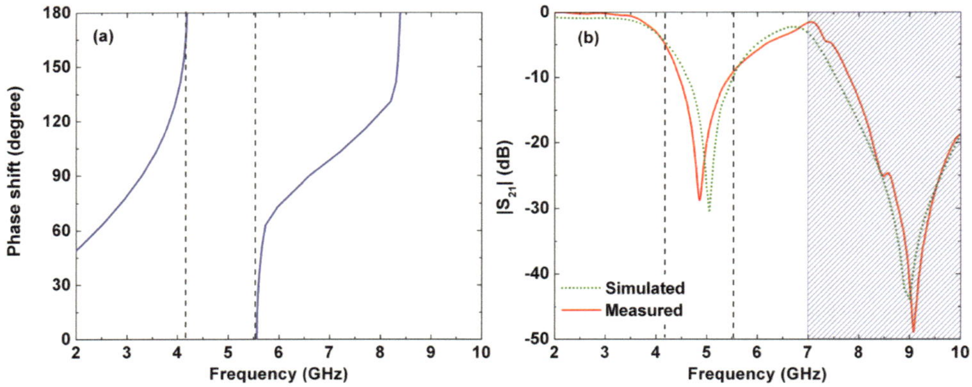

Fig. 10. (a) Dispersion analysis (b) simulated and measured insertion loss.

3.2 U-shaped resonators – Overcoming the spurious resonance

In the last section, we discussed the CPW-integrated CSRRs for use as a single metallization layer, high rejection stop band filter. Yet, the insertion loss of the CPW-CSRR suffers from a strong spurious stopband located on the high frequency side very close to the actual stopband. This spurious resonance results from a phenomenon called differentiating resonances (Ibraheem et al., 2008). At the core of this problem lies the slight difference in the length of the inner and the outer resonator arms. Thus, we proposed complementary u-shaped split resonator (CUSR) which is depicted in Fig. 11a (Al-Naib & Koch, 2008a). Here, the physical length of the u-shaped outer and inner rectangular resonator is identical. Thus, the spurious resonance present in case of the CSRR filters should disappear, resulting in a flat and low-loss passband response.
To provide a benchmark for the CUSR resonator performance we compare its properties to the ones of a CSRR. Figs. 11b & 11c illustrate the layout of the CSRR and CUSR cells integrated on a tapered CPW transmission line. The top layer contains the resonators, which are fabricated in a wet etching process. The lattice constant is 9.6 mm, the centre conductor width 9.15 mm and the slot width 0.45 mm. After the taper, the dimensions of the centre conductor and the slot width are 1 mm and 0.14 mm, respectively. The resulting characteristic impedance for the host line is Z0 = 50 Ohm. The CPW taper, which eliminates errors due to soldered connectors, was optimized for maximum matching between the two sides of the CPW. All structures are fabricated on an FR-4 substrate (dielectric constant ε_r = 4, loss tangent tanδ = 0.02, thickness h = 0.5 mm, and double sided copper clad of 35μm) using an in-house standard mask/photoetching technique.
In a next step we determined the S-parameters of the resonators within the frequency band of 1 to 12 GHz using an HP E8361A vector network analyzer (VNA) with a microstrip test

fixture (Wiltron 3680). As already explained in the previous section, a thru-short-line (TRL) calibration was performed for the CPW. The return loss was found to be better than 31 dB for the whole band of interest. Fig. 11a depicts the dimensions of the CUSR employed in this study. We have chosen an external radius of r = 3.6 mm, a separation of c = 0.4 mm, a width of d = 0.4 mm, and a metallic strip g = 0.4 mm for operation in the C-band. Figs. 11b & 11c show the fabricated structures for both CPW/CSRR & CPW/CUSR. In case of the CUSR, the resonator arm length for both u-shaped arms is identical in this design.

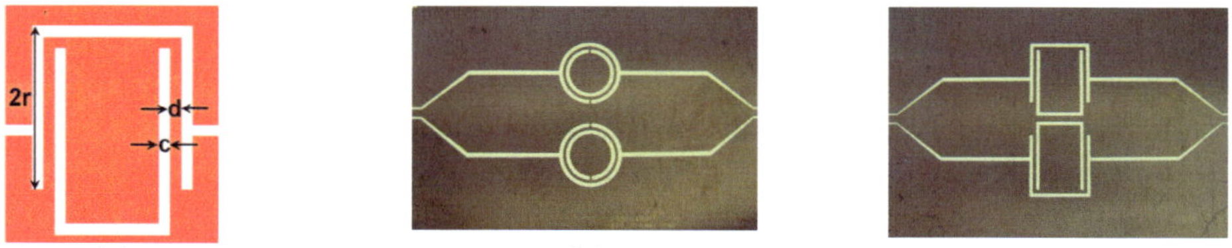

(a) (b) (c)

Fig. 11. (a) Schematic of CUSR with its dimensions and layout of CPW/CSRR structure (b) and CPW/CUSR (c).

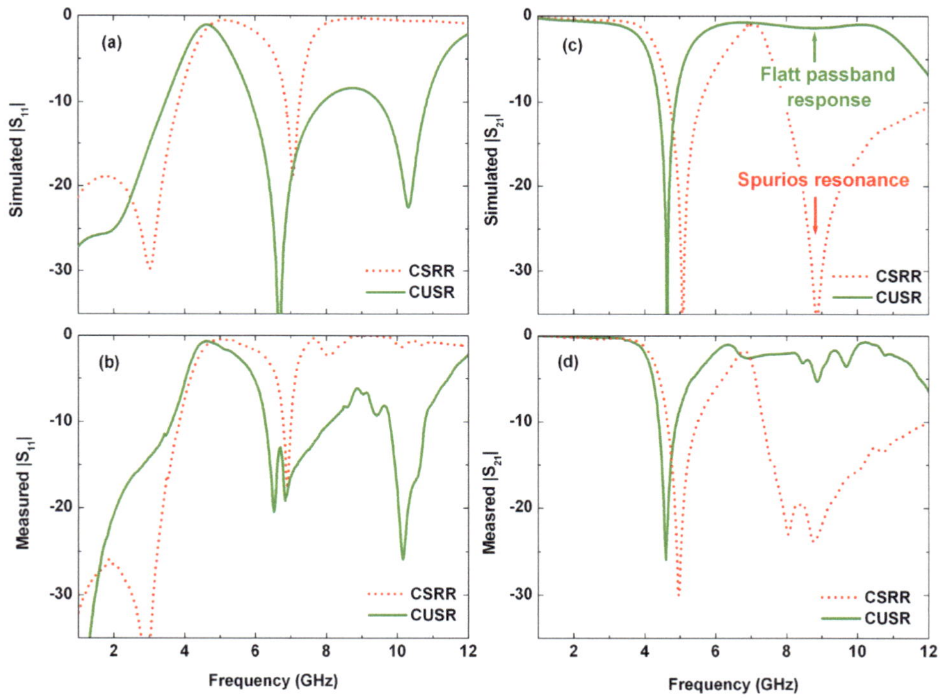

Fig. 12. Simulated (a) and measured (b) return loss for CPW-CSRR and CPW-CUSR structures. Simulated (c) and measured (d) insertion loss for CPW-CSRR and CPW-CUSR structures.

To simulate the transmission frequency response of the structures we use a commercially available 3D full-wave solver based on the finite element method (HFSS). Fig. 12 shows the simulated and measured magnitudes of the return and insertion losses of both CSRR (dashed line) and CUSR (solid line) structures. The primary stopband where the real part of the electric permittivity is expected to be negative is clearly pronounced in both cases. The simulated and measured return loss is shown in Figs. 12a and 12b. In case of the CSRR, the upper passband still has a considerable attenuation due to the spurious resonance induced by the differing length of the inner and the outer resonator as aforementioned in the previous section. In this upper passband the return loss of the CUSR is approximately 8 dB lower than that of the CSRR, offering a drastically improved stopband filter performance.

Figs. 12c and 12d depict the simulated and measured insertion loss. The stopbands are centred at approximately 4.6 GHz and 5.1 GHz for CUSR and CSRR, respectively. Although, both structures feature very high rejection levels and sharp transition edges, the previously mentioned spurious extended stopband centred around 8.8 GHz limits the applicability of the CSRR resonators. The measured data for the return and insertion losses agree well with the simulated results. The small discrepancies can be attributed to geometrical inhomogeneities introduced by the wet etching process. Please note that in addition to the improved filter performance the resonance frequency of the CUSRs is 10% less compared to the CSRRs, enabling a higher degree of miniaturization. For a thorough parametric study of the geometrical dimensions with regards to the resonance behaviour of the structures, the inclined reader is referred to (Al-Naib & Koch, 2008a).

3.3 Bandwidth modifying slots

We will now take a short excurse to an interesting alteration of the resonator geometry: It has been shown, e.g. in (Ibraheem & Koch, 2007), that slots inserted in the vicinity of a resonator can modify its bandwidth, enabling a simple method of custom tailoring filters to specific applications. To investigate this aspect for the CSRR and CUSR structures we fabricated structures as depicted in Fig. 13 with slots of 0.8 mm width and differing lengths close to the metamaterial resonators. The slot length *sl* varies from 0 to 3.6 mm in increments of 0.6 mm. Fig. 13b shows the bandwidth over the *sl* parameter. A continuous increase of the bandwidth with the length of the slots is observed for both structures. It is worth mentioning that the structures are still in the sub-wavelength range despite the presence of the slots. To conclude, slots in the vicinity of CSRRs or CUSRs allow easy custom tailoring of the filter bandwidth without the need for a time consuming redesign of the resonator itself.

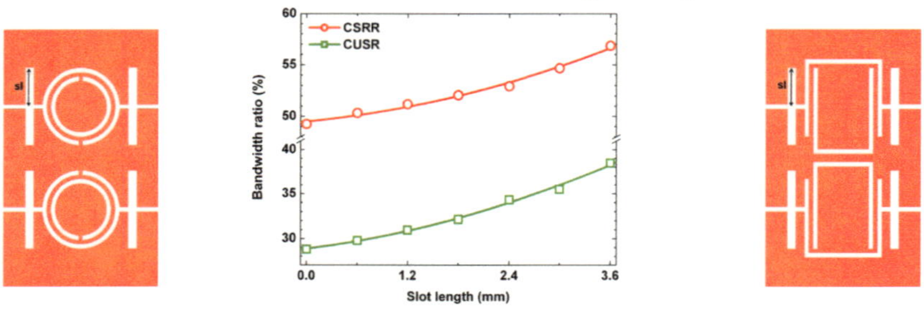

(a) (b) (c)

Fig. 13. (a) CPW-CSRR with nearby slots (b) bandwidth of the stop band filter vs. the slot length for CPW-CSRR and CPW-CSCR structures (c) CPW-CUSR with nearby slots.

3.4 Spiral CSRR/CPW

Many applications throughout microwave technology demand for a high degree of miniaturization. In order to meet this challenge, we will use this section to study the concept of spiral resonators in conjunction with the previously discussed CPW integrated CSRRs and demonstrate the high miniaturization potential which arises from this combination. The spiral resonator was proposed to further reduce the size of the SRRs (Baena et al., 2004). We utilize the complementary split rectangle resonators (CSCR) and the Complementary Spiral Resonator (CSR) with multi turns to obtain single layer bandstop filters with a high degree of miniaturization.

The structures of interest are depicted in Fig. 14. It shows the layout of the tapered CPW incorporating a unit cell of CSRRs, CSCRs, and CSRs, respectively. The resonators are symmetrically etched into the top layer. The lattice constant *a* is 8 mm. The FR-4 substrate is employed for all the structures (dielectric constant ε_r = 4, loss tangent tanδ = 0.02, and thickness *h* = 0.5 mm). They have an external radius of *r* = 3.6 mm, a separation of *c* = 0.2 mm, a width of *d* = 0.2 mm, and a metallic strip *g* = 0.2 mm (c.f. Fig. 9c).

To simulate the transmission frequency response, we use the commercial software package Ansoft HFSS (HFSS). Fig. 15 shows the simulated magnitude response for a CSRR, a CSCR and the new spiral CSRR. For all structures, there is a strong main stopband centred at 4.08 GHz, 3.36 GHz, and 1.38 GHz for the CSRR, CSCR, and the spiral CSRR, respectively. Please note that the resonance frequency of the spiral CSRR is only one-third compared to the resonance frequency of the CSRRs, revealing the high potential for miniaturization. Measurements, which are in good agreement with the simulations, are shown in Fig. 15b & 15d.

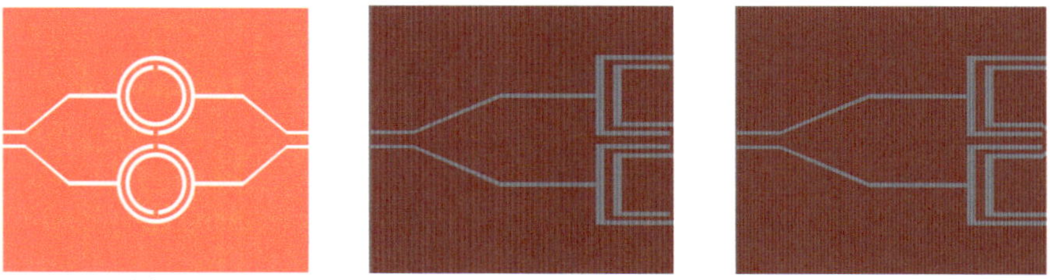

(a) (b) (c)

Fig. 14. Layout of the CPW/CSRR (a), CPW/CSCR (b) and CPW/CSR structures (c).

In order to achieve even higher degrees of miniaturization more turns can be added to the spiral resonators as shown in the insets of Fig. 16. We investigate two, four, six, and eight turn spiral resonators. However, increasing the number of turns not only reduces the resonance frequency (increasing the miniaturization) but also lowers the separation of the main stopband from the next higher frequency resonance. Fig. 16a depicts the dependence of the resonance frequency (Fig. 16a, left scale), the corresponding electrical size in terms of guided wavelength (Fig. 16a, right scale) and the frequency ratio of the second to the first resonance (Fig. 16b) on the number of spiral turns. Saturation in the change of the resonance frequency with increasing number of turns is revealed. This effect can be explained by the saturation of the resonator inductance and capacitance and has already been observed in (Bilotti et al., 2007). However, a very small electrical size of $\lambda g/50$ is achieved with an eight turn spiral resonator. The ratio between the fist and the second resonance, shown in Fig. 16b, decreases with the number of turns and also saturates at a value of approximately 2.2 for eight turns. Thus, in contrast to the CSRR structures discussed in the previous sections, the second resonance does not impact the excellent filter performance.

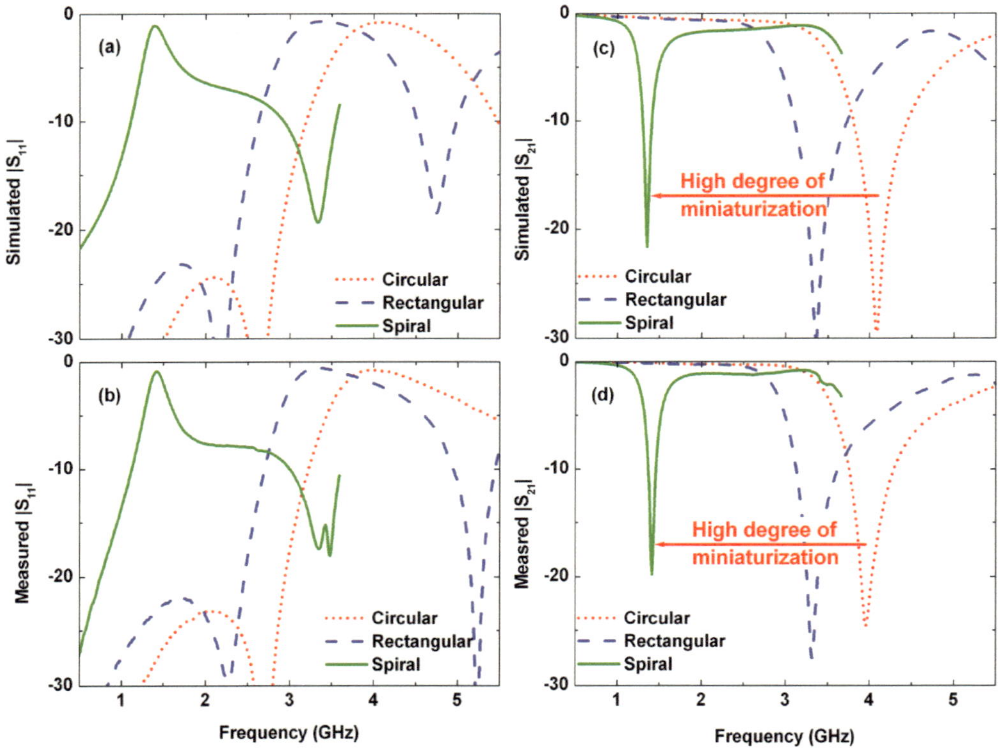

Fig. 15. Simulated (a) and measured (b) return loss for the CPW/CSRR (circular), CPW/CSCR (rectangular), and CPW/CSR (spiral) structures. Simulated (c) and measured (d) insertion loss for the CPW/CSRR (circular), CPW/CSCR (rectangular), and CPW/CSR (spiral) structures.

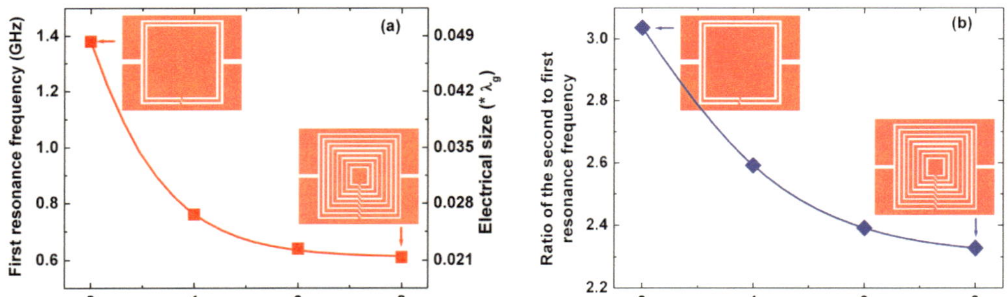

Fig. 16. (a) Resonance frequency versus number of spiral turns (b) ratio of the second to first resonance frequencies.

3.5 Bandpass Filter with CSRR/CPW

So far, we have presented how CSRRs can be employed to miniaturize conventional bandstop filters tremendously. Although bandstop filters are very important as mentioned earlier, many modern microwave applications, such as automotive radar and wireless communication systems, rely on bandpass filters (BPFs). Moreover, transceiver modules are based on bandpass filters to separate uplink from downlink. In order to succeed in the field of wireless applications, miniaturization of these versatile devices is mandatory to achieve a high integration density in the overall system.

Martin et al. proposed a left-handed materials which consists of CPW loaded with SRRs from the backside of a substrate, in combination with periodically aligned strip lines (SLs) on the upside, connecting the central conductor to the ground planes (Martin et al., 2003b). The frequency response of such structure exhibits low insertion losses and a sharp cutoff at the lower band edge. However, the attenuation level above the passband is usually not as high as required for practical filter applications. A workaround for this problem is to increase the number of unit cells to achieve the desired attenuation at the upper transition band edge. Unfortunately, this increase directly leads to a higher insertion loss at the centre frequency as well, so that a trade-off has to be found.

This section presents a BPF based on combining conventional SRRs with CSRRs resonators. It exhibits low insertion loss, sharp cut-off and high stopband attenuation. CSRRs feature low pass characteristics compared to SRRs which exhibit high pass behavior. Combining both SRRs and CSRRs allows a very flexible design of BPFs. The lower and higher band edges are defined by the resonance frequency of the SRR and CSRR, respectively.

Fig. 17a depicts one of the fabricated structures where CSRR cascaded with SRR. The dimensions of the SRRs are chosen such that the device operates in the C-Band around 4 GHz. Figs. 17b & 17c shows the dimensions of both SRRs and CSRRs which have the same width of c = 0.42 mm, a separation of d = 0.38 mm and a gap of g = 0.36 mm. The outer radius r is 3.2 mm and 3.6 mm for the SRRs and CSRRs, respectively. A commercial low cost FR-4 substrate (dielectric constant ε_r = 4, loss tangent tanδ = 0.02, thickness h = 0.5 mm) is used to fabricate the structures.

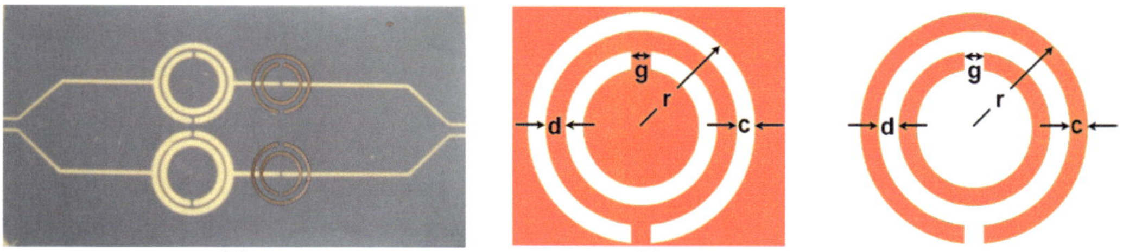

(a) (b) (c)

Fig. 17. (a) SRR/SL-CSRR loaded the CPW fabricated structure. Schematic of CSRR (b) and SRR (c) with relevant dimensions.

Three selected structures have been chosen for simulations and measurements. They include a unit cell of each, SRR/SL and CSRR as well as a combination of both. A commercially available 3D full-wave solver (HFSS) based on the finite element method is employed to calculate the S-parameters. Accompanying measurements were performed using a vector network analyzer (HP E8361A) with a microstrip test fixture (Wiltron 3680). The simulated phase for the three structures is given in Fig. 18. Examining the phase responses of the single elements and the overall structure shows the optimization options in terms of impedance matching. The phase response of the SRR/SL is mainly capacitive for the lower stopband. At the resonance frequency, a phase flip occurs, after which the SRR/SLs behave mainly inductive. The CSRR exhibits a gradually decreasing phase response over the whole frequency band of interest. The behavior of the CSRR changes from inductive to capacitive at the centre frequency which helps to compensate for the high inductive reactance of the SRR/SL after their phase flip resulting in lower insertion loss in the passband.

Figs. 19 & 20 shows simulated and measured return (S_{11}) and insertion (S_{21}) losses, respectively. The simulations qualitatively agree very well with the measurements. Simulation and measurement graphs reveal only small quantitative discrepancies, which can be attributed to manufacturing inaccuracies. At the centre frequency, the measured and simulated return loss for the combined SRR/SL-CSRR is 18.9 and 17.6 dB, respectively. The simulated and measured insertion loss at the centre frequency of f_c = 4 GHz is 2.7 dB and 3.4 dB, respectively. The out-of-band rejection is higher than 19 dB in the simulation and 17 dB in the measurement. Keeping in mind, that the whole structure has a geometrical outline of only 0.29 λg by 0.29 λg (with λg as guided wavelength) and is based on standard FR-4 substrate material, the performance is superior to previously demonstrated devices which employ cascaded SRR cells.

Another investigation has been performed with a more expensive microwave substrate (Rogers RO3003, dielectric constant ε_r= 3, tanδ = 0.0013, thickness h= 0.127 mm). The simulation predicts a reduction of insertion loss by 60% (1.6 dB) and an out-of-band rejection enhancement of 4 dB. In order to obtain the same centre frequency of the FR-4 sample, the radius of the split rings was reduced to 2.8 mm to compensate for the difference in substrate thickness and the dielectric constant. Experimentally, we found a reduction in the insertion loss at the centre frequency of 55% (1.5 dB), which is close to the predicted value. The demonstrated design allows very narrow passbands usually only available for filters constructed in more expensive technologies, e.g. based on high-temperature superconductor thin films.

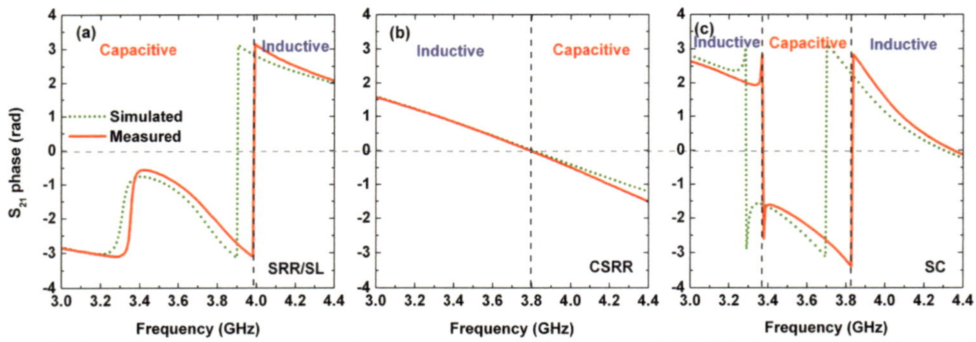

Fig. 18. Simulated and measured transmission phase for SRR/SL (a), CSRR (b) and both structures (c).

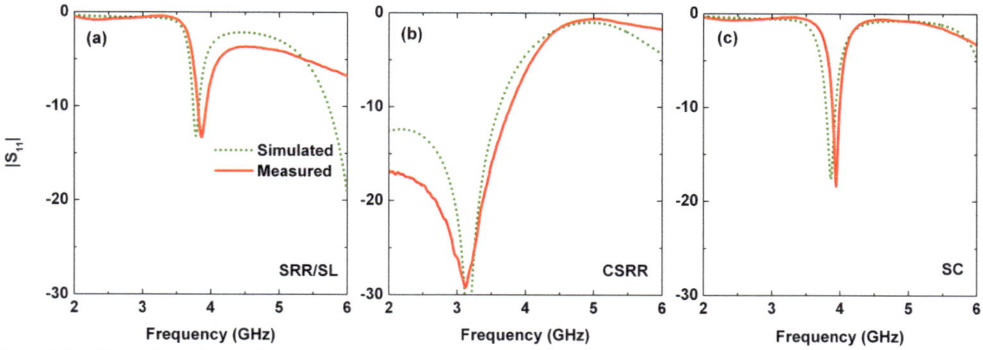

Fig. 19. Simulated and measured return loss for the SRR/SL (a), CSRR (b) and both structures (c).

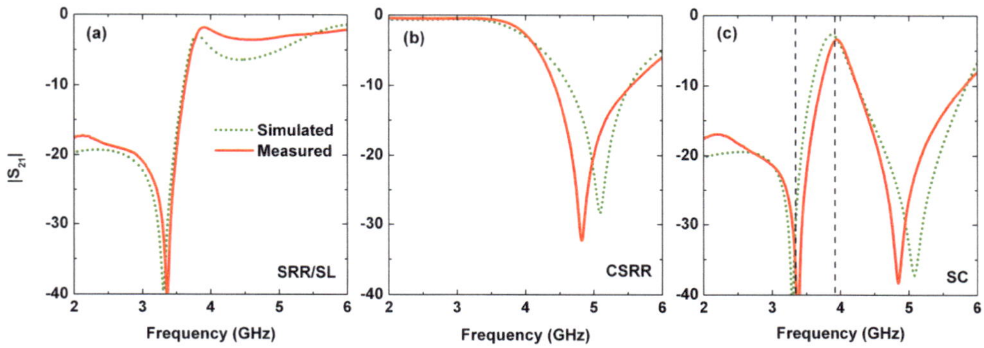

Fig. 20. Simulated and measured insertion loss for the SRR/SL (a), CSRR (b) and both structures (c).

Moreover, we analyzed the dependence of the insertion loss on the bandwidth ratio (for FR-4 and RO3003 substrates) in Fig. 21. The outer radius of the SRRs was swept from 3.0 mm to 3.6 mm and 2.6 mm to 3.2 mm in steps of 0.1 mm for the FR-4 and the RO3003, respectively. The resonance frequency of the SRRs is proportional to the inverse square root of the outer radius (Marques et al., 2002). Thus, the resonance frequency is decreased with an increase in the radius of the SRRs. This leads to a widening of the bandwidth of the overall structure. The bandwidth ratio is defined as the relation of the bandwidth over the centre frequency so that the bandwidth ratio is increased as well. As long as the centre frequency of the bandpass filter lies within the passband of both resonators, the influence of the insertion loss on the bandwidth ratio stays very small. While when the centre frequency of the bandpass filter lies in the transition region, reducing the bandwidth ratio a little further drastically increases the losses. For the current case, the losses remain low for bandwidth ratios of down to 10%. Hence, custom tailored filter designs can be achieved by varying only a single dimension of the structure, namely the outer radius of the SRRs. Additional SRR or CSRR cells could be added to improve the performance in case the required bandwidth ratio is below 10%. Moreover, multiple turn rectangular spiral resonators and its complementary could be employed to achieve a significant further miniaturization.

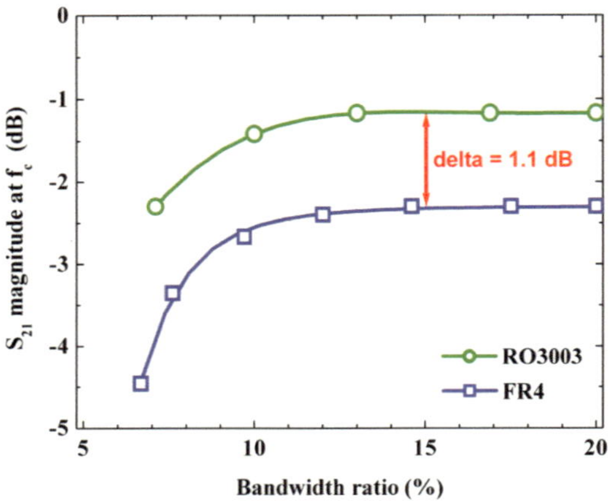

Fig. 21. Simulated S21 magnitude at the centre frequency versus the bandwidth ratio for a sweep of the outer radius of the SRR.

4. Conclusion

In conclusion we have discussed recent advances in the field of planar metamaterials. Especially, recently proposed planar metamaterial resonators for high performance, small footprint filters integrated in coplanar waveguide technology were considered. The following list, illuminating briefly the characteristic strengths and weaknesses of the previously discussed concepts, shall serve both as summary and reference to the inclined reader.

Complementary split ring resonator (CSRR) with slots
(Section 3.1, see also (Ibraheem & Koch, 2007))

+ Stopband filter
+ High stopband rejection
+ Low passband losses
+ Rather small outline dimensions
+ Slots allow custom-tailoring of the filter bandwidth
- Spurious resonance in the higher passband due to difference in inner and outer resonator arm length

U-shaped split resonator (CUSR)
(Section 3.2, see also (Al-Naib & Koch, 2008a))

+ Stopband filter
+ High stopband rejection
+ Low passband losses
+ Rather small outline dimensions
+ No spurious resonance in the higher passband due to equal resonator arm lengths

Combined CSRR/SRR with strip lines in series
(Section 3.5, see also (Al-Naib et al., 2008))

+ Bandpass filter
+ Low passband losses
+ High stopband rejection
+ Moderate outline dimensions
+ Well suited for frontend designs

Circular multiple turn complementary spiral resonators (CSRs)
(Section 3.4, see also (Al-Naib & Koch, 2008b))

+ Stopband filter
+ Extremely compact size
+ Low passband losses
+ High stopband rejection
+ Well suited for frequency selective surface designs with small resonator spacings
- Spurious resonance due to differing resonator arm lengths.
+ But: still a good separation between spurious resonance and main resonance is achieved

5. References

Al-Naib, I. A. I.; Jansen, C. & Koch, M. (2008). Miniaturized bandpass filter based on metamaterial resonators: a conceptual study, *Journal of Physics D: Applied Physics*, 41(20), 205002.

Al-Naib, I. & Koch, M. (2008a). Higher Degree of Miniaturization with Split Rectangle Resonators. *Proceedings of the 38th European Microwave Conference*, pp. 559-562, Amsterdam-The Netherlands.

Al-Naib, I. & Koch, M. (2008b). Miniaturization of Coplanar waveguide Bandstop Filter with Spiral Resonators, *Metamaterials'2008*, pp. 715-717, Pamplona-Spain.

Alu, A. & Engheta, N. (2005). Achieving transparency with plasmonic and metamaterial coatings, *Physical Review E*, 72(1), 016623.

Baena, J. D.; Marqués, R.; Medina, F. & Martel, J. (2004). Artificial magnetic metamaterial design by using spiral resonators, *Phys. Rev. B, Condens. Matter*, 69, pp. 014402(1)-014402(5).

Bilotti, F.; Toscano, A. & Vegni, L. (2007.) Design of spiral and multiple split-ring resonators for the realization of miniaturized metamaterial samples, *IEEE Trans. Microwave theory and technology*, 55, pp. 2258-2267.

Caloz, C. & Itoh, T. (2002). Application of the transmission line theory of left-handed (LH) materials to the realization of a microstrip "LH line". *IEEE Antennas and Propagation Society International Symposium*, 2, pp. 412-415.

Caloz, C. & Itoh, T. (2005). *Electromagnetic Metamaterials: Transmission Line Theory and Microwave Applications*, Wiley & Sons.

CST Microwave Studio®, http://www.cst.com.

Driscoll, T.; Basov, D. N.; Starr, A. F.; Rye, P. M.; Nemat-Nasser, S.; Schurig, D. & Smith, D. R. (2006). Free-space microwave focusing by a negative-index gradient lens. *Applied Physics Letters*, 88(8), 081101-3.

Eleftheriades, G. V. & Balmain, K. G. (2005). *Negative Refraction Metamaterials. Fundamental Principles and Applications*, Wiley & Sons.

Eleftheriades, G.; Grbic, A. & Antoniades, M. (2004). Negative-refractive-index transmission-line metamaterials and enabling electromagnetic applications. *IEEE Antennas and Propagation Society International Symposium*, 2, pp. 1399-1402.

Engheta, N. (2002). An idea for thin subwavelength cavity resonators using metamaterials with negative permittivity and permeability. *IEEE Antennas and Wireless Propagation Letters*, 1, pp. 10-13.

Engheta, N. & Ziolkowski, R. W. (2006). *Electromagnetic Metamaterials: Physics and Engineering Aspects*, Wiley & Sons.

Erentok, A. & Ziolkowski, R. (2005). Development of Epsilon-Negative (ENG) Metamaterials for Efficient Electrically Small Antenna Applications, *IEEE Antennas and Propagation Society International Symposium*, pp. 304-307, Washington, DC, USA.

Falcone, F.; Martin, F.; Bonache, J.; Marques, R. & Sorolla, M. (2004). Coplanar waveguide structures loaded with split-ring resonators, *Microwave and Optical Technology Letters*, 40(1), 3-6.

Feresidis, A.; Goussetis, G.; Shenhong, W. & Vardaxoglou, J. C. (2005). Artificial magnetic conductor surfaces and their application to low-profile high-gain planar antennas, *IEEE Transactions on Antennas and Propagation*, 53(1), 209-215.

Gay-Balmaz, P.; Maccio, C. & Martin, O. J. F. (2002). Microwire arrays with plasmonic response at microwave frequencies, *Applied Physics Letters*, 81(15), 2896-2898.

Greegor, R. B.; Parazzoli, C. G.; Nielsen, J. A.; Thompson, M. A.; Tanielian, M. H. & Smith, D. R. (2005). Simulation and testing of a graded negative index of refraction lens. *Applied Physics Letters*, 87(9), 091114-3.

HFSS™, available at http://www.ansoft.com/products/hf/hfss.

Hrabar, S.; Bartolic, J.; & Sipus, Z. (2005). Waveguide miniaturization using uniaxial negative permeability metamaterial, *IEEE Transactions on Antennas and Propagation*, 53(1), 110-119.

Ibraheem, I. A.; & Koch, M. (2007). Coplanar waveguide metamaterials: The role of bandwidth modifying slots, *Applied Physics Letters*, 91(11), 113517-3.

Ibraheem, I. A.; Schoebel, J. & Koch, M. (2008). Group delay characteristics in coplanar waveguide left-handed media, *Journal of Applied Physics*, 103(2), 024903-7.

Iyer, A.; & Eleftheriades, G. (2002). Negative refractive index metamaterials supporting 2-D waves, *IEEE MTT-S International Microwave Symposium Digest*, 2, pp. 1067-1070.

Lagarkov, A. & Kisel, V. (2001). Electrodynamic properties of simple bodies made of materials with negative permeability and negative permittivity. *Doklady Physics*, 46(3), 163-165.

Leonhardt, U. (2006). Optical Conformal Mapping, *Science*, 312(5781), 1777-1780.

Marques, R.; Medina, F. & Rafii-El-Idrissi, R. (2002). Role of bianisotropy in negative permeability and left-handed metamaterials, *Physical Review B*, 65(14), 144440.

Marques, R.; Martin, F. & Sorolla, M. (2008). *Metamaterials with Negative Parameters: Theory, Design, and Microwave Applications: Theory, Design and Microwave Applications*, John Wiley & Sons.

Martin, F.; Falcone, F.; Bonache, J.; Marques, R. & Sorolla, M. (2003a). Miniaturized coplanar waveguide stop band filters based on multiple tuned split ring resonators, *IEEE Microwave and Wireless Components Letters*, 13(12), 511-513.

Martin, F.; Bonache, J.; Falcone, F.; Sorolla, M. & Marques, R. (2003b). Split ring resonator-based left-handed coplanar waveguide, *Applied Physics Letters*, 83(22), 4652-4654.

Oliner, A. A. (2002). A periodic-structure negative-refractive-index medium without resonant elements, *IEEE AP-S/URSI Int. Symp. Dig.*, San Antonio, TX, 41-44.

Pacheco, J.; Grzegorczyk, T. M.; Wu, B.; Zhang, Y. & Kong, J. A. (2002). Power Propagation in Homogeneous Isotropic Frequency-Dispersive Left-Handed Media. *Physical Review Letters*, 89(25), 257401.

Pendry, J. B. (2000). Negative refraction makes a perfect lens, *Physical Review Letters*, 85(18), 3966–3969.

Pendry, J. B.; Holden, A. J.; Stewart, W. J. & Youngs, I. (1996). Extremely low frequency plasmons in metallic mesostructures, *Physical Review Letters*, 76(25), 4773-4776.

Pendry, J. B.; Holden, A. J.; Robbins, D. J. & Stewart, W. J. (1998). Low frequency plasmons in thin-wire structures, *Journal of Physics: Condensed Matter*, 10(22), 4785-4809.

Pendry, J.; Holden, A.; Robbins, D. & Stewart, W. J. (1999). Magnetism from conductors and enhanced nonlinear phenomena, *IEEE Transactions on Microwave Theory and Techniques*, 47(11), 2075–2084.

Pendry, J. B. & Smith, D. R. (2006). The quest for the superlens, *Scientific American*, 295(1), 60-7.

Ramakrishna, S. A.; & Grzegorczyk, T. M. (2008). *Physics and Applications of Negative Refractive Index Materials*, CRC.

Schurig, D.; Mock, J. J.; Justice, B. J.; Cummer, S. A.; Pendry, J. B.; Starr, A. F & Smith, D. R. (2006). Metamaterial Electromagnetic Cloak at Microwave Frequencies. *Science*, 314(5801), 977–980.

Shelby, R. A.; Smith, D. R.; & Schultz, S. (2001). Experimental Verification of a Negative Index of Refraction, *Science*, 292(5514), 77–79.

Simons, R. N. (2001). *Coplanar Waveguide Circuits, Components and Systems*. Wiley-IEEE, New York.

Smith, D. R.; Mock, J. J.; Starr, A. F. & Schurig, D. (2005). Gradient index metamaterials, *Physical Review E*, 71(3), 036609.

Smith, D. R.; Padilla, W. J.; Vier, D. C.; Nemat-Nasser, S. C. & Schultz, S. (2000). Composite Medium with Simultaneously Negative Permeability and Permittivity, *Physical Review Letters*, 84(18), 4184.

Wang, S.; Feresidis, A.; Goussetis, G. & Vardaxoglou, J. C. (2006). High-gain subwavelength resonant cavity antennas based on metamaterial ground planes, *IEE Proceedings-Microwaves, Antennas and Propagation*, 153(1), 1-6.

Wolff, I. (2006). *Coplanar Microwave Integrated Circuits*. John Wiley & Sons.

21

Passive Diplexers and Active Filters based on Metamaterial Particles

Daniel Segovia-Vargas[1], Vicente González-Posadas[2], José Luis Jiménez-Martín[2], Luis Enrique García-Muñoz[1] and Oscar García-Pérez [1]

[1] *Universidad Carlos III de Madrid*
Leganés-Madrid, Spain
[2] *Universidad Politécnica de Madrid*
Madrid, Spain

1. Introduction

Composite transmission lines are one of the main developments in the increasingly popular field of electromagnetic metamaterials, artificial electromagnetic structures with both negative electric permittivity and magnetic permeability (Caloz & Itoh, 2005). These structures present a backward wave or left-handed (LH) propagation instead of the conventional right-handed one. The first experimental microwave structures that presented this behavior were the result of combining thin wires and split-ring resonators (Shelby et al, 2001). Soon it was evident that narrow bandwidth operation and high losses were inherent to the resonant nature of this kind of metamaterials. In order to overcome the previous problems, some authors proposed the so-called metamaterial transmission lines or left-handed transmission lines LH TL (Sanada et al, 2004). The LH TL concept has been extended and generalized to the concept of composite right/left handed (CRLH) structures where mixed contributions of LH and RH cells occur in practice. More specifically, the CRLH transmission lines have become a very commonly used solution to obtain metamaterial properties with low losses and broader bandwidth. Then, below a certain frequency, a CRLH transmission line behaves as a left-handed transmission line while over higher frequencies it is basically a conventional right handed line. As a consequence of this combined behaviour, the phase response is not linear with respect to the frequency.
From the circuit application point of view, the two main characteristics of the CRLH transmission lines consist on obtaining miniaturized and/or dual band circuits. Then, it can be mentioned that dual band hybrid couplers (Lin et al, 2004), branch-line couplers (Keung & Cheng, 2004), dual-performance rat-race couplers (Castro-Galan et al, 2009) and enhanced rat-race couplers (Okabe, Caloz & Itoh 2004). The most critical aspect in the design of combiners, filters or diplexers with conventional CRLH transmission lines is the losses associated to them.
For the case of diplexers, using of the so called dual composite right left handed (D-CRLH) cells may overcome some of the previously stated problems. The D-CRLH transmission lines are the dual part of the CRLH transmission lines and were first proposed in (Caloz,

2006). Thus, this type of line presents a RH performance at low frequencies while LH at higher frequencies. From its equivalent circuit point of view, the D-CRLH section changes an equivalent bandpass of the CRLH section by a bandstop section. This is particularly useful in the design of diplexers. The design of a diplexer based on CRLH lines has shown drawbacks coming from the losses at the pass bands. A diplexer can be based either on allowing some frequency pass-bands or on rejecting the frequency stop-bands. The use of D-CRLH transmission lines can help to overcome the previous problems by working with the complementary rejection frequency bands. The design of microwave diplexers based on using of D-CRLH lines is presented in this chapter.

In addition, in order to avoid losses, the use of dual frequency active filters is also proposed. This dual-frequency performance can be achieved by making use of conventional CRLH transmission lines. The inclusion of this type of lines as feedback sections in a first order recursive topology can be used to generate a filtering response with two arbitrary pass bands. Additionally, dual-band couplers are also required. These may be implemented by means of CRLH structures or by shunted stub branch line structures. This second structure produces a strong rejection level at the central stop band what improves the overall response of the dual frequency active filter. Theoretical analysis and design procedures are verified by means of manufacturing and measurement of a prototype.

Then, passive diplexers based on D-CRLH lines and dual frequency active filters based on conventional CRLH transmission lines are presented in this chapter.

2. Dual Composite Right-Left Handed transmission line theory

The D-CRLH structure, shown in Fig. 1, behaves as the complementary structure of the conventional CRLH cell. This unit cell has a series parallel LC tank circuit and a shunt series LC tank. The D-CRLH indeed exhibits its left-handed band at high frequencies and its right-handed band at low frequencies and is of stop-band nature.

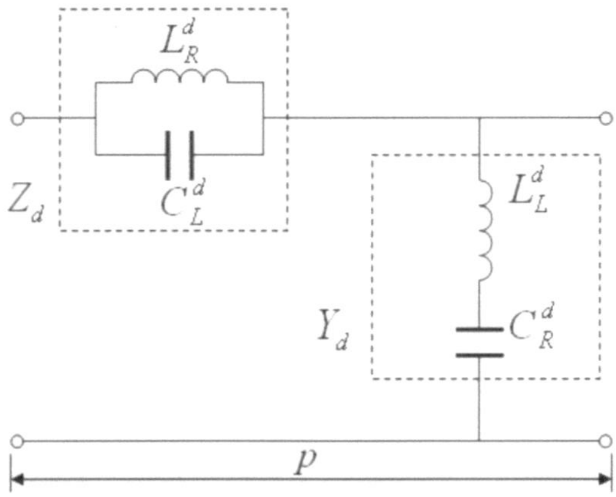

Fig. 1. Schematic of the D-CRLH Transmission Line

The equations running the performance of the D-CRLH balanced line are given in (1)-(3). If the resonance frequencies of the shunt LC tank and the series LC tank were w_{sh} and w_{se} a balanced condition could be achieved by satisfying (1)

$$\omega_0 = \sqrt{\omega_{sh} \cdot \omega_{se}} \tag{1}$$

Once the D-CRLH line is balanced, if a right and left cut-off frequencies are defined, then it can be written that

$$\omega_R = \sqrt{\frac{1}{L_R \cdot C_R}}; \omega_L = \sqrt{\frac{1}{L_L \cdot C_L}} \tag{2}$$

The characteristic impedance of the line is given by

$$Z_0 = Z_R = Z_L = \sqrt{\frac{L_R}{C_R}} = \sqrt{\frac{L_L}{C_L}} \tag{3}$$

Finally, the propagation constant is given by

$$\beta(\omega) = \frac{1}{\frac{1}{\beta_R} + \frac{1}{\beta_L}} = \frac{1}{\frac{1}{\omega/\omega_R} + \frac{1}{\omega_L/\omega}} = \frac{1}{\frac{\omega_R}{\omega} - \frac{\omega}{\omega_L}} = \frac{\omega_L \cdot \omega}{\omega_0^2 - \omega^2} \tag{4}$$

The stop-band frequencies can be computed from the local impedance by placing two cutting-off frequencies, one in the left-handed part and other at the right-handed one. These cutting-off frequencies are given by the following equation

$$\omega_{CL/R} = \omega_0 \cdot \sqrt{1 + \frac{\omega_L}{8 \cdot \omega_R} \pm \sqrt{\frac{\omega_L}{4 \cdot \omega_R} + \sqrt{1 + \frac{\omega_L}{16 \cdot \omega_R}}}} \tag{5}$$

where the sign + is for the left handed high-pass cutoff and the sign - is for the right handed low-pass cutoff. It must be noted that $\omega_{CL} > \omega_{CR}$.

Fig. 2 (taken from (González-Posadas et al, 2008)) shows the performance of a dual-CRLH line, both in amplitude and phase with the band-stop at 970 MHz. It can be seen that the structure presents a rejection bandwidth instead of a passing one (as it would be for the conventional CRLH case).

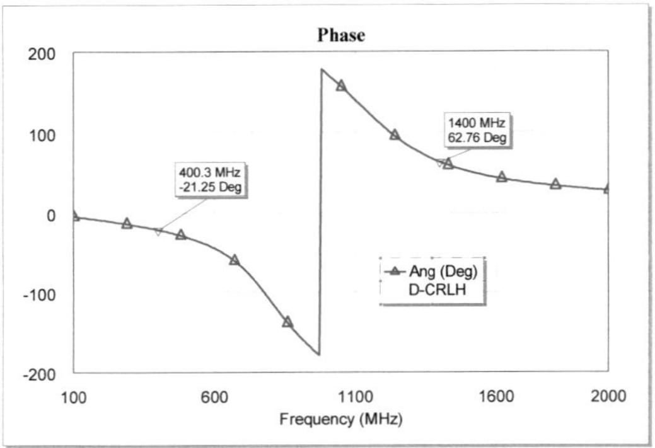

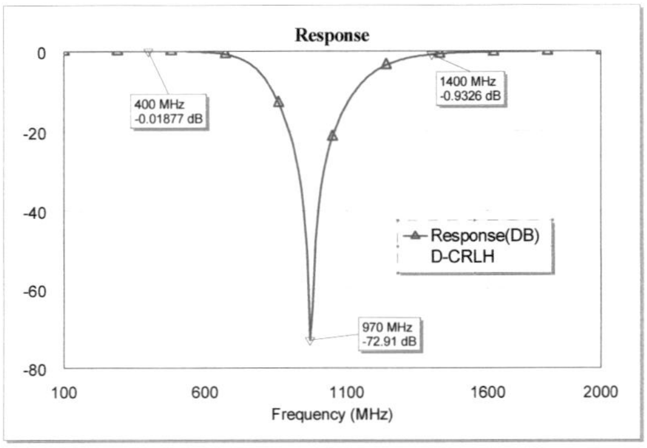

Fig. 2. D-CRLH Phase (left) and amplitude (right) response. The parameter are L_R=6 nH, C_R=4.452 pF, L_L=11.13 nH, C_L=2.4 pF.

3. Passive diplexer design

In this section the principles for the diplexer design based on D-CRLH lines will be presented. A generalization for a multistage diplexer design will be also given. Finally, some experimental results will be shown to prove the validity of the proposed design strategy.

3.1 Design principles

The realization of diplexers with CRLH lines has not shown a great development till now. Different topologies to build CRLH diplexers have been used. In all these cases the CRLH transmission line was used to allow the desired frequency band and reject the non-desired one. In this way, all the problems associated with the sensitivity and losses of the CRLH transmission lines were present. Then, (Bonache et al, 2005) used split-ring-resonators for a classical diplexer topology; (Horii, Caloz & Itoh, 2005) proposed a vertical topology to achieve a very compact diplexer at a price of large losses and a frequency shift; finally, (Wong, Balmain & Eleftheriades, 2006) proposed an original and non-compact planar topology for diplexer design.

The dual performance of the so called D-CRLH transmission lines allows a different design strategy for diplexers. The strategy is based on rejecting the non-desired frequency bands instead of allowing the desired ones. In this way, the diplexer design will be based on working with D-CRLH transmission lines tuned at the frequency that is not allowed. Thus, the band-stop performance of a D-CRLH line is used to design a diplexer. Then, for a diplexer separating the frequencies f_1 and f_2 (being $f_1 < f_2$), the first D-CRLH line, according to Fig. 3, has a stop-band at f_1 and a pass band at f_2; however, the second D-CRLH line presents a stop-band at f_2 and a pass band at f_1. From the left and right handed performance of the corresponding D-CRLH lines, it can be seen that the first D-CRLH line is left-handed at the higher passing frequency (f_2) while the second D-CRLH line is right-handed at the lower passing frequency (f_1), just the opposite as it would be for the conventional CRLH transmission line. In this way, any of these D-CRLH lines will be used in any of the branches of the proposed diplexer.

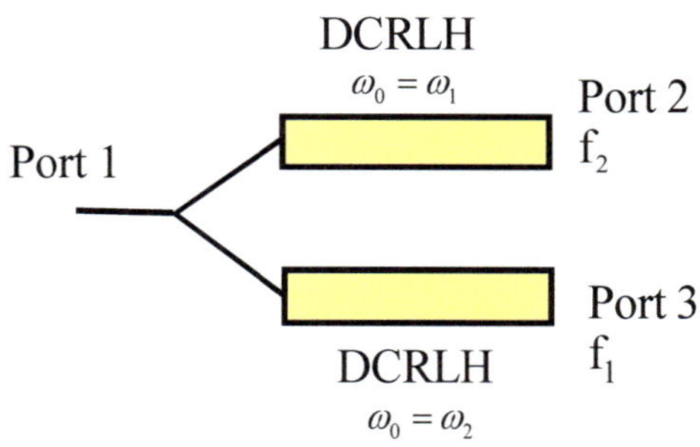

Fig. 3. Proposed D-CRLH diplexer structure

The design frequency has been chosen as the non-desired frequency (f_1 for the band allowing f_2 pass, say port 2 in the Fig. 3, and f_2 for the band allowing f_1 pass, say port 3 in Fig. 3). Let us introduce a factor *k(N)* , being *N* the number of cells of the D-CRLH transmission line, as the one that defines the ratio between the band stop frequency and the cutting-off frequencies of the D-CRLH lines. This factor depends on the number of cells in the D-CRLH line and on the frequency separation and will define two cutting-off frequencies: towards its right corresponding to the left-handed performance and towards its left corresponding to its right-handed performance. Without loss of generality, let us assume that the number of cells is 1 for the following design procedure.

First, choice of the band-stop frequencies for the two balanced D-CRLH structures. These frequencies are chosen in a way that the corresponding central frequency is the one allowing the other frequency pass

$$\omega_1 = \omega_{lower}; \omega_2 = \omega_{higher}; \omega_2 > \omega_1 \tag{6}$$

Secondly, define the right-handed cutting-off frequency of the D-CRLH section centred at ω_2 and the left-handed cutting-off frequency of the D-CRLH section centered at ω_1. Then, for the D-CRLH that allows passing the low frequencies the phase must be positive, that is right-handed, and left-handed for the line that allows passing the high frequency

$$\omega_{1cL} = \omega_{lower} \cdot k(N); \omega_{2cR} = \frac{\omega_{higher}}{k(N)} \tag{7}$$

The values for the right and left cutting off frequencies are, then, given as

$$\omega_{CL/R} = \omega_0 \cdot \sqrt{1 + \frac{[k(N)]^2}{8} \pm \sqrt{\frac{[k(N)]^2}{4} + \sqrt{1 + \frac{[k(N)]^2}{16}}}} \tag{8}$$

If the ratio between the higher and lower frequency is called *R*, a graphical representation between the value *R* and the factor *k(N)* can be obtained. This ratio is given in Fig. 4.

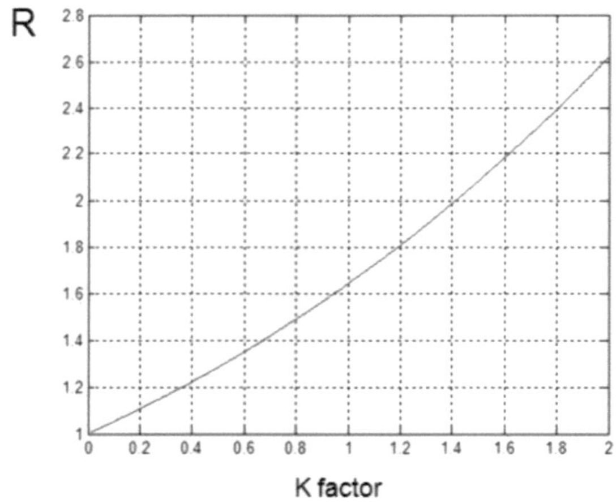

Fig. 4. *K* factor as a function of the ratio R between the higher and lower frequency

Once the value of the *k* factor and of the frequency ratio *R*, has been determined, the values for each branch left-handed and right-handed capacitors can be found out as follows. Thus for the branch rejecting the lowest frequency and allowing the highest one (port 2 in Fig. 3), the values for the components are given as

$$L_{1R} = \frac{Z_0 \cdot k(N)}{\omega_{lower}}; L_{1L} = \frac{Z_0}{\omega_{lower} \cdot k(N)}; C_{1L} = \frac{L_{1L}}{Z_0^2}; C_{1R} = \frac{L_{1R}}{Z_0^2} \tag{9}$$

And for the branch rejecting the highest frequency and allowing the lowest one (port 3 in Fig. 3), the values for the components are given as

$$L_{2R} = \frac{Z_0 \cdot k(N)}{\omega_{higher}}; L_{2L} = \frac{Z_0}{\omega_{higher} \cdot k(N)}; C_{2L} = \frac{L_{2L}}{Z_0^2}; C_{2R} = \frac{L_{2R}}{Z_0^2} \tag{10}$$

3.2 Multistage diplexer

For the general case where the number of cells is *N* (larger than 1), the values of the components are given by the following expressions

$$L_{1,2R,N} = \frac{L_{1,2R}}{N}; L_{1,2L,N} = L_{1,2L} \cdot N; C_{1,2R,N} = \frac{C_{1,2R}}{N}; C_{1,2L,N} = C_{1,2L} \cdot N \tag{11}$$

A discussion on the number of sections and on the D-CRLH performance can be done now. In (Caloz, 2006) the D-CRLH performance was proven for a 10-cell transmission line structure. If the number of cells is reduced the left-handed performance still maintains as it has been shown in (Herraiz et al, 2007). The only dependence on the number of cells to be implemented comes from the fact that the transient response between the low frequency right-handed region and the high-frequency left-handed one is more abrupt when the number of cells is larger. However, from the diplexer design point of view a trade-off between the desired losses and the rejection factor at the non desired frequency in the corresponding branch must be considered. In this way, for most single designs low orders (1 or 2) for the identical D-CRLH sections are enough to achieve the desired performance. Next section will show the design of a particular one-stage or multistage diplexer.

3.3 Experimental results

Since one of the most important features is its miniaturization, the presented design procedure will be applied to the low microwave frequency band, to separate frequencies working at TETRA-GSM bands (380 MHz and 960 MHz). As the working frequencies are far enough, initially a one stage design will be taken into account. The schematic of the proposed diplexer is shown in Fig. 5. (AWR ®). For this diplexer, the value for the factor *k(N)* is 2.01. The conventional transmission lines allow joining the two branches and soldering the SMA connectors and, at the same time they are optimized to achieve the lowest return losses. The proposed circuit has been implemented on an Arlon600 substrate with a relative permittivity of 6, height of 0.6mm.

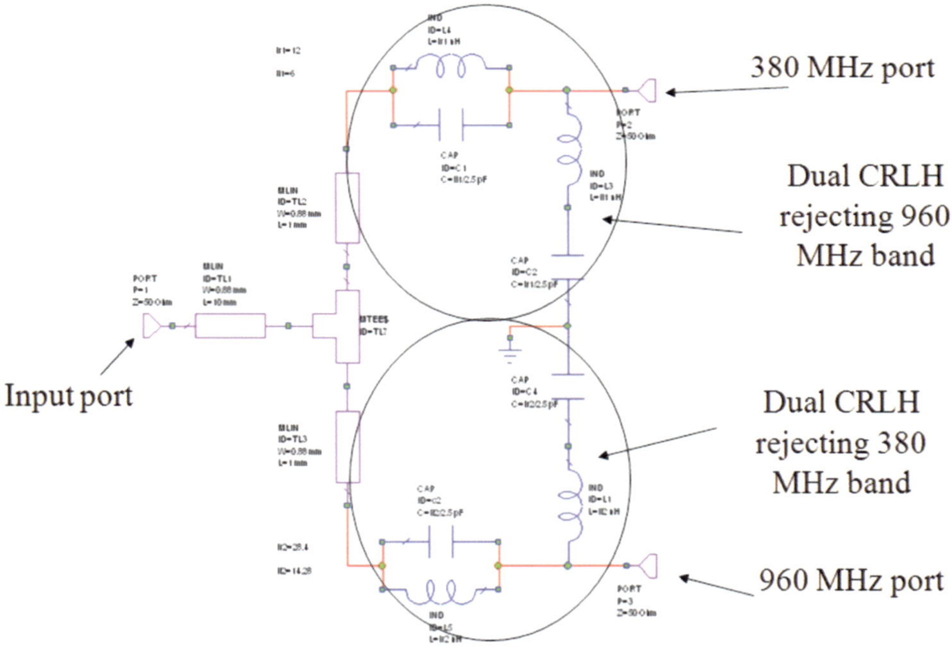

Fig. 5. Schematic of the one stage diplexer for separating TETRA and GSM frequencies.

The simulated transmission parameters (in dB) of any of the two branches are shown in Fig. 6. It can be seen that the upper part in the schematic of Fig. 5 corresponds to the s_{31} parameter in the transmission parameter of Fig. 6 since it rejects the 960 MHz band and allows the 380 MHz band. In the same way, the lower part corresponds to the s_{21} parameter.

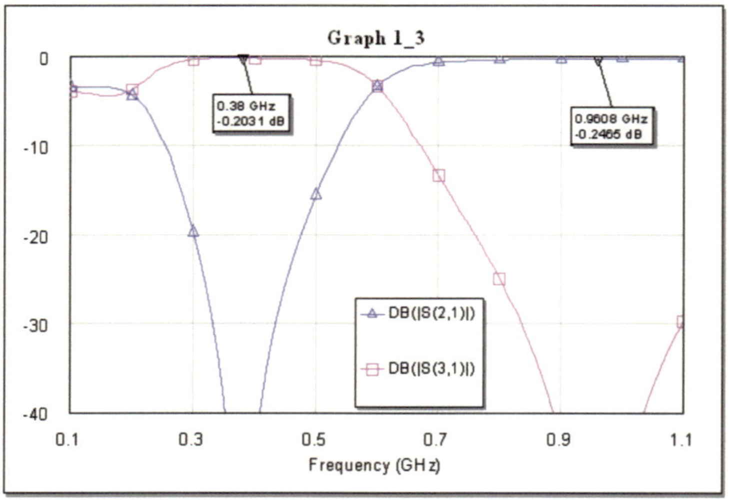

Fig. 6. Transmission parameters of any of the two diplexer branches

For the case presented before it can be seen that the simulated losses in the desired bandwidths (around 0.2 dB at any frequency) and the rejection parameter at the non desired frequency (lower than 40 dB) are enough for a good performance of the diplexer. If the diplexer frequencies were closer or a steeper slope would be needed a larger number of sections would be required. Fig. 7 shows a comparison between the transmission parameters (s_{31} and s_{21}) for a diplexer made with one D-CRLH cell (as in the previous case) and other diplexer made with four D-CRLH cells. It can be seen that the main difference is the slope of the rejection bandwidth. For the proposed example it can be seen that the number of sections can be done as low as possible, then the number of sections will be equal to 1.

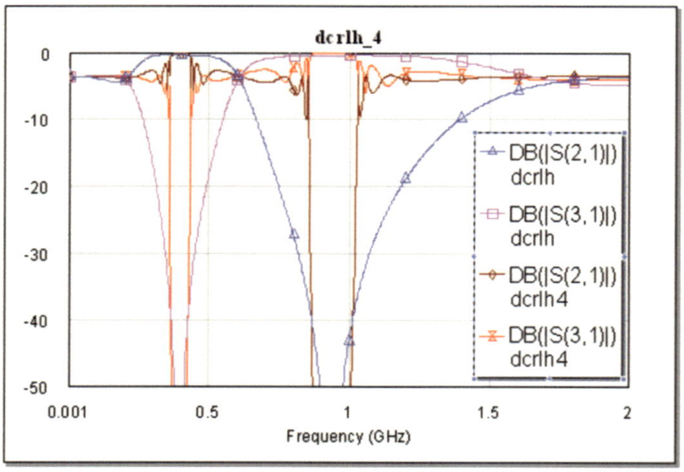

Fig. 7. Comparison between the transmission parameter for the previous one-D-CRLH-cell structure and a four-D-CRLH-cell structure.

Fig. 8. Photo of the proposed one-cell D-CRLH diplexer.

Finally, a prototype has been manufactured and measured. Fig. 8 shows a photo of the manufactured prototype. Fig. 9 shows the transmission and reflection measurement of the manufactured one-cell diplexer. The measurements show an excellent agreement with simulation. In addition, the insertion losses are lower than 0.4 dB at both frequency bands. The return losses are lower than -22 dB at each frequency band and at any diplexer port. Lastly the isolation between the two output ports is nearly 40 dB at both working frequencies. In addition, the compactness of the circuit is quite good and miniaturization has achieved for a diplexer circuit working in the UHF frequency bands.

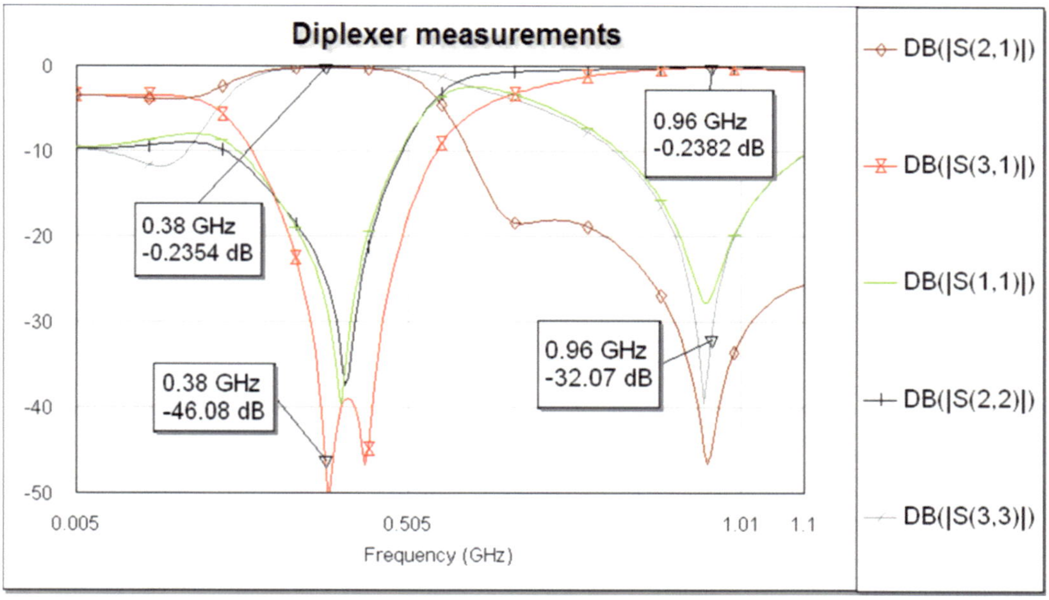

Fig. 9. Measurements of the transmission and reflection parameter for the one-cell D-CRLH diplexer.

4. Principles of active filters based on metamaterial cells.

One way of reducing the filtering losses is by means of active filters. The most usual microwave active filters can be classified into four main groups: filters based on negative resistance elements (Chang et Itoh, 1990), based on active inductors (Lucyszyn & Robertson, 1994), transversal, and recursive filters (Rauscher, 1985). In opposition to the first types, based on lumped elements, the two last ones are distributed filters, and therefore more adequate for high-frequency applications. The frequency selective response is generated by means of a combination of signal components that propagate along several electrical sections with different amplitudes and phase delays. The single difference between both schemes is that the transversal type only needs feed-forward branches, whereas the recursive one also includes feedback. This last approach is chosen as it is usually easier to design and produces more compact structures (the number of junction elements that occupy a large part of the total layout area, is in general smaller) (Rauscher, 1985). Of course, potential instability issues related to the feedback must be addressed (Billonet, Jarry & Guillon, 1995), but this is true for every microwave network that contains active elements, as they always show parasitic feedback.
The filtering characteristic of a first order recursive filter is generated by combination of the main path signal component with a properly weighted and delayed sample of the output of the network, which constitutes the feedback. The key of this process is therefore the phase response of the feedback section that must have certain value at the center frequency of the pass band. Thus, it can be implemented by means of a transmission line section. The objective is to extend the operation of recursive active filters to dual pass band frequency responses. In order to produce a dual-band response, the phase condition should be fulfilled at two different, controllable frequencies, and for this reason conventional transmission lines are not suitable. In fact, the solution proposed makes use of composite right/left-handed (CRLH) transmission lines (García-Pérez et al, 2009).

4.1 Theoretical design

Theoretical concepts of microwave active recursive filters derive from low frequency and discrete-time filtering techniques. An example of a first-order digital filter and its equivalent microwave recursive filter is shown in Fig. 10. In that figure *x(t)* and *y(t)* denote the time-dependent input and output signals respectively, a_0 and b_1 are gain weights, and τ represents a time delay. The frequency transfer function is given as

$$H(f) = \frac{Y(f)}{X(f)} = \frac{a_0}{1 - a_0 \cdot b_1 \cdot e^{-j \cdot 2\pi \cdot f \cdot \tau}} \tag{12}$$

The transfer function *H(f)* describes a periodic band-pass filter response. The center frequency of each band, f_0, is characterized by the total phase response of the loop Φ_{loop} being a multiple of 2π

$$\Phi_{loop} = -2\pi \cdot f \cdot \tau = -n \cdot 2\pi \Rightarrow f_0 = n \cdot 1/\tau, \tag{13}$$

where *n* denotes any integer number. Therefore, for those frequency values f_0, the signal coming from the feedback branch is combined in phase with the input signal, resulting in maximum gain values.

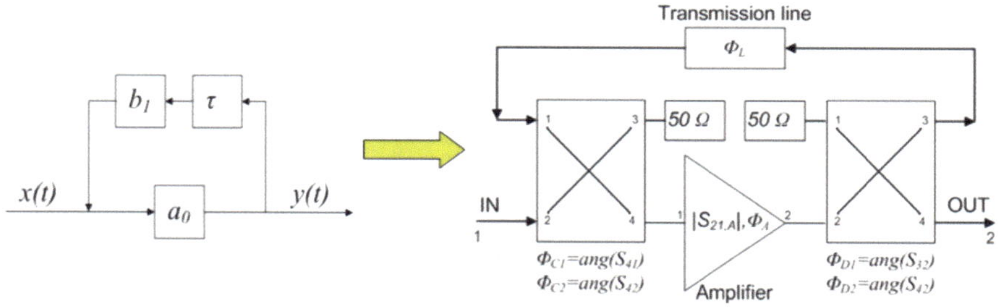

Fig. 10. Flow graph of a first order digital filter and equivalent microwave recursive active filter.

Although recursive filters of any order are possible, the most common cases are first-order topologies because of their simplicity. Of course, in order to fulfill stricter requirements of bandwidth or in-band ripple, higher order structures may be needed. However, the study presented here will be restricted to the simpler first order networks. The scheme presented in the left part of Fig. 10 can be transformed into a microwave filter circuit just by replacing all its parts with microwave components resulting in the right part of Fig. 10. For example, the time delay τ is implemented by means of a transmission line section while the weight coefficient a_0 corresponds to an amplifier. Also, the signals have been combined at the input/output ports by using branch-line couplers as combiners/dividers. Their power ratios can be absorbed into the term b_1.

In this context, the band-pass center frequencies from (13) are the values f_0 for which the loop phase condition is satisfied,

$$\Phi_{loop} = \Phi_A(f_0) + \Phi_{D1}(f_0) + \Phi_L(f_0) + \Phi_{C1}(f_0) = -n \cdot 2\pi \tag{14}$$

where $\Phi_A(f_0)$, $\Phi_{D1}(f_0)$, $\Phi_L(f_0)$ and $\Phi_{C1}(f_0)$ denote the phase responses of the amplifier, the power divider, the transmission line section and the power combiner respectively, all of them obtained at the band-pass frequency f_0, being n an integer number. Therefore, in the same way as the previous case, each operating frequency f_0 can be seen as the frequency at which the signal from the feedback transmission line is combined constructively (i.e.: with null relative phase shift) with the input power signal. This condition can be ensured by enforcing the loop phase to be multiple of 2π.

If the combiners were implemented with 3 dB combiners or power dividers then, if the filter gain function $|s_{21,F}(f_0)|$ were represented in front of the amplifier gain $|s_{21,A}(f_0)|$, two distinct operation zones could be observed (sub index $_F$ denotes filter while sub index $_A$ denotes amplifier). The first one corresponds to $|s_{21,A}(f_0)|$ under 6 dB, where $|s_{21,F}(f_0)|$ grows up with $|s_{21,A}(f_0)|$; in the second one $|s_{21,A}(f_0)|$ is over 6 dB and $|s_{21,F}(f_0)|$ decreases with it. There exists a value of the amplifier gain below which the circuit is unconditionally stable. That is $|s_{11,F}(f_0)|<0$ dB, at every frequency and for every combination of phases in the loop. This value corresponds to $|s_{21,A}(f_0)|=3.5$ dB when using 3 dB branch line couplers, but may take other values if different power combiners are used. From this value up to $|s_{21,A}(f_0)|=6$ dB the filter behaves as potentially unstable. For higher values ($|s_{21,A}(f_0)|>6$ dB) the circuit has been analytically demonstrated to be unconditionally unstable (Billonet, Jarry & Guillon, 1995) and (García-Pérez et al, 2009).

4.2 Requirements for dual band operation

Some important conditions must be addressed in order to design dual-band recursive filters. Firstly, the phase equation (14) must be simultaneously satisfied at both filtering frequencies. Since the phase delays introduced by the amplifier (Φ_A) and the power combiners (Φ_{D1}, Φ_{C1}) are fixed once those components are chosen, the only design variable is the phase delay of the feedback transmission line (Φ_L). For conventional single-band filters only one condition is imposed at the design frequency and, therefore, it can be fulfilled by choosing the correct length of transmission line. However, for dual-band responses two different conditions should be established and, in general, the linear phase delay exhibited by conventional transmission lines is not enough to simultaneously meet both of them. The inclusion of CRLH transmission lines as feedback lines is proposed, since their non-linear behavior provides more degrees of freedom. The second condition is that power combiners working simultaneously at both design frequencies are indispensable. They have a double purpose: obtain a good input/output match at the operating frequencies and isolate the input port with respect to the power coming from the feedback line avoiding stability problems, which are critical when working with feedback topologies.
The phase response exhibited by the whole CRLH transmission line at frequency f can be expressed as

$$\Phi_{CRLH}(f) = 2\cdot\Phi_{RH}(f) + N\cdot\frac{1}{2\pi\cdot f\cdot\sqrt{L_L\cdot C_L}} \tag{15}$$

where Φ_{RH} denotes the phase response of each transmission line section, N is the number of left handed T-cells, and L_L and C_L denotes the values of the lumped inductors and capacitors.
Usually, the phase delay given by a power combiner takes a fixed value at its operating frequencies, so (14) can be reduced to an expression only dependent with the phases of the transmission line section (Φ_L) and the amplifier (Φ_A). Since the phase delay of a branch line coupler is $\Phi_{C1}=\Phi_{D1}=\pi$ rad, the phase condition in (14) when using branch line couplers can be expressed as

$$\Phi_L(f)+\Phi_A(f) = -n\cdot 2\pi,\quad f = f_1, f_2 \tag{16}$$

where n can take any integer value, and f_1 and f_2 are the two desired band-pass frequencies. By substituting $\Phi_L(f)$ with the phase of a CRLH transmission line from (16), a system of two simultaneous equations is established. Its solutions (lengths of transmission line sections, number and values of lumped elements) constitute the design parameters used for the prototypes in next section.
Concerning the branch-line couplers two possible solutions could be available: dual frequency branch line couplers with CRLH lines (Lin et al, 2004) or dual frequency branch line couplers with parallel stubs fulfilling the branch line design equations (Cheng & Wong, 2004). The CRLH line-based solution has an important drawback with respect to the stub-loaded branch line. Apart from its complexity and the parasitic effects of the lumped elements, the CRLH line-based solution does not show transmission zeros at the intermediate band between the pass bands resulting in an overall spurious bands in the active filter (this will be shown in the next subsection of experimental results). For that reason this solution is discarded.

The second alternative shows a transmission zero located at the frequency halfway between the pass bands, for which the stubs behave as short circuits. As a result, rejection can be highly increased at the inner stop band. This is especially important if condition (14) is verified at the two pass bands for non-consecutive multiples of 2π, since spurious pass bands are also generated between them, in the stop band. The filtering characteristic of the branch line can be used to eliminate or at least mitigate these undesired transmission spikes. For that reason, this will be solution chosen.

4.3 Experimental results

In order to test the feasibility of the dual band recursive active filters described in the previous sections two prototypes have been designed and built: one with a CRLH branch line (that will be finally discarded) and other with parallel stubs branch line coupler.

4.3.1 Active filter with CRLH branch coupler

The first active filter has its pass bands centered at f_1=0.9GHz and f_2=1.6GHz, uses branch line couplers implemented with CRLH transmission lines as combiners. In order to reduce the total size of each combiner, a new modification has been introduced in their design that consists of placing two Schiffman lines sections at each horizontal branch of the hybrid, close to the left-handed sections. Each Schiffman section is a pair of coupled transmission lines which can be designed to be equivalent to a conventional transmission line, but needing less space. The result is a significant reduction of one of the two dimensions of each power combiner, and therefore of the whole circuit footprint. The active stage will be composed of the monolithic amplifier ERA-5+ of Mini-Circuits®. It is necessary to add an attenuator in series to maintain the gain under 6dB, in order to avoid instabilities.

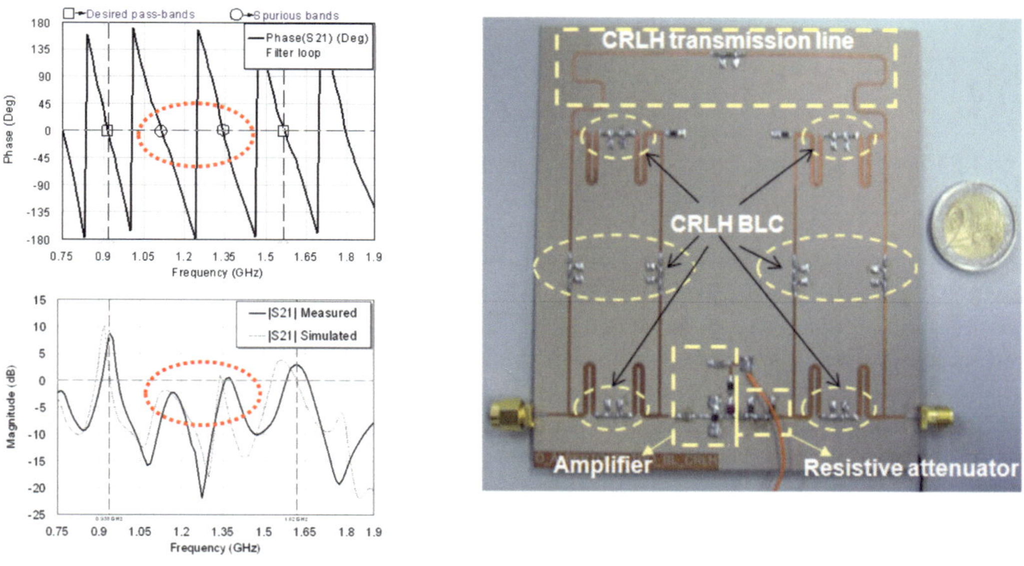

Fig. 11. Left part, phase response of the active filter showing the desired bands and the spurious ones. Right part, photo of the prototype with CRLH lines and Schiffman lines.

From Fig. 11 it can be seen that although the feedback CRLH line has been designed as electrically short as possible, the contributions in phase of the branch line coupler makes unavoidably the loop to be a section electrically long. This makes the phase condition to be fulfilled at undesired intermediate frequencies. The effect is the appearance of spurious pass bands as can be seen in Fig. 11. These undesired spurs at intermediate frequencies, between both desired pass bands, at 1175MHz and 1370MHz, are caused by the feedback loop phase response, that takes values near to a multiple of 2π rad. Then, the signal coming from the feedback line is constructively added at the input. These spurs, which appear out of the frequencies at which the hybrids are designed, are mitigated by the own transfer function of the hybrids, but are difficult to eliminate since the hybrid does not present a rejection band. For that reason this topology is discarded as a suitable one to achieve dual-band active filters.

4.3.2 Active filter with shunted stubs CRLH branch coupler

Another prototype has been designed and built with its pass bands centered at f_1=0.8 GHz and f_2=1.7 GHz. The shunted stubs branch line couplers present a more compact layout with very high rejection level between the pass bands. With regard to the active stage a single-stage distributed amplifier has been used to achieve a flat gain response over both band-pass frequencies. A resistive network has also been included to avoid instability problems. All these elements can be seen in Fig. 12.

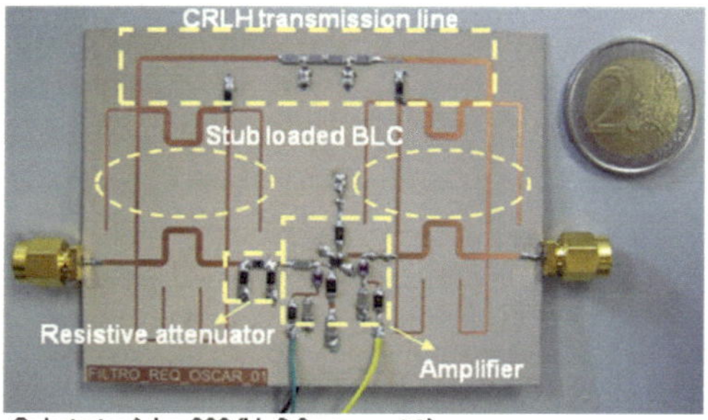

Substrate: Arlon 600 (H=0.6mm, εr=6.2)

Fig. 12. Photo of the prototype with shunted stubs branch line coupler

The feedback CRLH transmission line has been designed to fulfill two conditions: first, its phase response must satisfy (14) and, second, the equivalent characteristic impedance must be 50 Ω. Once again, although the feedback line has been designed as electrically short as possible, the complete phase loop including all the components delays is electrically long, and spurious bands appear unavoidably. Thus, two spike-shaped spurious pass bands appear at the inner stop band, corresponding to the non-desired solutions. Although non-desired solutions appear, they are cancelled out due to the high signal rejection introduced by the couplers in both the main path and the feedback loop. As stated before, this rejection is produced by the zero of the transmission and coupling coefficients associated with the branch-line port stubs. The amplitude response of the shunted branch line coupler can be seen in Fig. 13 where a strong rejection band appears between the two desired bands.
Due to this rejection band the two spikes closer to the pass bands are highly attenuated and have a reduced effect, with a transmission level of -8 dB at 1025 MHz and -6 dB at 1470 MHz. This can be seen in the overall response of the active filter that is shown in Fig. 14. The

upper part shows the phase response where the desired and spurious frequencies have been marked. It can be seen that two desired bands at 800 MHz and at 1700 MHz appear. However, three other spurious bands appear at 1025 MHz, 1175 MHz and 1470 MHz.

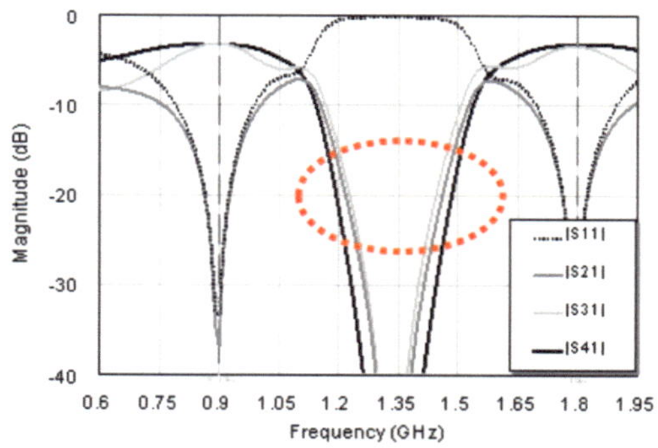

Fig. 13. Amplitude response of the shunted stubs branch line coupler.

The overall filter amplitude response is seen in the bottom part of Fig. 14. Due to the rejection of the branch line coupler the spurious band at 1175 MHz is completely rejected at levels lower than -30 dB. The other two spurious bands at 1025 MHz and 1470 MHz are also rejected presenting transmission levels lower than -6 dB. For that reason this topology for dual band active filters can be considered as suitable and will not be discarded as the previous one.

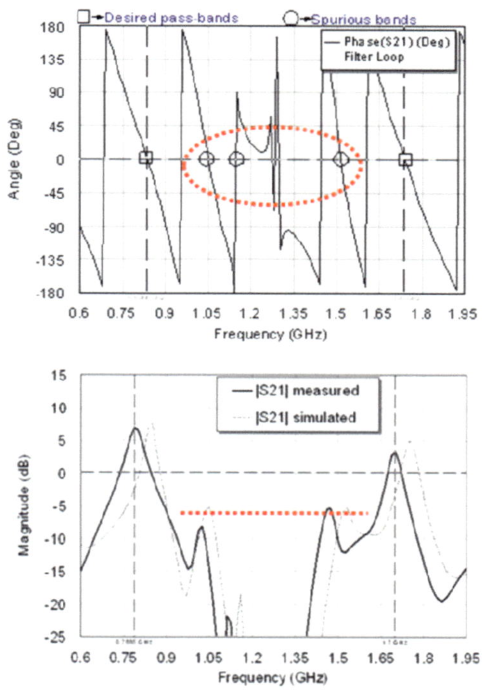

Fig. 14. Upper part, phase and amplitude response of the active filter showing the desired bands and the spurious ones (from García-Pérez et al, 2009).

5. Conclusion

This chapter is devoted to obtaining filtering and diplexing structures with losses as low as possible. First dual composite right/left handed cells have been proposed to reduce the losses; secondly active structures have been proposed to achieve dual band active filters.

The first part of the chapter is devoted to D-CRLH transmission lines as a good solution for the design of circuits that are alternatively used for rejecting or allowing different frequency bands. Then, a D-CRLH diplexer-circuit based on D-CRLH transmission lines has been proposed. A general design procedure for implementing arbitrary-frequency compact diplexers has been proposed. As the D-CRLH transmission lines de not present a frequency periodic performance, the proposed diplexer can be designed for every given frequency ratio f_2/f_1.

The second part of the chapter is devoted to the design of dual band active filters. A theoretical study describing the principles of this type of filters is developed through the text. Since the proposed scheme uses feedback sections, stability matters have to be taken into account. In this way, it may be necessary to limit the amplifier gain by adding attenuators in the active stage. Another important issue is the appearance of undesired spurious peaks at frequencies in the inner band. In order to mitigate their effect, branch-line couplers with shunted stubs are used to provide dual band operation and strong rejection between the pass-bands.

6. References

Caloz, C & Itoh, T. Electromagnetic Metamaterial Transmission Line Theory and Microwave Applications, IEEE Press and Wiley, ISBN 0471-66985-7, New York, , 2005.

R. A. Shelby, D. R. Smith & S. Schultz. Experimental verification of a negative index of refraction. Science, vol. 292, April 2001, pp. 77-79, ISSN 0036-8075.

Sanada, A.; Caloz, C. & Itoh, T., Characteristics of the composite right/left-handed transmission lines, IEEE Microwave and Wireless Component Letters, vol. 14, no. 2, Feb. 2004, pp. 68–70, ISSN 1531-1309.

I-Hsiang Lin, M. de Vicentis, C. Caloz, T. Itoh; Arbitrary Dual-Band Components Using Composite Right/Left-Handed Transmission Lines", IEEE Trans. On Microwave Theory and Techniques, vol. 52, no. 4, April 2004, pp. 1142-1149, ISSN 0018-9480.

K. Keung, M. Cheng; A Novel Approach to the design and implementation of dual-band compact planar 90 branch-line coupler", IEEE Trans. On Microwave, Theory and Techniques, vol. 52, no 11, November 2004, ISSN 0018-9480.

D. Castro-Galan, L. E. Garcia-Muñoz, D. Segovia-Vargas and V. Gonzalez-Posadas, Diversity monopulse antenna based on a dual frequency and dual mode CRLH rat-race coupler, Progress in Electromagnetic Research, PIERB, 2009, vol.14, pp. 87-106.

H. Okabe, C. Caloz and T. Itoh; A Compact Enhanced-Bandwidth Hybrid Ring Using an Artificial Lumped-Element Left-Handed Transmission-Line Section, IEEE Trans. On Microwave Theory and Techniques, vol. 52, no. 3, march 2004, pp. 798-804, ISSN 0018-9480.

Caloz C., Dual Composite Right/Left-Handed (D-CRLH) Transmission Line Metamaterial, IEEE Microwave and Wireless Component Letters, vol. 16, no. 11, Nov. 2006, pp. 585-587, ISSN 1531-1309.

V.González-Posadas, J.L. Jiménez, L.E. García-Muñoz, D. Segovia-Vargas, Novel Diplexer Made with Dual-Composite Right/Left-Handed Lines (D-CRLH), Proceedings on the Czech and Slovak Microwave and Radioelectronics Week, april 2008, Prague.

J. Bonache, I. Gil, J. García-García, F. Martín; Complementary splits ring resonators for microstrip diplexer design, Electronic Letters, vol. 41, no. 14, Jul. 2005, pp. 810-811.

Y. Horii, C. Caloz, T. Itoh; Super-compact multilayered Left-handed transmission line and diplexer application, IEEE Trans. Microw. Theory Tech., vol. 53, no. 4, Apr. 2005, pp. 1527-1534, ISSN 0018-9480.

J.K.H. Wong, K.G. Balmain and G.V. Eleftheriades, A diplexer based on the spatial filtering property of planar anisotropic transmission-line metamaterials, Proceedings IMS MTT 2006.

F. J. Herraiz-Martínez, V. González-Posadas, F. Iñigo-Villacorta and D. Segovia-Vargas, Low-cost Approach based on an Eigenfrequency Method to obtain the Dispersion Diagram in CRLH Structures, IEEE Microwave and Wireless Components Letters, vol. 17, no.1, January 2007, pp. 13-15, ISSN 1531-1309.

C.-Y. Chang, and T. Itoh, Microwave active filters based on coupled negative resistance method, IEEE Trans. Microw. Theory Tech., vol. 38, no. 12, Dec. 1990, pp. 1879-1884, ISSN 0018-9480.

S. Lucyszyn, and I. D. Robertson, Monolithic narrow-band filter using ultrahigh-Q tunable active inductors, IEEE Trans. Microw. Theory Tech., vol. 42, no. 12, Dec. 1994, pp. 2617-2622, ISSN 0018-9480.

C. Rauscher, Microwave active filters based on transversal and recursive principles, IEEE Trans. Microw. Theory Tech., vol. 33, no. 12, Dec. 1985, pp. 1350-1360, , ISSN 0018-9480.

L. Billonnet, B. Jarry, and P. Guillon, Stability diagnosis of microwave recursive structures using the NDF methodology, IEEE MTT-S International Microw. Symp Dig., vol. 3, May 1995, pp. 1419-1422.

O. García-Pérez, A. García-Lampérez, V. González-Posadas, M. Salazar-Palma, D. Segovia-Vargas, Dual band recursive active filters with composite right/left handed transmission lines, IEEE Trans. Microw. Theory Tech., vol.57, no 5, May2009, pp.1180-1187, ISSN 0018-9480.

K.-K. M. Cheng and F. L. Wong, A novel approach to the design and implementation of dual-band compact planar 90° branch-line coupler, IEEE Trans. Microw. Theory Tech., vol. 52, no. 11, Nov. 2004, pp. 2458-2463, ISSN 0018-9480.

Multifrequency and Multifunction Metamaterial-loaded Printed Antennas

Francisco Javier Herraiz-Martínez, Daniel Segovia-Vargas, Eduardo Ugarte-Muñoz, Luis Enrique García-Muñoz and Vicente González-Posadas
Radiofrequency Group, Carlos III University in Madrid
Spain

1. Introduction

Metamaterials can be broadly defined as electromagnetic structures engineered to achieve exotic or unusual properties (Caloz & Itoh, 2004; Eleftheriades & Balmain, 2005; Engheta & Ziolkowski, 2006; Marqués et al., 2007). Recently these features have been used in microwave and antenna engineering to develop devices with extraordinary properties. For example, microwave devices with extraordinary characteristics such as miniaturization or operation over multiple frequency bands have been developed (Caloz & Itoh, 2004; Eleftheriades & Balmain, 2005; Marqués et al., 2007). The effort in the antenna field has been put on the use of metamaterials for travelling-wave antennas and as substrates and superstrates to enhance the performance of the original antennas (Caloz & Itoh, 2004; Eleftheriades & Balmain, 2005; Engheta & Ziolkowski, 2006). One of the main applications of metamaterial structures in microwave engineering is the development of artificial Left-Handed (LH) Transmission Lines (TLs) (Caloz & Itoh, 2004; Eleftheriades & Balmain, 2005). These TLs are termed as LH because their behaviour is the dual of the conventional or Right-Handed (RH) ones. In the LH TLs, the electric field, magnetic field and propagation vectors form a LH triplet, allowing the propagation of backward-waves, contrary to the conventional case. If we consider a TL as the concatenation of infinite unit cells, the equivalent circuit model of a LH unit cell is a series capacitance and a shunt inductance which is the dual of a RH unit cell (a series inductance and a shunt capacitance) (Caloz & Itoh, 2004).

During the last years, wireless systems have achieved a great popularity and penetration in society. Cellular systems, positioning systems (GPS, Galileo), personal area networks (Bluetooth) and wireless local area networks (WiFi) are good examples. This fact has made that user terminals designed for two or more of these services are very common nowadays. From the antenna engineering point of view, the radiating elements for these terminals require challenging features. The first one is multifrequency, which means that the antennas must work at two or more arbitrary bands simultaneously. Another challenging feature is multifunctionality because in some cases different characteristics, such as polarization or radiation pattern, are required at each working band. Moreover, all these antennas must be small to integrate them into compact handheld devices attractive for the users. Furthermore, the use of cheap technologies is crucial for mass production. All these requirements cannot be easily achieved with conventional approaches. For that reason, novel technologies such as those based on metamaterial structures are being proposed to fulfil all of these requirements.

The proposed approach is called metamaterial-loaded printed antennas. It is based on conventional printed antennas loaded with a small number of metamaterial particles. Although metamaterial structures are ideally formed by an infinite number of unit cells, a small number of unit cells can be used to achieve devices with enhanced performance for practical purposes. For example, microwave devices such as filters or couplers have been developed with a small number of metamaterial cells (Caloz & Itoh, 2004; Eleftheriades & Balmain, 2005; Marqués et al., 2007). Furthermore, the use of a small number of particles does not increase the complexity and size of the antennas, providing easy design and manufacturing processes and allowing the antennas fit into modern devices. Finally, it is important to note that printed antennas have low profile, light weight, low cost and they are easy to integrate with circuitry and as elements of antenna arrays (Garg et al., 2001; James & Hall, 1989).

In this Chapter two kind of metamaterial-loaded printed antennas are presented. In Section 2 metamaterial-loaded printed dipoles are reviewed (Herraiz-Martínez et al., 2009). Basic theory and several examples are explained, showing the validity of this approach to develop multifrequency printed dipoles. Section 3 is devoted to microstrip patch antennas filled with LH structures (Herraiz-Martínez et al., 2008). A simple TL model of these antennas is used to demonstrate their multifrequency and multifunction features. Two different patch antennas filled with LH structures are studied, manufactured and measured, demonstrating the validity of the approach. Finally, the Chapter is concluded in Section 4.

2. Metamaterial-loaded Printed Dipoles

2.1 LC-loaded Printed Dipoles

The first experiment to show the validity of the metamaterial-loaded printed antennas consists of loading a simple printed wire antenna with LC parallel tanks. These cells have been chosen because most of the elemental metamaterial magnetic cells, such us Split Ring Resonators (SRRs) or spiral resonators, can be modelled as LC parallel tanks, as it was proposed in (Baena et al., 2005).

The reference antenna is an antipodal dipole printed on both sides of a dielectric substrate with height h (Fig. 1). Each half of the antipodal dipole is printed on one side of the substrate. The parameters of the dipole are the length L and the width W of each half. This configuration has been chosen because it avoids the use of a balun to feed the antenna. This is possible because the printed dipole is fed through a paired strips transmission line (Wadell, 1991) with a SMA connector soldered to the end of the line. The outer conductor of the SMA connector is soldered to one strip of the feeding line while the inner conductor is soldered to the other strip of the feeding line. The dimensions of the feeding line are the length L_f and the width W_f.

As an example, a dipole with L = 42.05 mm, W = 2.50 mm, L_f = 27.50 mm and W_f = 1.00 mm is considered. The substrate is the low-cost FR-4 (ε_r = 4.5, tan δ = 0.015 and h = 0.50 mm). These dimensions are chosen to obtain the resonant frequency of the fundamental mode (n = +1) at 1.5 GHz. When working at this mode, the current on the dipole has the $\lambda/2$ sinusoidal distribution, with minima at the edges of the dipole (open-circuit conditions) and maximum at the centre of the dipole (Fig. 2(a)). This current distribution provides the conventional dipolar-like radiation pattern (Fig. 2(b)).

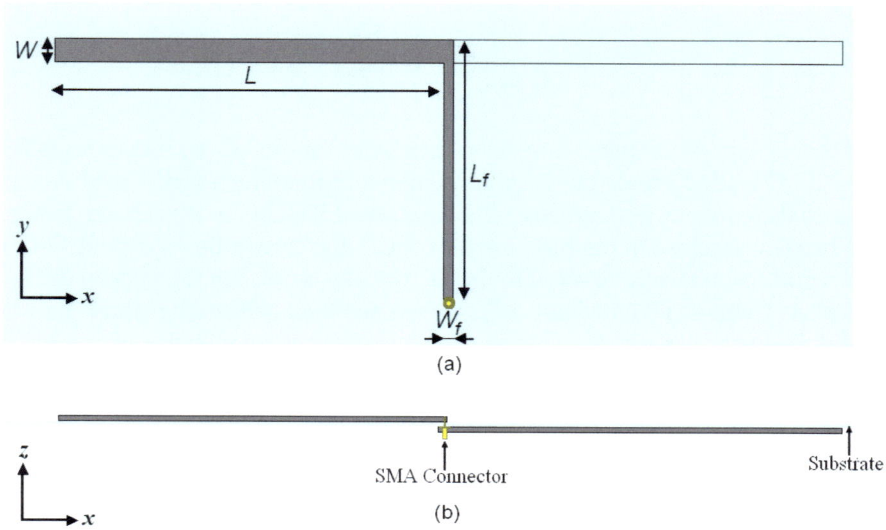

Fig. 1. Sketch of an antipodal printed dipole. (a) Top view. (b) Side view.

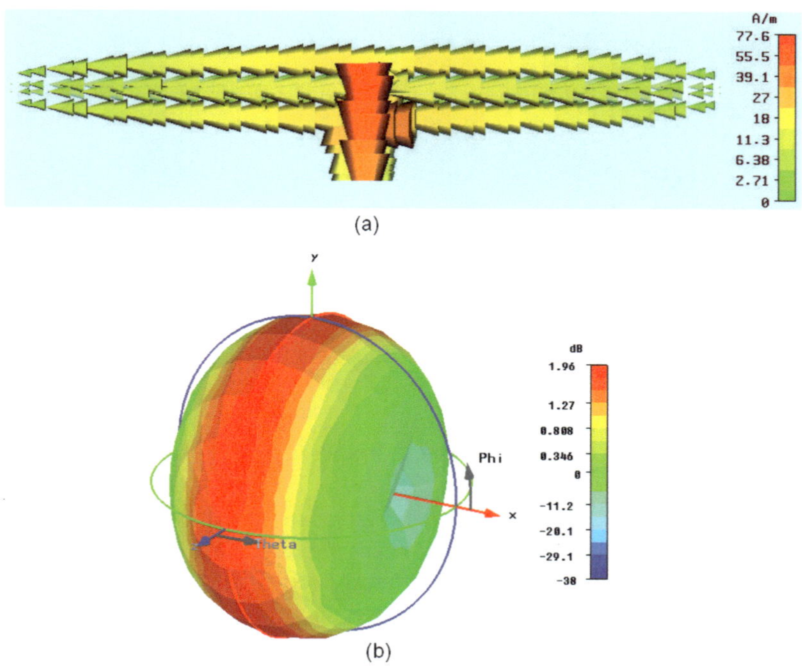

Fig. 2. Fundamental mode of an antipodal printed dipole. (a) Current distribution. (b) Radiation pattern.

To show the validity of the proposed approach, the reference printed dipole is loaded with one LC parallel tank per half, as it is shown in Fig. 3. The values of the LC components are L = 10 nH and C = 0.47 pF. These tanks are placed at a distance d = 35.00 mm from the centre of the dipole. The self-resonant frequency of the LC parallel tanks is computed as

$$f_{LC} = \frac{1}{2\pi\sqrt{LC}} \tag{1}$$

According to this expression the self-resonant frequency of the LC parallel tanks is 2.3 GHz. The proposed LC loaded dipole has been simulated and manufactured. This dipole presents an unusual dual-frequency performance (1.4 GHz and 1.9 GHz), as it is shown in Fig. 4. The first resonance is very close to the fundamental one of the conventional dipole. On the other hand, the second resonance is devoted to the LC parallel tanks, but it has been shifted down towards lower frequencies. It has been observed that when the ratio between the fundamental frequency of the dipole and the self-resonant frequencies of the LC tanks is larger than in this case, the frequency shift of the second resonance is considerably reduced. This is an important feature, because dual-frequency printed dipoles with arbitrary working frequencies can be designed.

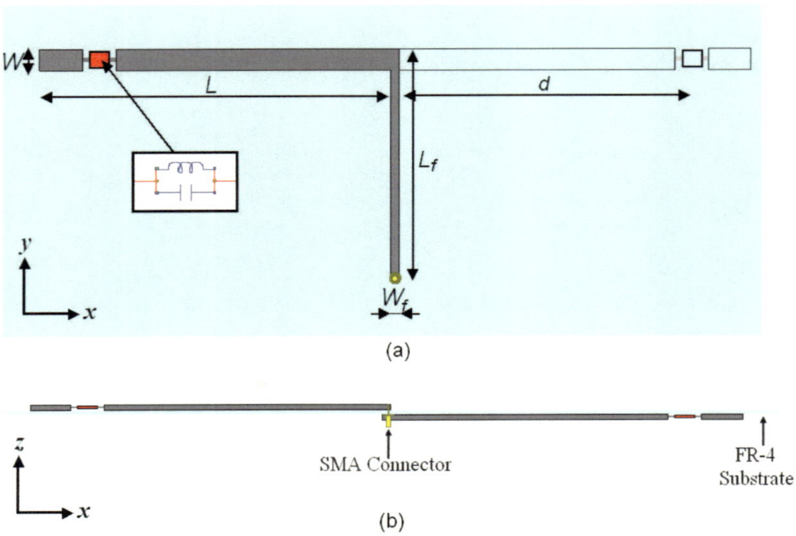

Fig. 3. Sketch of a LC-loaded antipodal printed dipole. (a) Top view. (b) Side view.

The currents on the dipole have a half-wavelength sinusoidal distribution (Fig. 5). At the first frequency they are similar to the unloaded dipole (Fig. 2(a)). On the other hand, the LC tanks impose a hard boundary condition (open-circuit) at the second frequency. This produces that there is almost no current between the tanks and the dipole edges. In this case, the minima are located at the tanks and the maximum is maintained at the feeding point. Both half-wavelength sinusoidal currents provide a dipolar-like radiation pattern (Fig. 6).

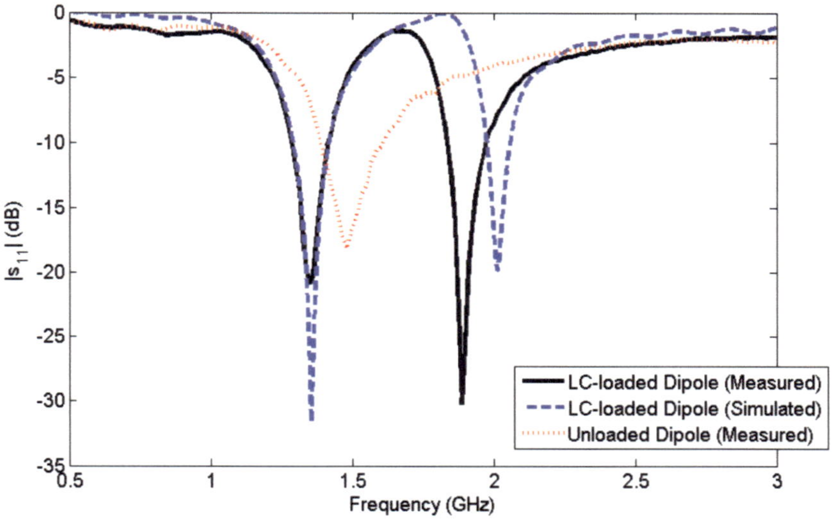

Fig. 4. Measured and simulated reflection coefficient of the LC-loaded dipole. The measured reflection coefficient of the unloaded dipole is also plotted.

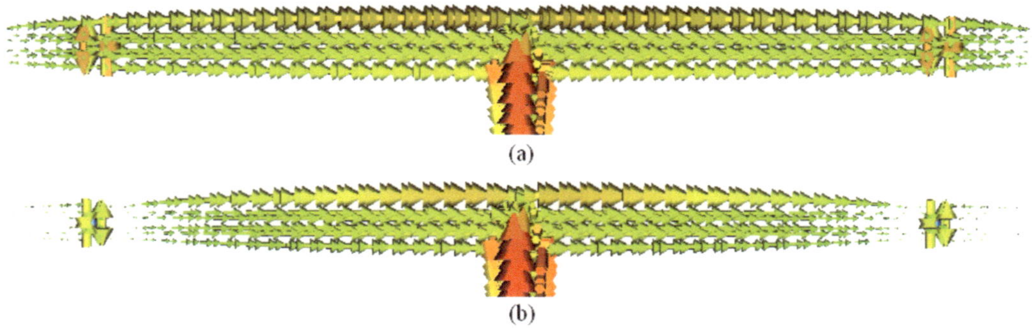

Fig. 5. Simulated currents on the LC-loaded printed dipole. (a) 1.4 GHz. (b) 1.9 GHz.

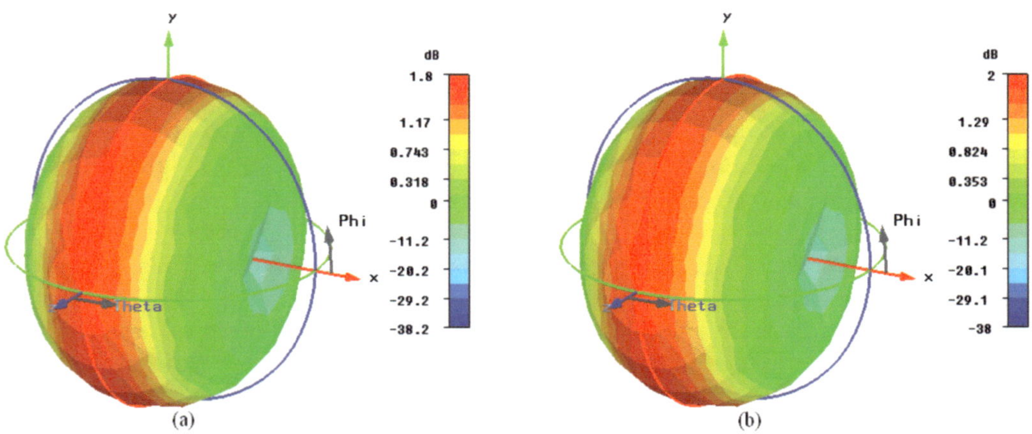

Fig. 6. Radiation patterns of the LC-loaded printed dipole. (a) 1.4 GHz. (b) 1.9 GHz.

2.2 Dual-frequency Metamaterial-loaded Printed Dipoles

The next step consists of loading the antipodal printed dipole with printed metamaterial particles in order to achieve a dual-frequency performance. One metamaterial particle or a set of them are coupled to each half of the dipole. In the example of Fig. 7, four SRRs are printed on the opposite side of each dipole half. This configuration has been chosen because it provides a proper matching within a wide range of ratios between the working frequencies. The SRRs parameters, according to Fig. 7(b), are the external radius R, the width of the strips W_{SRR} and the gap between strips g. The SRRs are placed at a distance c from the dipole centre. The separation between the centers of the SRRs is s. Moreover, other magnetic metamaterial particles can be used to obtain the desired multifrequency performance (spiral resonators, Omega particles ...). Regardless of the particular configuration, the metamaterial particles must be located in a way such the magnetic field generated by the currents on the dipole has an important component in the direction perpendicular to the plane of the particles. This is similar to the case of TLs loaded with metamaterial particles (Marqués et al., 2007).

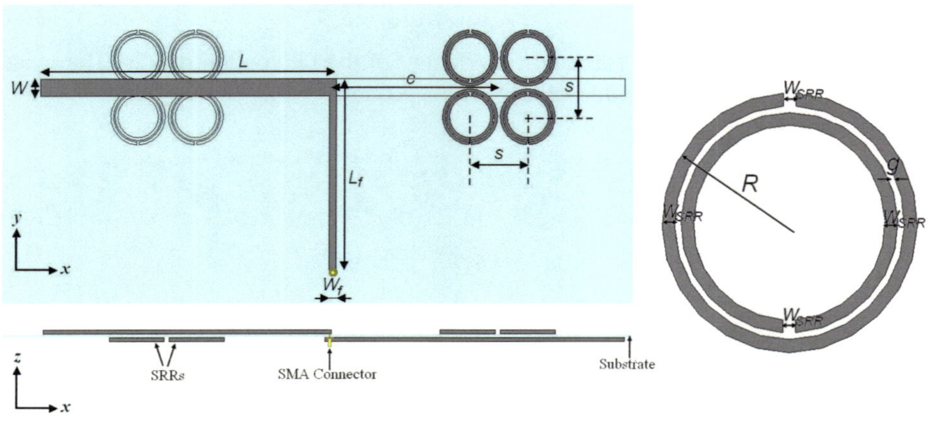

(a) (b)

Fig. 7. Sketch of a metamaterial-loaded printed dipole antenna: (a) Top and side views of the antenna with its design parameters. (b) Split Ring Resonator parameters.

A prototype of the proposed SRRs-loaded dipole antenna has been designed. The dimensions of the dipole are kept unchanged: L = 42.05 mm, W = 2.50 mm, L_f = 27.5 mm and W_f = 1.00 mm. The low-cost FR-4 substrate (h = 0.50 mm, ε_r = 4.50 and tan δ = 0.015) has been used. The prototype is loaded with a set of four SRRs per half, placed at a distance c =24.20 mm and the separation s is 8.40 mm. The parameters of the SRRs are the radius R = 4.00 mm, the width W_{SRR} = 0.40 mm and the gap between rings g = 0.20 mm. The theoretical resonant frequency of the unloaded antipodal dipole with these dimensions is 1.5 GHz, as showed in the previous Subsection. According to (Baena et al., 2005) the resonant frequency of the SRRs is 2.55 GHz.

The proposed configuration provides the desired dual-frequency performance. The first frequency (f_1) is close to the fundamental frequency of the dipole (n = +1 mode) while the additional frequency (f_{SRR}) occurs in the vicinity of the SRRs self-resonant frequency. Fig. 8 (obtained with Momentum) shows the current distributions on the proposed dipole at the two working frequencies. At f_1 (Fig. 8(a)) the current distribution is similar to the one obtained in the conventional $\lambda/2$ dipole (Fig. 2(a)). It can be appreciated that the effect of the SRRs at this frequency is negligible and nearly no current passes along them. This implies that the expected radiation pattern at this frequency is also similar to the reference dipole. In fact, the only modification in the antenna performance is a very slight frequency shift in the resonant frequency towards lower frequencies due to the capacitive parasitic effect of the SRRs. On the other hand, the working principle at f_{SRR} is somewhat different since the resonant frequency is imposed by the SRRs. At this frequency the SRRs are resonating, as can be seen in Fig. 8(b), where the currents through the SRRs are maximum. In this case, the SRRs are not radiating but imposing a hard boundary condition (an open circuit) where they are placed. In this way, the currents in the dipole are nearly zero from this position to the end of the overall dipole. Thus, the effect of the SRRs is similar to the one obtained with the LC parallel tanks in the previous Subsection. This means that the radiating element is formed by the current distribution on the dipole between the SRRs arrangements. It should be noted that this radiating element is a dipole with a length shorter than $\lambda/2$ sustaining a current between the edges where the SRRs are placed. Thus, it is expected that the proposed structure gives a dipolar-like radiation pattern at f_{SRR}. This is an important feature because the proposed dipoles not only present the desired dual-frequency performance but they keep the dipolar-like radiation pattern at both working frequencies.

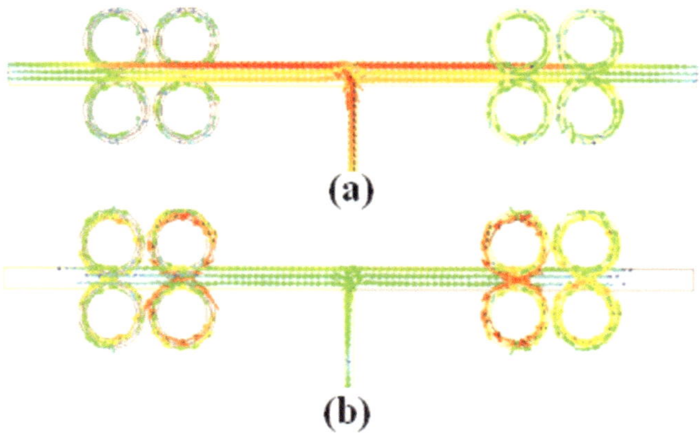

Fig. 8. (a) Currents on the SRRs-loaded dipole at f_1. (b) Currents on the SRRs-loaded dipole at f_{SRR}.

The proposed dual-frequency dipole and the conventional unloaded dipole have been manufactured (Fig. 9(a)). The measured reflection coefficients of both prototypes are shown in Fig 9(b). The reference dipole working frequency is 1.48 GHz, while the proposed SRRs-loaded dipole has the desired dual-frequency performance. The first resonance appears at f_1 = 1.32 GHz. The second working frequency (f_{SRR}) is 2.83 GHz. This frequency is shifted towards higher frequencies due to the overall coupling effects and the tolerances of the substrate and the manufacturing process. The bandwidth at the lower band is around 15% at the -10 dB level for both dipoles. On the other hand, the bandwidth at the additional band (at the -10 dB level) for the proposed dipole is much lower (1.27%). This is due to the high Q factor of the SRRs, what implies a much smaller bandwidth in the resonance imposed by them.

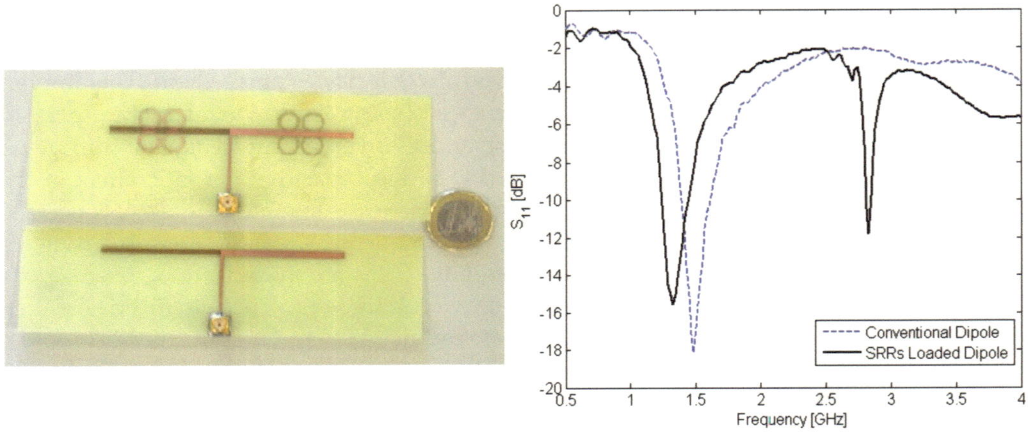

(a) (b)

Fig. 9. (a) Picture of the manufactured prototypes: SRRs-loaded antipodal dipole (top) and conventional antipodal dipole (bottom). (b) Measured reflection coefficient of both antennas.

Fig. 10 shows the measured radiation patterns of the proposed dual-frequency dipole at both working frequencies (1.32 GHz and 2.83GHz). Both of them are dipolar and similar to the conventional dipole. A ripple can be noticed especially at the YZ plane because of the presence of a metallic plate, part of the positioner, behind the antenna. The measured gain of the reference dipole is 1.99 dB while the gain of the dual-frequency prototype is 1.81 dB at f_1 and 0.67 dB at f_{SRR}.

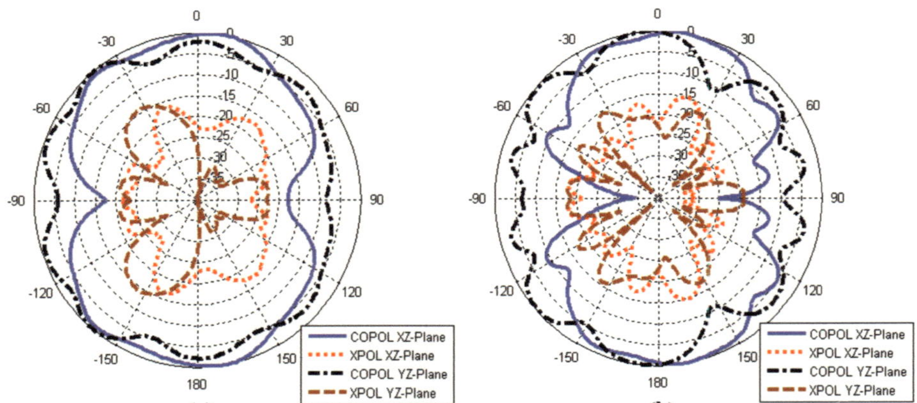

(a) (b)

Fig. 10. Measured radiation patterns of the proposed SRR-loaded dipole. (a) f_1 = 1.32 GHz (b) f_{SRR} = 2.83 GHz.

2.3 Multifrequency Metamaterial-loaded Printed Dipoles

The approach to obtain multifrequency printed dipoles (printed dipoles with three or more working frequencies) consists of exciting two or more additional resonances. These additional resonances are obtained with different pairs of metamaterial particles. At least a pair of metamaterial particles must resonate at each desired additional frequency. As an example, a triple-frequency printed dipole is proposed below.

According to Fig. 11, let us consider the case in which $R_2 = R_1 + 0.4$ mm. In this case, the resonant frequencies of the top and bottom pairs of SRRs are separated 0.22 GHz and a triple-frequency antenna can be obtained.

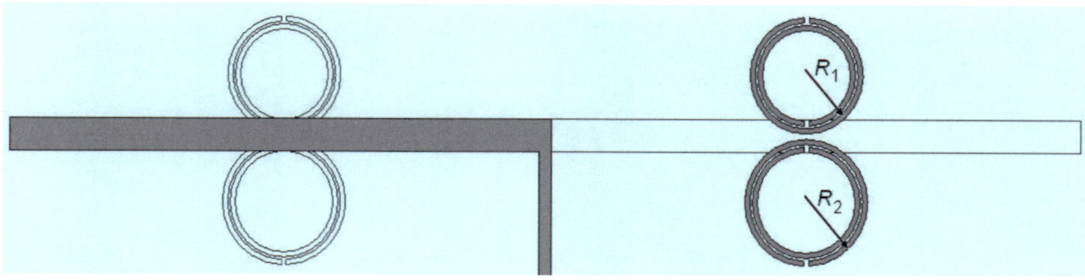

Fig. 11. Sketch of the proposed triple-frequency printed dipole loaded with SRRs.

Fig. 12 shows the simulated (CST Microwave Studio) reflection coefficient of the proposed triple-frequency antenna. The three working frequencies can be easily identified. The $|s_{11}|$ parameters of the dipoles loaded with both SRRs with $R = R_1$ and $R = R_2$ are also plotted. These plots show that the three resonances of the triple-frequency antenna are due to the fundamental frequency of the dipole and the self-resonant frequencies of the SRRs with different dimensions.

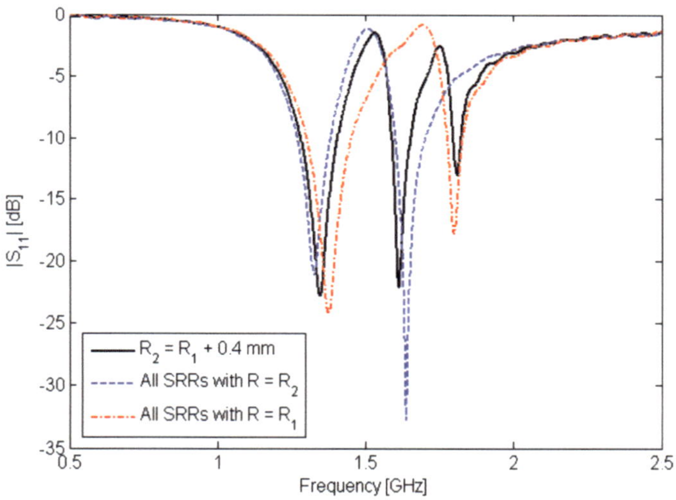

Fig. 12. $|s_{11}|$ parameter of the triple-frequency printed dipole antenna. The same information for the cases in which all the SRRs have the same dimensions is also plotted.

The simulated currents on the antenna at the three working frequencies are shown in Fig. 13. The distributions are similar to the one presented by the dual-frequency dipole (Fig. 8). The SRRs are not resonating at the first working frequency (f_1) and the currents on the dipole are similar to the unloaded dipole (Fig. 2(a)). The SRRs with R_2 are resonating at the second working frequency (f_2). Most of the current on the dipole is between the SRRs and there is almost no current between the SRRs and the edges of the dipole. At the third working frequency (f_3) the distribution is similar to the previous one (f_2) but the SRRs with

R_1 are resonating instead of R_2-SRRs. It is important to note that the radiation pattern is dipolar at the three working frequencies.

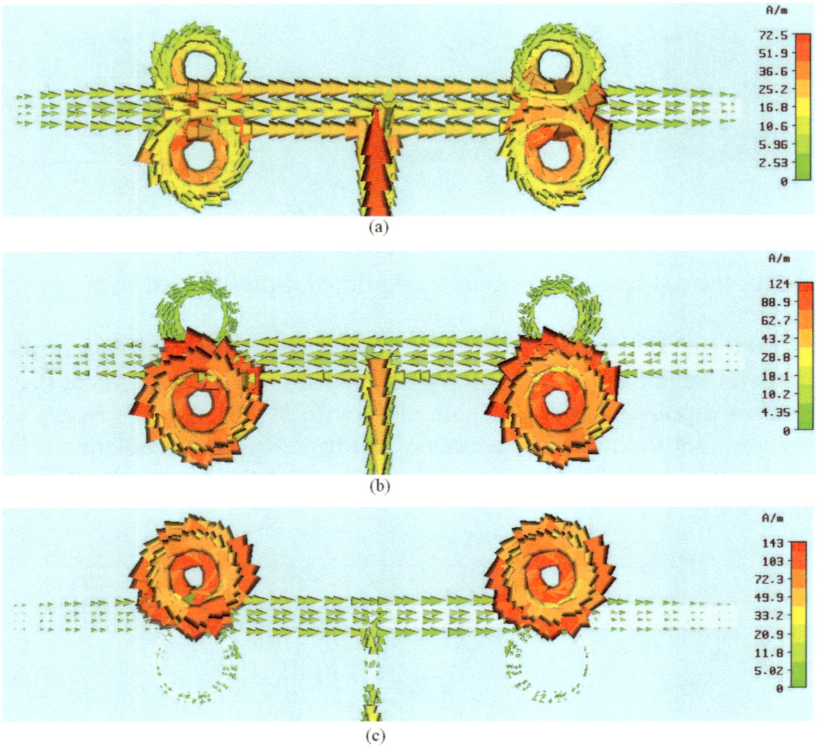

Fig. 13. Currents on the triple-frequency printed dipole loaded with SRRs. (a) f_1 (fundamental frequency of the unloaded dipole). (b) f_2 (resonant frequency of the SRRs with R_2). (c) f_3 (resonant frequency of the SRRs with R_1).

3. Microstrip Patch Antennas Filled with LH Structures

3.1 Simplified Transmission Line Model

A conventional microstrip patch antenna can be modelled as a RH TL. In this case, the resonant condition is given by the following equation:

$$\beta_n L = n\pi \tag{2}$$

where L is the equivalent TL length and n is the resonant index. As the TL has a RH behavior, the propagation constant β is always positive and linear with frequency. This means that all the modes have positive indices (n = +1, +2, +3, ...), all the resonant frequencies are harmonics of the fundamental one ($f_n = n\, f_{+1}$) and all the modes have $n\,\lambda/2$ electric field distribution, which means that all the modes different to the fundamental one have multiple lobes in the radiattion pattern. These characteristics make conventional patch antennas not suitable for multifrequency systems. The first reason is that arbitrary frequencies cannot be achieved because once the fundamental frequency is fixed, the other frequencies are always multiples of the fundamental one. The second reason is that radiation patterns with multiple lobes are not interesting for most of applications, since that implies a loss of directivity.

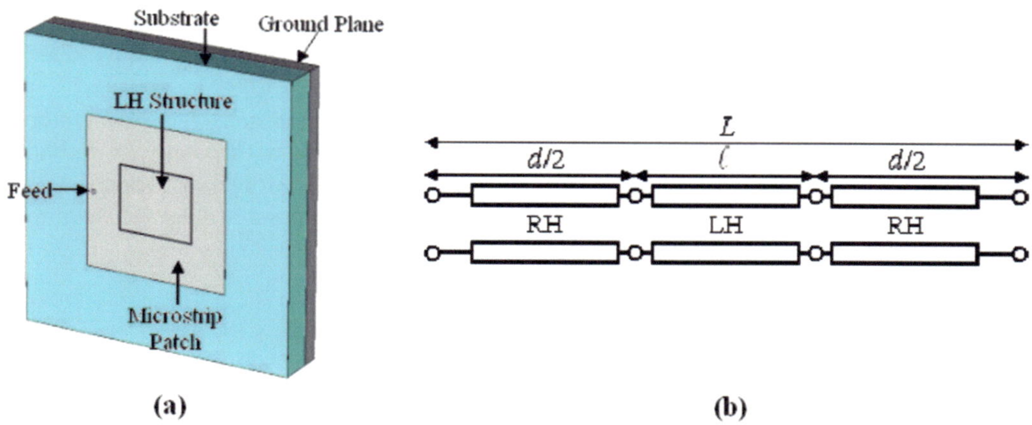

Fig. 14. Microstrip patch filled with LH structures. (a) Sketch of the antenna. (b) Equivalent TL model.

The proposed multifrequency patch antennas are based on a square microstrip patch filled with LH structures (Fig. 14(a)). For simplicity we can consider propagation along one main direction. In this case, the simplest equivalent antenna TL model is composed of a LH section between two RH sections (Fig. 14(b)).
The propagation constant is positive and linear with frequency in the RH sections while it is negative and proportional to $1/f$ in the LH sections. Then, in this case the resonant condition can be written as:

$$\beta_n L = \beta_n^{RH} d + \beta_n^{LH} \ell = k_1 f_n d - \frac{k_2}{f_n} \ell = n\pi \tag{3}$$

where k_1 and k_2 are positive constants; d and ℓ are the equivalent lengths of the RH and LH sections, respectively. In this case, it is possible to obtain modes with negative, zero or positive indices, contrary to the conventional case. Specifically, for a LH section composed of M unit cells, n takes values:

$$n = -M+1, -M+2, \ldots, -1, 0, +1, +2, \ldots \tag{4}$$

In particular if $M \geq 2$, two interesting modes are achieved below the mode equivalent to the fundamental one of the conventional patch antenna:
The first one is the n = -1 mode when the condition $\beta\ell = -\pi$ is satisfied. This mode has a half-wavelength electric field distribution similar to the fundamental mode of a conventional patch antenna. Thus, a patch-like radiation pattern is achieved at this mode.
The second one is the n = 0 mode when the condition $\beta\ell = 0$ is achieved. This mode has a uniform electric field distribution in amplitude and phase inside the patch antenna, which gives a monopolar radiation pattern (null at broadside). This type of radiation pattern cannot be achieved with conventional patches and only is present in short-circuited patch

antennas (González-Posadas et al., 2006). The possibility of using this mode provides a degree of multifunctionality due to the fact that radiation pattern diversity can be achieved with these antennas.

Moreover, the conventional (RH) modes ($n \geq +1$) are also present in the proposed antennas. It is important to note that the resonant frequencies are not forced to follow a harmonic ratio in this case. This is possible thanks to the non-linear behaviour introduced by the LH section.

In conclusion, two interesting kinds of multifrequency patch antennas can be developed with this approach. The first one is a dual-band patch antenna with patch-like radiation pattern at both bands by using the $n = \pm 1$ modes simultaneously. Moreover, when the $n = 0$ mode is also excited, a triple-frequency and dual-mode patch antenna is obtained. Examples of both antennas are presented below.

3.2 Triple-frequency and Dual-mode Microstrip Patch Antenna

The proposed antenna is shown in Fig. 15(a) where the patch dimensions (L x W) are 42 mm x 42 mm. The substrate is polypropylene (PP) with $\varepsilon_r = 2.2$ and $h = 10$ mm. The LH structure is implemented by using mushroom-type cells. These cells are based on microstrip patches grounded with vias and separation gaps between the cells. The vias provides the shunt inductances and the gaps the series capacitances of the LH section (Sanada et al., 2004). In particular, in this antenna the LH structure is based on a 2 x 1 mushroom-type cell array configuration and the dimensions of the mushrooms (L_m x W_m) are 10.6 mm x 17.8 mm. The diameter of the vias (d) is 0.7 mm, the gap between the two mushrooms (g_1) is 0.40 mm and the separation gap between the microstrip patch and the LH structure (g_2) is 0.20 mm. The antenna is fed through a coaxial probe placed 14 mm away from the centre. The dimensions of the ground plane are 80 mm x 80 mm. These dimensions were chosen to obtain the resonant frequencies at 1 GHz (GSM-900 band) for the $n = -1$ mode, 1.5 GHz (positioning systems) for the $n = 0$ mode and 2.2 GHz (UMTS) for the $n = +1$ mode.

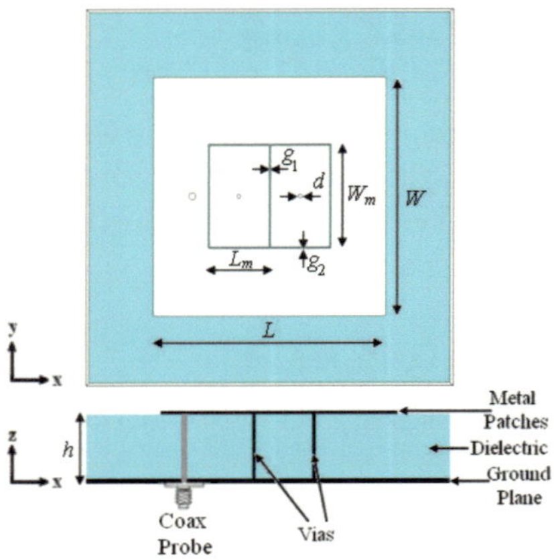

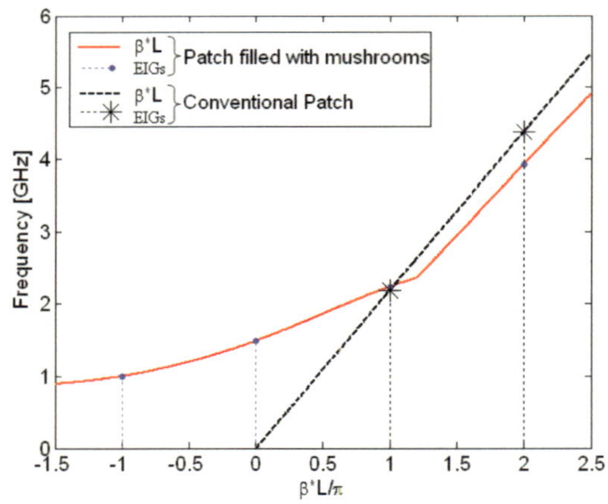

Fig. 15. Triple-frequency and dual-mode antenna based on a microstrip patch filled with LH cells. (a) Sketch of the antenna. (b) Electrical length (β L) and eigenfrequencies (EIGs) of the resonant modes of the proposed antenna (solid red line). The same information for the conventional patch antenna is also plotted in the dashed line.

Fig. 15(b) shows the electrical length of the proposed antenna versus frequency and the eigenfrequencies of the modes obtained through full-wave simulation. The same information for the conventional square patch antenna is also plotted. The electrical lengths have been computed by interpolating the discrete values of the eigenfrequencies. The linear ratio between the electrical length and frequency for the conventional patch antenna can be appreciated. On the other hand, the patch partially filled with the mushroom structures has the same relation for the modes with positive indices, but it has a LH behaviour at lower frequencies. This LH relation has been computed with the eigenfrequencies method (Herraiz-Martínez et al., 2007). The chart shows that for the index n = +1, the working frequency is very close to the fundamental frequency of the conventional patch (but not equal due to the residual LH effect of the metamaterial structure). As a first approximation for general design, the patch length is chosen to obtain the n = +1 frequency as the fundamental mode of the patch without mushrooms. On the other hand, the frequencies of the lower modes ($n \leq 0$) strongly depend on the mushroom structure selection. Large mushrooms lead to higher separation between the working frequencies and thus, higher slope in the LH region, while smaller mushrooms provide closer resonances and smaller slopes in the LH region of the electrical length chart. Then, the higher frequency (n = +1 mode) will mainly depend on the patch length (L), while the resonant frequencies of the other modes can be fixed by choosing the adequate mushroom parameters (basically L_m, W_m and the gaps, because the effect of the vias diameter d can be neglected). A detailed parametric study has been carried out in (Herraiz-Martínez et al., 2008).

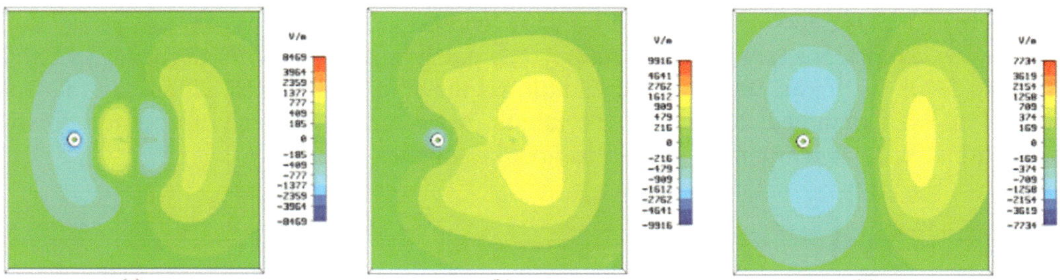

(a) (b) (c)

Fig. 16. Electric field distributions of the proposed antenna: (a) n = –1 mode (f_{-1} = 1.06 GHz), (b) n = 0 mode (f_0 = 1.45 GHz), (c) n = +1 mode (f_{+1} = 2.16 GHz)

The electric field distributions (CST Microwave Studio) for the modes with n = –1, n = 0 and n = +1 indices are shown in Fig. 16. The electric field distributions for the n = ±1 modes have half-wavelength electrical length (maximum in amplitude and 180° phase shift at the edges and null at the centre of the patch), similarly to the fundamental mode of a conventional patch antenna. A local 180° phase shift is observed inside the mushroom structure at f_{-1}, but this singularity does not affect the radiation behavior, as it will be shown in the radiation pattern of the experimental results. On the other hand, there is no local phase shift at f_{+1}. Lastly, the electric field distribution is almost uniform in phase and amplitude at f_0, as was predicted.

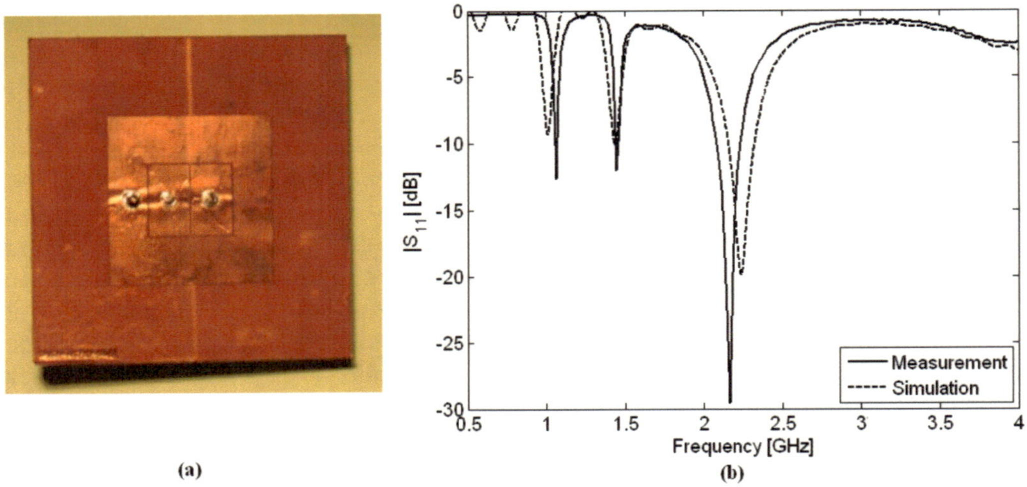

Fig. 17. (a) Picture of the triple-frequency and dual-mode patch antenna. (b) Simulated and measured reflection coefficient of the proposed antenna.

A prototype of this patch antenna has been manufactured (Fig. 17(a)). Fig. 17(b) shows the measured reflection coefficient. The return losses are -12.62 dB at f_{-1}= 1.06 GHz, -12.01 dB at f_0 = 1.45 GHz and -9.59 dB at f_{+1} = 2.16 GHz. The ratio between the resonant frequencies of the first and second modes is 1.37 and the ratio between the two dipolar modes is 2.04. The ratio between these modes can be arbitrarily chosen and depends on the patch and mushrooms dimensions, as explained before. The patch length is $\lambda_0/6.74$ at f_{-1}, $\lambda_0/4.92$ at f_0 and $\lambda_0/3.31$ at f_{+1}. It can be seen that a multifrequency antenna with different radiation modes has been obtained. Moreover, the length is strongly reduced with respect to the conventional $\lambda/2$ patch antennas at the additional frequencies.

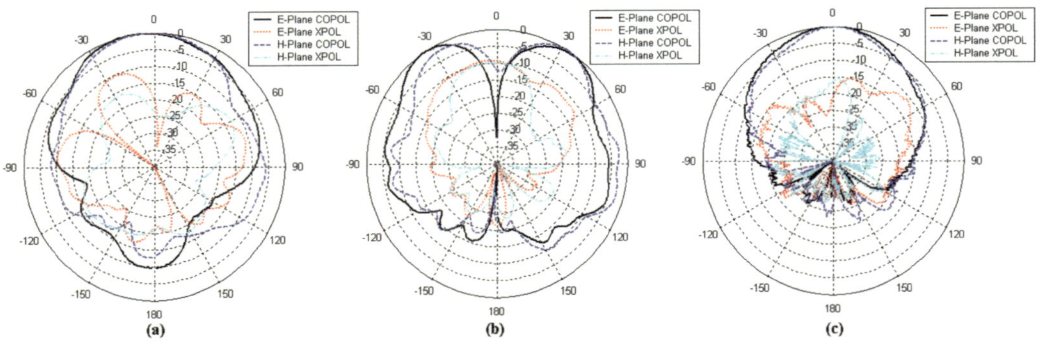

Fig. 18. Measured radiation patterns of the triple-frequency and dual-mode patch antenna. (a) n = -1 mode (1.06 GHz). (b) n = 0 mode (1.45 GHz). (c) n = +1 mode (2.16 GHz).

Finally, Fig. 18 shows the measured radiation patterns of the E-plane (x-z plane) and H-plane (y-z plane) and their corresponding cross-polar components. For the n = -1 mode (1.06 GHz) a patch-like radiation pattern can be seen in Fig. 18(a). The radiation pattern of the n = 0 mode (1.45 GHz) is monopolar as it can be seen in Fig. 18(b). A null in the broadside direction is appreciated. The depth of this null is 16 dB for the E-plane radiation pattern and

15 dB for the H-plane pattern. Two comments must be made concerning the n = 0 mode. First, the maximum of the radiation pattern that would be in the endfire direction is somewhat reduced due to the effect of the finite ground plane. Secondly, the n = 0 mode is excited in a weaker way than the n = 0 mode in a short circuited patch antenna (González-Posadas et al., 2006). This can be seen because of the higher level of the cross-polar component with respect to the short-circuited patch. The broadside radiation pattern is also achieved for the n = +1 mode (2.16 GHz, Fig. 18(c)). The measured gain of the antenna is -3 dB at f_{-1}, 1 dB at f_0 and 6.5 dB at f_{+1}. The gain of the additional modes (n = -1, 0) is reduced with respect to conventional patches because the electrical length of the patch antenna is also smaller due to the miniaturization achieved.

3.3 Dual-frequency Microstrip Patch Antenna

The proposed antenna is designed to work with similar radiation characteristics at 1.8 GHz (DCS band) and 2.2 GHz (UMTS band). Then, the first working frequency is obtained for the n = −1 mode while the second one is obtained for the n = +1 mode. Therefore, a patch-like radiation pattern is achieved at the two frequencies. The monopolar mode (n = 0) is not to be excited in this case. It has been seen that when the vias are aligned with respect to the probe, the ratio between the different working frequencies is higher and the n = 0 mode is excited. On the other hand, when the vias are not located in line with the probe, the frequency ratio between the patch-like modes (n = ±1) is reduced and the monopolar mode (n = 0) is weakly excited. For that reason, the number of cells in the y-direction has been doubled (resulting in a 2 x 2 LH structure instead of a 2 x 1) so the vias are not aligned with the feeding probe.

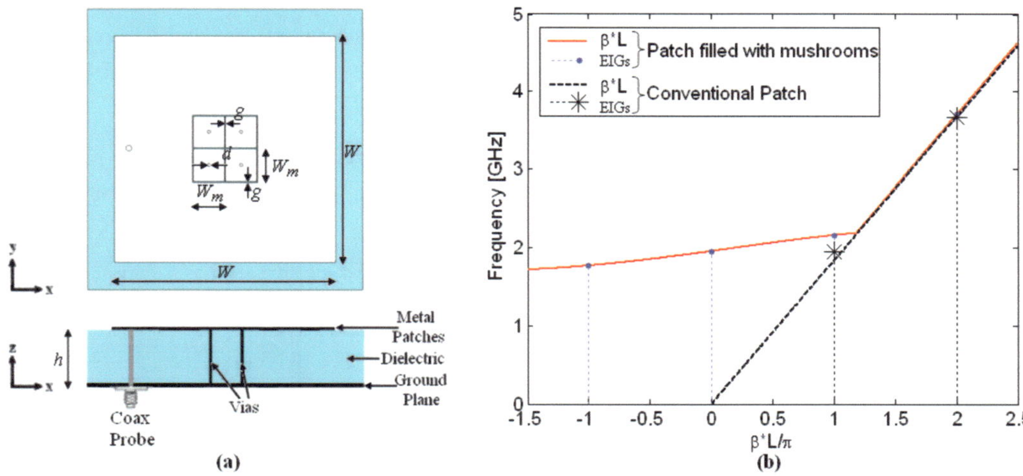

Fig. 19. Dual-frequency antenna based on a microstrip patch filled with LH cells. (a) Sketch of the antenna. (b) Electrical length (β L) and eigenfrequencies (EIGs) of the resonant modes of the proposed antenna (solid red line). The same information for the conventional patch antenna is also plotted in the dashed line.

Thus, the proposed design is shown in Fig. 19(a). It consists of a 48.2 mm square patch (W = 48.2 mm) partially filled with a 2 x 2 mushroom arrangement. The substrate is Polypropylene (PP) with ε_r = 2.2 and h = 8 mm. The mushrooms are squares of W_m = 6.8

mm, the vias diameter (*d*) is 0.70 mm and the separation gaps (*g*) are 0.2 mm. The patch is fed through a coaxial probe placed at 21 mm from the centre.

The modes of the proposed patch and its electrical length computed as in the previous case (Section 3.2) are plotted in Fig. 19(b). Once again, the antenna has a LH behavior at low frequencies and the conventional effect at higher frequencies. In this case, the slope in the LH region is lower than in the previous antenna (Fig. 15(b)) to produce the desired reduction in the frequency ratio between the patch-like modes.

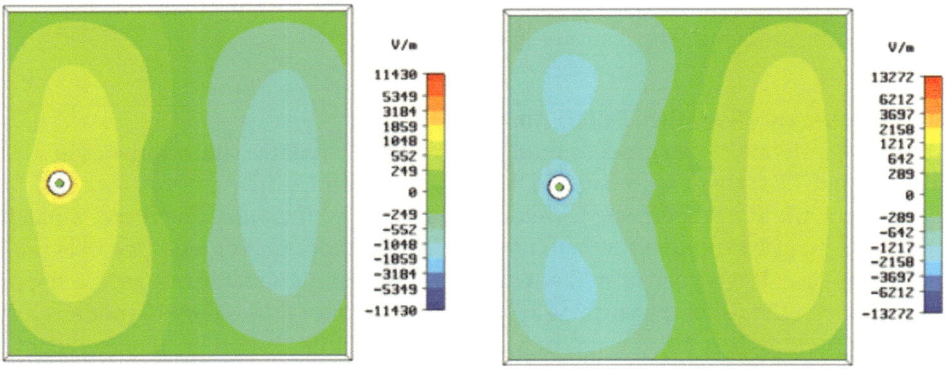

(a) (b)

Fig. 20. Electric field distributions for the proposed antenna: (a) n = –1 mode (f_{-1} = 1.81 GHz), (b) n = +1 mode (f_{+1} = 2.20 GHz)

Fig. 20 shows the electric field distributions (CST Microwave Studio) at the two working modes ($n = \pm 1$). Similar half-wavelength electric fields distributions are obtained at the two frequencies.

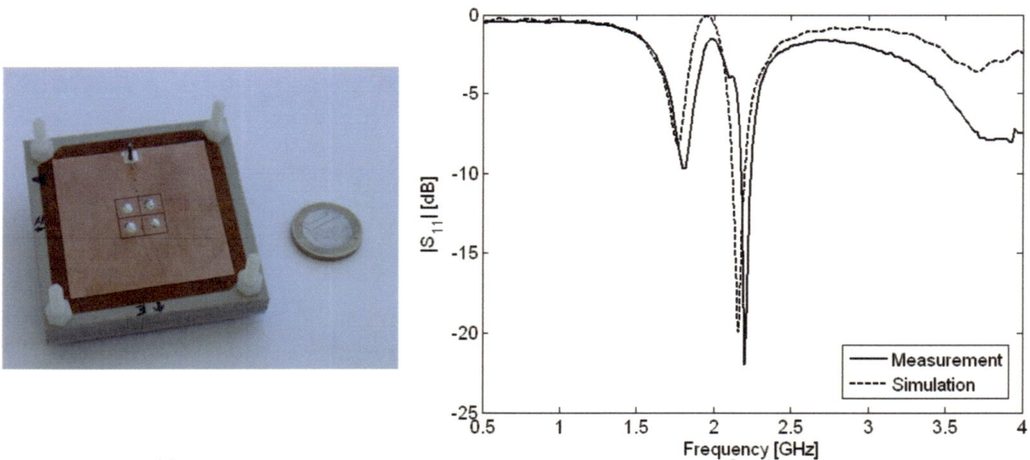

(a) (b)

Fig. 21. (a) Picture of the dual-frequency patch antenna. (b) Simulated and measured reflection coefficient of the proposed antenna.

Fig. 21(a) shows a picture of the dual-frequency antenna. The simulated and measured return losses are shown in Fig. 21(b). The measured return losses are −9.83 dB at f_{-1} = 1.81 GHz and −22.03 dB at f_{+1} = 2.20 GHz. The ratio between these two frequencies is 1.21, which is very difficult to achieve to achieve with another single-layer approach. In addition, a reduction factor in comparison with the conventional patch antenna has also been achieved for the n = −1 mode. In this case the patch length is $\lambda_0/3.44$ at f_{-1} and $\lambda_0/2.83$ at f_{+1}.

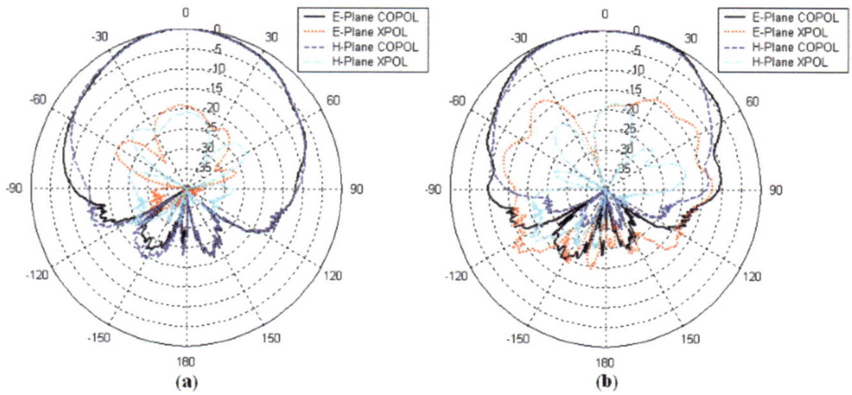

Fig. 22. Measured radiation patterns of the dual-frequency patch antenna. (a) n = −1 mode (1.81 GHz). (b) n = +1 mode (2.20 GHz).

The measured radiation patterns are shown in Fig. 22. The desired patch-like radiation pattern is obtained at both working frequencies. The XPOL component in the broadside direction is approximately −20 dB at both working frequencies. The measured gain of the antenna is 4.5 dB at the first working frequency and 6.8 dB at the second one.

4. Conclusion

Nowadays there is a huge demand on antennas with challenging requirements such us multifrequency, multifunctionality, miniaturization and low cost. All of these features cannot be achieved with conventional approaches. For that reason, novel approaches based on new technologies such us metamaterial structures are being developed. One of these approaches is the one presented in this Chapter: metamaterial-loaded printed antennas. In the present Chapter, it has been demonstrated the validity of this approach to develop antennas with such features. In particular, two kind of metamaterial-loaded antennas have been proposed. The first one is metamaterial-loaded printed dipoles and the second one is microstrip patch antennas filled with LH structures.

Regarding the first kind of antennas, three steps have been followed: initially the dipoles have been loaded with LC parallel tanks achieving dual-frequency performance. After that, the same characteristic has been achieved by loading the dipoles with metamaterial particles. Finally, metamaterials with different resonant frequencies have been used to achieve more than two working frequencies simultaneously. Several prototypes have been designed and manufactured showing good results.

For the microstrip patch antenna filled with LH structures, a simplified TL model has been presented to explain their multifrequency and multifunction behaviour. Moreover, two practical implementations have been proposed: a triple-frequency and dual-mode patch antenna and a dual-frequency patch antenna with reduced ratio between the working frequencies. All the results have been validated experimentally.

5. References

Baena, J. D.; Bonache, J.; Martín, F.; Marqués, R.; Falcone, F.; Lopetegui, T.; G. Laso, M. A.; García, J., Gil, I. & Sorolla, M. (2005). Equivalent circuit models for split ring resonators and complementary split ring resonators coupled to planar transmission lines. IEEE Transactions on Microwave Theory and Techniques, Vol. 53, No. 4, (April 2005), pp. 1451-1461, ISSN: 0018-9480

Caloz, C. & Itoh, T. (2004). Electromagnetic Metamaterials: Transmission Line Theory and Microwave Applications, Wiley-IEEE Press, ISBN: 0-471-669-857, New York

Eleftheriades, G. V. & Balmain, K. G. (2005). Negative-Refraction Metamaterials: fundamental principles and applications, Wiley-IEEE Press, ISBN: 0-471-601-462, New York

Engheta, N. & Ziolkowski, R. W. (2006). Metamaterials: Physics and Engineering Explorations, Wiley-IEEE Press, ISBN: 0-471-761-028, New York

Garg, R.; Bhartia, P.; Bahl, I.. & Ittipiboon, A. (2001). Microstrip Antenna Design Handbook, Artech House, ISBN: 0-89006-513-6, Norwood

González-Posadas, V.; Segovia-Vargas, D.; Rajo-Iglesias, E.; Vázquez-Roy, J. L. & Martín-Pascual, C. (2006). Approximate Analysis of Short Circuited Ring Patch Antenna Working at TM01 Mode. IEEE Transactions on Antennas and Propagation, Vol. 54, No. 6, (June 2006), pp. 1875-1879, ISSN: 0018-926X

Herraiz-Martínez, F. J.; González-Posadas, V.; Iñigo-Villacorta, F. & Segovia-Vargas, D. (2007). Low-cost Approach based on an Eigenfrequency Method to obtain the Dispersion Diagram in CRLH Structures. IEEE Microwave and Wireless Components Letters, Vol. 17, No.1, (January 2007), pp. 13-15, ISSN: 1531-1309

Herraiz-Martínez, F. J.; García-Muñoz, L. E.; González-Posadas, V. & Segovia-Vargas, D. (2008). Multi-frequency and dual mode patch antennas partially filled with Left-Handed structures. IEEE Transactions on Antennas and Propagation, Vol. 58, No. 8, Part 2, (August 2008), pp. 2527-2539, ISSN: 0018-926X

Herraiz-Martínez, F. J.; García-Muñoz, L. E.; González-Overjero, D.; González-Posadas V. & Segovia-Vargas, D. (2009). Dual-frequency printed dipole loaded with Split Ring Resonators. IEEE Antennas and Wireless Propagation Letters, Vol. 8, (2009), pp. 137-140, ISSN: 1536-1225

James, J. R. & Hall, P. S. (1989). Handbook of Microstrip Antennas, Peter Peregrinus, ISBN: 0-86341-150-9, London

Marqués, R.; Martín, F. & Sorolla, M. (2007). Metamaterials with Negative Parameters, John Wiley & Sons, ISBN: 978-0-471-74582-2, Hoboken, NJ

Sanada, A.; Caloz, C. & Itoh, T. (2004). Planar Distributed Structures with Negative Refractive Index. IEEE Transactions on Microwave Theroy and Techniques, Vol. 52, No. 4, (April 2004), pp. 1252-1263, ISSN: 0018-9480

Wadell, B.C. (1991). Transmission Line Design Handbook, Artech House, ISBN: 0-89006-436-9, Norwood, MA

23

Wideband planar plate monopole antenna

H. R. Hassani and S. M. Mazinani
Electrical & Electronic Eng. Dept., Shahed University
Tehran-IRAN

1. Introduction

The use of a single wideband antenna which covers a wide range of frequencies is very desirable for many applications including wireless and high data rate communication, position and tracking, sensing and imaging, and radar. Planar plate monopole antenna is a candidate. They are interesting due to their broad impedance bandwidth, linearly polarized omnidirectional azumuthal radiation pattern and are very cost effective to construct. They are planar structure, where a thin planar metal element can be used instead of the traditional wire element of a monopole antenna. It was first described by (Dubost & Zisler, 1976), who observed the wide impedance characteristics of this antenna. The antenna is capable of covering the 2-18GHz band with good radiating properties. This antenna can be used in various wireless communication applications, ranging from GSM1800, PCS1900, DCS, WCDMA/UMTS, the 2.45/5.2/5.8 GHz ISM bands, UNII, DECT, WLANs, the European Hiper LAN I, II, Bluetooth technology, and wireless local loop (WLL) 3.4-3.6 GHz and 10 GHz and UWB (3.1–10.6GHz).
In this chapter, important developments to the basic geometry of the planar plate monopole antenna are provided and discussed with respect to their impedance bandwidth, current distribution and their radiation patterns. Formulas that can be used to evaluate the frequency corresponding to the lower edge of the impedance bandwidth will also be provided. The basic antenna structures considered include the circular plate which yield very large impedance bandwidth and the rectangular (square) plate that provides a lower impedance bandwidth than the circular plate but its radiation pattern suffers less degradation over the bandwidth. Thus, most of the works reported in the literature on the subject of the planar plate monopole antennas are on ways of increasing the impedance bandwidth of the rectangular (square) monopole plate. This include: the bevelled square plate; the pin shorted square plate; dual fed and triple fed square elements and the rectangular monopole antenna loaded with small rectangular plates. Addition of slots to the planar monopole resulting in multiple frequency notch behaviour are also considered and discussed.

2. Planar Circular Monopole Antenna

One of the earliest monopole shapes whose properties were studied in the literature is the circular disc monopole antenna (Agrawall et al., 1998). Fig. 1 shows a metallic circular disc monopole, CDM, and an elliptical disc monopole, EDM, placed above a flat ground plane and fed through a coaxial feed via a narrow strip. VSWR results for an optimized disk, CDM and EDM, of thickness 1 mm above a ground plane of size 30 × 30 cm^2 are shown in Fig. 2. For the CDM antenna the height g = 1 mm gives the highest bandwidth ranging 1.17 to 12 GHz for a VSWR < 2. This range corresponds to bandwidth ratio of 1:10.2. The EDM antenna considered has an area equal to that of the CDM for comparison purposes. Fig. 2 also shows the VSWR of two EDM antennas with dimension a = 26 mm and b = 24 mm (with aspect ratio of 1.1) when fed either along the minor or the major axis. If the aspect ratio increases from 1.1 to 1.4, it can be shown that the bandwidth of the antenna decreases, Table 1.

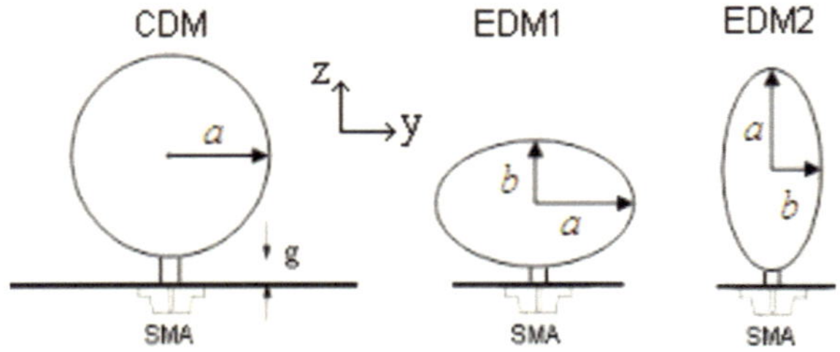

Fig. 1. Geometry of a circular and an elliptical disc planar monopole antenna. (Source: Agrawall et al., 1998).

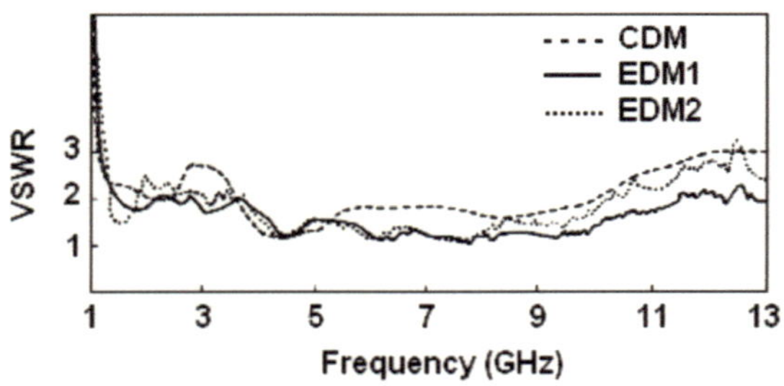

Fig. 2. VSWR of circular and elliptical disc planar monopole antenna. (Source: Agrawall et al., 1998).

Config.	a (mm)	b (mm)	Measured Freq. range for VSWR < 2 (GHz)	Theoretical Lower freq. for VSWR < 2 (GHz)	Bandwidth ratio
CDM	26	25	1.17 - 12.00	1.28	1:10.2
EDM1	26	24	1.21 - 13	1.31	1:10.7
EDM2			1.20 - 12.50	1.24	1:10.4
EDM1	27	23	1.38 - 11.49	1.37	1:8.3
EDM2			1.13 - 12.00	1.20	1:10.6
EDM1	28	22	1.37 - 11.30	1.41	1:8.2
EDM2			1.08 - 11.43	1.17	1:10.6
EDM1	29	21	1.58 - 10.45	1.46	1:6.6
EDM2			1.09 - 10.45	1.13	1:9.6

Table 1. VSWR bandwidth of CDM and EDM. (Source: Agrawall et al., 1998).

The frequency corresponding to the lower edge of the bandwidth of these monopole antennas can be determined approximately by equating the area of the planar configuration to that of a cylindrical wire of height l (which is same as that of planar disc height) with equivalent radius r given by $2\pi rl = \pi ab$. The length of a standard cylindrical monopole for real input impedance is given by (Balanis, 1982)

$$l = 0.24 \times \lambda \times F \tag{1}$$

Where $F = (\frac{l}{r})/(1+\frac{l}{r})$. From the above equations, the resonant frequency is given by

$$f = c/\lambda = (30\times 0.24) / (l+r) \quad \text{GHz} \tag{2}$$

where l and r are in centimetres. The theoretical frequencies calculated using the above equations for the discs considered in Table 1 shows agreement within ± 8 %.
From Table 1 it is seen that the simple circular disc can provide a very high impedance bandwidth. Fig. 3 shows the simulated E- and H-plane radiation patterns of the CDM antenna at three different frequencies over the bandwidth. From these results it can be seen that as the frequency increases from 2.5 to 9.0 GHz, the direction of maxima of the conical beam of the E-plane pattern varies from 30° to 60° from elevation, whereas the H-plane pattern remains nearly omnidirectional with maximum variation in azimuth increasing from 4 to 7 dB. The slight distortion in the patterns might be attributed to shape of the disc monopole, reflections from metallic surfaces and edge diffraction. These patterns are similar to that of a vertical linear monopole antenna of equivalent height on a finite ground plane, (Balanis, 1982).

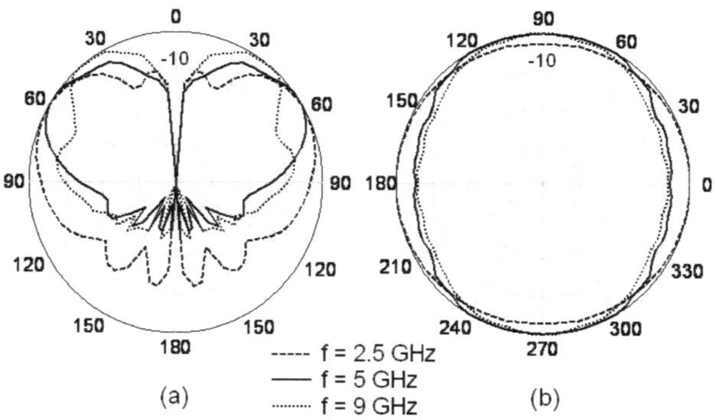

Fig. 3. The normalized radiation pattern of the CDM antenna (a) E-plane and (b) H-plane

Fig. 4 shows the distribution of current along the CDM antenna at two frequencies in the bandwidth. It can be seen that at the lower frequency, the current on the disc is uniform, but at the higher frequency, the current points in different directions. Thus, at higher frequency, we do not get a good omnidirectional pattern in the H-plane and the direction of the beam peak in the E-plane pattern varies from 30° to 60° from elevation.

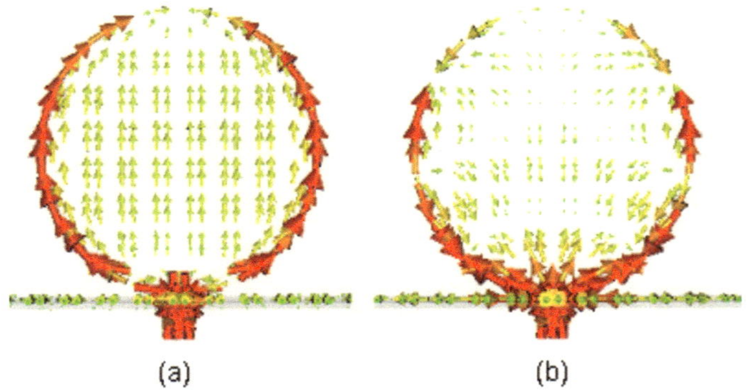

Fig. 4. The distribution of current along the circular planar monopole antenna at (a) 2.5 GHz and (b) 9 GHz

3. Square Planar Monopole Antenna

One of the simple planar monopole (PM) antenna shapes that have received a lot of attentions is the square (or rectangular) plate, Fig. 5. The square element is easier to deal with during the optimization process. The planar element is located a distance g above a ground plane, and is fed by a narrow strip through a SMA connector. The bandwidth of the PM antenna is set mainly by the radiating element dimensions, L and to obtain the maximum impedance bandwidth a suitable feed gap separation, g is required.

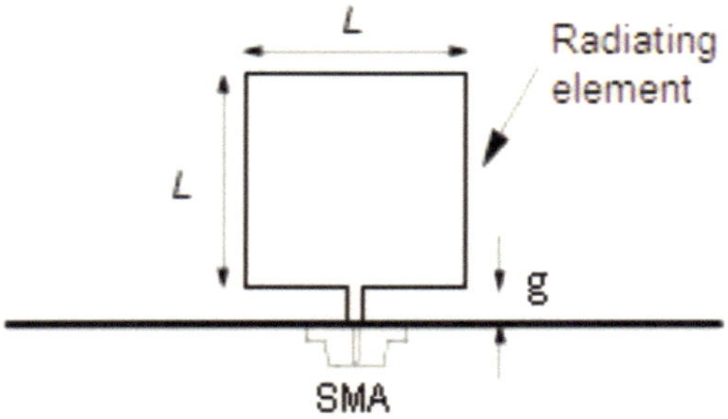

Fig. 5. The square planar monopole plate antenna above a square ground plane

The effect that the square PM antenna dimension and its separation from ground have on the impedance bandwidth can be obtained through the study of the return loss results, (Ammann, 1999); (Ammann & Chen, 2003a). In all cases the thickness of the planar plate monopole is 0.5 mm copper sheet and the square ground plane considered is of side 100 mm and the SMA connector has a feed-probe diameter of 1.2 mm. Table 2 gives for a square plate planar monopole of various dimensions, L, the lower and upper frequency limits, and hence, the bandwidth based on 10 dB return loss. In each case the feed gap has been optimized for the highest bandwidth.

As is typical for monopole antennas, the lower edge of the impedance bandwidth is inversely proportional to the overall length of the element. In the case of PM antenna, the overall length also includes the feed gap (i.e. $L + g$). Typically, the length of the square PM corresponds to about 0.21 of a free space wavelength at the lower-edge frequency; this is shorter than a quarter-wave monopole due to a reduced length-to-radius factor.

Square Size L (mm)	Frequency Limits (GHz)	Bandwidth (MHz)	Optimum Feed Gap (mm)
60	1.16 - 2.08	920	3
55	1.23 - 2.19	960	3
50	1.34 - 2.35	1010	3
45	1.44 - 2.59	1150	2.5
40	1.59 - 2.96	1370	2.5
35	1.86 - 3.53	1670	2.5
30	1.98 - 4.05	2090	2.5
25	2.38 - 5.20	2820	2.5
20	2.68 - 6.50	3820	2.2

Table 2. The impedance bandwidth for the square element of various dimensions, *L*. (Source: Ammann & Chen, 2003a).

The lower frequency limit of a rectangular radiating element, size *L*×*W*, can be determined from the following simple formula:

$$f_l = \frac{7.2}{\left(L + \frac{W}{\pi} + g\right)} \tag{3}$$

The above results shows that the impedance bandwidth is dependent on the feed gap, *g* and this gap must he optimized for maximum bandwidth. The frequency corresponding to the lower edge of the bandwidth is fairly independent of the feed gap, *g*, but the upper frequency is heavily dependent on it. This can be seen in Fig. 6, which shows the return loss for a 30 × 30 mm^2 square monopole with feed gaps of 0.8, 1.6, and 2.5 mm.

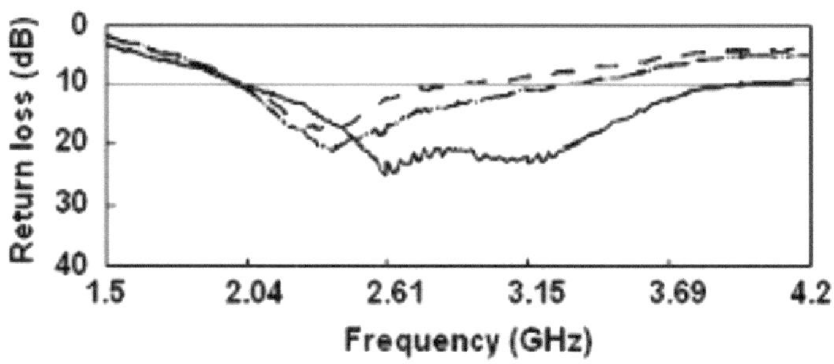

Fig. 6. The return loss of a 30 mm square monopole with feed gaps of 0.8 (dashed), 1.6 (dot-dashed), and 2.5 mm (solid). (Source: Ammann & Chen, 2003a).

Fig. 7 shows the E- and H-plane radiation patterns of the square PM antenna at various frequencies over the bandwidth.

Upon comparison of the square PM antenna with the circular PM antenna one can see that the circular plate provides higher impedance bandwidth but, unlike the circular PM antenna the patterns of the square (or rectangular) plate monopole are fairly stable with frequency. As such, majority of the works, reported in the literature, carried out on the PM antennas use the square (or rectangular) shaped plates.

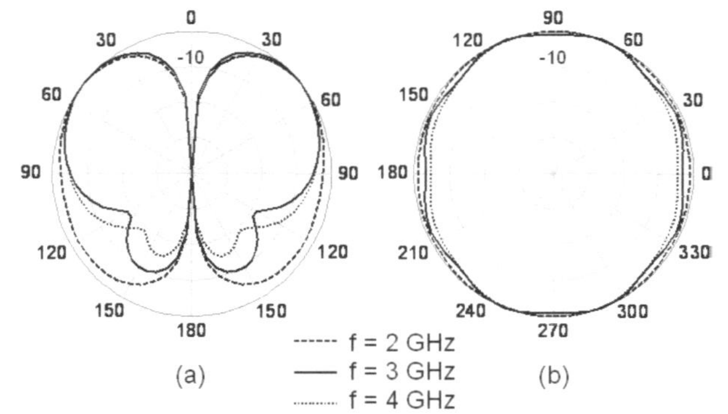

Fig. 7. The radiation pattern of the square monopole antenna at different frequencies, (a) E- (b) H-plane

To increase the impedance bandwidth of the square PM antenna, various techniques that change the basic shape of the monopole plate has been reported. These techniques include: cutting one or both edges of the monopole plate near the ground plane, so called bevelling; shorting the monopole plate to the ground; exciting the monopole plate antenna at two or three feeding points; and loading the monopole plate at its radiating edges with small rectangular plates. In the following these techniques are given.

4. Rectangular planar monopole antenna with bevel

A significant increase in impedance bandwidth can be achieved by cutting, or bevelling, the edge of the square monopole near the ground plane on one or both sides of the feed probe (Ammann, 2001), as shown in Fig. 8.

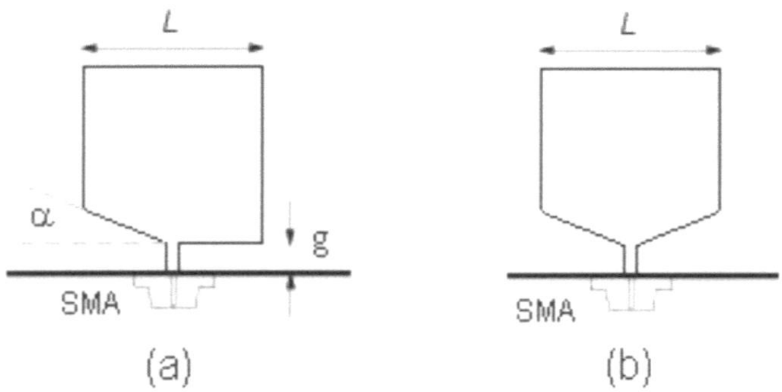

Fig. 8. Planar square monopole antenna with (a) asymmetrical and (b) symmetrical bevel

The basic antenna structure considered here is square shaped and is of side 25 mm, and thickness 0.2 mm, placed above a 150 mm square ground plane and fed via an SMA connector. The upper and lower edge frequencies of this simple square antenna are 2.35 and 4.95 GHz, representing an impedance bandwidth ratio of 2.1:1. One of the edges of the PM antenna can be cut, asymmetrically bevelled, and fine control of the impedance bandwidth can be achieved by varying the angle of the bevel. If the square element is bevelled by $\alpha = 10^{o}$ on one side of the feed probe, the upper edge frequency increases to 5.3 GHz. If the bevel is increased to $\alpha = 40^{o}$, the upper edge frequency increases to 6.0 GHz, while the lower edge frequency drops to 2.175 GHz. This represents an impedance bandwidth ratio of 2.75:1. If the planar element is symmetrically bevelled on both sides of the feed probe, the upper edge frequency is increased further. For a symmetrical bevel of $\alpha = 40^{o}$ a significant increase in the upper edge frequency, 12.5 GHz, can be achieved, representing an impedance bandwidth ratio of 5.75:1. Further increases in bevel do not increase the impedance bandwidth. The upper and lower edge frequencies for both asymmetrically and symmetrically bevelled PM antennas are given in Table 3.

Bevel α (degrees)	Bandwidth (GHz) Asymmetrically Bevelled	Bandwidth (GHz) Symmetrically Bevelled
0	2.35 - 4.95	2.35 - 4.95
10	2.20 - 5.30	2.12 - 5.95
20	2.19 - 5.75	2.11 - 6.75
30	2.17 - 5.97	2.10 - 7.25
40	2.17 - 6.00	2.10 - 12.50

Table 3. The impedance bandwidth of the bevelled square PM antenna. (Source: Ammann, 2001).

The radiation patterns of the asymmetric and symmetric bevelled monopole antenna are nearly constant with frequency over the bandwidth. The radiation patterns are quasi omnidirectional over the impedance bandwidth, the bevel has no noticeable effect on radiation patterns (less than a decibel). Fig. 9(a) shows the E-plane patterns of the symmetric bevelled monopole antenna at two frequencies. Fig. 9(b) shows the H-plane pattern, where the pattern is omnidirectional to within ±1.6 and ±2.9 dB at 2.4 and 5.8 GHz, respectively.

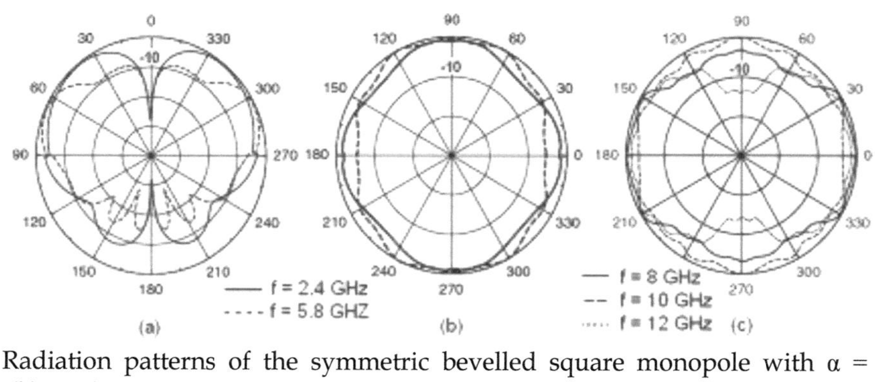

Fig. 9. Radiation patterns of the symmetric bevelled square monopole with $\alpha = 40^{o}$ (a) E-plane (b) H-plane (c) H-plane at higher frequencies. (Source: Ammann, 2001).

The maximum gain at 2.4 GHz is 2.7 dBi at θ = 52° but is 1.2 dBi at θ = 90°. At 5.8 GHz, the maximum gain is 4.6 dBi at θ = 65°, and is only 2.4 dBi at θ = 90°. The pattern of the 40° symmetrically bevelled element suffers some degradation at higher frequencies, Fig. 9(c). At these frequencies, the dimensions of this antenna are no longer small compared to a wavelength, and the radiation pattern exhibits some directivity.
The increase in impedance bandwidth of the PM antenna through bevelling can be described through Transmission Line Modelling (TLM) (Valderas et al., 2006). The study of the current flow on a PM antenna reveals that it is mostly concentrated in the vertical and horizontal edges. Fig. 10 shows the current amplitude distribution for a simple square PM antenna. The horizontal current distribution is observed to be focused on the bottom edge of the PM antenna, near the ground, where it is greater in amplitude than the vertical component. Thus, this edge will hardly contribute to the radiation. As such, the structure could be modelled as a transmission line loaded with the radiation resistance of the antenna Fig. 11a. Transmission line broadband matching techniques could then be applied to the antenna. The relations between the geometrical parameters of the monopole and the transmission lines are shown in Table 4. From the TLM point of view, bevelling technique is equivalent to forcing the horizontal transmission line to exhibit characteristic impedance that is a function of the distance from the feed point. On the other hand, continuously tapered lines are suitable techniques to achieve broadband matching. Both of these aspects are joined in the TLM applied to the PM antenna. A bevel, indeed, would act as a tapered line since a continuously increasing height above the ground plane means that the characteristic impedance will also increase continuously in a tapered line fashion, Fig. 11b.

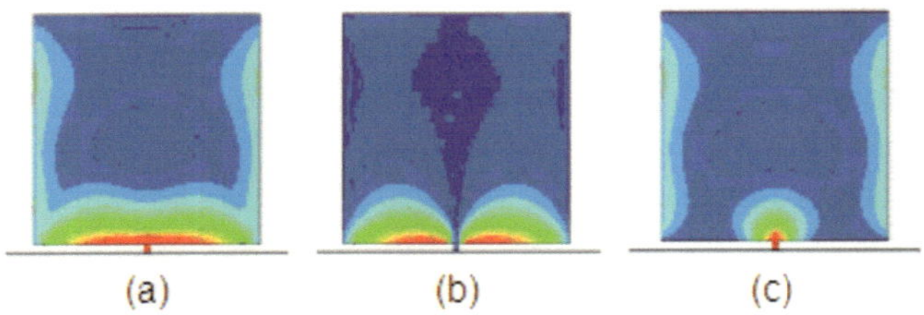

Fig. 10. The current amplitude distribution (a) absolute, (b) horizontal and vertical components. (Source: Valderas et al., 2006).

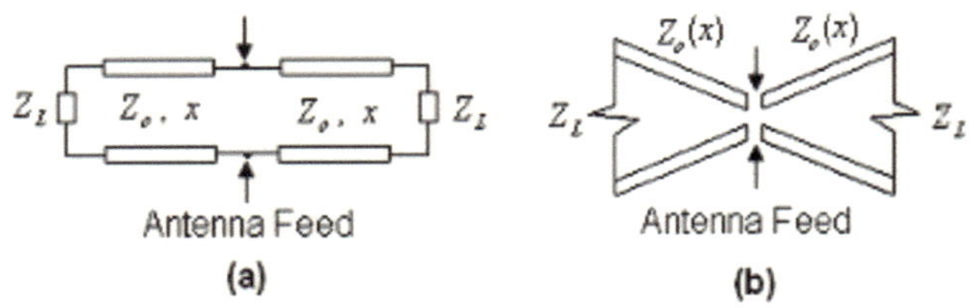

Fig. 11. Qualitative model of the (a) PM antenna (b) PM antenna with bevel based on TLM

PM antenna	TLM
g	Z_o
$\frac{L}{2}$	x
Radiating edge	Z_L

Table. 4. Relation between the planar monopole and transmission line parameters. (Source: Valderas et al., 2006).

5. Shorted planar monopole antenna

The monopole antenna with shorting post is shown in Fig. 12. The shorting post is located at one comer of the planar element, and could be cylindrical with about 1 mm diameter, or a strip with width 2 mm.

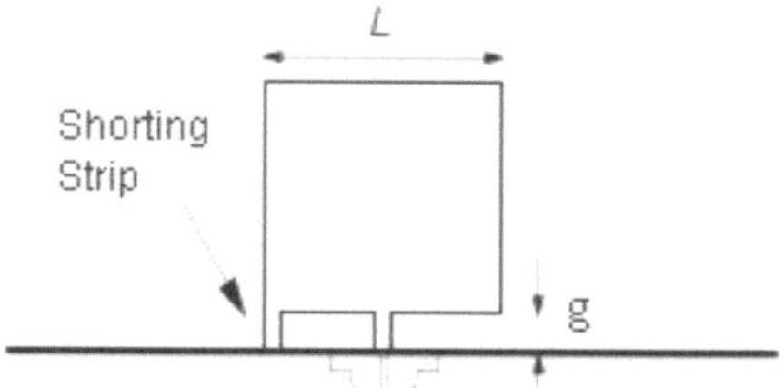

Fig. 12. Square planar monopole plate antenna with shorting post at one corner

Compared to the simple monopole antenna the feed-gap separation in the present structure needs to be reduced for optimum impedance bandwidth. Similar to other antennas, for a given planar monopole antenna, the use of shorting post has been shown to reduce the lower-edge frequency by introducing an extra mode, (Lee et al., 1999); (Ammann & Chen, 2003a). The presence of the post increases the antenna bandwidth and makes it smaller in height. Fig. 13 shows the return loss for a 25 × 25 mm^2 PM antenna with and without the shorting strip, which shows an impedance bandwidth of 114% and 76%, respectively. The asymmetry in the structure produces some distortion in the radiation pattern, particularly at the higher frequencies, Fig. 14.

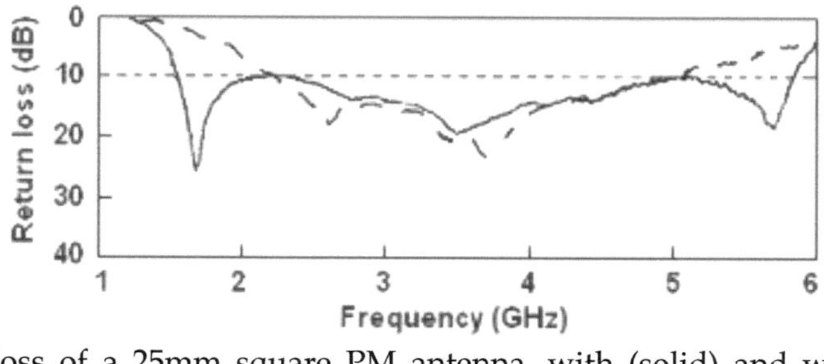

Fig. 13. Return loss of a 25mm square PM antenna, with (solid) and without (dash) the shorting post. (Source: Ammann & Chen, 2003a).

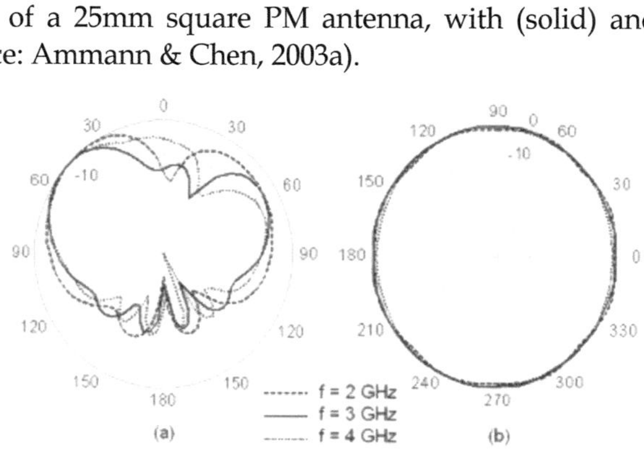

Fig. 14. Radiation patterns of the shorted square monopole antenna (a) E-plane (b) H-plane.

Fig. 15 shows the current distribution on the monopole antenna with and without the shorting post at one corner. It can be seen that for the simple PM antenna currents distribute symmetrically around the feed point. When the shorting post is added to the bottom edge, current distribution on the antenna becomes asymmetric and current is strong near the shorted edge and ground plane. Thus, the radiation from this edge is lower than the opposite edge. Because of this asymmetric, the radiation pattern becomes asymmetric and degraded in the side of shorting.

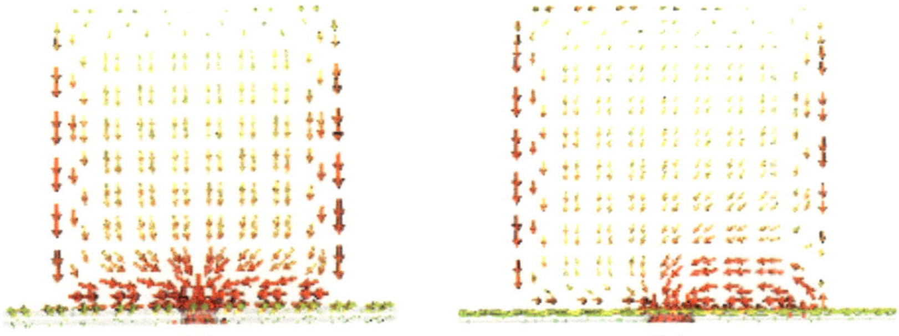

Fig. 15. Current distribution on a square monopole plate (a) without and (b) with shorting post

6. Shorted planar monopole antenna with bevel

It was shown earlier that addition of a bevel to one side of a square planar monopole, SPM antenna increases the upper edge frequency significantly and increases the bandwidth of the antenna. Control of this upper frequency is possible by adjusting the bevel angle. The addition of a shorting post to one side of the planar monopole also increases the bandwidth. A combination of shorting post and bevel can yield an impedance bandwidth of 800 MHz to 11 GHz, which is suitable for combinations of cellular and UWB systems (Ammann & Chen, 2003b). Fig. 16 shows a shorted PM antenna with bevel. The antenna is constructed using 0.2 mm thick copper sheet on a 200 mm square ground plane and fed via an SMA connector. The square plate dimension, L, is 60 mm, the shorting strip is 2 mm wide and the feed gap g is 1.2 mm. The monopole plate is bevelled on the side opposite the shorting strip.

The VSWR for the SPM antenna and SPM with shorting strip, SHPM, and SHPM with bevel, SHPMB, are shown in Fig. 17. As can be seen the shorting strip reduces the lower-edge frequency and the bevel raises the upper-edge frequency significantly. The plot shows the effect of introducing the bevel and shorting strip on the impedance bandwidth. The feed gaps are optimized for maximum bandwidth in each case. The 3 : 1 VSWR impedance bandwidths are 890–2250 MHz for the SPM antenna, 730–2500 MHz for the SHPM antenna and 850 MHz to >10.5 GHz for the SHPMB antenna.

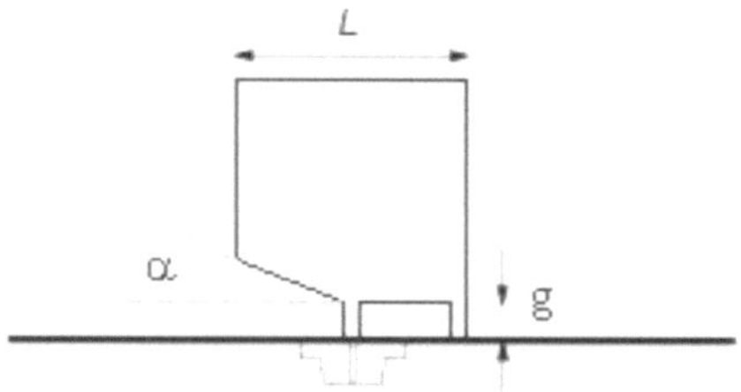

Fig. 16. The geometry of the shorted PM antenna with bevel

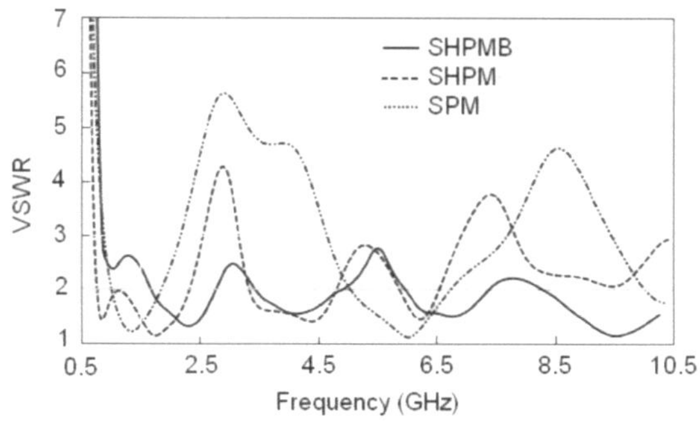

Fig. 17. The VSWR for the SPM, SHPM and SHPMB antennas. (Source: Ammann & Chen, 2003b)

VSWR plots over the range 0.5–10.5 GHz are shown in Fig. 18(a) for the SPM antenna with bevel angles of $\alpha = 0^{o}$ to $\alpha = 40^{o}$. It can be seen that the VSWR variation with frequency is reduced by the addition of the bevel. Fig. 18(b) shows the VSWR for the SPM antenna with both bevel and shorting strip. It can be observed that the VSWR remains below 3: 1 over most of the frequency range when $\alpha = 10^{o}$ and 20°. The bevel tends to increase the VSWR slightly at the lower frequencies but reduces the VSWR significantly at higher frequencies. For $\alpha = 10^{o}$, the lower and upper-edge frequencies (3: 1 VSWR) are 790 MHz and >10.5 GHz, respectively. The VSWR is slightly greater than 3: 1 only in the regions from 1.05–1.22 GHz and from 7.5–8.2 GHz.

The radiation patterns are shown in Fig. 19. The plots indicate nearly omnidirectional patterns in the H plane and typical monopole patterns in the E plane cuts at the lower frequencies. The plots are given at 900 MHz, 2.4, and 5.8 GHz.

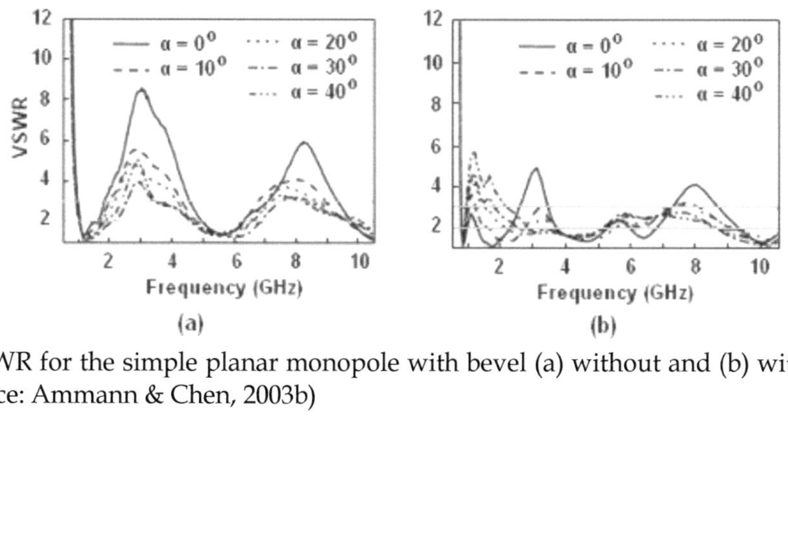

Fig. 18. VSWR for the simple planar monopole with bevel (a) without and (b) with shorting strip. (Source: Ammann & Chen, 2003b)

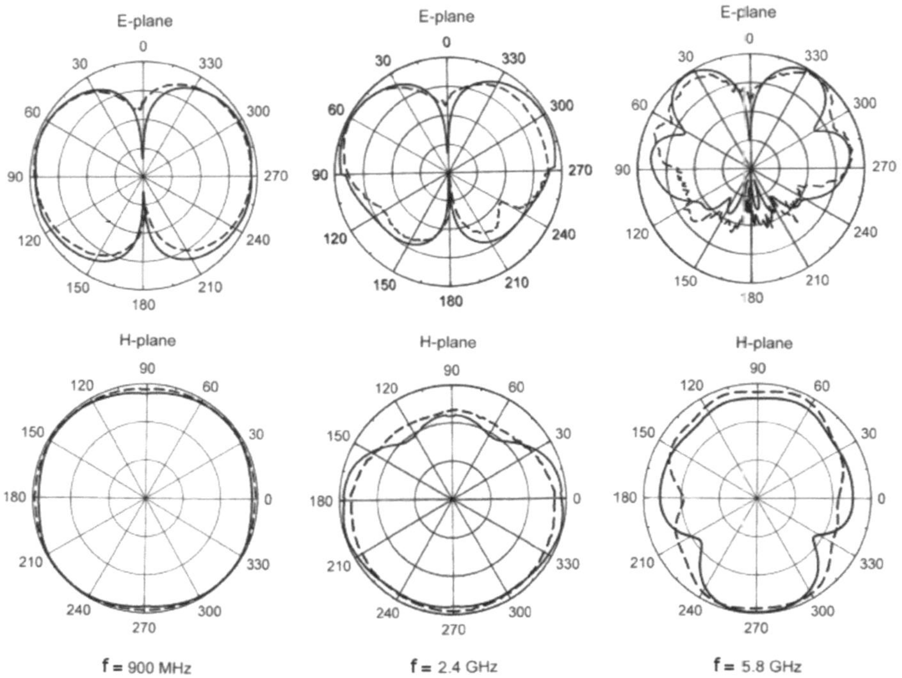

Fig. 19. The E and H-plane radiation patterns of the SHPMB antenna. (Source: Ammann & Chen, 2003b)

7. Monopole Antenna with a dual and triple Feeding Strip

The theory of characteristic modes (Harrington & Mautz, 1971) has been used by (Antonino-Daviu et al., 2003) to investigate the modes associated to a square PM antenna. From the knowledge of mode behaviour, it is shown in that paper that the performance of the antenna could be improved by using a double feed configuration. Such a double feed generates a pure and intense vertical current distribution in the whole structure and avoids horizontal currents, which degrade the polarization properties and the impedance bandwidth performance of the antenna. However, to generate the double feed, an additional microstrip feeding network under the ground plane of the PM antenna was used that complicates the total antenna configuration and increases the fabrication cost of the antenna.

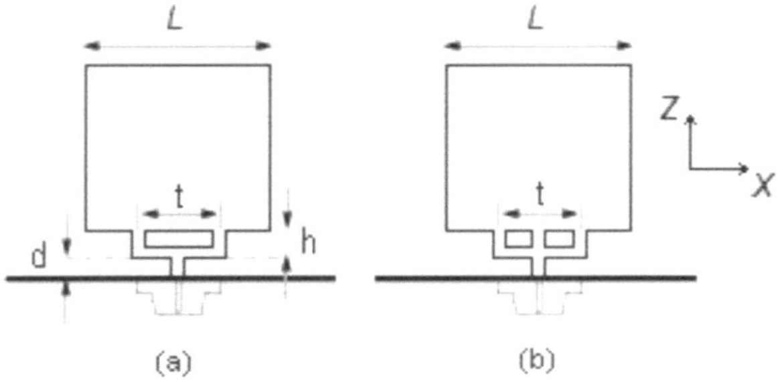

Fig. 20. Geometry of the square planar monopole antenna with (a) two-branch feeding strip and (b) trident-shaped feeding strip.

Following the above work, (Wong et al., 2005) suggested a simple dual and trident-shaped feeding strip suitable to achieve bandwidth enhancement of a square PM antenna. The planar monopole with the simple feeding strip is easily fabricated using a single metal plate and fed using a 50 ohm SMA connector and, in addition, no external feeding network is required, Fig. 20.
In Fig. 20, a 0.2 mm thick square planar monopole antenna of side length L (40 mm) mounted above a ground plane of size 150 mm x150 mm is shown. The feeding strips are uniform and set to 2 mm. By adjusting the three parameters d, t and h, increase in the impedance bandwidth of the planar monopole antenna can be achieved. Parameter d controls the coupling between the ground plane and the lower edge of the planar monopole, which effectively varies the input reactance of the antenna. Thus impedance matching of the antenna can be fine-tuned, and optimized impedance bandwidth can be obtained for the antenna. The return loss for the simple planar monopole antenna with single feed, double branch feed and triple branch feed are shown in Fig. 21. With the size of the square plate set as above, the impedance bandwidth (10 dB return loss) would have a lower edge frequency of less than 1.5 GHz. For the double and triple branch feed structures by selecting the parameters t, h and d of the feeding strip to be 15, 3.5, and 1.0 mm, respectively, the upper edge frequency of the impedance bandwidth, for the double feed would be around 10.2 GHz and for the triple feed would be larger than 11.4 GHz. For the single feed PM antenna, to obtain the highest bandwidth, requires a feeding strip with d = 2.5 mm. It is seen that the triple branch feeding strip shows an impedance bandwidth of 1.276-11.448 GHz (a frequency ratio of 8.32), while the two-branch feeding strip shows a bandwidth of 1.354-10.182 GHz (a frequency ratio of 7.52) and that of the simple feeding strip is 1.455-3.286 GHz (a ratio of 2.26). The wide impedance bandwidth of the triple fed PM antenna makes the antenna suitable for application in the new broadband wireless metropolitan area network system using the IEEE 802.16a (2–11 GHz) standard.
The current distribution behaviour on the planar monopole antenna for the three feeding structures is shown in Fig. 22. It is seen that the triple fed PM antenna provides a more uniform current distribution compared to a dual-feed design and a single-feed design. This leads to a much improved impedance bandwidth for the square planar monopole antenna.
Radiation patterns of the triple fed planar monopole antenna are shown in Fig. 23 at three different frequencies of 2, 6 and 10 GHz.

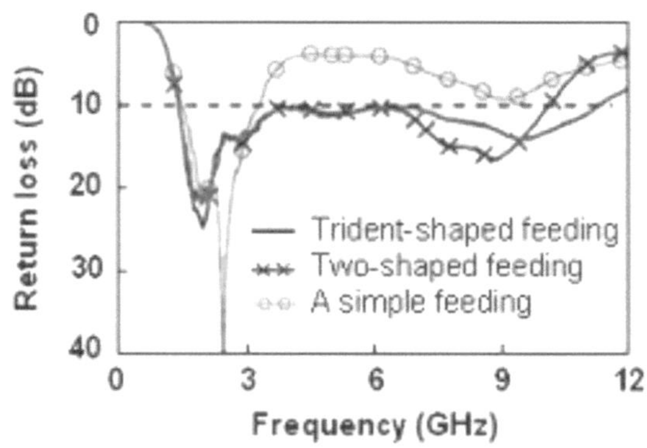

Fig. 21. The return loss of the planar monopole antennas with a trident-shaped feeding strip, a two-branch feeding strip and a simple feeding strip. (Source: Wong et al., 2005).

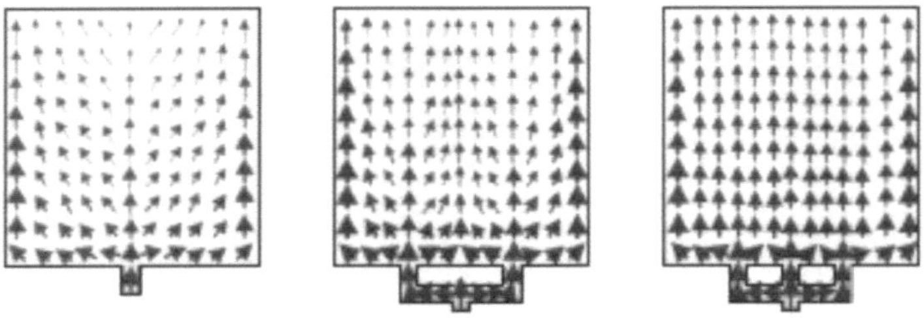

Fig. 22. The current distributions for the three antennas at f = 2.5 GHz. (Source: Wong et al., 2005).

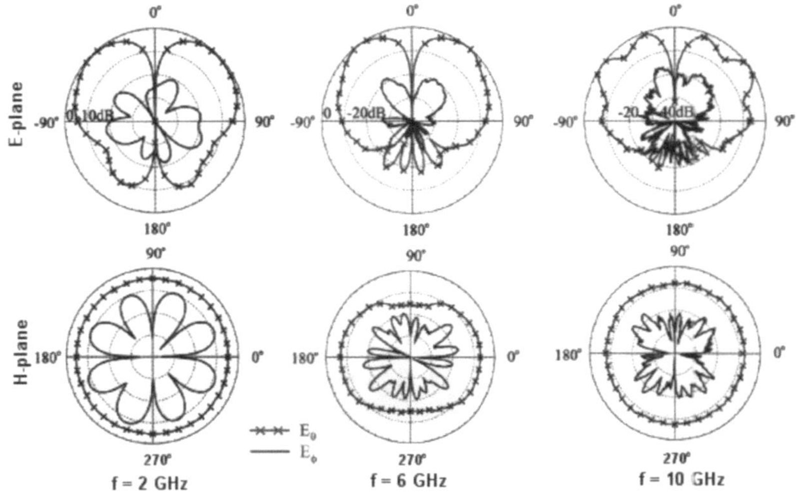

Fig. 23. Radiation pattern of the triple fed planar monopole antenna at three different frequencies. (Source: Wong et al., 2005).

Fig. 24 shows the antenna gain for frequencies across the impedance bandwidth. For frequencies up to about 6 GHz, it is seen that the antenna gain monotonically increases from about 4.0 to 7.0 dBi. For the higher frequency portion of the impedance bandwidth, however, the antenna gain varies relatively slightly in the range of 6.5–7.0dBi.

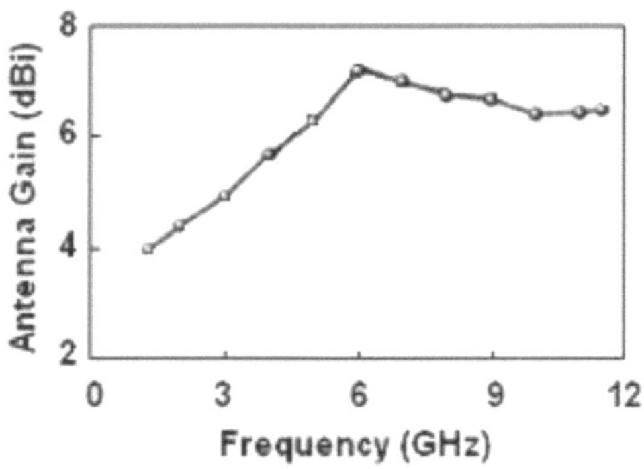

Fig. 24. Gain of the triple fed planar monopole antenna. (Source: Wong et al., 2005).

8. Plate loaded planar monopole antenna

The planar monopole antenna structures considered so far can provide at most a bandwidth over the range 1.2-11.5 GHz. In PM antennas the lower limit of the bandwidth is set by the monopole plate dimensions. To increase the upper frequency limit even further the plate loading can be employed (Mazinani & Hassani, 2009a). The plate loaded planar monopole antenna, PLPM, and its parameters are shown in Fig. 25. The antenna is constructed using copper sheet of thickness 0.2 mm, and dimension L = 20 mm and W = 12 mm, placed on a small circular ground plane of radius 50 mm. Based on obtaining the widest bandwidth, the feed gap parameter, g, is set at 1mm. A 50Ω coaxial probe feeds the bottom of the antenna through the ground plane via a 1.2 mm connector. This PM antenna is loaded at its two radiating edges by small rectangular plates. The parallel plate placed on the two sides of the radiating element is determined by two parameters L_s and W_s.

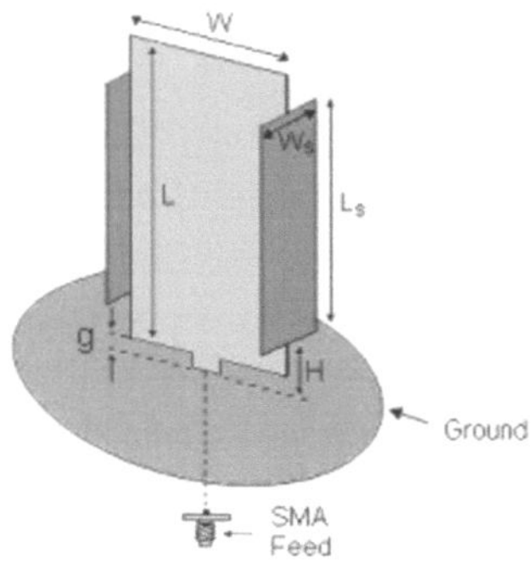

Fig. 25. Planar monopole antenna loaded with a pair of rectangular plate. (Source: Mazinani & Hassani, 2009a).

Based on a return loss of 10 dB, the effect of the plate loading dimension on frequency and bandwidth are listed in Table 5. It is obvious that the size of the plate loading has a pronounce effect on the upper resonant frequency and thus on the impedance bandwidth of the monopole antenna. For the rectangular plate loading, with small values of W_s the return loss shows multiband behavior and with a good choice for L_s a wideband behavior can be obtained. From this Table, it is seen that for the rectangular loading plates with W_s = 6 mm and L_s = 12 mm the upper frequency limit is 16.7 GHz. Fig. 26 shows the return loss for the plate loaded PM antenna. It can be seen that the addition of the plates increases the upper edge frequency significantly resulting in a bandwidth of 2.9-16.7GHz.

W_s (mm)	L_s (mm)	Bandwidth (GHz)
4	12	2.9 - 6.2 , 8.9 - 13.5 , 14.7 - 17.5
6	12	2.9 - 16.7
8	12	2.9 - 15.6
6	10	2.9 - 6.7 , 9.2 - 18.5
6	14	2.9 - 12.5

Table 5. Dimensions and impedance bandwidth of rectangular loading plates. (Source: Mazinani & Hassani, 2009a).

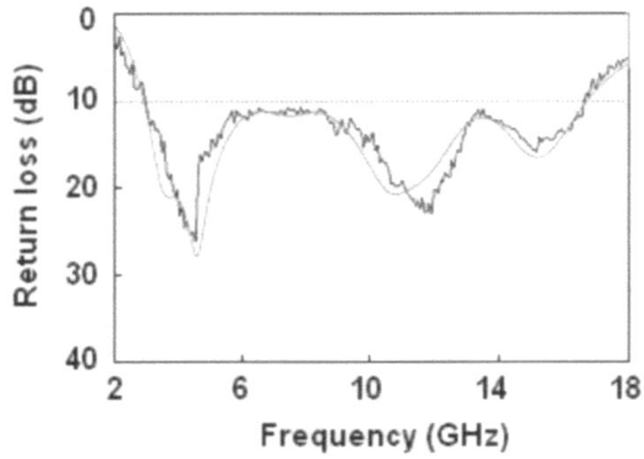

Fig. 26. Simulated (gray) and measured (black) return loss of plate loaded PM antenna. $W_s = 6$ mm and $L_s = 12$ mm. (Source: Mazinani & Hassani, 2009a).

The effect of the presence of the loading plates can be seen through application of the TLM and the current flow on the PM antenna. In section 4, TLM was applied to a beveled PM antenna and through tapering the equivalent transmission line increase in the impedance bandwidth was obtained. Another way to increase the antenna bandwidth is to match the radiation impedance of the antenna to the characteristic impedance of the equivalent transmission line of the PM antenna which is assumed to be matched to the antenna feed. To do so, one can add a shunt impedance loaded stub transmission line (line with Z'_s terminated in Z'_{LS}) to the radiation impedance Z_{Lo}, as shown in Fig. 27(a). It is assumed that the monopole antenna is at a height above the ground plane that leads to being impedance matched to the antenna feed.

In the same way that a rectangular planar monopole antenna can be modeled as a transmission line terminated with an impedance Z_L, the proposed shunt transmission line terminated in an impedance Z'_{LS} can be modeled as a rectangular plate. This leads to a new planar monopole antenna structure. The TLM model shown in Fig. 27(a) is equivalent to a rectangular planar monopole antenna loaded with a pair of parallel plates, placed on the two radiating edges of the antenna. To make the final antenna structure symmetric, one can divide the shunt transmission line into two equal halves, Fig 27(b). In this way, the equivalent plate can be attached to the radiating edge at its symmetric line. The relation between the geometrical parameters of the proposed antenna and its equivalent transmission lines are shown in Table 6.

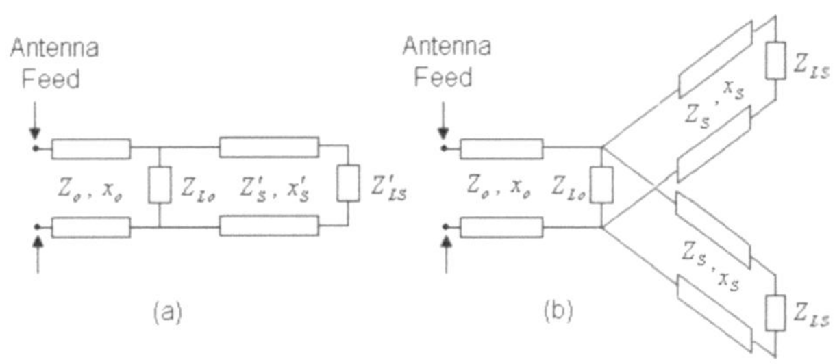

Fig. 27. The TLM of the halved PLPM antenna (a) PM antenna with single shunt impedance loaded stub transmission line and (b) equivalent two section shunt impedance loaded stub transmission line. (Source: Mazinani & Hassani, 2009a).

PLPM	g	$\frac{W}{2}$	Radiating edge	H	$\frac{W_S}{2}$	Loading plate radiating edge
TLM	Z_o	x_o	Z_{Lo}	Z_S	x_S	Z_{LS}

Table 6. Relation between the PLPM antenna and transmission line parameters

Fig. 28 shows the amplitude current distribution on a simple PM and on the PLPM antenna at frequency of 4 GHz. It is obvious that a similar current distribution that takes place on the simple PM antenna also takes place on the loading plates of the PLPM antenna

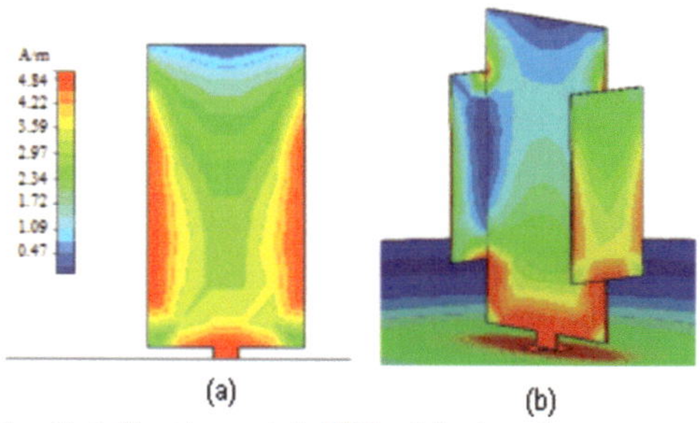

Fig. 28. Current amplitude distribution at 4 GHz (a) simple rectangular PM antenna, (b) the plate loaded PM antenna. (Source: Mazinani & Hassani, 2009a).

The measured normalized E and H-plane radiation patterns of the proposed PLPM antenna at 4, 10, and 16 GHz, are shown in Fig. 29. The H-plane pattern of the antenna shows a good acceptable omnidirectional behavior at all frequencies.

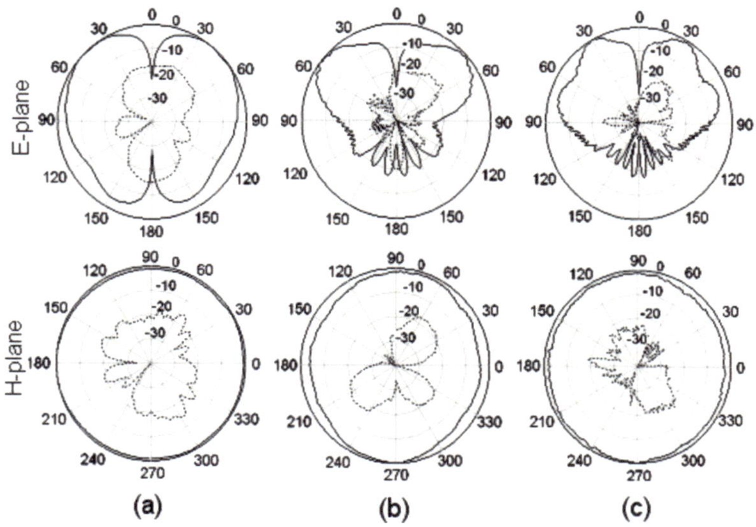

Fig. 29. The co-polar (solid) and cross-polar (dash) pattern of the PLMP antenna at (a) 4 GHz, (b) 10 GHz and (c) 16 GHz. (Source: Mazinani & Hassani, 2009a).

This is in contrast to the usual PM antennas where due to the asymmetry in the configuration of the antenna in the two orthogonal planes good omnidirectional pattern at higher frequencies is not achievable. By adding the loading plates to the PM antenna, as shown in Fig. 28(b), the currents on the two edges of the loading plate results in constructive interference, resulting in a stronger radiation pattern in the direction normal to the plane of the loading plate. Thus, the overall radiation pattern of the antenna becomes more omnidirectional than the case of the simple PM antenna. The E-plane radiation pattern of the PLMP antenna is similar to those of the simple PM antenna. With increase in the frequency of operation, a dip in the main beam in the E-plane pattern is in evidence, due to the large electrical size of the antenna. In all cases the cross polarization level is lower than -15dB, similar to that of a simple PM antenna.

Fig. 30 shows the measured PLPM antenna gain variation against frequency. It is known that a single PM antenna has some 2 - 4.5 dB of gain. Thus, loading plate has increased the gain, especially at higher frequencies where gain of 7 dB is noted.

The above PLPM antenna with its very wide impedance bandwidth can also be used as an internal antenna for mobile handset (Mazinani & Hassani, 2009b). Also due to its higher gain, such antenna is useful for superdirective arrays.

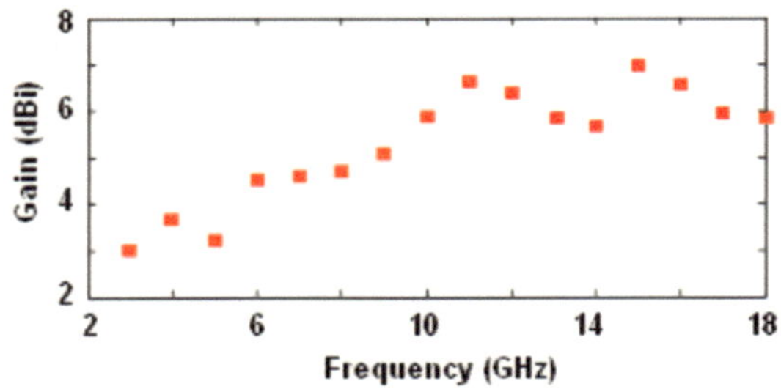

Fig. 30. Gain of PM antenna loaded with rectangular plates. (Source: Mazinani & Hassani, 2009a).

9. Planar Monopole Antenna with Band-Notch Characteristic

A single planar monopole antenna can cover the 2-18 GHz frequency band with good performance. This frequency range has interference to the existing narrower communication systems frequency bands. There are a few frequency bands that are reserved for narrowband wireless technologies (such as WLAN, HYPERLAN/2, IEEE802.11a ...), thus, there is a need in the wideband device to provide filtering to avoid interference from or causing interference to narrowband devices. However, the use of a filter will increase the complexity of the wide band system. Rather than using filtering electronics, it has been shown that by creating a slot in the interior of the radiating element, a planar monopole can exhibit a single or multiple narrow frequency notch bands while maintaining the wideband performance. The shape, size and position of the slot on the antenna surface play an important role in the determination of the frequency center and the bandwidth of the notch. Most of the works reported in the literature include U-shaped vertical slots and simple rectangular horizontal slots cut from the edge of the antenna. In the following sections, these structures are described.

9.1 U shaped vertical slots

A technique to create single or double band notch behavior is by creating single or multiple half wavelength U-shaped slots placed vertically along the surface of the monopole antenna, [Lee et al. 2005] and [Lee et al., 2006]. Fig. 31 shows the structure of the PM antenna with a combination of U and ∩-slots. Other combinations of such slots are also possible. A copper planar element of thickness 0.2 mm, size 20mm x 27 mm and beveling angle 12°, is vertically placed at spacing of 1 mm over the circular ground plane of radius 75mm and fed via a 50 ohm SMA connector.

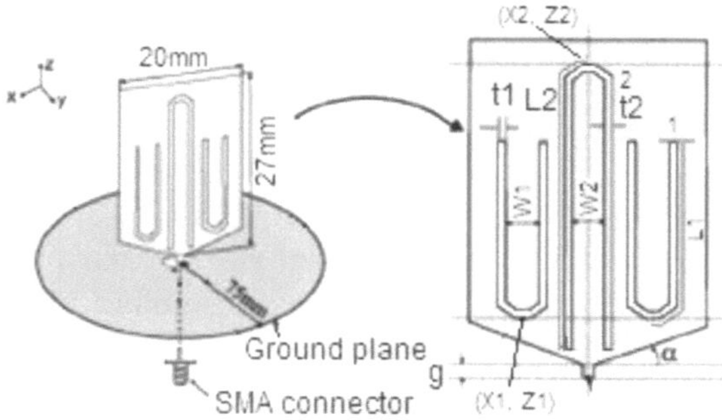

Fig. 31. Geometry of the planar monopole antenna with U-slots. (Source: Lee et al., 2006).

Fig. 32 shows the return loss results for the PM antenna with various U-slot shapes placed on the antenna. The size and position of the parameters used for slots 1 and 2 are (t_1, L_1, X_1, Z_1, W_1) = (1, 15.6, 0, 2, 3) mm and (t_2, L_2, X_2, Z_2, W_2) = (1, 24.6, 0, 25.7, 10.5) mm. Also shown in this Figure is return loss of the PM antenna without any slots, the reference antenna. Fig. 32(b) shows the planar antenna with single ∩-shaped slot, where creation of one notch band at 2.96 GHz is noticeable, and Fig. 32(c) shows the return loss for a U-shaped slot that makes a notch band at 4.77 GHz. The bandwidth of the notches can be controlled by adjusting W1 and W2.

Fig. 32. The return loss of the PM antenna (a) without slot, (b) with ∩-slot, (c) with U-slot and (d) combination of ∩ and U-slots. - - - - simulated, -------- Measured. (Source: Lee et al., 2006).

Combining these two U and ∩-shaped slots, Fig. 32(d), the return loss results show two notches at the same frequencies as those of Fig. 32(b) and (c), i.e. when the two slots are placed beside each other results shows little mutual coupling between the two.
The centre frequency of the notch bands for either the U-slot or the ∩-slot can be accurately predicted from the following formula:

$$f_{notch} \approx \frac{c}{4L_n} \tag{4}$$

Where c is the speed of light and L_n is the half length of a U or ∩-slot, Fig. 31.
By varying the slot position and the distance between the slot arms one can control the notch bandwidth and its centre frequency.
For the structure of Fig. 32(d), the normalized surface current distribution at three different frequencies, passband 2.4GHz, at lower notch centre frequency 2.96GHz, and at upper notch centre frequency 4.81GHz are shown in Fig. 33. One can see that at the passband, Fig. 33(a), there are more current distributions near the feeding point. At the lower notch centre frequency, Fig. 33(b), the current is concentrated around the bottom edge of the ∩-shaped slot and is oppositely directed between the interior and exterior of the slot. Thus, the antenna operates in a transmission-line-like mode, which transforms the nearly high impedance (open circuit) at the top of the slot to nearly zero impedance (short circuit) at the antenna feeding. This zero impedance at the feeding point leads to the desired high attenuation and impedance mismatching near the notch frequency. Fig. 33(c) shows that at the upper notch frequency, current is concentrated around the top edge of the U-shaped slot and is oppositely directed between the interior and exterior of the slot. As in previous case, the antenna operates in a transmission-line-like mode, which transforms the nearly zero impedance at the top of the slot to nearly high impedance at the antenna feeding, leading to the desired high attenuation near the notch frequency.

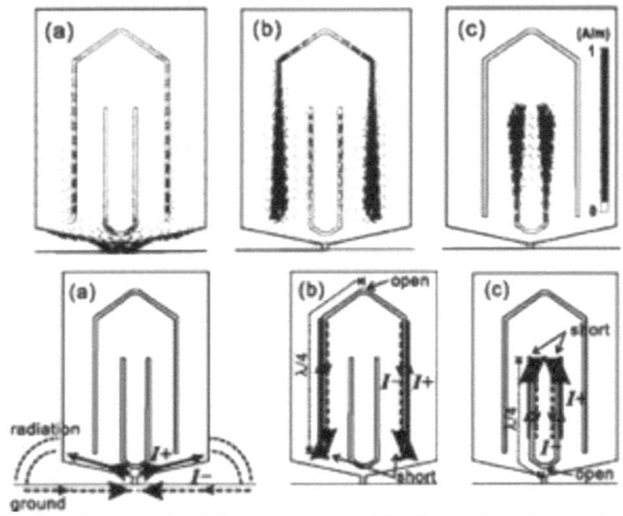

Fig. 33. Current distribution on the PM antenna with slots. At (a) passband, 2.4 GHz, (b) at Lower notch frequency, 2.96 GHz and (c) at higher notch frequency 4.81GHz. (Source: Lee et al., 2006).

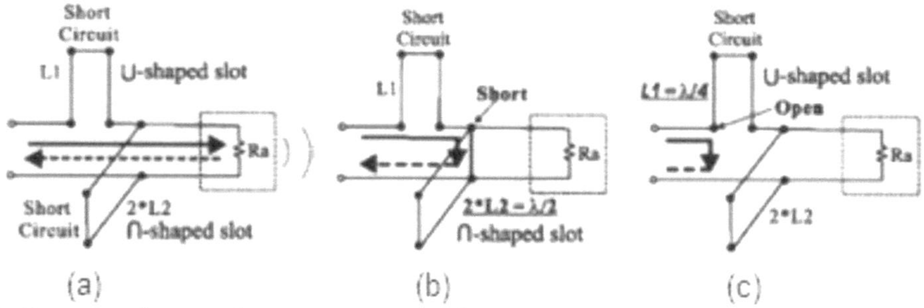

Fig. 34. Conceptual equivalent-circuit model for antenna with U and ∩-shaped slots. At (a) passband (b) first notch frequency, (c) second notch frequency. (Source: Lee et al., 2006).

Fig. 34 shows the conceptual equivalent-circuit model for the antenna, that includes a series stub, a shunt stub, and antenna resistance R_a. At passband the stubs are non operational and radiation takes place. The stubs are a short-circuit stub with L1 = 15.6 mm and an open-circuit stub with L2=24.6 mm. When L2 is equal to λ/4 in Fig. 34(b), at 2.96 GHz, the input impedance at the feeding point is zero (short circuit). Also, when L1 is equal to λ/4 in Fig. 34(c), at 4.78 GHz, the input impedance at the feeding point is high (open circuit) due to the quarter-wave transformer. In these two cases, destructive interference for the excited surface currents in the antenna will occur, which causes the antenna to be non responsive at those frequencies.

9.2. Simple horizontal slots

Simple horizontal slots cut from the edges of the wideband monopole planar antenna can also create notch bands (Rahmati & Hassani, 2009). Such slots cut from the edges of the monopole antenna are easier to create as compared to U-slots cut from the centre of the antenna. Fig. 35 shows the antenna structure where a copper planar element of thickness, 0.2 mm, size 22mm × 25 mm and beveling angle of 17°(or h_b=4mm), is mounted 0.5 mm over the circular ground plane of radius 40mm. Through placement of single horizontal slot, a single and a double pair of horizontal symmetrical slots placed on the edges of the monopole antenna single and double tunable notch characteristics can be obtained.

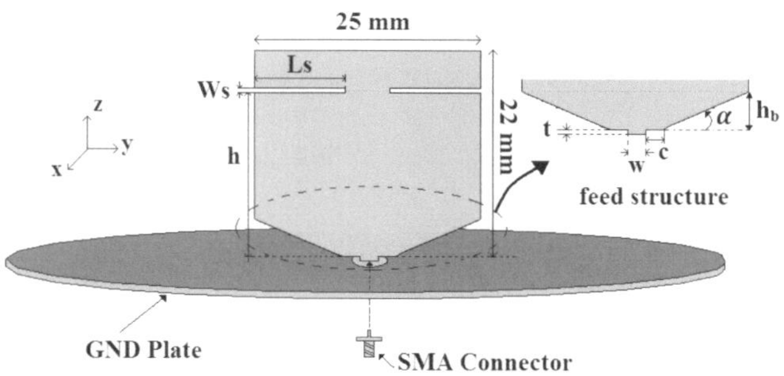

Fig. 35. Geometry of the planar monopole antenna with simple horizontal cuts. (Source: Rahmati & Hassani, 2009).

The effect that a single horizontal slot cut from one edge of a monopole antenna has on the antenna are shown in Fig. 36. There are three parameters, *Ls*, *Ws* and *h* that affect the notch band. Fig. 36(a) shows the return loss of the antenna for various position of the slot, *h*. The results show that when *h* is varied from 11mm to 20mm the notch bandwidth decreases from 20% to 4%, the return loss of the notch at centre frequency decreases and only a small shift in the notch centre frequency takes place. Fig. 36(b) shows the return loss as slot length, *Ls*, is varied. The results show that with an increase in *Ls*, the notch centre frequency decreases with an improved notch behavior. Although not shown, results on variation of slot width, *Ws*, shows that as *Ws* is increased, the notched bandwidth increases from 9% to 20% and only a small shift in notch centre frequency takes place. From these results, the centre frequency of the notch can be obtained through the following approximate formula:

$$f_{\text{notch}} \cong \frac{c}{4 \times L_s + 0.5 \times h} \qquad (5)$$

where *c* is the speed of light. It can be seen that the notch frequency is controlled mostly by the slot length *Ls*.

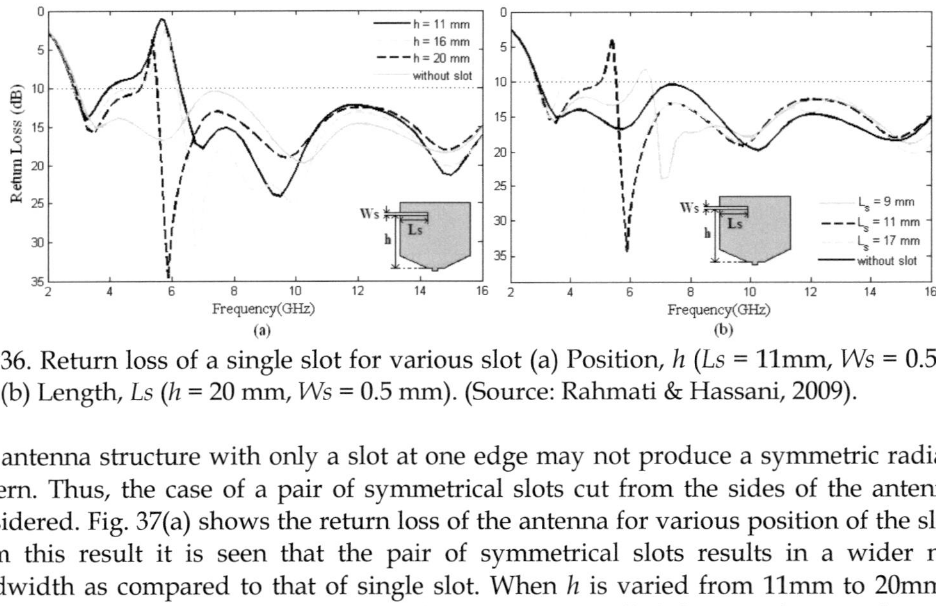

Fig. 36. Return loss of a single slot for various slot (a) Position, *h* (*Ls* = 11mm, *Ws* = 0.5mm) and (b) Length, *Ls* (*h* = 20 mm, *Ws* = 0.5 mm). (Source: Rahmati & Hassani, 2009).

The antenna structure with only a slot at one edge may not produce a symmetric radiation pattern. Thus, the case of a pair of symmetrical slots cut from the sides of the antenna is considered. Fig. 37(a) shows the return loss of the antenna for various position of the slot, *h*. From this result it is seen that the pair of symmetrical slots results in a wider notch bandwidth as compared to that of single slot. When *h* is varied from 11mm to 20mm the notch bandwidth reduces from 55% to 12% and a very small shift in notch centre frequency takes place. The single pair of symmetrical slots can be modeled as two resonances connected in series to the initial monopole antenna. Slots of equal length result in equal resonant frequency and higher notch bandwidth.

The effect of various slot length, *Ls*, is shown in Fig. 37(b). It is seen that the length of the slot determines the centre frequency of the notched band. As *Ls* is increased the notch centre frequency shifts toward the lower frequency with an increase in return loss level. Results on variation of *Ws*, similar to the previous single slot case, shows that as *Ws* is varied from 0.3mm to 1mm, the notch bandwidth increases from 9% to 20% and only a small shift in notch centre frequency takes place. From the results of Fig. 37 it is seen that the centre frequency of the notch is more dependent on the length and position of the slot while the slot position also affects the bandwidth of the notch. From these results one can state that for slot position h <16mm the centre frequency of the notch can be obtained through the following approximate formula:

$$f_{\text{notch}} \cong \frac{c}{4 \times L_s + h} \tag{6}$$

while for h>16mm, the effect of slot position becomes less, thus the formula in equation (5) would be more suitable.

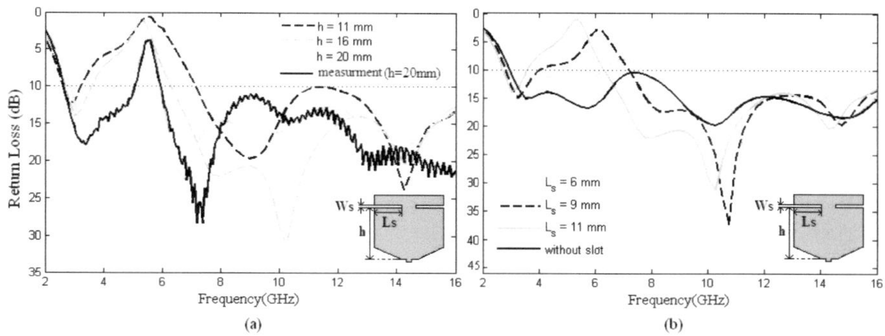

Fig. 37. Return loss of a single pair of symmetric slots for various slot (a) height, h. (Ls = 11 mm, Ws = 0.5 mm) and (b) length, Ls. (h = 16 mm, Ws = 0.5 mm). (Source: Rahmati & Hassani, 2009).

Fig. 38 shows the normalized surface current distribution over the planar monopole antenna for the single and for a pair of slots at various frequencies. In Fig. 38(a) and (c), where the antenna operates at pass band frequencies 3.5 and 13GHz, there are more current distributions near the feeding point(i.e. slot does not resonate and has little effect). At notch frequency, 5.7 GHz, as shown in Fig. 38(b), current is concentrated around the edge of the slot while there is almost no current at the feeding point. The high concentration of current around the slot can be represented as a short circuited stub. The edge loaded slot monopole antenna can then be modeled as a short circuited stub in series with the unloaded monopole radiation resistance, R, Fig. 38(d, e). For the case of single pair of symmetric slots, there would be two short circuited stubs connected in series with R. At notch frequency, the slot is almost $\lambda/4$ and thus transforms short circuit at the slot to open circuit at the antenna feeding point. This leads to the desired high attenuation and impedance mismatching around the notch frequency.

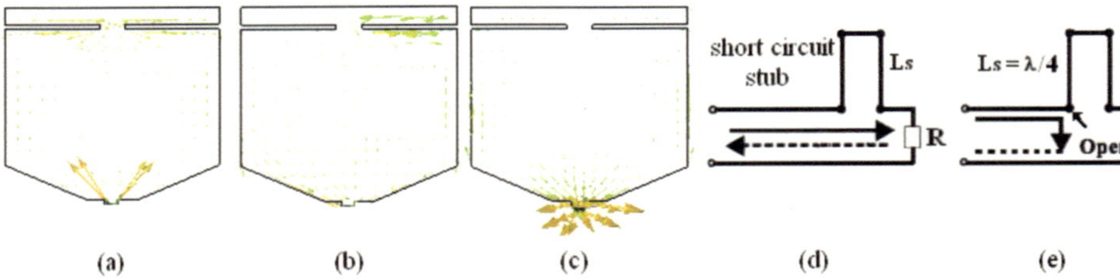

Fig. 38. Surface current of antenna with a pair of slots, at the (a) pass band 3.5 GHz, (b) notch frequency 5.7 GHz, (c) pass band 13 GHz. The equivalent-circuit model for slot loaded planar monopole antenna at (d), pass band frequency (e) notch frequency. (Source: Rahmati & Hassani, 2009).

To create multiple notch bands multiple horizontal slots placed on top of each other are required. Here, the case of the planar monopole antenna with two pairs of horizontal symmetrical slots is presented. The presence of the second pair of slots creates an extra resonance (notch), whose centre frequency is dependent on the parameters of this pair of slots. In the cases studied the dimensions of the upper pair of slots are kept fixed while those of the lower pair of slots are varied.

Fig. 39(a) shows the return loss results for various lengths of the lower pair of slots, *Ls2*. It is seen that increasing *Ls2* towards *Ls1* the centre frequency of the second resonance approaches that of the first, thus increasing the bandwidth of the notch.

Fig. 39(b) shows the return loss for various heights of the lower pair of slots. From Fig. 35, beveling height, h_b is 4mm, thus, *h2*=3mm is for a slot which is placed in the beveling region of the monopole antenna. As the length of this slot in such a region is small, the upper centre frequency would be quite high. As the slot is moved up from this region (i.e. *h2* >4mm) there is not much change in upper centre frequency while its return loss level reduces. Similar to the two previous cases, as the width of the slot, *Ws2*, is increased the bandwidth of the upper resonance increases while its centre frequency changes slightly.

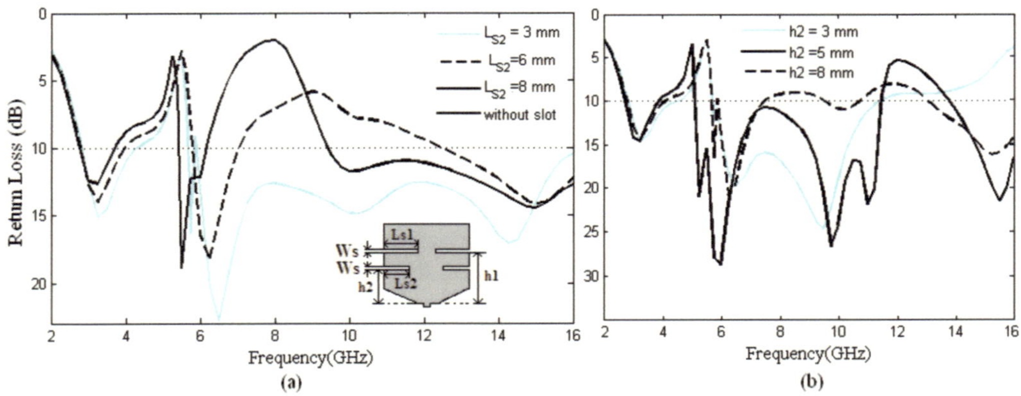

Fig. 39. Return loss of two pairs of symmetrical slots for various lower pair of slots (a) lengths *Ls2*, (*h1*= 20 mm, *Ls1*= 11 mm, *h2* = 8 mm, *Ws* = 0.5 mm) and (b) heights *h2*, (*h1* = 20 mm, *Ls1* = 11 mm, *Ls2* = 5 mm, *Ws* = 0.5 mm). (Source: Rahmati & Hassani, 2009).

10. References

Agrawall, P. A.; Kumar, G. & Ray, K. (1998). Wideband planar monopole antennas, *IEEETransactions on Antennas Propagation*, vol. 46, no. 2, Feb. 1998, pp. 294–295.

Ammann, M. J. (1999). Square Planar Monopole Antenna, Proc. IEE Nat. Antennas Propag. Conf., York, U.K., IEE Publication no. 461, 1999, pp. 37-40.

Ammann, M. J. (2001). Control of the impedance bandwidth of wideband planar monopole antennas using a beveling technique. *Microwave and Optical Tech. Letters*, vol. 30, no. 4, August 2001, pp. 229-232.

Ammann, M. J. & Chen, Z. N. (2003a). Wideband Monopole Antennas for Multi-Band Wireless Systems, *IEEE Antennas and Propagation Magazine*, vol. 45, no. 2, April 2003, pp. 146-150.

Ammann, M. J. & Chen, Z. N. (2003b). A Wide-Band Shorted Planar Monopole With Bevel, *IEEE Transactions on Antennas and Propagation*, vol. 51, no. 4, April 2003, pp. 901-903.

Antonino-Daviu, E.; Cabedo-Fabre´s, M.; Ferrando-Bataller, M. & Valero-Nogueira, A. (2003). Wideband double-fed planar monopole antennas, *Electronics Letters*, vol. 39, no. 23, 13th Nov. 2003, pp. 1635-1636.

Balanis, C. A. (1982). Antenna Theory: Analysis and Design, New York, Harper and Row, 1982.

Dubost, G. & Zisler, S. (1976). *Antennas a Large Bande*, Masson, Paris, New York, 1976, pp. 128–129.

Harrington, R.F. & Mautz, J. R. (1971). Theory of characteristic modes for conducting bodies, *IEEE Transactions on Antennas Propagation*, vol. 19, no. 5, Sept. 1971, pp. 622–628.

Lee, E.; Hall, P. S. & Gardner, P. (1999). Compact wideband planar monopole Antenna, *Electronics Letters*, vol. 35, no. 25, 9th Dec. 1999, pp. 2157-2158.

Lee, Wang-Sang; Lim, Won-Gyu & Yu, Jong-Won (2005). Multiple Band-Notched Planar Monopole Antenna for Multiband Wireless Systems, *IEEE Microwave and Wireless Components Letters*, vol. 15, no. 9, Sept. 2005, pp. 576-578.

Lee, Wang-Sang; Kim, Dong-Zo; Kim, Ki-Jin & Yu, Jong-Won (2006). Wideband Planar Monopole Antennas With Dual Band-Notched Characteristics, *IEEE Transaction on Microwave Theory and Techniques*, vol. 54, no. 6, June 2006, pp. 2800-2806.

Mazinani, S. M. & Hassani, H.R. (2009a). A Novel Broadband Plate-Loaded Planar Monopole Antenna, *IEEE Antennas and Wireless Propagation Letters*, vol. 8, 2009, pp. 1123-1126.

Mazinani, S. M. & Hassani, H.R. (2009b). A wideband internal plate loaded planar monopole antenna for mobile handset, *J. of Electromagn. Waves and Appl.*, vol. 23, 2009, pp. 1273–1282.

Rahmati, B. & Hassani, H.R. (2009). Frequency notched wide band planar monopole antennas, *Progress In Electromagnetics Research C*, vol. 9, 2009, pp. 131-143.

Valderas, D.; Legarda, J.; Gutiérrez, I. & Sancho, J. I. (2006). Design of UWB Folded-Plate Monopole Antennas Based on TLM, *IEEE Transactions on Antennas & Propagation*, vol. 54, no. 6, June 2006, pp. 1676-1687.

Wong, Kin-Lu; Wu, Chih-Hsien & Su, Saou-Wen (Stephen) (2005). UltrawideBand Square Planar Metal-Plate Monopole Antenna With a Trident-Shaped Feeding Strip, *IEEE Transactions on Antennas & Propagation*, vol. 53, no. 4, April 2005, pp. 1262-1269.

24

Collinear Microstrip Patch Antennas

Alois Holub and Milan Polívka
Czech Technical University in Prague
Czech Republic

1. Introduction

The original idea of the collinear principle in the antenna design comes from Franklin (Franklin, 1925). He faced the problem of resonant long wire antennas. In principle, the standing wave current distribution on the long straight wire produces *n* radiation lobes of the same level, depending on the number *n* of half-wave antenna sections. Employing non-radiating quarter-wave stubs Franklin converted the original out-phase current distribution into an in-phased distribution of currents on collinear segments (represented by solid red arrows in Fig. 1), thus producing only one major radiation beam. A key advantage of such arrangement is represented by the high gain of the antenna with the properties of series antenna array, whereas the simplicity of the single feeding point is maintained. All antenna structures based on this principle are known as collinear arrays (CoA). The latter are composed of in-phase fed radiating elements that lie in the straight line. Their radiation is typically broadside and perpendicular to the axis of collinear elements. Since Franklin's times many collinear antenna structures have been proposed. The principle representatives of the CoA are described later on.

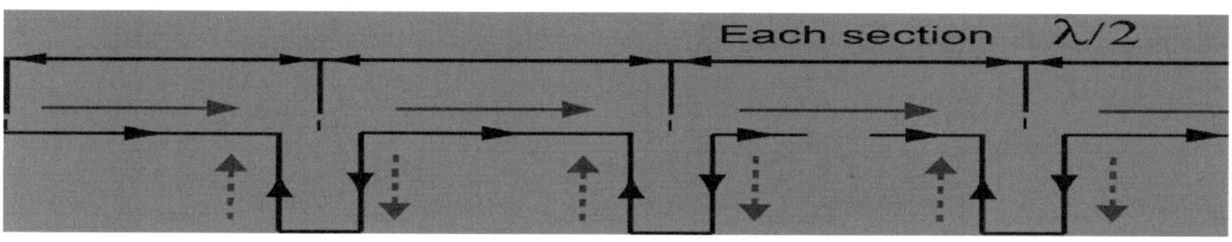

Fig. 1. Sketch of vector current distributed on original Franklin collinear wire dipole. Only collinear segments provide in-phase current distribution and contribute to radiation.

The first coaxial collinear (CoCo) antenna was proposed in 1972 (Balsley & Ecklund, 1972). It is constructed of series of half-wavelengths of the coaxial cable connected together by an electrically interchanging of the inner-and outer-conductors at each junction, see Fig. 2. From the physical point of view, the resulting antenna takes form of a one single long section of flexible coaxial line. Nevertheless, from the electrical point of view, it is composed of a number of collinear half-wave dipoles fed in phase. Although the principle of operation of the CoCo antenna is based on the Franklin's idea, the concept of radiating coaxial is far more complex. Due to the Ampere's circuital law, the currents in the inner line conductor and on the inside surface of the outer line conductor must be equal and opposite. Referring to Fig. 2, it is obvious that one feeder feeds two coaxial lines only. Consequently, there can be no current over the outer cable conductor. This current represents the antenna radiating current. In fact, the generator excites two antennas (one starting with the outer line conductor, whereas the other with the inner line conductor). Therefore, both of these two antennas are fed 180° out of phase. Consequently, there will be a non-zero total current along the coaxial-line sections, i.e., a radiating current. A frequently presented variation of the described CoCo antenna is the monopole concept over the ground plane. Further electromagnetically coupled coaxial dipole array antenna has been proposed as a modification of the CoCo (Miyashita et al., 1999).

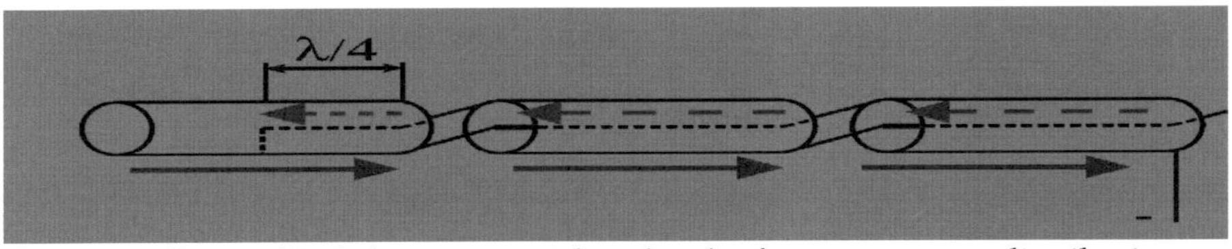

Fig. 2. Coaxial collinear (CoCo) antenna with a sketch of vector current distribution

Franklin-type microstrip line antenna (Nishimura et al, 1979) was probably one of the first microstrip line type antennas based on the collinear principle. The antenna configuration and the current distribution on the radiating microstrip line are depicted in Fig. 3a. The operational principle is based on a presumption that, in general, a non-radiating microstrip line can be manufactured and used in order to radiate by means of the suitable bending of the strip conductor. The aforementioned bending perturbs transmission characteristics of the microstrip line periodically; thereby it operates as a linear array antenna. The presented shape of the microstrip turns out to be very effective, because even the phasing stubs contribute to the radiation. The microstrip is designed for 50 Ω. The end of the microstrip line is terminated by an open-circuit. When the lengths of the straight section and the bending section (see Fig. 3a) are set at the level of about one-half of guide wavelength, the eventual current distribution can be represented by the arrow. As a result, the direction of the radiation is broadside of the array and the polarization is parallel to the straight section.

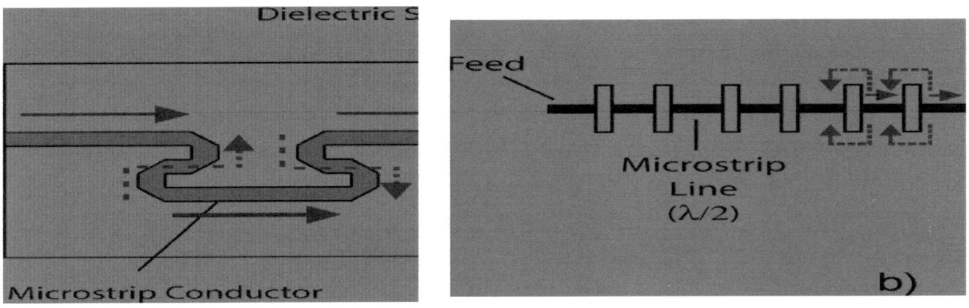

Fig. 3. Franklin-type microstrip line antenna a) and Microstrip-Franklin antenna b) with a vector current distribution

Another microstrip antenna derived from the Franklin's collinear idea was the Microstrip-Franklin antenna (Solbach, 1982). The layout of the structure is shown in Fig. 3b). Unlike the

CoCo antenna, the principle of the operation is more similar to the Franklin's folded wire. Solbach started with the endeavour to suppress the radiation of the half-wavelength connecting lines situated between the patch radiators in microstrip array. In comparison with Franklin, he proposed two 100 Ω quarter-wave phasing stubs, producing a 180° phase shift between the terminals of the microstrip transmission lines. The stubs were designed symmetrically in order to maintain the parasitic discontinuity effects at the junction of stub and microstrip patch on the low levels. The currents on the phasing stubs are mutually opposite in direction so that the radiation produced by the stubs is cancelled. Nevertheless, the electrical fringe fields of the terminating microstrip lines superimpose the phase in the slot between the lines. The resulting electric field in the slot was revealed as a prevailing source of radiation in the arrangement; see Fig. 3b). Such phasing stub can be described as a slot radiator embedded into the microstrip line and employing the radiators in question. It is possible to design the antenna array with a low spurious radiation and also a low surface wave excitation.

One of the latest structures with applied omnidirectional principle to the microstrip structure is represented by the Omnidirectional Planar Microstrip Antenna (OMA) (Bancroft and Bateman, 2004). The geometry of the OMA is presented in Fig. 4. The antenna consists of top and bottom traces. The top layer traces range from wide to narrow, while complementing the narrow to wide traces on the bottom layer. The antenna is fed with a probe at the junction of the first narrow line and the next wide section connection. The principle of operation is similar to the CoCo antenna, since all wide half-wavelength parts of the antenna radiate. The radiation is omnidirectional, because the odd and even half-wavelength wide section radiate in opposite directions. The impedance matching is achieved by variations of the value of W_2.

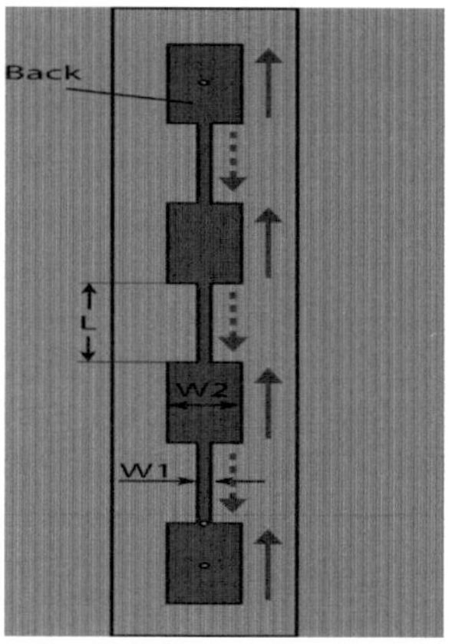

Fig. 4. Omnidirectional microstrip line antenna with a sketch of vector current distribution

Another collinear type antenna is represented by an arrangement of folded slot analogues to the wire type, e.g. (Chen et al., 2007).

The next chapters are going to deal with the development and efficient analysis of a novel arrangement of collinear antenna in the microstrip technology providing hemispherical pattern called Collinear Microstrip Patch Antenna (CoMPA) (Polívka & Holub, 2005).

As it is going to be demonstrated, the latter can also be modularly extended in area for higher gain (Polívka & Holub, 2006).

2. Development of Collinear Microstrip Patch Antennas

This section is based on the explanation presented for the first time in papers (Polívka & Holub, 2005) and (Polívka & Holub, 2006). It describes the operational principle of a novel type of collinear antenna array, designed and realized in the microstrip patch antenna technology. The operational principle is explained via surface current distribution on the patch in the way analogical to the case of Franklin wire CoA. The performance of the antenna principle is verified on the realized prototypes for 869 MHz RFID and 2.4 GHz ISM bands, which are employed in the real applications.

2.1 Principle of CoMPA operation

Firstly, let us explain the meaning of the following three expressions frequently used in the microstrip patch antenna technology: radiator, patch and motif. The term 'radiator' stand for a complete radiating element that includes the ground plane. 'Patch' is a conductive part of a radiator that is situated in the height h above the ground plane. The term 'motif' is used for a particular geometrical shape of the patch.

The principle of the operation of CoMPA is based on the application of geometrical 'perturbation elements', i.e. slots and notches introduced in the patch that, itself, operates on higher order modes. In our case, the antenna resonates dominantly on the TM_{0X} mode, where X determines the number of current half-wavelengths in the resonant longitudinal dimension of the patch (in figures indicated on the y-axis). The impact of the perturbation elements can be explained as the way how to eliminate the radiation from even out-phase electric current distribution, which is forced to flow round these elements. This approach is analogous to the application of $\lambda/4$ curved sections in the Franklin wire antenna.

From the point of view of the radiation pattern calculation, the slots (and notches) can be put together with outer radiation edges that are considered as a radiation source in case that the equivalent method of magnetic currents is used. The E-field distribution along the external perpendicular edges is nearly constant. On the contrary the E-field distribution along the inner slots reaches its maximum in the center, but shows degressive trend in its value as long as it approaches the slot edges, which corresponds to the currents flowing around. A detailed analysis of the E-field distribution along the edges of the CoMPA represents the objective of Chapter 3. Vector surface current distributions on the CoMPAs, operating on the mode TM_{03} and TM_{05} are demonstrated in Fig. 5.

The CoMPA operating with TM_{03} mode ($CoMPA_{03}$) with one central narrow slot constitutes the simplest example of the implementation of the above-described principle.. The slot of the length of approx. $\lambda_g/2$ and the width of a fragment of λ_g makes the second (even) current wavelength to flow around (see Fig. 5a). The same effect can be explained in case of CoMPA operating on the mode TM_{05} ($CoMPA_{05}$) (see Fig. 5b). The optimized dimensions of $CoMPA_{05}$ scaled to the wavelength are depicted in Fig. 5c. The solid red and dash blue arrows represent the in-phase (approx. $\lambda_g/2$ long) and out-phase (approx. $\lambda_g/5$ long) source areas, respectively.

Physical dimensions of the realized antenna prototype (see Fig. 6a) are listed below: patch size 268 × 643 mm, slot length 172.5 mm and ground plane size 298 × 680 mm. The patch is carried by plastic distance posts located over the ground plane at the height of 10 mm (approx. 0.03 λ_0). The distance between the coaxial feed placed on the y-axis and the inner edge of one of the slots equals approx. 0.047 λ_g. All the structure was modeled in the IE3D method of moment simulation tool with a finite ground plane.

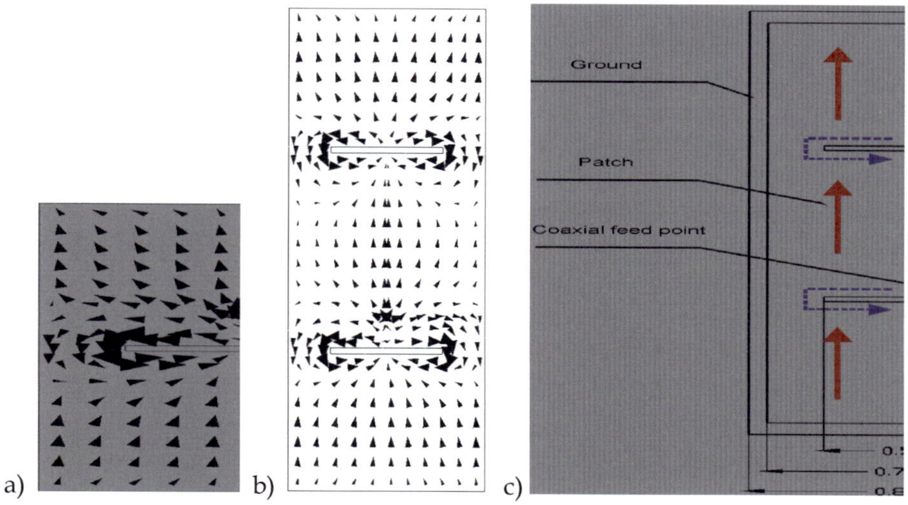

Fig. 5. Vector surface current distribution represented by black arrows (simulated by IE3D) on the patch of the CoMPA operating on mode a) TM_{03}, and b) TM_{05}. c) Scheme of the $CoMPA_{05}$ with dimensions related to the wavelength with schematic current distribution.

The measured reflection coefficients of the realized prototype with and without a hardened polystyren (HPS) radom of 3 mm thickness placed at the height of 40 mm over the patch is illustrated in Fig. 6b.

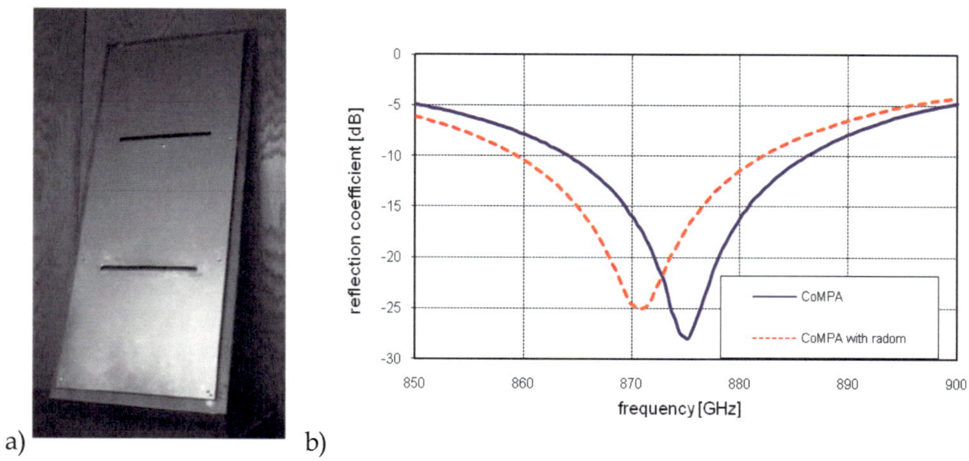

Fig. 6. a) Photograph of $CoMPA_{05}$ prototype for 869 MHz band, b) measured reflection coeffcient of realized prototype with depicted influence of HPS radom.

The measured radiation patterns are presented in Fig. 7. The distance between the center of odd in-phase rectangular parts is equal to approx. 0.65 λ_0, which explains the sidelobe level of about -13 dB similarly to the 0.5 λ_0 spaced uniform array. The front-back ratio is approx. 19 dB in the E-plane, although the ground plane dimensions exceed the size of the patch itself by only some 0.1 λ_0 (!). The antenna gain without cover has been measured by means of the substitution method at 869.5 MHz in the antenna anechoic chamber (Mazánek et al., 2000). The simulated values of the directivity 12.6 dBi and gain 12.4 dBi (the corresponding efficiency amounts to 95 %) are slightly higher than the measured gain 11.7 dBi. This phenomenon can result from the fabrication tolerance on one hand and the presupposed gain measurement error, which equals at least +/-0.5 dBi on the other hand.. The radiation of the antenna is directional in the E-plane and wider in the H-plane, which corresponds to the linear array of radiators in the y-axis. The measured 3dB beamwidth of 25° in the E-plane and of 65° in the H-plane confirms the gain enhancement (when compared to the gain of the standard rectangular microstrip patch operating on TM_{01} mode). In fact, the aforementioned gain reaches approx. 6-9 dBi, indeed depending on the heigth and the substrate used. The value of the impedance bandwidth BW = 2.8 % (related to VSWR = 2) is relatively low. It arises from the resonant character of the structure as it is expected in case of the patch-type antennas. In case of a particular RFID application at 869 MHz, where this prototype was used (Švanda et al., 2007), the impedance bandwidth is sufficient, because merely the 250 kHz band was required.

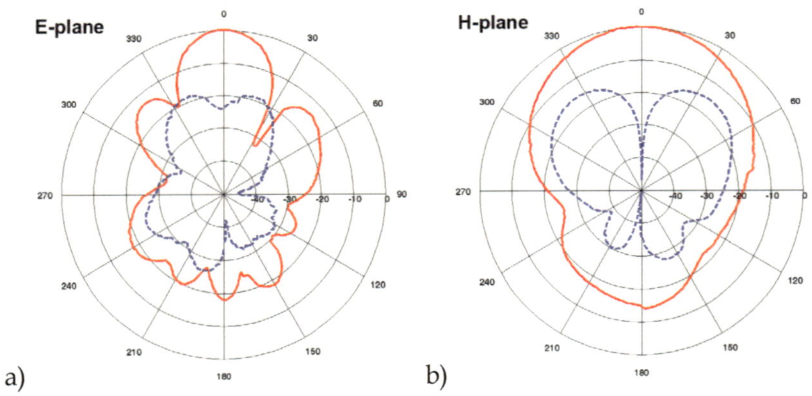

Fig. 7. Measured co-polar (E_{co}) and cross-polar (E_x) radiation patterns of $CoMPA_{05}$ for 869 MHz band in a) E-plane, and b) H-plane.

2.2 Planar extension of CoMPA

All examples of collinear antennas introduced in the first state-of-the-art chapter have one quality in common - each design can be considered as a linear antenna array. However, the principle of CoMPA in the microstrip patch technology enables the extension of the structure perpendicularly to the longitudinal axis of the array. This measure is introduced in the following text. The essence of this lateral extension of currents is similar to the principle used in the grid flat-panel array (Kraus & Marhefka, 2002).

The first stage of the explanation has to be dedicated to the structure operating on the TM_{03} mode, i.e. $CoMPA_{03}$. The latter can be considered as a linear array of radiators, which can be laterally extended. Yet in order to preserve the surface current distribution, it is indispensable to add a pair of lateral notches. These notches (of the length of approx. $\lambda_g/4$) are placed perpendicularly to the patch border, at the same y-coordinates as the slots. In consequence, the current distribution of the TM_{03} mode remains the same as on CoMPA. In addition, the similar phenomenon of currents that flow around the notches is maintained (see Fig. 8a). As a result, the J_y component plays a dominant role on the surface of the patch and the radiator exhibits a broadside hemispherical radiation with an enhanced gain. The domination of the J_y component on the majority of the patch surface is a crucial condition for maintainance of a reasonably low cross-polar level. Due to the fact that the area of radiating sources was extended, a higher level of directivity is presumed. By combining longitudinal and lateral extensions of $CoMPA_{03}$, a motif with two central slots and two pairs of lateral notches operating on TM_{05} mode (called planar CoMPA; hereinafter referred to as $PCOMPA_{05}$) is realized (see Fig. 8b). Fig. 8c shows $PCoMPA_{05}$ motif divided by vertical and horizontal dashed lines into basic modules that form building blocks of the previously elaborated less complex versions of discussed PCoMPAs. The horizontal dashed lines divide the motif into areas with opposite orientation of the surface currents. These currents are denoted in the same fashion as in Fig. 5c, i.e. by solid red and dash blue arrows. The solid red arrows represent the in-phase regions, where the surface currents are nearly straight and oriented towards the y-axis (they are approximately $\lambda_g/2$ high). On the contrary, the dash blue arrows illustrate the out-phase regions, where the surface currents flow around the slots and notches. As a consequence, in the latter type the currents are oriented mostly towards the x-axis (their high is of approximately $\lambda_g/5$).

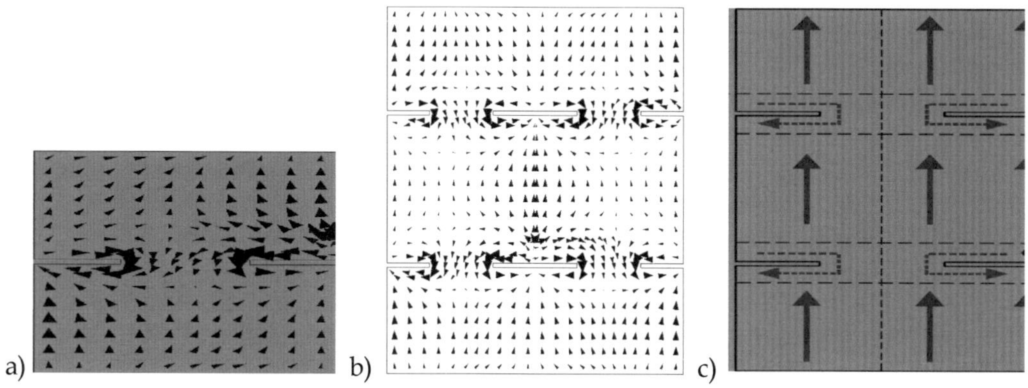

Fig. 8. Vector surface current distribution represented by black arrows on PCoMPA with a) TM_{03} and b) TM_{05} modes (simulated by IE3D). c) Scheme of $PCoMPA_{05}$ showing separation to basic segments and line demarking in-phase and out-phase current source areas

The initial design of the $PCoMPA_{05}$ antenna prototype that is developed for 2.4 GHz band is based on the principle described above. The center design frequency was set to 2.44 GHz and the antenna initial dimensions were optimized by means of build in procedures of IE3D simulator, where the criteria of impedance matching and maximum gain were followed. Physical dimensions of the final antenna prototype follow: patch size 189 × 222 mm, slot length 56 mm, notches length 28 mm and ground plane size 240 × 260 mm. The patch is carried over the ground plane by plastic distance posts (in the simulation, this fact was neglected) at the height of 5 mm (approx. 0.04 λ_0). The patch is fed by a coaxial probe placed on the y-axis at the distance of around 0.047 λ_g above one of the slots. Measured and simulated reflection coefficients of the realized prototype are depicted in Fig. 9b.

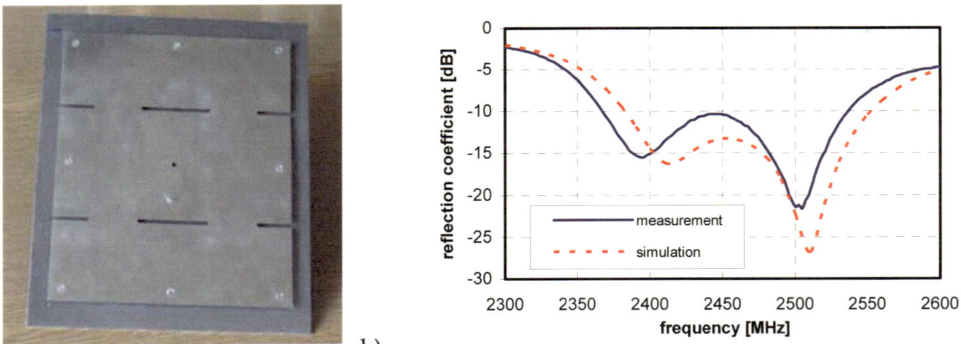

a) b)

Fig. 9. a) Photograph of realized PCoMPA$_{05}$ prototype, b) measured and simulated reflection coefficient

The distance between the centers of the in-phase source current areas equal 0.68 λ_0. Consequently, similarly to the case of 0.5 λ_0 element spacing in the linear uniform array, the sidelobe level is equal to about -10 dB. The front-back ratio is approximately 20 dB in both, the E plane and H plane. However, the ground plane dimensions exceed at all sides the size of the patch itself by merely some 0.2 λ_0.. The measured radiation patterns are presented in Fig. 10. The simulated (15.8 dBi) and the measured gains (15.4 dBi) result in 92 % efficiency. The impedance bandwidth 173 MHz, i.e. BW = 7.1 % (for VSWR = 2) is sufficient for example for the 2.4 GHz ISM applications.

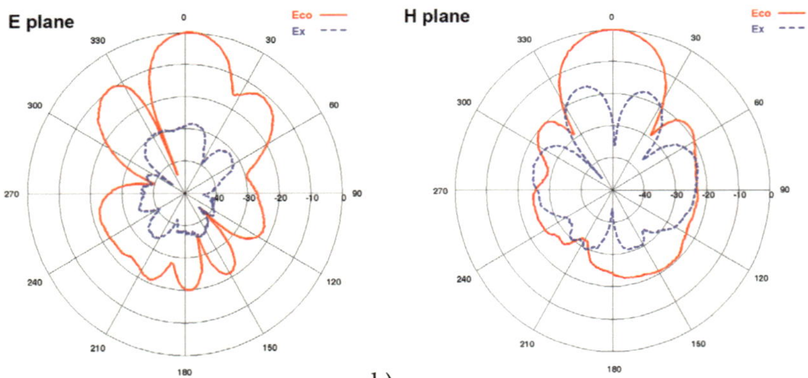

a) b)

Fig. 10. Measured co-polar (E_{co}) and cross-polar (E_x) radiation patterns of PCoMPA$_{05}$ prototype in a) E-plane, and b) H-plane.

3. Efficient Analysis of Collinear Microstrip Patch Antennas

An accurate and reliable characterization of both microwave and millimeter-wave antennas and circuits is one of the basic prerequisites for a successful computer-aided design (CAD), which constitutes a key prerequisite for a fast and cheap production process. Accordingly, our attention within the analysis of the CoMPA is concentrated mainly on a fast initial design showing sufficient accuracy that would not require the use of an expensive electromagnetic simulator. The selected analysis approach should be able to use the CoMPA rectangular building blocks for the effective implementation of the method and would also be suitable for more complex types of CoMPAs. All these requirements fulfill the multiport network model (MNM) (Gupta et al., 1981), which, together with innovations implemented by the authors, is going to be subject to a brief recapitulation here.

3.1 Multiport network model of patch antennas

The MNM is a method based on the Green's function approach (Okoshi, 1985) that is restricted to planar circuit components with regular canonical shapes. It can be considered as an extension of the cavity model (Lo et al., 1979), (Richards et al., 1981). The patch antenna is analyzed as a two-dimensional planar network, whereas the electromagnetic fields underneath the patch and outside the patch are modeled separately as networks that are then connected together via edge ports (Gupta & Hall, 2000); see Fig. 11.

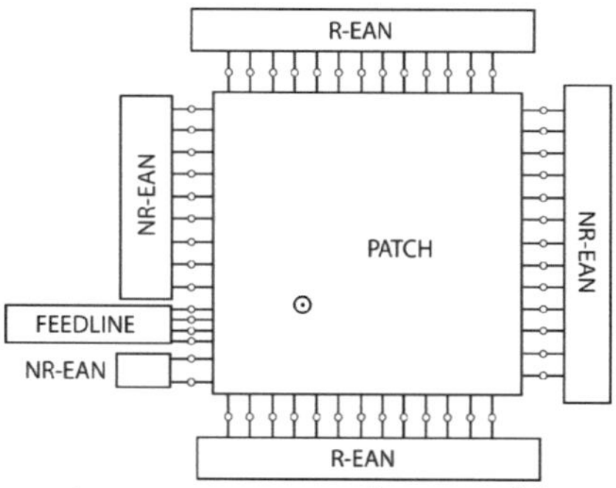

Fig. 11. Rectangular patch antenna represented as multiport network, connected with radiating and non-radiating edge admittance networks (R-EAN and NR-EAN) according to (Gupta & Hall, 2000)

The evaluation of the field underneath the patch is derived from the Green's function and can be expressed in terms of Z-matrix of multiport network in the following way:

$$Z_{pq} = \frac{j\omega\mu h}{LW} \cdot \frac{\sum_{m=0}^{\infty}\sum_{n=0}^{\infty} \sigma_m \sigma_n (x_p, y_p) \phi_{mn}(x_q, y_q)}{k_x + k_y - k^2} \tag{1}$$

where eigenfunction ϕ_{mn}, for ports oriented along the y-direction is:

$$\phi_{mn}(x, y) = \cos(k_x x)\cos(k_y y)\, sinc\left(\frac{k_y W}{2}\right) \tag{2}$$

and for ports oriented along the x-direction the following equation applies:

$$\phi_{mn}(x,y) = \cos(k_x x)\cos(k_y y)sinc\left(\frac{k_x L}{2}\right) \tag{3}$$

where the function sinc(z) is defined as sin(z)/z and, at the same time

$$k_x = \frac{m\pi}{L}, \quad k_y = \frac{n\pi}{W}$$

$\sigma_m = 1$ if m = 0, resp. $\sigma_m = 2$ if m $\neq 0$

$$k^2 = \omega^2 \mu \varepsilon_0 \varepsilon_r (1 - j\delta)$$

with δ being the loss tangent of the dielectric, L and W rectangle's length and width, and h the substrate's height. Points (x_p, y_p) and (x_q, y_q) denote the locations of the ports p and q, respectively.

The outer fields are modeled by means of so-called edge admitance networks (EAN), which might be considered as either radiating or non-radiating, depending on the shape of the voltage distribution along the edge. The non-radiating EAN (NR-EAN) are multiport networks consisting merely of the capacitance C (representing the energy stored in the fringing field). On the contrary, the radiating EAN (R-EAN) consists of parallel combination of the capacitance C and the conductances G (representing the power carried away by radiation and surface waves). The formulae for G and C can be found in (James & Hall, 1989).

The segmentation and desegmentation methods (Gupta et al., 1981) are used in order to identify the Z-matrix of non-regular shaped components, composed of the elementary segments, for which Green's functions are available. This technique enables to connect these segments into the complex planar shape via external ports. The voltage distribution and further s-parameters can be derived easily from the Z-matrix. The mathematical description of the technique is presented in several antenna handbooks, e.g. (James & Hall, 1989).

3.2 MNM of CoMPA$_{03}$

The very first MN model of CoMPA$_{03}$ was originally developed in order to apply an MNM method on a patch antenna with inner slot (Holub & Polívka, 2007a) and to compare the results with IE3D full-wave method of moments based simulation, which was assumed referential. Firstly, the antenna geometry was designed and optimized in IE3D simulator. The design frequency equalled $f = 2.44$ GHz. The optimized structure dimensions are listed below: patch length L = 147.5 mm, patch width W = 86.0 mm, slot length L_s = 56.0 mm, slot width W_s = 2.5 mm, space between the slot and the coaxial feeding probe L_f = 29.0 mm and the air substrate height h = 5 mm. The segmentation method, instead of desegmentation one, was used as the latter produces small numbers in Z-matrix. It addition, it gives rise to a consequential error during the process of desegmentation resulting from a substantial subtility of the inner slot(s) as well as from a small distance between the additional inner and external ports of the slot. The complete MN model of CoMPA$_{03}$ is made up of four segments; see Fig. 12a. The collateral segments are identical and the port distribution along the edges is uniform except of the central part with the ports No 11 and 35. The width of these two ports is equal to the width of the slot, W_S. The Z-matrixes of two central parts are nearly identical, because their dimensions and the port layout are the same. Nevertheless, the upper central Z-matrix involves the feeding port No. 1 representing a coaxial feeding probe. The Z-matrix of the complete segmented structure is composed of 97 × 97 elements. The distribution of EAN's is presented in Fig. 12b. The collateral edges are considered non-radiating. The EAN's contains the edge capacitance C only. The radiating EAN's (R-EAN's) are connected to the top and the bottom external ports of the patch.

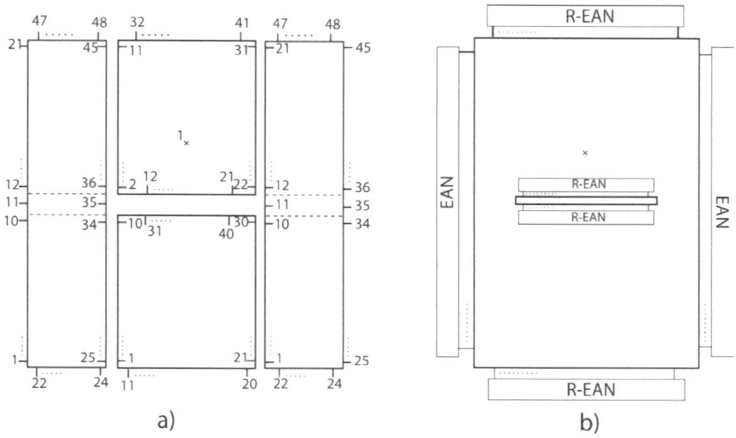

Fig. 12. a) Segmentation of MN model of CoMPA$_{03}$, b) MN model of CoMPA$_{03}$ with connected EANs and R-EAN

The situation (i.e. EAN) slightly varies along the edges of the inner slot. The capacitance at the edge of the slot is different from the capacitance at the external edges of the antenna and thus should be considered in a way shown in Fig. 13a. The aforementioned configuration matches the capacitance of the gap in the microstrip line (Gupta et al, 1981). This approach requires inclusion of the capacitance C_g between the opposite ports at the slot edges; see Fig. 13a. However, from the comparison of MNM with IE3D simulation results we have learned that the approach based on the consideration of the slot edges as just two external edges (see Fig. 13b) leads to results that are sufficiently accurate. The accuracy of the MN model depends on the number N of eigenfunctions in the calculation of Z-matrices of individual segments. The influence of the numer N on the results is demonstrated in the graph in Fig. 14. For the simplest CoMPA$_{03}$ radiator, the required accuracy can be achieved provided that the number N is of at least $N = 20$. The higher is the N, the higher is the accuracy. Yet, indeed, the duration of the calculation increases accordingly.

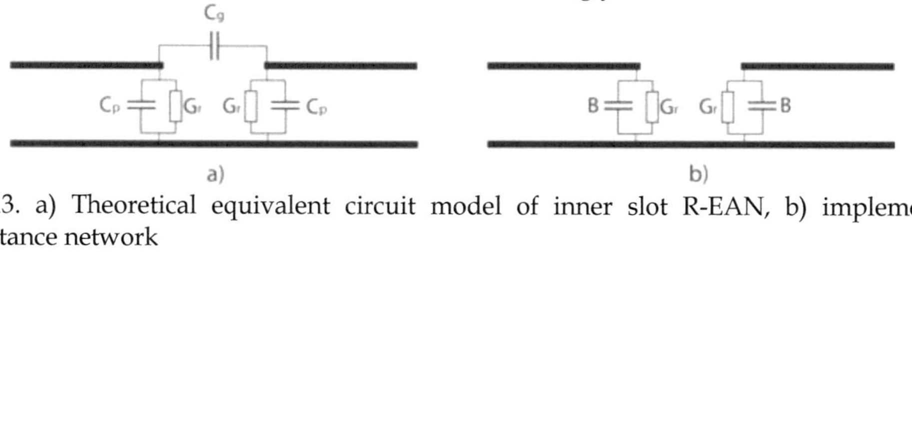

Fig. 13. a) Theoretical equivalent circuit model of inner slot R-EAN, b) implemented admittance network

The same structure was simulated in IE3D (for comparison see results indicated in Fig. 14b. A slight difference in the orientation of the curves in Smith diagram between the IE3D and MNM results is caused by an inductance of the coaxial probe, which is not included in the MN model.

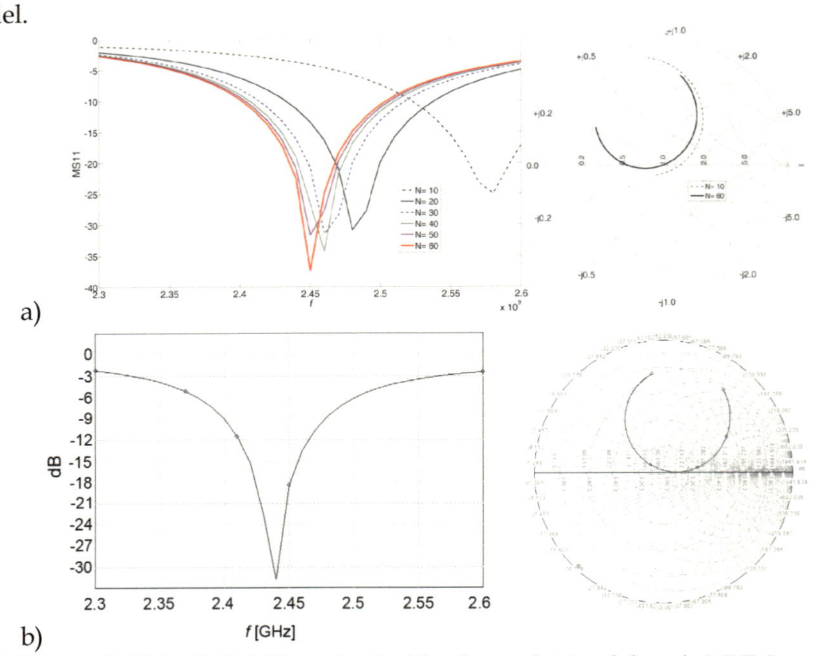

Fig. 14. MS11 and S11 of CoMPA$_{03}$ in Smith chart obtained by a) MNM, and b) IE3D simulation

In order to get an idea of the physical principle of the radiation structure, it is advisable to visualize the surface current distribution (see Fig. 15). Similar characteristic can be achieved in MNM by a display of the voltage distribution along the edges. Given the fact that the currents flow against the direction of voltage gradient, we can sketch the arrows representing the principal directions of the currents at the edges. Although the visualization of the voltage distribution cannot comprehend the direction of the currents inside the structure, it represents a useful instrument for the analysis of the radiation structure.

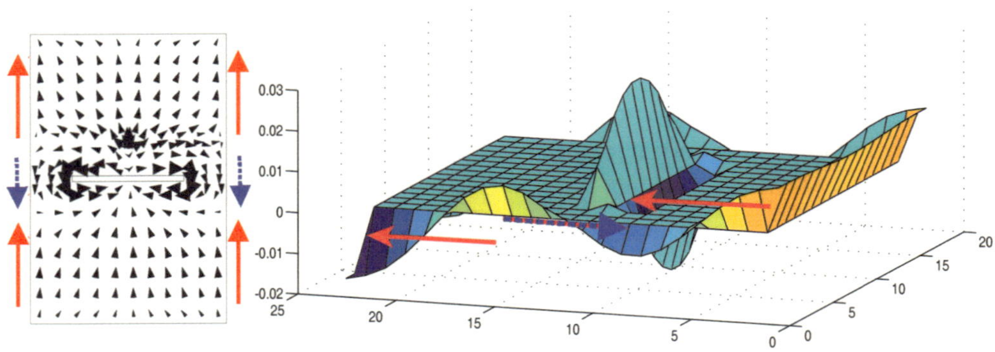

Fig. 15. Comparison of vector surface current distribution visualized by IE3D (left) and edge voltage distribution obtained by MN modelling (right)

A more complex MN model of the CoMPA$_{05}$, optimized for the frequency of 869 MHz, has been presented (Holub & Polívka, 2007b). The following rule applies: the higher is the structure complexity, the higher is the number of possible variants of segmentation. For instance, the segmentation of the CoMPA$_{05}$ can be derived from the CoMPA$_{03}$ segmentation by connecting additional rectangular segments to the shape of the CoMPA$_{03}$. However, such approach would be ineffective as it would contain too many segments with dissimilar proportions.

For edges with variable voltage distribution (non-radiating edges and inner slots), the number of segments per each half-wavelength should be considered from 8 to 10. In case of the uniform distribution, this number amounting to around 4-5 per each half-wavelength is, in general, sufficient.

3.3 Modeling of zero thickness of inner slots

The complexity of models (Holub & Polívka, 2007a), (Holub & Polívka, 2007b) results from a relatively complicated segmentation. The latter has to be used, due to a very limited width of the inner slots. This disadvantage can be eliminated by an effective MN modeling of the CoMPA, where the slot width is considered zero (instead of e.g. W_s = 0.5 mm); see Fig. 16. This approach has been first presented in (Holub & Polívka, 2008). Merely two unequal matrices have to be computed: the central matrix with a coaxial feeding and the top-bottom matrices. After the process of segmentation is accomplished, the ports along the opposite edges of the slots are located at the same xy-coordinates (this step has only an insignificant impact on the results). Due to the shift of the slot edge ports towards the center of the slots, the length of the CoMPA sections adjoining with the slots is extended. The ports along the slot edges are connected to the modified radiating EANs (MR-EANs); see Fig. 17c.

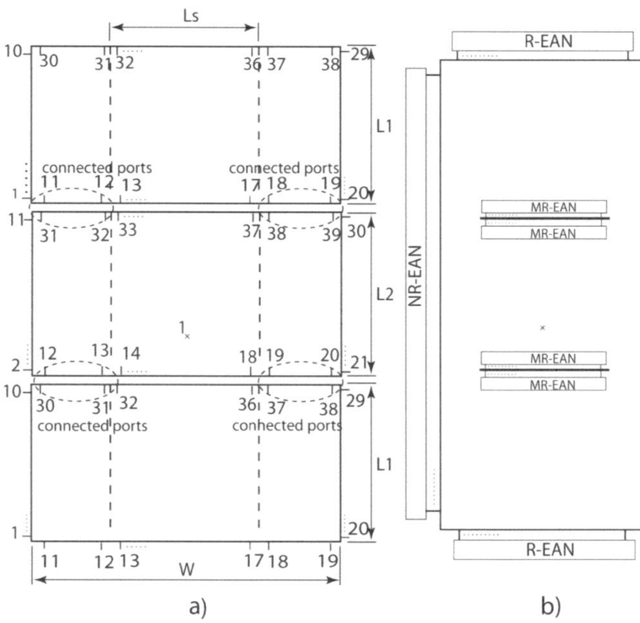

Fig. 16. Effective segmentation of MN model of CoMPA$_{05}$ considering zero thickness of slots, b) MN model of CoMPA$_{05}$ connected with R-EANs and MR-EANs

The results of the IE3D simulation confirmed that the MR-EANs should not contain the capacitances representing fringing fields at the edges of the slots. These fields are minor, since the opposite edge of the slot and its influence is partly included in the model itself by means of the shift of the slot ports towards the axis of the slot. As for the external fields, they can be incorporated either by the connection of the susceptance B or by a short extension of ΔL. The elements of the Y-matrix characterizing the EANs are computed from the equivalent circuits indicated in Fig. 16.

The distribution of the EANs for the $CoMPA_{05}$ is depicted in Fig. 17. The main advantage of the presented solution consists in its compactness and simplicity. Compared to the previous models, the presented solution requires fewer steps during the segmentation, which does not save much of the calculation time (approx. 5 %) though. But still, the code is simpler and easier for implementation.

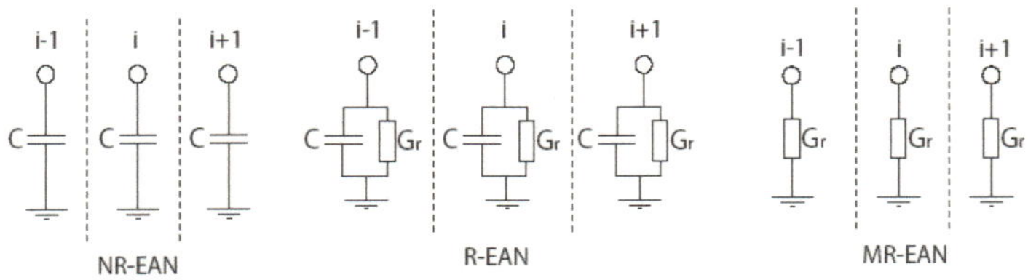

Fig. 17. Elements of Y-matrices characterizing non-radiating edge admittance network (NR-EAN), radiating edge admittance network (R-EAN) and modified radiating EAN (MR-EAN)

3.4 Efficient implementation of MNM for $CoMPA_{0X}$ analysis

Although the amount of the saved computational time is insignificant, the main feature of the above-described approach lies in the difference in segmentation. In case we divide the CoMPA antenna into the physically logical blocks with similar current distribution (thatwould be bounded by radiating slots), these modules match the segments in MNM segmentations. The complexity of the structure can be determined by the excited TM_{xy} mode and/or the number of patch modules (i.e. the areas between the slots). In the effective MNM method modelling we can take advantage of the periodicity of the structure. The development of a universal code for the previous models would be substantially complicated (especially the part dedicated to the algorithm of segmentation). Besides, the obtained model would be inefficient and nearly unusable. Owing to the presented novel effective segmentation, the situation is considerably simplified.

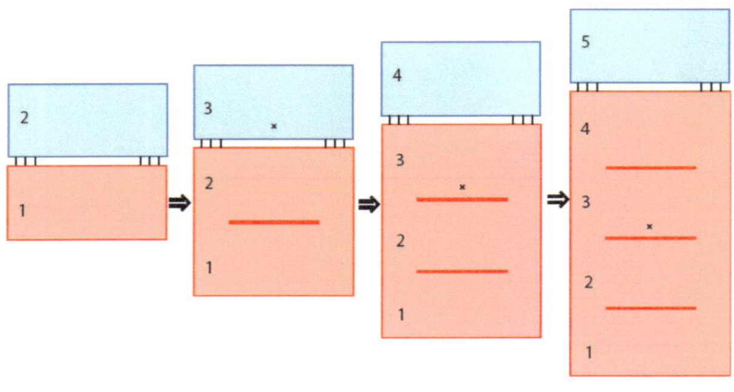

Fig. 18. Sequential assembly of 5 segments in $CoMPA_{09}$ MN modeling with feeding probe position according to eq. (4)

From the given geometry dimensions of the internal and external modules as well as the number of ports along particular edges of the segments, it is necessary to allocate the positions of the individual ports and their widths. The next step is represented by the calculation of the Z-matrices for the internal and external CoMPA modules.

The first port of the calculated internal matrix is the coaxial feeding port. Such type of the matrix is utilized uniquely for the central module with feeding port. The location of the port in question is determined by its distance from the central slot. When the first row and column of the matrix, representing the first feeding port, is erased, the matrix for all other internal modules, without any calculation, is obtained.

Subsequently, the connection of the individual ports to the whole structure is initiated. Fig. 18 depicts the sequential assembly of the five modules on the example of the $CoMPA_{09}$ excited by the mode TM_{09}. Red segments represent the modules already connected to the structure; the coax feeding is marked as ×.

After the connection of the further segment, the ports along the edges are disarranged. Before the next assembly, the port distribution has to be rearranged. The first and the last segments stand for peripheral modules. The position of the feeding segment is determined by the following relation:

$$feed = \text{floor}\left(\frac{CoMPA_{\text{level}}}{2}\right) + 1 \tag{4}$$

where *feed* is the order number of feeding module, $CoMPA_{\text{level}}$ represents the total number of antenna modules (MNM segments) and 'floor' stands for the Matlab function *round toward minus infinity*.

3.5 MNM and IE3D computational time comparison

To evaluate the effectiveness and virtues of the MN method, it is necessary to compare its computational time with the one of another method. Here we use IE3D as a reference. This EM simulator is well-suited for planar structures as is CoMPA utilized in our case.

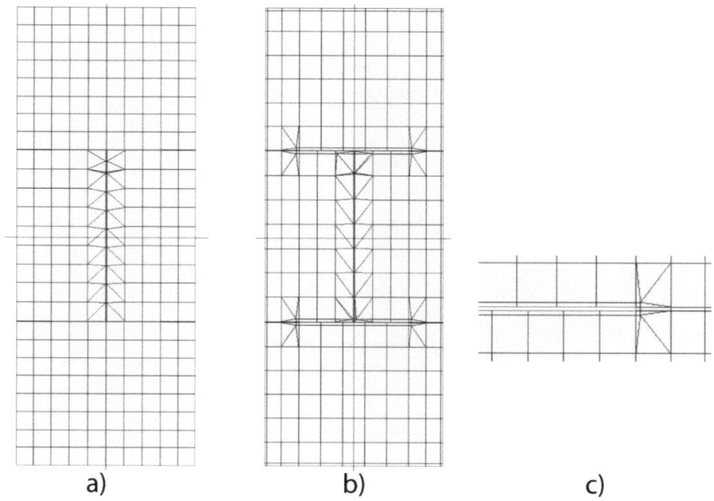

Fig. 19. IE3D mesh of $CoMPA_{05}$ with 8 cells per wavelength, a) without AEC, b) with AEC

As for the accuracy of the results, IE3D enables to use so-called automatic edge cells (AEC), which are narrow edge cells employed for a precise modeling of current distribution; see Fig. 19b, c. Usually it is recommended to set the mesh density to 20 cells per wavelength. As it is shown, the application of 8 cells per wavelenght, together with the AEC, leads to sufficiently accurate results, provided that a symmetric matrix solver (SMS) is used. The frequency shift of the reflection coefficient minimum when a number of 8 and 20 of cells per wavelength with AEC is used is equal to just about 0.8%. While the difference between MNM and IE3D optimal results is about 2.5% which is acceptable for initial design; see Fig. 20.

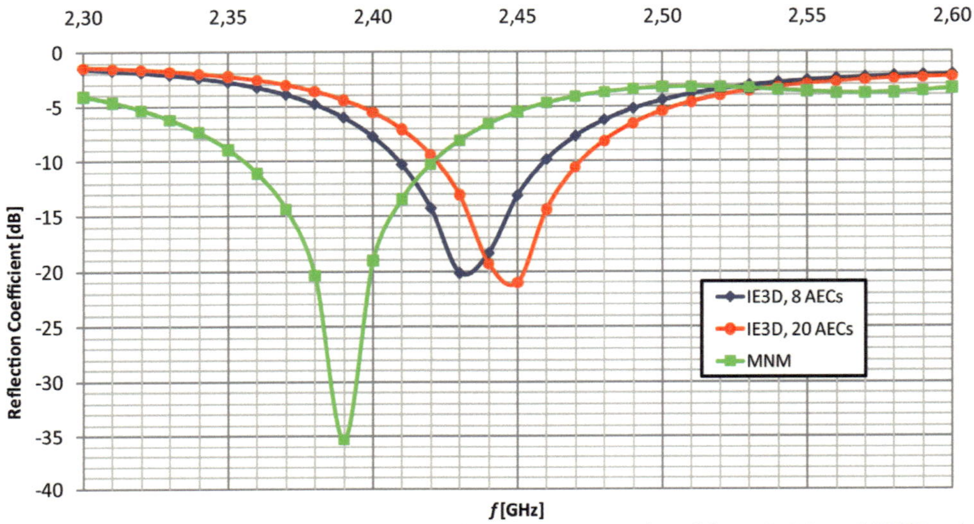

Fig. 20. Reflection coefficient comparison of $CoMPA_{05}$ simulated by MNM and IE3D with 8 and 20 cells per wavelength (with AECs)

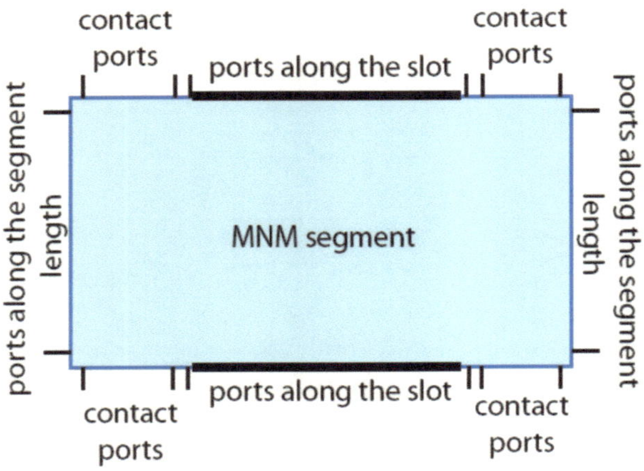

Fig. 21. Port distribution along the edges of MNM segment

A low number of mesh cells with such a good accuracy can be explained by the surface current distribution on the CoMPA, where the gradient of the currents attains the highest level around the slots, for due to the employment of the AEC, the meshing is fine at the edges.
The distribution of the ports along the edges in the used MNM is depicted in Fig. 21, where the number of ports along the particular sections utilized for comparison with IE3D are 8, 7, and 6 for the segment length, slot length and contact port length, respectively. The number of eigenfunctions in summation of the Z-matrix is equal $N = 40$.
Because of a short computational time per frequency (that is less than one second), the antennas were analyzed at 31 frequency points within 2.3 - 2.6 GHz frequency band. The obtained results are summarized in the diagram in Fig. 22. The principal difference between IE3D and MNM consists in the rule that the computational time in IE3D rapidly rises with the increasing number N of CoMPA segments. However, the MNM calculation stays nearly constant, yet there is an exception - the calculation of the $CoMPA_{03}$ (two CoMPA modules) requires half of the time, because the model is composed of two segments derived from one matrix, representing peripheral elements. The slight growth of the computational time (see Fig. 22) is attributable to the rising number of segmentation cycles. On the contrary, the main time consuming part - the computation of MNM Z-matrices - remains unchanged. To state the key virtue of the proposed MNM approach, it is necessary to point out that we can calculate the CoMPA of any level without the significant rise in the computational time.

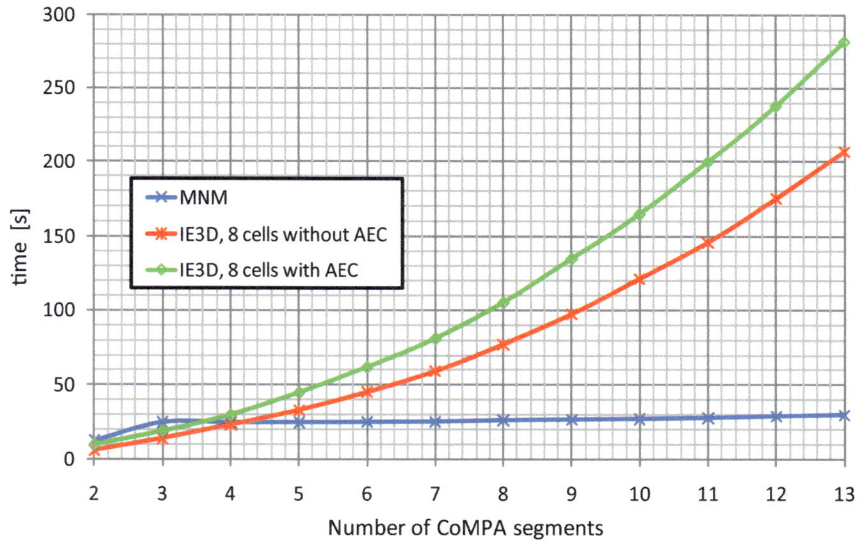

Fig. 22. Comparison of computational time of MNM method and IE3D simulation; IE3D (8 cells), IE3D-optimal (8 cells with AEC)

4. Conclusion

A sort of novel collinear microstrip patch antennas with a hemispherical radiation pattern, showing an increased gain of approx. 12 ÷ 15 dBi has been comprehensively introduced.

The essence of the operation has been explained via surface current distribution of operational modes of the antenna, which can be described as slots and notches loaded microstrip patch operating with higher TM_{0X} modes. Due to the collinear arrangement of the in-phase source current areas, the directivity can be enhanced by an increase in the order of the operational mode together with the enlargement of the patch longitudinal and lateral dimensions. The advantage of such arrangement, when compared to a classical patch array, is represented by a very simple structure without the need for any feeding network. The drawback, however, is given by the limited impedance bandwidth and also the impossibility to control the amplitude distribution as well as the phase of source currents on the structure. This structure is namely suitable for applications, where the gain ranging from approx. 12 to 20 dBi is required. Typical applications are terminal antennas destined for communication purposes or RFID reader antennas.
Subsequently, the effective multiport network model has been implemented for the CoMPA fast initial design, based on the presumtion of zero slot width. The results then match very sufficiently the results obtained by the IE3D simulator. The comparison of the computational times of the MNM and the IE3D shows that in case the number N of CoMPA modules increases, the IE3D computational time is approximately proportional to the N^2, while the MNM time remains nearly constant or increases very slowly. When stressing the advantages of the novel type of segmentation introduced in this chapter, it is crucial to mention also the compactness and simplicity of the MNM algorithm, indeed when compared to the original implementation.

5. Acknowledgement

This work has been performed at the Department of Electromagnetic Field of the Czech Technical University in Prague. It was supported through the project of the Czech Science Foundation No. 102/08/1282 called "Artificial electromagnetic structures for miniaturization of high-frequency and microwave radiation and circuit elements". In addition, it was financed also by the Czech Ministry of Education, Youth and Sports within the Research Project in the Area of the Prospective Information and Navigation Technologies MSM 6840770014, and also the COST project IC0603 "Antenna Systems & Sensors for Information Society Technologies".

6. References

Balsley, B. B. & Ecklund, W. L. (1972). A Portable Coaxial Collinear Antenna, *IEEE Trans. on Antennas and Propagation*, Vol. 20, No. 1972, pp. 513-516, ISSN 0018-926X

Bancroft, R. & Bateman, B. (2004). An Omnidirectional Planar Microstrip Antenna, *IEEE Trans. on Antennas and Propagation*, Vol. 52, No.11, November 2004, pp. 3151-3153, ISSN 0018-926X

Chen, S., Lan, I. & Hsu, P. (2007). In-Line Series-Feed Collinear Slot Array Fed by a Coplanar Waveguide, *IEEE Trans. on Antennas and Propagation*, Vol. 55, No. 6, (June 2007), pp. 1739-1744, ISSN 0018-926X

Franklin, C. S. (1925). Improvements in wireless telegraph and telephone aerials, *U.K. British Patent GB242342*, Nov. 5, 1925

Holub, A. & Polívka, M. (2007a). Application of MNM on Collinear Microstrip Patch Antenna, *Proceedings on Antennas and Propagation Symposium [CD-ROM]*, pp. 61 - 64, Hawaii, June 2007, Honolulu

Holub, A. & Polívka, M. (2007b). Multiport Network Modeling of a Complex Canonicaly Shaped Patch Antenna, *Proceedings on European Conference on Antennas and Propagation [CD-ROM]*, pp. 1-5, ISBN 978-0-86341-842-6, Edinburgh, Scotland, November 2007, Stevenage, Herts: The Institution of Engineering and Technology (IET), Edinburgh

Holub, A. & Polívka, M. (2008). Effective segmentation in Multiport Network Model method for analysis of planar antennas with thin slots, *Proceedings of International Symposium on Antennas and Propagation 2008 [CD-ROM]*, Taipei, Taiwan, October 2008, Yuan Ze University and Oriental Institute of Technology, Taipei

Gupta, K. C., Greg, R. & Chadha, R. (1981). *Computer-Aided Design of Microwave Circuits*, Artech House, ISBN 0890061068

Gupta, K. C. & Hall, P. S. (2000). *Analysis and Design of Integrated Circuit Antenna Modules*, John Wiley & Sons. ISBN 0471190446, New York

James, J. R. & Hall, P. S. (1989). *Handbook of Microstrip Antennas*, Peter Peregrinus Ltd., ISBN 0-86341-150-9, London, UK

Kraus, J. D., & Marhefka, R. J. (2002). *Antennas for all Applications*, McGraw-Hill, ISBN 0-07-232103-2, New York

Lo, Y. T., et al. (1979). Theory and experiment on microstrip antennas. *IEEE Trans. on Antennas and Propagation*, Vol. 27, No. 2, (March 1979), pp. 137-145, ISSN 0018-926X

Mazánek, M., Klepal, M., Pechač, P., Polívka, M., Bartík, H. (2000). Anechoic and EMC Chambers - Modelling, Design, Testing, *Proceedings of Millennium Conference on Antennas and Propagation*, pp. 156-160, The Netherlands, 2000, European Space Agency, Noordwijk

Miyashita, H., Ohmine, H., Nishizawa, K., Makino, S. & Urasaki, S. (1999). Electromagnetically Coupled Coaxial Dipole Array Antenna. *IEEE Trans. on Antennas and Propagation*, Vol. 47, No.11, November 1999, pp. 1716-1725, ISSN 0018-926X

Nishimura, S., Nakanano, K. & Makimoto, T. (1979). Franklin-type Microstrip Line Antenna, *International Symposium Digest, Antennas and Propagation*, pp. 134-137, Washington, 1979, IEEE, Seattle

Okoshi, T. (1985). *Planar Circuits for Microwave and Lightwaves*, Springer-Verlag, ISBN 0387138536, New York

Polívka, M., Holub, A. & Mazánek, M. (2005). Collinear Microstrip Patch Antenna. *Radioengineering*, Vol. 14, No. 4, (December 2005), p. 40-42. ISSN 1210-2512

Polívka, M. & Holub, A. (2006). Planar Version of Collinear Microstrip Patch Antenna, *Conference Proceedings MIKON 2006*, pp. 959-962, ISBN 83-906662-7-8, Poland, May 2006, Telecommunication Research Institute, Warsaw

Richards, W. F., et al. (1981). An improved theory for microstrip antennas and applications. *IEEE Trans. on Antennas and Propagation*, Vol. 29, No. 1, (January 1981), pp. 38-46, ISSN 0018-926X

Solbach, K. (1982). Microstrip-Franklin Antenna. *IEEE Trans.on Antennas and Propagation*, Vol. 30, No. 4, (July 1982), pp. 773-775, ISSN 0018-926X

Švanda, M., Polívka, M. & Hudec, P. (2007). Application of the UHF RFID system for the identification of sportsmen in mass races, In: *Proceedings of the European Microwave Association*. Vol. 3, No. 4 (December, 2007), pp. 295-301, Edizioni Plus - Universita di Pisa, ISBN 88-8492-324-7, Pisa

Design of Non-Uniformly Excited Linear Slot Arrays Fed by Coplanar Waveguide

JP Jacobs, J Joubert and JW Odendaal
University of Pretoria
South Africa

1. Introduction

Slot dipole antennas fed by coplanar waveguide (CPW) have a number of appealing characteristics that include relatively low feed line dispersion and losses at millimeter-wave frequencies, ease of integration with circuit components due to the uniplanar configuration, and considerably wider impedance bandwidth than microstrip patch antennas on comparable substrates, *e.g.*, (Miao et al., 2000). In the sub-millimeter and millimeter-wave ranges, CPW-fed slots have been used to feed dielectric lens antennas (Focardi et al., 2002); arrays of CPW-fed slots are also expected to be used for space applications (Neto et al., 2003).

While a variety of CPW-fed slot arrays have been reported (see section 3), rigorous design procedures for CPW-fed slot arrays that incorporate the effects of element interactions in the form of mutual coupling have only rarely been presented (Huang et al., 1999; Jacobs & Joubert, 2009b). An iterative design typically involves the *a priori* generation of self-admittance data for representative slot dimensions, and requires the calculation of mutual coupling between all possible slot pairs in the array; this is especially pertinent if a non-uniform aperture distribution aimed at achieving reduced sidelobe levels is to be realized.

In this chapter an overview of the current state-of-the-art in the design of non-uniform linear CPW-fed antennas, as well as its subsidiary calculations, is presented. In section 2, the field distribution in an isolated CPW-fed is shown; this is important information for the calculation of mutual coupling. Section 3 focuses on the calculation of mutual coupling using a reciprocity-based paradigm. This is done in the context of two important practical cases. The first involves mutual coupling between slots on electrically thin substrates, where a simplified yet accurate method can be used that obviates use of the substrate Green's function. Arrays of slots on such substrates exhibit bi-directional radiation. The second involves an extended reciprocity approach for calculating the mutual admittance between slots on a conductor-backed two-layer substrate, a configuration that would be useful when unidirectional radiation is required. Section 4 gives details of two types of CPW-fed linear array design. First, the iterative design using an approximate procedure of a uniform array on a conductor-backed two-layer substrate is summarized; measured results are presented. The procedure doesn't require the explicit calculation of the mutual admittance between pairs of slots, but does take such interactions into account through full-wave calculations of the input impedance of the array as a whole. Second, the implementation of a rigorous iterative design procedure for a non-uniform array with reduced sidelobe levels on an electrically thin substrate is described; measured sidelobe levels of about 16 dB below the main beam were achieved. Due to the electrically thin substrate, the design procedure could draw on the simplified method for calculating mutual admittance outlined in section 3. Section 5 briefly summarizes some of the main findings.

2. Field properties of isolated CPW-fed slots

CPW-fed slots are normally operated in the vicinity of their second resonances because of the favourable impedance bandwidth properties here (compared to the first-resonant region). A CPW-fed slot on a single dielectric layer is shown in Fig. 1. In order to more accurately account for its use in a linear array environment (see Fig. 4), the slot is terminated in a perfect short-circuit realized by a section of transmission line of length l_s.

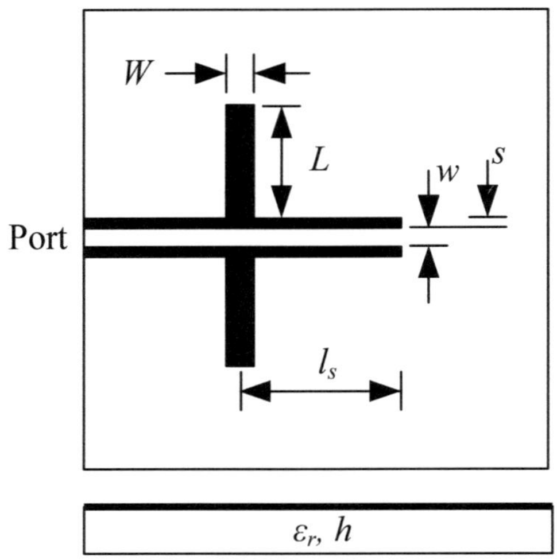

Fig. 1. Top and side views of CPW-fed slot on single-layer substrate. L ≡ half-length and W ≡ width of radiating slot; s ≡ slot width and w ≡ center strip width of CPW feed line; h ≡ dielectric layer height; ε_r ≡ relative dielectric constant; l_s ≡ length of CPW stub implementing short-circuit.

When evaluating the mutual admittance between pairs of slots, the aperture electric field distribution is required (see section 3). IE3D (Zeland Software, 2001), a full-wave electromagnetic simulator that employs magnetic current modeling and assumes laterally infinite ground planes and dielectrics, can be used to obtain this information. Figs. 2 and 3 (Jacobs & Joubert, 2009a) show magnitudes and phases at 6 GHz of the tangential electric fields along longitudinal slot centers of three isolated slots on an electrically thin substrate with h = 1.575 mm and ε_r = 2.33 (at 6 GHz, h = $0.048\lambda_d$ = $0.0315\lambda_0$, with λ_d and λ_0 the wavelengths in the dielectric and free space respectively). The slots' widths W was 1 mm, their half-lengths L were in the vicinity of the second-resonant half-length L_{res} = 21. 75 mm, and the length of the short-circuit stub l_s was $0.48\lambda_{CPW}$ = 19.4 mm. The slots were fed by a 87 Ω CPW feed line that had dimensions w = 3 mm and s = 1 mm. (The same radiating slot and feed line dimensions were used in obtaining the mutual admittance results of Figs. 6–8.) The aperture field vector component shown is the one directed across the width of the slot; the orthogonal component was negligible by comparison.

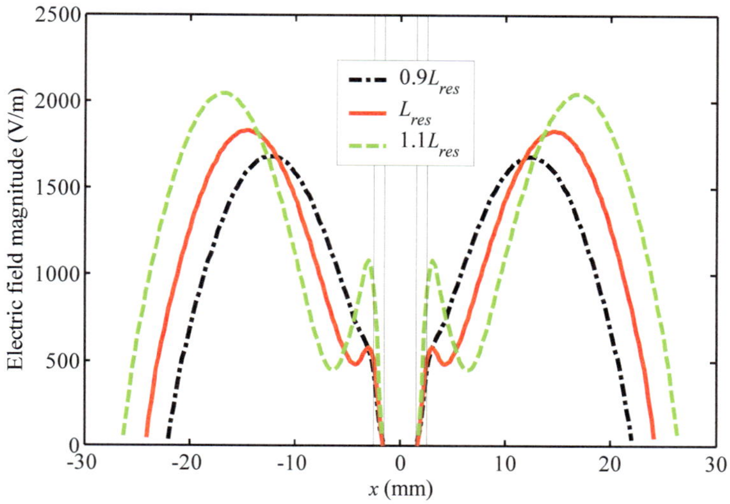

Fig. 2. Magnitude of electric field of isolated slots with half-lengths in vicinity of second-resonance half-length L_{res}. L_{res} = 21.75 mm, W = 1 mm, l_s = 19.4 mm = 0.48 λ_{CPW}, h = 1.575 mm, ε_r = 2.33, w = 3 mm, s = 1 mm. Vertical lines correspond to position of CPW.

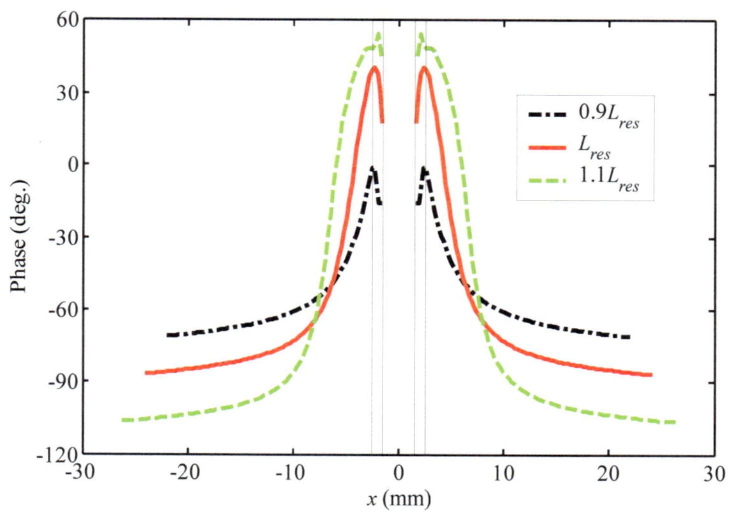

Fig. 3. Phase of electric field of isolated slots with half-lengths in vicinity of second-resonance half-length L_{res}. L_{res} = 21.75 mm, W = 1 mm, l_s = 19.4 mm = 0.48 λ_{CPW}, h = 1.575 mm, ε_r = 2.33, w = 3 mm, s = 1 mm. Vertical lines correspond to position of CPW.

Contrary to the nearly constant phases of CPW-fed slots with half-lengths in the vicinity of the first-resonant half-length (Jacobs, 2007), the phases of these second-resonant slots exhibit a sharp rise close to the CPW feed line, while changing little in the outer reaches of the slots. Increases in slot length result in offsets of phases with respect to phases of preceding lengths. Similar graphs were obtained for slots on conductor-backed two-layer substrates (Jacobs, 2007).

3. Mutual coupling between CPW-fed slots

The focus of this section is the calculation of mutual coupling between slot pairs in linear arrays of slots fed in series by CPW. The geometry of such an array, designed for a broadside main lobe, is shown in Fig. 4. While a variety of types of CPW-fed slot arrays have been reported, for instance wideband linear CPW-fed log-periodic dumb-bell slot arrays (Kim et al., 2006), amplifier arrays using CPW-fed folded slot antennas (Tsai et al., 1994), and CPW-fed planar (two-dimensional) slot arrays in multi-chip module-deposition (MCM-D) technology (Soliman et al., 1999), the rigorous calculation of mutual coupling between slot array elements in the context of iterative array design procedures has only rarely been addressed, *e.g.*, Huang et al. (1999). For certain applications, such calculations seem unnecessary: the design of *uniform* CPW-fed arrays on both single-layer substrates (Qui et al., 2002) and conductor-backed two-layer substrates (Jacobs et al., 2003) have been accomplished without explicitly calculating the mutual coupling between individual pairs of slots, as noted earlier. On the other hand, the design of *non-uniform* arrays subject to rigorous sidelobe-level specifications generally requires explicit, accurate accounts of the mutual coupling between all possible slot pairs in the array.

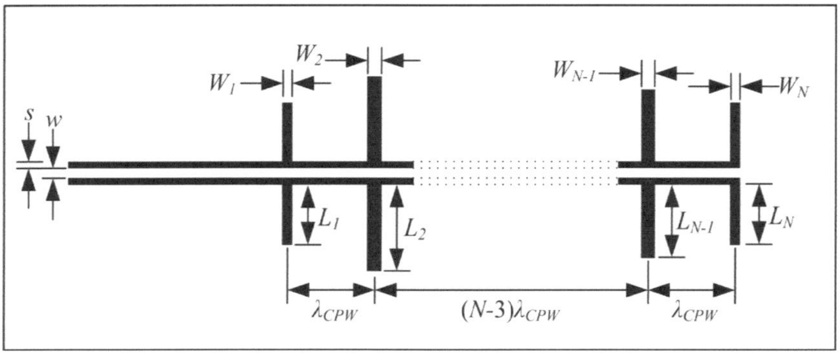

Fig. 4. Top view of N-element CPW-fed linear slot array with equiphase element excitations. $L_m \equiv$ half-length and $W_m \equiv$ width of radiating slot m, where $m = 1, 2 .. N$; $s \equiv$ slot width and $w \equiv$ center strip width of CPW feed lines; $\lambda_{CPW} \equiv$ CPW wavelength.

3.1 Mutual coupling between slots on electrically thin single-layer dielectric substrate

For electrically thin substrates a simplified method based on a well-known reciprocity-based expression can be used to find the mutual admittance between two broadside CPW-fed slots (Jacobs & Joubert, 2009a). The method is eminently suitable for easy incorporation into iterative array design algorithms, and for fast evaluation. The geometry is shown in Fig. 5; slot orientations correspond to their orientation in a linear array such as that of Fig. 4. The slots in Fig. 5 are intended to be accurate models of slots in an array environment - hence their termination in CPW sections implementing short-circuits that extend beyond their radiating portions (this will be more fully described below). The kind of mutual admittance calculation performed here is required when adopting a so-called first-order interaction approach (Amitay et al., 1972) towards finding the mutual admittance between any two slots in an array: the mutual admittance between a specific pair of slots is, for the sake of simplicity, assumed to be identical to the mutual admittance between them when the other slots in the array are removed.

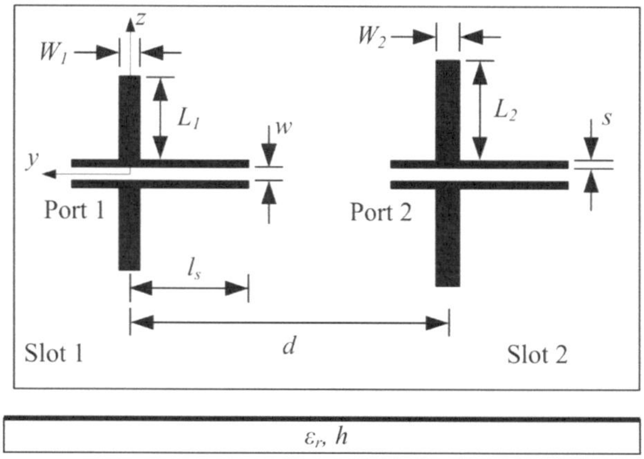

Fig. 5. Top and side views of broadside CPW-fed slots on single-layer dielectric substrate. $L_1 = L_2 \equiv$ half-lengths and $W_1 = W_2 \equiv$ widths of radiating slots; $d \equiv$ distance between radiating slots; $s \equiv$ slot width and $w \equiv$ center strip width of CPW feed lines; $h \equiv$ dielectric layer height; $\varepsilon_r \equiv$ relative dielectric constant; $l_s \equiv$ length of CPW stub implementing short-circuit.

The simplified method assumes that, for mutual admittance calculations between CPW-fed slots on electrically thin single-layer substrates, the inhomogeneous air-dielectric layer-air medium can be sufficiently accurately approximated by a homogenous free space (this is different from the approach taken by Huang et al. (1999)). The mutual admittance between CPW-fed slots *1* and *2* of Fig. 5 can then be found by adapting results originally derived for wire dipoles radiating in free space in (Balanis, 1996) to slots in an infinite ground plane radiating in free space. If it is assumed that CPW feed lines have negligible effect on mutual coupling, only the radiating portions of the CPW-fed slots need to be considered. Hence,

$$Y_{21} = -\frac{1}{V_1 V_2}\left[\int_{-(w/2+s+L_2)}^{-(w/2+s)} H_{z21}(z_2) I_{m2}(z_2) dz_2 + \int_{w/2+s}^{w/2+s+L_2} H_{z21}(z_2) I_{m2}(z_2) dz_2\right] \tag{1a}$$

where

$$H_{z21}(z_2) = H_r(z_2)\cos\theta - H_\theta(z_2)\sin\theta \tag{1b}$$

with

$$H_r(z_2) = \int_{-(w/2+s+L_1)}^{-(w/2+s)} I_{m1}(z_1)\eta_0 \frac{L_1 \cos\theta}{2\pi r^2}\left[1+\frac{1}{jk_0 r}\right]e^{-jk_0 r}dz_1 + \int_{w/2+s}^{w/2+s+L_1} I_{m1}(z_1)\eta_0 \frac{L_1 \cos\theta}{2\pi r^2}\left[1+\frac{1}{jk_0 r}\right]e^{-jk_0 r}dz_1 \quad (1c)$$

and

$$H_\theta(z_2) = \int_{-(w/2+s+L_1)}^{-(w/2+s)} I_{m1}(z_1) j\eta_0 \frac{k_0 L_1 \sin\theta}{4\pi r}\left[1+\frac{1}{jk_0 r}-\frac{1}{(k_0 r)^2}\right]e^{-jk_0 r}dz_1 + \int_{w/2+s}^{w/2+s+L_1} I_{m1}(z_1) j\eta_0 \frac{k_0 L_1 \sin\theta}{4\pi r}\left[1+\frac{1}{jk_0 r}-\frac{1}{(k_0 r)^2}\right]e^{-jk_0 r}dz_1 \quad (1d)$$

In the above, H_{z21} is the magnetic field of equivalent magnetic current I_{m1} at the position of slot *2* when slot *1* radiates in isolation (I_{m1} represents slot *1* when radiating in isolation), I_{m2} is the equivalent magnetic current representing slot *2* when radiating in isolation, V_1 is the terminal voltage of slot *1* when radiating in isolation, V_2 is the terminal voltage of slot *2* when radiating in isolation, L_1 and L_2 are the half-lengths of slots *1* and *2* respectively, s is the slot width and w the center strip width of the CPW feed line, $r = \sqrt{(z_2 - z_1)^2 + d^2}$, $k_0 = \omega\sqrt{\mu_0 \varepsilon_0}$ is the free-space wavenumber, and $\eta_0 = \sqrt{\mu_0 / \varepsilon_0}$ the free-space intrinsic impedance. r and θ are spherical coordinates and the dimensions constituting the integration boundaries are shown in Fig. 5.

The above equations can be implemented as follows to find Y_{21} against slot spacing d for two CPW-fed slots on an electrically thin substrate:

First, each CPW-fed slot of Fig. 5 is simulated in isolation using IE3D (in the case of twin slots, only one slot needs to be simulated). As noted earlier, in order to accurately model a CPW-fed linear array environment, each slot is terminated in a perfect short-circuit realized by a section of transmission line of length l_s, with l_s measured from the center of the radiating portion of the slot. For composite slot (feed line, radiating slot and short-circuit termination) k, where k = 1 or 2, the simulation yields the tangential electric field along the center of the radiating slot (see Figs. 2 & 3), from which its equivalent magnetic current density I_{mk} can be obtained. It is assumed that the radiating slot electric field only has a vector component across the width of the slot (in other words, the small longitudinal component predicted by IE3D is neglected), and that the field is constant across the width of the slot. The latter assumption, which enables calculation of I_{m1} and I_{m2} by simply multiplying the corresponding equivalent magnetic current density by the slot width, can be enforced in IE3D by adopting a discretization that allows for only one cell across the slot width. In addition to the slot tangential field, the isolated-slot moment-method analysis also gives the terminal voltages V_k, where k = 1 or 2. Excitation ports in IE3D are defined at the ends of feed lines (*cf.* Fig. 5). The port voltage computed by IE3D at the end of the feed line is in fact the terminal voltage, and is found by integrating the transverse electric field in one of the two CPW slots over the slot width. In the mutual admittance calculations described below, terminal voltages (and isolated slot self-admittances) were referred to centers of radiating slots.

Second, Equation (1) was evaluated for each instance of slot separation d. In accordance with the definition of the quantities that constitute equation (1), I_{m1}, I_{m2}, V_1 and V_2 were kept the same for all values of d. Curves of Y_{21} against d obtained with the reciprocity-based method outlined above were compared with curves computed using IE3D. Using a moment-method approach towards this end implies that the entire two-slot structure of Fig. 5 needs to be solved for each instance of d. Ports were defined at the ends of the CPW feed lines, and from the full-wave solution IE3D calculated the two-port Y parameters of the structure with respect to the above ports. Y parameters (including Y_{21}) were afterwards referred to centers of radiating slots.

Y_{21} against d with $0.9\lambda_{CPW} \leq d \leq 2\lambda_{CPW}$ was computed for three pairs of identical broadside (twin) slots on the electrically thin substrate with h = 1.575 mm and ε_r = 2.33 (at 6 GHz, $h = 0.048\lambda_d = 0.0315\lambda_0$, with λ_d and λ_0 the wavelengths in the dielectric and free space respectively). The slots' widths W was 1 mm, their half-lengths $L_1 = L_2$ were $0.9L_{res}$, L_{res} and $1.1L_{res}$ respectively, with L_{res} = 21. 75 mm, and the length of the short-circuit stub l_s was $0.48\lambda_{CPW}$ = 19.4 mm. The slots were fed by a 87 Ω CPW feed line that had dimensions w = 3 mm and s = 1 mm. The resonant slot had a self-impedance of 17 Ω, and hence can be considered a 'typical' slot from a linear array perspective, given that the input impedance of a broadside CPW-fed linear array is simply the sum of the slots' active impedances, and since slot self-impedance is often considered a first-order approximation to slot active impedance (Elliott, 1981). The self-impedances of the $0.9L_{res}$ and $1.1L_{res}$ slots were 20-j16 Ω and 16+j19 Ω respectively.

The real and imaginary parts of Y_{21} against normalized slot separation d/λ_{CPW} for the twin slots with $L_1 = L_2 = L_{res}$ are shown in Fig. 6; results from both the reciprocity-based method (the curves labeled 'rec.') and IE3D are given. Likewise, Y_{21} against d/λ_{CPW} for twin slots with $L_1 = L_2 = 0.9L_{res}$ and $L_1 = L_2 = 1.1L_{res}$ are shown in Figs. 7 and 8 respectively. The third set of curves (labeled 'improved rec.') in each of Figs. 6–8 correspond to an improved version of the reciprocity-based method that used, instead of the terminal voltages V_k obtained from IE3D, new terminal voltages $V_{k,\,new}$ that were computed as follows. The most accurate results for Y_{21} in a linear array with broadside main lobe are required when the slot spacing $d = \lambda_{CPW}$; this is the minimum spacing between slot elements, and the effect of mutual coupling would be the greatest here. Thus, for each of the twin slot cases represented in Figs. 6–8, the value of Y_{21} was computed at $d = \lambda_{CPW}$ using IE3D; so was the reaction integral in equation (1). The product $V_{1,\,new}\, V_{2,\,new}$ was then obtained from the quotient of IE3D's Y_{21} and the reaction integral value. Since twin slots were involved, $V_{1,\,new}$ and $V_{2,\,new}$ are equal. These voltages were then used in a new evaluation of equation (1) over the range of d.

Figs. 6–8 reveal generally good agreement between Y_{21} computed using the original reciprocity-based method and Y_{21} computed using IE3D, with agreement being best for the $0.9L_{res}$ twin slots and least good for the longest (*i.e.*, $1.1L_{res}$) twin slots. For each of the three cases, the magnitude of the difference between the Y_{21}-against-d curves obtained with the original reciprocity-based method and IE3D were computed, and normalized to the magnitude of the isolated slot self-admittance, $|Y_{self}|$. Averaged over the range of d, the normalized magnitude of the difference was found to be 10.2% for the L_{res} slots, 1.8% for the

$0.9L_{res}$ slots, and 12% for the $1.1L_{res}$ slots. These average errors decreased to 3%, 1.4%, and 2.4% respectively when the improved reciprocity-based method was used. Hence excellent results can be obtained at the cost of further pre-processing required to compute the terminal voltages $V_{1,\ new}$ and $V_{2,\ new}$ in the manner described above. In an actual linear array design, this would need to be done once only, presumably for the same matrix of slot lengths and widths that self-admittance data will be generated for (Huang et al., 1999).
In order to verify the accuracy of IE3D's calculations, Y_{21} against frequency was computed using that simulator for a twin slot configuration with a fixed layout identical to that shown in Fig. 5, except for one of the slots being rotated by 180° in order to enable feeding it from the edge of the substrate via a coaxial launcher. The twin slots were designed to be at their second resonances at 6 GHz, and had $L_1 = L_2 = 20.95$ mm = $L_{res,\ 6\ GHz}$, and $W_1 = W_2 = 1$ mm on an electrically thin substrate with $h = 0.813$ mm, $\varepsilon_r = 3.38 \pm 0.05$ and $\tan\delta = 0.0027$ (*i.e.*, Rogers RO4003C laminate). The feed line characteristic impedance was 83 Ω ($w = 3$ mm and $s = 1$ mm). Fig. 9 shows good agreement between measured and simulated Y_{21} (referenced to the centers of the slots) data.

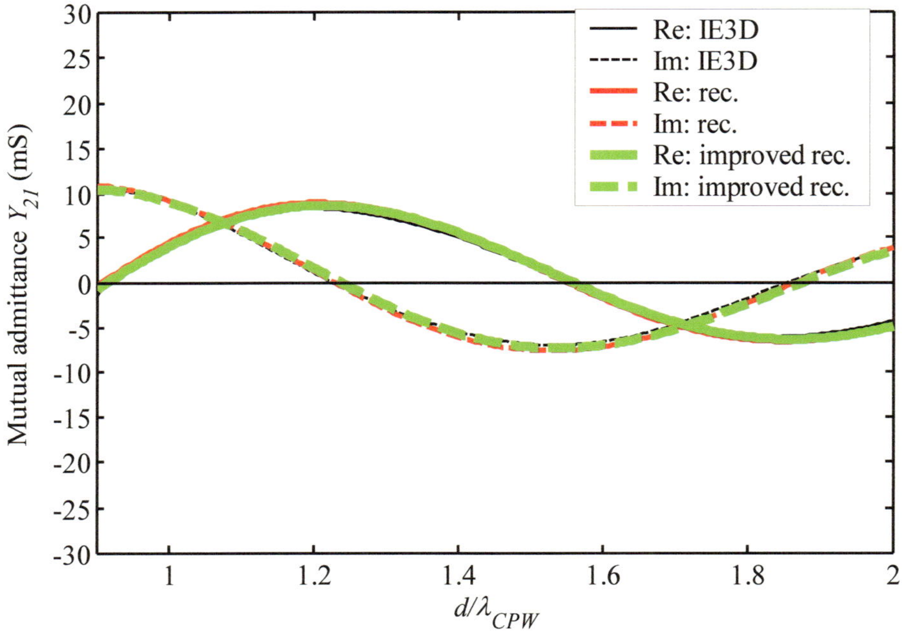

Fig. 6. Mutual admittance Y_{21} against broadside distance d/λ_{CPW} at 6 GHz for CPW-fed twin slots with $L_1 = L_2 = L_{res} = 21.75$ mm. $W_1 = W_2 = 1$ mm, $l_s = 19.4$ mm $= 0.48\ \lambda_{CPW}$, $h = 1.575$ mm, $\varepsilon_r = 2.33$, $w = 3$ mm, $s = 1$ mm.

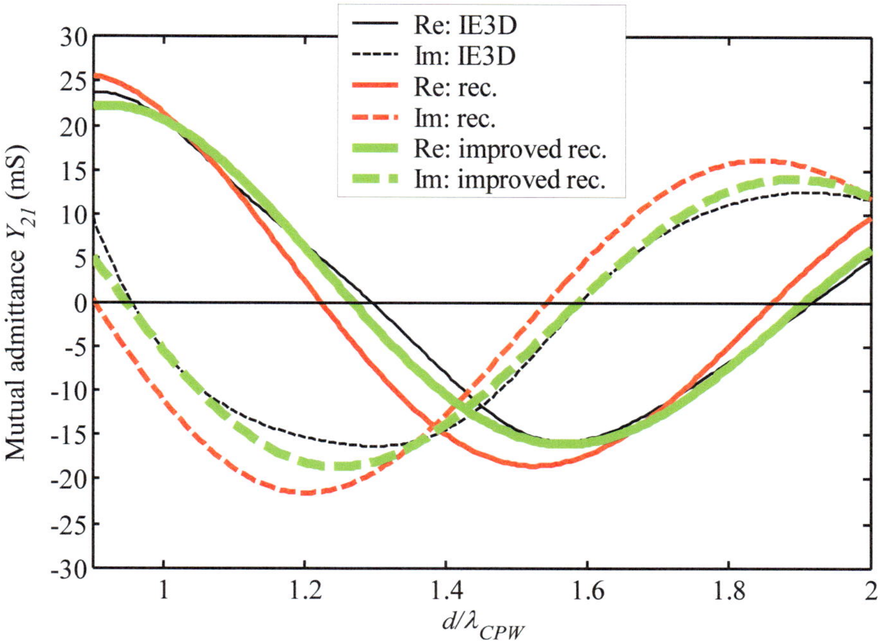

Fig. 7. Mutual admittance Y_{21} against broadside distance d/λ_{CPW} at 6 GHz for CPW-fed twin slots with $L_1 = L_2 = 0.9L_{res} = 19.575$ mm. $W_1 = W_2 = 1$ mm, $l_s = 19.4$ mm $= 0.48\ \lambda_{CPW}$, $h = 1.575$ mm, $\varepsilon_r = 2.33$, $w = 3$ mm, $s = 1$ mm.

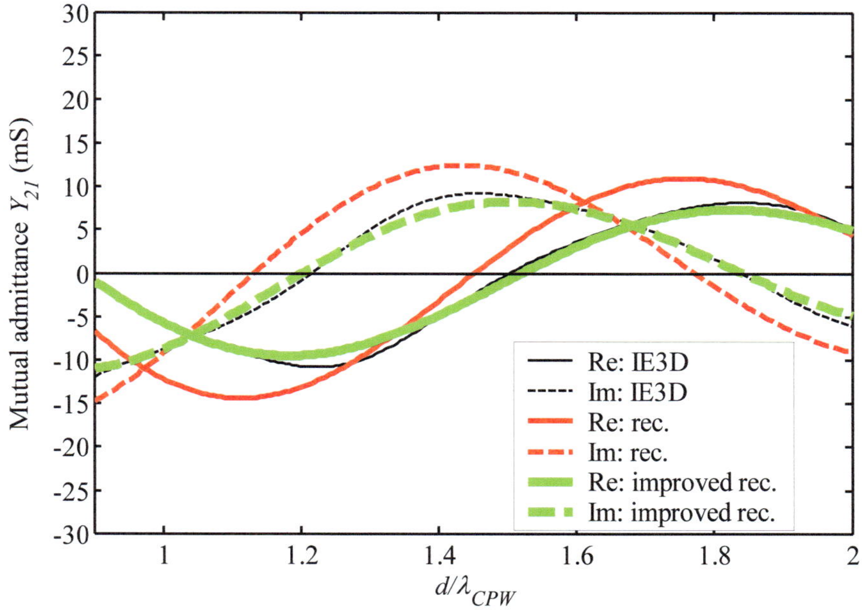

Fig. 8. Mutual admittance Y_{21} against broadside distance d/λ_{CPW} at 6 GHz for CPW-fed twin slots with $L_1 = L_2 = 1.1L_{res} = 23.925$ mm. $W_1 = W_2 = 1$ mm, $l_s = 19.4$ mm $= 0.48\ \lambda_{CPW}$, $h = 1.575$ mm, $\varepsilon_r = 2.33$, $w = 3$ mm, $s = 1$ mm.

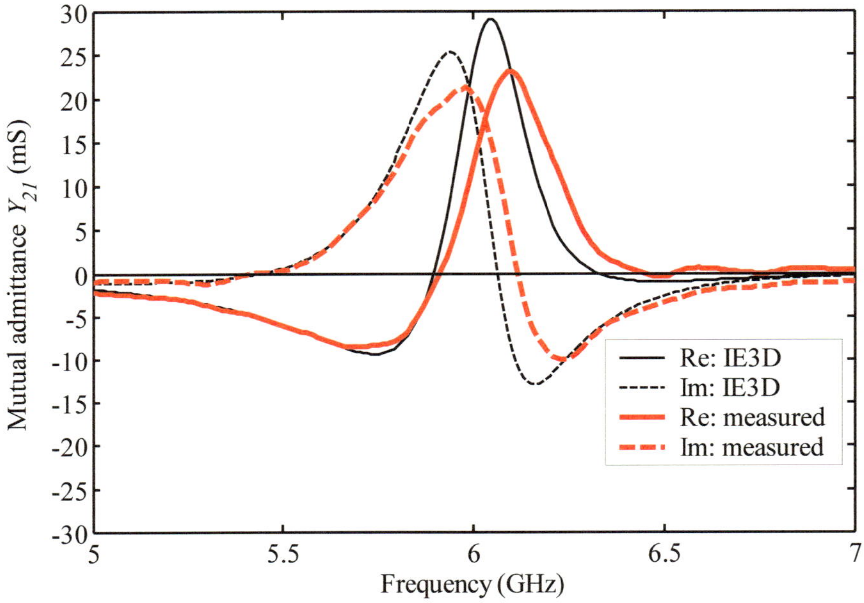

Fig. 9. Computed and measured mutual admittance Y_{21} against frequency for CPW-fed twin slots. $L_1 = L_2 = L_{res,\,6\,GHz}$ = 20.95 mm, $W_1 = W_2$ = 1 mm, l_s = 18.3 mm = $\lambda_{CPW,\,6\,GHz}$, $d = \lambda_{CPW,\,6\,GHz}$ = 38.1 mm, h = 0.813 mm, ε_r = 3.38±0.05, tan δ = 0.0027, w = 3 mm, s = 1 mm.

3.2 Slots on conductor-backed two-layer substrate

Fig. 10 shows twin CPW-fed slots on a two-layer conductor-backed substrate with an air bottom layer (i.e., ε_{r2} = 1). For applications that require unidirectional radiation, this particular structure is of interest as the radiation efficiency attainable with appropriately-spaced twin slots on such a substrate is comparable to that of twin slots on $\lambda_d/4$ single-layer substrates with a reflector positioned $\lambda_o/4$ away from the CPW ground planes side (Qiu et al., 2000) (λ_d is the wavelength in the dielectric; λ_o is the free-space wavelength). This is true even though the substrate of Fig. 10 is of lesser electrical height than the twin slots on a single-layer substrate - potentially an advantage at microwave frequencies.

3.2.1 Effect of conducting back plane

The antenna input impedance of a linear slot array on $\lambda_d/4$ single dielectric substrate is unaffected by a back reflector placed $\lambda_o/4$ on the side of the CPW ground planes (Qiu et al., 2002). This is an indication that the back reflector has minimal influence on mutual coupling between the array slots. However, in this section the presence of a back reflector placed below a two-layer substrate (Fig. 10) is shown to change the mutual coupling between slots. This implies a change in internal coupling (i.e., coupling on the dielectric side) due to the presence of the back plane. Thus, the appropriate parallel-plate two-layer Green's function would be needed in order to account for mutual coupling with sufficient accuracy in an iterative array design procedure (Jacobs et al., 2005).

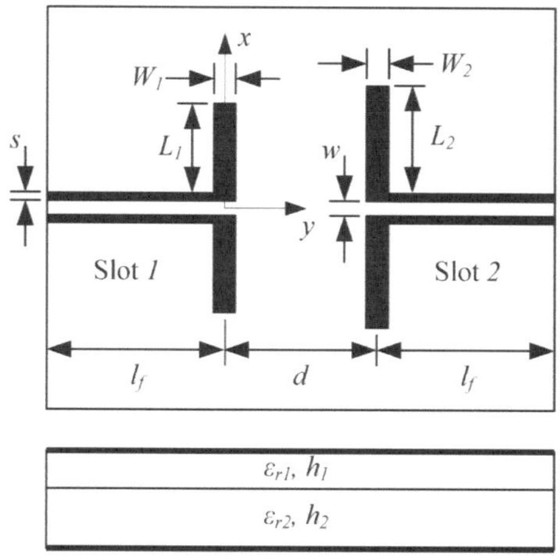

Fig. 10. Top and side views of broadside CPW-fed twin slot antennas on conductor-backed two-layer substrate. $L \equiv$ half-length and $W \equiv$ width of radiating slots; $d \equiv$ distance between radiating slots; $s \equiv$ slot width and $w \equiv$ centre strip width of feed lines; $l_f \equiv$ length of feed lines; h_1 and $h_2 \equiv$ dielectric layer heights; ε_{r1} and $\varepsilon_{r2} \equiv$ relative dielectric constants.

Assuming infinite top and bottom conducting planes and dielectric layers with dielectric constants $\varepsilon_{r1} = 3.38$ and $\varepsilon_{r2} = 1$, with the top substrate layer height $h_1 = 0.813$ mm, simulations were carried out at 10 GHz to demonstrate the effect of back plane height on mutual coupling between the twin slots of Fig. 10 (Jacobs et al., 2005). Three values of bottom layer height, or back plane distance, were considered, namely $h_2 = \infty$, $\lambda_o/4$, and $\lambda_o/6$ (at 10 GHz, $\lambda_o/4 = 7.5$ mm and $\lambda_o/6 = 5$ mm). The case $h_2 = \infty$ is equivalent to the absence of a back plane. A CPW feed line with characteristic impedance of about 50 Ω for the case $h_2 = \infty$ is used (w = 3.7 mm and s = 0.2 mm). For each of the cases $h_2 = \infty$, $\lambda_o/4$, and $\lambda_o/6$, an isolated (radiating) slot with a width W = 0.4 mm was designed to be resonant at 10 GHz by adjusting its half-length L. The resulting resonant slot half-lengths were 10.36 mm, 10.60 mm and 10.67 mm respectively, with corresponding self-impedances of 12.5 Ω, 11.2 Ω, and 13.9 Ω.

Fig. 11 shows the real and imaginary parts g_{21} and b_{21} of the normalized mutual admittance y_{21} as a function of normalized distance d/λ_{CPW} for each of the three cases $h_2 = \infty$, $\lambda_o/4$, and $\lambda_o/6$. y_{21} is the mutual admittance Y_{21} normalized with respect to the relevant isolated resonant slot self-admittance. The reason for normalizing Y_{21} is that, in a linear array context, the relative size of mutual admittance magnitudes with respect to the slots' self-admittance magnitudes is an indicator of the extent of the effect of mutual coupling on the array input impedance.

The normalized curves indicate that, in the absence of a back plane, the maximum values of the magnitudes of the real part (G_{21}) and imaginary part (B_{21}) of the mutual admittance Y_{21} are about 40% and 27% respectively of the resonant slot self-admittance; the maximum of $|g_{21}|$ occurs at $d = \lambda_{CPW}$ and the maximum of $|b_{21}|$ at $d = 1.4\,\lambda_{CPW}$.

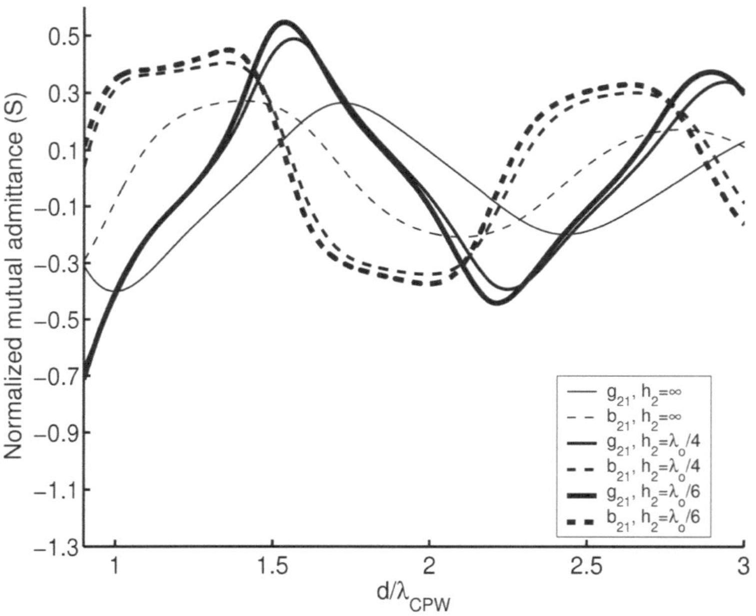

Fig. 11. Real and imaginary parts g_{21} and b_{21} of normalized mutual admittance y_{21} vs. interslot distance d/λ_{CPW} at 10 GHz.

Adding a back plane at $h_2 = \lambda_o/4$ results in significantly higher relative maximum values of about 68% and 40% for $|G_{21}|$ and $|B_{21}|$ respectively; other extrema also show notable increases compared to the $h_2 = \infty$ case. The back plane has the effect of shifting the curves for g_{21} and b_{21} so that their extrema in general are not aligned with those of the curves for $h_2 = \infty$. These effects can be ascribed to internal mutual coupling (the external equivalent problem, concerned with fields in the half-space adjacent to the CPW ground planes, is unchanged by the addition of a back plane), indicating that the input impedance of an array on the single-layer substrate would be significantly affected when a back plane is placed $\lambda_o/4$ away. Decreasing the back plane distance to $h_2 = \lambda_o/6$ results in curves quite similar to, and more or less "in phase" with, the curves for the case $h_2 = \lambda_o/4$ with somewhat higher maximum values for g_{21} and b_{21} (0.71 and 0.45 respectively vs. 0.68 and 0.4), suggesting that the effect of mutual coupling increases as back plane distance decreases. At $d = \lambda_{CPW}$ the magnitude of the real part of the mutual admittance was about 40% of the relevant resonant slot self-admittance for all cases of h_2.

3.2.2 An extended reciprocity-based method for calculating mutual admittance

In this section, a computational strategy based on the well-known expression derived from reciprocity is proposed to calculate the mutual admittance between two CPW-fed slots on conductor-backed two-layer substrates (Jacobs et al, 2009). The formulation iteratively updates the field distribution in both slots to account for the coupling interaction between the slots. The classical reciprocity-based approximation assumes that the field in the slot is the same whether the slot is radiating in isolation or in the presence of a second short-circuited slot. In cases where mutual coupling is high, such as for resonant-length twin slots, this assumption becomes invalid.

The two CPW-fed slots on a conductor-backed two-layer substrate, as shown in Fig. 10, are center-fed by an infinitesimal voltage source that is placed in the centre of a short section of slotline connecting the two CPW feed line slots at their ends. Employing the standard formulation derived from reciprocity the mutual admittance Y_{12} between them is (Nauwelaers & Van de Capelle, 1988)

$$Y_{12} = -\frac{1}{V_1 V_2} \int_{S_2} \mathbf{H}_{21} \cdot \mathbf{M}_2 dS_2 \tag{2}$$

where $\mathbf{M}_2$ is the equivalent magnetic current density representing slot *2*, V_1 and V_2 are the terminal voltages of slots *1* and *2*, respectively and, assuming that the effects of mutual coupling between CPW feed lines, and between feed lines and radiating slots are negligible, S_2 is the surface area of slot *2*. $\mathbf{H}_{21}$ is the magnetic field at the position of slot *2* due to the equivalent magnetic current density, $\mathbf{M}_1$, representing slot *1*. $\mathbf{H}_{21}$ is calculated as the sum of the internal field (using the Green's function for the conductor-backed two-layer substrate), and the external field (using the free-space Green's function).

The initial estimates for $\mathbf{M}_1$ and $\mathbf{M}_2$ are determined by calculating the electric field distributions in the isolated slots using a moment-method solver e.g. IE3D. The isolated-slot moment-method analysis also yielded the terminal voltages V_1 and V_2.

A schematic representation of CPW-fed twin slots is shown in Fig. 12. In situation *a*, slot *1* is assumed to be excited such that its terminal voltage V_1^a equals V_1 (*i.e.*, the terminal voltage of slot *1* when radiating in isolation), while slot *2* is short-circuited. Hence

$$I_1^a = Y_{11} V_1^a \approx Y_{self} V_1 \tag{3}$$

where I_1^a is the terminal current entering port *1*. The current I_2^a entering port *2* can then be found using (Balanis, 1982) adapted for slots (from its original form for wire dipoles):

$$I_2^a = -\frac{1}{V_2} \int_{S_2} H_{21x} M_{2x} dS_2 \tag{4}$$

where all quantities on the right-hand side are similar in meaning to the corresponding quantities in equation (2). For the first interaction, H_{21x} is the longitudinal component of the magnetic field at the position of slot *2* when slot *1* radiates in isolation with terminal voltage V_1 (and terminal current I_1^a), M_{2x} and V_2 are the equivalent magnetic current density and terminal voltage respectively of slot *2* when radiating in isolation, and S_2 is the surface area of slot 2. From equations (2) and (4) it follows that

$$I_2^a = V_1 Y_{12} \tag{5}$$

where Y_{12} is the first iteration of the mutual admittance, computed using equation (2) with the field distribution for the slots radiating in isolation.

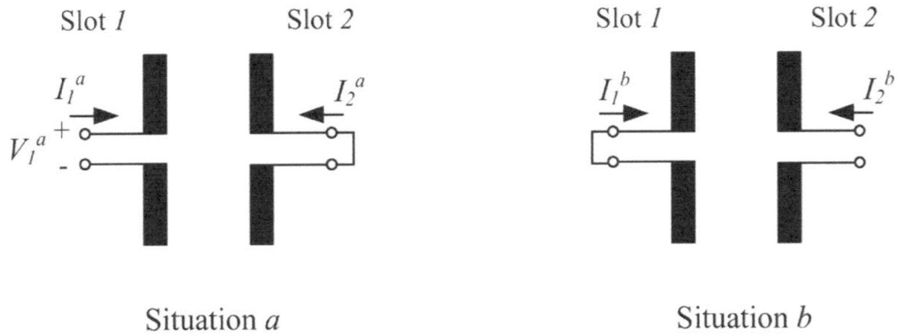

Fig. 12. Schematic representation of CPW-fed twin slots with terminal voltages and currents, and port terminations used in higher-order iterative procedure.

Situation *b* is set up to account for the effect that the current I_2^a induced by slot *1* at the terminals of slot *2* (*cf.* situation *a*) in turn has on slot *1*. Hence slot 2 is assumed to be excited by a terminal current

$$I_2^b = I_2^a \tag{6}$$

while slot *1* is short-circuited. The short-circuit current I_1^b at port *1* can then be found from

$$I_1^b = -\frac{1}{V_1} \int_{S_1} H_{12x}^b M_{1x} dS_1 \tag{7}$$

In the above, H_{12x}^b is the magnetic field at the position of slot *1* when slot *2* radiates in isolation (its terminal current is I_2^b), while M_{1x} and V_1 are the equivalent magnetic current density and terminal voltage respectively of slot *1* when radiating in isolation. Since $H_{12x}^b / I_2^b = -H_{21x} / I_1^a$, it follows from equations (2) and (7) that

$$I_1^b = -\frac{I_2^b}{I_1^a} V_2 Y_{21} \tag{8}$$

with $Y_{21} = Y_{12}$.

The final step in the iterative procedure is to revisit situation *a* and compute updated versions $I_1^{a,update}$ and $V_1^{a,update}$ of the current I_1^a and voltage V_1^a at port *1*:

$$I_1^{a,update} = I_1^a + I_1^b \tag{9}$$

$$V_1^{a,update} \approx \frac{I_1^{a,update}}{Y_{self}} \tag{10}$$

An updated version of the mutual admittance $Y_{12} = Y_{21}$ that includes the effect of the above higher-order interactions between the slots can then be found from

$$Y_{12,update} = Y_{21,update} = \frac{I_2^a}{V_1^{a,update}} \tag{11}$$

while an estimate $Y_{11,est}$ of the two-port self-admittance Y_{11} (at this point only Y_{self} is available) can be computed as follows:

$$Y_{11,est} = \frac{I_1^a}{V_1^{a,update}} \tag{12}$$

The process of updating the two-port admittance parameters can be repeated iteratively until the effect of the next higher-order interaction is negligible.

Fig. 13 shows the Y_{12}-against-*d* curves for the classical reciprocity approach, the iterative reciprocity-based formulations and a moment-method solver (IE3D) for the case of two identical (twin) resonant slots. Y_{12} against broadside slot separation *d* was computed at 10 GHz for CPW-fed slots on a conductor-backed two-layer substrate configured for high radiation efficiency. The substrate had h_1 = 0.813 mm = $0.05\lambda_d$, h_2 = 5 mm = $\lambda_0/6$, ε_{r1} = 3.38, and ε_{r2} = 1 (λ_d is the dielectric wavelength in the top layer and λ_0 the free-space wavelength). The substrate allowed for propagation of the TM_0 two-layer parallel-plate mode only. Slot dimensions were *W* = 0.4 mm and $L = L_{res}$ = 10.87 mm, yielding a resonant self-resistance of 14 Ω; dimensions of the 50 Ω feed line were *w* = 3.7 mm and *s* = 0.2 mm.

The extended reciprocity-based curves resemble the shape of the moment-method curves much closer than the classical reciprocity-based curves. The reciprocity-based approach for mutual admittance calculations between CPW-fed slots on conductor-backed two-layer substrates – modified to account for higher-order interactions – is a viable alternative to a moment-method-based approach, offering comparable accuracy and the advantages of simplicity of implementation within iterative array design procedures.

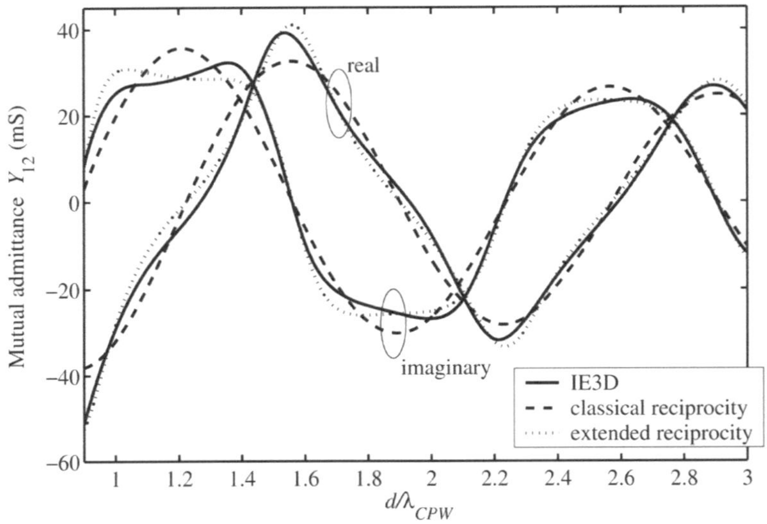

Fig. 13. Mutual admittance Y_{12} against broadside distance d/λ_{CPW} for CPW-fed twin slots with $L_1 = L_2 = L_{res} = 10.87$ mm computed using the classical and extended reciprocity approaches, and IE3D. $W_1 = W_2 = 0.4$ mm; $h_1 = 0.813$ mm; $h_2 = 5$ mm; $\varepsilon_{r1} = 3.38$; $\varepsilon_{r2} = 1$; $l_f = 0.5\lambda_{CPW}$.

4. Linear arrays of CPW-fed slots

4.1 Uniform array on conductor-backed two-layer substrate

This section describes how an approximate iterative design procedure for uniformly-excited CPW-fed linear slot arrays on single-layer substrates (Qiu et al., 2002) can be extended to the case of an 8-element CPW-fed slot array on a conductor-backed higher-lower permittivity two-layer substrate (Jacobs et al., 2003). When designing series-fed broadside linear slot arrays, it is desirable to have slots excited in-phase and with equal magnitude for maximum gain. This implies that slots should be spaced λ_{CPW} rather than $\lambda_g/2$ apart, even though the latter the spacing would seem preferable due to the ensuing phase cancellation of the substrate mode which would result in improved radiation efficiency (λ_{CPW} is the wavelength of the CPW feed line; λ_g is the wavelength of the dominant substrate mode). Previously reported investigations of linear arrays on single-layer substrates, and $\lambda_d/4$ substrates with a back reflector positioned $\lambda_o/4$ away, however indicated that a λ_{CPW} spacing (as opposed to $\lambda_g/2$ spacing) does not affect gain adversely (Qiu et al., 2002).

The top view layout of a generic CPW-fed linear array is shown in Fig. 4; given the present context, all slot half-lengths were assumed to be equal and likewise for slot widths. For simulation purposes, substrate layer heights and dielectric constants were selected (as before) to ensure that a non-leaky CPW feed line would result - this requires a higher-lower permittivity substrate, the higher permittivity layer being adjacent to the slots (Huang & Kuo, 1998). The selected values were $h_1 = 0.813$ mm, $\varepsilon_{r1} = 3.38$, $h_2 = 5$ mm ($\lambda_o/6$ at 10 GHz), and $\varepsilon_{r2} = 1.1$ (see side view in Fig. 10). Using IE3D, a 50 Ω CPW feed line was designed that had $w = 3.7$ mm and $s = 0.2$ mm.

The array (N = 8) was designed for operation at 10 GHz using the following steps in an iterative fashion; the procedure relies on the assumption that slot active impedances are largely determined by the corresponding slot self-impedances (and to a lesser extent by mutual coupling).

1. The self-impedance $Z_{in,slot}$ of a single slot dipole with half-length L and (radiating slot) width W was computed using IE3D (infinite ground planes were assumed throughout). For the first iteration, convenient values of W and L were chosen.
2. An 8-element array was constructed of identical slots spaced λ_{CPW} apart, each with the W and L determined in the previous step. The self-impedance $Z_{in,array}$ of the array was determined using IE3D.
3. The difference Z_{diff} between the array input impedance and the desired input impedance of 50 Ω was found.
4. Using IE3D, new L and W values were found that would yield a new slot self-impedance of $Z_{in,slot} - Z_{diff}/N$.
5. Steps (2) and onwards were repeated until $Z_{in,array}$ was within an acceptable margin from 50 Ω.

The iterative procedure converged to L and W values of 12.04 mm and 1.17 mm respectively. In Fig. 14 measured return loss against frequency is shown for the above design manufactured on a 12"x18" Rogers RO4003C substrate. The deviation of the measured operating frequency of 9.7 GHz from the design frequency of 10 GHz can be attributed to manufacturing and material considerations that likely included deviations from specified dielectric constants, and inconsistencies in the thickness of the polystyrene bottom layer.
The E-plane co-polarized radiation pattern measured at 9.7 GHz is shown in Fig. 15. Good agreement with the predicted pattern was exhibited, indicating insignificant leakage of the dominant two-layer parallel-plate mode from the sides of the antenna. Sidelobe levels were below -13 dB as expected except for a sidelobe at 85° that had a level of about -10 dB; this could have been caused by an inadvertently-excited higher-order mode on the CPW feed.

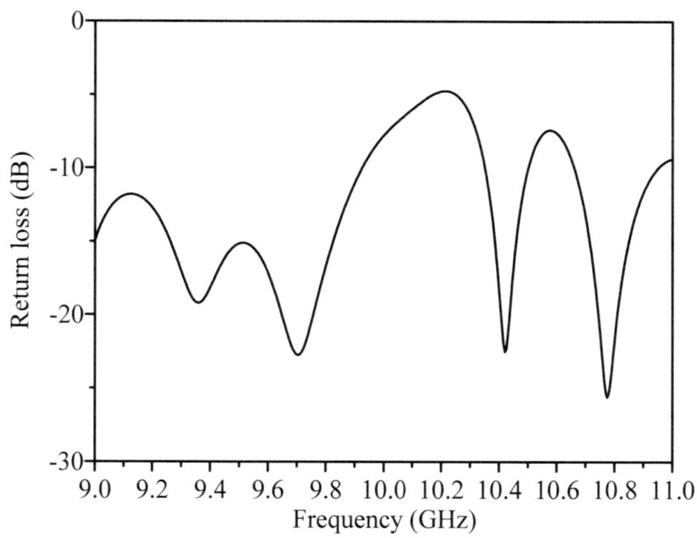

Fig. 14. Measured return loss against frequency for 8-element CPW-fed slot array on conductor-backed two-layer substrate.

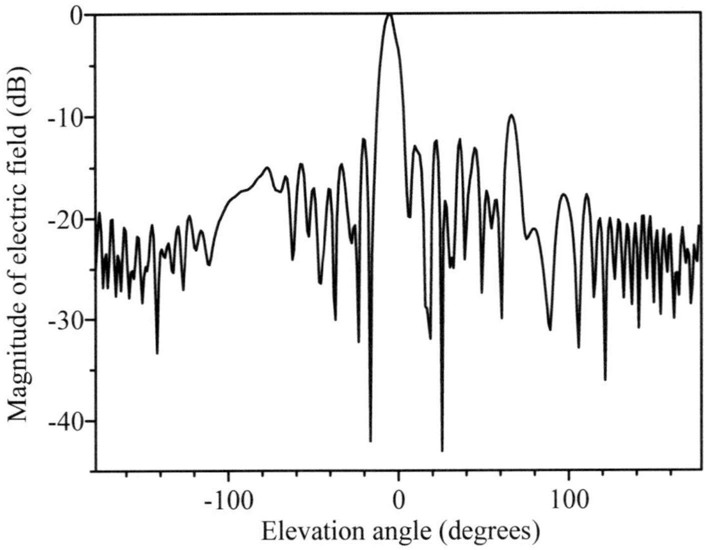

Fig. 15. E-plane co-polarization pattern measured at 9.7 GHz for 8-element CPW-fed slot array on conductor-backed two-layer substrate.

4.2 Non-uniform array on single-layer substrate

This section describes design of a linear non-uniform slot array series-fed by CPW, on an electrically thin substrate for computational simplification, with sidelobe levels substantially reduced beyond that of a uniform array (Jacobs & Joubert, 2009b). The geometry of such a linear array is as shown in Fig. 4. A low-sidelobe level specification typically requires explicit, accurate accounts of the mutual coupling between all possible slot pairs in the array, and hence the ability to enforce a suitably tapered slot excitation. In what follows, the iterative design of a 5-element non-uniform linear CPW-fed array with reduced sidelobe levels on an electrically thin substrate is described. Because of the electrically thin substrate, mutual coupling can be accounted for using the reciprocity-based method involving the assumption of a homogeneous free-space medium, as discussed in section 3.1 (Jacobs & Joubert, 2009a).

4.2.1 Array design procedure

The first step was to synthesize a 5-element Dolph-Tschebychev array of isotropic radiators to have a maximum sidelobe level of -20 dB below the main beam maximum (Balanis, 1996). The spacing between the radiators was $d = 0.78\lambda_0$, with λ_0 the free-space wavelength at the design frequency of 5.6 GHz (this frequency belongs to the upper WLAN band). $0.78\lambda_0$ equals λ_{CPW}, the wavelength of a 70 Ω CPW feed line on a substrate with height h = 1.575 mm and relative permittivity ε_r = 2.33, the substrate of choice for the physical array; the CPW centre strip and slot widths were w = 3 mm and s = 0.5 mm respectively (see Fig. 14). The aim of this particular spacing was to ensure in-phase excitation of the slots. The resulting excitation values were $[V_1\ V_2\ V_3\ V_4\ V_5]$ = [0.52 0.83 1.00 0.83 0.52].

The second step entailed setting up a slot self-admittance database, as well as corresponding field distribution and terminal voltage databases for the calculation of mutual admittances. In particular, self-admittance data was generated using IE3D for a grid of half-lengths and widths of isolated CPW-fed slots: half-length values were between 18.5 and 27.5 mm (at a 0.75 mm interval), and widths were between 0.25 and 3.25 mm (at a 0.25 mm interval). Aperture field and terminal voltage data was collected concurrently with the self-admittance data.

The steps followed in iteratively designing the array are now outlined (see also Huang et al. (1999)) - consider in this regard the generic linear CPW-fed slot array of Fig. 4 that has N slots spaced λ_{CPW} apart with slot half-lengths and widths L_m and W_m ($m = 1, 2, ..., N$). In network terms, the array can be viewed as an equivalent transmission line circuit with the slots represented by series active impedances separated by λ_{CPW}-long sections of transmission line (Huang et al., 1999; Meide et al., 2002). An active voltage V_m exists across the series active impedance representing slot m, while a current I is common to all slot impedances due to the series nature of the equivalent circuit. The corresponding active admittance of slot m is $Y_m^a = I / V_m$. For an N-element array, the iterative design proceeds as follows:

1. A desired input impedance $Z_{in,des}$ is chosen (50 Ω for the purposes of this letter). The desired slot active admittances $Y_{m,des}^a$ can then be found using the following two equations:

$$Z_{in,des} = \sum_{m=1}^{N} \frac{1}{Y_{m,des}^a} \tag{13}$$

$$Y_{1,des}^a V_1 = ... = Y_{N,des}^a V_N \tag{14}$$

2. Starting values for the lengths and widths (L_m, W_m) can then be assumed or calculated. Good starting values can be obtained by initially assuming zero mutual coupling between the slots and then calculating slot dimensions (L_m, W_m) from the self-admittance database that would realize $Y_{mm} = Y_{m,des}^a$.
3. These dimensions (L_m, W_m) are then used to calculate the mutual admittances Y_{mn} between all possible slot pairs m and n according to section 3.1 (Jacobs & Joubert, 2009a). A linear interpolation scheme is used to calculate the updated mutual admittances Y_{mn} during this step for specific (L_m, W_m) dimensions (from the discreet field distribution and terminal voltage database compiled previously).
4. A set of $2N$ non-linear equations can then be solved to enforce the desired relative slot excitations, the chosen matching criteria, and of course resonance of the N slots in the array. The mutual admittances Y_{mn} are kept constant as calculated in step (3), and a new set of (L_m, W_m) are sought (in effect a new set of self-admittances Y_{mm} are determined) that will satisfy the design equations. The equations are of the form:

$$\frac{Y_m^a}{Y_N^a} = \frac{V_N}{V_m}\text{, for } m = 1, 2, \ldots, N-1 \tag{15}$$

$$\sum_{m=1}^{N} \frac{1}{Y_m^a} = Z_{in,des} \tag{16}$$

$$\text{Imag}(Y_m^a) = 0\text{, for } m = 1, \ldots, N \tag{17}$$

Throughout the non-linear equation solving process, a bivariate spline interpolation scheme (McNamara & Joubert, 1994) is used to calculate new self-admittances Y_{mm} for arbitary (L_m, W_m) dimensions, using the previously compiled discreet self-admittance database.

5. Updated mutual admittances for the new slot dimensions can then be calculated, and the true active admittances determined using $Y_m^a = \sum_{n=1}^{N} Y_{mn} \frac{V_n}{V_m}$, from which the actual input impedance of the array can then be found. If Z_{in} is not close enough to the desired value of $Z_{in,des}$, steps 4 and onwards are repeated until convergence is obtained.

4.2.2 Results and discussion

Application of the above procedure yielded the following dimensions for the $N = 5$ array: $L_1 = L_5 = 23.9$ mm, $W_1 = W_5 = 0.3$ mm, $L_2 = L_4 = 23.1$ mm, $W_2 = W_4 = 1.15$ mm, and $L_3 = 23.1$ mm, $W_3 = 2.2$ mm. For the purpose of simulating the full array in IE3D, a short-circuit stub termination of length $l_s = 21$ mm $\approx \lambda_{CPW}/2$ was added (see Figure 4), as well as a front-end CPW feed section of length $\lambda_{CPW}/2$. The simulated array was etched on a Rogers RT/duroid 5870 laminate ($h = 1.575$ mm, $\varepsilon_r = 2.33\pm0.02$, $\tan\delta = 0.0009$) of dimensions 230 mm × 135 mm (see Figure 16), and its reflection coefficient, radiation patterns and gain were measured.

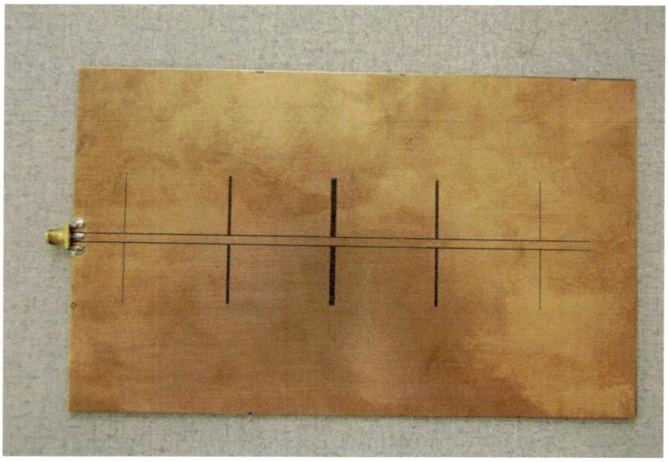

Fig. 16. The manufactured 5-element CPW-fed array.

Fig. 17 shows the simulated and measured array reflection coefficient against frequency. The simulated and measured resonant frequencies were 5.61 GHz and 5.59 GHz respectively, which are both very close to the design frequency of 5.6 GHz. The simulated fractional bandwidth was 1.5% (VSWR < 2), while the corresponding measured bandwidth was 2.1%, confirming the inherently narrow-band nature of the feeding scheme. Discrepancies between simulation and measurement can likely be attributed to manufacturing errors.

Fig. 18 displays the simulated and measured normalized E-plane radiation patterns (the array lies in the xy plane). The simulated co-polarization pattern has a maximum sidelobe level of about -17 dB, which is higher than the designed-for Dolph-Tschebychev sidelobe level of -20 dB. This can be attributed to the accuracy of mutual admittance calculations. The measured co-polarized pattern on the whole agrees well with the simulated pattern (the discontinuity in the simulated pattern at Theta = ±90° is due to IE3D setting the electric field component perpendicular to the infinite ground plane, *i.e.*, the co-polarized component, to zero at far-field observation points coinciding with the ground plane; this apparently is an artifact of the simulator). The somewhat higher maximum sidelobe levels of about -16 dB in the measurement can possibly be attributed to the finite substrate (simulations were carried out for a laterally infinite substrate). Fig. 18 also gives the measured cross-polarization which is significantly below co-polarization levels (simulated cross-polarization was too small to show on this graph).

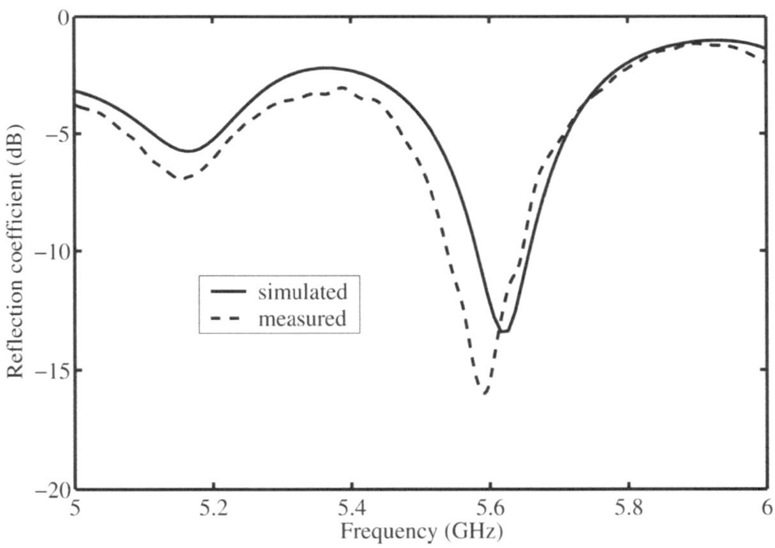

Fig. 17. Simulated and measured reflection coefficient vs. frequency.

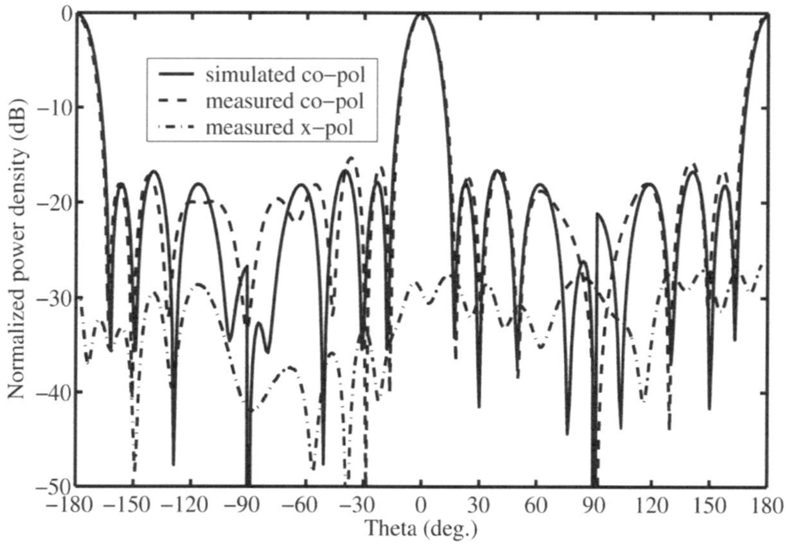

Fig. 18. Simulated and measured E-plane radiation patterns.

The maximum boresight gain of the array is at 5.62 GHz, *i.e.*, 14 dBi (predicted) and 13.5 dBi (measured). The corresponding radiation efficiency was calculated using IE3D to be 97%, indicating negligible power loss to substrate modes due to the electrically thin substrate.

5. Final remarks

The design of non-uniform linear CPW-fed slot arrays with sidelobe levels significantly reduced below that of uniform arrays was presented; efficient methods for accomplishing important parts of the design, such as the calculation of mutual coupling subject to representative substrate configurations, were summarized. Results of a practical implementation on an electrically thin single-layer dielectric substrate was shown; measured sidelobe levels of -16 dB below could be achieved.

6. References

Amitay, N.; Galindo, V. & Wu, C. P. (1972). *Theory and analysis of phased array antennas*, Wiley-Interscience, ISBN 10: 0471025534, New York.

Balanis, C. A. (1996). *Antenna theory: analysis and design*, 2nd ed., ISBN-10: 0471592684, Wiley, New York.

Chen, S.-Y.; Lan, I.-C. & Hsu, P. (2007). In-Line Series-Feed Collinear Slot Array Fed by a Coplanar Waveguide, *IEEE Transactions on Antennas and Propagation*, Vol. 55, No. 6, pp. 1739-1744, ISSN: 1350-2417.

Chen, W.-S. & Wong, K.-L. (2000). A dual-frequency coplanar waveguide-fed slot antenna, *Microwave and Optical Technology Letters*, Vol. 25, No. 3, pp. 226–228, ISSN: 0895-2477.

Elliott, R. S. (1981). *Antenna theory and design*, Prentice-Hall, ISBN: 0130383562, Englewood Cliffs, New Jersey.

Focardi, P.; Neto, A. & McGrath, W. R. (2002). Coplanar-waveguide-based tetraherts hot-electron-bolometer mixers - improved embedding circuit description, *IEEE Transactions on Microwave Theory and Techniques,* Vol. 50, pp. 2374–2383, ISSN: 0018-9480.

Huang, J.-F. & Kuo, C.-W. (1998). More investigations of leakage and nonleakage conductor-backed coplanar waveguide. *IEEE Transactions on Electromagnetic Compatibility*, Vol. 40, No. 3, pp. 257-261, ISSN: 0018-9375.

Huang, T.-F.; Lu, S.-W. & Hsu, P. (1999). Analysis and design of coplanar waveguide-fed slot antenna array, *IEEE Transactions Antennas and Propagation*, Vol. 47, No. 10, pp. 1560-1565, ISSN: 0018-926X.

Jacobs, J. P. (2007). Self and mutual admittance of CPW-fed slots on conductor-backed two-layer substrates, *Microwave and Optical Technology Letters*, Vol. 49, No. 11, pp. 2798–2802, ISSN: 0895-2477.

Jacobs J. P. & Joubert, J. (2009a). Fast and efficient calculation of mutual admittance between CPW-fed slots on electrically thin substrates, *International Journal of RF and Microwave Computer-Aided Engineering*, Vol. 19, No. 2, pp. 277-284, ISSN: 1096-4290.

Jacobs J. P. & Joubert, J. (2009b). Design of a linear nonuniform CPW-fed slot array with reduced sidelobe levels, *Microwave and Optical Technology Letters*, Vol. 51, No. 9, pp. 2175-2178, ISSN: 0895-2477.

Jacobs, J. P.; Joubert, J. & Odendaal, J. W. (2003). Conductor-backed CPW-fed broadside linear slot antenna array on two-layer dielectric substrate, *Proceedings of International ITG-Conference on Antennas*, pp. 61–63, Berlin, Sep. 2003, Germany.

Jacobs, J. P.; Joubert, J. & Odendaal, J. W. (2005). Effect of back plane distance on mutual coupling between CPW-fed slots on conductor-backed two-layer substrates. *Microwave and Optical Technology Letters,* Vol. 47, No. 5, pp. 407-409, ISSN: 0895-2477.

Jacobs, J. P.; Joubert, J. & Odendaal, J. W. (2009). Extended reciprocity-based computation of mutual admittance between CPW-fed slots on conductor-backed two-layer substrates. *Microwave and Optical Technology Letters*, Vol. 51, No. 1, pp. 91-94, ISSN: 0895-2477.

Kim, S. H.; Choi, J. H.; Baik, J.W. & Kim, Y. S. (2006). CPW-fed log-periodic dumb-bell slot antenna array, *Electronics Letters*, Vol. 42, No. 8, pp. 436–438, ISSN: 0013-5194.

McNamara D. A. & Joubert, J. (1994). On the use of bivariate spline interpolation of slot data in the design of slotted waveguide arrays, *ACES Journal*, Vol. 9, No. 1, pp. 6-9, ISSN: 1054-4887.

Miao, M.; Ooi, B. L. & Kooi, P. S. (2000). Broadband CPW-fed wide slot antenna, *Microwave and Optical Technology Letters,* Vol. 25, No. 3, pp. 206–211, ISSN: 0895-2477.

Nauwelaers, B. K. J. C. & Van de Capelle, A. R. (1988). Integrals for the mutual coupling between dipoles or between slots: with or without complex conjugate. *IEEE Transactions on Antennas and Propagation,* Vol. 36, No., 10, pp. 1375-1381, ISSN: 0018-926X.

Neto, A.; De Maagt, P. & Maci, S. (2003). Optimized basis functions for slot antennas excited by coplanar waveguides, *IEEE Transactions on Antennas and Propagation*, Vol. 51, No. 7, pp. 1638–1646, ISNN: 1350-2417.

Qiu, M.; Simcoe, M. & Eleftheriades, G. V. (2000). Radiation efficiency of printed slot antennas backed by a ground reflector, *Proceedings of 2000 IEEE Antennas and Propagation Society International Symposium*, pp. 1612-1615, ISBN: 0-7803-6372-8, Salt Lake City, July 2000, USA.

Qiu, M.; Simcoe, M. & Eleftheriades, G. V. (2002). High-gain meanderless slot arrays on electrically thick substrates at millimeter-wave frequencies, *IEEE Transactions Microwave Theory and Techniques*, Vol. 50, No. 2, pp. 517–528, ISSN: 0018-9480.

Soliman, E. A.; Brebels, S.; Beyne, E. & Vandenbosch, G. A. E. (1999). 2x2 and 4x4 arrays of annular slot antennas in MCM-D technology fed by coplanar CPW networks, *IEE Proceedings Microwaves, Antennas and Propagation*, Vol. 146, No. 5, pp. 335–338, ISSN: 1350-2417.

Tsai, H. S.; Rodwell, M. J. W. & York, R. A. (1994). Planar amplifier array with improved bandwidth using folded-slots, *IEEE Microwave and Guided Wave Letters*, Vol. 4, No. 4 pp. 112–114, ISSN: 1051-8207.

Zeland Software (2001). IE3D User's Manual, Release 8.